A DICTIONARY OF CIVIL & ENVIRONMENTAL ENGINEERING

A DICTIONARY OF CIVIL & ENVIRONMENTAL ENGINEERING

Dictionary for Principles and Practice of Engineering (PE) Examination

---

Harry C. Friebel, Ph.D., P.E., D.CE

GOLDEN RATIO
PUBLISHING

Published by Golden Ratio Publishing, LLC, Berlin, New Jersey.

Permissions Department
Golden Ratio Publishing, LLC
Email: info@goldenratiopublishing.com

Library of Congress Control Number: 2020918807

Library of Congress subject headings:
Civil engineering – Dictionaries
Hydraulic engineering – Dictionaries
Environmental engineering – Dictionaries
Highway engineering – Dictionaries
Traffic engineering – Dictionaries
Structural engineering – Dictionaries
Geotechnical engineering – Dictionaries
Building – Dictionaries

To my children, Nathaniel and Juliet - your bravery, strength and perseverance are my inspiration.

TESTIMONIALS

Here's what former examination takers are saying about the need to bring Dr. Friebel's dictionary to the PE examination:

A DICTIONARY OF CIVIL, WATER RESOURCES & ENVIRONMENTAL ENGINEERING

"***This is an ABSOLUTE must!*** *- If you are taking the PE Civil-Water Resources exam, you need this book! The people that make the test (evil NCEES) know what is in your Lindeburg book, so they aren't going to give you freebie questions directly from that resource. I got a few answers right out of this book for the October 2015 exam, none of those answers were in the Lindeburg nor the Goswami books...."* Amazon verified purchase by Predadactyl on December 13, 2015.

"***Don't miss easy questions on the PE exam*** *- Because you don't know the definition of a word! You are sure to come across terms that you have never heard before, and the answer will be apparent as long as you can figure out what the problem statement is asking.* Amazon verified purchase by M.J.D. on November 9, 2015.

"***A MUST HAVE for the Civil PE Exam*** *- This is a MUST HAVE for the Civil PE Exam. There are so many topics covered in the AM, breadth portion of the exam, that you are sure to come across some terms that you have not yet encountered or are not familiar with. Due to the specialized content, you are unlikely to find most of these terms in a standard dictionary. I tried a handful of other "Scientific" and "Engineering" dictionaries but this is hands down the best for the PE Exam.* Amazon verified purchase by Dan B on March 6, 2015.

"***Very useful!*** *- I used this book for the PE exam, and I am sure it helped me get 2-3 problems correct that I otherwise would have had to guess on. Very useful!* Amazon verified purchase by Bb on December 16, 2014.

"***Great book for my engineering library*** *- I bought the book to help me in defining terms that I was unfamiliar with while taking the PE exam. I used the book a few* *times during my exam and quite possibly helped in getting a few questions correct. This book is also a great addition to my engineering library at my office and would recommend it to any practicing water resource engineer.* Amazon verified purchase by Kellen A. Lopes on May 27, 2014.

*"**Great for the PE exam** - I was skeptical when told to purchase this for studying the PE exam but have been pleasantly surprised with how handy it's come in already. It's a great reference and though I was originally planning to sell it back after my exam, I will keep it. I'm happy to have a quick reference at my desk for Civil Engineering.* Amazon verified purchase by Chantal on October 31, 2013.

*"**A must have when taking the PE exam** - I took the PE exam for the first time last October, many years after graduation, and came up short. Fortunately, when I took the exam for the second time in April I purchased Dr. Friebel's Dictionary of Civil, Water Resources & Environmental Engineering which proved to be extremely helpful, I passed! The dictionary provided quick access to at least 4 or 5 problems, both AM and PM variety. I strongly urge anyone taking the Water Resources depth exam to add this book to your arsenal. Ultimately, I passed using only the CERM, Metcalf & Eddy, and Dr. Friebel's dictionary.* Amazon verified purchase by PAL on June 11, 2013.

*"**Great Resource** - I had 2 dictionaries for the P.E. exam before learning of Dr. Friebel's WRE dictionary. Neither had much in the way of water resources or environmental. I purchased a copy here (fast service) after taking a class from Dr. Friebel on the hunch that if he put the same effort into the book I wouldn't be disappointed. A scan across all 3 quickly showed that it is very expansive regarding environmental terms and very different from the typical civil dictionary. I used it for a few practice exams but didn't need it for the April 13 Civil/Transportation exam. Lucky I guess. I do know some people who did use it for the exam and scored a few points they wouldn't have. I rank it between a must have and cheap insurance. You won't find a better WRE dictionary near this cost. Buy it for the exam and when you pass take it to work.* Amazon verified purchase by New P.E. on June 2, 2013.

ACKNOWLEDGMENTS

I credit my knowledge and understanding of Engineering to my colleagues and former academic mentors. I am very grateful for all their patience, advice and guidance I have received over the years. Since I began helping others pass the Principles and Practice of Engineering (PE) Examination, I have also benefited enormously from discussions with former students, fellow PE review course instructors and readers of my website (www.goldenratiopublishing.com). Special acknowledgement is made to my former review instructor colleagues who helped me formulate and refine my teaching methods to assist others in passing the PE examination. I have been taught so much, by so many people - you have all inspired me! I again owe my sincerest gratitude to both Amy Hendrickson (texnology.com) for assisting me with my book deign layouts and Robert Anthony for designing my company logo and first book cover.

Over the years, I have been inspired and supported by my immediate family and sincerest friends. In addition to my sisters Victoria and Tara and my brother Damian, I would also like to thank: Peter Banfe, Thomas Hurley, Mike Scott, Randall Wise, Peter Smith, Laura Bittner, Scott Hansen and Doug Erwin - I cannot put into words the kindness you have shown me and my family over the past few years. Finally, I would especially like to acknowledge: Susan Bergen, Rebecca Hansen and Susan Hall for having the courage to speak up and tell the truth - protecting those most precious to me that didn't have the ability to protect themselves.

To my children Nathaniel and Juliet, thank you for your unconditional love and patience - while I can't promise to be here for the rest of your life, I can promise to always love you for the rest of mine.

H. C. F.

ABOUT THE AUTHOR

Harry C. Friebel, Ph.D., P.E., D.CE is the founder and president of the publishing company Golden Ratio Publishing, LLC. Dr. Friebel is a hydraulic engineer with the U.S. Army Corps of Engineers (views expressed herein are the authors own and do not represent the views of the U.S. Army Corps of Engineers or Federal government).

Dr. Friebel's first publication, ***A Dictionary of Civil, Water Resources & Environmental Engineering***, has sold over 10,000 copies as of the date of this publication.

Dr. Friebel is a former Civil, Water Resources and Environmental Engineering P.E. review course instructor. Dr. Friebel formerly taught Hydraulic Design to both undergraduate and graduate students as an adjunct professor at Rowan University, in Glassboro, New Jersey. He earned both a B.S. and M.S. engineering degree from Florida Institute of Technology and a Ph.D. from Stevens Institute of Technology. He is a registered Professional Engineer (P.E.), Diplomate of Coastal Engineering (D.CE) and has served as principal investigator on numerous coastal engineering and water research projects sponsored by federal, state and local government agencies.

For more information about Dr. Friebel or Golden Ratio Publishing, please visit:

www.goldenratiopublishing.com

If you have any comments regarding this book or Golden Ratio Publishing, please contact:

info@goldenratiopublishing.com

AFTER YOU PASS, WRITE ME

Please, after your pass, write me your success story! I'd love to post it on my website (www.goldenratiopublishing.com) or even feature it in my next book.

I welcome your feedback and suggestions. If this book helps, please let me know. If I can improve it in any way or there is a term you would like to see included in a future edition, let me know that too. I only write and teach because I love it - I experienced some exhilarating teachers in my career and now it is time for me to give back.

I must admit - I do enjoy the positive feedback I receive from students. Through their comments I was inspired to write my first dictionary. Now, with the additional feedback I have received from *A Dictionary of Civil, Water Resources & Environmental Engineering*, I have produced a broader and even more comprehensive dictionary. Your input will help me improve future editions, so please send me your thoughts:

info@goldenratiopublishing.com

www.goldenratiopublishing.com

There you will find updates to this book and announcements of future publications. I built it for you, so please take advantage of it.

PREFACE

Think you don't need a dictionary, think again!

This dictionary was written with the Principles and Practice of Engineering (PE) Civil and Environmental examinations in mind. Typically, a PE examinee spends well over a thousand dollars on review books, courses and sample examinations. Many previous examinees believe they did not pass the examination the first time due to missing a single problem or two. What if one of those questions could have been answered correctly with a dictionary? Do you really want to chance not understanding what a word means in any of your examination questions? What is it worth to increase your chances of getting additional problems correct?

Think of this dictionary as insurance! Hopefully you won't need it, but what if you do? Be wary of anyone stating a dictionary is not necessary for the exam. If you don't think the good people of National Council of Examiners for Engineering and Surveying (NCEES) that put the examination together are not aware of the exact terms defined (and NOT defined) in your review manual glossary, you're kidding yourself. I have received testimonial after testimonial from former students thanking me for recommending that they bring a dictionary to the exam. Don't believe me? Go on Amazon and read the reviews for yourself. Terms can (and do) appear on the examination that you may not be familiar with. What if they are not defined in your reference books, then what? Everyone's vocabulary is different and what's familiar to you, may not be familiar to the person sitting next to you and vice versa. This dictionary may very well be the difference in you passing your examination.

I sincerely hope you find this dictionary an invaluable resource. I spent many years gathering, writing, and re-writing the terms found in this dictionary to provide a treasure of material for passing the PE Civil and Environmental examinations. I wrote this dictionary with only one objective, to help you pass the PE examination!

H. C. FRIEBEL

Berlin, New Jersey
October, 2020

ABBREVIATIONS

Δ CHANGE.

Θ_E EQUIVALENT POTENTIAL TEMPERATURE.

° DEGREES.

7Q10 SEVEN-DAY, 10-YEAR LOW FLOW.

A AIRWORTHY.

A AMS ARCTIC AIR MASS.

A/C ASBESTOS-CEMENT.

A/E ARCHITECTURAL AND ENGINEERING.

AAAS AMERICAN ASSOCIATION FOR THE ADVANCEMENT OF SCIENCE.

AAC ALL AMERICAN CANAL.

AADT AVERAGE ANNUAL DAILY TRAFFIC.

AAHU AVERAGE ANNUAL HABITAT UNITS.

AAM ANNUAL ARITHMETIC MEAN.

AAR ALKALI AGGREGATE REACTION.

AASHTO AMERICAN ASSOCIATION OF STATE HIGHWAY AND TRANSPORTATION OFFICIALS.

AAWDT ANNUAL AVERAGE WEEKDAY TRAFFIC.

AAWU ALASKAN AVIATION WEATHER UNIT.

AB 1. AHARONOVBOHM; 2. ASSEMBLY BILL.

ABC ASSOCIATION OF BOARDS OF CERTIFICATION.

ABNDT ABUNDANT.

ABS 1. ACRYLONITRILE-BUTADIENE-STYRENE; 2. ALKYL BENZENE SULFONATE.

ABT ABOUT.

ABV ABOVE.

AC 1. ACRE; 2. ALTERNATING CURRENT.

AC-FT 1. ACRE-FOOT; 2. ACRE-FEET.

ACCAS ALTOCUMULUS CASTELLANUS.

ACCUMS ACCUMULATION.

ACEC AREA OF CRITICAL ENVIRONMENTAL CONCERN.

ACER ASSISTANT COMMISSIONER - ENGINEERING AND RESEARCH.

ACES AUTOMATED COASTAL ENGINEERING SYSTEM.

ACGIH AMERICAN CONFERENCE OF GOVERNMENTAL INDUSTRIAL HYGIENIST.

ACH AIR CHANGES PER HOUR.

ACHP ADVISORY COUNCIL ON HISTORIC PRESERVATION.

ACLD ABOVE CLOUD LEVEL.

ACM ASBESTOS-CONTAINING MATERIAL.

ACN AIRCRAFT CLASSIFICATION NUMBER.

ACPY ACCOMPANY.

ACR 1. ALKALI-CARBONATE REACTION; 2. ACUTE-CHRONIC RATIO.

ACRE-FT 1. ACRE-FOOT; 2. ACRE-FEET.

ACRM ASSISTANT COMMISSIONER-RESOURCES MANAGEMENT.

ACRS ACROSS.

ACT AUTOMATED GUIDEWAY TRANSIT.

ACTS ACTION CORRESPONDENCE TRACKING SYSTEM.

ACWA ASSOCIATION OF CALIFORNIA WATER AGENCIES.

ACWM ASBESTOS-CONTAINING WASTE MATERIALS.

ACYC ANTICYCLONE.

AD ACTIVITY DESCRIPTION.

ADAS AUTOMATED DATA ACQUISITION SYSTEM.

ADDS AVIATION DIGITAL DATA SERVICE.

ADF 1. AUTOMATIC DIRECTION FINDER; 2. ACTIVE DARK FILAMENT.

ADI ACCEPTABLE DAILY INTAKE.

ADJ ADJACENT.

ADMS ACCESSIBILITY DATA MANAGEMENT SYSTEM.

ADP 1. AHERA DESIGNATED PERSON; 2. AUTOMATED DATA PROCESSING; 3. ACOUSTIC DOPPLER PROFILER; 4. AVERAGE WAVE PERIOD (SECONDS) OF ALL WAVES DURING THE 20-MINUTE PERIOD.

ADPC ACOUSTIC DOPPLER CURRENT PROFILER.

ADR ALTERNATIVE DISPUTE RESOLUTION.

ADSS ADVANCED DECISION SUPPORT SYSTEM.

ADT AVERAGE DAILY TRAFFIC.

ADVIS 1. ADVISORY PROGRAM; 2. ADVISORY.

AF 1. ACRE-FOOT; 2. ACRE-FEET.

AF/YR ACRE-FEET PER YEAR.

AFC AUTOMATIC FARE COLLECTION.

AFO ANIMAL FEEDING OPERATION.

AFOS AUTOMATION OF FIELD OPERATIONS AND SERVICES.

AFS ARCH FILAMENT SYSTEM.

AFTN AFTERNOON.

AFTR AFTER.

AFWA AIR FORCE WEATHER AGENCY.

AFY ACRE-FEET PER YEAR.

AFZ AUTO-FREE ZONE.

AG 1. AGRICULTURE; 2. ASSOCIATION OF GOVERNMENTS.

AGC 1. AUTOMATIC GENERATION CONTROL; 2. AUTOMATIC GAIN CONTROL.

AGFS AVIATION GRIDDED FORECAST SYSTEM.

AGL ABOVE GROUND LEVEL.

AGM ANNUAL GEOMETRIC MEAN.

AGN AGAIN.

AGP ALGAL GROWTH POTENTIAL.

AHD AHEAD.

AHERA ASBESTOS HAZARD EMERGENCY RESPONSE ACT.

AHOS AUTOMATIC HYDROLOGIC OBSERVING SYSTEM.

AHOS-S AUTOMATIC HYDROLOGIC OBSERVING SYSTEM - SATELLITE.

AHOS-T AUTOMATIC HYDROLOGIC OBSERVING SYSTEM - TELEPHONE.

AHS ARCHAEOLOGICAL AND HISTORICAL SERVICES.

AI ANNUAL INSPECTION.

AIDS ACQUIRED IMMUNE DEFICIENCY SYNDROME.

AIPC AMERICAN INDIAN PROGRAM COUNCIL.

AIRFA AMERICAN INDIAN RELIGIOUS FREEDOM ACT.

AIRMET AIRMAN'S METEOROLOGICAL ADVISORY.

AIS AUTOMATED INFORMATION SYSTEM.

AISC AMERICAN INSTITUTE OF STEEL CONSTRUCTION.

AIV AVIATION IMPACT VARIABLES.

ALC AGENCY LOCATION CODE.

ALERT AUTOMATED LOCAL EVALUATION IN REAL TIME.

ALP AIRPORT LAYOUT PLAN.

ALS APPROACH LIGHT SYSTEM.

ALUM ALUMINUM SULFATE.

AMA ADAPTIVE MANAGEMENT AREA.

AMCR ALTERNATIVE MANAGEMENT CONTROL REVIEW.

AMP 1. AMPERE; 2. ADAPTIVE MANAGEMENT PLAN.

AMR ACQUISITION MANAGEMENT REVIEW.

AMS AMERICAN METEOROLOGICAL SOCIETY.

AMVER AUTOMATED MUTUAL ASSISTANCE VESSEL RESCUE SYSTEM.

ANBURS ALPHANUMERIC BACKUP REPLACEMENT SYSTEM.

ANC ACID NEUTRALIZING CAPACITY.

ANSI AMERICAN NATIONAL STANDARDS INSTITUTE.

AO 1. ARCTIC OSCILLATION; 2. AREA OFFICE.

AOP 1. ANNUAL OPERATING PLAN; 2. AQUATIC ORGANISM PASSAGE.

AOR AREA OF REVIEW.

AP 1. ANALYSIS PLAN; 2. ANOMALOUS PROPAGATION.

APCD AIR POLLUTION CONTROL DISTRICT.

APE AREA OF POTENTIAL EFFECT.

API ANTECEDENT PRECIPITATION INDEX.

APR ACTIVE PROMINENCE REGION.

APRCH APPROACH.

APRCHG APPROACHING.

APRNT APPARENT.

APS ACCESSIBLE PEDESTRIAN SIGNAL.

APST AVIATION PRODUCTS AND SERVICES TEAM.

APTS ADVANCED PUBLIC TRANSPORTATION SYSTEMS.

AQIA AIR QUALITY IMPACT ANALYSIS.

AQMD AIR QUALITY MANAGEMENT DISTRICT.

AQMP AIR QUALITY MANAGEMENT PLAN.

AR ACTIVE REGION.

ARAM AVIATION, RANGE, AND AEROSPACE METEOROLOGY.

ARAR APPLICABLE OR RELEVANT AND APPROPRIATE REQUIREMENT.

ARARS APPLICABLE OR RELEVANT AND APPROPRIATE REQUIREMENTS.

ARMS AUTOMATED RECORDS MANAGEMENT SYSTEM.

ARP ADDRESS RESOLUTION PROTOCOL.

ARSR AIR ROUTE SURVEILLANCE RADAR.

ART AUTOMATIC RADIO-THEODOLITE.

ARTCC AIR ROUTE TRAFFIC CONTROL CENTER.

ARTS 1. ADVANCED RURAL TRANSPORTATION SYSTEM; 2. AUTOMATED RESOURCES TRAINING SYSTEM.

ARWRI AMERICAN RIVER WATER RESOURCES INVESTIGATION.

AS ASSISTANT SECRETARY.

ASAP 1. AHOS SHEF AUTOMATIC PROCESSING SYSTEM; 2. AS SOON AS POSSIBLE.

ASAPTRAN SOFTWARE COMPONENT OF ASAP.

ASC ADMINISTRATIVE SERVICE CENTER.

ASCE AMERICAN SOCIETY OF CIVIL ENGINEERS.

ASCII AMERICAN STANDARD CODE FOR INFORMATION INTERCHANGE.

ASCMP ADMINISTRATIVE SUPPORT CAREER MANAGEMENT PROGRAM.

ASCS AGRICULTURAL STABILIZATION AND CONSERVATION SERVICE.

ASL ABOVE SEA LEVEL.

ASOS AUTOMATED SURFACE OBSERVING SYSTEM.

ASOS IDS FOUR CHARACTER IDENTIFIER ASSIGNED TO EACH AUTOMATED SURFACE OBSERVING SYSTEM.

ASR 1. AIRPORT SURVEILLANCE RADAR; 2. ACTIVE SURGE REGION; 3. ALKALI-SILICA REACTION.

ASR-9 AIRPORT SURVEILLANCE RADAR.

ASSE AMERICAN SOCIETY OF SAFETY ENGINEERS.

ASSR ALKALI-SILICATE/SILICA REACTION.

ASTM AMERICAN SOCIETY FOR TESTING AND MATERIALS.

ASTS ABOVE GROUND STORAGE TANKS.

ATC 1. AIR TRAFFIC CONTROL; 2. AUTOMATIC TRAIN CONTROL.

ATCRBS AIR TRAFFIC CONTROL RADAR BEACON SYSTEM.

ATDTDCS AUTOMATED TONE DIAL TELEPHONE DATA COLLECTION SYSTEM.

ATIS 1. ADVANCED TRAVELER INFORMATION SYSTEM; 2. AUTOMATIC TERMINAL INFORMATION SERVICE.

ATM 1. ATMOSPHERIC; 2. ACTIVE TRAFFIC MANAGEMENT.

ATMP AIR TEMPERATURE.

ATMS ADVANCED TRAFFIC MANAGEMENT SYSTEM.

ATP ADENOSINE TRIPHOSPHATE.

ATTM AT THIS TIME.

ATU AEROBIC TREATMENT UNIT.

ATV ALL-TERRAIN VEHICLE.

AU 1. ANIMAL UNIT; 2. ASTRONOMICAL UNIT.

AUM ANIMAL UNIT MONTHS.

AUO ADMINISTRATIVELY UNCONTROLLABLE OVERTIME.

AVADS AUTOMATED VACANCY ANNOUNCEMENT DISTRIBUTION SYSTEM.

AVC AUTOMATIC VEHICLE CLASSIFICATION.

AVCSS ADVANCED VEHICLE CONTROL AND SAFETY SYSTEMS.

AVHRR ADVANCED VERY HIGH RESOLUTION RADIOMETER.

AVI AUTOMATIC VEHICLE IDENTIFICATION.

AVL AUTOMATIC VEHICLE LOCATION.

AVMT ANNUAL VEHICLE MILES TRAVELED.

AVN AVIATION MODEL.

AVO AVERAGE VEHICLE OCCUPANCY.

AVP AVERAGE WAVE PERIOD.

AWARDS AGRICULTURAL WATER RESOURCES DECISION SUPPORT.

AWHPS AREA WIDE HYDROLOGIC PREDICTION SYSTEM.

AWIPS ADVANCED WEATHER INTERACTIVE PROCESSING SYSTEM.

AWOL ABSENCE WITHOUT LEAVE.

AWSC ALL-WAY STOP-CONTROLLED.

AWTR ADVANCED WATER TREATMENT RESEARCH.

AWWA AMERICAN WATER WORKS ASSOCIATION.

A&E ARCHITECTURAL AND ENGINEERING.

B/C BENEFIT-COST RATIO.

BA BIOLOGICAL ASSESSMENT.

BACM BEST AVAILABLE CONTROL MEASURES.

BACT BEST AVAILABLE CONTROL TECHNOLOGY.

BADA BAY AREA DISCHARGERS ASSOCIATION.

BAF BIOACCUMULATION FACTOR.

BAM BINARY ANGULAR MEASURE.

BAS BUILDING AUTOMATION SYSTEM.

BAT 1. BEST AVAILABLE TECHNOLOGY ECONOMICALLY ACHIEVABLE; 2. BEST AVAILABLE TREATMENT TECHNOLOGIES.

BCF BIOCONCENTRATION FACTOR.

BCKG BACKING.

BCM BECOME.

BCMNG BECOMING.

BCT BEST CONVENTIONAL POLLUTANT CONTROL TECHNOLOGY.

BD BLOWING DUST.

BDAT BEST DEMONSTRATED AVAILABLE TECHNOLOGY.

BEA U.S. BUREAU OF ECONOMIC ANALYSIS.

BEST BUDGET ESTIMATE SYSTEM TOOLS.

BFR BEFORE.

BGN BEGIN.

BHND BEHIND.

BIA U.S. BUREAU OF INDIAN AFFAIRS.

BINOVC BREAKS IN OVERCAST.

BIP BUDGET INPUT PROGRAM.

BKN BROKEN.

BLD BUILD.

BLDUP BUILDUP.

BLM U.S. BUREAU OF LAND MANAGEMENT.

BLO BELOW.

BLV BEFORE.

BM BENCHMARK.

BMP BEST MANAGEMENT PRACTICE.

BMPS BEST MANAGEMENT PRACTICES.

BNDRY BOUNDARY.

BO BIOLOGICAL OPINION.

BOD 1. BIOCHEMICAL OXYGEN DEMAND; 2. BIOLOGICAL OXYGEN DEMAND.

BOM U.S. BUREAU OF MINES.

BON BASIS OF NEGOTIATION.

BOR U.S. BUREAU OF RECLAMATION.

BOVC BASE OF OVERCAST.

BPA BONNEVILLE POWER ADMINISTRATION.

BPJ BEST PROFESSIONAL JUDGMENT.

BPT BEST PRACTICABLE CONTROL TECHNOLOGY CURRENTLY AVAILABLE.

BR MIST.

BRAIN BUREAU OF RECLAMATION ACQUISITION INFORMATION NETWORK.

BRC BUDGET REVIEW COMMITTEE.

BRD USGS BIOLOGICAL RESOURCES DIVISION.

BRF BRIEF.

BRK BREAK.

BRM BUSINESS REPLY MAIL.

BRN BULK RICHARDSON NUMBER.

BRT BUS RAPID TRANSIT.

BS BLOWING SNOW.

BSC BASE SAFETY CONDITION.

BSD BRIGHT SURGE ON THE DISK.

BSL BRIGHT SURGE ON THE LIMB.

BSO BUSINESS SERVICES OFFICE.

BTR BETTER.

BTU BRITISH THERMAL UNITS.

BTWN BETWEEN.

BUDS BUSINESS UTILIZATION DEVELOPMENT SPECIALIST.

BUS BUSWORK.

BW BODY WEIGHT.

BWER BOUNDED WEAK ECHO REGION.

C 1. DRAG COEFFICIENT; 2. THERMAL CONDUCTANCE; 3. CENTIGRADE; 4. CELSIUS

C-D COLLECTOR-DISTRIBUTOR.

C. PERFRINGENS CLOSTRIDIUM PERFRINGENS.

CA CLOUD-TO-AIR LIGHTNING.

CAA COLD AIR ADVECTION.

CAD COLD AIR DAMMING.

CADAS CENTRALIZED AUTOMATED DATA ACQUISITION SYSTEM.

CAFO CONCENTRATED ANIMAL FEEDING OPERATION.

CAPE CONVECTIVE AVAILABLE POTENTIAL ENERGY.

CAPS CENTER FOR ANALYSIS AND PREDICTION OF STORMS.

CAS CHEMICAL ABSTRACT SERVICE.

CAT CLEAR AIR TURBULENCE.

CAVU 1. CEILING AND VISIBILITY UNLIMITED; 2. CLEAR OR SCATTERED CLOUDS (VISIBILITY GREATER THAN 10 MI).

CB CUMULONIMBUS.

CBA COST-BENEFIT ANALYSIS.

CBD CENTRAL BUSINESS DISTRICT.

CBI CONFIDENTIAL BUSINESS INFORMATION.

CBMAM CUMULONIMBUS MAMMA.

CC CLOUD-TO-CLOUD LIGHTNING.

CCC CRITERIA CONTINUOUS CONCENTRATION.

CCITT CONSULTATIVE COMMITTEE FOR INTERNATIONAL TELEPHONE AND TELEGRAPH.

CCL CONVECTIVE CONDENSATION LEVEL.

CCN CONTEMPLATED CHANGE NOTICE.

CCS CARBON CAPTURE AND STORAGE.

CCTV CLOSED-CIRCUIT TELEVISION.

CD 1. CADMIUM; 2. COLD.

CDB 1. CLIMATE DIAGNOSTICS BULLETIN; 2. COMPUTING DEVELOPMENT BRANCH.

CDC CLIMATE DIAGNOSTIC CENTER.

CDD COOLING DEGREE DAYS.

CDF CONFINED DISPOSAL FACILITY.

CDT CENTRAL DAYLIGHT TIME.

CE CONDITIONALLY EXEMPT GENERATORS.

CEM CIVIL EMERGENCY MESSAGE.

CERCLA COMPREHENSIVE ENVIRONMENTAL RESPONSE, COMPENSATION, AND LIABILITY ACT.

CERCLIS FEDERAL COMPREHENSIVE ENVIRONMENTAL RESPONSE, COMPENSATION, AND LIABILITY INFORMATION SYSTEM.

CFC CHLOROFLUOROCARBON.

CFC'S CHLOROFLUOROCARBONS.

CFCS CHLOROFLUOROCARBONS.

CFI COMPREHENSIVE FLARE INDEX.

CFL COMPACT FLUORESCENT LAMP.

CFM CUBIC FEET PER MINUTE.

CFP COLD FRONT PASSAGE.

CFR 1. CODE OF FEDERAL REGULATIONS; 2. COMPREHENSIVE FACILITY REVIEW.

CFS 1. CUBIC FEET PER SECOND; 2. CONTAINER FREIGHT STATION.

CFS-DAY CUBIC FEET PER SECOND-DAY.

CFSM CUBIC FEET PER SECOND PER SQUARE MILE.

CG CLOUD-TO-GROUND LIGHTNING.

CHC CHANCE.

CHEMTREC CHEMICAL TRANSPORTATION EMERGENCY CENTER.

CHG CHANGE.

CHGS CHANGES.

CHP COMBINED HEAT AND POWER.

CI 1. CIRRUS CLOUDS; 2. CURIE; 3. CONFORMAL INDEX.

CIA CRITICAL IMPACT ANGLE.

CIG CEILING.

CIN OR B- CONVECTIVE INHIBITION.

CIO CHIEF INFORMATION OFFICER.

CIP CRITICAL IMPACT POINT.

CIU CATEGORICAL INDUSTRIAL USER.

CLD CLOUD.

CLIMO CLIMATOLOGY/CLIMATOLOGICAL.

CLR CLEAR.

CLRG CLEARING.

CLSM CONTROLLED LOW STRENGTH MATERIAL.

CM 1. CENTIMETER; 2. COMBINED MOMENT.

CM^2 SQUARE CENTIMETER.

CMC CRITERIA MAXIMUM CONCENTRATION.

CMP 1. CENTRAL MERIDIAN PASSAGE; 2. CORRUGATED METAL PIPE.

CMPLT COMPLETE.

CMPLX COMPLEX.

CNG COMPRESSED NATURAL GAS.

CNIF CALIBRATION NETWORK INFORMATION FILES.

CNS CENTRAL NERVOUS SYSTEM.

CNTR CENTER.

CNTRL CENTRAL.

CNVG CONVERGE.

CNVTV CONVECTIVE.

COD CHEMICAL OXYGEN DEMAND.

COE U.S. ARMY CORPS OF ENGINEERS.

COH COEFFICIENT OF HAZE.

CONUS CONTINENTAL UNITED STATES.

COR CORRECTION.

CORE COMMONLY RECOGNIZED STRUCTURAL ELEMENTS.

CORPS OF ENGINEERS U.S. ARMY CORPS OF ENGINEERS.

CPC CLIMATE PREDICTION CENTER.

CPM CRITICAL PATH METHOD.

CPT CONE PENTEROMETER TESTING.

CPVC CHLORINATED POLYVINYL CHLORIDE.

CRCP CONTINUOUSLY REINFORCED CONCRETE PAVEMENT.

CRN CORONAL RAIN.

CRP 1. CHILD RESISTANT PACKAGING; 2. CRITICAL RAINFALL PROBABILITY.

CRS CONSOLE REPLACEMENT SYSTEM.

CSF CONFIDENTIAL STATEMENT OF FORMULA.

CSI CONDITIONAL SYMMETRIC INSTABILITY.

CSO COMBINED SEWER OVERFLOW.

CSS COMBINED SEWER SYSTEM.

CST CENTRAL STANDARD TIME.

CSTL COASTAL.

CTC CENTRALIZED TRAFFIC CONTROL.

CTG CONTROL TECHNIQUE GUIDELINES.

CTM CONTINUUM STORM.

CTR CENTERLINE TURNING RADIUS.

CTY CITY.

CU CUMULUS CLOUDS.

CUFRA CUMULUS FRACTUS.

CVO COMMERCIAL VEHICLE OPERATION.

CVR COVER.

CWA 1. CLEAN WATER ACT; 2. COUNTY WARNING AREA.

CWF COASTAL WATERS FORECAST.

CWFA COUNTY WARNING AND FORECAST AREA.

CWLM CUMULATIVE WORKING LEVEL MONTHS.

CWR CONTINUOUS WELDED RAIL.

CYC CYCLONE.

CYCLGN CYCLOGENESIS.

DALR DRY ADIABATIC LAPSE RATE.

DAMBRK DAM BREAK FORECASTING MODEL.

DAPM DATA ACQUISITION PROGRAM MANAGER.

DATACOL SOFTWARE SYSTEM THAT SUPPORTS RFC GATEWAY FUNCTIONS.

DATANET HYDROLOGIC DATA NETWORK ANALYSIS SOFTWARE.

DB DECIBEL.

DBE DESIGN BASIS EARTHQUAKE.

DC DIRECT CURRENT.

DCP DATA COLLECTION PLATFORM.

DDC DIRECT DIGITAL CONTROLS.

DDS DATA DISTRIBUTION SYSTEM.

DDT 1. DICHLORO-DIPHENYL-TRICHLOROETHANE; 2. DICHLOROCLIPHENYLTRI-CHLOROETHANE.

DEWP DEW POINT.

DFUS DIFFUSE.

DHRO DIVERSITY AND HUMAN RESOURCES OFFICE.

DI 1. DEIONIZE; 2. DURABILITY INDEX.

DIR DIRECTION.

DME DISTANCE MEASURING EQUIPMENT.

DMNT DOMINANT.

DMR DISCHARGE MONITORING REPORT.

DMSH DIMINISH.

DNAPL DENSE NON-AQUEOUS PHASE LIQUID.

DNR DEPARTMENT OF NATURAL RESOURCES.

DNS DENSE.

DNSTRM DOWNSTREAM.

DO DISSOLVED OXYGEN.

DOC 1. DEPARTMENT OF COMMERCE; 2. DESIGNERS" OPERATING CRITERIA.

DOH DEVELOPMENT AND OPERATIONS HYDROLOGIST.

DP 1. DEEP; 2. DEW POINT; 3. DIRECT PUSH.

DPD DIETHYL-PHENYLENE-DIAMINE.

DPNG DEEPENING.

DPTH DEPTH.

DPTR DEPARTURE.

DPVA DIFFERENTIAL POSITIVE VORTICITY ADVECTION.

DQOS DATA QUALITY OBJECTIVES.

DR DIRECTION.

DRCTN DIRECTION.

DRE DESTRUCTION AND REMOVAL EFFICIENCY.

DRFT DRIFT.

DSA SPECIAL TROPICAL DISTURBANCE STATEMENT.

DSD DARK SURGE ON DISK.

DSF DISAPPEARING SOLAR FILAMENT.

DSIPT DISSIPATE.

DTRT DETERIORATE.

DU DOBSON UNIT.

DURG DURING.

DURN DURATION.

DVA DEFICIENCY VERIFICATION ANALYSIS.

DVLP DEVELOP.

DVMT DAILY VEHICLE MILES TRAVELED.

DVV DOWNWARD VERTICAL VELOCITY.

DWNDFT DOWNDRAFT.

DWNSLP DOWNSLOPE.

DWOPER DYNAMIC WAVE ROUTING MODEL.

DWPT DEW POINT.

DWT DEADWEIGHT TONNAGE.

DZ DRIZZLE.

E 1. EAST; 2. ESTIMATED; 3. VOLTAGE.

E. COLI COLIFORM BACTERIUM ESCHERICHIA COIL.

EA ENVIRONMENTAL ASSESSMENT.

EAP EMERGENCY ACTION PLAN.

EBND EASTBOUND.

EBS EMERGENCY BROADCAST SYSTEM.

EC EFFECTIVE CONCENTRATION.

ECMF EUROPEAN CENTER FOR METEOROLOGY FORECAST MODEL.

ECMWF EUROPEAN CENTER FOR MEDIUM-RANGE WEATHER FORECASTS.

EDB ETHYLENE DIBROMIDE.

EDD ENFORCEMENT DECISION DOCUMENT.

EDT EASTERN DAYLIGHT TIME.

EEO EQUAL EMPLOYMENT OPPORTUNITY.

EFCT EFFECT.

EFR EMERGING FLUX REGION.

EGL ENERGY GRADE LINE.

EHS EXTREMELY HAZARDOUS SUBSTANCE.

EIF ENHANCED IFLOWS FORMAT.

EIS ENVIRONMENTAL IMPACT STATEMENT.

EL EQUILIBRIUM LEVEL.

ELEV ELEVATION.

ELF EXTREMELY LOW FREQUENCY.

ELG EFFLUENT LIMITATIONS GUIDELINES.

ELSW ELSEWHERE.

ELT EMERGENCY LOCATOR TRANSMITTER.

ELY EASTERLY.

EMBDD EMBEDDED.

EMC ENVIRONMENTAL MODELING CENTER.

EMF ELECTROMOTIVE FORCE.

EML ELEVATED MIXED LAYER.

EMU ELECTROMAGNETIC UNIT.

ENDG ENDING.

ENHNCD ENHANCED.

ENSO EL NIÑO-SOUTHERN OSCILLATION.

ENTR ENTIRE.

EOC EMERGENCY OPERATIONS CENTER.

EOP EMERGENCY OPERATIONS PLAN.

EOX EXTRACTABLE ORGANIC HALIDES.

EP TOXIC EXTRACTION PROCEDURE.

EPA ENVIRONMENTAL PROTECTION AGENCY.

EPACT ENERGY POLICY ACT.

EPCTG EXPECTING.

EPL ERUPTIVE PROMINENCE ON LIMB.

EPM EQUIVALENT PER MILLION.

EPP EMERGENCY PREPAREDNESS PLAN.

EPV EQUIVALENT POTENTIAL VORTICITY.

ERLY EARLY.

ERN EASTERN.

ESA ENDANGERED SPECIES ACT.

ESAL EQUIVALENT SINGLE-AXLE LOAD.

ESP 1. ELECTROSTATIC PRECIPITATOR; 2. EXTENDED STREAMFLOW PREDICTION.

ESPINIT ESP INITIALIZATION PROGRAM.

EST EASTERN STANDARD TIME.

ET EVAPOTRANSPIRATION.

ETA ESTIMATED TIME OF ARRIVAL.

EUV EXTREME ULTRAVIOLET.

EVE EVENING.

EWD EASTWARD.

EWI EQUAL-WIDTH INCREMENT.

EWS EARLY WARNING SYSTEM.

EWW EXTREME WIND WARNING.

EXCLD EXCLUDE.

EXTD EXTEND/EXTENDED.

EXTRM EXTREME.

EXTSV EXTENSIVE.

E_H REDOX POTENTIAL.

F 1. FOG; 2. FUJITA; 3. NATURAL FREQUENCY; 4. FAHRENHEIT.

F SCALE FUJITA SCALE.

FA FORECAST AREA.

FAA FEDERAL AVIATION ADMINISTRATION.

FASTST FASTEST.

FAV FINAL ACUTE VALUE.

FAWS FLIGHT ADVISORY WEATHER SERVICE.

FBO GREAT LAKES FREEZE-UP/BREAK-UP OUTLOOK.

FCM FRACTURE-CRITICAL MEMBER.

FCS FLUOROCARBONS.

FCST 1. FORECAST; 2. NWSRFS FORECAST PROGRAM TO PRODUCE OPERATIONAL FORECASTS.

FDF FUNDAMENTALLY DIFFERENT FACTORS.

FDM FINITE DIFFERENCE METHOD.

FEM FINITE ELEMENT METHOD.

FEMA FEDERAL EMERGENCY MANAGEMENT AGENCY.

FERC FEDERAL ENERGY REGULATORY COMMISSION.

FFA FLASH FLOOD WATCH.

FFG FLASH FLOOD GUIDANCE.

FFS FLASH FLOOD STATEMENT.

FFW FLASH FLOOD WARNING.

FG FOG.

FID FLAME IONIZATION DETECTOR.

FIFO FIRST IN, FIRST OUT.

FIFRA FEDERAL INSECTICIDE, FUNGICIDE AND RODENTICIDE ACT.

FLG FALLING.

FLRY FLURRY.

FLS FLOOD STATEMENT.

FLW 1. FLOOD WARNING; 2. FOLLOW; 3. FLOW-WIND.

FM 1. FROM; 2. FATHOM; 3. FINENESS MODULUS.

FNSI FINDING OF NO SIGNIFICANT IMPACT.

FNT FRONT.

FNTGNS FRONTOGENESIS.

FNTLYS FRONTOLYSIS.

FONSI FINDING OF NO SIGNIFICANT IMPACT.

FOPS FALLING OBJECT PROTECTION STRUCTURE.

FOUS FORECAST OUTPUT UNITED STATES.

FPO FLOOD POTENTIAL OUTLOOK.

FPS FUJITA-PEARSON SCALE.

FQT FREQUENT.

FRA FEDERAL RAILROAD ADMINISTRATION.

FRC FIBER-REINFORCED COMPOSITE.

FRMG FORMING.

FROPA FRONTAL PASSAGE.

FROSFC FRONTAL SURFACE.

FRST FROST.

FRZ FREEZE.

FRZN FROZEN.

FS 1. U.S. FOREST SERVICE; 2. FEASIBILITY STUDY; 3. FACTOR OF SAFETY.

FSCBG SPECIFIC AERIAL SPRAY DISPERSION MODEL.

FSS FLIGHT SERVICE STATION.

FT 1. FEET; 2. FOOT.

FT^2 SQUARE FEET.

FT^3 CUBIC FEET.

FTP FILE TRANSFER PROTOCOL.

FVT FORECAST VERIFICATION TOOL.

FWD FORWARD.

FWS FISH & WILDLIFE SERVICE.

FZRA FREEZING RAIN.

G/KG GRAMS PER KILOGRAM.

GAL GALLON.

GC GAS CHROMATOGRAPHY.

GCMS GENERAL CIRCULATION MODELS.

GEMPAK GENERAL ENVIRONMENTAL METEOROLOGICAL PACKAGE.

GEN GENERAL.

GEOSEQUESTRATION GEOLOGIC SEQUESTRATION.

GF GROUND FOG.

GFCI GROUND-FAULT CIRCUIT INTERRUPTER.

GFS GLOBAL FORECAST SYSTEM.

GH GAUGE HEIGHT.

GIS GEOGRAPHIC INFORMATION SYSTEM.

GLF OPEN LAKES FORECAST.

GLS GREAT LAKES STORM SUMMARY.

GMDSS GLOBAL MARITIME DISTRESS AND SAFETY SYSTEM.

GMT 1. GREENWICH MEAN TIME; 2. UNIVERSAL COORDINATED TIME.

GND GROUND.

GNRL GENERAL.

GOES GEOSTATIONARY ORBITING ENVIRONMENTAL SATELLITE.

GOES DCS GEOSTATIONARY ORBITING ENVIRONMENTAL SATELLITE DATA COLLECTION SYSTEM.

GOMOOS GULF OF MAINE OCEAN OBSERVING SYSTEM.

GPS GLOBAL POSITIONING SYSTEMS.

GRAS GENERALLY RECOGNIZED AS SAFE.

GRDL GRADUAL.

GRT 1. GROSS REGISTERED TONNAGE; 2. GREAT.

GRTST GREATEST.

GSHP GROUND SOURCE HEAT PUMP.

GSTY GUSTY.

GTR GREATER.

GW GIGAWATT.

GWH GIGAWATT-HOUR.

H DIVERSITY INDEX.

H_s SIGNIFICANT WAVE HEIGHT.

HADS HYDROMETEOROLOGICAL AUTOMATED DATA SYSTEM.

HAR HIGHWAY ADVISORY RADIO.

HAUNCHES HAUNCH.

HBI HILSENHOFF"S BIOTIC INDEX.

HC HYDROCARBONS.

HCFCS HYDROCHLOROFLUOROCARBONS.

HDD HEATING DEGREE DAYS.

HEPA HIGH-EFFICIENCY PARTICULATE AIR.

HF HIGH FREQUENCY.

HFCS HYDROFLUOROCARBONS.

HG MERCURY.

HGL HYDRAULIC GRADE LINE.

HI 1. HEAT INDEX; 2. HIGH.

HIC 1. HYDROLOGIST IN CHARGE; 2. HEAD INJURY CRITERIA.

HLRW HIGH-LEVEL RADIOACTIVE WASTE.

HLS HURRICANE LOCAL STATEMENT.

HMA HOT MIX ASPHALT.

HMD HEMISPHERIC MAP DISCUSSION.

HMR HYDROMETEOROLOGICAL REPORT.

HMRT HAZARDOUS MATERIALS RESPONSE TEAM.

HOD HYDROLOGIST ON DUTY.

HOT HIGH-OCCUPANCY TOLL.

HOTL HIGH-OCCUPANCY TOLL LANE.

HOV HIGH-OCCUPANCY VEHICLE.

HP HORSEPOWER.

HP STORM HIGH-PRECIPITATION STORM.

HP SUPERCELL HIGH-PRECIPITATION SUPERCELL.

HPA HECTOPASCAL.

HPC HYDROMETEOROLOGICAL PREDICTION CENTER.

HPD HOURLY PRECIPITATION DATA.

HR HOUR.

HRL HYDROLOGICAL RESEARCH LABORATORY AT THE OFFICE OF HYDROLOGY.

HRS 1. HOURS; 2. HAZARD RANKING SYSTEM.

HSA HYDROLOGIC SERVICE AREA.

HSB HYDROLOGIC SYSTEMS BRANCH IN THE OFFICE OF HYDROLOGY.

HSF HIGH SEAS FORECAST.

HTC HYDROMETEOROLOGICAL TRAINING COUNCIL.

HV HAVE.

HVAC HEATING, VENTILATION, AND AIR CONDITIONING.

HVY HEAVY.

HWMC HAZARDOUS WASTE MANAGEMENT COORDINATOR.

HWRF HURRICANE WEATHER RESEARCH AND FORECASTING MODEL.

HWVR HOWEVER.

HYDROUNIT HYDROELECTRIC UNIT.

HZ 1. HAZARD; 2. HAZE; 3. HERTZ.

I IMPORTANCE FACTOR.

I.D. IDENTIFICATION.

I/M INSPECTION AND MAINTENANCE.

IAP INSTRUMENT APPROACH PROCEDURE.

IBI INDEX OF BIOTIC INTEGRITY.

IC 1. IN-CLOUD LIGHTNING; 2. INHIBITION CONCENTRATION.

ICR INFORMATION COLLECTION REQUEST.

ICS INCIDENT COMMAND SYSTEM.

IDF INFLOW DESIGN FLOOD.

IDLH IMMEDIATELY DANGEROUS TO LIFE OR HEALTH.

IFLOWS INTEGRATED FLOOD OBSERVING AND WARNING SYSTEM.

IFR INSTRUMENT FLIGHT RULES.

ILEV INHERENTLY LOW EMISSION VEHICLE.

ILS INSTRUMENT LANDING SYSTEM.

IMF INTERPLANETARY MAGNETIC FIELD.

IMPL IMPULSE.

IMPT IMPORTANT.

IN HG INCHES OF MERCURY.

IN^2 SQUARE INCH.

IN^3 CUBIC INCHES.

INCR INCREASE.

INDC INDICATE.

INLD INLAND.

INTERNATIONAL SYSTEM OF UNITS SI SYSTEM.

INTS INTENSE.

INTSFY INTENSIFY.

INTST INTENSIFY.

IOU INVESTOR OWNED UTILITY.

IP ICE PELLETS.

IPM INTEGRATED PEST MANAGEMENT.

IPP INDEPENDENT POWER PRODUCER.

IPV IMPROVE.

IR INFRARED.

IRI INTERNATIONAL ROUGHNESS INDEX.

IRIS INTEGRATED RISK INFORMATION SYSTEM.

IRP INTEGRATED RESOURCE PLANNING.

IS IMPACT SEVERITY.

ISOLD ISOLATED.

ITCZ INTERTROPICAL CONVERGENCE ZONE.

ITS INTELLIGENT TRANSPORTATION SYSTEM.

I&M ENHANCED INSPECTION AND MAINTENANCE.

J/KG JOULES PER KILOGRAM.

JCP JOINTED CONCRETE PAVEMENT.

JHA JOB HAZARD ANALYSIS.

JMF JOB-MIX FORMULA.

JRCP JOINTED REINFORCED CONCRETE PAVEMENT.

JSTR JET STREAM.

JTWC JOINT TYPHOON WARNING CENTER.

K SMOKE.

K AMS COLD AIR MASS.

KG KILOGRAM.

KM^2 SQUARE KILOMETERS.

KT KNOT.

KTS KNOTS.

KV KILOVOLT.

KVA KILOVOLT-AMPERE.

KW KILOWATT.

KW-HR KILOWATT-HOUR.

KWH KILOWATT-HOUR.

L LITER.

LA LOAD ALLOCATIONS.

LALS LIGHTNING ACTIVITY LEVELS.

LARC LIMITED AUTOMATIC REMOTE COLLECTOR.

LAS LINEAR ALKYLATE SULFONATE.

LASH LIGHTER ABOARD SHIP.

LAT LATITUDE.

LAWEB GREAT LAKES WEATHER BROADCAST.

LC_{50} LETHAL CONCENTRATION 50.

LCD LOCAL CLIMATOLOGICAL DATA.

LCL 1. LIFTING CONDENSATION LEVEL; 2. LESS THAN CARLOAD; 3. LESS THAN CONTAINER LOAD.

LD_{50} LETHAL DOSE 50.

LDLO LETHAL DOSE LOW.

LDS LIGHTNING DETECTION SYSTEM.

LDT LOCAL DAYLIGHT TIME.

LEA LOCAL EDUCATION AGENCY.

LEL LOWER EXPLOSIVE LIMIT.

LEOP LOCAL EMERGENCY OPERATIONS PLAN.

LEPC LOCAL EMERGENCY PLANNING COMMITTEE.

LEWP LINE ECHO WAVE PATTERN.

LFC LEVEL OF FREE CONVECTION.

LFT LIFT.

LFWS LOCAL FLOOD WARNING SYSTEM.

LGT LIGHT.

LGWV LONG WAVE.

LI 1. LANGELIER INDEX; 2. LIFTED INDEX.

LID LOW IMPACT DEVELOPMENT.

LIFO LAST IN, FIRST OUT.

LIFR LOW INSTRUMENT FLIGHT RULES.

LIVV LIFTED INDEX VERTICAL VELOCITY.

LKLY LIKELY.

LL LIQUID LIMIT.

LLJ LOW LEVEL JET.

LLRW LOW-LEVEL RADIOACTIVE WASTE.

LLWS LOW LEVEL WIND SHEAR.

LM LUMEN.

LMTD LIMITED.

LN LINE.

LNAPL LIGHT NON-AQUEOUS PHASE LIQUID.

LOA 1. LETTER OF AGREEMENT; 2. LENGTH OVERALL.

LOAEL LOWEST OBSERVED ADVERSE EFFECT LEVEL.

LOC LEVEL OF CONCERN.

LOD LIMIT OF DETECTION.

LOL LOSS OF LIFE.

LON LENGTH-OF NEED.

LOP LEVEL OF PROTECTION.

LOPE LOW DENSITY POLYETHYLENE.

LOPRES LOW PRESSURE.

LORAN LONG RANGE NAVIGATION.

LOS LEVEL OF SERVICE.

LOS SCORE LEVEL OF SERVICE SCORE.

LOW-E LOW EMISSIVITY.

LP STORM LOW-PRECIPITATION STORM.

LPS LOOP PROMINENCE SYSTEM.

LRG LARGE.

LRL LABORATORY REPORTING LEVEL.

LRV LIGHT RAIL VEHICLE.

LSD LAND-SURFACE DATUM.

LSR LOCAL STORM REPORT.

LST LOCAL STANDARD TIME.

LT-MDL LONG-TERM METHOD DETECTION LEVEL.

LTD LIMITED.

LWFS LOCAL FLOOD WARNING SYSTEM.

LWL LENGTH ON THE WATERLINE.

M 1. METER; 2. MILLION.

M^2 SQUARE METER.

M^3 CUBIC METERS.

M&C MATERIALS AND CONSTRUCTION.

MAC MAXIMUM ACCEPTABLE CONCENTRATION.

MACT MAXIMUM AVAILABLE CONTROL TECHNOLOGY.

MAF MILLION ACRE-FEET.

MAFOR GREAT LAKES MARINE FORECAST.

MALR MOIST ADIABATIC LAPSE RATE.

MAP 1. MAXIMUM ACCEPTABLE PRESSURE; 2. MEAN AREAL PRECIPITATION.

MAREP MARINE REPORT.

MAX MAXIMUM.

MAXT HIGH TEMPERATURE FOR THE DAY.

MB MILLIBAR.

MBAS METHYLENE BLUE ACTIVE SUBSTANCES.

MC MOISTURE CONTENT.

MCC MESOSCALE CONVECTIVE COMPLEX.

MCE MAXIMUM CREDIBLE EARTHQUAKE.

MCG MICROGRAM.

MCG/G MICROGRAMS PER GRAM.

MCG/KG MICROGRAMS PER KILOGRAM.

MCL MAXIMUM CONTAMINANT LEVEL.

MCLG MAXIMUM CONTAMINANT LEVEL GOAL.

MCS MESOSCALE CONVECTIVE SYSTEM.

MCV MESOSCALE CYCLONIC VORTICES.

MD MESOSCALE DISCUSSION.

MDA MODIFICATION DECISION ANALYSIS.

MDE MAXIMUM DESIGN EARTHQUAKE.

MDL METHOD DETECTION LIMIT.

MDSS MAINTENANCE DECISION SUPPORT SYSTEM.

MDT 1. MODERATE; 2. MOUNTAIN DAYLIGHT TIME.

MEGG MERGING.

MEI MAXIMALLY EXPOSED INDIVIDUAL.

MESO MESOCYCLONE.

MESOLOW MESOSCALE LOW-PRESSURE CENTER.

METRO METROPOLITAN.

MEV 1. MEGAELECTRON VOLT; 2. MILLION ELECTRON VOLTS.

MF MEDIUM FREQUENCY.

MG MILLIGRAM.

MG/L MILLIGRAMS PER LITER.

MGD MILLION GALLONS PER DAY.

MI MILE.

MI^2 SQUARE MILE.

MIC METEOROLOGIST IN CHARGE.

MID MIDDLE.

MIN 1. MINIMUM; 2. MINUTE.

MINT MINIMUM TEMPERATURE.

MIS MANAGEMENT INFORMATION SYSTEM.

MISC MISCELLANEOUS.

MISG MISSING.

MJO MADDEN-JULIAN OSCILLATION.

ML MINIMUM LEVEL.

MLCAPE MEAN LAYER CAPE.

MLLI MEAN LAYER LIFTED INDEX.

MLLW MEAN LOWER LOW WATER.

MLS MICROWAVE LANDING SYSTEM.

MLW MEAN LOW WATER.

MO MOMENT.

MOA MILITARY OPERATIONS AREA.

MOBY MARINE OPTICAL BUOY.

MOE MARGIN OF EXPOSURE.

MOR MODULUS OF RUPTURE.

MOS MODEL OUTPUT STATISTICS.

MOU MEMORANDUM OF UNDERSTANDING.

MOV MOVE.

MOVG MOVING.

MP 1. MARITIME POLAR; 2. MEASURING POINT.

MPL MAX PARCEL LEVEL.

MPN MOST PROBABLE NUMBER.

MPO METROPOLITAN PLANNING ORGANIZATION.

MPR MEAN PANEL RATING.

MR MORE.

MRF MATERIALS RECOVERY FACILITY.

MRGL MARGINAL.

MRI MEAN ROUGHNESS INDEX.

MRL MINIMUM REPORTING LEVEL.

MRNG MORNING.

MS/CM MICROSIEMENS PER CENTIMETER.

MSA METROPOLITAN STATISTICAL AREA.

MSDS MATERIAL SAFETY DATA SHEET.

MSG MESSAGE.

MSGP MULTI-SECTOR GENERAL PERMIT.

MSL MEAN SEA LEVEL.

MSLP MEAN SEA LEVEL PRESSURE.

MST MOUNTAIN STANDARD TIME.

MSTR MOISTURE.

MT 1. MARITIME TROPICAL; 2. MOUNTAIN.

MTC MANUAL TRAFFIC CONTROL.

MTN MOUNTAIN.

MTNS MOUNTAINS.

MTS MOUNTAINS.

MULI MOST UNSTABLE LIFTED INDEX.

MVFR MARGINAL VISUAL FLIGHT RULES.

MVS MOVES.

MW MEGAWATT.

MWD MEAN WAVE DIRECTION.

MWH MEGAWATT-HOUR.

MWS MARINE WEATHER STATEMENT.

N 1. MANNING'S ROUGHNESS COEFFICIENT; 2. NORTH.

NAAQS NATIONAL AMBIENT AIR QUALITY STANDARDS.

NAD 27 NORTH AMERICAN DATUM OF 1927.

NAM NORTH AMERICAN MESO.

NAO NORTH ATLANTIC OSCILLATION.

NAPL NON-AQUEOUS PHASE LIQUID.

NAV NAVTEX FORECAST.

NAVAID NAVIGATION AID.

NAVD 88 NORTH AMERICAN VERTICAL DATUM OF 1988.

NAWAS NATIONAL WARNING SYSTEM.

NAWQA NATIONAL WATER QUALITY ASSESSMENT.

NBAR NON-BINDING ALLOCATIONS OF RESPONSIBILITY.

NC NO CHANGE.

NCAR NATIONAL CENTER FOR ATMOSPHERIC RESEARCH.

NCCF NOAA CENTRAL COMPUTER FACILITY.

NCDC NATIONAL CLIMATIC DATA CENTER.

NCEP NATIONAL CENTERS FOR ENVIRONMENTAL PREDICTION.

NDB NON-DIRECTIONAL BEACON.

NDBC NATIONAL DATA BUOY CENTER.

NDFD NATIONAL DIGITAL FORECAST DATABASE.

NE NORTHEAST.

NEC NECESSARY.

NECP SUIT NON-ENCAPSULATING CHEMICAL PROTECTIVE SUIT.

NEG NEGATIVE.

NELY NORTHEASTERLY.

NEPA NATIONAL ENVIRONMENTAL POLICY ACT.

NERC NORTH AMERICAN ELECTRIC RELIABILITY COUNCIL.

NERN NORTHEASTERN.

NESDIS NATIONAL ENVIRONMENTAL SATELLITE, DATA, AND INFORMATION SERVICE.

NESHAPS NATIONAL EMISSIONS STANDARDS FOR HAZARDOUS AIR POLLUTANTS.

NEWD NORTHEASTWARD.

NEXRAD NEXT GENERATION RADAR.

NG/L NANOGRAMS PER LITER.

NGT NIGHT.

NGVD 29 NATIONAL GEODETIC VERTICAL DATUM OF 1929.

NGVD NATIONAL GEODETIC VERTICAL DATUM.

NHC NATIONAL HURRICANE CENTER.

NHOP NATIONAL HURRICANE OPERATIONS PLAN.

NIFC NATIONAL INTER-AGENCY FIRE CENTER.

NIL NONE.

NIOSH NATIONAL INSTITUTE FOR OCCUPATIONAL SAFETY AND HEALTH.

NLY NORTHERLY.

NM NAUTICAL MILE.

NMBR NUMBER.

NMC NATIONAL METEOROLOGICAL CENTER.

NMHC NON-METHANE HYDROCARBON.

NML NORMAL.

NMOG NON-METHANE ORGANIC GASES.

NMRS NUMEROUS.

NO NITRIC OXIDE.

NOAA NATIONAL OCEANIC AND ATMOSPHERIC ADMINISTRATION.

NOAEL NO OBSERVABLE ADVERSE EFFECT LEVEL.

NOEC NO-OBSERVED-EFFECT-CONCENTRATION.

NOEL NO-OBSERVED-EFFECT-LEVEL.

NOGAPS NAVY OPERATIONAL GLOBAL ATMOSPHERIC PREDICTION SYSTEM.

NOHRSC NATIONAL OPERATIONAL HYDROLOGIC REMOTE SENSING CENTER.

NOHSCP NATIONAL OIL AND HAZARDOUS SUBSTANCES CONTINGENCY PLAN.

NOTAM NOTICE TO AIRMEN.

NPDES NATIONAL POLLUTANT DISCHARGE ELIMINATION SYSTEM.

NPL NATIONAL PRIORITIES LIST.

NPS 1. NATIONAL PARK SERVICE; 2. NON-POINT SOURCE.

NR NEAR.

NRC NATIONAL RESPONSE CENTER.

NRCS NATIONAL RESOURCES CONSERVATION SERVICES.

NRI NATIONWIDE RIVERS INVENTORY.

NRN NORTHERN.

NRT 1. NET REGISTERED TONNAGE; 2. NATIONAL RESPONSE TEAM.

NRW NARROW.

NS NIMBOSTRATUS.

NSDWR NATIONAL SECONDARY DRINKING WATER REGULATIONS.

NSH NEARSHORE FORECAST.

NSOM NATIONAL WEATHER SERVICE OPERATIONS MANUAL.

NSPS NEW SOURCE PERFORMANCE STANDARDS.

NSR NEW SOURCE REVIEW.

NSSFC NATIONAL SEVERE STORM FORECAST CENTER.

NSSL NATIONAL SEVERE STORMS LABORATORY.

NT NANOTESLA.

NTA NITRILOTRIACETIC ACID.

NTFO NEXT IN, FIRST OUT.

NTU NEPHELOMETRIC TURBIDITY UNIT.

NUG"S NON-UTILITY GENERATORS.

NVA NEGATIVE VORTICITY ADVECTION.

NW NORTHWEST.

NWD NORTHWARD.

NWIS NATIONAL WATER INFORMATION SYSTEM.

NWLY NORTHWESTERLY.

NWP NUMERICAL WEATHER PREDICTION.

NWRD NORTHWESTWARD.

NWRN NORTHWESTERN.

NWS NATIONAL WEATHER SERVICE.

NWSH NATIONAL WEATHER SERVICE HEADQUARTERS.

NWSO NATIONAL WEATHER SERVICE OFFICE.

NWSRFS V5.0 NATIONAL WEATHER SERVICE RIVER FORECAST MODEL VERSION 5.

NWWS NATIONAL WEATHER WIRE SERVICE.

NXT NEXT.

OASIS OPEN ACCESS SAME-TIME INFORMATION SYSTEM.

OBE OPERATING BASIS EARTHQUAKE.

OBS OBSERVATION.

OBSC OBSCURE.

OC 1. ORGANIC CARBON; 2. OPERATING CHARACTERISTICS.

OCED ORGANIZATION OF ECONOMIC AND COOPERATIVE DEVELOPMENT.

OCFNT OCCLUDED FRONT.

OCNL OCCASIONAL.

OCR OVER-CONSOLIDATION RATIO.

OEODM OFFICE OF EQUAL OPPORTUNITY AND DIVERSITY MANAGEMENT.

OFF OFFSHORE WATERS FORECAST.

OFSHR OFFSHORE.

OGP OFFICE OF GLOBAL PROGRAMS.

OH OFFICE OF HYDROLOGY.

OHD OVERHEAD.

OLR OUTGOING LONGWAVE RADIATION.

OML OPERATIONS MANUAL LETTER.

OPC OCEAN PREDICTION CENTER.

OPP OFFICE OF PESTICIDE PROGRAMS.

ORIG ORIGINAL.

OSC ON-SCENE COORDINATOR.

OSHA 1. OCCUPATIONAL SAFETY AND HEALTH ADMINISTRATION; 2. OCCUPATIONAL SAFETY AND HEALTH ACT.

OS&Y OUTSIDE SCREW-AND-YOKE.

OTLK OUTLOOK.

OTR OTHER.

OTRW OTHERWISE.

OVC OVERCAST.

OVNGT OVERNIGHT.

OVR OVER.

OVRN OVERRUN.

OVRNGT OVERNIGHT.

O&M OPERATIONS AND MAINTENANCE.

P.T. PIPE THREAD.

PA PASCAL.

PAC PACIFIC.

PAH POLYCYCLIC AROMATIC HYDROCARBON.

PAR 1. POPULATION AT RISK; 2. PRECISION APPROACH RADAR.

PAT PATTERN.

PB 1. LEAD; 2. POLYBUTYLENE.

PBL 1. PLANETARY BOUNDARY LAYER; 2. PROBABLE.

PC PICOCURIE.

PCA POLAR CAP ABSORPTION.

PCBS POLYCHLORINATED BIPHENYLS.

PCI 1. PICOCURIE; 2. PAVEMENT CLASSIFICATION INDEX.

PCNS POLYCHLORINATED NAPHTHALENES.

PCPN PRECIPITATION.

PCT PERCENT.

PD 1. PERIOD; 2. PERCENT DEFECTIVE.

PDE PARTIAL DIFFERENTIAL EQUATION.

PDI PALMER DROUGHT INDEX.

PDM PRECEDENCE DIAGRAMMING METHOD.

PDMT PREDOMINANT.

PDO PACIFIC DECADAL OSCILLATION.

PDS PARTICULARLY DANGEROUS SITUATION.

PDSI PALMER DROUGHT SEVERITY INDEX.

PDT PACIFIC DAYLIGHT TIME.

PE POLYETHYLENE.

PEIS PROGRAMMATIC ENVIRONMENTAL IMPACT STATEMENT.

PEL PERMISSIBLE EXPOSURE LIMIT.

PERT PROJECT EVALUATION AND REVIEW TECHNIQUE.

PETE POLYETHYLENE TEREPTHALATE.

PETN PENTAERYTHRITOL TETRANITRATE.

PFCS PERFLUOROCARBONS.

PFR PERIODIC FACILITY REVIEW.

PGA PEAK GROUND ACCELERATION.

PHF PEAK HOUR FACTOR.

PHFO HONOLULU NATIONAL WEATHER SERVICE FORECAST OFFICE.

PHWC PROJECT HAZARDOUS WASTE COORDINATOR.

PI PLASTICITY INDEX.

PIBAL PILOT BALLOON.

PICS PRODUCTS OF INCOMPLETE COMBUSTION.

PIREP PILOT REPORT.

PIX PICTURE.

PK PEAK.

PL PLASTIC LIMIT.

PM PARTICULATE MATTER.

PMD PROGNOSTIC DISCUSSION.

PMF PROBABLE MAXIMUM FLOOD.

PMI PROJECT MANAGEMENT INSTITUTE.

PMO PORT METEOROLOGICAL OFFICER.

PMP PROBABLE MAXIMUM PRECIPITATION.

PMR PROPORTIONATE MORTALITY RATIO.

PMS PAVEMENT MANAGEMENT SYSTEM.

PNA PACIFIC NORTH AMERICAN TELECONNECTION.

PNHDL PANHANDLE.

POH PROBABILITY OF HAIL.

POHCS PRINCIPAL ORGANIC HAZARDOUS CONSTITUENTS.

POP PROBABILITY OF PRECIPITATION.

POS POSITIVE.

POTWS PUBLICLY OWNED TREATMENT WORKS.

PP POLYPROPYLENE.

PPB PARTS PER BILLION.

PPI PLAN POSITION INDICATOR.

PPINE PLAN POSITION INDICATES NO ECHOES.

PPM PARTS PER MILLION.

PQL PRACTICAL QUANTIFICATION LIMIT.

PQPF PROBABILISTIC QUANTITATIVE PRECIPITATION FORECAST.

PRCP PRECIPITATION.

PRECD PRECEDE.

PRES PRESSURE.

PRF PULSE REPETITION FREQUENCY.

PRIM PRIMARY.

PRIN PRINCIPAL.

PROPA PROPAGATION.

PRP POTENTIALLY RESPONSIBLE PARTY.

PRST PERSIST.

PRT 1. PERSONAL RAPID TRANSIT; 2. PULSE REPETITION TIME.

PRVD PROVIDE.

PSBL POSSIBLE.

PSBLY POSSIBLY.

PSG PASSAGE.

PSI 1. POUNDS PER SQUARE INCH; 2. POLLUTANT STANDARD INDEX; 3. PRESENT SERVICEABILITY INDEX.

PSIA POUNDS PER SQUARE INCH, ABSOLUTE.

PSIG POUNDS PER SQUARE INCH, GAUGE.

PSR PRESENT SERVICEABILITY RATING.

PST PACIFIC STANDARD TIME.

PTCHY PATCHY.

PTCLDY PARTLY CLOUDY.

PTDY PRESSURE TENDENCY.

PTLY PARTLY.

PTN PORTION.

PTTN PATTERN.

PTWC PACIFIC TSUNAMI WARNING CENTER.

PTYPE PRECIPITATION TYPE.

PUHCA PUBLIC UTILITY HOLDING COMPANY ACT.

PURPA PUBLIC UTILITY REGULATORY POLICIES ACT.

PV PHOTOVOLTAIC.

PVA POSITIVE VORTICITY ADVECTION.

PVC POLYVINYL CHLORIDE.

PVL PREVAIL.

PWL PERCENT WITHIN LIMITS.

PWO PUBLIC SEVERE WEATHER OUTLOOK.

PWP PERMANENT WILTING POINT.

PWS PUBLIC WATER SYSTEM.

Q QUALITY INDEX.

QA QUALITY ASSURANCE.

QC QUALITY CONTROL.

QG QUASIGEOSTROPHIC.

QLA QUALITY LEVEL ANALYSIS.

QLCS QUASI-LINEAR CONVECTIVE SYSTEM.

QN QUESTION.

QPF QUANTITATIVE PRECIPITATION FORECAST.

QSTNRY QUASI-STATIONARY.

QTR QUARTER.

QUAD 1. QUADRANT; 2. ONE QUADRILLION.

R 1. RAIN; 2. RATIO.

RA 1. RAIN; 2. REMEDIAL ACTION.

RAC REFERENCE AMBIENT CONCENTRATION.

RACM 1. REASONABLY AVAILABLE CONTROL MEASURES; 2. REGULATED ASBESTOS-CONTAINING MATERIAL.

RACT REASONABLY AVAILABLE CONTROL TECHNOLOGY.

RADAR RADIO DETECTION AND RANGING

RADIOFAX ABBREVIATION FOR RADIO FACSIMILE.

RAFC REGIONAL AREA FORECAST CENTER.

RAFS REGIONAL ANALYSIS AND FORECASTING SYSTEM.

RAOB RADIOSONDE OBSERVATION.

RAWS REMOTE AUTOMATED WEATHER STATIONS.

RCC ROLLER COMPACTED CONCRETE.

RCKY ROCKY MOUNTAINS.

RCMD RECOMMEND.

RCRA RESOURCE CONSERVATION AND RECOVERY ACT.

RCV RECEIVE.

RDA RADAR DATA ACQUISITION.

RDG RIDGE.

RDS RADIUS.

REBAR REINFORCING BAR.

RECLAMATION BUREAU OF RECLAMATION.

RECS RENEWABLE ENERGY CERTIFICATES.

REF REFERENCE.

RELBL RELIABLE.

REP 1. REPRESENT; 2. REPRESENTATIVE.

REPI RENEWABLE ENERGY PRODUCTION INCENTIVE.

RFC RIVER FORECAST CENTER.

RFD REFERENCE DOSE.

RFP REQUEST FOR PROPOSALS.

RFQ REQUEST FOR QUALIFICATIONS.

RGD RAGGED.

RGN REGION.

RH RELATIVE HUMIDITY.

RHI RANGE-HEIGHT INDICATOR.

RHWMC REGIONAL HAZARDOUS WASTE MANAGEMENT COORDINATOR.

RI RIDEABILITY INDEX.

RIOGD RIO GRANDE.

RIOMETER RELATIVE IONOSPHERIC OPACITY METER.

RLS RELEASE.

RLTV RELATIVE.

RM RIVER MILE.

RMCL RECOMMENDED MAXIMUM CONTAMINANT LEVEL.

RMN REMAIN.

RMTN REGIONAL METEOROLOGICAL TELECOMMUNICATIONS NETWORK.

RMV REMOVE.

RN RIDE NUMBER.

RNA RIBONUCLEIC ACID.

RNAV AREA NAVIGATION.

RNFL RAINFALL.

RO REVERSE OSMOSIS.

RO/RO ROLL-ON/ROLL-OFF.

ROD RECORD OF DECISION.

ROSA REMOTE OBSERVATION SYSTEM AUTOMATION.

ROTG ROTATING.

ROW RIVER OUTLET WORKS.

RO&M REVIEW OF OPERATION AND MAINTENANCE.

RPD RAPID.

RPG RADAR PRODUCT GENERATOR.

RPLC REPLACE.

RPM 1. REVOLUTIONS PER MINUTE; 2. REMEDIAL PROJECT MANAGER.

RPRT REPORT.

RQL REJECTABLE QUALITY LEVEL.

RQR REQUIRE.

RRA RECLAMATION REFORM ACT OF 1982.

RRQ RIGHT REAR QUADRANT.

RRT REGIONAL RESPONSE TEAM.

RSG RISING.

RSN REASON.

RTC REFERENCE TISSUE CONCENTRATION.

RTE ROUTE.

RTG REGIONAL TRANSMISSION GROUP.

RTOR RIGHT TURN ON RED.

RTRD RETARD.

RTRN RETURN.

RTVS REAL TIME VERIFICATION SYSTEM.

RUC RAPID UPDATE CYCLE.

RVA RIVER SUMMARY.

RVF RIVER FORECAST.

RVI RIVER ICE STATEMENT.

RVR 1. RIVER RECREATION STATEMENT; 2. RUNWAY VISUAL RANGE.

RVS RIVER STATEMENT.

RWIS ROAD WEATHER INFORMATION SYSTEM.

S 1. SOUTH; 2. SNOW; 3. SECOND.

S/W SHORTWAVE.

S/WV SHORTWAVE.

SAB 1. SATELLITE ANALYSIS BRANCH; 2. SCIENCE ADVISORY BOARD.

SAL SALINITY.

SAME SPECIFIC AREA MESSAGE ENCODING.

SAMEX STORM AND MESOSCALE ENSEMBLE EXPERIMENT.

SAR SODIUM ADSORPTION RATIO.

SARA SUPERFUND AMENDMENTS AND REAUTHORIZATION ACT OF 1986.

SATL SATELLITE.

SAV SUBMERGED AQUATIC VEGETATION.

SBCAPE SURFACE BASED CAPE.

SBND SOUTHBOUND.

SBSD SUBSIDE.

SC 1. STRATOCUMULUS; 2. SUDDEN COMMENCEMENT.

SCA SMALL CRAFT ADVISORY.

SCAHS SMALL CRAFT ADVISORY FOR HAZARDOUS SEAS.

SCARB SMALL CRAFT ADVISORY FOR ROUGH BAR.

SCAW SMALL CRAFT ADVISORY FOR WINDS.

SCS SOIL CONSERVATION SERVICE.

SCT SCATTERED.

SDM STATION DUTY MANUAL.

SDWA SAFE DRINKING WATER ACT.

SE SOUTHEAST.

SEC SPACE ENVIRONMENT CENTER.

SEE SAFETY EVALUATION EARTHQUAKE.

SEED SAFETY EVALUATION OF EXISTING DAMS.

SEF SAFETY EVALUATION FLOOD.

SEISMO PERTAINS TO EARTHQUAKES.

SELS SEVERE LOCAL STORM.

SELY SOUTHEASTERLY.

SERC STATE EMERGENCY RESPONSE COMMISSION.

SERN SOUTHEASTERN.

SESC SPACE ENVIRONMENT SERVICES CENTER.

SEWD SOUTHEASTWARD.

SFC SURFACE.

SFL SEQUENCE FLASHING LIGHTS.

SG SNOW GRAINS.

SGFNT SIGNIFICANT.

SH SHOWER.

SHARS SUBTLE HEAVY RAINFALL SIGNATURE.

SHEF STANDARD HYDROLOGIC EXCHANGE FORMAT.

SHEFPARS SOFTWARE DECODER FOR SHEF DATA.

SHFT SHIFT.

SHLW SHALLOW.

SHRA RAIN SHOWERS.

SHRAS SHOWERS.

SHRT SHORT.

SHRTWV SHORTWAVE.

SHSN SNOW SHOWERS.

SHWR SHOWER.

SI 1. SHRINKAGE INDEX; 2. SUDDEN IMPULSE; 3. SEVERITY INDEX.

SI SYSTEM INTERNATIONAL SYSTEM OF UNITS.

SIC STANDARD INDUSTRIAL CLASSIFICATION.

SID 1. STANDARD INSTRUMENT DEPARTURE; 2. SUDDEN IONOSPHERIC DISTURBANCE.

SIGMET SIGNIFICANT METEOROLOGICAL ADVISORY.

SIGWX SIGNIFICANT WEATHER.

SIP STATE IMPLEMENTATION PLANS.

SITE SUPERFUND INNOVATIVE TECHNOLOGY EVALUATION.

SITOR SIMPLEX TELETYPE OVER RADIO.

SIU SIGNIFICANT INDUSTRIAL USER.

SL 1. SEA LEVEL; 2. SHRINKAGE LIMIT.

SLD SOLID.

SLGT SLIGHT.

SLO SLOW.

SLOSH SEA, LAKE AND OVERLAND SURGES FROM HURRICANES.

SLP SEA LEVEL PRESSURE.

SLT SHUTTLE LOOP TRANSIT.

SLY SOUTHERLY.

SM 1. STATUTE MILES; 2. SUM TOTAL FOR MONTH.

SMA 1. SOIL MOISTURE ACCOUNTING; 2. STONE MATRIX ASPHALT.

SMCL SECONDARY MAXIMUM CONTAMINANT LEVEL.

SMPDBK SIMPLIFIED DAM BREAK MODEL.

SMW SPECIAL MARINE WARNING.

SN STRUCTURAL NUMBER.

SNOTEL SNOW TELEMETRY.

SNR SIGNAL-TO-NOISE RATIO.

SNW SNOW.

SNWFL SNOWFALL.

SO SOUTHERN OSCILLATION.

SOCS SYNTHETIC ORGANIC CHEMICALS.

SOD SAFETY OF DAMS.

SOO SCIENCE AND OPERATIONS OFFICER.

SOP STANDING OPERATING PROCEDURES.

SOQ STATEMENT OF QUALIFICATIONS.

SOV SINGLE-OCCUPANT VEHICLE.

SPC STORM PREDICTION CENTER.

SPCC SPILL PREVENTION CONTROL AND COUNTERMEASURE PLAN.

SPCLY ESPECIALLY.

SPCP SPILL PREVENTION, CONTAINMENT, AND COUNTERMEASURES PLAN.

SPENES NESDIS SATELLITE PRECIPITATION ESTIMATES.

SPMD SEMIPERMEABLE MEMBRANE DEVICE.

SPP SMALL POWER PRODUCER.

SPS 1. STANDARD PROJECT STORM; 2. SEVERE WEATHER POTENTIAL STATEMENT.

SPUI SINGLE-POINT URBAN INTERCHANGE.

SQG SMALL QUANTITY GENERATOR.

SQLN SQUALL LINE.

SRF SURF ZONE FORECAST.

SRH STORM-RELATIVE HELICITY.

SRM STORM RELATIVE MEAN.

SRN SOUTHERN.

SS SANDSTORM.

SSB SOLAR SECTOR BOUNDARY.

SSHP SITE SPECIFIC HYDROLOGIC PREDICTION SYSTEM.

SSHS SAFFIR/SIMPSON HURRICANE SCALE.

SSO SANITARY SEWER OVERFLOWS.

SST SEA SURFACE TEMPERATURE.

ST STRATUS.

STAR STANDARD TERMINAL ARRIVAL ROUT.

STFR STRATUS FRACTUS.

STG STRONG.

STJ SUBTROPICAL JET.

STORET STORAGE AND RETRIEVAL.

STRFM STRATIFORM.

SV SLOPE VARIANCE.

SVOC SEMIVOLATILE ORGANIC COMPOUND.

SVR SEVERE.

SVRL SEVERAL.

SW 1. SOUTHWEST; 2. SNOW SHOWERS.

SWD SWELL DIRECTION.

SWE SNOW WATER EQUIVALENT.

SWEAT SEVERE WEATHER THREAT.

SWF SHORT WAVE FADE.

SWH SWELL HEIGHT.

SWI SHOWALTER INDEX.

SWLY SOUTHWESTERLY.

SWP SWELL PERIOD.

SWRN SOUTHWESTERN.

SWS SEVERE WEATHER STATEMENT.

SWWD SOUTHWESTWARD.

SX STABILITY INDEX.

SXN SECTION.

SYNOP SYNOPTIC.

T 1. TEMPERATURE; 2. NATURAL PERIOD OF VIBRATION; 3. THUNDERSTORM.

T ROLLS TRANSVERSE ROLLS.

TACAN TACTICAL AIR NAVIGATION.

TAF TERMINAL AERODROME FORECAST.

TAFB TROPICAL ANALYSIS AND FORECAST BRANCH.

TCE TRICHLOROETHYLENE.

TCMS TRANSPORTATION CONTROL MEASURES.

TCP TRADITIONAL CULTURAL PROPERTY.

TCU TOWERING CUMULUS.

TD TROPICAL DEPRESSION.

TDA TODAY.

TDH TOTAL DYNAMIC HEAD.

TDM TRANSPORTATION DEMAND MANAGEMENT.

TDS TOTAL DISSOLVED SOLIDS.

TDWR TERMINAL DOPPLER WEATHER RADAR.

TECP TOTALLY ENCAPSULATING CHEMICAL PROTECTIVE.

TEMP TEMPERATURE.

TEU TWENTY-FOOT EQUIVALENT UNIT.

TGA TECHNICAL-GRADE ACTIVE INGREDIENT.

THE "ACT" CLEAN WATER ACT.

THETA-E EQUIVALENT POTENTIAL TEMPERATURE.

THK THICK/THICKNESS.

THM TRIHALOMETHANE.

THN THIN.

THRFTR THEREAFTER.

THRU THROUGH.

THRUT THROUGHOUT.

THSD THOUSAND.

TIL UNTIL.

TIP TRANSPORTATION IMPROVEMENT PROGRAM.

TKE TURBULENT KINETIC ENERGY.

TLC TOTAL LATERAL CLEARANCE.

TLV THRESHOLD LIMIT VALUE.

TMA TRUCK-MOUNTED ATTENUATOR.

TMDL TOTAL MAXIMUM DAILY LOAD.

TMW TOMORROW.

TNDCY TENDENCY.

TNGT TONIGHT.

TOC TOTAL ORGANIC CARBON.

TOP CLOUD TOP.

TOR 1. TORNADO; 2. TORNADO WARNING.

TPC TROPICAL PREDICTION CENTER.

TPH TOTAL PETROLEUM HYDROCARBONS.

TPW TOTAL PRECIPITABLE WATER.

TRE TOXICITY REDUCTION EVALUATION.

TROF TROUGH.

TROP TROPOPAUSE.

TROWAL TROUGH OF WARM AIR ALOFT.

TRPCL TROPICAL.

TRRN TERRAIN.

TS TROPICAL STORM.

TSD TECHNICAL SUPPORT DOCUMENT.

TSDF TREATMENT, STORAGE AND DISPOSAL FACILITY.

TSI THERMAL SYSTEM INSULATION.

TSM 1. TRANSPORTATION SYSTEM MANAGEMENT; 2. TOTAL STORM MANAGEMENT.

TSP TOTAL SUSPENDED PARTICLES.

TSRA THUNDERSTORMS WITH RAIN.

TSS TOTAL SUSPENDED SOLIDS.

TSTM THUNDERSTORM.

TUA TOXIC UNIT ACUTE.

TUC TOXIC UNIT CHRONIC.

TUS TOXIC UNITS.

TUTT LOW TROPICAL UPPER-TROPOSPHERIC TROUGH.

TVS TORNADO VORTEX SIGNATURE.

TWA TIME-WEIGHTED AVERAGE.

TWEB TRANSCRIBED WEATHER BROADCASTS.

TWTDS TREATMENT WORKS TREATING DOMESTIC SEWAGE.

U URANIUM.

UAA USE ATTAINABILITY ANALYSIS.

UCC ULTRA CLEAN COAL.

UCP UNIT CONTROL POSITION.

UDI UNDER DIRECT INFLUENCE.

UGC UNIVERSAL GEOGRAPHIC CODE.

UHF ULTRA HIGH FREQUENCY.

UIC UNDERGROUND INJECTION CONTROL.

ULJ UPPER LEVEL JET.

UNSBL UNSEASONABLE.

UNSTBL UNSTABLE.

UPR UPPER.

UPSLP UPSLOPE.

UPSTRM UPSTREAM.

UPWP UNIFIED PLANNING WORK PROGRAM.

USACE U.S. ARMY CORPS OF ENGINEERS.

USBR U.S. BUREAU OF RECLAMATION.

USCS UNIFIED SOIL CLASSIFICATION SYSTEM.

USDW UNDERGROUND SOURCE OF DRINKING WATER.

USFS U.S. FOREST SERVICE.

USGS U.S. GEOLOGICAL SURVEY.

UST UNDERGROUND STORAGE TANK.

UT UNIVERSAL TIME.

UTC UNIVERSAL COORDINATED TIME.

UV ULTRAVIOLET

UVM UPWARD VERTICAL MOTION.

UVV UPWARD VERTICAL VELOCITY.

UWNDS UPPER WINDS.

UZA URBANIZED AREA.

V 1. VOLT; 2. WIND SPEED.

V/C VOLUME-TO-CAPACITY.

VA VOLT-AMPERE.

VAAC VOLCANIC ASH ADVISORY CENTERS.

VAD VELOCITY AZIMUTH DISPLAY.

VARS VOLT-AMPERES REACTIVE.

VASI VISUAL APPROACH SLOPE INDICATOR.

VAV VARIABLE AIR VOLUME.

VCNTY VICINITY.

VCP VOLUME COVERAGE PATTERN.

VFD VARIABLE FREQUENCY DRIVE.

VFR VISUAL FLIGHT RULES.

VHF VERY HIGH FREQUENCY.

VIL VERTICALLY-INTEGRATED LIQUID.

VIS 1. VISIBLE SATELLITE IMAGERY; 2. VISIBLE; 3. VISIBILITY.

VKMT VEHICLE MILES OF TRAVEL.

VLCTY VELOCITY.

VLF VERY LOW FREQUENCY.

VLY VALLEY.

VMD VOLUME MEDIAN DIAMETER.

VMS VARIABLE MESSAGE SIGN.

VMT VEHICLE MILES TRAVELLED.

VOC VOLATILE ORGANIC COMPOUND.

VOR VHF OMNIDIRECTIONAL RANGE.

VORT MAX VORTICITY MAXIMUM.

VOS VOLUNTARY OBSERVING SHIP PROGRAM.

VOT VORTICITY.

VR VEER.

VRT MOTN VERTICAL MOTION.

VSB VISIBLE SATELLITE IMAGERY.

VSBY VISIBILITY.

VTEC VALID TIME EVENT CODE.

VVP VOLUME VELOCITY PROCESSING.

VWP VAD WIND PROFILE.

W WEST.

WAA WARM AIR ADVECTION.

WAPA WESTERN AREA POWER ADMINISTRATION.

WBND WESTBOUND.

WBZ WET BULB ZERO.

WC/ATWC WEST COAST AND ALASKA TSUNAMI WARNING CENTER.

WCM WARNING COORDINATION METEOROLOGIST.

WDIR WIND DIRECTION (THE DIRECTION THE WIND IS COMING FROM IN DEGREES CLOCKWISE FROM TRUE N).

WDLY WIDELY.

WDSPRD WIDESPREAD.

WER 1. WATER-EFFECT RATIO; 2. WEAK ECHO REGION.

WET WHOLE-EFFLUENT TOXICITY.

WFO WEATHER FORECAST OFFICE.

WFP WARM FRONT PASSAGE.

WH WATT-HOUR.

WHMIS WORKPLACE HAZARDOUS MATERIALS INFORMATION SYSTEM.

WIM WEIGH-IN-MOTION.

WINT WINTER.

WL 1. WHITE LIGHT; 2. WORKING LEVEL.

WLA WASTE LOAD ALLOCATION.

WLM WORKING LEVEL MONTH.

WLY WESTERLY.

WMC WORLD METEOROLOGICAL CENTER.

WMO WORLD METEOROLOGICAL ORGANIZATION.

WQBEL WATER QUALITY-BASED EFFLUENT LIMIT.

WQS WATER QUALITY STANDARD.

WRCC WESTERN REGIONAL CLIMATE CENTER.

WRN WESTERN.

WRNG WARNING.

WSCC WESTERN SYSTEMS COORDINATING COUNCIL.

WSFO WEATHER SERVICE FORECAST OFFICE.

WSHFT WIND SHIFT.

WSR-57 NATIONAL WEATHER SERVICE WEATHER SURVEILLANCE RADAR DESIGNED IN 1957 THAT WAS REPLACED BY WSR-88D UNITS.

WSR-74 NATIONAL WEATHER SERVICE WEATHER SURVEILLANCE RADAR DESIGNED IN 1974 THAT WAS REPLACED BY WSR-88D UNITS.

WSR-88D 1. WEATHER SURVEILLANCE RADAR - 1988 DOPPLER; 2. NEXRAD UNIT.

WSW WINTER STORM MESSAGE.

WT WARNING TIME.

WTMP SEA SURFACE TEMPERATURE (CELSIUS).

WTR WATER EQUIVALENT.

WVHT SIGNIFICANT WAVE HEIGHT.

WWH WIND WAVE HEIGHT.

WWP WIND WAVE PERIOD.

WX WEATHER.

WY WATER YEAR.

XBT EXPENDABLE BATHYTHERMOGRAPH.

XCITED EXCITED.

XCPT EXPECTING.

XPC EXPECT.

XSEC CROSS SECTION.

YD^2 SQUARE YARD.

YD^3 CUBIC YARD.

Z 1. RADAR REFLECTIVITY FACTOR; 2. ZULU TIME; 3. ZULU.

ZFP ZONE FORECAST PRODUCT.

ZL FREEZING DRIZZLE.

ZNS ZONES.

ZOI ZONE OF INTRUSION.

ZR FREEZING RAIN.

A

A INDEX Daily average level for geomagnetic activity.

A_k Area of every natural draft opening (k) in complete enclosure, in m^2.

A-HORIZON Uppermost of three layers (A-, B-, C-) of soil, up to 2 inches (6 feet 6 inches) deep that collects organic matter from plant roots and leaches its fine grains and soluble components into B-horizon, beneath.

A-SCALE Sound level measurement scale that discriminates against low frequencies and approximates human ear.

A-SCALE SOUND LEVEL Measurement of sound approximating sensitivity of human ear used to note intensity or annoyance level of sounds.

AASHTO SOIL CLASSIFICATION SYSTEM Soil classification system based on field performance of soils for highway construction: A-1-a (gravel with or without well-graded fines,) A-1-b (sand with or without well-graded fines,) A-2-4 (granular material with silty fines,) A-2-6 (granular material with clayey fines,) A-3 (poorly graded sand with almost no fines or gravel,) A-4 (mostly silty fines,) A-5 (soil with silty fines,) A-6 (silty or clayey fines with low liquid limit,) A-7-5 (plastic clays and silts).

AB Prefix denoting practical quantification of theoretical electromagnetic unit.

ABAMPERE Unit of electric current equal to 10 amperes that's equal to current, flowing through two infinitely long parallel conductors, 1 cm apart, that produces force of 2 dynes/cm.

ABANDONED AREA Abandoned place where work has stopped and additional work isn't intended.

ABANDONED MINE Mine where operations have happened before and either appropriate reclamation bond or financial assurance was released or forfeited or, if no reclamation bond or another financial guarantee was posted, no mining operations have happened for five years or more.

ABANDONED WELL 1. Well no longer being used or cannot be used because of its poor condition; 2. Well whose use has been permanently discontinued or in state of such disrepair that it cannot be used for its intended purpose.

ABANDONMENT 1. Relinquishing of public interest in right to use water for any purpose or approach to utilizing such water or of right of way or action thereon without any intent to retrieve or use again for highway and/or transportation purposes; 2. Cessation of work with contractor and removal of equipment and uninstalled materials from site.

ABATE Reduce in quantity, degree or intensity.

ABATEMENT Reducing degree or intensity of environmental pollutants.

ABATEMENT DEBRIS Waste from remediation activities.

ABATSON Baffle that deflects airborne sound waves down.

ABATVENT Apparatus that reduces force of or prevents entry of wind.

ABATVOIX Reflector of airborne sound waves.

ABAXIAL Surface of leaf that faces away from stem.

ABC EXTINGUISHER Fire extinguisher that can be used on fires involving ordinary combustibles, flammable liquids and energized electrical equipment.

ABERRANT 1. Atypical; 2. Departing from normal type or structure.

ABFARAD 1. Unit of electric capacitance equal to 10^9 farads; 2. Capacitance of capacitor carrying charge of 1 abcoulomb with potential difference of 1 abvolt between its plates.

ABHENRY 1. Unit of electrical inductance equal to equal to 10^{-9} henry; 2. Inductance produced when rate of change of current of 1 ampere/s generates induced electromotive force of 1 abvolt.

ABIOCEN Non-living components of environment.

ABIOSESTON Non-living matter floating in water.

ABIOTIC Absence of living organisms.

ABLATION 1. Depletion of snow and ice from any mass of frozen water through processes of melting, sublimation and evaporation, wind erosion and calving; 2. Wasting and removal from rock mass of material by physical processes (i.e., wind erosion or by chemical processes).

ABLUTION FITTING Sanitary washing fitting made for communal use, generally with water sprays controlled by touchless or foot valves.

ABNORMAL EROSION Erosion speeds at variance with all geological standard, as result of human activity.

ABOHM Unit of electric resistance equal to 10^{-9} ohm.

ABOVE GROUND HEIGHT Maximum height from natural ground surface to top of dam.

ABOVEGROUND BIOMASS Aboveground part of plant, excluding root system.

ABOVEGROUND RELEASE Discharge to surface of land or to surface water.

ABOVEGROUND STORAGE FACILITY Tank or alternative container, with bottom on plane not more than 150 mm (6 in.) below surrounding surface.

ABOVEGROUND TANK Tank situated such that its whole surface area is entirely over plane of adjacent surrounding surface.

ABRASION Reduction of rock fragments or surfaces by wearing, friction, grinding or rubbing action of other rock particles or of transport medium.

ABRASIVE WEAR Removal of substances from surfaces in relative movement with cutting or abrasive actions of tough particle.

ABRIDGED REPORT ON RIVER GAUGE STATION (E-19A) Abridged version of E-19, E-19a was used to update E-19 as additional information or changes occur at station during intervening five year period.

ABRUPT WAVE Transitionary wave or increase in depth of water within open channel as result of abrupt change in states of flow.

ABS Kind of surface active agent that, in waste treatment processes, tends to cause foaming and resist biological degradation.

ABS PIPE Stiff plastic conduit manufactured from acrylonitrile butadiene styrene.

ABSCISSA Number whose absolute value (modulus) is perpendicular distance of point from vertical axis (i.e., Y-axis).

ABSOLUTE 1. Chemical substance free, for all practical purposes, of impurities; 2. Separate of any arbitrary standard (i.e., non-relative).

ABSOLUTE BLOCK Block that allows only one train to occupy defined section of track.

ABSOLUTE COMPACTION Density quantify on basis of utmost theoretical measure of voidless combination as ascertained by recognized test method.

ABSOLUTE ERROR Mistakes of position relative to some grid or graticule.

ABSOLUTE FILTER RATING Diameter of biggest hard spherical particle that will pass via filter under specified test conditions.

ABSOLUTE FILTRATION RATING Diameter of biggest spherical particle that will pass via filter under specified test conditions.

ABSOLUTE HUMIDITY Mass of water vapor per unit volume of air.

ABSOLUTE MANOMETER Device whose calibration may be determined from its quantifiable physical constants that's exact same for all vapors and gases.

ABSOLUTE PERMISSIVE Length of track where no train is permitted to enter while it is occupied by another train.

ABSOLUTE POUNDS PER SQUARE INCH Relative to vacuum, atmospheric pressure plus gauge pressure.

ABSOLUTE PRESSURE Sum of atmospheric pressure plus gauge pressure.

ABSOLUTE SPECIFIC GRAVITY Ratio of mass (referred into vacuum) of certain volume of solid or liquid in stated temperature (e.g., 50 $^{\circ}$F) to mass (referred to vacuum) of equivalent volume of gas-free distilled water at stated temperature.

ABSOLUTE TEMPERATURE Temperature measured or computed on absolute scale where absolute zero is minimum and scale units equivalent in magnitude to centigrade degrees.

ABSOLUTE VELOCITY Speed of body in regard to fixed point on earth's surface.

ABSOLUTE VISCOSITY Measure of fluid's inclination to resist flow without consideration to its density.

ABSOLUTE VOLUME 1. In instance of solids, displacement volume of particles themselves, including their permeable and impermeable voids, but excluding any space between them; 2. In instance of fluids, their volume.

ABSOLUTE ZERO Hypothetical temperature defined by complete lack of heat, equivalent to about -460 $^{\circ}$F.

ABSOLUTELY STABLE AIR Atmospheric condition that exists when environmental lapse rate is less than moist adiabatic lapse rate.

ABSOLUTELY UNSTABLE AIR Atmospheric condition that exists when environmental lapse rate is greater than dry adiabatic lapse rate.

ABSORBED DOSE Amount of substance that enters body of exposed organism through physical or biological processes.

ABSORBED MOISTURE Moisture that has entered solid by absorption and contains physical properties not substantially different from ordinary water in exact same temperature and pressure.

ABSORBED WATER Water held mechanically in soil or rock mass and having physical properties not substantially different from ordinary water at same temperature and pressure.

ABSORBENCY Property of material to imbibe liquids.

ABSORBENT Substance effective at taking in water, moisture or alternative fluid.

ABSORBENT SEPARATOR Apparatus that retains specific soluble and insoluble contaminants by molecular adhesion.

ABSORBER Substance or apparatus for removing certain fluid or gas from material to constructed level, either chemically or mechanically.

ABSORBER PLATE In solar energy system, that part of collector system that transfers solar heat to working medium.

ABSORBING WELL Well dug or drilled via impermeable strata right into dry strata and used to drain away water.

ABSORPTION 1. Entrance of water into soil or rocks by all natural processes including infiltration of precipitation or snowmelt, gravity flow of streams into valley alluvium into sinkholes or other large openings and movement of atmospheric moisture; 2. Uptake of water, other fluids or dissolved chemicals by cell or organism; 3. Taking in of fluids or other substances through cells or tissues; 4. Passing of substance into circulatory system of body; 5. Entry of toxicants through skin; 6. Dissolving or mixing of substance in gaseous, liquid or solid form with groundwater; 7. Process where incident radiant energy is retained by substance by conversion to some other form of energy.

ABSORPTION BARRIER Any of exchange sites of body (e.g., skin, lung tissue, gastrointestinal-tract wall) that permit uptake of various substances at different rates.

ABSORPTION BED Comparatively big pit tilled with rough aggregate, used to get effluent from waste treatment system.

ABSORPTION CHILLER Cooler that works from evaporation of fluid in low pressure side of closed system; refrigerant being recovered by consuming activity of another fluid.

ABSORPTION COEFFICIENT Volume of gas that's dissolved (referred to 0 $^{\circ}$C and 760 mm Hg pressure).

ABSORPTION FACTOR Fraction of chemical making contact with organism absorbed by organism.

ABSORPTION FIELD Effluent disposal system, especially from septic tanks, comprising network of perforated pipes placed in rough gravel in trenches that slope away from tank exit.

ABSORPTION HYGROMETER Device used to quantify amount of water vapor in atmosphere by measuring upsurge in weight of drying agent over that sample of air is drawn.

ABSORPTION LOSS Water lost to adjacent soil.

ABSORPTION RATE Weight of fluid taken in with substance following immersion for specific period in liquid of known properties and temperature.

ABSORPTION TEST Evaluation made to ascertain skill of substances to absorb vapors or fluids under given conditions.

ABSORPTION TRENCH Trench including perforated conduit or distribution tile surrounded by rough aggregate, where effluent from wastewater treatment process is released.

ABSORPTIVE CAPACITY Volume of waste materials that can be deposited in special surroundings without environmental damage or aesthetic impact.

ABSORPTIVE TERRACE Terrace of changing width formed on either side of ridge to be able to induce moisture to propagate and be kept over broadest possible region.

ABSORPTIVITY 1. Ability of substance to consume (light, heat, moisture, gases); 2. Fraction of radiant energy consumed by top it hits.

ABSTRACT Brief background of all information along with brief history of transactions.

ABSTRACT OF BIDS Overview of unit prices provided by owner, used as cornerstone for choosing low bid on tendered contract.

ABSTRACT OF TITLE Record revealing condensed history of title to property, including parts of all conveyances or other pertinent instruments relating to estate or interest in property and all liens, charges, encumbrances and releases.

ABSTRACTING Procedure of assembling data of identical character under item prepared for charging; primary phase of working up bill of materials.

ABSTRACTION Temporary or permanent withdrawal of water from source.

ABSTRACTIVE USE Use of resource that's briefly taken from its natural environment.

ABUTMENT 1. Part of valley or canyon wall against which dam is constructed where left and right abutments of dams are defined with observer viewing dam looking in downstream direction; 2. Part of dam that contacts riverbank; 3. Structure that supports ends of dam or bridge; 4. Part of structure terminal point or receives thrust or pressure; 5. Action or place of abutting.

ABUTMENT FLASHING Flashing that shields head and sides of abutment that's level at head and stepped in at inclines.

ABUTMENT PIECE Lowest structural member in framing that receives and distributes loads via upright; soleplate.

ABUTMENT SEEPAGE Reservoir water that moves through seams or pores in natural abutment material and exits as seepage.

ABUTTAL Components where property abuts on other land; borders.

ABVOLT 1. Unit of electromotive force or potential difference equal to 10^{-8} volt; 2. Potential that exists between two points when 1 erg of work must be done to transfer 1 abcoulomb of charge from one point to other.

ABWATT 1. Unit of power equal to 10^{-7} watt; 2. Power dissipated when current of 1 abampere flows across potential difference of 1 abvolt.

ABYSSAL Water depth where light will penetrate.

ABYSSINIAN WELL Perforated tube driven into earth where groundwater is pumped.

ABYSSOBENTHOS Ocean floor below water depth where light will penetrate.

ABYSSOPELAGIC Organisms that exist in water at depths greater than 3,000 m.

ACCELERATE STOP DISTANCE Runway distance required to accelerate aircraft to specified speed and bring aircraft to complete stop.

ACCELERATED ABSORPTION TEST Evaluation where outcome is hastened by testing at conditions more severe than those expected in typical service.

ACCELERATED COMPLETION Condition to get contractor to finish work at earlier date than was scheduled in contract files.

ACCELERATED DESIGN AND CONSTRUCTION Technique whereby construction is started before completion of final detail drawings.

ACCELERATED LIFE TEST Process of designing to approximate in brief time deteriorating effects obtained under normal service conditions.

ACCELERATED SERVICE TEST Test where some service condition is exaggerated in order to obtain result in shorter time.

ACCELERATED WEATHERING Testing of material samples to cycles of weather to test materials durability.

ACCELERATION Time rate of change of velocity vector, either of magnitude or direction or both.

ACCELERATION LANE Auxiliary or speed-change lane that allows vehicles to accelerate to highway speeds before entering through-traffic lanes of highway.

ACCELERATION/DECELERATION DELAY Vehicle delay from slowing and then speeding up to running speed.

ACCELEROGRAM Record from accelerometer showing acceleration as function of time.

ACCEPTABLE Acceptable to authority having jurisdiction.

ACCEPTABLE DAILY INTAKE Estimate of largest amount of chemical that person can be exposed on daily basis not anticipated to result in adverse effects.

ACCEPTANCE 1. Written sign that goods and/or materials furnished and/or installed are as stipulated; 2. Consignee's receipt of dispatch that terminates common carrier contract for transport.

ACCEPTANCE CONSTANT Quality index minimum allowed.

ACCEPTANCE INSPECTION Review by owner's representative of work scheduled for total or partial payment.

ACCEPTANCE LIMIT Limiting value placed on quality measure that permits acceptance.

ACCEPTANCE NUMBER Maximum number of units in sample considered defective or nonconforming that permits acceptance.

ACCEPTANCE OF BATCH SEQUENCE Condition where quantity of rejected batches in sequence is less than or equivalent to sequence satisfactory amount as dependent on right sampling strategy.

ACCEPTANCE OF WORK Handing over of completed construction project, excepting possibly any agreed-upon deficiencies or make-goods, with obligation for payment less any holdback.

ACCEPTANCE PLAN Recommended procedure of sampling, measuring and testing, together with standards for acceptability of lot of material or construction.

ACCEPTANCE SAMPLING AND TESTING Testing done to determine if quality of production is acceptable in terms of specifications.

ACCEPTANCE TEST Review that affirms if substance meets approval criteria.

ACCEPTED Item is "accepted" if it has been inspected and found to be safe by person or laboratory qualified to make such inspection.

ACCEPTED BID Offer taken by owner or his representative and that forms basis of contract.

ACCEPTED ENGINEERING PRACTICES Practices that are compatible with standards of practice required by registered professional engineer.

ACCESS Method or way of approach.

ACCESS CHAMBER Space big enough for tradesperson to enter.

ACCESS CHARGE 1. Charge levied on power supplied or its customer, for access to utility's transmission or distribution system; 2. Charge for right to send electricity over another's wires.

ACCESS CLEANOUT Pipe fitting with removable plate.

ACCESS CONTROL POINT Location staffed to restrict entry of unauthorized personnel into risk area during emergency and/or disaster events involving use of vehicles, barricades or other measures to deny access.

ACCESS DOOR Door giving access to machine rooms, machine and equipment spaces or other generally closed areas.

ACCESS FLOOR Flooring, typically modular and demountable, above structural floor, beneath where cables and other services are located.

ACCESS FLOORING Modular flooring parts that can be readily removed to offer access to underfloor ducting.

ACCESS FOR PHYSICALLY CHALLENGED Specific measures taken in design and construction to accommodate special needs of physically disabled.

ACCESS HOLE Opening big enough for planned use.

ACCESS POINT 1. Driveway, intersection or opening that provides roadway access; 2. Device that permits wireless devices to connect to network.

ACCESS ROAD Paved or unpaved road that provides connection between public road and utility.

ACCESS SHAFT Concrete portion of outlet works between shaft house and gate chamber that provides vertical access to gates.

ACCESS SWITCH Keyed switch where circuit could be activated.

ACCESS TIME Average travel time from point of origin to departure terminal or vehicle.

ACCESS-POINT DENSITY Total number of roadway access points divided by segment length.

ACCESSIBILITY 1. Percentage of people that complete trip within specified time; 2. Design of products, devices, services or environments for people with disabilities.

ACCESSIBLE 1. Wiring capable of being removed or exposed without damaging structure and/or finish or not permanently closed in; 2. Gear not guarded by locked doors, elevation or other alternative means.

ACCESSIBLE ENVIRONMENT Land surfaces, surface waters, oceans, atmosphere and all of lithosphere beyond restricted region.

ACCESSIBLE PEDESTRIAN SIGNAL Device that communicates both audible and/or vibrating indications of pedestrian signal timing.

ACCESSIBLE PEDESTRIAN SIGNAL DETECTOR Device designated to assist pedestrian who has visual or physical disabilities in activating pedestrian phase.

ACCESSION Additions to property caused by annexation of fixtures or of alluvial deposits.

ACCESSORY 1. Any auxiliary apparatus that can't be categorized as attachment; secondary component or assemblage of components that lends to general function and utility of machine; 2. Parts used to join, attach or link structural components or services.

ACCESSORY BOX Weatherproof rectangular carton, mourned on or inside wall, used to house electrical junctions and parts.

ACCESSORY CLOUD Cloud dependent on larger cloud system for development and continuance (e.g., roll clouds, shelf clouds, wall clouds).

ACCESSORY SEAL Seal useful for sealing accessory gear.

ACCESSORY SPECIES Species of tree that represents between 25% and 50% of stand.

ACCIDENT 1. Sudden, unwanted event that adversely affects man or environment; 2. Happening by chance; surprising, unforeseen occasion identifiable as to time and location.

ACCIDENT ASSESSMENT Evaluation of nature, severity and impact of accident.

ACCIDENT INSURANCE Insurance indemnity for loss of time, lack of facility, medical expenses, as well as other defined risks and expenses as result of injury.

ACCIDENT PREVENTION Preparation and mindful actions taken to decrease chance for incidence of injuries.

ACCIDENT SITE Location of unexpected occurrence, failure or loss, either at plant or along transportation route, resulting in release of hazardous materials.

ACCIDENTAL ERROR Unbiased, small and inevitable errors in observations.

ACCIDENTAL OCCURRENCE accident, including continuous or repeated exposure to conditions that results in physical harm or property damage.

ACCIDENTAL RELEASE Unexpected or non-sudden release of oil from underground storage tank that results in dependence on corrective actions and damages for physical harm or property damage.

ACCIDENTAL SPECIES Species of tree that represents less than 25% of stand.

ACCLIMATION Adjustment of organism to new habitat or environment.

ACCLIMATIZATION Physiological and behavioral adjustments of organism to changes in its environment.

ACCLIVITY Steepness or upslope of line or plane relative to horizontal.

ACCOMMODATION BRIDGE Temporary bridge.

ACCOUNT CODE STRUCTURE Assignment of alphanumeric identification to work parts within individual price bundles.

ACCOUNT NUMBER Identification of work package to ensure all costs incurred because of its conclusion could be correctly assigned and allocated.

ACCOUNT PAYABLE Amount owing to your lender, normally resulting in purchase of goods or services.

ACCOUNT RECEIVABLE Amount asserted against debtor, generally resulting from your selling of products or services.

ACCOUNTABILITY Being answerable for exercise of one's ability as well as performance of one's responsibilities.

ACCREDIT Authorize; grant qualifications to; give power to; certify.

ACCREDITED Individual or lab recognized by way of competent authority as qualified and able to do predetermined duty or job.

ACCRETION Process of growth or addition.

ACCRUAL DATE Date of event causing loss or damage or date that loss or damage needs to have been discovered via exercise of reasonable care.

ACCRUAL METHOD Bookkeeping process that recognizes revenues and expenses when they occur regardless of when cash is received or paid.

ACCRUED ASSET Collecting but not yet enforceable claim caused by supply of services or cost of cash (as during building or development) that's been only partially performed, just isn't yet billable and is not paid for.

ACCRUED DEPRECIATION Difference between reproduction cost new or replacement cost new and present worth of those improvements, plant or utilities, as measured on date of appraisal.

ACCRUED EXPENSE Incurred expense where there presently is not any enforceable claim as result of job or contract not having been finished to your billable period.

ACCRUED INTEREST Interest that has been earned but is not due or payable.

ACCRUED LIABILITY Accumulating but not yet enforceable claim by way of consultant or contractor resulting from partially finished work not yet billable.

ACCRUED REVENUE Brought in sales resulting from partly finished work that cannot yet be charged.

ACCUMULATED DEFICIENCY Total deficiency from established normal condition to date from start.

ACCUMULATED DEPRECIATION Total that asset continues to be depreciated since being obtained or put in service.

ACCUMULATED EXCESS Total excess from established normal condition to date from start.

ACCUMULATED SPECULATIVELY Stuff collected before being recycled.

ACCURACY 1. Degree of conformity of measure to standard or true value; 2. How close predicted or measured value is to true value or benchmark.

ACEQUIA Gravity-driven waterways that can be from simple ditches with dirt banks to lined with concrete.

ACID 1. Corrosive solution with pH value between 0 and 7; 2. Hydrogen-containing compound that reacts with metals to form salts; 3. pH modifier used in U.S. Fish and Wildlife Service wetland classification system.

ACID AEROSOL Acidic liquid or solid particles small enough to become airborne.

ACID COPPER CHROMATE Waterborne wood preservative comprising copper sulfate, sodium dichromate and chronic acid.

ACID DEPOSITION Complex chemical and atmospheric phenomenon that occurs when emissions of sulfur and nitrogen compounds and other substances are transformed by chemical processes into atmosphere and then deposited on Earth in either wet forms through rain, "acid rain," snow or fog or dry forms through acidic gases or particulates.

ACID DEWPOINT Temperature where dilute sulfuric acid appears as condensate when flue gas containing sulfur trioxide and water vapor is cooled below saturation temperature.

ACID GAS 1. Gas flow of hydrogen sulfide (H2S) and carbon dioxide (CO2) that continues to be separated from sour natural gas with sweetening unit. 2. By-product of incomplete combustion having pH value of less than 6.5.

ACID LEAD Totally processed, 99.9% pure lead featuring modest quantity of copper.

ACID MINE DRAINAGE 1. Drainage of water from areas that have been mined for coal or other mineral ores; 2. Water with low pH because of its contact with sulfur-bearing material and is harmful to aquatic organisms.

ACID MINE WATER Water present in or releasing from mine that contains minerals in solution and having low pH, usually brought on by oxidation of iron sulfide to sulfuric acid.

ACID MIST Sulfuric acid mist.

ACID NEUTRALIZER Apparatus in drainage system where acid discharge is likely, comprising chamber filled with limestone where flow is conducted to create neutral effluent.

ACID NEUTRALIZING CAPACITY Measure of ability of base (e.g., water, soil) to resist changes in pH, equivalent sum of all bases or base-producing materials, solutes plus particulates, in aqueous system that can be titrated with acid to equivalence point.

ACID OR FERRUGINOUS MINE DRAINAGE Mine drainage that, before any treatment, either has pH of less than 6.0 or total iron concentration equivalent to or greater than 10 mg/l.

ACID PRECIPITATION Precipitation containing relatively high concentrations of acid-forming chemicals that have been released into atmosphere and combined with water vapor harmful to environment.

ACID RAIN 1. Precipitation that has been rendered acidic by airborne pollutants harmful to environment; 2. Result of sulfur dioxide and nitrogen oxides reacting in atmosphere with water and returning to Earth.

ACID RECOVERY Sulfuric acid pickling operations offering procedures for reclaiming unreacted acid from spent pickling acid solutions.

ACID REGENERATION Those hydrochloric acid pickling operations offering procedures for regenerating acid from spent pickling acid solutions.

ACID RESISTANCE Extent where surface will resist attack by acids.

ACID SOOT Carbon particles, 1-3 mm in diameter, resulting from incomplete combustion, held together by water made acid due to combination with sulfur trioxide.

ACIDIC Condition of water or soil that contains sufficient amount of acid substances to lower pH below 7.0.

ACIDIC DEPOSITION Transfer by either by wet-deposition processes (e.g., rain, snow, dew, fog, frost, hail) or by dry deposition (e.g., gases, aerosols, fine to coarse particles) of acidic or acidifying substances from atmosphere to surface of Earth or to objects on its surface.

ACIDIFICATION Forced or accelerated leaching by exposing substance into higher concentration of solvents (e.g., acids) than would normally be encountered.

ACIDIFIED Addition of acid to sample to lower pH below 2.0.

ACIDITY Quantitative ability of aqueous solution to react with hydroxyl ions.

ACOUSTIC Linking to sound and hearing.

ACOUSTIC ABSORPTION COEFFICIENT Acoustic energy absorbed by surface, divided by total energy incident on it, subject to frequency of particular sound.

ACOUSTIC CLIP Flooring clip having flexible rubber pad, used to lessen transmission of sound and vibrations.

ACOUSTIC CONSTRUCTION Building techniques aimed at measurably: reducing or attenuating sound transmission.

ACOUSTIC DESCRIPTOR Numeric, symbolic or narrative information describing product's acoustic properties as they are determined according to prescribed test methodology.

ACOUSTIC FINISHINGS Materials, usually porous, inside that sound energy is adsorbed.

ACOUSTIC PLASTER Gypsum plaster that, when applied to wall or ceiling, has high sound absorbency.

ACOUSTIC REFLEX Muscular action of mechanical mechanism of human ear to protect against loud and sudden sound.

ACOUSTIC TILE Rectangle of acoustic board, usually interlocking for direct attachment into subbase or square edged and mounted in suspended ceiling system.

ACOUSTIC TRACE Line on vibration record that records sound level.

ACOUSTICAL ASSURANCE PERIOD Given amount of time where exhaust system, correctly used and maintained, must continue in conformity using recommended standard.

ACOUSTICAL BARRIER Baffle or obstacle designed and set to reduce or stop passing of sound waves.

ACOUSTICAL BOARD Any type of special material used in control or prevent passage of sound.

ACOUSTICAL CEILING Ceiling system built to decrease transmission of sound: may consist of layer of absorbent material or suspended system supporting acoustical plank.

ACOUSTICAL CORRECTION Measures taken to boost acoustical performance of finished room or auditorium.

ACOUSTICAL IMPEDANCE Mathematical expression for qualifying substance regarding its energy transport properties.

ACOUSTICAL MATERIALS Sound consuming or attenuating materials.

ACOUSTICAL TILE Wall and ceiling tiles composed of substances and of design meant to attenuate sound waves.

ACOUSTICAL TRANSMISSION FACTOR Ratio of original sound volume transmitted through/beyond any obstruction.

ACOUSTICS Study of sound; quality of sound.

ACQUIESCENCE Concurrence by owners of adjoining property that resolves boundary or rights dispute.

ACQUIRED CHARACTER Change in organism in response to environmental change.

ACQUIRED IMMUNE DEFICIENCY SYNDROME Condition in people where immune system endures advancing failure.

ACQUISITION Procedure of receiving property or right of way.

ACQUISITION OR TAKING Obtaining property for government use.

ACRE Unit of area equal to 43,560 ft^2 or 4,840 yd^2.

ACRE-FOOT 1. Unit of volume equivalent to volume of water required to cover one acre to depth of 1 foot that equals 43,560 ft^3, 326,851 gallons.

ACRE-INCH Unit of volume, equal to amount required to cover 1 acre to depth of 1 in.

ACRE-INCH-DAY Unit of volume equal to flow required to cover 1 acre to depth of 1 in. in 24-hour period.

ACRYLIC FIBER Manufactured synthetic fiber where fiber-forming substance is any long-chain synthetic polymer composed of at least 85% by weight of acrylonitrile units.

ACT Laws or change to laws, authorized into law with governing body.

ACT OF GOD Circumstance supposed to be beyond human control.

ACTINIDES Group of elements, having atomic numbers that range from 89 to 103, created in nuclear reactions or from radioactive decay.

ACTINOMETER Device used to assess heat absorbed by blackened disc or chamber.

ACTINOMYCETES Big group of mold-like microorganisms giving off odor characteristic of rich earth and that are important in stabilization of solid waste by composting.

ACTION LEVEL 1. Level of lead or copper that, if exceeded, triggers treatment or other requirements that water system must follow; 2. Regulatory levels recommended by EPA for enforcement by FDA and USDA when pesticide residues occur in food or feed commodities for reasons other than direct application of pesticide; 3. Airborne concentration of asbestos of 0.1 fibers/cm^3 of atmosphere computed as 8-hour time-weighted average; 4. Exposure amount (i.e., concentration of substance in air) where specific OSHA regulations to safeguard workers take effect.

ACTION STAGE Stage that, when reached by rising stream, represents level where National Weather Service or partner/user needs to take some type of mitigation action in preparation for possible significant hydrologic activity.

ACTIVATED ALUMINA Granular, porous form of aluminum trioxide effective at consuming water along with other materials, used in pollution management.

ACTIVATED CARBON 1. Highly adsorbent form of carbon used to remove odors and toxic substances from liquid or gaseous emissions used to remove dissolved organic matter from waste drinking water; 2. Adsorptive particles or granules of carbon where particles or granules have high capacity to selectively remove certain trace and soluble materials from water.

ACTIVATED CARBON PROCESS Technique for removing sulfur dioxide from flue gases.

ACTIVATED CHARCOAL Charcoal made by carbonizing organic substance, normally in lack of oxygen, used in granular or powder form to absorb odors.

ACTIVATED MANGANESE DIOXIDE PROCESS Technique for removing sulfur dioxide from flue gases by dry absorption to create ammonium sulfate.

ACTIVATED SILICA Negatively charged colloidal particle formed by reaction of dilute sodium silicate solution using dilute solution of acidic substance or another activant; used mainly as coagulant aid.

ACTIVATED SLUDGE Sludge floc that results when primary effluent is mixed with bacteria-laden sludge and then agitated and aerated to promote biological treatment, speeding breakdown of organic matter in raw sewage undergoing secondary waste treatment.

ACTIVATED SLUDGE LOADING Weight of biochemical oxygen demand (BOD) in applied liquor per unit volume of aeration capacity or per weight of activated sludge per day.

ACTIVATED SLUDGE PROCESS Aerobic sewage treatment method where air is introduced to raw sewage to promote mixing and bacterial digestion.

ACTIVATED SOLIDS Mix of organisms in wastewater solids generated in existence of dissolved oxygen.

ACTIVATION 1. Creation, under aerobic conditions or organisms effective at consuming organic material from water in activated sludge process; 2. Process of making material radioactive by bombardment with neutrons, protons or alternative nuclear particles; 2. Process of moving plant or gear from dormant state to your working state.

ACTIVATION PRODUCT Material made radioactive as result of irradiation, especially by neutrons in nuclear reactor.

ACTIVATOR 1. Chemical added to pesticide to increase its activity; 2. Compound ingredient used to boost effectiveness of accelerator.

ACTIVE Solar activity levels with at least one geophysical event or several larger radio events (10 cm) per day (i.e., Class M Flares).

ACTIVE BED Layer of material between bed surface and hypothetical depth at which no transport will occur for given gradation of bed material and flow conditions.

ACTIVE CAPACITY 1. Reservoir capacity normally usable for storage and regulation of reservoir inflows to meet established reservoir operating requirements (e.g., irrigation, power, municipal and industrial use, fish and wildlife, recreation, water quality) that extends from highest of either top of exclusive flood control capacity, top of joint use capacity or top of active conservation capacity, to top of inactive capacity; 2. Total capacity less sum of inactive and dead capacities.

ACTIVE CONSERVATION CAPACITY 1. Total amount of reservoir capacity normally available for release from reservoir below maximum storage level; 2. Total or reservoir capacity minus inactive storage capacity; 3. Volume of water between outlet works and spillway crest.

ACTIVE CONSERVATION STORAGE 1. Portion of water stored in reservoir that can be released for all useful purposes (e.g., municipal water supply, power, irrigation, recreation, fish, wildlife); 2. Volume of water stored between inactive pool elevation and flood control stage.

ACTIVE DARK FILAMENT Active prominence seen on disk.

ACTIVE EARTH PRESSURE Minimum value of earth pressure that exists when soil mass is permitted to yield sufficiently to cause its internal shearing resistance along potential failure surface to be completely mobilized.

ACTIVE FACTORS Supplies energy and nutrients for natural processes in plants.

ACTIVE FAULT Fault which, because of its present tectonic setting, can undergo movement from time to time in immediate geologic future.

ACTIVE FIRE PROTECTION System that features electrical and mechanical detectors and equipment to warn of and/or extinguish fires, typically within enclosed spaces and structures.

ACTIVE GRADE CROSSING WARNING SYSTEM Flashing-light signals to inform road users of approach or presence of trains at highway crossings.

ACTIVE INGREDIENT 1. Component that kills or otherwise controls, target pests in pesticide product; 2. Defoliant that may cause leaves or foliage to drop from plant; 3. Desiccant that will artificially accelerate drying of plant tissue; 4. Plant regulator that, through physical, biochemical activity, will accelerate or retard rate of growth or rate of maturation or otherwise change behavior of ornamental or crop plants or their merchandise.

ACTIVE INSTITUTIONAL CONTROL 1. Controlling access to disposal site by any means besides passive institutional controls; 2. Performing maintenance operations or remedial activities at site; 3. Controlling or cleaning up releases from site;

4. Tracking parameters associated with disposal system operation.

ACTIVE LAYER 1. Depth from bed surface to equilibrium depth continually mixed by flow; 2. Area above permafrost that thaws.

ACTIVE LIFE OF HAZARDOUS WASTE FACILITY Interval from first receipt of hazardous waste at facility until receipt of certification with qualified authority of final closure.

ACTIVE LIFE OF SOLID WASTE FACILITY Span of operation starting with all first receipt of solid waste and ending at completion of closure actions.

ACTIVE LONGITUDE Approximate center of range of heliographic longitudes in that "active regions" are more numerous and more flare-active than average.

ACTIVE MINING AREA Area, on and beneath land, used or affected in action associated with extraction, removal or recovery of coal, ore, minerals or aggregate from its natural deposits.

ACTIVE PASSINGS When bicyclist passes another bicyclist moving in same direction.

ACTIVE PORTION Part of facility where treatment, storage or disposal operations are or have already been ran.

ACTIVE PRESSURE Minimum earth pressure force from soil tending to overturn free retaining wall.

ACTIVE PRIORITY Traffic signal that adjusts signal timing priority due to bus arrival.

ACTIVE PROMINENCE Prominence displaying material motion and changes in appearance over few minutes of time.

ACTIVE PROMINENCE REGION Portion of solar limb displaying active prominences.

ACTIVE REGION Localized, transient volume of solar atmosphere that plages, sunspots, faculae, flares, etc. may be observed.

ACTIVE REPAIR TIME Time where repair work is performed on equipment.

ACTIVE SERVICE Drain that's receiving wastewater from process unit that can always keep water seal.

ACTIVE SOLAR SYSTEM Solar system that uses mechanical devices to collect, store and distribute useful energy.

ACTIVE SOUND ATTENUATOR Sound generator that creates waves asynchronous to many created by fans in heating or ventilating or air-conditioning system.

ACTIVE STORAGE Volume of storage facility accessible to be used.

ACTIVE STORAGE CAPACITY 1. Total amount of reservoir capacity normally available for release from reservoir below maximum storage level (volume of water between outlet works and spillway crest); 2. Total or reservoir capacity minus inactive storage capacity.

ACTIVE SURGE REGION Active Region that exhibits group or series of spike-like surges that rise above limb.

ACTIVE TRAFFIC MANAGEMENT Increases peak capacity and eases traffic flow on roadways.

ACTIVE TRANSPORT Energy-expending mechanism by which cell moves chemical across cell membrane from point of lower concentration to point of higher concentration, against diffusion gradient.

ACTIVE WATER Water having corrosive qualities.

ACTIVITY 1. Ratio of plasticity index to percent by dry mass of soil particles finer than 0.002 mm (2 microns) in size; 2. Work performed over interval of time; 3. Total flow of energy through system in unit of time; 4. Measure of transformation speed of radioactive nuclei in particular time; 5. Number of radioactive material.

ACTIVITY DEFINITION Identifying unique tasks that have to be performed to create various project deliverables.

ACTIVITY DESCRIPTION Brief phrase or label used in project network diagram that typically describes extent of work of task.

ACTIVITY DURATION Real time or estimated time needed to achieve something.

ACTIVITY DURATION ESTIMATING Estimate of amount of work periods that will be necessary to complete individual tasks.

ACTIVITY OF CLAY Ratio of plasticity index to percent by weight of total sample of clay smaller than 0.002 mm (0.00008 in.) in grain size.

ACTIVITY ORIENTED SYSTEMS Systems that present information in terms of activities (i.e., periods of time).

ACTUAL AGE OF PROPERTY Number of years that have passed since construction.

ACTUAL CASH VALUE Actual market value of something insured at time of its loss or damage.

ACTUAL COST Actual costs accrued during process of completing all work.

ACTUAL COST OF WORK PERFORMED Total costs incurred (direct and indirect) in executing work within specified period of time.

ACTUAL CREDITS Emission credits according to real production volumes as contained in reports submitted to regulatory authority.

ACTUAL DIMENSION Measurement as measured or computed.

ACTUAL EMISSIONS True rate pollution emissions from emissions unit.

ACTUAL FINISH DATE Point in time that work really finished on action.

ACTUAL GROUNDWATER VELOCITY Effective rate of motion of groundwater percolating through water-bearing substance, as quantified from volume of groundwater passing by means of unit cross-sectional area in unit of time divided by effective porosity.

ACTUAL PRODUCTIVE TIME Time spent employing machine to execute job.

ACTUAL SLOPE Incline where excavation face is excavated.

ACTUAL START DATE Point in time that work really began on action.

ACTUATE Set in operation.

ACTUATED CONTROL Detects all phases of traffic movement via detector.

ACTUATED OPERATION Traffic control signal operation where all signal phases function on basis of actuation.

ACTUATION Change or extension of traffic signal phase via detector.

ACTUATOR 1. Mechanical linkage that conducts motion.Mechanical linkage that conducts motion; 2. Linear or rotary apparatus that converts hydraulic or fluid energy into mechanical energy.

ACULTURAL VIBRATION Shaking unexpected and unknown to onlooker.

ACUTE 1. Occurring over short period of time used to describe brief exposures and effects that appear promptly after exposure; 2. Stimulus severe enough to rapidly induce effect; 3. Effect observed in 96- hours or less is typically considered acute in aquatic toxicity tests; 4. Disease or medical condition at or approaching crisis.

ACUTE ANGLE Angle between 0° and 90°.

ACUTE DERMAL LD_{50} Mathematically derived approximation of single dermal dose of material that could cause 50% mortality to your test population under certain conditions.

ACUTE EFFECT Adverse effect on any living organism that results in severe symptoms that develop rapidly and often subside after exposure stops.

ACUTE EXPOSURE 1. Exposure to toxic substance that occurs in short or single time period; 2. Single exposure to toxic substance that results in severe biological harm or death; 3. Usually characterized as lasting no longer than day.

ACUTE HEALTH EFFECT Immediate (i.e., within hours, days) effect that may result from exposure to certain drinking water contaminants (e.g., pathogens).

ACUTE INHALATION LC_{50} Mathematically derived approximation of concentration of material that could cause 50% mortality to your test population under given conditions.

ACUTE INHALATION TOXICITY Adverse effects caused by substance following single uninterrupted exposure by inhalation over short period of time (24 hours or less).

ACUTE LC_{50} Concentration of material, expressed as parts per million of medium, fatal to 50% of test population of animals under test conditions.

ACUTE LETHAL TOXICITY Fatal effect created on organism in just brief period of time of exposure to your substance.

ACUTE ORAL LD_{50} Mathematically derived approximation of single oral dose of material that could cause 50% mortality to your test population under given conditions.

ACUTE ORAL TOXICITY Adverse effects occurring within quick time of oral administration of one dose of material or multiple doses given within 24 hours.

ACUTE TOXICITY 1. Ability of substance to cause poisonous effects resulting in severe biological harm or death soon after single exposure or dose; 2. Any severe poisonous effect resulting from single short-term exposure to toxic substance; 3. LD_{50} of substance (lethal dose at which 50% of test animals succumb to toxicity of chemicals) is typically used as measure of its acute toxicity.

ACUTE TOXICITY TEST 1. Chemical substance that generates intensely hazardous effects if it kills within limited period of time; 2. Process used to look for concentration of material that creates toxic effect (death) on stated portion of test organisms in short span of time.

ACUTE-CHRONIC RATIO Ratio of acute toxicity of effluent or toxicant to its chronic toxicity.

ACUTELY TOXIC CONDITIONS Those acutely toxic to aquatic organisms following their short-term exposure within affected area.

ADAPT Make suitable for particular purpose by modification or change.

ADAPTATION 1. Adjustment to environmental conditions; 2. Changes in organism's physiological structure or function or habits that allow it to survive in new surroundings.

ADAPTATION PARAMETER Data related to specific WSR-88D unit that may consist of meteorological or hydrological parameters or of geographic boundaries, political boundaries, system configuration or other like data.

ADAPTER Fitting that joins devices (e.g., pipes) of different materials, sizes or threads.

ADAXIAL Surface of leaf that faces toward stem.

ADD-ON CONTROL DEVICE Air pollution control device (e.g., carbon absorber, incinerator) that reduces pollution in exhaust gas.

ADDENDA Additions or supplements issued before bid opening that clarifies, corrects or changes bidding documents or requirements of contract documents.

ADDENDUM 1. Something to be added (e.g., text added to contract); 2. Height by which tooth of gear projects beyond standard pitch circle or line.

ADDITIONAL WORK Rise in range of work initially contracted for, leading to variation.

ADDITIVE Chemical substance deliberately added to food or beverage to improve its stability or impart another desired quality.

ADDITIVE DATA Group of coded remarks that includes pressure tendency, amount of precipitation and maximum/minimum temperature during specified periods of time.

ADDITIVE EFFECT Combined effect of two or more chemicals equal to sum of their individual effects.

ADDITIVITY Characteristic property of mixture of toxicants that exhibits total toxic effect equal to arithmetic sum of effects of individual toxicants.

ADDRESSABLE SYSTEM SMOKE DETECTOR Smoke detection system where individual sensors are programmed to signal their location when activated.

ADDS Summated quantities that bring together all work and materials of identical nature.

ADENOSINE TRIPHOSPHATE 1. Organic, phosphate rich compound important in transfer of energy in organisms; 2. Excellent indicator of presence of living material in water providing sensitive and rapid estimate of biomass reported in micrograms per liter.

ADEQUATE SO_2 EMISSION LIMITATION Discharge limit approved or promulgated as sufficient to achieve and take care of atmospheric quality standard in regions impacted by smokestack discharges without usage of any unauthorized dispersion technique.

ADEQUATE STORAGE Setting of pesticides in appropriate containers as well as in safe areas in order to minimize chance for escape that could cause unreasonable adverse impacts on environmental surroundings.

ADEQUATELY WET Material sufficiently mixed or penetrated with liquid to prevent release of particulates.

ADEQUATELY WETTED Sufficiently mixed or coated with water or aqueous solution to prevent dust emissions.

ADHESION 1. Shearing resistance between soil and another material under zero externally applied pressure; 2. Land quality of sticking to metal components of mechanical equipment; 3. State where two surfaces are held together by interfacial effects that could include molecular forces and interlocking actions.

ADHESIVE WATER Film of water encompassing grain of water-carrying substance after gravity water has been drained off.

ADHESIVE WEAR Removal of substances from surfaces in relative motion due to surface contact.

ADIABAT Line on thermodynamic chart relating pressure and temperature of substance (e.g., air) undergoing transformation where no heat is exchanged with its environment.

ADIABATIC 1. Changes in temperature caused by expansion (cooling) or compression (warming) of body of air as it rises or descends in atmosphere, with no exchange of heat with surrounding air; 2. Condition where heat neither enters nor leaves system.

ADIABATIC LAPSE RATE Rate of decrease of temperature experienced by parcel of air when it is lifted in atmosphere under restriction that it cannot exchange heat with its environment.

ADIABATIC PROCESS Process which occurs with no exchange of heat between system and its environment.

ADIRONDACK TYPE SNOW SAMPLING SET Snow sampler consisting of 5-ft. fiberglass tube, 3 in. in diameter, 60-in. snow depth capacity, with serrated-edge steel cutter at one end and twisting handle at other.

ADIT 1. Nearly horizontal underground excavation in abutment having opening in only one end; 2. Opening in face of dam for access to galleries or operating chambers.

ADJACENT Bordering, contiguous or neighboring.

ADJUSTABLE TIMBER SINGLE-POST SHORE Individual timber used with fabricated damp to obtain adjustment.

ADJUSTED BASE COST Base construction cost adjusted for alternate components, plus or minus.

ADJUSTED DISCHARGE Discharge data that have been mathematically adjusted (i.e., to remove effects of daily tide cycle or reservoir storage).

ADJUSTING NUT Threaded nut that, when turned, will alter some already recognized allowance.

ADJUSTING SCREW 1. leveling apparatus or jack composed of threaded screw and adjusting handle, usually employed for vertical adjustment of shoring and formwork; 2. Threaded screw that, when turned, will induce two linked faces to go farther apart or closer together.

ADJUSTMENT Behavioral response to environmental conditions.

ADJUSTMENT FACTORS Numerical value added or subtracted from parameter.

ADJUTAGE TUBE Tube fitted into orifice; efflux tube.

ADMINISTERED DOSE Amount of substance given to test subject to determine dose-response relationships.

ADMINISTRATIVE AUTHORITY Public officials, bodies, commissions or institutions authorized under law to enforce provisions of policy, code and regulation.

ADMINISTRATIVE CHARGES Overhead prices caused by management.

ADMINISTRATIVE CLOSURE Creating, collecting and disseminating information to formalize project completion.

ADMINISTRATIVE ORDER Legal document directing individual, business or other entity to take corrective action or refrain from activity, describes violations and actions to be taken and can be enforced in court.

ADMINISTRATIVE ORDER ON CONSENT 1. Legal agreement and individual, business or other entity where violator agrees to pay for correction of violations, take required corrective or cleanup actions or refrain from activity; 2. Describes actions to be taken, may be subject to comment period, applies to civil actions and can be enforced in court.

ADMINISTRATIVE PROCEDURES ACT Law that spells out procedures and requirements related to promulgation of regulations.

ADMINISTRATIVE RECORD All documents considered or relied on in selecting response action.

ADMIXTURE Mixing in something.

ADOBE Fine clay or loess deposits caused by ice abrasion during glacation.

ADOLESCENT RIVER River in second stage of new drainage system; one that has largely obliterated lakes and waterfalls of its youthful stage that possesses well-cut channel that may reach base level at its mouth.

ADSORBATE Material being removed by adsorption process.

ADSORBED WATER Water in soil or rock mass, held by physicochemical forces, having physical properties substantially different from absorbed water or chemically combined water, at same temperature and pressure.

ADSORBENT Material (e.g., activated carbon) responsible for removing undesirable substance in adsorption process.

ADSORBER Substance or apparatus used to adsorb another substance.

ADSORPTION 1. Removal of pollutant from air or water by collecting pollutant on surface of solid material (e.g., activated carbon removes organic matter from wastewater); 2. Bonding of chemicals to soil particles or other surfaces; 3. Adherence of gas, liquid or dissolved material on surface of solids; 4. Increase in concentration of gas or solute at interface of two-phase system.

ADSORPTION ISOTHERM Graphical representation of relationship between solute concentration and mass of solute species adsorbed on aquifer sediment or rock.

ADSORPTION RATIO Amount of test substance adsorbed by solid phase (i.e., soil) divided by amount of test substance in solution phase that's in equilibrium with solid phase.

ADSORPTION WATER Water retained on surface of solid particles by molecular forces with emission of heat.

ADULTERANTS Chemical impurities or substances that by law do not belong in food or pesticide.

ADULTERATED 1. Any pesticide whose strength or purity falls below quality stated on its label; 2. Food, feed or product that contains illegal pesticide residues.

ADVANCE 1. Establish something to happen ahead of set mark or time; 2. Payment made on account of, but before end of, contract or before receipt of goods or services.

ADVANCE GRADING Area vehicle traverses prior to contacting end terminal.

ADVANCE PREEMPTION Notification of approaching train forwarded to highway traffic signal controller unit or assembly by railroad equipment in advance of activation of railroad warning devices.

ADVANCE PREEMPTION TIME Additional time requested to preempt traffic signal controller from railroad company equipment beyond what is provided to railroad active warning devices.

ADVANCED AIR EMISSION CONTROL DEVICES Air pollution control equipment used to treat air discharge that has been initially treated.

ADVANCED DECISION SUPPORT SYSTEM Computer software designed to provide easy access to and allow efficient use of methods of analysis and information management.

ADVANCED GAS-COOLED REACTOR Nuclear reactor working at 675 °C and higher that's fuelled with enriched uranium dioxide, cooled by gaseous carbon dioxide that uses graphite as moderator.

ADVANCED PUBLIC TRANSPORTATION SYSTEMS Application of transportation management and information technologies to public transit systems to increase their efficiency of operation and improve safety.

ADVANCED RURAL TRANSPORTATION SYSTEM Intelligent transportation system for travelers in and through rural areas as well as of agencies responsible for operation and maintenance of rural transportation systems.

ADVANCED SEWAGE TREATMENT Treatment necessary to remove suspended and dissolved substances not normally removed during secondary treatment.

ADVANCED TRAFFIC MANAGEMENT SYSTEM System that integrates technology primarily to improve safety and vehicle traffic flow.

ADVANCED TRAVELER INFORMATION SYSTEM Traffic system to assist surface transportation travellers in moving from starting location to their desired destination.

ADVANCED TREATMENT Level of wastewater treatment more stringent than secondary treatment that requires 85% reduction in conventional pollutant concentration or significant reduction in non-conventional pollutants.

ADVANCED VEHICLE CONTROL AND SAFETY SYSTEMS Safety system to avoid crashing through improved drive performance.

ADVANCED VERY HIGH RESOLUTION RADIOMETER Main sensor on U.S. polar orbiting satellites.

ADVANCED WASTEWATER TREATMENT Any treatment of sewage that goes beyond secondary or biological water treatment stage and includes removal of nutrients (e.g., phosphorus, nitrogen) and high percentage of suspended solids.

ADVANCED WEATHER INTERACTIVE PROCESSING SYSTEM Allows operator to overlay meteorological data from variety of sources.

ADVECTION 1. Transport of matter or energy in horizontal stream of air; 2. Process where solutes are transported by moving groundwater.

ADVECTION FOG Fog that forms when warm air flows over cold surface and cools from below until saturation is reached.

ADVENTIVE PLANT Introduced or alien plant that grows without aid but is not permanently established.

ADVERSE CONSEQUENCES Negative impacts that may result from failure of dam where primary concerns are loss of human life, economic loss (including property damage), lifeline disruption and environmental impact.

ADVERSE EFFECT Action that has apparent direct or indirect negative effect on conservation and recovery of ecosystem component listed as threatened or endangered.

ADVERSE EFFECTS DATA FIFRA requires pesticide registrant to submit data to EPA on any studies or other information regarding unreasonable adverse effects of pesticide at any time after its registration.

ADVERSE POSSESSION Right of occupant to obtain title in defiance of another person's lawful title following occupancy to get mandatory statutory period in genuine, open, notorious, exclusive use.

ADVERTISEMENT Public announcement for work bids.

ADVIS Program that combines API method of estimating runoff with unit hydrograph theory to estimate streamflow for headwater basin.

ADVISORY 1. Highlights special weather conditions that are less serious than warning; 2. Non-regulatory document that communicates risk information to those who may have to make risk management decisions.

ADVISORY COUNCIL ON HISTORIC PRESERVATION Promotes preservation, enhancement and sustainable use of our nation's diverse historic resources that advises President/Congress on national historic preservation policy.

ADVISORY SPEED Recommendation speed.

AEOLIAN 1. Materials carried, deposited, produced or eroded by wind; 2. Windblown.

AEOLIAN DEPOSIT Carried or blown by wind.

AEOLIAN DEPOSITS Wind-deposited material (e.g., dune sands, loess deposits).

AERATE 1. Impregnate with gas (e.g., air); 2. Act of combining with air; 3. To supply air to water, soil or other media.

AERATED LAGOON Holding and/or treatment pond that speeds up natural process of biological decomposition of organic waste by stimulating growth and activity of bacteria that degrade organic waste.

AERATED POND Natural or man-made wastewater treatment pond where mechanical or diffused-air aeration supplements oxygen supply.

AERATION 1. Process of adding air to water by either passing air through water or passing water through air and may be passive (i.e., waste is exposed to air) or active (i.e., mixing or bubbling device introduces air); 2. Process that promotes biological degradation of organic matter in water.

AERATION PERIOD 1. Time when water is subjected to mechanical aeration; 2. Time when combined liquor is subjected to mechanical aeration while getting activated sludge treatment.

AERATION TANK Open-topped chamber where air is added under pressure to fluid (e.g., sewage, wastewater) to oxidize organic fraction.

AERATION ZONE Depth of soil below ground to top of groundwater level; divided into three layers, upper soil belt, intermediate belt and lower capillary fringe.

AERATOR Apparatus that adds air to water.

AERIAL ATTACK Utilization of aircraft in fighting forest fires, mostly to drop fire retarding or extinguishing solutions on or in course of, conflagration.

AERIAL PHOTOGRAMMETRY Interpreting information from aerial photographs.

AERIAL PHOTOGRAPHY Photos taken from air at regular times and used in photo interpretation to offer information regarding land development, plant life and land forms.

AERIAL PHOTOMAP Aerial photograph or photomosaic where fundamental mapping data was added.

AERIAL PHOTOMOSAIC Composite of aerial photographs depicting some of earth's surface.

AERIAL PLANKTON Spores, bacteria along with other microorganisms floating in air.

AERIAL SURVEYING Studying done through photo-interpretation of pictures taken from aircraft flying grid pattern in fixed elevation within target area.

AEROALLERGEN Airborne pollen and organic debris irritant to many people, causing allergenic states.

AEROALLERGENS Any of variety of allergens (e.g., pollens, grasses, dust) carried by winds.

AEROBE Organism that uses oxygen to support life, used in water purification and sewage treatment.

AEROBIC 1. Condition where free (atmospheric) or dissolved oxygen is present in water; 2. Opposite of anaerobic; 3. Life or processes that require or are not destroyed by, presence of oxygen.

AEROBIC BACTERIA Bacteria that require free elemental oxygen for growth.

AEROBIC DECOMPOSITION Rot or breaking down of organic substance in existence of free or dissolved oxygen.

AEROBIC DIGESTION Breakdown of organic parts by microbial activity in existence of oxygen.

AEROBIC PROCESS Process that relies on existence of dissolved oxygen.

AEROBIC RESPIRATION Oxidation of organic compounds by oxygen.

AEROBIC TREATMENT Process by which microbes decompose complex organic compounds in presence of oxygen and use liberated energy for reproduction and growth.

AEROBIC TREATMENT UNIT Mechanical wastewater treatment unit that provides secondary wastewater treatment for single home, cluster of homes or commercial establishment by mixing air (oxygen) and aerobic and facultative microbes with wastewater.

AEROBIC WASTE TREATMENT PROCESS Breakdown of wastes by microorganisms in existence of free or dissolved oxygen.

AEROBIOSIS Biological processes that must have oxygen.

AEROCHLORINATION Injection of combination of chlorine gas and compressed air into wastewater for removing grease.

AERODYNAMIC DIAMETER Diameter of sphere of unit density that acts aerodynamically like particles of test material.

AERODYNAMIC DRAG Resistance of solid body to move through air.

AERODYNAMIC ROUGHNESS Turbulence in moving air caused by unevenness of surface that it passes over.

AERODYNAMICS Branch of physics that deals with all different forces related to air or other gases in movement.

AEROELASTIC VIBRATION Structure vibration due to wind.

AEROFILTER Layer of coarse aggregate utilized for high-speed filtering of sewage effluent.

AEROGENERATOR Apparatus that relies on motion of air to turn rotor blades to produce electricity.

AEROHYDROUS Of both water and air commonly applied to minerals containing water in pores or cavities.

AERONAUTICAL INFORMATION PUBLICATION Operational manual for flight planning.

AEROSOL 1. Finely divided material (solid or liquid) suspended in air or other gaseous environment; 2. Pressurized gas used to propel substances out of container; 3. Larger than molecule and can be filtered from air.

AEROSOL PROPELLANT Inert liquid having low boiling point that vaporizes instantaneously at room temperature on release of pressure; used to carry and disperse compound.

AEROSOL SPRAY Container where propellant is combined using material held under pressure that discharges its contents as fine mist.

AEROSPHERE Lower part of atmosphere surrounding earth, consisting of troposphere and stratosphere.

AESTHETIC OBJECTIVE Established parameters or features of drinking water that can impair its approval by consumers or interfere with practices for supplying great-quality water.

AESTHETIC VALUE Worth of item based on performance or appearance.

AESTHETICS Study or philosophy of item value based on appearance.

AESTIDURILIGNOSA Mixed evergreen and deciduous forest.

AESTILIGNOSA Deciduous forest and brush.

AETIOLOGY 1. Study of causes or causation; 2. Study or investigation of causes of disease.

AFFECTED ENVIRONMENT Existing biological, physical, social and economic conditions of area subject to change, both directly and indirectly, as result of proposed human action.

AFFECTED LANDFILL Under Clean Air Act, landfills that meet criteria for capacity, age and emissions rates set by EPA and are required to collect and combust their gas emissions.

AFFECTED PUBLIC 1. People who live and/or work near hazardous waste site; 2. Human population adversely impacted following exposure to toxic pollutant in food, water, air or soil.

AFFLUENT STREAM Stream or river that flows into another or other body of water; tributary.

AFFLUX 1. Difference in elevation between high flood levels downstream and upstream of weir; 2. Upstream increase of water level above normal caused by obstruction in channel or constriction of bridge or weir or by regulation.

AFTER-FLUSH Water bled from stream refilling cistern following flushing of toilet and redirected to bowl to make certain seal is intact.

AFTERBAY 1. Body of water immediately downstream from power plant or pumping plant that regulates fluctuating discharges; 2. Short stretch of stream or conduit or to pond or reservoir; 3. Tailrace of hydroelectric generating station, with restricted outflow at discharge of turbines; 4. Initial section of stream or conduit forming discharge from storage pond or reservoir.

AFTERBAY DAM Dam located downstream from large hydroelectric power plant used to regulate discharges downstream.

AFTERBURNER Burner located so that combustion gases are made to pass through its flame in order to remove smoke and odors that may be attached to or separated from incinerator proper.

AFTERFILTER High-efficiency filter located near terminal unit in air-conditioning system.

AFTERSHOCK Little seismic tremor following earthquake of considerable magnitude and happening close to focus of primary shock.

AFTFRPRECIPITATION Deposit of colloidal calcium carbonate on individual grains of sand filter and/or walls of conduits of related supply system following treatment of water with lime.

AGDISP Particular atmospheric dispersion model used for treating transport and diffusion of aerially sprayed pest control agents in agricultural applications.

AGE/STRENGTH RELATIONSHIP Increase in durability of substance as result of maturing process and/or chemical reaction.

AGENCY 1. Relationship between principal and agent whereby agent is employed to undertake certain acts on principal's behalf when dealing with third party; 2. Administrative branch of government.

AGENT Any physical, chemical or biological entity that can be harmful to organism.

AGENT ORANGE Toxic herbicide and defoliant.

AGEOSTROPHIC Atmosphere movement on account of asymmetry involving horizontal pressure gradient force as well as deviating force because of wind speed that's related to vertical motion, cloud formation and weather change.

AGGLOMERATE 1. Ice cover of floe formed by freezing together of various forms of ice; 2. Collecting together into floc of suspended colloidal particles; 3. Coarse-grained pyroclastic rock.

AGGLOMERATION Particle build-up.

AGGRADATION 1. Geologic process wherein streambeds, floodplains, sandbars and bottom of water bodies are raised in elevation by deposition of sediment; 2. Opposite of degradation; 3. Building up of portion of land toward uniformity of grade or gradient.

AGGRADING RIVER River in process of building up its valley bottom through deposition of materials carried forward by flow.

AGGREGATE 1. Crushed rock or gravel screened to sizes for use in road surfaces, concrete or bituminous mixes; 2. Mass or cluster of soil particles, often having characteristic shape.

AGGREGATE BLENDING Process of intermixing two or more aggregates to create distinct pair of properties, usually, but not entirely, to enhance grading.

AGGREGATE DELAY Summation of delays for all vehicles passing through signals, multiple lanes or groups.

AGGREGATE FACILITY Individual drain system together with ancillary downstream sewer lines and oil-water separators, down to and including any secondary oil-water separator.

AGGRESSIVE CONDITIONS Environmental conditions that require special consideration in selection of materials and/or construction.

AGGRESSIVE SOLUTION Any chemical that attacks material where it is spread.

AGING Process where substance or complex changes with time.

AGITATION Shaking or stirring so as to mix various constituents.

AGITATOR Mechanical equipment for blending and/or aerating.

AGONIC LINE Line on chart joining points of zero magnetic declination.

AGONISTIC BEHAVIOR Competitive or aggressive behavior displayed by animal toward another.

AGREEMENT Written agreement between two parties.

AGRIC HORIZON Depositional soil horizon resulting from cultivation.

AGRICULTURAL DRAIN Tiles, typically 4 inches in diameter, placed end to end without shutting joints so as to empty subsoil.

AGRICULTURAL DRAINAGE 1. Process of directing excess water away from root zones by natural or artificial means; 2. Water drained away from irrigated farmland.

AGRICULTURAL DRAINAGE WELLS Class V wells that receive agricultural runoff including improved sinkholes, abandoned drinking water wells, wells that recharge aquifers with agricultural tail waters and wells that drain flood irrigation.

AGRICULTURAL LIME Hydrated lime, used to condition soil.

AGRICULTURAL POLLUTION 1. Farming wastes, including runoff and leaching of pesticides and fertilizers; 2. Erosion and dust from plowing; 3. Improper disposal of animal manure and carcasses; 4. Crop residues and debris.

AGRICULTURAL SOLID WASTE Solid waste that's created by rearing of animals, as well as creating and harvesting of crops or trees.

AGRICULTURAL WASTE Poultry and livestock manure and residual materials in liquid or solid form generated from production and marketing of poultry, livestock or fur-bearing animals that also includes grain, vegetable and fruit harvest residue.

AGROCHEMICAL Synthetic chemicals (i.e., pesticides, fertilizers) used in agricultural production.

AGROECOSYSTEM 1. Land used for crops, pasture and livestock and adjacent uncultivated land that supports other vegetation and wildlife and associated atmosphere, underlying soils, groundwater and drainage networks; 2. Community of microorganisms, plants as well as creatures, with their abiotic environment, that exists on farmed acreage.

AGROFORESTRY Practice of agricultural incorporating cultivation and conservation of trees.

AGRONOMIC RATES Digested sewage sludge application rates that provide amount of nitrogen needed by specific crop or vegetation being grown while minimizing amount that passes below root zone.

AGRONOMY Study and application of soil and plant sciences to agriculture.

AHERA DESIGNATED PERSON Person designated by Local Education Agency to ensure that AHERA requirements for asbestos management and abatement are properly implemented.

AIMLESS DRAINAGE Stream or flow pattern common to swampy ground.

AIR Mixture of gases comprising Earth's atmosphere.

AIR BALANCING Chain of alterations made to air-conditioning system to make sure that conditioned air is equally distributed to any or all rooms and places.

AIR BARRIER Material used in house envelope to retard passage of air.

AIR BINDING 1. Decrease in rate of infiltration into soil resulting from increase in pressure of air trapped in interstices between soil grains; 2. Reduction in efficiency of filter, conduit or pump due to entrained air.

AIR BOTTLE Thick-walled, steel or composite metal cylinder equipped with control valve, pressure gauge and nipple outlet for high-pressure hose connection, used to contain compressed air or oxygen as emergency breathing supply.

AIR BREAK 1. Brake operated by piston driven by compressed air; 2. Piping arrangement where waste discharges into open connection and then into receptacle/interceptor to prevent backflow or siphonage.

AIR CHAMBER Closed pipe chamber on discharge of reciprocating pump that absorbs irregularities in hydraulic conditions and helps induce uniform flow in suction and discharge lines and protect pump against shocks caused by pulsating flow.

AIR CHAMBER PUMP Displacement pump designed using air chamber where air is alternately compressed and expanded from water displaced by pump.

AIR CHANGE Measure of how many times air within defined space (e.g., building) is replaced.

AIR CHANGES PER HOUR Movement of volume of air in given period of time (e.g., if building has one air change per hour, means that air in building will be replaced in one-hour period).

AIR CIRCULATION Movement of air, either through convection or under pressure in circulation system, from one place to another.

AIR CLASSIFICATION Process employing air stream to separate materials by difference in density and aerodynamic properties.

AIR CLASSIFIER Mechanical device using air currents to separate solid components into "light" or "heavy" fractions.

AIR CLEANER FILTER ELEMENT Apparatus to get rid of particulates from primary air that enters air induction system of engine.

AIR COLLECTOR Solar collector that uses air as heat transfer medium.

AIR COMPRESSOR Machine used to take in air at ambient pressure and compress it to pressures up to about 8.5 atmospheres.

AIR CONDITIONER Mechanical device capable of adjusting temperature and humidity of collected air and delivering it back to rooms and areas contained within system.

AIR CONDITIONING Process of colling, heating , humidifying, dehumidifying, cleaning and/or circulating air.

AIR CONDITIONING SYSTEM Assembly of gear, detector and controls required for conditioning and circulation of air to defined standards.

AIR CONDITIONING UNIT Compact, independent, self-contained device that conditions and circulates air.

AIR CONTAMINANT Particulate matter, gas or combination thereof, other than water vapor.

AIR CONTAMINANT CONTROL Process that neutralizes air contaminants and odors.

AIR CONTAMINANTS Airborne materials, solid, liquid or gaseous, that are believed to cause harm to life forms or damaging to operation of machinery.

AIR COOLER Apparatus that lowers temperature of air between its inlet and outlet stages.

AIR CURTAIN 1. Air bubbling through perforated pipe causes upward water flow that slows spread of oil used to stop fish from entering polluted water; 2. Use of bubbles rising from perforated pipe placed across bottom of tank to prevent horizontal mixing of tank contents.

AIR DEFICIENCY Not enough adequate air within air-fuel mixture, to provide amount of oxygen needed to fully oxidize fuel.

AIR DIFFUSER Device that converts stream of air into diffuse and irregular flow to promote mixing of primary air with secondary air, typically within MOM or space.

AIR DISPLACEMENT PUMP Displacement pump where compressed air is used as displacement agent in cylinders.

AIR DISTRIBUTION Supply of conditioned air from central plant under pressure through ductwork to exit louvres and vents.

AIR DUCT Pipe, tube or passageway for conveying air, normally associated with heat, jug, ventilating and air conditioning.

AIR ELIMINATOR Pipe fitting for automatic release of trapped air and gas from supply system.

AIR EMISSIONS Airborne solid particulates, gaseous pollutants or odors emanating from source.

AIR ENTRAINING Procedure of injecting air into substance in way such that it forms tiny bubbles of consistent size and dispersion.

AIR ESCAPE Mechanism that permits excess air to discharge from water pipe.

AIR EXCAVATION EQUIPMENT Gear to force compressed air by means of wand to be able to remove soil from around conduit, cable or another buried services in form of jetting.

AIR EXCHANGE RATE Rate at which outside air replaces indoor air in given space.

AIR FILTER Apparatus that immobilizes solid or gaseous pollutants from air.

AIR GAP Open vertical gap or empty space that protects and separates drinking water supply from contamination by backflow or back siphonage by another water system in treatment plant.

AIR HANDLING EQUIPMENT Variety of devices designed to move air under controlled conditions.

AIR HANDLING UNIT Type of heating and/or cooling distribution that channels warm/cool air to different parts of building.

AIR HEATER Heat exchanger where air passes and is heated by medium of higher temperature.

AIR INLET Opening where air is drawn.

AIR INTAKE Apparatus where air is drawn.

AIR LIFT Means of lifting fluids by pushing air to underside of open discharge pipe submerged in well or tank.

AIR LOCK 1. Bubble of air, located in pipe or in pump that has not been properly primed, that prohibits movement of liquid; 2. Chamber that can be made airtight, separating two other rooms, chambers or areas.

AIR MAINTENANCE DEVICE Apparatus that automatically keeps necessary air pressure in sprinkler system.

AIR MASS Body of air covering large area that possesses horizontally uniform, distinct characteristics.

AIR MASS THUNDERSTORM Thunderstorm not associated with front or other type of synoptic-scale forcing mechanism.

AIR MIXING PLENUM Chamber in air-conditioning system where recirculating air is combined with fresh air.

AIR MONITORING Means taken to measure quality of air and its constituents.

AIR NAVIGATION FACILITY Any facility used, available for use or designated for use in aid of air navigation.

AIR OUTLET Apparatus that allows air to pass from within conditioning or ducting system.

AIR OXIDATION REACTOR Device or process vessel where one or more organic reactants are combined with air or combination of air and oxygen, to produce one or more organic compounds, including ammoxidation and oxychlorination reactions.

AIR PACK Self-contained breathing gear providing you with air supply from tanks by way of regulator.

AIR PADDING Pumping dry air into container to assist with withdrawal of liquid or to force liquefied gas out of container.

AIR PARCEL Theory involving body of air whose properties and transformations may be determined from laws of physics and mechanics.

AIR PERMEABILITY Property that allows passage of air through mass.

AIR PLENUM Any space used to convey air in building, furnace or structure.

AIR POLLUTION Presence of contaminants or pollutant substances in air that interfere with human health or welfare or produce other harmful environmental effects.

AIR POLLUTION CONTROL AGENCY Any agency charged with responsibility and has authority to monitor air quality and enforce means that to maintain defined quality.

AIR POLLUTION CONTROL DEVICE Mechanism or equipment that cleans emissions generated by source by removing pollutants that would otherwise be released to atmosphere.

AIR POLLUTION INDEX Arbitrary expression of concentration of one or more pollutants within air sample, used to scale intensity of air pollution in location of sampling.

AIR POLLUTION POTENTIAL Meteorological potential for air pollution problems, considered without regard to presence or absence of actual pollution sources.

AIR PRESSURE SPRINKLER SYSTEM Sprinkler system where air pressure is used to force water from storage tank into system.

AIR PUMP Air pump that compresses, exhausts or forces it through something.

AIR PURGE VALVE Valve that removes trapped air from piping system.

AIR QUALITY 1. Measure of health-related and visual characteristics of air; 2. Concentration in measured sample of air of one or more pollutants.

AIR QUALITY ACT Laws that creates minimal acceptable air quality.

AIR QUALITY CRITERIA Levels of pollution and lengths of exposure above which adverse health and welfare effects may occur.

AIR QUALITY MANAGEMENT Principles and strategy where Air Quality Act or similar laws will be managed.

AIR QUALITY MODEL Mathematical or conceptual model used to estimate present or future air quality.

AIR QUALITY RESTRICTED OPERATION OF SPRAY TOWER
Process using formulas that need high speed of wet scrubbing to keep desired quality of smokestack gases that creates greater amount of waste water than can be recycled.

AIR QUALITY STANDARDS Level of pollutants prescribed by regulations that are not be exceeded during given time in defined area.

AIR REHEATER Apparatus that adds heat to air being circulated through system.

AIR RELEASE VALVE Valve used to release air from pipe or fitting under pressure.

AIR RELIEF VALVE Air apparatus set in apex of pipeline to mechanically vent air and prevent from becoming air-bound.

AIR RIGHTS Property rights for management or particular usage of designated airspace.

AIR ROUTE SURVEILLANCE RADAR Long-range radar to control airspace within and around borders of United States.

AIR ROUTE TRAFFIC CONTROL CENTER Facility for controlling aircraft flying in particular airspace at high altitudes between approaches and departures at airports.

AIR SCOURING Method to clean up to 8 inches in diameter water mains by introducing short, high-pressure air bursts that remove most deposits from inside walls.

AIR SIDE ECONOMIZER Air economizer cycle reduces load on chilled water system by increasing flow of outside air above minimum required for ventilation when outside air temperature is favorable in comparison to return air temperature.

AIR SLAKED Wetted by exposure to moisture in atmosphere.

AIR SLAKING Process of breaking up or sloughing when indurated soil is exposed to air.

AIR SPACE Landfills remaining volume or capacity for waste disposal at given time, not including volume occupied by cap, cover and liner.

AIR SPARGING Injecting air or oxygen into aquifer to strip or flush volatile contaminants as air bubbles up through groundwater and is captured by vapor extraction system.

AIR STAGNATION Meteorological situation where there is major buildup of air pollution in atmosphere.

AIR STAGNATION ADVISORY National Weather Service product issued when major buildups of air pollution, smoke, dust or industrial gases are expected near ground for period of time.

AIR STRIPPING Treatment system that removes volatile organic compounds from contaminated groundwater or surface water by forcing airstream through water and causing compounds to evaporate.

AIR SUPPLY UNIT Machine to refill, on-site, exhausted compressed self-contained breathing apparatus air tanks.

AIR TEST Evaluation of integrity of drain where segment of system is sealed and subjected to positive air pressure that shouldn't drop below predetermined percentage of initial pressure over defined amount of time.

AIR TOXICS Any air pollutant for which national ambient air quality standard does not exist (i.e., excluding ozone, carbon monoxide, PM-10, sulfur dioxide, nitrogen oxide) that may reasonably be anticipated to cause cancer, respiratory, cardiovascular or developmental effects, reproductive dysfunctions, neurological disorders, heritable gene mutations or other serious or irreversible chronic or acute health effects in humans.

AIR TRAFFIC CONTROL System designed to give assistance to aircraft, to prevent collisions and to manage aircraft flow.

AIR TRAFFIC CONTROL RADAR BEACON SYSTEM Air traffic control system used to separate air traffic and enhance surveillance radar monitoring.

AIR TRANSPORTABLE MOBILE UNIT Modularized transportable unit containing communications and observational equipment necessary to support meteorologist preparing on-site forecasts at wildfire or other incident.

AIR TREATMENT Any process that changes properties of air.

AIR VENT 1. Method of releasing trapped air, at highest location within water distribution system; 2. Opening, possibly covered by grate or grill that permits air to circulate.

AIR VENTILATION Amount of clean air needed to keep desired quality of air in space.

AIR VESSEL Vessel, tank or duct containing or conducting air under pressure.

AIR WASH Procedure where compressed air and water wash medium of rapid sand filter.

AIR WASHER System that cleans air typically using water spray system.

AIR WAVES Airborne oscillations or pulsations.

AIR-ADMITTANCE VALVE Device that lets air into pipe system to ease small differences in pressure.

AIR-BOUND State where air obstructs free flow of water due to being trapped at high point of pipeline or pump.

AIR-DISTRIBUTING ACOUSTICAL CEILING Hung acoustical ceiling that allows passing of measured volume of air from plenum pressurized above.

AIR-EXHAUST VENTILATOR Fan ducted to outside that removes odors and/or fumes.

AIR-HANDLING SYSTEM System installed in building to circulate and recondition air.

AIR-SLACK Clay that will swell after absorbing moisture and being subjected to air.

AIR-SPACE RATIO Ratio of volume of water that can be drained from saturated soil or rock under action of force of gravity to total volume of voids.

AIR-TO-AIR RESISTANCE Resistance to flow or mixing of air caused by temperature or pressure differentials, laminar flow, etc.

AIR-VOID RATIO Ratio of volume of airspace to total volume of voids in soil mass.

AIR-WATER STORAGE TANK Water storage container that contains air, under pressure, above water level.

AIR/OIL TABLE Surface between vadose zone and ambient oil where pressure of oil in porous medium is equal to atmospheric pressure.

AIRBLAST 1. Sudden rush of air, in confined space, following disturbance (e.g., explosion); 2. Pressure wave of sound caused by disturbance travelling through atmosphere.

AIRBLAST FOCUSING Sound energy concentrated in small, ground area as result of refraction of sound waves.

AIRBORNE PARTICULATE MATTER Suspended tiny particles of solid matter.

AIRBORNE PARTICULATES Total suspended particulate matter found in atmosphere as solid particles or liquid droplets.

AIRBORNE RELEASE Release of any pollutant into air.

AIRBORNE SNOW SURVEY PROGRAM Center that makes airborne snow water equivalent and soil moisture measurements over large areas of country that are subject to severe and chronic snow melt flooding.

AIRBORNE SNOW WATER EQUIVALENT MEASUREMENT THEORY Theory based on fact that natural terrestrial gamma radiation is emitted from potassium, uranium and thorium radioisotopes in upper eight inches of soil.

AIRBORNE SOUND Noise and/or sound passing through air.

AIRCRAFT CLASSIFICATION NUMBER Number expressing relative effect of aircraft on runway pavement for specified standard subgrade category.

AIRFIELD SOIL CLASSIFICATION System based on sieve analyses and consistency limits of soils where cohesive soils below liquid limit of 50% are generally classified as silts and above liquid limit of 50% are generally classified as clays.

AIRLIFT PUMP Pump consisting of two pipes that passes compressed air from smaller pipe that in turn increases pressure, forcing fluid into larger pipe, typically used in tank or well.

AIRLINE MASK Outside contaminated area, breathing apparatus linked to air supply.

AIRLINE PRESSURE REGULATOR Device that transforms fluctuating air pressure supply to continuous, lower pressure flow.

AIRPLANE Powered heavier-than-air aircraft with fixed wings.

AIRPORT Airfield including one or more runways and one or more passenger terminals for commercial airports.

AIRPORT ADVISORY AREA Area within ten miles of airport without control tower or where tower is not in operation but where flight service station is located.

AIRPORT ADVISORY SERVICE Provides pilots with updates on local weather and information about known traffic operating on and around airport.

AIRPORT CAPACITY Measure of maximum number of aircraft operations that can be accommodated at airport within given period of time.

AIRPORT LAYOUT PLAN Depicts existing facilities and planned development for airport.

AIRPORT SURVEILLANCE RADAR Airport radar system used to detect and display presence and position of aircraft in airspace around airports.

AIRSHIP Lighter-than-air aircraft that can be propelled forward through air.

AIRSPEED Speed of aircraft relative to air through which it is flying.

AIRTIGHT At designed pressures, impermeable to passage of air or other gases.

AIRTIGHT INSPECTION COVER Sealable, non-ventilating drain cover.

AIRTIGHT WOOD STOVE Closed container that suffocates fire and prevents rapid burning of wood, making it burn more evenly and longer that controls both air input and exhaust.

AIRTIGHTNESS 1. Quality of being airtight; 2. Ability of building envelope to resist infiltration and exfiltration of air.

AIRWAY Flight path used by airplanes.

AIRWORTHY Meeting standards or suitable for safe flight.

ALACHLOR Herbicide, marketed under trade name Lasso, used mainly to control weeds in corn and soybean fields.

ALANINE Non-essential amino acid found in most animal proteins.

ALAR Trade name for daminozide, pesticide that makes fruits color deeper, firmer and less likely to drop off before harvesting.

ALARM 1. Sound and/or information intended to give notice of approaching danger; 2. To disturb.

ALARM BELL Bell that rings after detecting out-of-normal condition.

ALARM CIRCUIT Electrical circuit that links two parts of alarm system.

ALARM CONTACT Switch that works for preset high, low or abnormal conditions.

ALARM SIGNAL Signal indicating emergency or warning of danger.

ALARM SWITCH Device that triggers alarm to sound.

ALARM VERIFICATION Automatic fire system that must report alarm status for minimum time after being reset.

ALASKA CURRENT 1. North Pacific Ocean current flowing counterclockwise in Gulf of Alaska; 2. Northward flowing (warm) division of Aleutian Current.

ALBEDO 1. Reflectivity; 2. Fraction of radiation striking surface reflected by that surface.

ALBERTA CLIPPER Fast moving low pressure system that moves southeast out of Canadian Province of Alberta (southwest Canada) through Plains, Midwest and Great Lakes region usually during winter that's usually accompanied by light snow, strong winds and colder temperatures that's sometimes referred to as "Saskatchewan Screamer."

ALBIC Soil light in color due to absence of clay and iron oxides.

ALCOHOLIC FERMENTATION Ethanol created by fermentation of sugar.

ALDEHYDES Air pollutant, that can be irritant, consisting of organic compounds.

ALDICARB Insecticide sold under trade name Temik, made from ethyl isocyanate.

ALDRIN + DIELDRIN Organochlorine insecticides.

ALERT 1. Flood warning system consisting of remote sensors, data transmission by radio and computer software package developed by National Weather Service; 2. Generic term used for decision making software package.

ALERT FLOOD WARNING SYSTEM Cooperative, community-operated flood warning system.

ALERT STAGE Elevation or stage, of stream at which need-to-know officials are notified of threat of possible flooding that's synonymous with monitor stage or caution stage.

ALEUTIAN CURRENT Eastward flowing North Pacific Ocean current which lies north of North Pacific Current.

ALEUTIAN LOW Semi-permanent, subpolar area of low pressure located in Gulf of Alaska near Aleutian Islands.

ALEVIN Young fish which has not yet absorbed its yolk sac.

ALGAE 1. Simple plants containing chlorophyll that produce oxygen during sunlight hours and use oxygen during night hours; 2. Microscopic plants that contain chlorophyll and live floating or suspended in water.

ALGAL BLOOM Rapid and flourishing growth of algae (e.g., green, bluegreen) in and on body of water as result of high phosphate concentration from farm fertilizers and detergents.

ALGAL GROWTH POTENTIAL Maximum algal dry weight biomass that can be produced in natural water sample under standardized laboratory conditions.

ALGICIDAL Acts as algicide.

ALGICIDE Substance or chemical used specifically to kill or control algae.

ALGISTATIC Inhibits algal growth.

ALGORITHM 1. Computer program (or set of programs) designed to systematically solve certain kind of problem; 2. Step by step procedure for solving problem.

ALIASING Process by which frequencies too high to be analyzed with given sampling interval appear at frequency less than Nyquist frequency.

ALIGHT To get off or exit vehicle.

ALIGNMENT 1. Process of adjusting mechanism such that its parts are aligned; 2. Precise route or course taken by linear way (e.g., road, railway, footpath) between two points.

ALIPHATIC Class of organic compounds where carbon atoms are arranged in open chain.

ALIQUOT Portion of sample where one or more aliquots make up sample.

ALKALI 1. Soluble salt obtained from ashes of plants; 2. Base that dissolves in water creating hydroxide ions and turns reddened litmus to blue.

ALKALI ACCUMULATION In upper layers of soil, concentration of carbonates and hydroxides of alkali metals due to evaporation of water.

ALKALI AGGREGATE REACTION Deterioration of concrete by which alkali in cement paste in concrete reacts chemically with silica or carbonate present in some aggregates.

ALKALI CARBONATE REACTION Reaction of alkali's which occurs between certain argillaceous dolomitic limestones and alkaline pore solution in concrete and causes expansion and extensive cracking.

ALKALI FLAT 1. Dry lake; 2. Remnant of flat lake consisting of fine-grained sediments infused with alkali salts.

ALKALI RESISTANCE When substance resists reaction with alkali.

ALKALI SOIL Soil where development of majority of plant harvests is reduced due to alkalinity and/or sodium content.

ALKALI-SILICA REACTION Reaction of alkali's with aggregate with various forms of poorly crystalline reactive silica: opal, chert, flint and chalcedony and also tridymite, crystoblite and volcanic glasses.

ALKALI-SILICATE/SILICA REACTION Reaction of alkali's with strained quartz is thought to be one reactive component of aggregates causing this reaction.

ALKALINE 1. Having pH greater than 7.0; 2. Water, wastewater or soil containing sufficient amount of alkali substances to raise pH above 7.0; 3. Quality of being bitter due to alkaline content, one of class of caustic bases.

ALKALINE CLEANING Using of solution to remove oil along with similar compounds from metal surface.

ALKALINE CLEANING BATH Bath, consisting of alkaline cleaning solution, to process workpiece.

ALKALINE CLEANING RINSE Rinse following alkaline cleaning tub.

ALKALINE MINE DRAINAGE Mine drainage that has total iron concentration of less than 10 mg/l and pH greater than or equal to 6.0.

ALKALINE SOIL Soil, having pH range of 7.0 and 8.5, containing soluble salts.

ALKALINE WATER 1. Water with pH greater than 7.0; 2. Water having comparatively low dissolved solids, but higher sodium content.

ALKALINITY Capacity of bases, water or solutes to neutralize acids.

ALKYL BENZENE SULFONATE Surface active agent that causes foaming and withstands biological degradation in wastewater treatment.

ALKYL MERCURY Hazardous mercurial compound that can affect body if inhaled, comes in contact with eye or skin or inhaled that's utilized as seed dressing and fungicide.

ALKYL SULFONATES 1. Esters of alkane sulfonic acids; 2. Surface-active (alkylating) agents utilized in artificial detergents.

ALL-AIR System where conditioned air is passed to atmosphere following single circulation cycle.

ALL-WAY STOP-CONTROLLED INTERSECTION Intersection requiring vehicle to stop before proceeding through.

ALL-WEATHER AIRPORT Airport equipped with facilities to permit landing of qualified aircraft and aircrewmen without regard to operational weather limits.

ALLELOCHEMISTRY Effect of organic compound produced and discharged from one organism on another organism separated from its source.

ALLELOPATHY Influence of plants upon each other caused by products of metabolism.

ALLERGEN Substance that causes allergic reaction in individuals sensitive to it.

ALLEY Narrow street or passageway.

ALLEY COLLECTION Group of solid waste containers placed next to or within street.

ALLIGATOR CRACKING Distress in asphalt pavement where cracks form pattern that look like reptile scales.

ALLOCHTHONOUS 1. Deposits of material that originated elsewhere; 2. Originating in place other than where it is found.

ALLOPATRIC Having separate and mutually exclusive areas of geographical distribution.

ALLOTMENT Allowance of specific amount of money or other credit of particular thing to particular person.

ALLOWABLE BEARING CAPACITY Maximum pressure that can be permitted on foundation soil, giving consideration to all pertinent factors, with adequate safety against rupture of soil mass or movement of foundation of such magnitude that structure is impaired.

ALLOWABLE COSTS Prices that are assigned to job that are considered necessary and reasonable.

ALLOWABLE EMISSIONS Emission rate of source calculated using maximum rated capacity of source, subject to enforceable permit conditions or other enforceable limits and any applicable emission standards.

ALLOWABLE HEADWATER Ponded water depth permitted at culvert entrance.

ALLOWABLE HEADWATER DEPTH Depth measured from culvert invert to water surface elevation.

ALLOWABLE PILE BEARING LOAD Maximum load that can be permitted on pile with adequate safety against movement of such magnitude that structure is endangered.

ALLOWABLE STATIC DESIGN STRESS Stress permissible for calculations involving unfactored loading.

ALLOWABLE-CUT EFFECT Immediate increase in today's allowable cut of forest timber attributable to expected future increases in yields.

ALLOWANCE 1. Customary deduction from gross weight of goods; 2. Minimum time operator is paid even if not worked.

ALLOY PIPE Conduit (steel) fabricated with variety of elements including carbon that provides greater durability and corrosion resistance than common, carbon steel conduit.

ALLUVIAL 1. Relating to and/or sand deposited by flowing water (e.g., riverbed, flood plain, delta); 2. Alluvium deposited by stream or flowing water.

ALLUVIAL AQUIFER Water-bearing deposit of unconsolidated material (e.g., sand, gravel) left behind by river or other flowing water.

ALLUVIAL CHANNEL Water channel where bed and banks are made up of loose sediment.

ALLUVIAL CLAY Clay substance deposited by streams and/or rivers.

ALLUVIAL CONE Steep, narrow, cone-shaped alluvial fan formed where swift stream suddenly slows.

ALLUVIAL DEPOSIT Clay, silt, sand, gravel or other sediment deposited by action of streams or rivers.

ALLUVIAL FAN Fan-shaped accumulation of sediment deposited as stream and/or river flow decreases in velocity.

ALLUVIAL PLAIN Plain formed by deposition of alluvial material eroded from periodic flooding of stream and/or river.

ALLUVIAL REACH Reach of river with sediment bed composed of same type of sediment material as that moving in stream.

ALLUVIAL RIVER 1. River where bed and banks are made up of mobile sediment and/or soil; 2. River that has formed its channel through aggradation.

ALLUVIAL SOIL Fine-grained soil deposited by flowing water.

ALLUVIAL STREAM Stream whose channel boundary is composed of appreciable quantities of sediments transported by flow and which generally changes its bed forms as rate of flow changes.

ALLUVIAL-SLOPE SPRING Spring occurring where watertable slope and surface gradient are equal on lower slope of alluvial cone.

ALLUVIATION Deposition of alluvium by stream and/or river.

ALLUVIUM 1. Material transported and deposited by flowing water; 2. Sediments deposited by erosional processes usually in beds of rivers and streams, on flood plain, on delta or at base of mountain.

ALONG-SLOPE WIND SYSTEM Closed, thermally driven diurnal mountain wind circulation whose lower branch blows up or down sloping sidewalls of valley or mountain and upper branch blows in opposite direction closing circulation.

ALPHA DECAY Radioactive decay by emitting alpha particle that reduces atom's atomic number by two.

ALPHA PARTICLE Positively charged nucleus of helium-4 atom (consisting of two protons and two neutrons) emitted as consequence of radioactivity.

ALPHANUMERIC DISPLAY Digital readout that shows both letters and numbers.

ALPINE Land elevations above timber line and below snow line.

ALPINE OROGENY Orogenic phase in Late Mesozoic (Eoalpine) and current Cenozoic period that produced mountain ranges of Alpide belt (i.e., Alps and Carpathians in southern Europe and Atlas Mountains in northwestern Africa).

ALPINE SNOW GLADE Marshy clearing between slopes above timberline in mountains.

ALTERED DISCHARGE Discharge other than usual current and/or improved discharge.

ALTERED SPEED ZONE Engineering study determined speed limit.

ALTERNATE BAR Wave patterns of alluvial deposit where crest and trough alternate between banks of channel.

ALTERNATE BARS Bars formed in staggered pattern near banks of channels.

ALTERNATE BID Bid that states amount to be added to or subtracted from quoted price if alternate methods and materials are chosen.

ALTERNATE DEPTH Open channel flow depths with same discharge where specific energy is same.

ALTERNATE METHOD Any method of sampling and analyzing that's not reference or equivalent method but that has been demonstrated to produce adequate results.

ALTERNATE PROPOSAL Submission that different from what's described in bid documents.

ALTERNATING CURRENT Electric current that reverses its direction (i.e., positive/negative values) at regular intervals.

ALTERNATING DEVICE In accordance with predetermined sequence, device to or where stream may be automatically or manually diverted.

ALTERNATING SPRINKLER SYSTEM Fire protection sprinkler system that can have pipes full of water for summer period and empty for winter.

ALTERNATIVE COMPLIANCE 1. Policy that allows facilities to choose among methods for achieving emission-reduction or risk-reduction instead of command-and control regulations that specify standards and how to meet them; 2. Use of theoretical emissions bubble over facility to cap amount of pollution emitted while allowing company to choose where and how (within facility) it complies.

ALTERNATIVE EFFLUENT LIMITATIONS Regulation, limitation or standard that controls thermal component of discharge.

ALTERNATIVE ENERGY Energy generated in ways that do not deplete natural resources or harm environment.

ALTERNATIVE FUELS Substitutes for traditional liquid, oil-derived motor vehicle fuels like gasoline and diesel including mixtures of alcohol-based fuels with gasoline, methanol, ethanol, compressed natural gas and others.

ALTERNATIVE METHOD Any method of air pollution sampling and/or analyzing that's not reference and/or equivalent method but continues to produce adequate results.

ALTERNATIVE REMEDIAL CONTRACT STRATEGY CONTRACTORS Government contractors who provide project management and technical services to support remedial response activities at National Priorities List sites.

ALTERNATIVE TECHNOLOGY Procedure for wastewater treatment that provides for reuse of water, recycle of wastewater, eliminate discharge of pollutants or recover energy.

ALTERNATIVE TO CONVENTIONAL TREATMENT WORKS FOR SMALL COMMUNITY Alternative technology used by treatment works in small communities include specific defined alternative technologies as well as individual and onsite systems; small diameter gravity, pressure or vacuum sewers conveying treated or partially treated wastewater.

ALTERNATIVE TOOL Analysis procedure that computes performance of transportation system.

ALTERNATIVE WATER SUPPLIES Include seawater, brackish groundwater, surface water, stormwater, reclaimed water, aquifer storage and recovery projects and any other non-traditional supply source identified in regional water supply plan.

ALTERNATIVES Courses of action that may meet objectives of proposal at varying levels of accomplishment, including most likely future conditions without project or action.

ALTIMETER Instrument that indicates altitude of object above fixed level.

ALTIMETER SETTING Correction of station pressure to sea level used by aviation.

ALTITUDE Absolute height of location above reference level, usually measured from sea level.

ALTITUDE CONTROL VALVE Valve that controls level of tank by remaining open while tank is not full and closes when tanks reaches its maximum level.

ALTOCUMULUS Cloud characterized by globular masses or rolls of large whitish or greyish cloudlets with shaded portions, in layers or patches, with individual elements being larger and darker than those of cirrocumulus and smaller than those of stratocumulus at altitude of approximately 8,000 to 20,000 feet.

ALTOCUMULUS CASTELLANUS Mid-level clouds (bases generally 8,000 to 15,000 feet) often taller than wide, of which at least fraction of their upper parts show cumulus-type development indicating instability aloft that may precede rapid development of thunderstorms.

ALTOSTRATUS Cloud of class characterized by generally uniform gray or bluish sheet or layer of striated, fibrous or uniform appearance, lighter in color than nimbostratus and darker than cirrostratus at altitude of approximately 8,000 to 20,000 feet.

ALUM astringent salt (aluminum sulfate) used as coagulant in potable and wastewater treatment.

ALUMINA Aluminum oxide, used as desiccant, especially in mining, material sciences and ceramics.

ALUMINIZED STEEL PIPE Steel conduit that combines strength of steel with corrosion resistance of aluminum.

ALUMINOUS CEMENT Cements consisting predominantly of hydraulic calcium aluminate.

ALUMINUM Metallic chemical element with atomic number of 13.

ALUMINUM HYDRAULIC SHORING Pre-engineered shoring system designed to support sidewalls of excavation, comprised of aluminum hydraulic cylinders (cross braces) with vertical (uprights) or horizontal (walers) rails.

ALUMINUM PIPE Conduit fabricated of aluminum alloy used for all types of fabrication projects where lightweight and corrosion resistance is primary concern.

ALUMINUM SULFATE Chemical compound soluble in water and is mainly used as coagulating agent in purification of drinking water and wastewater treatment plants.

AMALGAMATION Dissolving or blending of metal (e.g., gold and silver) in mercury to separate it from its parent material.

AMBIENT 1. Surrounding natural conditions or environment at given place and time; 2. Environmental or surrounding conditions; 3. Encircling.

AMBIENT AIR Any unconfined portion of atmosphere (i.e., open air, surrounding air).

AMBIENT CONDITIONS 1. Weather or atmospheric conditions in place; 2. Set of parameters used for dimensioning ventilation and cooling systems.

AMBIENT MEDIUM Material surrounding or contacting organism (e.g., outdoor air, indoor air, water, soil) where chemicals or pollutants can reach organism.

AMBIENT MOISTURE Moisture in immediate surroundings of something.

AMBIENT MONITORING All forms of monitoring conducted beyond immediate influence of discharge pipe or injection well and may include sampling of sediments and living resources.

AMBIENT TEMPERATURE Temperature of surrounding air or other medium.

AMBIENT TOXICITY Measured by toxicity test on sample collected from water body.

AMBIENT WATER CRITERION Concentration of toxic pollutant in navigable water that will not result in adverse impact on important aquatic life or on consumers of such aquatic life, after exposure of that aquatic life for periods of time exceeding 96 hours and continuing at least through one reproductive cycle.

AMBURSEN DAM Buttress dam where upstream part is relatively thin flat slab usually made of reinforced concrete.

AMEBA single-celled, protozoan organism occurring in soil and water and parasitic in animals.

AMENDED WATER Water that surfactant (wetting agent) has been added.

AMENDMENT Substance added to improve.

AMENITY AREA 1. Regarding residential, space provided for active or passive recreation and enjoyment of occupants of residential development; 2. Regarding nOn-residential, space provided for active or passive recreation and enjoyment of public, during hours that development is open to public, that shall be owned and maintained by owners of development.

AMERICAN STANDARD FITTINGS Standards published by American Standards Institute regarding types and dimensions of pipe fittings.

AMERICAN WHEEL Water turbine where flow is inward first and then becomes axial.

AMIDSHIPS Middle of ship, either longitudinally or laterally.

AMINO ACID Any organic compound containing both amino and carboxylic acid functional group.

AMMONIA Gaseous compound of nitrogen and hydrogen that's common by-product of animal waste that readily converts to nitrate in soils and streams and is used in potable water and wastewater engineering.

AMMONIA SLIP Amount of unreacted ammonia emitted from control equipment such as electrostatic precipitator, selective catalytic reduction (SCR) or selective non-catalytic reduction process.

AMMONIA-N At pH 9.5, result obtained by manual distillation followed by Nesslerization method.

AMMONIACAL NITROGEN Measure of nitrogen present as ammonia, toxic pollutant often found in landfill leachate and in waste products, such as sewage, liquid manure and other liquid organic waste products.

AMMONIATOR 1. To combine or impregnate with ammonia or ammonium compound in potable water or wastewater treatment; 2. To subject to ammonification.

AMMONIFICATION 1. Formation of ammonia or its compounds from nitrogenous compounds; 2. Treatment or impregnation of something with ammonia.

AMMONIUM One form of nitrogen usable by plants.

AMORTIZATION 1. Reduction of loan principal over series of payments; 2. Distribution of cost of intangible asset, such as intellectual property right, over projected useful life of asset.

AMOUNT OF PESTICIDE OR ACTIVE INGREDIENT Ingredient used in producing pesticide expressed as weight for solid and/or semisolid products and as weight and/or volume of liquid products.

AMPERAGE Measurement of amount of electricity flowing past given point in conductor per second.

AMPERE 1. Unit of electric current or rate of flow of electrons equal to flow of coulombs; 2. One volt across 1 ohm of resistance causes current flow of 1 ampere.

AMPERE PER METER Unit of magnetic field strength equal to 1 ampere/meter.

AMPERE TURN Unit of magnetomotive force equal to 1 ampere flowing through 1 turn of magnetizing coil.

AMPERE-HOUR Unit of electric charge that allows 1 ampere of current to flow for 1 hour.

AMPEROMETRIC Based on electric current that flows between two electrodes in solution.

AMPEROMETRIC TITRATION Class of titrations where equivalence point is determined through measurement of electric current produced by titration reaction.

AMPHIBIAN Vertebrate animals that have life stages both in water and on land (e.g., salamanders, frogs, toads).

AMPHIBIAN AND REPTILE POISONS AND REPELLENTS Includes all substances or mixtures intended for preventing, destroying, repelling or mitigating amphibians and reptiles declared to be pests.

AMPHOTERIC 1. Having characteristics of both acid and base and capable of reacting as either; 2. Amphiprotic.

AMPLIFICATION 1. Modification of input bedrock ground motion by overlying unconsolidation materials; 2. Independently increasing quantity.

AMPLIFIER Device used to increase strength of analog signal.

AMPLITUDE 1. Maximum magnitude of quantity; 2. Maximum absolute value of vertical component of curve or function; 3. Maximum height of wave.

AMPLITUDE OF TIDE Semi-range of component tide.

AMPROMETRIC TITRATION Way of measuring concentrations of certain substances in water using electric current that flows during chemical reaction.

ANA-FRONT Frontal boundary where main shield of clouds and precipitation is located behind actual frontal boundary.

ANABAENA Genus of filamentous cyanobacteria that exist as plankton, commonly occurring in masses and often contaminating drinking water, giving it fishy odor and taste.

ANABATIC WIND 1. Upslope wind generated as result of local surface heating and apart from effects of larger scale circulation; 2. Opposite of katabatic wind.

ANABRANCH Diverging branch of river, creek or stream that re-enters main stream.

ANACLINAL STREAM Stream whose gradient is in opposing sense to dip of underlying strata.

ANADROMOUS 1. Fish that migrate from salt water to freshwater to breed; 2. Going up rivers to spawn.

ANADROMOUS FISH Migratory species that are born in freshwater, live mostly in estuaries and ocean water and return to freshwater to spawn.

ANAEROBE 1. One that does not require oxygen to sustain its metabolic processes; 2. Anaerobic organism.

ANAEROBIC 1. Without oxygen; 2. Condition where free (atmospheric) or dissolved oxygen is not present in water; 3. Life or process that occurs in or is not destroyed by, absence of oxygen; 4. Opposite of aerobic.

ANAEROBIC BACTERIA Bacteria that do not live or grow when oxygen is present.

ANAEROBIC BIODEGRADATION Decomposition of organic matter by microorganisms in absence or near absence of free oxygen where decomposition end-products are enriched in carbon.

ANAEROBIC CONTACT PROCESS 1. Anaerobic digester; 2. Wastewater treatment procedure where waste stabilization microorganisms are moved upstream to boost treatment speed.

ANAEROBIC DIGESTION Series of biological processes where microorganisms break down biodegradable material in absence of oxygen.

ANAEROBIC RESPIRATION Chemical process where energy is produced from food without using oxygen.

ANAEROBIC WASTE TREATMENT Bio-treatment (i.e., microorganisms) without oxygen.

ANALOG 1. Class of measuring devices where output varies continuously as function of input (non-digital); 2. Historical instance of given meteorological scenario or feature used for comparison with another scenario or feature; 3. Organ or structure similar in function to one in another kind of organism but is of dissimilar evolutionary origin; 4. Structural derivative of parent compound that often differs from it by single element.

ANALOG DETECTOR Generates electronic signals to represent its observations.

ANALOG DISPLAY Screen that yields constant indicator relative to reference.

ANALOG PROCESSING Procedure of processing information utilizing representative value of physical variable.

ANALOG SIGNAL Signal that varies in continuous manner.

ANALOGOUS ARTICLES Articles not detailed in classification but having characteristics similar to those classified.

ANALOGY Relationship of resemblance or equivalence between two situations, people or objects, especially when used as basis for explanation or extrapolation.

ANALYSIS 1. process of breaking down substance into its constituent parts or result of this process; 2. Record of evaluation of potable water or wastewater.

ANALYSIS HOUR Hour where capacity analysis on element is performed.

ANALYSIS PERIOD Period of time where economic analysis is made.

ANALYTICAL MODEL Mathematical model generally assuming homogeneous aquifer properties, uniform flow direction and hydraulic gradient, uniform aquifer thickness, with simple upper and lower boundaries and lateral boundaries are placed at infinite distance.

ANALYZER Apparatus that runs regular or continuous measurement of some variable.

ANCHOR BLOCK 1. Massive block built to withstand forces (i.e., thrust or pull); 2. Mass of material appropriately placed around pipe to prevent movement when pipe is carrying fluid.

ANCHOR BOLT Fastener to connect objects or structures to concrete.

ANCHOR GATE Gate, used to close off channel or canal, that's heavy and supported at its upper bearing.

ANCHOR ICE 1. Submerged frazil ice attached or anchored to river bottom, irrespective of its formation; 2. Ice in bed of stream or upon submerged body or structure.

ANCHOR ICE DAM Accumulation of anchor ice which acts as dam and raises water level.

ANCHORAGE Block installed along pipeline to withstand forces caused by fluids passing through conduit or changes in temperature of conduit; 2. Harbor, river or offshore area that can accommodate ship at anchor.

ANCHORAGE ZONE Region of structure where prestress force is transferred from prestressing steel to concrete and distributed more widely to member.

ANCHORED WALL Retaining wall that uses cables and anchors in rock or soil behind it.

ANCHORS 1. Device use to resist movement of transportation rails; 2. Ties holding adjoining parts together.

ANCIENT MONUMENT Old or ancient building or structure or site of historical interest protected by regulation.

ANCIENT SOIL Layer of buried, old, weathered subsurface soil.

ANCILLARY DATA Other categories of data critical to interpreting water-quality data and formulating courses of action.

ANCILLARY EQUIPMENT Machines and other technical things that are used with main items of equipment to create complete system.

ANCILLARY SERVICES Other energy-related services that are required to control system frequency, to meet changing scheduling requirements, to react to changing loads and unexpected contingencies and to ensure system stability (i.e., preventing blackouts).

ANDESITE Fine-grained, medium gray volcanic rock of intermediate composition between rhyolite and basalt.

ANECHOIC ROOM Room designed to completely absorb reflections of either sound or electromagnetic waves.

ANEMOMETER Instrument for measuring speed of wind.

ANEMOPHILY Form of pollination whereby pollen is distributed by wind.

ANEROID BAROMETER Instrument for measuring atmospheric pressure where needle, attached to top of evacuated box, is deflected as changes in atmospheric pressure cause top of box to bend in or out.

ANESTHETIC EFFECT Loss of sensation with or without loss of consciousness.

ANGELS Radar echoes caused by birds, insects and localized refractive index discontinuities.

ANGER SCRAPER Hydraulically-powered auger, found in middle of scraper bowl, that's self-loading.

ANGLE DOZER Bulldozer with its blade set at angle to push earth to one side.

ANGLE GATE VALVE Gate valve with elbow on one end.

ANGLE GLOBE VALVE Valve with inlet and outlet set at 90° to each other.

ANGLE OF ATTACK Angle between chord line of airfoil and airflow over it.

ANGLE OF CURRENT Regarding stream gauging, angular difference between angle made by current and 90°.

ANGLE OF EXTERNAL FRICTION 1. Angle between abscissa and tangent of curve representing relationship of shearing resistance to normal stress acting between soil and surface of another material; 2. Angle of wall friction.

ANGLE OF FLARE Angle between culvert barrel centerline and direction of wing wall.

ANGLE OF INCIDENCE Angle that straight line, ray of light, etc., meeting surface makes perpendicular to surface at point of impact.

ANGLE OF INDRAFT Angle that actual wind direction crosses isobars.

ANGLE OF INTERNAL FRICTION 1. Angle between axis of normal stress and tangent to Mohr envelope at point representing given failure-stress condition for solid material; 2. Angle of shear resistance.

ANGLE OF NATURAL REPOSE Steepest angle of descent or dip relative to horizontal plane that material can be piled without slumping by gravity.

ANGLE OF OBLIQUITY Angle between direction of resultant stress or force acting on given plane and normal to that plane.

ANGLE OF REFLECTION Angle at which reflected ray of energy leaves reflecting surface measured between outgoing ray and perpendicular to surface at point of incidence (i.e., where ray strikes).

ANGLE OF REPOSE Angle between horizontal and maximum slope that particular soil or geologic material assumes through natural processes.

ANGLE OF SHEAR RESISTANCE 1. Angle between axis of normal stress and tangent to Mohr envelope at point representing given failure-stress condition for solid material; 2. Angle of internal friction.

ANGLE OF THREAD Angle between sides of screw thread measured in axial plane.

ANGLE OF TURN Measured change of vehicle direction.

ANGLE OF WALL FRICTION 1. Angle between abscissa and tangent of curve representing relationship of shearing resistance to normal stress acting between soil and surface of another material; 2. Angle of external friction.

ANGLE STOP VALVE Shutoff valve used to stop flow of liquid in pipe.

ANGLE VALVE Manually operated valve, used for regulating flow of fluid in pipe, with its outlet opening oriented at right angles to its inlet opening.

ANGLED TEE T-shaped pipe fitting where main inlet is at shallow angle.

ANGLER-DAY Time spent fishing by one person for any part of day.

ANGSTROM Unit of length equal to 0.1 nanometer.

ANGULAR FREQUENCY 1. Scalar measure of rate of rotation; 2. Frequency of sound (i.e., radians/second).

ANGULAR INCONFORMITY Angular discordance between younger and older rocks.

ANGULAR MISALIGNMENT Small angle, on same horizontal plane, formed between intersecting axes of two conduits.

ANGULAR PARTICLE Angular shape of aggregate particles.

ANGULARITY 1. Property or characteristic of possessing angles; 2. Angle between centerline of stream or conduit and line normal to centerline of obstruction; 3. Sharp point or angle.

ANGULARITY CORRECTION Correction to be made to observed velocity when direction of current is not at right angles to discharge section line.

ANHYDRITE Saline evaporite consisting of anhydrous calcium sulfate (gypsum).

ANHYDROUS 1. Having little or no water; 2. Having no water of crystallization.

ANHYDROUS PRODUCT Product that would exist after all water is removed.

ANICUT Dam made in stream for maintaining and regulating irrigation.

ANIMAL DANDER Tiny scales of animal skin commonly considered indoor air pollutant.

ANIMAL FEED Food given to domestic animals.

ANIMAL FEEDING OPERATION Lot or facility (other than aquatic animal production facility) where following conditions are met: animals (other than aquatic animals) have been, are or will be stabled or confined and fed or maintained for total of 45 days or more in any 12-month period and crops, vegetation, forage growth or post-harvest residues are not sustained in normal growing season over any portion of lot or facility.

ANIMAL SPECIALTIES Water use associated with production of fish in captivity except fish hatcheries, fur-bearing animals in captivity, horses, rabbits and pets.

ANIMAL STUDIES Investigations using animals as surrogates for humans with expectation that results are pertinent to humans.

ANIMAL UNIT Standard unit used in calculating relative grazing impact of different kinds and classes of livestock.

ANIMALS Living organisms that feed on organic matter, typically having specialized sense organs and nervous system and able to respond rapidly to stimuli.

ANION Negatively charged ion in electrolyte solution, attracted to anode under influence of difference in electrical potential.

ANIONIC SURFACTANT Surface active substance that contains wash active and degreasing abilities on surfaces of metals.

ANISOTROPIC Having properties that differ according to direction of measurement.

ANISOTROPIC MASS Mass having different properties in different directions at any given point.

ANISOTROPIC SOIL Soil mass having different properties in different directions at any given point.

ANISOTROPY 1. Flow conditions vary with direction; 2. Conditions under which one or more hydraulic properties (e.g., hydraulic conductivity) of aquifer vary from reference point.

ANKERITE Any of group of mixed carbonate minerals.

ANNEALING WITH OIL Use of oil to quench workpiece as it passes from annealing furnace.

ANNEALING WITH WATER Use of water to quench workpiece as it passes from annealing furnace.

ANNEX Building joined or associated with main building.

ANNEXATION Legal merging of territory into another body.

ANNUAL 7-DAY MINIMUM Lowest mean value for any 7-consecutive-day period in year whereas year may be either calendar year or water year (October 1 through September 30).

ANNUAL 1. Happening once each year; 2. Corporate fiscal year.

ANNUAL ALLOWABLE HARVEST Amount of lumber scheduled to be eliminated from certain unit in single year.

ANNUAL AVERAGE Maximum allowable for entire processing season or calendar year by applicable annual average limitation.

ANNUAL AVERAGE DAILY TRAFFIC Total volume of vehicle traffic of roadway for year divided by 365 days.

ANNUAL AVERAGE WEEKDAY TRAFFIC Average traffic during weekdays (i.e., Monday through Friday).

ANNUAL CAPACITY FACTOR Unitless ratio of actual electrical energy output over year to maximum possible electrical energy output over that year.

ANNUAL CAPITAL Annual plant and equipment costs, plus taxes and interest.

ANNUAL COKE PRODUCTION Coke produced in batteries connected to coke byproduct recovery plant over 1 year.

ANNUAL CYCLE Set of changes or events that uniformly or consistently, take place at same time of year.

ANNUAL DOCUMENT LOG Detailed information maintained at facility on PCB waste handling at facility.

ANNUAL ENERGY COST Variable costs relating to energy production in year.

ANNUAL FAILURE PROBABILITY Probability of load multiplied by probability of failure.

ANNUAL FLOOD Maximum discharge peak or highest flow at point on stream during any particular calendar/water year (e.g., October 1 - September 30).

ANNUAL FLOOD SERIES List of annual floods.

ANNUAL GROWTH Change in value of measurement over period of year.

ANNUAL GROWTH RING Layer of wood formed in plant during single period of growth.

ANNUAL INSPECTION Yearly inspections of dam and appurtenant facilities conducted by local operating office that address both O&M and dam safety issues.

ANNUAL LOAD FACTOR Factor equal to energy generated in year divided by product of peak demand for that year and number of total hours in year.

ANNUAL OPERATING COST Includes all annual operation and maintenance expense, wheeling, purchased power, etc.

ANNUAL RUNOFF Total quantity of water runoff discharged from drainage basin in year that may be reported as volume, discharge per unit of drainage area or depth of water on drainage basin.

ANNUAL VARIATION Yearly change in star's mean right ascension or declination produced by precession of equinoxes and proper motion of star.

ANNUAL VEHICLE MILES TRAVELED Number of miles traveled by vehicles on given portion of road during year.

ANNUAL WORK PLAN Annual budget document that describes proposed work to be performed and details amount of funds required.

ANNUALIZED COST Yearly payments for determined period of time, equal in value to onetime payment based on chosen interest rate.

ANNUALIZED LOSS OF LIFE Sum of probability of dam failure multiplied by annual probability of loading and estimated number of lives that would be lost for each dam failure scenario under particular loading category.

ANNUALIZED METHOD Converts short-term costs or rate into annual costs or rate.

ANNULAR Ring-shaped.

ANNULAR SPACE Ring-shaped space located between two circular objects (e.g., two pipes).

ANNULUS 1. Space between casing of well and well bore; 2. Space between tubing and casing of well; 3. Ring of sun not covered by moon in annular solar eclipse; 4. Structure in fern that consists of differentially thick-walled cells on sporangium that bend and distort as result of drying; 5. Membranous remnants of partial veil that leaves ring on stem of mushroom.

ANNULUS GROUTING Grouting opening between two ring-shaped objects (e.g., between two conduits).

ANNUNCIATOR 1. Signalling device that shows which of several electrical circuits is active; 2. Anything that announces something.

ANODE 1. Positive pole or electrode of electrolytic system; 2. Attracts negatively charged particles or ions (anions).

ANODIZE To coat surface of metal electrolytically with oxide, either as protection or decoration.

ANOMALIES Externally visible skin or subcutaneous disorders, including deformities, eroded fins, lesions and tumors.

ANOMALOUS PROPAGATION Radar term for false (i.e., non-precipitation) echoes resulting from non-standard propagation of radar beam under certain atmospheric conditions.

ANOMALY Deviation of measurable unit over period in given region from long-term average.

ANORTHOSITE Phaneritic, intrusive igneous rock characterized by predominance of plagioclase feldspar.

ANOXIC Without oxygen.

ANOXIC DENITRIFICATION Biological nitrogen removal process where nitrate nitrogen is converted by microorganisms to nitrogen gas in absence of dissolved oxygen.

ANTAGONISM 1. Situation where two chemicals interfere with each others actions or one chemical interferes with action of other; 2. Characteristic property of mixture of toxicants that exhibits less-than-additive total toxic effect.

ANTARCTIC "OZONE HOLE" Seasonal depletion of ozone in upper atmosphere above large area of Antarctica.

ANTECEDENT Earlier, either in time or order.

ANTECEDENT CONDITIONS Watershed conditions prevailing prior to event that's normally used to characterize basin wetness (e.g., soil moisture).

ANTECEDENT DRAINAGE Drainage system that maintains its original course and pattern despite changes in underlying topography.

ANTECEDENT FLOOD Flood or series of floods assumed to occur prior to occurrence of inflow design flood.

ANTECEDENT MOISTURE 1. Relative wetness or dryness of watershed or sanitary sewershed; 2. Amount of wetness of soil at beginning of runoff.

ANTECEDENT PRECIPITATION Rainfall that occurred prior to event being consideration.

ANTECEDENT PRECIPITATION INDEX Index of moisture stored within drainage basin before storm.

ANTECEDENT STREAM Stream that maintains its original course and pattern despite changes in underlying rock topography.

ANTECEDENT-PRECIPITATION INDEX Weighted summation of daily precipitation amounts, used as index of soil moisture.

ANTENNA GAIN Measure of effectiveness of directional antenna as compared to isotropic radiator where maximum value is called antenna gain by convention.

ANTHELION Luminous white spot that appears on parhelic circle at same altitude as Sun and 180° from it in azimuth.

ANTHRACITE FILTER Water treatment plant filter that consists of multiple layers of anthracite ordered in decreasing size to trap particulate matter.

ANTHROPIC Of or pertaining to mankind or humans or period of humanity's existence.

ANTHROPOGENIC 1. Man made; 2. Created, occurring because of or influenced by human activity.

ANTHROPOGENIC BETA PARTICLE AND PHOTON EMITTERS All radionuclides emitting beta particles and/or photons listed in Maximum Permissible Body Burdens and Maximum Permissible Concentrations of Radionuclides in Air and Water for Occupational Exposure.

ANTHROPOGENIC SOURCE Pollutant source caused or produced by humans.

ANTI-BACKSLIDING Provision in U.S. Federal Regulations that requires reissued permit to be as stringent as previous permit with some exceptions.

ANTI-EXTRUSION RING Thin ring installed on low pressure side of seal to prevent elastomer extrusion into clearance gap.

ANTI-FLOOD VALVE Check valve at flood level to stop reverse flow of sewage and wastewater from drainage network.

ANTI-INTRUDER FENCING Woven wire fencing barrier to prevent intrusion into restricted areas or retain personnel or goods within restricted areas.

ANTI-MICROBIAL Agent that kills microbes.

ANTI-WIND Upper or return branch of along-valley wind system, as confined within valley and blowing in direction opposite to winds in lower altitudes of valley.

ANTIBIOSIS Association between organisms detrimental (harmful) to one or more of them.

ANTIBIOTIC Any substance that can destroy or inhibit growth of bacteria and similar microorganisms.

ANTICHLORS Reagents used to remove excess chlorine residuals or bleaching liquor from potable water or treated wastewater effluent.

ANTICLINAL SPRING Contact spring occurring along surface outcrop of anticline, from pervious stratum overlying one less pervious.

ANTICLINE 1. Fold in Earth's crust that curves upward (i.e., convex upward) whose core contains stratigraphically older rock; 2. Upward fold in rock layers that creates arched or dome-like uplift of sedimentary layers.

ANTICORROSION TREATMENT Treatment aimed at eliminating corrosion-producing qualities of substance.

ANTICYCLOGENESIS Formation or intensification of anticyclone or high pressure center.

ANTICYCLONE Large-scale circulation of winds around central region of high atmospheric pressure, clockwise in Northern Hemisphere, counterclockwise in Southern Hemisphere.

ANTICYCLONIC ROTATION 1. Rotation in opposite sense as Earth's rotation; 2. Opposite of cyclonic rotation.

ANTIDEGRADATION 1. Policies which ensure protection of water quality for particular water body where water quality exceeds levels necessary to protect fish and wildlife propagation and recreation on and in water; 2. Special protection of waters designated as outstanding natural resource waters that's typically adopted by each state to minimize adverse effects on water.

ANTIDEGRADATION CLAUSE Part of federal air/water quality requirements prohibiting deterioration where pollution levels are above legal limit.

ANTIDUNE Bedform found in fluvial environments, opposing direction of flow.

ANTIDUNE MOVEMENT Occurrence in water flowing on sand deposition where sand dune moves opposite current.

ANTIFER Kind of armored layer protection using precast concrete blocks shaped as altered cubes.

ANTIFREEZE SPRINKLER SYSTEM Wet pipe sprinkler system employing automatic sprinklers that are attached to piping system that contains antifreeze solution and that are connected to water supply.

ANTIGRAFFITI TREATMENT Finish applied to discourage markings.

ANTIHAMMER DEVICE Chamber designed to absorb shock caused by rapidly closed valve.

ANTIHEAVE MEASURE Precaution taken to avoid soil heave.

ANTILLES CURRENT Current that originates in vicinity of Leeward Islands as part of Atlantic North Equatorial Current.

ANTIMICROBIAL AGENTS Drugs, chemicals or other substances that kills microorganisms or stops their growth.

ANTINODE Region of maximum amplitude situated between adjacent nodes of vibrating body.

ANTIOZONANT Any chemical compound that prevents or slows down degradation of material by ozone gas in air.

ANTISIPHON PIPE 1. Pipe with ventilation; 2. Branch vent.

ANTISIPHON TRAP Drainage system designed to preserve water seal by preventing siphonage.

ANTISIPHON VALVE check valve that allows water to only flow in one direction.

ANTISTATIC AGENT Compound used for treatment of materials or their surfaces in order to reduce or eliminate buildup of static electricity.

ANTIVIBRATION MOUNTING Flexible mounting for engine, gearbox or other mechanical item that reduces amount of noise and vibration passed from item.

ANVIL Flat, spreading top of cumulonimbus cloud, often shaped like anvil.

ANVIL CRAWLER Lightning discharge occurring within anvil of thunderstorm, characterized by one or more channels that appear to crawl along underside of anvil.

ANVIL DOME Large overshooting top or penetrating top.

ANVIL ROLLOVER Slang for circular or semicircular lip of clouds along underside of upwind part of back-sheared anvil, indicating rapid expansion of anvil.

ANVIL ZITS Slang for frequent (often continuous or nearly continuous), localized lightning discharges occurring from within thunderstorm anvil.

AP INDEX Averaged Planetary Index based on data from set of specific stations.

APERIODIC That does not recur periodically or have periodic vibration.

APEX 1. Highest or greatest point of something; 2. Topmost vertex of cone or pyramid.

APHELION 1. Point on annual orbit of body (e.g., about Sun) farthest from Sun; 2. Opposite of perihelion.

APHOTIC ZONE 1. Part of deep body of ocean or lake that receives insufficient sunlight for photosynthesis to be possible; 2. Depth in water where light doesn't penetrate.

API METHOD Statistical method to estimate amount of surface runoff which will occur from basin from given rainstorm based on antecedent precipitation index, physical characteristics of basin, time of year, storm duration, rainfall amount and rainfall intensity.

API SCALE American Petroleum Institute gravity scale is universally accepted scale of relative density of fluids used in fuel technology.

APOGEAN RANGE Average range occurring at time of apogean tides.

APOGEAN TIDES Tide of reduced range when moon is near apogee (i.e., farthest point from earth in its elliptical orbit).

APOGEE 1. Farthest distance between Moon and Earth or Earth and Sun; 2. Point, in orbit about any planet, farthest from planet.

APPARATUS 1. Entirety of means whereby specific production is made existent or task accomplished; 2. Complex machine or instrument; 3. Firefighting vehicle used for emergency response.

APPARENT COHESION Resistance of particles to being pulled apart, due to surface tension of moisture film surrounding each particle.

APPARENT DENSITY Mass per unit volume (or weight per unit volume) of material, including voids that are inherent in material.

APPARENT DIRT CAPACITY In fluid filter evaluation, amount of dirt that can be added to filter test system before terminal differential pressure is reached.

APPARENT GROUNDWATER VELOCITY Volume of groundwater passing through unit area of porous medium, perpendicular to direction of flow, in unit of time.

APPARENT SPECIFIC GRAVITY Ratio of weight of volume of substance to weight of equal volume of reference substance.

APPARENT TEMPERATURE Measure of human discomfort due to combined heat and humidity (e.g., heat index).

APPARENT VISCOSITY Shear stress applied to fluid divided by shear rate.

APPARENT WIND Speed and true direction from which wind appears to blow with reference to moving point (i.e., relative wind).

APPENDIX Emergency operations plan element attached to functional annex to provide information on special approaches or requirements generated by unique characteristics of specified hazards of particular concern to jurisdiction.

APPLETON LAYER Highest region of ionosphere (from 90 to 600 miles up) that contains highest concentration of free electrons and is most useful for long-range radio transmission.

APPLICABLE OR RELEVANT AND APPROPRIATE REQUIREMENTS State or federal statute that pertains to protection of human life and environment in addressing specific conditions or use of particular cleanup technology at Superfund site.

APPLICABLE REQUIREMENTS Cleanup standards, standards of control and other substantive requirements, criteria or limitations that specifically address hazardous substance, pollutant, contaminant, remedial action, location or other circumstance.

APPLICABLE STANDARD Any requirement relating to quality of water containing or potentially containing pollutants that may affect groundwater, discharge of sludge or disposal practice.

APPLICABLE WATER QUALITY STANDARDS Water quality standards adopted by governing body.

APPLICATION FACTOR Ratio between that concentration of substance that produces particular chronic response in organism and that that causes death in 50% of population within given time.

APPLICATION FOR PAYMENT Written request by contractor for payment under terms of contract.

APPLICATOR Tool or device used to apply fluid or semi-fluid substance to surface.

APPLIED CLIMATOLOGY Scientific analysis of climatic data in light of useful application for operational purpose.

APPLIED COST Term used in cost accounting to denote cost assigned to something, that may be different from actual cost.

APPLIED DOSE Amount of substance in contact with primary absorption boundaries of organism (e.g., skin, lung tissue, gastrointestinal track) and available for absorption.

APPLIED WATER 1. Water delivered to user; 2. May be used for either inside uses or outside watering, does not include precipitation or distribution losses and may apply to metered or unmetered deliveries.

APPORTIONMENT Division of income and expenses and sharing carried out according to plan or formula.

APPRAISAL 1. Act or process of developing opinion of value; 2. Judgment or assessment of value of something, especially formal one.

APPRAISAL APPROACH Method used in estimating market value of business, business ownership interest, intangible asset, property or security.

APPRAISAL ESTIMATE Estimate used in appraisal study as aid in selecting most economical plan by comparing alternative features or for determining whether more detailed investigations of potential project are economically justified used to obtain approximate costs in short period of time with inadequate data that's not to be used for project authorization.

APPRAISAL INCREASE Upward change in estimate of market value of item originating from reappraisal by certified professional.

APPRAISAL LEVEL OF DETAIL Level of detail necessary to facilitate making decisions on whether or not to proceed with detailed study and evaluation of any alternative.

APPRAISAL REPORT Report incorporating approximation value of something.

APPRAISAL STUDY Study incorporating appraisal level of detail.

APPRAISED INVENTORY Comprehensive listing of things making up constructed property, with or without unit costs or prices appended.

APPRAISED PRICE Evaluation of item's price based on given point in time performed by professional.

APPRAISED VALUE Evaluation of property's value based on given point in time performed by professional.

APPRAISER One who performs assessment of real property.

APPROACH 1. Lanes at intersection to accommodate turn movements; 2. Way aircraft comes in to land at airport.

APPROACH CHANNEL 1. Channel upstream from that portion of spillway having concrete lining or concrete structure or upstream from intake structure of outlet works; 2. Channel generally unlined, excavated in rock or soil, with or without riprap, soil cement or other types of erosion protection.

APPROACH CONTROL Responsible for controlling all instrument flights operating by airports.

APPROACH DELAY Stopped-time delay for approach.

APPROACH GRADE Average slope along traffic approach.

APPROACH LIGHT SYSTEM Lighting system installed on approach end of airport runway.

APPROACH SECTION Cross-section of stream channel that's normal to current.

APPROACH SIGNAL Fixed signal to govern approach to other signal(s).

APPROACH SLOPE Path airplane follows on its final approach to land on runway.

APPROPRIATE RIGHT Right of beneficial use.

APPROPRIATE SENSITIVE BENTHIC MARINE ORGANISMS Two or more species that together represent filter-feeding, deposit-feeding and burrowing characteristics.

APPROPRIATE SENSITIVE MARINE ORGANISM At least one species each representative of phytoplankton or zooplankton, crustacean or mollusk and fish species chosen from among most sensitive species documented in scientific literature or accepted by EPA as being reliable test organisms to determine anticipated impact of wastes on ecosystem at disposal site.

APPROPRIATE TECHNOLOGY Ideological movement that involves small-scale, labor-intensive, energy efficient, environmentally sound, people-centered and locally controlled projects.

APPROPRIATION 1. Public funds set aside for specific purpose; 2. Amount of something legally set apart or assigned to particular purpose or use.

APPROPRIATION DOCTRINE System for allocating water to private individuals used in most Western states whereas doctrine of Prior Appropriation was in common use throughout arid west as early settlers and miners began to develop land that was based on concept of "First in Time, First in Right."

APPROPRIATIVE Water rights to or ownership of water supply that is acquired for beneficial use of water by following specific legal procedure.

APPROPRIATOR OF WATER RIGHTS person who redirects and places to valuable use that water of stream or alternative body of water under water right got through appropriation.

APPROVAL 1. Expression granting permission; 2. Indication of agreement with proposal; 3. Acknowledgement that person, thing or event meets requirements.

APPROVED Officially agree to or accept as satisfactory.

APPROVED EQUAL Material, equipment or method approved by architect or engineer for use in work as being acceptable as equivalent to those originally specified in contract documents.

APPROVING AUTHORITY Official authorized under statute to enforce specified requirements.

APPROXIMATE QUANTITIES Development of elemental cost plan to finish.

APPURTENANCE 1. Thing to that another pertains; 2. Appendage added to something else; 3. Minor property (e.g., greenhouse) that passes with main property when it is sold.

APPURTENANCES Equipment used for some specific task.

APPURTENANT STRUCTURE Ancillary features of dam (e.g., outlets, spillways, power plants, tunnels, bridges, drain systems, tunnels, towers).

APPURTENANT WORKS Structures such as spillways, either in or separate from dam; reservoir and its rim; low-level outlet works; and water conduits including, but not limited to, tunnels, pipelines or penstocks, either through dam or its abutments.

APRON 1. Hard surface bordering structure or area; 2. Section of concrete or riprap constructed upstream or downstream from control structure to prevent undercutting of structure; 3. Floor or lining of concrete, timber or other suitable material at toe of dam, discharge side of spillway, chute or other discharge structure, to protect waterway from erosion from falling water or turbulent flow; 4. Short ramp with slight pitch; 5. Sides of tree's canopy.

AQUACULTURE Science of farming organisms (e.g., plants, animals, fish, shellfish, algae) that live in water.

AQUACULTURE WELLS Class V wells that dispose of water used for cultivation of marine and freshwater animals and plants.

AQUAFALFA Ground having high water table.

AQUATIC Living, growing or occurring in or on water.

AQUATIC ALGAE Microscopic plants that grow in sunlit water containing phosphates, nitrates and other nutrients.

AQUATIC ANIMALS Animal, either vertebrate or invertebrate, that lives in water for most or all of its lifetime.

AQUATIC ECOSYSTEM Stream channel, lake or estuary bed, water, and/or biotic communities and habitat features that occur therein.

AQUATIC ENVIRONMENT AND AQUATIC ECOSYSTEM Waters that include wetlands that serve as habitat for interrelated and interacting communities and populations of plants and animals.

AQUATIC FLORA Plants that have adapted to living in aquatic environments.

AQUATIC GROWTH Floating, drifting or attached organisms in body of water.

AQUATIC GUIDELINES 1. Specific levels of water quality which, if reached, may adversely affect aquatic life; 2. Non-enforceable guidelines issued by governmental agency or other institution.

AQUATIC HABITAT Environments characterized by presence of standing or flowing water.

AQUATIC LIFE CRITERIA Water-quality guidelines for protection of aquatic life/organisms.

AQUATIC ORGANISM PASSAGE Modification or removal of barriers that restrict or impede movement of aquatic organisms.

AQUEDUCT 1. Pipe, man-made canal, conduit or channel designed to transport water from remote source, usually by gravity; 2. Structure carrying water over river or depression.

AQUEOUS 1. Something made up of, similar to or containing water; 2. Watery.

AQUEOUS METAMORPHISM Change in feel and composition of stone by solution and redeposition of mineral matter through influence of water that they comprise.

AQUEOUS SOLUBILITY Maximum concentration of chemical that will dissolve in pure water at reference temperature.

AQUEOUS VAPOR Gaseous form of water.

AQUICLUDE 1. Solid, impermeable area (e.g., clay layer) that limits movement of groundwater; 2. Formation that contains water but cannot transmit it rapidly enough to furnish significant supply to well or spring.

AQUIFER 1. Natural underground bed or layer (i.e., soil, sand, gravel, rock or porous storage) that contains water usually capable of yielding large amount (i.e., supply) of water; 2. Water-bearing stratum of permeable rock, sand or gravel; 3. Underground water-bearing geologic formation or structure that provides groundwater reservoir.

AQUIFER RECHARGE/RECOVERY WELLS Class V wells that are used to inject fluids to recharge aquifer and may have secondary purposes (e.g., saline intrusion prevention, subsidence control, aquifer storage and recovery).

AQUIFER REMEDIATION WELLS Class V wells that are used to clean up, treat or prevent contamination of underground sources of drinking water.

AQUIFER SERVICE AREA Place above aquifer such as region where whole population served by aquifer resides.

AQUIFER TEST Test to determine hydraulic properties (e.g., transmissivity, hydraulic conductivity, storage coefficient) of aquifer.

AQUIFER, CONFINED 1. Aquifer bounded above and below by confining beds (e.g., impermeable material) where hydraulic head is above top of aquifer; 2. Aquifer under pressure so that when aquifer is penetrated by well, water will rise to level associated with pressure.

AQUIFER, UNCONFINED Aquifer whose hydraulic head surface (i.e., water table) is at equilibrium with atmospheric pressure.

AQUIFUGE Geological formation which has no interconnected openings and cannot hold or transmit water.

AQUIHERBOSA Plant community of herbaceous plants that exists in abundantly wet areas.

AQUITARD Geological formation that may contain groundwater but is not capable of transmitting significant quantities of it under normal hydraulic gradients.

ARABLE Having soil or topographic features suitable for cultivation (e.g., farming).

ARABLE AREA Land capable of being ploughed and used to grow crops.

ARABLE LAND Land which when farmed that provides reasonable return to farm's expenses.

ARBITRATION Process where two or more parties use impartial third party to resolve dispute.

ARBORETUM Place where many varieties of tree are grown for research, educational and ornamental purposes.

ARBORICULTURE Branch of horticulture concerned with planting and growth of trees.

ARCH CLOUDS Clouds that can stretch large distances along mountain range.

ARCH DAM Dam, curved upstream in plan view, that directs water against walls to provide force to compress dam.

ARCH FILAMENT SYSTEM Bright, compact plage crossed by system of small, arched filaments, that is often sign of rapid or continued growth in Active Region.

ARCH-BUTTRESS DAM Buttress dam that is curved in plan.

ARCH-GRAVITY DAM Arch dam that is only slightly thinner than gravity dam.

ARCHAIC Cultural stage following earliest known human occupation in New World (about 5,500 B.C. to A.D. 100) characterized by generalized hunting and gathering lifestyle and seasonal movement to take advantage of variety of resources.

ARCHEAN Relating to or denoting eon that constitutes earlier (or middle) part of Precambrian, where there was no life on earth.

ARCHEOLOGY Study of human cultures through recovery and analysis of their material relics.

ARCHIMEDEAN SCREW 1. Screw machine, open on all sides, that transfers water from low-lying body of water; 2. Pump consisting of inclined, revolving, corkscrew-shaped shaft tightly enclosed in pipe.

ARCHIMEDES PRINCIPLE Upward buoyant force exerted on body immersed in fluid, whether fully or partially submerged, is equal to weight of fluid that body displaces and acts in upward direction at center of mass of displaced fluid.

ARCHING Transfer of stress from yielding part of soil or rock mass to adjoining less-yielding or restrained parts of mass.

ARCHING ACTION Loads transmitted due to lateral confinement provided by surrounding concrete.

ARCHITECTURAL COATINGS Coverings (e.g., paint, roof tar) that are used on exteriors of buildings.

ARCTIC Region within Arctic Circle, or, loosely, northern regions in general, characterized by very low temperatures.

ARCTIC FRONT Boundary or front separating deep, cold arctic air from shallower, relatively less cold polar air.

ARCTIC SEA SMOKE Steam fog, but often specifically applied to steam fog rising from small open water within sea ice.

ARCTIC SMOKE Fog that forms in arctic regions when very cold air flows over warmer waters.

ARCUS Low, horizontal cloud formation associated with leading edge of thunderstorm outflow (i.e., gust front).

ARE Unit of area equal to 100 m^2.

AREA 1. Measure of extent of surface measured in square units; 2. Particular extent of surface defined by boundaries.

AREA COVERAGE Geographic area where station can communicate.

AREA CURVE Curve expressing connection between area and another variant.

AREA DIFFERENTIAL SYSTEM Valve controlled by way of opposing pistons of distinct displacement.

AREA DRAIN Drain for surface areas immediately adjacent.

AREA ESTIMATING METHOD Process of estimating prices by multiplying adjusted gross floor area with unit cost.

AREA FORECAST DISCUSSION National Weather Service product intended to provide well-reasoned discussion of meteorological thinking which went into preparation of Zone Forecast Product that at end of discussion includes list of all advisories, non-convective watches and non-convective warnings.

AREA LANDFILLING Depositing waste in layers and forming terraces over identified landfilling area.

AREA NAVIGATION Instrument flight rules navigation that allows aircraft to choose any course within network of navigation beacons.

AREA OF CONCERN Condition or object along roadside that requires precaution.

AREA OF DIVERSION Defines limiting flow lines along which water moves into pumping depression, normal to potentiometric contours.

AREA OF INFLUENCE Area (i.e., area) covered by drawdown curves of given pumping well or combination of wells at particular time.

AREA OF INFLUENCE OF WELL Area surrounding well within which piezometric surface has been lowered when pumping has produced maximum steady rate of flow (i.e., zone of influence).

AREA OF REVIEW Area around deep injection well that must be checked for artificial penetrations(e.g., other wells) to determine if flow between aquifers will be induced by injection operation before permit is issued.

AREA RAINFALL Average rainfall over area.

AREA SAMPLING Total area under investigation is divided into small sub-areas that are sampled randomly.

AREA SOURCE 1. Any source of air pollution released over relatively small area but which cannot be classified as point source (e.g., vehicles, small engines, small businesses, household activities, biogenic sources); 2. Array of pollutant sources, so widely dispersed and uniform in strength that they can be treated in dispersion model as aggregate pollutant release from defined area at uniform rate.

AREA TYPE Environment description where system element is located.

AREA WIDE HYDROLOGIC PREDICTION SYSTEM Computer system which automatically ingests areal flash flood guidance values and WSR-88D products and displays this data and other hydrologic information on map background.

AREA-CAPACITY CURVE Graph showing relation between surface area of water in reservoir, corresponding volume and elevation.

AREA-CAPACITY TABLE Table giving reservoir storage capacity and sometimes surface areas, in terms of elevation increments.

AREAL Adjective version of noun “area.”

ARENACEOUS 1. Sandy; 2. Growing in sandy soil.

ARGILLACEOUS 1. Pertaining to clay; 2. Made of, containing or resembling clay.

ARGILLITE Fine-grained sedimentary rock, intermediate between shale and slate, sometimes used as building material.

ARGON Chemical element that's colorless, odorless, inert gas (Ar) with atomic number of 18.

ARID 1. Very dry; 2. Term describing climate or region where precipitation is so deficient in quantity or occurs so infrequently (i.e., less than 25 cm or 10 inches of rainfall annually) that intensive agricultural production is not possible without irrigation.

ARITHMETIC MEAN Measure of central tendency of set of values computed by dividing sum of values by their number.

ARM 1. Long, narrow, rigid section of object extending from main body; 2. One of two parts of chromosome.

ARMOR Naturally occurring surface of pebbles, rocks or boulders that line bed of waterway or beach and provide protection against erosion.

ARMORED PIPE Big diameter plastic pipe augmented by steel wire embedded in its wall.

ARMORING 1. Process of progressive coarsening of bed layer by removal of fine particles until it becomes resistant to scour; 2. Formation of resistant layer of relatively large particles resulting from removal of finer particles by erosion.

AROCLOR Registered trademark for group of polychlorinated biphenyls that were manufactured by Monsanto Company prior to 1976.

AROMATIC 1. Derived from benzene; 2. Having closed ring of alternate single and double bonds with delocalized electrons; 3. Fragrant or spicy.

AROMATICS Type of hydrocarbon (e.g., benzene, toluene) with specific type of ring structure.

ARRAY Orderly series, arrangement or sequence.

ARRIVAL RATE Average of vehicles arriving at travel lane, road or location.

ARRIVAL TYPE Six categories that determine progression quality at intersection that's signalized.

ARROW 1. Sign or symbol used to indicate direction; 2. Projectile consisting of shaft, point and tail with stabilizing fins shot from bow; 3. Inflorescence or tassel of mature sugar cane plant.

ARROW BOARD/PANEL Lighted board that flashes arrows to temporarily control traffic.

ARROW DIAGRAMMING METHOD Network diagramming technique where tasks are represented by arrows.

ARROYO Small, deep, flat-floored channel or gully of ephemeral or intermittent stream, usually with nearly vertical banks cut, into unconsolidated material due to infrequent rainfall and depth of cut which does not penetrate below level of permanent groundwater.

ARSENIC Non-metallic chemical element (As) with atomic number of 33.

ARSENICALS Pesticides containing arsenic.

ARTERIAL DRAINAGE Drainage system, where different number of drains flow collectively into one main channel.

ARTERIAL HIGHWAY High capacity urban road.

ARTERIAL SERVICE Bus route for arterial roads.

ARTERIAL STREET Major thoroughfare, particularly one used to provide large capacity for travelling distances.

ARTERIAL VENT Vent serving building drain and public sewer.

ARTESIAN Groundwater, held under pressure in porous rock or soil confined by impermeable geological formations, that rises to surface under its own hydrostatic pressure.

ARTESIAN AQUIFER Aquifer containing water, under positive pressure, between two relatively impermeable boundaries where water level can rise above ground surface yielding flowing well.

ARTESIAN BASIN Subsurface bowl full of water under adequate pressure that it is going to rise over underparts of overlying impervious strata if given chance.

ARTESIAN CAPACITY Speed where well will produce water at surface as result of artesian pressure.

ARTESIAN DISCHARGE Rate of flow from artesian well.

ARTESIAN FLOW Flow from aquifer where pressure within aquifer forces groundwater to rise above land surface naturally without using pump.

ARTESIAN HEAD Space above (or below) ground level to which water within artesian aquifer would rise if free to achieve that.

ARTESIAN PRESSURE Pressure applied by groundwater against overlying impermeable or less permeable formation, bottom of which is at lesser level that that of free water surface of groundwater.

ARTESIAN WATER Groundwater under pressure when tapped by well and is able to rise above level at which it is first encountered.

ARTESIAN WATER POWER Energy of artesian well supplied by pressure of water discharged from well.

ARTESIAN WELL 1. Well drilled into confined aquifer with enough hydraulic pressure for water to flow to surface without pumping; 2. Free flowing well.

ARTESIAN-FLOW AREA Surface area where water in underlying aquifer is under sufficient hydrostatic pressure to rise to surface.

ARTICULATED BUS Bus fitted with joint to increase its manoeuvrability.

ARTICULATED CAR Vehicle that has permanent or semi-permanent pivot joint in its construction, allowing vehicle to turn sharply.

ARTICULATED CONCRETE MATTRESS Concrete blocks used for scour protection and stability.

ARTICULATED DUMP TRUCK Very large heavy duty type of dump truck used to transport loads over rough terrain.

ARTIFACT Any human-made or used object, intact or in pieces, 50 years or older.

ARTIFICIAL 1. Man-made; 2. Unnatural; 3. False, misleading.

ARTIFICIAL CONTAMINANT Sample of known composition and particle size distribution introduced into system for test purposes.

ARTIFICIAL CONTROL Weir or other man-made structure which serves as control for stream-gaging station.

ARTIFICIAL DRAINS Man-made or constructed drains.

ARTIFICIAL FUEL Fuel consisting of small particles consolidated into lumps or blocks.

ARTIFICIAL HARBOR Safe place produced by deliberately constructed breakwaters, sea walls or jetty's or that can be constructed by dredging.

ARTIFICIAL ISLAND Fill placed that has been constructed by people rather than formed by natural means.

ARTIFICIAL LAKE Man-made basin used to store water for community.

ARTIFICIAL NAVIGABLE WATERWAY Stream, channel or canal made navigable by dredging and building of controlling works.

ARTIFICIAL OBSTRUCTION Any obstacle other when compared to natural obstacle that impedes or prevents passage or flow.

ARTIFICIAL RAIN Human induced weather precipitation.

ARTIFICIAL RECHARGE Process where water is put back into groundwater reservoir from surface-water supplies (e.g., irrigation, induced infiltration from streams/wells, pumping water directly into aquifer).

ARTIFICIAL REEF Use of man-made underwater structure to provide cover and promote life for marine organisms.

ARTIFICIAL REGENERATION Forest sites regenerated by planting seedlings or by direct seeding.

ARTIFICIAL RESUSCITATION Means of assisting or stimulating respiration.

ARTIFICIAL SLUDGE Substitutes for return sludge in activated sludge process.

ARTIFICIAL SUBSTRATE Device that purposely is placed in stream or lake for colonization of organisms.

ARTISAN SPRING Spring whose water flows from crack in cap rock of aquifer.

AS-BUILT DRAWING Original design drawings revised to reflect any changes made in field submitted by contractor upon completion of construction project.

ASBESTOS Mineral fiber (i.e., magnesium silicate) that can pollute air or water and cause cancer or asbestosis when inhaled.

ASBESTOS ABATEMENT Procedures to control fiber release from asbestos-containing materials in building or to remove them entirely, including removal, encapsulation, repair, enclosure, encasement and operations and maintenance programs.

ASBESTOS ACTIVITY PLANS Written procedures in school's asbestos-management plan that detail steps Local Education Agency (LEA) will follow in performing initial and additional cleaning, operation and maintenance-program tasks, periodic surveillance and re-inspection required by Asbestos Hazard Emergency Response Act.

ASBESTOS ASSESSMENT Evaluation of physical condition and potential for damage of all friable asbestos containing materials and thermal insulation systems.

ASBESTOS BLANKET Asbestos-filled wrap placed around pipes to reduce heat loss.

ASBESTOS CEMENT Fire-resisting weatherproof building material made from cement and asbestos.

ASBESTOS DEBRIS Sections of asbestos-cement building materials.

ASBESTOS DIAPHRAGM Percolating diaphragm that allows through flow of brine from anode to cathode that separates chlorine and hydrogen gas spaces.

ASBESTOS ENCAPSULATION Treatment of asbestos-containing materials with sealant material that surrounds or embeds asbestos fibers in adhesive matrix to prevent release of fibers.

ASBESTOS FELT Material produced by combining felted asbestos fiber with asphalt or elastomer base.

ASBESTOS FIBER . Mineral found in natural settings such as caves and outdoor deposits.

ASBESTOS JOINT RUNNER Asbestos rope, wrapped around pipe and then clamped in position used to hold molten lead poured in caulked joint.

ASBESTOS MIXTURE Mixture that contains bulk asbestos or another asbestos mixture as intentional component.

ASBESTOS PLASTER Non-flammable pipe insulation containing asbestos fibers, diatomaceous earth and gypsum plaster.

ASBESTOS PROGRAM MANAGER Building owner or designated representative who supervises all aspects of facility asbestos management and control program.

ASBESTOS REMOVAL Removal of asbestos from building and structures.

ASBESTOS SHINGLES Roof or wall shingles made with asbestos cement board.

ASBESTOS TAILINGS Any solid waste that contains asbestos and is product of asbestos mining or milling operations.

ASBESTOS-CEMENT BOARD Sheet produced from asbestos cement that's fire-resistant.

ASBESTOS-CEMENT PIPE Circular conduit fabricated from concrete comprising number of asbestos fibers.

ASBESTOS-CEMENT PRODUCTS Products made in stiff substance composed basically of asbestos fiber and portland cement.

ASBESTOS-CEMENT SHEETS Plain or corrugated sheets, used as roof or walk cladding, made from asbestos cement.

ASBESTOS-CONTAINING WASTE MATERIALS Mill tailings or any waste that contains commercial asbestos from asbestos mills by manufacturing, fabricating and spraying operations.

ASBESTOS-DIATOMITE Mixture of asbestos fibers, diatomaceous earth and portland cement used to make insulating, fire-resistant building products.

ASBESTOS-FREE PRODUCTS Variety of products that perform exactly same functions as ones that have historically included asbestos fibers but that are made with substitute substances.

ASBESTOSIS Disease associated with inhalation of asbestos fibers that makes breathing progressively more difficult and can be fatal.

ASEISMIC DESIGN Procedure for earthquake resistant design to limit potential damage of buildings to tolerable level.

ASEISMIC REGION 1. Not subject to earthquakes; 2. Resisting destructive forces of earthquakes.

ASEPTIC Free of disease-causing microbes.

ASEXUAL REPRODUCTION Reproduction by which offspring arise from single organism.

ASH 1. Solid remains after complete combustion; 2. Non-aqueous remains of material subjected to any complete oxidation process; 3. Fine particles from volcano, volcanic ash.

ASH CONTENT Measure of deposit that remains after combustion under given conditions.

ASH DUMP Small door inside fireplace in center of fireplace floor or inner hearth that leads to ash pit.

ASH MASS Mass or amount of residue present after residue from dry-mass determination has been ashed in muffle furnace at temperature of 500 $^{\circ}$C for 1 hour.

ASH SLUICE Trench or channel where water flushes deposit from ash pit into storage or disposal point.

ASHFALL ADVISORY Advisory issued for conditions associated with airborne ash plume resulting in ongoing deposition at surface that may originate directly from volcanic eruption or indirectly by wind suspending ash.

ASIATIC CLOSET Wash closet designed to be used in squatting position and comprises elongated bowl positioned either at or just above floor level.

ASKAREL Mixture of polychlorinated biphenyls, formerly used in electrical apparatus.

ASME-TYPE LP GAS CYLINDER Fuel container for liquefied petroleum gas.

ASPECT 1. Compass direction where slope or building, faces; 2. Any specific feature, part or element of something; 3. Way something appears when viewed from certain direction or perspective.

ASPECT RATIO Ratio of object's longest dimension to its next-longest dimension.

ASPHALT 1. Sticky, black and highly viscous liquid or semi-solid, composed almost entirely of bitumen, present in most crude petroleum's and in some natural deposits; 2. Hard ground covering used for roads and walkways.

ASPHALT PROCESSING Storage and blowing of asphalt.

ASPHALT PROCESSING PLANT Plant that blows asphalt to be used in fabrication of asphalt products.

ASPHALT SOIL STABILIZATION Treatment of naturally occurring, non-plastic or moderately plastic soil with with cutback or liquid asphalt at ambient temperatures to improve load-bearing qualities of soil.

ASPHALT STABILIZATION Mixing of asphalt with sand and/or soil to improve granular cohesion, waterproofing or alternative variable.

ASPHALT STORAGE TANK Apparatus to store asphalt at asphalt roofing plants, petroleum refineries and asphalt processing plants.

ASPHYXIANTS Chemicals that starve cells of individual from life-giving oxygen needed to sustain metabolism.

ASPHYXIATING POLLUTANTS Air pollution that displace oxygen being transferred to hemoglobin molecules.

ASPHYXIATION 1. Death due to lack of oxygen; 2. Acute lack of oxygen.

ASPIRATE To remove liquid or gas by means of suction.

ASPIRATION Drawing of air into combustion chamber.

ASPIRATOR Pump for removing gases or liquids.

ASPIRATOR MIXER Gas-air proportioning device that causes flow of combustion air to induce proper amount of gas into airstream.

ASSAY Test for specific chemical, microbe or effect.

ASSEMBLAGE Process of assembling or bringing together.

ASSEMBLY 1. Act of putting together set of pieces, fragments or elements; 2. Congregation of people in one place for purpose.

ASSEMBLY BOLT Threaded bolt used to hold together briefly several parts during installation and/or construction.

ASSEMBLY DRAWING Shows how parts of product fit together.

ASSEMBLY LANGUAGE Programming language where source code of programs is composed of mnemonic instructions, each of which corresponds directly to machine instruction for particular processor.

ASSEMBLY ROD Rod or bolt, used in machine assembly, that has external threads.

ASSEMBLY SEQUENCE Aligning and merging fragments from longer sequence in order to reconstruct original sequence.

ASSEMBLY TIME 1. Elapsed time after application of adhesive until its strength becomes effective; 2. Time elapsed in performing assembly or sub-assembly operation.

ASSEMBLY TORQUE Torque applied at final assembly.

ASSESSED VALUE Dollar value assigned to property to measure applicable taxes.

ASSESSED WATERS Water bodies for which state is able to make use-support decisions based on actual information.

ASSESSMENT Act of assessing or amount (of tax, levy or duty etc) assessed.

ASSESSMENT ENDPOINT Explicit expression of environmental value to be protected.

ASSESSMENT RATIO Ratio of assessed value to market value to determine property tax liability.

ASSESSMENT ROLL Record of taxable persons and property in taxing jurisdiction.

ASSESSOR One who discovers worth of property for tax or insurance evaluation.

ASSET MANAGEMENT Process of managing, monitoring and maintaining value of commercial entity.

ASSETS Any property or object of value that one possesses.

ASSIGNABLE CAUSE Variation where specific event is linked to inconsistency in system.

ASSIGNMENT SHEET List of operators and regular scheduled bus runs.

ASSIMILATION Ability of body of water to purify itself of pollutants.

ASSIMILATIVE CAPACITY Capacity of natural body of water to receive wastewaters or toxic materials without deleterious effects and without damage to aquatic life or humans who consume water.

ASSOCIATED FACILITY Term used to describe those facilities examined by respective regional or area office including most carriage, distribution and drainage systems, small diversion works, small pumping plants and power plants, open and closed conduits, tunnels, siphons, small regulating reservoirs, waterways and class B bridges.

ASSOCIATED PRINCIPAL USER User with dedicated communications to WSR-88D unit.

ASSOCIATION 1. Environmental unit comprising two or maybe more species in closer proximity than would happen by chance; 2. Level of reliance or independence between couple of variants.

ASSOCIATION OF BOARDS OF CERTIFICATION International organization representing boards which certify operators of waterworks and wastewater facilities.

ASTHENOSPHERE Zone of Earth's upper mantle, below lithosphere.

ASTRONOMICAL DAWN 1. Time at which Sun is 18° below horizon in morning; 2. Point in time at which Sun starts lightening sky where prior to this time, sky is completely dark.

ASTRONOMICAL DUSK 1. Time at which Sun is 18° below horizon in evening; 2. Point in time at which Sun no longer illuminates sky.

ASTRONOMICAL TWILIGHT Period where sun is between 12° and 18° below horizon in morning and evening.

ASTRONOMICAL UNIT Mean Earth-Sun distance that's equal to 1.496×10^{13} cm or 214.94 solar radii.

ASYMMETRIC Not similar in size, shape, form or arrangement of parts on opposite sides of line, point or plane.

AT JACKING Pre-stressing tendons during tensioning.

AT REST PRESSURE Thrust from earth on to fused retaining wall.

AT TRANSFER Passing prestressing force to concrete.

AT-GRADE INTERSECTION Where roadway intersects highway at same grade.

AT-REST EARTH PRESSURE Value of earth pressure when soil mass is in its natural state without having been permitted to yield or without having been compressed.

AT-THE-SOURCE At or prior to commingling of delacquering scrubber liquor blowdown with other wastewaters.

ATMOMETER Instrument that measures rate of evaporation of liquid.

ATMOSPHERE 1. Air surrounding and bound to Earth; 2. Gases surrounding any astronomical body; 3. Unit of pressure equal to air pressure at sea level.

ATMOSPHERE FOR TESTING Atmosphere at ambient humidity and temperature.

ATMOSPHERIC BOUNDARY LAYER Layer of air adjacent to bounding surface where effects of friction are significant that's roughly lowest one or two kilometers of Earth's atmosphere.

ATMOSPHERIC CEILING Height in atmosphere where heated column ceases to rise.

ATMOSPHERIC CIRCULATION MODEL Mathematical model for quantitatively describing, simulating and analyzing structure of circulation in atmosphere and underlying causes.

ATMOSPHERIC CORRECTION Process of removing effects of atmosphere on reflectance values of images taken by satellite or airborne sensors.

ATMOSPHERIC DEPOSITION Transfer of substances from air to surface of Earth either in wet form (e.g., rain, fog, snow, dew, frost, hail) or in dry form (e.g., gases, aerosols, particles).

ATMOSPHERIC DISPERSION Disperse of air pollutants in ambient atmosphere.

ATMOSPHERIC DISPLACEMENT System or process of using water fog in superheated region causing water to be converted into steam that enlarges and displaces atmosphere in burning room or building.

ATMOSPHERIC INVERSION Horizontal layers of air that increase in temperature with height.

ATMOSPHERIC MOISTURE Actual amount of water in atmosphere.

ATMOSPHERIC PRESSURE Pressure exerted by Earth's atmosphere at any given point above sea level.

ATMOSPHERIC RADIATION Infrared radiation (energy in wavelength interval of 3-80 micrometer) emitted by or being propagated through atmosphere that consists of both upwelling and downwelling components.

ATMOSPHERIC STORAGE TANK Tanks are designed for pressures that are approximately equal to that of atmosphere.

ATMOSPHERIC-TYPE VACUUM BREAKER Mechanical types of backflow prevention where float, kept elevated by regular water flow, drops into test seal in event flow is reversed.

ATOMIC ENERGY Energy released by atom in nuclear reaction (e.g., fission or fusion).

ATOMIC NUMBER 1. Number of protons in atom; 2. Number of electrons surrounding nucleus of neutral atom of element.

ATOMIC SPECTRUM Spectrum of frequencies of electromagnetic radiation emitted or absorbed during transitions of electrons between energy levels within atom.

ATOMIC WEIGHT Term used to represent mean relative atomic mass of element in nature.

ATOMIZATION Procedure where flow of water or gas impinges upon molten metal stream, breaking it into droplets that solidify as powder particles.

ATOMIZING HUMIDIFIER Apparatus that adds tiny particles of water into flow of air.

ATTACHED GROWTH PROCESSES Wastewater treatment processes where microorganisms and bacteria treating wastes are attached to media in reactor.

ATTACHMENT Device attached to piece of equipment or tool.

ATTAINMENT AREA Area considered to have air quality as good as or better than national ambient air quality standards as defined in Clean Air Act.

ATTENDED LANE Toll lane adjacent to human occupied toll booth.

ATTENUATION 1. Reduction in peak of hydrograph resulting in more broad, flat hydrograph; 2. Decrease in amplitude of flood wave due to channel geometry and energy loss as it progresses downstream; 3. Process by which compound is reduced in concentration over time (i.e., absorption, adsorption, degradation, dilution, and/or transformation); 4. Decrease with distance of sight caused by particulate pollution; 5. Reduction of radar beam power due to reflection or absorption of energy when it strikes target; 6. Decrease in amplitude of seismic waves with distance due to geometric spreading, energy absorption and scattering.

ATTENUATOR Any device that attenuates signal, but especially electronic device that reduces amplitude of signal.

ATTERBERG LIMITS Boundaries (determined by laboratory tests) of moisture content in soil between liquid state and plastic state (i.e., liquid limit), between plastic state and semisolid state (i.e., plastic limit) and between semisolid state and solid state (i.e., shrinkage limit).

ATTERBERG TEST Procedure for ascertaining plasticity of soils.

ATTIC FAN Ventilation fan that regulates heat level of building's attic by exhausting hot air.

ATTITUDE Orientation of vehicle relative to horizon, direction of motion, etc.

ATTO Prefix symbolizing 10^{-18}.

ATTRACTANT Chemical or agent that lures insects or other pests by stimulating their sense of smell.

ATTRACTANTS Materials or mixtures that, through their property of bringing specific creatures, are utilized as pesticides to mitigate or ruin vertebrate or invertebrate animals.

ATTRACTING GROIN Structure that fixes channel nearby by transferring water toward itself.

ATTRACTIONS Daily trips, from visitors and local residents, to area.

ATTRIBUTE ACCEPTANCE PLAN Procedure that grades construction material as conforming and nonconforming, without looking at quantitative measurements.

ATTRIBUTE SURVEY Survey to determine important components of recreational experience.

ATTRITION Wearing or grinding down of substance by friction where dust from such processes contributes to air pollution.

ATTRITION ARRANGEMENT Using deaths, retirements and voluntary resignations to reduce labor force.

ATYPICAL Not typical.

AUDIBLE ALARM Alarm that sets off one or more of sirens of security system.

AUDIBLE SIGNAGE Mechanical annunciation of advice or directions.

AUDIBLE SOUND Property of sound that most determines pitch between approximately frequencies of 20 to 20,000 Hz.

AUDIO FREQUENCY Frequency or range of frequencies of sound waves, that are audible.

AUDIOGRAPH Instrument used to measure person's hearing that determines rate where sound levels diminish over distance.

AUDIOMETRIC ROOM Room constructed and finished of materials that will absorb over 99% of sound generated within room.

AUDIOVISUAL WARNING SYSTEM Alarm apparatus that indicates operator of gear of out-of-tolerance conditions or potentially dangerous circumstances.

AUDIT 1. Result of such examination or account as adjusted by auditors; 2. Final account.

AUFEIS Sheet-like layered mass of ice.

AUGER Rotating drill, used to take core samples, having screw thread that carries cuttings away from face.

AUGMENTED FLOW Increased volume of water entering channel or running overland.

AUGMENTED REPORT Meteorological report prepared by automated surface weather observing system for transmission with certified observers signed on to system to add information to report.

AURORA Atmospheric phenomenon created by charged particles from sun striking upper atmosphere, creating colored lights in sky that occurs mainly in high-latitude night sky.

AURORA AUSTRALIS 1. Luminous, radiant emission from upper atmosphere over middle and high latitudes and centered around Earth's magnetic poles; 2. Same as Aurora Borealis, but in Southern Hemisphere, known as southern lights.

AURORA BOREALIS Luminous, radiant emission from upper atmosphere over middle and high latitudes in northern hemisphere.

AURORAL OVAL Oval band around each geomagnetic pole that is locus of structured aurorae.

AUTHOR'S SIGNATURE Signature of person(s) with primary responsibility for writing document.

AUTHORITY Power to enforce rules or give orders.

AUTHORITY HAVING JURISDICTION Person or office charged with enforcing.

AUTHORIZED PERSON Person assigned to perform specific type of duty or duties or to be at specific location or locations who is trained and has demonstrated ability to perform such duty or duties safely and effectively.

AUTHORIZED PROGRAM OR STATE State, Territorial, Tribal or interstate program which has been approved or authorized by EPA.

AUTO-FREE ZONE Area where automobile traffic is prohibited.

AUTO-SUPPRESSION SYSTEM Part of active fire protection system where detectors automatically activate fire extinguishers to suppress fire.

AUTOCHTHONOUS 1. Native to place where found; 2. Indigenous; 3. Buried in place.

AUTOCLAVE Device that uses steam to sterilize equipment and deactivate bacteria, viruses, fungi and spores.

AUTOGAMY Self-fertilization, fertilizing pollen being derived from same blossom as pistil acted upon.

AUTOGENETIC DRAINAGE Drainage by streams whose courses have been determined solely by conditions of land surface over which they flow.

AUTOGENIC SUCCESSION Succession driven by biotic components of ecosystem.

AUTOGYRO Aircraft where lift is provided by unpowered rotating wings and thrust is provided by conventional propeller.

AUTOIMMUNITY Condition where one's immune system attacks one's own tissues (i.e., autoimmune disorder).

AUTOLYSIS Destruction of organism's cells by enzymes produced by organism itself.

AUTOMATED EVENT-REPORTING GAUGE For river stage gauges, pressure transducer type gauge that can be programmed to report if water surface rises or falls by predetermined amount.

AUTOMATED GUIDEWAY TRANSIT Automated transit system where vehicles are automatically guided along guideway.

AUTOMATED LOCAL EVALUATION IN REAL TIME Local flood warning system where river and rainfall data are collected via radio signals in real-time at ALERT base station.

AUTOMATED METHOD Procedure for quantifying concentrations of ambient air pollutant where sample collection, evaluation and measurement are performed mechanically.

AUTOMATED MUTUAL ASSISTANCE VESSEL RESCUE SYSTEM System operated by U.S. Coast Guard which computes nearest available rescue vessels for vessels in distress using vessel track and position reports supplied by participating vessels.

AUTOMATED REPORT Meteorological report prepared by automated surface weather observing system for transmission and with no certified weather observers signed on to system.

AUTOMATED SURFACE OBSERVING SYSTEM Joint effort of National Weather Service, Federal Aviation Administration and Department of Defense to serve as nation's primary surface weather observing network.

AUTOMATIC ALARM Alarm actuated when preset of predetermined state is surpassed or invalidated.

AUTOMATIC BLOCK SIGNAL CONTROL SYSTEM System where signals are automatically actuated by switch, rail, train or signal circuit.

AUTOMATIC CONTROL VALVE Flow control valve that regulates flow or pressure of fluid using fluid pressure itself.

AUTOMATIC DIRECTION FINDER Device that calculates relative bearing of aircraft to radio station.

AUTOMATIC FARE COLLECTION Components that automate fare ticketing.

AUTOMATIC FIRE ALARM SYSTEM System to detect occurrence, alert control panel and proper authorities and notify occupants to take action.

AUTOMATIC FIRE PUMP Pump that provides required water pressure in fire standpipe or sprinkler system should system pressure fall below preset value.

AUTOMATIC FIRE VENT Device in roof of building that operates automatically to control fire or smoke.

AUTOMATIC GAIN CONTROL Method of automatically controlling gain of receiver, particularly one that holds output level constant regardless of input level.

AUTOMATIC GATE Mechanical device used to open and close gate without human assistance.

AUTOMATIC GENERATION CONTROL Computerized power system regulation to maintain scheduled generation within prescribed area in response to changes in transmission system operational characteristics.

AUTOMATIC LANE Roadway toll lane that automatically bills.

AUTOMATIC OPERATION Functioning of systems, equipment or processes in desired manner at proper time under control of mechanical or electronic devices that operate without human intervention.

AUTOMATIC PROGRESSION Advances in pay rates on periodic basis.

AUTOMATIC RADIOTHEODOLITE Ground-based radio direction finder that automatically tracks balloon-borne radiosonde.

AUTOMATIC RECORDING GAUGE Automatic instrument for continuously measuring and recording graphically.

AUTOMATIC SEAL Seal that's activated by pressure of fluid it seals.

AUTOMATIC SMOKE ALARM Apparatus that activates alarm upon detection of combustion.

AUTOMATIC SPILLWAY Apparatus for releasing water into drainage paths constructed along canal sides.

AUTOMATIC SPRINKLER Nozzle that spreads water from water supply system using flow control apparatus.

AUTOMATIC SPRINKLER KIT Set of tools used to start and/or stop flow of water from sprinkler.

AUTOMATIC SPRINKLER SYSTEM built-in fire protection system of underground and overhead piping.

AUTOMATIC TEMPERATURE COMPENSATOR Apparatus that measures temperature of fluid through device and automatically adjusts measured volume to equivalent volume.

AUTOMATIC TERMINAL INFORMATION SERVICE Broadcast of recorded aeronautical information continuously.

AUTOMATIC TRAIN CONTROL Train protection system that uses speed control mechanisms via computer.

AUTOMATIC VALVE Valve that opens or closes without human assistance.

AUTOMATIC VEHICLE CLASSIFICATION System to identify vehicles without driver or human observer interaction.

AUTOMATIC VEHICLE IDENTIFICATION System to identify vehicles in various situations.

AUTOMATIC VEHICLE LOCATION Determining and transmitting geographic location of vehicle.

AUTOMATION OF FIELD OPERATIONS AND SERVICES Computer system linking National Weather Service offices for transmission of weather data.

AUTOMOBILE MODE Roadway that provides service to all road users, excluding transit buses.

AUTOTROPHY Organisms that thrive by using inorganic materials for energy and grout.

AUTUMN 1. Season of year that's transition period from summer to winter, occurring as Sun approaches winter solstice; 2. Meteorological autumn (different from standard/astronomical autumn) begins September 1 and ends November 30.

AUTUMNAL EQUINOX Equinox at which Sun approaches Southern Hemisphere, marking start of astronomical autumn in Northern Hemisphere.

AUXILIARY 1. Supplementary or subsidiary; 2. Helping; giving assistance or support.

AUXILIARY EQUIPMENT Accessory equipment necessary for operation of generating station.

AUXILIARY LANE Roadway added between interchanges.

AUXILIARY SPILLWAY Any secondary spillway designed to be operated infrequently, possibly in anticipation of some degree of structural damage or erosion to spillway that would occur during operation.

AVAILABILITY FACTOR Amount of time that machine is able to produce over certain period, divided by amount of time in period.

AVAILABILITY SESSION Informal meeting at public location where interested citizens can talk with public officials on one-to-one basis.

AVAILABLE CAPACITY Amount of water held in soil available to plants.

AVAILABLE CHLORINE Measure of amount of chlorine available in chlorinated lime, hypochlorite compounds and other materials used as source of chlorine when compared with that of liquid or gaseous chlorines.

AVAILABLE DILUTION Ratio of amount of untreated wastewater to diluting flow at point of discharge.

AVAILABLE MOISTURE 1. Moisture content at permanent wilting point; 2. Difference in water content of soil between field capacity and permanent wilting.

AVAILABLE MOISTURE CAPACITY Difference between amount of soil-pore water at field moisture capacity and amount at wilting point.

AVAILABLE OXYGEN Amount of dissolved oxygen readily available for oxidation of organic matter in body of water.

AVAILABLE TIME-SPACE Total time-space available for circulating pedestrians.

AVAILABLE WATER Difference between field capacity maximum amount of water soil can hold and wilting point where plant can no longer extract water from soil.

AVAILABLE WATER SUPPLY Source, means or process of supplying water in excess of quantity necessary to supply valid prior rights and demands.

AVALANCHE Mass of snow, rock, and/or ice falling down mountain or incline.

AVALANCHE ADVISORY Preliminary notification that conditions may be favorable for development of avalanches in mountain regions.

AVERAGE 1. Arithmetic mean; 2. Arithmetic mean obtained by adding quantities and dividing sum by number of quantities; 3. Measure of central tendency.

AVERAGE ANNUAL DAILY TRAFFIC Total volume of traffic along segment or point for one year, divided by 365 days.

AVERAGE ANNUAL FLOOD Flood discharge equal to mean of discharged of all of maximum yearly floods during period of record.

AVERAGE ANNUAL RUNOFF Average value of annual runoff amounts calculated for selected period of record that represents average hydrologic conditions.

AVERAGE BUS STOP SPACING Mean distance between bus stops.

AVERAGE CONCENTRATION Average of analyses made over time and/or space.

AVERAGE COST FLOW Cost of goods available for sale is divided by number of units produced in period to arrive at average cost per unit.

AVERAGE DAILY FLOW Total quantity of liquid to point, divided by number of days of flow measurement.

AVERAGE DAILY TRAFFIC Average number of vehicles passing specific point during period of time, divided by days in that period of time.

AVERAGE DAY Daily traffic volumes along segment or past point.

AVERAGE DEGREE OF CONSOLIDATION Ratio of total volume change in soil mass at given time to total volume change anticipated in soil mass due to primary consolidation.

AVERAGE DIMENSION Average of corresponding measurements of random units.

AVERAGE DISCHARGE Arithmetic average of all complete water years of record of surface water discharge whether consecutive or not.

AVERAGE EFFICIENCY Efficacy of equipment over range of load where equipment works.

AVERAGE END AREA Procedure of computing volume of earthwork between two cross-sections where volume equals typical cross-sectional areas multiplied by distance between two cross-sections.

AVERAGE END CONCEPT Averaging of two end cross-sections of reach in order to smooth numerical results.

AVERAGE ENERGY Total power generation produced by power plant during all of years of its actual or simulated operation divided by number of years of actual or simulated operation.

AVERAGE FARE Mean of passenger fares.

AVERAGE FLOW Arithmetic average of flows measured at fixed point.

AVERAGE FREQUENCY OF OCCURRENCE Average number of years between similar events.

AVERAGE FROST PENETRATION Average based on historical data of depth of frost penetration.

AVERAGE GRADE Average of levels of varied earth locations throughout construction site.

AVERAGE GROUNDWATER VELOCITY 1. Average distance travelled per unit of time by body of groundwater; 2. Total volume of groundwater passing through unit cross-sectional area divided by porosity of water-transmitting material.

AVERAGE HAUL Area of mass diagram representing number of cubic yard stations of haul between balance points divided by ordinate of mass that yardage is hauled.

AVERAGE INTERSTITIAL VELOCITY Average rate of groundwater flow to interstices expressed as product of hydraulic conductivity and hydraulic gradient divided by effective porosity.

AVERAGE MONTHLY DISCHARGE LIMITATION Highest allowable average of daily discharges over calendar month, calculated as sum of all daily discharges measured during calendar month divided by number of daily discharges measured during that month.

AVERAGE MONTHLY DISCHARGE LIMITATIONS Highest allowable average of daily discharges over calendar month, calculated as sum of all daily discharges measured during that month divided by number of days on which monitoring was performed (except in case of fecal coliform).

AVERAGE RELATIVE HUMIDITY Average of measured relative humidities.

AVERAGE RUNNING SPEED Length of road segment divided by mean time of vehicles to pass that segment.

AVERAGE SEA LEVEL Average level of surface of one or more of Earth's oceans from which heights such as elevations may be measured.

AVERAGE SPEED Vehicle speeds divided by number of vehicles observed.

AVERAGE SPOT SPEED Mean spot speed of several vehicles measured at given spot.

AVERAGE STREAMFLOW Average rate of discharge of stream during given interval.

AVERAGE STREAMFLOW VELOCITY Typical speed of stream flowing in channel or conduit at given cross-section or in certain reach.

AVERAGE TRAVEL SPEED Distance of road segment divided by travel time of vehicle.

AVERAGE VEHICLE OCCUPANCY Mean number of people divided by number of vehicles traveling past specific point.

AVERAGE VELOCITY 1. Mean discharge divided by cross-sectional area; 2. Displacement divided by time.

AVERAGE WAVE PERIOD Average period (seconds) of highest one-third of wave observed during 20 minute sampling period.

AVERAGE WEEKLY DISCHARGE LIMITATION Highest allowable average of daily discharges over calendar week, calculated as sum of all daily discharges measured during calendar week divided by number of daily discharges measured during that week.

AVERAGE YEAR Year where observed quantities approximate mathematical mean of such observed quantifies for considerably longer period.

AVERAGE YEAR SUPPLY Average annual supply of water development system over long period.

AVERAGE YEAR WATER DEMAND Demand for water under average hydrologic conditions for defined level of development.

AVERAGING PERIOD Period of time over which receiving water concentration is averaged for comparison with criteria concentrations.

AVIAN Having to do with birds.

AVIATION MODEL 120-hour numerical model of atmosphere.

AVOIDED COST Cost utility would incur to generate next increment of electric capacity using its own resources.

AVOIRDUPOIS WEIGHT English and American system of weights based on pound of 16 ounces.

AVONGARD CRACK MONITORS Measure movements transverse and along joint or crack.

AVULSION Rapid change in channel direction and form that occurs during catastrophic, rare, floods.

AVULSION CUTOFF 1. Place of property circumscribed by old and also newest channels when river changes its path; 2. River's actions when avulsion happens.

AWARD To determine, make or grant.

AWARD OF CONTRACT After contract has been executed, act of sending signed agreement and other contract documents to lowest responsible bidder.

AWASH Washed by waves or tide when flush with surface of water.

AXENIC 1. Containing only single species of microorganism; 2. Isolated from organisms of any other species.

AXIAL FLOW Movement in direction of axis.

AXIAL FLOW COMPRESSOR Rotating, airfoil-based compressor where gas or working fluid principally flows parallel to axis of rotation or axially.

AXIAL FLOW PUMP Type of pump that essentially consists of propeller (an axial impeller) in pipe.

AXIAL FORCE DIAGRAM Graphical rendering of axial load acting at each segment of structural member, plotted to scale and with proper sign as ordinate at each point of member and along reference line representing length of member.

AXIAL LOADING Force applied along length of member (i.e., along lines of axis).

AXIAL MOTION Rotary motion of object around its own axis.

AXIAL SEAL Shaft seal where squeeze is on both top and bottom of O-ring's cross-section.

AXIAL TO IMPELLER direction of discharge of flow, parallel to impeller shaft, with propeller pumps.

AXIAL-FLOW FAN Type of compressor fan that increases pressure of air flowing through it by forcing air to move parallel to shaft about which blades rotate.

AXIAL-PISTON PUMP Positive displacement pump that has number of pistons in circular array within cylinder block.

AXIS 1. Straight line around which shaft or body revolves or is symmetrically arranged; 2. Fixed one-dimensional figure (e.g., line); 3. Centerline of tunnel or shaft; 4. Main stem or central part about which organs or plant parts such as branches are arranged.

AXIS OF DAM Vertical plane or curved surface, appearing as line in plan or cross-section, that horizontal dimensions can be referred, coincident with upstream face at top of dam.

AXIS OF IMPELLER Imaginary line running along center of shaft (i.e., impeller shaft).

AXIS OF ROTATION 1. Axis about which any body rotates; 2. Line joining North Pole and South Pole about which Earth rotates daily.

AXIS OF SYMMETRY Any of one or more lines about which geometric figure is symmetric.

AXIS OF WELD Imaginary line along weld used to describe positions of localized welds.

AXLE Spindle where wheel revolves or revolves with wheel.

AXLE CORRECTION Factors that adjust axle counts into vehicle counts.

AXLE COUNT Number of axles having contact with ground or rail.

AXLE GROUP At least 2 adjacent axles.

AXLE LOAD Total weight transferred by one axle and its wheels to road or railway surface.

AXLE SPACING Horizontal distance between axles.

AXLE UNIT Single or tandem axle.

AXLE-GROUP Two or more parallel axles on single vehicle or series of vehicles.

AZIMUTH Direction in terms of 360° compass where north is 0°, east is 90°, south is 180° and west is 270°.

AZIMUTH ANGLE Direction or bearing toward which sloping surface faces.

AZINPHOS-METHYL Acetylcholinesterase inhibitor formerly used as organophosphate insecticide.

AZODRIN Organophosphate insecticide.

AZOLLA Mosquito fern (any of several species of small floating aquatic ferns in genus Azolla) that forms symbiotic relationship with Anabaena (i.e., blue-green algae) that can fix nitrogen.

AZORES CURRENT One of currents of North Atlantic subtropical gyre.

AZORES HIGH Semi-permanent, subtropical area of high pressure in North Atlantic Ocean off East Coast of North America that migrates east and west with varying central pressure that's alternate term for Bermuda High.

AZOTOBACTER Genus consisting of most critical of aerobic, nitrogen-fixing bacteria.

B

B Abbreviation used on long-term climate outlooks issued by CPC to indicate areas that are likely to be below normal for given parameter (e.g., temperature, precipitation).

B BLOCK SIGNAL Fixed signal controlling use of block for railways.

B-HORIZON Commonly referred to as 'subsoil', between A-horizon and C-horizon, that consists of mineral layers which may contain concentrations of clay or minerals such as iron or aluminium oxides or organic material moved there by leaching.

BACILLARIOPHYTA Single-celled (unicellular), colonial or filamentous algae with siliceous cell walls constructed of two overlapping parts.

BACILLUS Any of various rod-shaped, spore-forming aerobic bacteria in genus Bacillus.

BACK 1. Support or resource in reserve; 2. Keel and keelson of ship; 3. Top surface of flat or inclined timber; 4. Roof or top of underground opening; 5. Ore found between certain amount along with surface or between two amounts.

BACK BLADING Dropping blade into rough surface and driving backwards to smooth out surface.

BACK DOOR COLD FRONT Cold front moving south or southwest along Atlantic seaboard and Great Lakes.

BACK DRAIN Perforated drain pipe set behind base of wall to lessen water pressure acting on wall's backside.

BACK FURROW First cut of plow, from which slice is laid on undisturbed soil.

BACK HAUL Return trip that truck or cargo ship makes after making delivery.

BACK OF LEVEE Levee side that faces secure region (i.e., opposite river it contains).

BACK OF QUEUE Distance between signalized intersection stop line and last queued vehicle.

BACK PRESSURE 1. Pressure that can cause water to backflow into water supply when user's water system is at higher pressure than public water system; 2. Pressure opposed to desired flow in confined places (e.g., conduits); 3. Measured loss in efficiency in engine exhaust system due to friction and/or pumping losses; 4. Earth pressure exerted on vertical walls of abutment by fill; 5. Lateral loading of piling.

BACK PRESSURE VALVE Control valve designed to hold pressure on pressurized production vessels and release upstream pressure when designated set point is reached.

BACK PROP Raking strut used to transfer bulk of lumber to ground in deep trenches.

BACK RIPPER Teeth fitted to dig in when dozer is reversed but float when moving forward.

BACK SHORE Area of shore that lies between average high tide mark and vegetation.

BACK SIPHONAGE Reverse flow condition created by difference in water pressures that causes water to flow back into distribution pipes of drinking water supply from any source other than intended one.

BACK VENT Vent for plumbing fixture located on downstream (sewer) side of trap to protect trap against siphonage.

BACK-BUILDING THUNDERSTORM Thunderstorm where new development takes place on upwind side, such that storm seems to remain stationary or propagate in backward direction.

BACK-END MATERIALS RECOVERY Engineered system that provides for collection of discrete reusable materials from mixed wastes that have been burned or treated.

BACK-END SYSTEM Mix of system elements that alters chemical properties of waste or converts its parts into energy or compost.

BACK-INLET GULLY Trapped receptacle at surface level that waste or storm water can pass through easily but gases, odours and rodents from drainage and sewer system can not pass back other way.

BACK-SHEARED ANVIL Slang for thunderstorm anvil which spreads upwind, against flow aloft.

BACKBLOW Reversal of conventional flow of water, under pressure, in well to free clogging stuff from screen or strainer.

BACKBOARD 1. Temporary board on exterior of scaffold; 2. Port or larboard side of ship.

BACKCROSS To cross hybrid with one of its parents.

BACKFALL Incline opposite to intended direction of flow that can lead to backflow or ponding.

BACKFILL Refill of material previously excavated.

BACKFILL BLADE Dozing blade fitted to machine to permit operator to replace material excavated from trench or pit.

BACKFILL CONCRETE Concrete used in refilling excavation in lieu of earth material.

BACKFILL DENSITY Percent compaction needed or anticipated for backfill.

BACKFIRE 1. To fire in opposite direction; 2. Fire started upwind of existing brush fire, to burn against and cut off its spread; 3. Experience premature ignition of fuel or ignition of exhaust gases, making popping sound.

BACKFLOW 1. Backing up of water through conduit or channel in direction opposite to normal flow; 2. Reverse flow condition created by difference in water pressures that causes water to flow back into distribution pipes of drinking water supply from any source other than intended one.

BACKFLOW CONNECTION Any arrangement whereby backflow can occur.

BACKFLOW PREVENTER Apparatus in potable water supply system that prevents backflow of water in connections on its outlet.

BACKFLOW PREVENTER VALVE Device designed to prevent potable water from being contaminated with non-potable water during condition of backflow.

BACKGROUND CONCENTRATION Concentration of substance in particular environment indicative of minimal influence by human (anthropogenic) sources.

BACKGROUND CONCENTRATION OF POLLUTANTS Concentration of unwanted materials generally or naturally present in place, independent of any local or special generator or source of pollutant.

BACKGROUND CONTAMINATION Total of extraneous particles which can be introduced in procedure for obtaining, storing, transferring and assessing fluid sample.

BACKGROUND LEVEL 1. Concentration of substance in environmental media (e.g., air, water or soil) that occurs naturally or is not result of human activities; 2. In exposure assessment, concentration of substance in defined control area, during fixed period of time before, during or after data-gathering operation.

BACKGROUND RADIATION Ionizing radiation naturally present in environment.

BACKHOE 1. Equipment consisting of digging bucket or scoop on end of articulated arm, drawn backwards to move earth; 2. Multi-purpose tractor with front-mounted loading bucket and rear-mounted digging bucket.

BACKHOE TAMPING Processing measure, frequently found in direct-dump transport systems, where standard backhoe is utilized to compact waste within open-top transport trailer.

BACKHOE/LOADER Heavy equipment vehicle that consists of tractor like unit fitted with loader-style shovel or bucket on front and backhoe on back.

BACKING Counterclockwise shift in wind direction.

BACKING WIND Wind that can shift in counterclockwise direction with time at given location or change direction in counterclockwise sense with height.

BACKPACK 1. Gear used to fight brush or forest fire; 2. Knapsack, supported by straps, worn on person's back for purpose of carrying things.

BACKSCATTER Portion of particles and/or radiation scattered back in incident direction.

BACKSIGHT Rod reading taken on point of known elevation, benchmark or turning point that's added to known elevation to arrive at height of instrument.

BACKSIPHON PREVENTER Plumbing fixture designed to prevent backflow of water into potable supply.

BACKSIPHONAGE Form of backflow caused by negative or below atmospheric pressure within water system.

BACKSLOPE 1. Downward slope away from bank of river, mountain, etc.; 2. Portion beyond ditch sides of earth embankment for highway cut that rejoins original ground.

BACKUP 1. Copy of file or record, stored separately from original, that can be used to recover original if it is destroyed or damaged; 2. Accumulation of material caused by (partial) obstruction or (complete) blockage of flow or movement of material or accumulation of material that causes overflow due to flow being greater than maximum possible flow.

BACKWARD PASS Calculation of latest finish and latest start dates within Critical Path Method (CPM).

BACKWASH 1. Reversal of flow through rapid sand filter to free material clogging filter; 2. Seaward return of water following uprush of waves across beach or shore; 3. Water, spray or waves thrown back by obstruction.

BACKWASHING Reversing flow of water back through filter media to remove entrapped solids.

BACKWATER 1. Water backed up or retarded in its course as compared with its normal or natural condition of flow; 2. Body of water where flow is slowed or turned back by obstruction (e.g., bridge, dam, opposing current, movement of tide); 3. Small, generally shallow body of water with little or no current of its own; 4. Stagnant water in small stream or inlet.

BACKWATER AREA Lands, adjacent to stream, flooded by backwater.

BACKWATER CURVE Longitudinal profile of water surface in open channel where depth of flow has been increased by obstruction, increase in channel roughness, decrease in channel width or flattening of bed slope.

BACKWATER EFFECT Effect which dam or other obstruction or construction has in raising surface of water upstream from it.

BACKWATER FLOODING Upstream flooding caused by downstream conditions (e.g., channel restriction, high flow in downstream confluence stream).

BACKWATER FUNCTION Ratio for determining shape of backwater curve or increased depth, of water in waterway at various distances upstream of dam or other obstruction raising upstream elevation of water surface.

BACKWATER GATE Gate installed at end of drain or outlet pipe to prevent backward flow of water or wastewater.

BACKWATER PROFILE Longitudinal profile of water surface in stream where water surface is raised above its normal level by natural or artificial obstruction.

BACKWATER RATIO Ratio of backwater to water-surface drop through constriction.

BACKWATER VALVE Valve installed to allow water or sewage to flow only one way.

BACKYARD COMPOSTING Diversion of organic food waste and yard trimmings from municipal waste stream by composting them in one's yard through controlled decomposition of organic matter by bacteria and fungi into humus-like product.

BACTERIA 1. Single-celled microscopic organisms typically spherical, rodlike or spiral and threadlike in shape, often clumped into colonies; 2. Microscopic living organisms that can aid in pollution control by metabolizing organic matter in sewage, oil spills or other pollutants; 3. Microscopic living organisms that can cause human, animal and plant health problems.

BACTERIAL ANALYSIS Evaluation of potable water, raw water and wastewater to learn existence, amount as well as identification of bacteria.

BACTERIAL CORROSION Corrosion caused or promoted by microorganisms.

BACTERIAL TREATMENT Treatment of wastewater including bacteria.

BACTERICIDE Pesticide used to control or destroy bacteria.

BACTERIOCHLOROPHYLLS Bacterial pigments that trap light energy.

BACTERIOLOGICAL AFTER GROWTH Growth of bacterial development following treatment of water supply.

BACTERIOLOGICAL COUNT Means for quantifying amounts of organisms from sample.

BACTERIOPHAGES Virus that parasitizes bacterium by infecting it and reproducing inside it.

BACTERIOSTAT Biological or chemical agent that causes bacteriostasis.

BAD AIR Polluted air or gas or air including inadequate oxygen that makes up danger.

BADGER Tool used to get rid of excess mortar in interior joints of drain tile as it's set.

BADLANDS Arid terrain characterized by severe erosion of sedimentary rocks.

BAFFLE 1. Flat board or plate, deflector, guide or similar device constructed or placed in flowing water or slurry systems to cause more uniform flow velocities to absorb energy and to divert, guide or agitate liquids; 2. Surface that is placed inside open area to inhibit direct motion from one part to another, without preventing motion altogether; 3. Lever for operating throttle valve of winding engine.

BAFFLE AERATOR Aerator where baffles cause turbulence and minimize short-circuiting.

BAFFLE BLOCK 1. Block, usually of concrete, constructed in channel or stilling basin to dissipate energy of water flowing at high velocity; 2. One of series of upright obstructions designed to dissipate energy as in case of stilling basin or drop structure.

BAFFLE CHAMBER Chamber designed to promote settling of fly ash and coarse particulate matter by changing direction and/or reducing velocity of gases produced by combustion of refuse/sludge.

BAFFLE PIER Protective structure in course of fast-flowing water to dissipate energy.

BAG DAM Temporary dam constructed of bags filled using blend of soil or sand and cement.

BAG FILTER Filter inside housing that fills with water or water, it seeps through filter's tiny holes, leaving behind unwanted contaminants.

BAG HOUSE Air pollution control device that removes particulates out of air or gas released from commercial processes or combustion for electricity generation.

BAG PLUG Inflatable drain stopper usually placed at lowest point in piping system; used during testing of system's integrity.

BAG TRAP S-shaped trap in plumbing where vertical inlet and outlet pipes are in alignment.

BAGGING OPERATION Mechanical procedure where bags are full of non-metallic minerals.

BAGHOUSE FILTER Large fabric bag, usually made of glass fibers used to eliminate intermediate and large particles.

BAGWORK Revetment, consisting of heavy material sewn into bags, for protecting embankments against scour.

BAILER 1. Pipe equipped with valve at lower end used to remove slurry from bottom or side of well as it is being drilled or to collect groundwater samples from wells or open boreholes; 2. Machine for bailing water out of pit; 3. Tube of varying length.

BAILEY BRIDGE Portable, pre-fabricated, truss bridge.

BAILING BUCKET Pail-like tool for removing water in hole during drilling or in preparation for concrete placement.

BAKE OVEN Apparatus which uses heat to dry or cure coatings.

BALANCE 1. To make credits and debits of (an account) correspond; 2. Pair of scales for measuring weight.

BALANCE BAR Large, heavy beam extending along upper edge of lock gate used to open and close gate when water level is equal on both sides of gate.

BALANCE PIPE Pipe between two points used to equalize pressure.

BALANCE POINT 1. Point where all excavated material continues to be utilized as fill; 2. When place's heat output equals heat input; 3. When weight is equally distributed when resting on single spot.

BALANCED BID Each bid item includes its own direct cost plus it pro rata share of project overhead, markup, bond and tax.

BALANCED CUT-AND-FILL When volume of material to be excavated equals volume of material needed as fill of earthwork.

BALANCED EARTHWORKS When volume of material to be excavated equals volume of material needed as fill of earthwork.

BALANCED GATE To discharge water, gate that operates automatically according to upstream head.

BALANCED HEAD CONDITION Condition where water pressure on upstream and downstream sides of object are equal (e.g., emergency, regulating gate).

BALANCED VALVE Value where internal forces from pressure are equal.

BALANCED, INDIGENOUS COMMUNITY Biotic community typically characterized by diversity, capacity to sustain itself through cyclic seasonal changes and presence of necessary food chain species.

BALANCING 1. Adjustment of air or water by means of system; 2. Utilization of long-term or temporary storage areas to attenuate stormwater runoff.

BALANCING RESERVOIR Surface or elevated water storage facility that acts to control flow within system.

BALANCING VALVE Discharge valve that ensures that design flow rate is achieved at all times irrespective of any pressure changes within system.

BALER Machine that compacts materials into high density bales.

BALING Compacting solid waste into blocks to reduce volume and simplify handling.

BALL CLAY Type of clay that commonly consists of 20-80% kaolinite, 10-25% mica and 6-65% quartz.

BALL COCK Valve controlling flow by means of ball pierced with opening in one direction and fixed to rotate on spindle at right angles to opening.

BALL JOINT 1. Consists of bearing stud and socket enclosed in casing; 2. Flexible conduit joint formed just like ball or sphere.

BALL LIGHTNING Short-lived glowing ball sometimes observed to float in air and thought to consist of ionized gas associated with thunderstorms.

BALL NOZZLE Sprinkler nozzle having cup at its end along with sphere within cup.

BALL TEST Drain test where bail is flushed through drain from one manhole to next to detect any partial blockage.

BALL VALVE Contrivance where ball, placed in circular cup with hole in its bottom, operates as valve.

BALL-MILLING Repeated churning action of cobbles, gravel and sand caused by force of water in stilling basin or other structure by which severe concrete abrasion can occur.

BALLAST 1. Heavy material placed in hold of ship to provide stability; 2. Coarse gravel or similar material laid to form bed for roads or railroads or in making concrete.

BALLAST VOYAGE Journey with no cargo on board to get ship in position for next loading port or docking.

BALLED AND BURLAPPED Of or relating to tree, shrub or other plant prepared for transplanting by allowing roots to remain covered by ball of soil around which canvas or burlap is tied.

BALLISTIC SEPARATOR Machine that sorts organic from inorganic matter for composting.

BAND APPLICATION Spreading of chemicals over or next to, each row of plants in field.

BAND SCREEN Infinite moving belt of wire mesh that traps big solids in intake into power station, water or wastewater treatment plant.

BANDED 0.02 ft to 0.1 ft.

BANDED STEEL PIPE Steel pipe for which strength has been increased by use of bands shrunk around shell.

BANDPASS FILTER Filter whose frequencies are between given upper and lower cutoff values, while substantially attenuating all frequencies outside these values.

BANDWIDTH Frequency range between lowest and highest frequencies that are passed through component, circuit or system with acceptable attenuation.

BANK 1. Margins of channel; 2. Incline; 3. In excavation, mass of soil rising above digging or trucking level; 4. Deposit of earth worked by excavations above water level; 5. Edge of river, lake or other watercourse; 6. Elevation or rising ground, under sea; 7. Shallow area of shifting sand, gravel, mud, etc.

BANK CUBIC METER 1 m^3 of material as it lies naturally.

BANK CUBIC YARD 1 yd^3 of material as it lies naturally.

BANK FULL Established river stage at given location along river that is intended to represent maximum safe water level that will not overflow river banks or cause any significant damage within river reach.

BANK GRAVEL Naturally deposited gravel intermixed with sand or clay found in and next to rivers and streams.

BANK MATERIAL Earth material in natural state before excavation or blasting.

BANK MEASURE Measure of volume of mass of soil or rock, before excavation, in its natural position.

BANK MIGRATION Lateral or horizontal movement of banks of stream.

BANK PROTECTION Measures (e.g., vegetation) taken to prevent or reduce erosion of earthen sloping face.

BANK REVETMENT Form of bank protection that covers whole wetted face, from deepest area of river bed to over level of greatest anticipated flooding or wave.

BANK SAND Fine-textured that generally contains percentage of clay and silt along with greatest percentage of sand having sharp edges.

BANK SEDIMENT RESERVOIR Portion of alluvium on sides of channel.

BANK SLOPING Sloping land with equipment.

BANK STORAGE Water that has infiltrated from reservoir into surrounding land where it remains in storage until water level in reservoir is lowered.

BANK YARD Cubic yard of soil or rock measured in situ, before excavation.

BANK-RUN GRAVEL Unprocessed gravel excavated from naturally occurring deposit between 6 mm (0.25 in.) to 150 mm (6 in.).

BANKFULL Water level or stage, at which stream, river or lake is at top of its banks and any further rise would result in water moving into flood plain.

BANKFULL DISCHARGE Flow discharge prior to river spilling onto its floodplain.

BANKFULL STAGE Established river stage at certain point along river that is intended to represent maximum safe water level which will not overflow river banks or cause any significant damage within reach of river.

BANKING System for recording qualified air emission reductions for later use in bubble, offset or netting transactions.

BANNER CLOUD Cloud plume often observed to extend downwind behind isolated mountain peaks, even on otherwise cloud-free days.

BANQUETTE Bench or berm built on land side of lever to improve width of its lower potion.

BAR 1. Unit of pressure, equal to 14.504 psi; 2. Silt, sand and/or gravel deposited at mouth of estuary of river or bay.

BAR CHART Graph in form of boxes of different heights, with each box representing different value or category of data and heights representing frequencies.

BAR RACK Display composed of vertical, horizontal or inclined parallel bars placed in channel or waterway to trap floating and partially-submerged debris.

BAR SCREEN Device used to remove large solids in wastewater treatment.

BARBER POLE Slang for thunderstorm updraft with visual appearance including cloud striations that are curved in manner similar to stripes of barber pole.

BARCHAN Arc-shaped sand ridge comprised of well-sorted sand.

BARE SITE Designated section of land cleared of debris, structures and vegetation.

BARE SOIL Soil or sand not covered by grass, sod, other live ground covers, wood chips, gravel, artificial turf or similar covering.

BARGE Large, flat-bottomed towed or self-propelled boat used mainly for river and canal transport of heavy goods or bulk cargo.

BARGE BED Natural or made mud bottom near river bank where barges can rest at low tide while being loaded or unloaded.

BARITE Mineral, barium sulfate, used in either pure or impure form as concrete aggregate.

BARIUM Silvery-white, soft metallic chemical element with atomic number of 56.

BARIUM SULFATE TEST Fast chemical color evaluation to ascertain pH value of soil.

BARK Rough outer covering of stalks and roots of woody plants.

BARMINUTOR Grinder used to reduce size of influent solids to wastewater plant so that they can be more easily treated or removed in subsequent processes.

BARN WASTEWATER Water contaminated from animal enclosures.

BARNES' FORMULA Means for computing flow in slimy sewers.

BAROCLINIC Atmospheric system, where isobars are at angle to isopycnals or isotherms, where density of air depends primarily upon pressure.

BAROCLINIC LEAF SHIELD Cloud pattern on satellite images.

BAROCLINIC ZONE Region where temperature gradient exists on constant pressure surface.

BAROCLINITY Measure of state of stratification in fluid where surfaces of constant pressure (i.e., isobaric) intersect surfaces of constant density (i.e., isosteric).

BAROGRAM Analog record of pressure produced by barograph.

BAROGRAPH Barometer that records air pressure observations continuously.

BAROMETER Instrument that measures atmospheric pressure.

BAROMETRIC PRESSURE 1. Downward pressure of atmosphere onto Earth; 2. At sea level, unit of air pressure is one atmosphere; 3. Pressure of atmosphere as indicated by barometer.

BAROTHERMOGRAPH Instrument for recording both pressure and temperature, as of atmosphere.

BAROTROPIC where pressure of atmosphere is dependent upon its density only.

BAROTROPIC SYSTEM Weather system where temperature and pressure surfaces are coincident (i.e., temperature is uniform or no temperature gradient) on constant pressure surface.

BAROTROPY 1. State of fluid where surfaces of constant density or temperature are coincident with surfaces of constant pressure (i.e., state of zero baroclinity).

BARRAGE Artificial obstruction placed in water to increase water level or divert it.

BARRANCA Steep-sided canyon or ravine.

BARREL 1. Water passage in culvert; 2. Wave that breaks with hollow compartment; 3. Cylindrical container with sides that bulge outward and having flat ends; 4. That part of pipe where bore and wall thickness remain uniform; 5. Solid drum; 6. Volumetric unit of measure for crude oil and petroleum products equivalent to 42 U.S. gallons.

BARREL NIPPLE Short length of conduit threaded outside on each end.

BARREL SAMPLER Open-ended steel tube used to collect soil samples.

BARREL WIDTH Inside diameter of drainage conduit.

BARRICADE Obstacle, barrier or bulwark.

BARRIER 1. Obstacle or impediment; 2. Dam or sill across little stream, generally on steep inclines, constructed to retard flow and decrease cutting action of stream.

BARRIER BAR Elongate offshore ridge, submerged at least at high tide, built up by action of waves or currents.

BARRIER BEACH Narrow, elongate sandy ridge rising slightly above high-tide level and extending generally parallel with mainland shore, but separated from it by lagoon.

BARRIER COATING Layer of material that obstructs or prevents passage of something through surface to be protected (e.g., grout, caulk, various sealing compounds).

BARRIER EFFECT Surface film that forms and shields substance from further corrosion.

BARRIER FILTER Filter that removes particles via porous structure.

BARRIER FOREST Treed inclines that hold back snow and so prevent avalanches.

BARRIER ISLAND Island close to and parallel to, main coastline.

BARRIER JET Jet-like wind current that forms when stably-stratified low-level airflow approaches mountain barrier and turns to left to blow parallel to longitudinal axis of barrier.

BARRIER PAIR Pair of functions.

BARRIER PLANTING Tactical use of vegetation to form physical or visual display.

BARRIER SPIT Depositional formations caused by lateral movement of water along shoreline known as littoral drift.

BARRIER SPRING Spring formed above fault between raised bedrock block and thick deposit of alluvium; 2. Spring that occurs when raising of confining bed forces groundwater to rise to ground surface.

BARRIER-SEPARATED LANE Lane separated from adjacent lane(s) via barrier.

BARRING AND WEDGING Removal of stone by prying with bars and insertion of wedges.

BARTEL'S ROTATION NUMBER Serial number assigned to 27-day rotation periods of solar and geophysical parameters where Rotation 1 in this sequence was assigned arbitrarily by Bartel to begin in January 1833.

BASALT 1. Fine-grained, dark-colored volcanic rock rich in iron-bearing minerals; 2. Consistent year-round energy use of facility; 3. Minimum amount of electricity supplied continually to facility.

BASCULE Bridge having one end that rises as other lowers.

BASCULE GATE 1. Gate hinged along one edge, usually either top or bottom edge; 2. Something that restrains flood or outpouring.

BASE 1. Substance that has pH value between 7 and 14; 2. Less free hydrogen ions than hydroxyl ions; 3. Foundation or starting point; 4. Any of class of generally water-soluble compounds, having bitter taste, that turn red litmus blue and react with acids to form salts; 5. Lowermost part of column, between shaft and pedestal or pavement; 6. End of leaf, petal or similar organ where it is attached to its support; 7. Number raised to power of exponent.

BASE BID Amount of money stated in bid as sum for which bidder offers to perform work, not including that work for which alternate bids are also submitted.

BASE CONDITIONS Standard conditions adjusted to prevailing conditions.

BASE COURSE Layer of specified or selected material of planned thickness constructed on subgrade or subbase for purpose of serving one or more functions (e.g., distributing load, providing drainage, minimizing frost action).

BASE DATA Digital fields of reflectivity, mean radial velocity and spectrum width data in spherical coordinates provided at finest resolution available from radar.

BASE DISCHARGE Discharge value, determined for selected stations, above which peak discharge data are published and selected so that average of about three peak flows per year will be published.

BASE ELBOW Pipe bend having flange or pad cast on it as seat for supporting column or bracket.

BASE EXCHANGE Physicochemical process whereby one species of ions adsorbed on soil particles is replaced by another species.

BASE FLOOD 1. Flood having 1% chance of being equaled or exceeded in any given year; 2. National standard for floodplain management is base or 1% chance flood that has at least one chance in 100 of occurring in any given year (100-year flood).

BASE FLOOD ELEVATION Computed elevation to which floodwater is anticipated to rise during flood having 1% probability of being equaled or exceeded in given year.

BASE FLOOD PLAIN Flood plain inundated by 100-year flood.

BASE FLOODPLAIN Land area covered by floodwaters of base flood.

BASE FLOW Portion of stream flow that's not runoff and results from seepage of water from ground into channel slowly over time, long-term lake storage or snowmelt.

BASE LEG Short, descending flight path at right angles to approach end extended centerline of landing runway.

BASE LENGTH Distance measured from points where travel lanes, both merging and diverging, converge.

BASE LEVEL 1. Lowest point river or stream can flow; 2. Lower limit for erosion process; 3. Profile where stream neither degrades nor aggrades its course.

BASE LOAD Minimum level of demand over span of time.

BASE LOG First log in face log resting directly on ground.

BASE PERIOD Time that flow in stream exceeds base flow due to isolated storm event.

BASE PERIOD FLEET Buses needed for base schedule.

BASE PERIOD SERVICE Vehicle operations during base schedule.

BASE PROCESS Production procedure where raw materials are introduced in batches at manufacturing plant.

BASE PRODUCTS Products that present some representation of base data.

BASE RATE Pay rate per unit of time, unit or work performed.

BASE REFLECTIVITY One of three fundamental quantities (along with base (radial) velocity and spectrum width) that Doppler radar measures.

BASE RESISTANCE Resistance to current of bulk component of cell.

BASE RUN Transit route that's usually eight hours long with unpaid breaks.

BASE RUNOFF Sustained or fair weather runoff composed largely of groundwater effluent.

BASE SAFETY CONDITION Level of loading above which dam failure does not contribute incremental loss of life.

BASE SATURATION Percentage of soil.

BASE SATURATION FLOW RATE Maximum rate of flow for vehicular traffic.

BASE STATION Computer that accepts radio signals from ALERT gaging sites, decodes data, places data in database and makes data available to other users.

BASE STOCK COST FLOW System where price of items sold and/or consumed is dependent upon assuming predetermined minimum amount is carried in stock forever and in fixed price.

BASE SUBSISTENCE DENSITY Density above which survival of human population is impossible.

BASE TEMPERATURE Reference temperature for ascertaining liquid densities or correcting measured volume of liquid.

BASE THICKNESS 1. Maximum thickness or width of dam measured horizontally between upstream and downstream faces and normal to axis of dam, but excluding projections for outlets or other appurtenant structures; 2. Thickness is used for gravity or arch dams and width is used for other dams.

BASE TIME Transit vehicle time during off-peak time.

BASE WIDTH Time duration of unit hydrograph.

BASE-EXCHANGE PROCESS Procedure enabling exchange of positive ions on prepared medium.

BASE-FLOW DEPRECIATION CURVE Feature rate where base flow of stream reduces until supplemented by precipitation or runoff.

BASED FLOOD Flood that has 1% or greater chance of recurring in any year or flood of magnitude equalled or exceeded once in 100 years on average.

BASEFLOW 1. Sustained low flow of stream in absence of direct runoff that includes natural (groundwater inflow) and human-induced streamflows; 2. Portion of stream discharge derived from natural storage.

BASELINE 1. Line that's base for measurement or for construction; 2. Datum used as basis for calculation or for comparison; 3. Fortitude amount of organism to your unique concentration of material.

BASELINE ALTERNATIVE Conditions that would prevail if no actions were taken.

BASELINE CONDITION Alternatives that would prevail if no actions were taken.

BASELINE PROFILE Used for survey of environmental conditions and organisms existing in region prior to unnatural disturbances.

BASELINE UNIFORM DELAY Average uniform traffic delay without initial queue.

BASELOAD Minimum constant amount of load connected to power system over given time period.

BASELOAD PLANT Power plant that is normally operated to take all or part of minimum load of system and which consequently produces electricity at essentially constant rate and runs continuously that's operated to maximize system mechanical and thermal efficiency and minimize system operating costs.

BASELOADING Running water through power plant at roughly steady rate, thereby producing power at steady rate.

BASIC 1. Opposite of acidic; 2. Water that has pH of greater than 7.

BASIC DATA Records of events and states excluding anything developed by way of computation or approximation.

BASIC FARE Fare class that offers lowest price in exchange for no amenities.

BASIC FIXED SITES Sites on streams at which streamflow is measured and samples are collected for temperature, salinity, suspended sediment, major ions and metals, nutrients and organic carbon to assess broad-scale spatial and temporal character and transport of inorganic constituents of streamwater in relation to hydrologic conditions and environmental settings.

BASIC FREEWAY SEGMENT Freeway with two or more lanes for traffic use in each direction.

BASIC HYDROLOGIC DATA Includes inventories of features of land and water that vary spatially (e.g., topographic and geologic maps) and records of processes that vary with both place and time.

BASIC HYDROLOGIC INFORMATION Includes surveys of water resources of particular areas and study of their physical and related economic processes, interrelations and mechanisms.

BASIC ROCK Dark colored igneous rock comparatively low in silica (45-55%).

BASIC WIND SPEED Three-second wind gust speed at 33 ft.

BASIC-STAGE FLOOD Arbitrary rate chosen of flow of stream used as lower limit in choosing floods to be examined.

BASIC-STAGE FLOOD SERIES List of all flood peaks that exceed chosen base stage or discharge, regardless of number of peaks occurring in year.

BASIDIOMYCETES Group of fungi having external spores that develop in basidia.

BASIN 1. Area having common outlet for its surface runoff; 2. Depression, natural or artificial, containing water; 3. Drainage basin; 4. Rock formation scooped out by water erosion; 5. Wide bowl for washing, sometimes affixed to wall.

BASIN AND RANGE PHYSIOGRAPHY Region characterized by series of generally north-trending mountain ranges separated by alluvial valleys.

BASIN BOUNDARY Topographic dividing line around perimeter of basin, beyond which overland flow (i.e., runoff) drains away into another basin.

BASIN LAG Time it takes from centroid of rainfall for hydrograph to peak.

BASIN MIXER Mixing faucet where hot water, cold water and rate of flow might be fixed using one control.

BASIN RECHARGE Rainfall that adds to residual moisture of basin in order to help recharge water deficit (i.e., water absorbed into soil that does not take form of direct runoff).

BASIN RUNOFF MODEL Any one of computer programs that mathematically models basin characteristics to forecast reservoir inflow from rainfall and/or streamflow data.

BASIN STATES Arizona, California, Colorado, New Mexico, Utah and Wyoming.

BATCH FED INCINERATOR Incinerator that's occasionally permitted to burn down or burn out solid waste.

BATCH LOADER REFUSE TRUCK Compactor truck built using loading hopper and mechanized panel that embroils solid wastes to body.

BATCH PLANT Mixing equipment that combines various ingredients to form concrete.

BATCH PROCESS 1. Process not continuous, but is carried out with discrete amount of material.

BATCH WATER HEATER Water storage tank, within insulated container using glazed cover, used to heat and store water.

BATCH-TYPE FURNACE Furnace used for heat treatment of materials, with or without direct firing, shut down periodically.

BATESIAN MIMICRY Form of biological resemblance where noxious or dangerous, organism, equipped with warning system such as conspicuous coloration, is mimicked by harmless organism.

BATH MIXER Blending faucet where hot and cold water, rate of flow and shower control are combined for bathtub.

BATHOLITH Large irregular mass of intrusive igneous rock that has melted or forced itself into surrounding strata.

BATHYAL Relating to ocean bottom, between sublittoral and abyssal zones.

BATHYMETER Instrument that measures depth of ocean water.

BATHYMETRY Science of measuring depths of oceans, lakes, seas, etc.

BATHYPELAGIC Of or pertaining to parts of oceans at depths between 1,000 and 4,000 meters (3,280 and 13,123 feet) deep.

BATILLARIOPHYTA 1. Green or brownish unicellular alga including chlorophyll and having siliceous walls; 2. Diatoms.

BATTER 1. Inclination from vertical; 2. Pile driven at angle to widen area of support and resist thrust.

BATTER PILE Angular driven pile to resist lateral forces.

BATTERY OF WELLS Interconnected wells where water is drawn with individual pump.

BATTURE Sea bed or river bed that has been raised or elevated.

BAUME′ SCALE Instrument to measure density of various liquids.

BAY 1. Body of water more or less three-quarters surrounded by land; 2. Bank or dam to keep back water; 3. Leaf of this or certain other species of tree or shrub.

BAY AND PROMONTORY Bay that's dominant upland characteristic is ridge line that shapes it.

BAYHEAD BARRIER Barrier beach protecting head of bay, but separated from it by lagoon.

BAYMOUTH BARRIER Barrier that partially encloses bay at its entrance.

BEACH 1. Shore of body of water, especially when sandy or pebbly; 2. Face of hydraulically filled earth dam; 3. To run aground on beach.

BEACH DRIFT Drifting of sediments, especially marine sediments, in patterns parallel to contours of beach, due to action of waves and currents.

BEACH EROSION Movement of beach materials by some combination of high waves, currents and tides or wind.

BEACH PIPE Hydraulic fill conduit used to build earth dam with slurry of earth and water.

BEACHING 1. Action of water waves by which beach materials settle into water because of removal of finer materials; 2. Loose-graded stones, from 0.07 to 0.2 m (2.75 to 8 in.) in diameter, used in layer 03 to 0.6 m (12 to 24 in.) thick as revetment; 3. Event where marine animal beaches itself.

BEACON Electronic traffic signal that broadcasts flashing signal.

BEAM 1. Principal horizontal piece of timber or iron in building 2. To emit light; 3. Maximum width of vessel.

BEAM FILLING Measure of variation of hydrometeor density throughout radar sampling volume.

BEAM WIDTH Angular width of antenna pattern.

BEAN SHEET Common term for pesticide data package record.

BEAR'S CAGE Slang for region of storm-scale rotation, in thunderstorm, that is wrapped in heavy precipitation.

BEAR-TRAP GATE Any of family of crest gates consisting of two leaves, upstream leaf hinged and sealed at upstream edge and downstream leaf hinged and sealed on its downstream edge.

BEARER Horizontal scaffolding member that platform rests on.

BEARING Beam, column or element carrying weight or load.

BEARING JOINT Joint where bolt limits lateral movement of elements by shank of bolt bearing upon sides of holes in clamped elements.

BEARING PILE Pile with bearing load through point bearing or friction.

BEAT 1. Interference between two tones of almost equal frequency; 2. Pulsation or throb.

BEAUFORT SCALE Empirical measure that relates wind speed to observed conditions between calm and hurricane that classifies conditions between 1 and 12.

BEAVER TAIL Slang for particular type of inflow band with relatively broad, flat appearance suggestive of beaver's tail.

BECQUEREL Unit of radioactivity corresponding to one disintegration per second.

BED 1. Place or flat surface or layer, on which something else rests or is laid; 2. Bottom of watercourse or body of water.

BED AND HAUNCHING Weak concrete, partially encircling pipe placed in trench, not less than 150 mm (6 in.) thick.

BED ELEVATION Height of streambed above specified level.

BED FORM Feature at interface of fluid and moveable bed due to bed material being moved by fluid flow.

BED FORMS Irregularities found on bottom of stream that are related to flow characteristics (e.g., dunes, ripples, antidunes).

BED LAYER Flow layer, several grain diameters thick (usually taken as two grain diameters thick), immediately above bed.

BED LOAD 1. Bed material that moves in continuous contact (i.e., rolling or sliding) with bed layer; 2. Large particles with density or grain size that preclude movement far above or for long distance with stream bed under natural conditions of flow.

BED LOAD DISCHARGE Quantity of bed load passing cross-section in unit of time (i.e., rate).

BED MATERIAL Sediment mixture of which streambed, lake, pond, reservoir or estuary bottom is composed.

BED MATERIAL DISCHARGE Total rate (e.g., tons/day) at which bed material is transported by given flow at given location on stream.

BED MATERIAL LOAD Total rate (e.g., tons/day) at which bed material is transported by given location on stream.

BED ROCK General term for rock, usually solid, that underlies soil or other unconsolidated, superficial material.

BED SEDIMENT Material that temporarily is stationary in bottom of stream or other watercourse.

BED SEDIMENT CONTROL VOLUME Source-sink component of sediment sources in river system (other component is suspended sediment in inflowing discharge) where user-defined dimensions are movable bed width and depth and average reach length.

BED SHEAR STATIONARY Unit area force exerted by fluid flowing past boundary.

BED SLOPE Inclination of channel bottom.

BED-LOAD DISCHARGE Quantity of bed load passing cross-section of stream in unit of time.

BED-MATERIAL DISCHARGE Part of total sediment discharge that is composed of grain sizes found in bed that's assumed equal to transport capability of flow.

BEDDING 1. Ground or layer of such, placed beneath and beside buried conduit to support load; 2. Layer of material laid over surface to fill any irregularities between it and something placed on top of it; 3. Any material used by or provided to animals to lie on.

BEDDING PLANE Separation or weakness between two layers of rock, caused by changes during building up of rock-forming material.

BEDLOAD Sediment that moves on or near bottom of streambed and is in almost continuous contact with bed.

BEDROCK 1. Solid rock at surface or underlying other surface materials; 2. Rock of relatively great thickness and extent in its native location; 3. Any solid rock, not exhibiting soil-like properties, that underlies soil or other unconsolidated, bed material.

BEDROCK MOTION PARAMETERS Numerical values representing vibratory ground motion (e.g., particle acceleration, velocity and displacement, frequency content, predominant period, spectral intensity, duration) that define design earthquake.

BEGIN ACTUAL CONSTRUCTION Construction activities that may occur or are prohibited prior to issuance of permit.

BEGINNING EVENT Start of first action in critical path program.

BEGINNING OF BREAKUP Date of definite breaking, movement or melting of ice cover or significant rise of water level.

BEGINNING OF FREEZUP Date on which ice forming stable winter ice cover is first observed on water surface.

BELANGER'S CRITICAL VELOCITY Condition of fluid flow in open channels for which velocity head equals one-half mean depth.

BELL Expanded or enlarged, end of pipe section, into which next pipe fits.

BELL AND SPIGOT Pipe joint where pipe ending in bell-like shape is joined to pipe ending in spigot like shape.

BELL DOLPHIN Used to moor boats, big bell shaped fender supported on piles in open water.

BELL TILE Clay pipe segments with one end enlarged to loosely join with another conduit segment.

BELLMOUTH Bell-shaped extension at end of conduit.

BELLMOUTH OVERFLOW Overflow structure situated in reservoir, with opening in shape of inverted bell.

BELLMOUTHED ORIFICE Orifice using short conduit externally.

BELLOWS SEAL Seal that incorporates metal bellows as spring-loading mechanism to maintain face contact when seal is not pressurized and eliminates traditional secondary sealing requirements.

BELLY DUMP Hauler with clam shell type dump gate that discharges its load from underside.

BELLY PAN Sheet of metal that covers and protects all or portion of underside.

BELOWGROUND RELEASE Any release to subsurface of land or to groundwater.

BELOWGROUND STORAGE FACILITY Tank or other container located other than as defined as aboveground.

BELT CONVEYOR Carrying apparatus that moves material by way of endless belt.

BELT LINE RAILROAD Railroad that operates within or around city that connects with larger railroads.

BELT OF CEMENTATION Portion of zone of fracture where mineral matter is commonly deposited in cracks, fissures, intergranular spaces and other openings of earth.

BELT OF PHREATIC FLUCTUATION Belt of watertable changes.

BELT OF SOIL WATER Upper subdivision of zone of aeration limited above by land surface and below by intermediate belt that contains plant roots and water available for plant growth.

BELT OF WATERTABLE FLUCTUATION Part of lithosphere that, because of fluctuations of water table, lies part of time in zone of saturation and part of time in overlying zone of aeration.

BELT OF WEATHERING Stretching from bottom surface to groundwater table where physical and chemical processes are harmful and result in breaking down of substance where belt is composed.

BEN EPA's computer model for analyzing violator's economic gain from not complying with law.

BENCH Working level or step in cut.

BENCH FLUME Conduit on bench, cut on sloping ground.

BENCH SCALE ANALYSIS Small-scale lab investigation.

BENCH TERRACE Series of level or virtually level strips running across slope at vertical intervals, supported by steep banks or risers.

BENCH-SCALE TESTS Laboratory testing of potential cleanup technologies.

BENCHING 1. Concrete around half-round drainage channel in manhole, sloping upwards in tile border to manhole wall; 2. Berm above ditch; 3. Procedure of excavating whereby terraces or ledges are operated in stepped contour.

BENCHMARK 1. Permanent or temporary point of known or assumed elevation above sea level; 2. Permanent reference monument (elevation) whose known elevation is tied to national network used in survey.

BEND 1. Change of direction in piping; 2. Short part of curved conduit.

BENDERGAIN Allowance included when quantifying amount substance that has to be of exact span when bent.

BENDWAY Waterway location where flow changes direction at curvature.

BENDWAY SCOUR Scour that occurs on river bed near outside of bend due to variable velocity distribution.

BENEDICTION PROCESS Dressing or processing of gold-bearing ores.

BENEFICATION Development of chemical and/or physical properties of materials or products via removal or alteration of unwanted elements or impurities.

BENEFICATION SYSTEM 1. Treatment of substance to enhance its form or properties; 2. Mechanical procedure for removing contaminants and cleaning rubbish.

BENEFICIAL ORGANISM Any organism that benefits growing process, including insects, arachnids, other animals, plants, bacteria, fungi, viruses and nematodes.

BENEFICIAL USE 1. Legal term describing person's right to enjoy benefits of specific property; 2. Water loss through use for betterment of society (e.g., irrigation, municipal use).

BENEFICIAL USES Management objectives.

BENEFICIARY Any individual, entity or governmental agency (local, state or federal) that benefits from project.

BENEFICIATION Reduction in size of particles of ore before further processing to recover metal.

BENEFIT-COST ANALYSIS Economic method for assessing benefits and costs of achieving alternative health-based standards at given levels of health protection.

BENEFIT-COST RATIO Ratio of present value of project benefits to present value of project costs, used in economic analysis.

BENT 1. State of being curved, crooked or inclined from straight line; 2. Transverse frame of framed structure.

BENTGRASS GRASS Large and very nearly cosmopolitan genus of plants in grass family, found in nearly all countries in world.

BENTHAL DEMAND Dissolved oxygen demand of water overlying benthal deposits caused by upward diffusion of products of decomposition.

BENTHAL DEPOSITS Mud, sand or gravel deposited by moving water.

BENTHIC 1. Plants or animals that live on bottom of lakes, streams or oceans; 2. Bottom of rivers, lakes or oceans; 3. Bottom- or depth-inhabiting.

BENTHIC FAUNA Organisms attached to, resting on bottom or living in bottom sediments of water body.

BENTHIC ORGANISM Form of aquatic life (organisms) that inhabits bottom of aquatic environment including bacteria, fungi, insect larvae and nymphs, snails, clams and crayfish that's useful as indicator of water quality.

BENTHOS 1. Organisms, flora and fauna at bottom of body of water; 2. Sea or lake bottom.

BENTONITE Colloidal clay that's expansible when moist and used to provide tight seal around well casing.

BENTONITE CLAY Clay with high content of mineral montmorillonite, usually characterized by high swelling on wetting and shrinkage on drying.

BENTONITE SLURRY TRENCH System that accomplishes vertical cutoff wall with low permeability characteristics that satisfy most requirements for watertight barrier.

BENTONITIC CLAY Clay with high content of mineral montmorillonite, usually characterized by high swelling on wetting and shrinkage on drying.

BENZENE Aromatic hydrocarbon, colorless, volatile and flammable liquid, whose structure consists of ring of alternate single and double bonds.

BERGERON PROCESS Process by which ice crystals in cloud grow at expense of supercooled liquid water droplets.

BERGY BIT Piece of ice which has broken away from iceberg, extending 1-5 meters above sea surface and 100-300 m^2 in area.

BERM 1. Horizontal narrow ledge or shelf, as along top or bottom of slope; 2. Terrace formed by wave action along beach; 3. Mound or bank of earth.

BERMUDA HIGH Semi-permanent, subtropical area of high pressure in North Atlantic Ocean off East Coast of North America that migrates east and west with varying central pressure.

BERNOULLI'S THEORY States that increase in speed of fluid occurs simultaneously with decrease in pressure or decrease in fluid's potential energy.

BERTH Space for ship to moor or vehicle to park.

BERTHING IMPACT Forces impacting piers during vessel's docking.

BERYLLIUM Metal hazardous to human health when inhaled as airborne pollutant discharged by machine shops, ceramic and propellant plants and foundries.

BEST ANGLE-OF-CLIMB AIRSPEED Angle where most excess thrust is available to ascend altitude.

BEST AVAILABLE CONTROL MEASURES Term used to refer to most effective measures (according to EPA guidance) for controlling small or dispersed particulates and other emissions from sources (e.g., roadway dust, soot and ash from wood stoves and open burning of rush, timber, grasslands, trash).

BEST AVAILABLE CONTROL TECHNOLOGY Currently available technology producing greatest reduction of air pollutant emissions, taking into account energy, environmental, economic and other costs.

BEST AVAILABLE TECHNOLOGY Water treatment that EPA certifies to be most effective for removing contaminant.

BEST AVAILABLE TECHNOLOGY ECONOMICALLY ACHIEVABLE Technology-based standard established by Clean Water Act as most appropriate means available on national basis for controlling direct discharge of toxic and non-conventional pollutants to navigable waters.

BEST BID/TENDER Bid to finish work or supply services and/or materials that most closely matches requirements of owner.

BEST DEMONSTRATED AVAILABLE TECHNOLOGY Most effective commercially available means of treating specific types of hazardous waste.

BEST GATE Gate opening where peak efficiency of turbine occurs at particular head.

BEST MANAGEMENT PRACTICE Agricultural practice that has been determined to be effective, practical means of preventing or reducing non-point-source pollution.

BEST MANAGEMENT PRACTICES Schedules of activities, prohibitions of practices, maintenance procedures and other management practices to prevent or reduce discharge of pollutants to waters of United States.

BEST PRACTICABLE CONTROL TECHNOLOGY CURRENTLY AVAILABLE First level of technology-based standards established by CWA to control pollutants discharged to waters of U.S.

BEST PRACTICABLE WASTE TREATMENT TECHNOLOGY Cost-effective technology that can treat wastewater, combined sewer overflows and non-excessive infiltration and inflow in publicly owned or individual wastewater treatment works.

BEST PROFESSIONAL JUDGMENT Method used by permit writers to develop technology-based NPDES permit conditions on case-by-case basis using all reasonably available and relevant data.

BEST RATE-OF-CLIMB AIRSPEED Airspeed where most excess thrust is available to ascend altitude.

BEST TRACK Subjectively-smoothed representation of tropical cyclone's location and intensity over its lifetime.

BEST-VALUE SELECTION Contract selection process where award is based on combination of quality and cost factors.

BETA PARTICLE Energetic electron or positron produced as result of nuclear reaction or nuclear decay.

BETA RADIATION High-speed electron or positron emitted in radioactive decay of atomic nuclei.

BETA RATIO Number of particles in upstream (before filtration) divided by number of particles downstream (after filtration).

BETA X Filtration ratio obtained during multi-pass evaluation.

BETHELL PROCESS Process of impregnating wood with preservatives.

BETTERMENT Improvement that adds to its value.

BEVEL GEAR Gear where two wheels working together lie in different planes and have their teeth cut at right angles to surfaces of two cones whose pieces coincide with point where axes of wheels would meet.

BI-DIRECTIONAL Moving in two directions.

BI-LEVEL RAIL CAR Rail car with two levels of seating.

BIAS Systematic error introduced into sampling or testing by selecting or encouraging one outcome or answer over others.

BIAT Timber support or beam in shaft.

BIB Tap for connection of hose.

BICARBONATE ALKALINITY Most prominent form of alkalinity in water composed of carbon atom in center of formation with three oxygen atoms attached to it.

BICYCLE Vehicle that has two wheels and propelled by rider pedaling.

BICYCLE FACILITIES Lanes, parking and storage to encourage bicycling.

BICYCLE FACILITY Lane, improvement or provision to encourage bicycling.

BICYCLE LANE Lane designated for use by bicyclists.

BICYCLE MODE Using bicycle to travel on roadway.

BICYCLE PATH Lane for use of bicycle riders.

BICYCLE ROUTE SYSTEM National long-distance cycling route network of United States.

BICYCLE SPEED Distance per time traveled by bicycle.

BICYCLE TRACK Separated path for bicycles.

BID Offer of price.

BID ABSTRACT On given project, compilation of bidders and their respective bids, usually separated into individual items.

BID CALL On given project, compilation of bidders and their respective bids, usually separated into individual items.

BID DATE Prescribed date and time where bid must be submitted to be considered.

BID DOCUMENTATION Documents (e.g., request for proposal, plans and specifications, contract, rules and regulations, insurance requirements, list of approved subcontractors, scope narrative) to be shared with bidders.

BID DOCUMENTATION ESCROW Saving successful bid documents.

BID DOCUMENTS Documents that are available to all contractors when submitting job bid.

BID FORM Form provided to bidder that becomes part of bid documents.

BID GUARANTEE Security assuring that bidder will not withdraw bid within period specified for acceptance and will execute written contract and furnish required bonds within time specified in bid after receipt of specified forms.

BID OPENING Stage in bidding process where received bid-envelopes are opened and examined by advertiser of invitation-to-bid.

BID PACKAGE Full set of bidding documents.

BID PRICE Price offered by potential buyer when bidding for goods.

BID QUANTITIES Quantities for unit-price items in bid schedule.

BID SECURITY Amount of money that may be calculated as percentage of budget estimate of procurement requirement or percentage of bidder's bid price.

BID SHOPPING Practice of divulging contractor's or subcontractor's bid to other prospective contractor(s) or subcontractor(s) before award of contract in order to secure lower bid.

BID/TENDER NOTICE Advertisement calling for bids.

BIDDER Someone who bids.

BIDDER'S LIST List of those who've required records so that you can organize bid.

BIDDING DOCUMENTS Documents required to be submitted in response to invitation to bid.

BIDDING OR NEGOTIATION PHASE Fourth phase of architect's basic services, when competitive bids or negotiated proposals are sought as basis for awarding contract.

BIDDING PERIOD Number of days or weeks up to bid date that are available to prospective bidder to prepare and lodge bid.

BIDDING REQUIREMENTS Document that includes bid form, basic and supplementary instructions to bidders and minimum acceptable standards set by originator (owner) of project for which bids are invited.

BIDET Low-mounted plumbing fixture or type of sink intended for washing perineal area of body.

BIENNIAL Plant which produces seeds during its second year and then dies.

BIENNIAL PLANT Flowering plant that takes two years to complete its biological life cycle.

BIFURCATE Divided or forked into two.

BIFURCATION 1. Division into two branches; 2. Place where two roads, tributaries, etc. part or meet; 3. Point where channel divides, when proceeding from seaward, into two distinct channels.

BIFURCATION GATE Gate positioned where conduit is divided into two branches to divert flow into or out of, one or other.

BIG BASIN Large retention, detention basins or ponds that detain and slow stormwater, allowing sediment, chemicals and trash to be filtered out before water is released into receiving waters.

BIG LINE Fire hose with minimum diameter of 65 mm (2.5 in.).

BIG-CONE DOUGLAS FIR TREE Douglas fir of California having cones 4 to 8 inches long.

BIGHT 1. Bend or curve in coastline, river or other geographical feature; 2. Area of sea lying between two promontories, larger than bay, wider than gulf.

BIKEWAY Bicycle lane or path.

BILL OF LADING Document where master of carrier (e.g., ship) acknowledges receipt of goods for transport.

BILL OF MATERIALS List of materials and components used in manufactured item, sometimes appended to engineering diagram of item.

BILLOW CLOUD Cloud consisting of broad parallel bands oriented perpendicular to wind.

BILTMORE STICK Tool used to measure various tree dimensions, such as diameter at breast height and height.

BIMETAL Beverage containers with steel bodies and aluminum tops that's handled differently from pure aluminum in recycling.

BIMETALLIC CORROSION Electrochemical process where one metal corrodes preferentially when it is in electrical contact with another, in presence of electrolyte.

BIN Radar sample volume.

BINAURAL Of, relating to, affecting or designed for use with two ears.

BIND To exert strong chemical attraction.

BINDER Portion of soil passing No. 40 United States Standard sieve.

BINDER COURSE Intermediate course between base and surfacing material that consists of bituminous material.

BINDER SOIL Portion of soil passing No. 40 (0.425 mm) United States Standard sieve.

BINDING ARBITRATION Judgment made by third party to settle dispute between two other parties, which is obligatory.

BINOCULAR VISION Vision system where two eyes work together to produce unified field of view which is wider and stereoscopic and where objects can be more readily discerned.

BIOACCUMULANTS Substances that increase in concentration in living organisms as they take in contaminated air, water or food because substances are very slowly metabolized or excreted.

BIOACCUMULATE Net uptake of material by organism from food, water, and/or respiration that results in elevated internal concentrations.

BIOACCUMULATION 1. Process where substance enters through gills, epithelial tissues, dietary or other sources and accumulates in tissues of living organisms; 2. Biological sequestering of substance at higher concentration than that at which it occurs in surrounding environment or medium, intake and retention of non-food substances by living organism from its environment, resulting in build-up of substances in organism.

BIOACCUMULATION FACTOR Ratio of substance's concentration in tissue versus its concentration in ambient water, in situations where organism and food chain are exposed.

BIOAERATION Form of activated sludge process where air is blown through raw sewage.

BIOASSAY Test which determines effect of chemical on living organism.

BIOASSIMILATION Accumulation of substance within habitat.

BIOAVAILABILITV Relative amount of administered test substance that reaches systemic circulation.

BIOAVAILABILITY 1. Capacity of chemical constituent to be taken up by living organisms either through physical contact or by ingestion; 2. Degree of ability to be absorbed and ready to interact in organism metabolism; 3. Measure of physicochemical access that toxicant has to biological processes of organism.

BIOCHEMICAL Chemical processes that occur inside or are mediated by living organisms.

BIOCHEMICAL ACTION Chemical change caused by metabolism of living organisms.

BIOCHEMICAL OXIDATION 1. Oxidation caused by biological activity; 2. Chemical mix of oxygen and organic matter.

BIOCHEMICAL OXYGEN DEMAND 1. Amount of oxygen needed by aerobic microorganisms to decompose all organic matter in sample of water; 2. Measurement of amount of oxygen utilized by decomposition of organic material, over specified time period (usually 5 days) in wastewater sample where greater BOD, greater degree of pollution.

BIOCHEMICAL OXYGEN DEMAND TEST Process that quantifies rate of oxygen use under controlled conditions of time and temperature.

BIOCHEMICAL OXYGEN DEMAND, FIRST STAGE That part associated with biochemical oxidation of carbonaceous (as distinct from nitrogenous) material.

BIOCHEMICAL OXYGEN DEMAND, IMMEDIATE STAGE First volume of oxygen employed by polluted liquid promptly upon being inserted into water containing dissolved oxygen.

BIOCHEMICAL OXYGEN DEMAND, SECOND STAGE Oxygen demand related to biochemical oxidation of nitrogenous substances.

BIOCHEMICAL OXYGEN DEMAND, STANDARD STAGE Biochemical oxygen demand as determined under normal laboratory processes for 5 days at 20 °C, expressed in mg/l.

BIOCHEMICAL OXYGEN DEMAND, ULTIMATE STAGE Complete quantity of oxygen necessary to meet biochemical oxygen demand of sample.

BIOCIDES Substances that kill living organisms.

BIOCLASTIC 1. Describing sedimentary rock composed of remains of small organisms; 2. Describing form of erosion or formation caused by action of living organisms.

BIOCLIMATOLOGY Interdisciplinary field of science that studies interactions between biosphere and Earth's atmosphere on time scales of order of seasons or longer.

BIOCOEN All living parts of surroundings.

BIOCONCENTRATION Accumulation of chemical in tissues of fish or other organism to levels greater than in surrounding medium or environment.

BIOCONCENTRATION FACTOR Ratio of substance's concentration in tissue versus its concentration in water, in situations where food chain is not exposed or contaminated.

BIOCONVERSION Conversion of biological material into usable products; but especially conversion of biomass into biofuel.

BIOCRITERIA Numerical values or narrative expressions that describe reference biological integrity of aquatic communities that inhabit water of given designated aquatic life use.

BIODEGRADABLE Capable of decomposing quickly through action of microorganisms under natural conditions.

BIODEGRADATION Transformation of substance into new compounds through biochemical reactions or actions of microorganisms (e.g., bacteria).

BIODEGRADATION, ANAEROBIC Decomposition of organic matter by microorganisms in absence or near absence of free oxygen where decomposition end-products are enriched in carbon.

BIODIESEL Fuel for diesel engines made of renewable organic raw materials as opposed to fossil hydrocarbons.

BIOENGINEERING Applications of principles of engineering to any of biological or medical sciences.

BIOFILTRATION Technique for pollution control using living material to capture and biologically degrade process pollutants.

BIOFLOCCULATION Clumping together of fine, dispersed organic particles by action of bacteria and algae.

BIOFUEL Any fuel obtained from renewable biological resource, especially from biomass.

BIOGAS Mixture of methane and carbon dioxide, produced by anaerobic digestion of organic waste matter, used as fuel.

BIOGASIFICATION Process of converting biomass to biogas, which can then be used as fuel.

BIOGEOCLIMATIC CLASSIFICATION Method to classify and manage sites on ecosystem-specific basis.

BIOGEOSPHERE Outer portion of Earth's crust as far below surface as life exists.

BIOLOGICAL ADDITIVES Microbiological cultures, enzymes or nutrient additives which are deliberately introduced into oil discharge for particular reason for supporting biodegradation to mitigate impacts of discharge.

BIOLOGICAL AMPLIFICATION Increasing concentration of substance, such as toxic chemical, in tissues of organisms at successively higher levels in food chain.

BIOLOGICAL ASSESSMENT Evaluation of biological condition of water body by using biological surveys and other direct measurements of resident biota in surface water.

BIOLOGICAL BENCHMARK Population level or fitness of plant or animal species used as benchmark for measuring pollution in natural systems (habitats).

BIOLOGICAL CONCENTRATOR Filter feeders (e.g., shellfish, plant, bacterium) that concentrate heavy metals and other stable compounds present in dilute concentrations of water or land.

BIOLOGICAL CONTAMINANTS Living organisms or derivates (e.g., viruses, bacteria, fungi, mammal and bird antigens) that can cause harmful health effects when inhaled, swallowed or otherwise taken into body.

BIOLOGICAL CONTROL Use of animals and organisms that eat or otherwise kill or out-compete pests.

BIOLOGICAL CONTROL AGENT Method of controlling pests such as insects, mites, weeds and plant diseases using other organisms.

BIOLOGICAL CRITERIA Numerical values or narrative expressions that describe reference biological integrity of aquatic communities that inhabit water of given designated aquatic life use.

BIOLOGICAL DIVERSITY Number and kinds of organisms or species per unit area or volume at given time.

BIOLOGICAL FILTRATION Action of bacteria in tank breaking down dangerous ammonia, converting them to nitrites and then nitrites to less toxic nitrates.

BIOLOGICAL GROWTH Activity and growth of any and all living organisms.

BIOLOGICAL INDICATOR Provides information on whether necessary conditions were met to kill specified number of microorganisms for given sterilization process, providing level of confidence in process.

BIOLOGICAL INTEGRITY Ability to support and maintain balanced, integrated, functionality in natural habitat of given region where functionally is defined as condition of aquatic community that inhabits unimpaired water bodies of specified habitat as measured by community structure and function.

BIOLOGICAL MAGNIFICATION Process whereby certain substances (e.g., pesticides, heavy metals) move up food chain, work their way into rivers or lakes and are eaten by aquatic organisms (e.g., fish) that in turn are eaten by large birds, animals or humans.

BIOLOGICAL MEASUREMENT Measurement taken in biological medium.

BIOLOGICAL MEDIUM One of major components of organism (e.g., blood, fatty tissue, lymph nodes, breath) where chemicals can be stored or transformed.

BIOLOGICAL MONITORING Method of assessing chemical exposures by measuring chemical or its breakdown products in biological sample (e.g., urine, blood).

BIOLOGICAL NUTRIENT REMOVAL Process used for nitrogen and phosphorus removal from wastewater before it is discharged into surface or groundwater.

BIOLOGICAL OPINION Document stating U.S. Fish and Wildlife Service and National Marine Fisheries Service opinion as to whether federal action is likely to jeopardize continued existence of threatened or endangered species or result in destruction or adverse modification of critical habitat.

BIOLOGICAL OXIDATION Decomposition of complex organic materials by microorganisms that occurs in self-purification of water bodies and in activated sludge wastewater treatment.

BIOLOGICAL OXYGEN DEMAND Measure of amount of oxygen consumed in biological processes that break down organic matter in water, over specified time period (usually 5 days) in wastewater sample where greater BOD, greater degree of pollution.

BIOLOGICAL PESTICIDES Certain microorganism (e.g., bacteria, fungi, viruses, protozoa) that are effective in controlling pests that usually do not have toxic effects on animals and people and do not leave toxic or persistent chemical residues in environment.

BIOLOGICAL PROCESSES Processes characteristic of or resulting from, activities of living organisms.

BIOLOGICAL SHIELDING 1. Structure designed to prevent migration of living organisms from one part of system to another; 2. Shielding provided to attenuate or absorb nuclear radiation.

BIOLOGICAL STRESSORS Organisms accidentally or intentionally dropped into habitats where they do not evolve naturally (e.g., gypsy moths, Dutch elm disease, certain types of algae, bacteria).

BIOLOGICAL SURVEY Collecting, processing and analyzing representative portion of resident aquatic community to determine community structure and function.

BIOLOGICAL TREATMENT Treatment technology that uses bacteria to consume organic waste.

BIOLOGICALLY EFFECTIVE DOSE Amount of deposited or absorbed compound reaching cells or target sites where adverse effect occur or where chemical interacts with membrane.

BIOLOGICALLY-ACTIVE FLOC Floc caused by activity of biological activities; activated sludge.

BIOLOGICALS Medical product made from living organisms and their products derived from biological sources.

BIOLOGY Scientific study of life.

BIOLUMINESCENCE Emission of light by living organism.

BIOMAGNIFICATION Process whereby certain substances (e.g., pesticides, heavy metals) move up food chain, work their way into rivers or lakes and are eaten by aquatic organisms (e.g., fish) that in turn are eaten by large birds, animals or humans where organisms eventually contain higher concentrations of substance than their food sources.

BIOMASS 1. Total mass or amount of living organisms in particular area or environment; 2. Amount of living matter, in form of organisms, present in particular habitat; 3. Vegetation used as fuel or source of energy.

BIOMASS FUEL Fuel developed from organic materials.

BIOMASS HARVESTING Removal of more woody material from site than would be removed under typical roundwood harvest.

BIOME Entire community of living organisms in single major ecological area.

BIOMETALLURGY Describes biotechnological processes that involve interactions between microorganisms and metals or metal-bearing minerals.

BIOMETEOROLOGY Study of relationship between atmospheric conditions (weather) and living organisms.

BIOMONITORING 1. Use of living organisms in water quality surveillance to indicate compliance with water quality standards or effluent limits and to document water quality trends; 2. Use of biological entity as detector and its response as measure to determine environmental conditions.

BIOPHOTOLYSIS Photolysis of biochemical compound; but especially production of hydrogen by action of light on algae or any other organism.

BIOPLEX System where wastes generated at each phase of sequence of phases are utilized as raw materials to get succeeding period.

BIOREACTOR Fermentation vat, containing microorganisms or biochemically active substances, used for waste recycling or for making drugs.

BIOREMEDIATION 1. Use of living organisms to clean up oil spills or remove other pollutants from soil, water or wastewater; 2. Use of organisms (e.g., non-harmful insects) to remove agricultural pests or counteract diseases of trees, plants and garden soil.

BIOSCRUBBER Odor removal filter consisting of gas scrubber and biological reactor.

BIOSENSOR Analytical device comprising biological recognition element (e.g., enzyme, receptor, DNA, antibody, microorganism) in intimate contact with electrochemical, optical, thermal or acoustic signal transducer that together permit analyses of chemical properties or quantities.

BIOSESTON Living particulate matter suspended in bodies of water such as lakes and seas.

BIOSOLIDS Solid, semi-solid or liquid untreated residue generated during treatment of domestic sewage in treatment facility that's mostly organic solid product.

BIOSPHERE Portion of Earth and its atmosphere that can support life.

BIOSPHERIC CYCLES Concept of cyclic mass exchange and/or recycling as mechanism of biosphere's functioning is developed.

BIOSTABILIZER Machine that converts solid waste into compost by grinding and aeration.

BIOSTIMULANT Any material of biological origin that enhances plant growth or development when applied to soil.

BIOSURVEY Collecting, processing and analyzing representative portion of resident aquatic community to determine community structure and function.

BIOSYSTEMATICS Taxonomy based upon statistical data of evolution of organisms.

BIOTA All living organisms of area.

BIOTECHNOLOGY 1. Application of principles and practices of engineering and technology to life sciences; 2. Techniques that use living organisms or parts of organisms to produce variety of products (from medicines to industrial enzymes) to improve plants or animals or to develop microorganisms to remove toxics from bodies of water or act as pesticides.

BIOTIC Of, pertaining to or produced by life or living organisms.

BIOTIC COMMUNITY Naturally occurring assemblage of plants and animals that live in same environment and are mutually sustaining and interdependent.

BIOTIC FACTORS Living parts of ecosystem.

BIOTIC INDEX Scale for showing quality of environment by indicating types of organisms present in it.

BIOTIC INFLUENCE Biological influence on vegetable life, in contrast to climatic influence.

BIOTIC POTENTIAL Inherent capacity of organism to reproduce and survive, that is pitted against limiting influences of environment.

BIOTIC PROVINCE Geographic region characterized by presence of one or more ecological associations that differ at least quantitatively from those of adjoining provinces and marked by tendency to act as center of ecological dispersion.

BIOTOPE Smallest geographical unit of habitat, characterized by high degree of uniformity in environment and its plant and animal life.

BIOTRANSFORMATION Conversion of substance into other compounds by organisms that includes biodegredation.

BIOTYPE 1. Genetically homogeneous population composed only of closely similar individuals; 2. Genotypic race or group of organisms.

BIOVOLUME DETERMINATION 1. Used to estimate biomass of algae in aquatic systems; 2. Number of cells of any organism counted by using microscope and grid or counting cell.

BITTERNS Drenched brine solution staying after precipitation of sodium chloride in solar evaporation process.

BITUMINOUS BLEEDING Excess bitumen on pavement surface.

BITUMINOUS CONCRETE Aggregate with bitumen as binder.

BITUMINOUS MACADAM Densely packed layer binded with bitumen.

BITUMINOUS PAVEMENT Pavement using bituminous binder for aggregate layers.

BLACK ALUM Used in water treatment, aluminum sulfate and tiny portion of activated carbon.

BLACK CHERRY TREE Deciduous woody plant species belonging to genus Prunus.

BLACK ICE Transparent ice formed in rivers and lakes or on roads and bridges.

BLACK IRON PIPE Steel conduit used to transport gas in homes and businesses that hasn't been galvanized.

BLACK LIQUOR OXIDATION SYSTEM Air or oxygen that oxidizes black liquor and associated material in storage tanks.

BLACK LIQUOR SOLIDS Dry weight of solids that enter recovery furnace in black liquor.

BLACK SMOKE Smoke composed of carbon particles that indicate incomplete combustion.

BLACK SPRUCE TREE North American species of spruce tree in pine family.

BLACKBODY 1. Hypothetical "body" that absorbs all of electromagnetic radiation striking it (does not reflect or transmit any incident radiation); 2. Not only absorbs all wavelengths, but emits at all wavelengths with maximum possible intensity for any given temperature.

BLACKBODY RADIATION 1. Electromagnetic radiation emitted by ideal blackbody adhering to radiation laws; 2. Theoretical maximum amount of electromagnetic radiation of all wavelengths that can be emitted by body at given temperature.

BLACKOUT Disconnection of source of electricity from all electrical loads in certain geographical area brought about by emergency forced outage or other fault in generation, transmission or distribution system serving area.

BLACKWATER Water that contains animal, human or food waste and toilet usage water.

BLADE 1. Usually part of excavator which digs and pushes dirt but does not carry it; 2. Thin, flat part of plant leaf, attached to stem.

BLADE MIXING Mixing of materials, on site, using blade of motor grader.

BLANK 1. Sample for control experiment that does not contain any of analyte of interest; 2. Empty space.

BLANK FLANGE Flange not yet drilled.

BLANKET 1. Covering or encompassing everything; 2. Rich material encircling quick reactor core to trap and retain neutrons and create more fissile material.

BLANKET BOG Area of peatland, forming where there is climate of high rainfall and low level of evapotranspiration, allowing peat to develop not only in wet hollows but over large expanses of undulating ground.

BLANKET DRAIN Layer of pervious material placed to facilitate drainage of foundation and/or embankment.

BLANKET INSULATION Glass fiber or other compressible fibrous insulation, generally in roll form.

BLAST Loosen or move rock or soil by means of explosives or explosion.

BLAST FENCE Device that redirects engine exhaust to prevent damage.

BLASTHOLE DRILL Air percussion, cord, fusion-kind or rotary drill made to drill blasthole.

BLASTING MATS Blanket usually composed of woven cable or interlocked rings placed over blast to reduce flyrock.

BLEACH 1. Chemical, such as sodium hypochlorite or hydrogen peroxide or preparation of such chemical, used for disinfecting or whitening; 2. To lose color due to stress-induced expulsion of symbiotic unicellular algae.

BLEACHING POWDER White powder, mixture of calcium hypochlorite and calcium chloride, prepared by action of chlorine on calcium hydroxide, used as strong bleaching agent and disinfectant.

BLEED 1. Removal of unwanted air or fluid; 2. Percolation, seepage or exudation of liquid through surface.

BLEED OFF To circumvent or pass by.

BLEED VALVE Valve for running off liquid or for allowing accumulations of gas in liquid to blow off.

BLEEDER 1. Valve designed to release small amount of excess pressure from system; 2. Anything that saps resource produced by something else.

BLEEDER PIPE 1. Conduit placed to allow water from outside basement retaining wall to pass through foundation into drains within building; 2. Pipe inserted in seal to relieve gas pressure from sealed area.

BLEEDER TILE Terra cotta pipe for conveying water from drainage tile to sewer or drain.

BLEEDING Flow or loss from something.

BLENDED WATER Mixture of hot and cold water, generally created and controlled by means of blending valve.

BLENDING VALVE Three-way valve that permits entering liquid to be mixed with liquid that circulates through valve.

BLIND 1. To conceal some covert deed or design; 2. Continuous wall around storage tank containing potentially hazard liquid.

BLIND DRAIN 1. Trench filled with pervious materials such as crushed stone where water flows toward outlet; 2. Ground drain without apparent outlet.

BLIND FLANGE Plate for covering or closing end of pipe.

BLIND NIPPLE 1. Short piece of piping or tubing having one end closed off; 2. Nipple, one end of which is capped.

BLINDING 1. Thin coat of sand or gravel used to fill holes in new road surface; 2. Filling or stopping up of openings in screen or sieve by substance being divided; 3. Clogging of filter media.

BLIZZARD 1. Large snowstorm accompanied by strong winds and greatly reduced visibility caused by blowing snow; 2. Conditions are expected to prevail for period of 3 hours or longer: Sustained wind or frequent gusts to 35 miles hour or greater and considerable falling and/or blowing snow (i.e., reducing visibility frequently to less than 1/4 mile).

BLIZZARD WARNING Issued for winter storms with sustained or frequent winds of 35 mph or higher with considerable falling and/or blowing snow that frequently reduces visibility to 1/4 of mile or less where these conditions are expected to prevail for minimum of 3 hours.

BLOATED CLAY Clay which has expanded during firing, owing to entrapped air or breakdown of sulfides or other ingredients.

BLOCK Small dam or obstruction adequate to stop flow in channel.

BLOCK CRACKING Pavement crack pattern caused by shrinkage of asphalt concrete and daily temperature cycling.

BLOCK SHEAR RUPTURE Combination of tension and shear that tears connection members away from connection.

BLOCKED FLOW Flow approaching mountain barrier too weak or too stable to be carried over barrier.

BLOCKING 1. Blocks used to support (e.g., building) temporarily; 2. State where anticyclone stays standing and cyclones move around it.

BLOCKLOADING Providing consistent amount of electrical power in stated period of time.

BLOCKOUT Means to physically stop operation of gear.

BLOOD PRODUCTS Any product derived from human blood (e.g., blood plasma, platelets, red or white corpuscles, derived licensed products).

BLOOM 1. Proliferation of algae and/or higher aquatic plants in body of water; 2. Flower (blossom) of plant.

BLOW SAND Wind-driven or drifted sand.

BLOWBY Material that moves under pressure past seal or gasket.

BLOWDOWN 1. Cooling fluid discharged from plant at end of its cycle; 2. Removal of liquid and solid hydrocarbons from refinery vessel by use of pressure; 3. Stand of trees uprooted or broken by wind.

BLOWING 1. Descriptor used to amplify observed weather phenomena whenever phenomena are raised to height of 6 feet or more above ground; 2. Up motion of earth material in bottom of cofferdam or excavation as result of groundwater pressure.

BLOWING DUST OR SAND 1. Strong winds over dry ground, that has little or no vegetation, can lift particles of dust or sand into air; 2. Airborne particles that can reduce visibility, cause respiratory problems and have abrasive effect on machinery.

BLOWING SNOW ADVISORY Issued when wind driven snow reduces surface visibility, possibly hampering traveling.

BLOWING WELL Well where air disperses.

BLOWOFF 1. Blowing off of steam, water, etc.; 2. Vent to atmosphere of fluid or vapor held under pressure; 3. Outlet for discharge of sediment or water front sewer.

BLOWOFF VALVE Dump valve or compressor bypass valve is pressure release system present in most turbocharged engines.

BLOWOUT 1. Sudden release of oil and gas from well; 2. Small saucer- or trough-shaped hollow or depression formed by wind erosion on pre-existing dune or other sand deposit.

BLOWUP Inflation (i.e., upward movement) of incompressible materials into joint space of pavement.

BLUE STAIN Dark-colored microscopic fungi cause bluish or grayish discoloration in sapwood of tree.

BLUE STAIN FUNGUS Vague term including various fungi that cause dark staining in sapwood.

BLUE TOPS Grade stakes whose tops indicate finish grade level.

BLUE VITRIOL Copper sulfate.

BLUE WATCH OR BLUE BOX Slang for severe thunderstorm watch.

BLUE-BABY SYNDROME Condition most common in young infants and certain elderly people that can be caused by ingestion of high amounts of nitrate, which results in blood losing its ability to effectively carry oxygen.

BLUE-GREEN ALGAE 1. Group of phytoplankton and periphyton organisms with blue pigment in addition to green pigment called chlorophyll; 2. Cause nuisance water-quality conditions in lakes, streams and slow-flowing rivers; 3. Cyanophyta.

BLUE-GREEN ALGAE OR BACTERIA Algae containing red and blue pigments in addition to chlorophylls.

BLUEGRASS GRASS Grass that spreads by underground stems and makes dense sod, favoring bright places and well-drained soils.

BLUFF High, steep bank (e.g., by river or sea or beside ravine or plain).

BLUSTERY 1. Same as Breezy; 2. 15 to 25 mph winds.

BOARD FOOT Piece of lumber 1 ft^2 and approximately 1 in. thick.

BOARD MEASURE 1. Measurement system for lumber; 2. Piece of lumber 1 ft square and approximately 1 in. thick is one board foot.

BOD Measure of quantity of oxygen utilized by microorganisms to break down organic waste materials in water, where greater BOD, greater degree of pollution.

BOD_5 Amount of dissolved oxygen consumed in five days by biological processes breaking down organic matter.

BOD_7 Biochemical oxygen demand of liquid as determined by incubation at 20 °C for period of 7 days.

BODILY INJURY Part of your insurance policy that pays for costs associated with injuries to other person or people involved.

BODILY INJURY AND PROPERTY DAMAGE Liability coverage can help pay for medical expenses and damaged property.

BODY Principal section of valve or similar pipe fitting.

BODY BURDEN Amount of chemical stored in body at given time.

BODY ELLIPSE Area of human body standing.

BODY FLUIDS Liquid emanating or derived from humans.

BODY WAVE Waves propagated in interior of Earth (i.e., compression and shear waves of earthquake).

BODY WAVE MAGNITUDE Magnitude of earthquake measured as common logarithm of maximum displacement amplitude (microns) and period (seconds) of body waves.

BOG 1. Type of wetland that accumulates appreciable peat deposits; 2. Nutrient-poor, acidic wetland dominated by waterlogged, spongy mat of sphagnum moss and decomposing vegetable matter that ultimately forms thick layer of acidic peat, primarily fed by rain water.

BOIL 1. Point where fluid begins to change to vapor; 2. Uplift of piece of water surface brought on by churning up movement of water; 3. Perpendicular flow of water and land to underside of excavation.

BOILER 1. Apparatus for heating circulating water or other heat transferring liquid; 2. Vessel designed to transfer heat produced by combustion or electric resistance to water that provide hot water or steam.

BOILER FEED WATER Input water utilized by boiler.

BOILER FEEDWATER Water driven right into boiler to replace that evaporated in generation of steam.

BOILER HORSEPOWER 1. Measure of boiler's capacity to produce steam; 2. Equivalent of heat required to change 34.5 lb (16 kg) per hour of water at 212 °F to steam.

BOILER OPERATING DAY 24-hour period where fossil fuel is combusted in steam generating unit.

BOILER SCALE Deposit that forms directly on heat transfer surfaces when constituent solubility limits are exceeded and resulting compounds precipitate onto tube surfaces.

BOILING MANGE Temperature spread between initial boiling point and final boiling point.

BOILING POINT Temperature where liquid will start to become gas, boil and vapor pressure equals given external pressure.

BOLSON 1. Relatively flat and arid valley surrounded by mountains and that drains into central basin; 2. Term used in desert regions of Southwestern United States.

BOLSTER 1. Beam in middle of railway truck, supporting body of car; 2. Short, horizontal structural timber between post and beam for enlarging bearing area of post and/or reducing span of beam.

BOLT ASSEMBLY Bolts, nuts and washers.

BOMB Popular expression of rapid intensification of cyclone (i.e., low pressure) with surface pressure expected to fall by at least 24 millibars in 24 hour.

BOND 1. Binding agreement; 2. Documentary obligation to pay sum or to perform contract; 3. Link or force between neighbouring atoms in molecule.

BOND RELEASE Time where proper regulatory authority returns bond based upon requirements have been satisfactorily completed.

BONDED TENDON Tendon bar continuously embedded in concrete.

BONNET 1. Upper portion or cover of gate valve body where disk rises when valve is opened; 2. Wire mesh, usually in shape of sphere, used to cover top of vent pipes.

BONNET PACKING Material that prevents leaks around valve stems.

BONNEVILLE POWER ADMINISTRATION One of five federal power marketing administrations that sell low cost electric power produced by federal hydroelectric dams to agricultural and municipal users that serves Idaho, Oregon and Washington as well as parts of Nevada and Wyoming.

BONUS FOR EARLY COMPLETION When Owner stands to receive tangible benefit from early completion of project.

BONUS-AND-PENALTY CLAUSE Contract clause assigning payment of bonus to contractor when completing work before stipulated date.

BOOK INVENTORY Cost of inventory on hand, as stated in organization's accounting records.

BOOK VALUE Capital sum where property or equipment is revealed in accounts.

BOOM 1. Floating device used to contain oil on body of water; 2. Piece of equipment used to apply pesticides from tractor or truck.

BOOMING 1. Experiencing period of prosperity or rapid economic growth; 2. Abrupt discharge of sizable amount of collected water in placer mining.

BOOSTER COMPRESSOR Compressor that discharges into another compressor's suction line.

BOOSTER CYCLE Interval where added hydraulic pressure is applied move solid waste into transport trailer or stationary compactor.

BOOSTER FAN Auxiliary fan that increases air pressure in system (e.g., HVAC) during certain peak requirement times.

BOOSTER PUMP Machine which will increase pressure of fluid.

BOOSTER RELAY VALVE Valve used to accelerate application and release of pressurized air.

BOOSTER STATION Pumping station that raises pressure on discharge side of pump or pumps in water distribution system.

BORA Regional downslope wind whose source is so cold that it is experienced as cold wind, despite compression warming as it descends lee slope of mountain range.

BORDER Earth berm developed to carry irrigation water in field.

BORDER ICE 1. Ice sheet in form of long border attached to bank or shore; 2. Shore ice.

BORDER IRRIGATION Process of irrigation by flushing large volume of water over relatively flat field surface in short period of time.

BORDERLINE SOILS Soils that have characteristics of two of classification groups in Unified Soil Classification System.

BORE 1. Tool, such as auger, for making hole by boring; 2. Internal diameter of tank, hose or pipe; 3. Make hole through something; 4. Capped well drilled to tap artesian water; 5. Sudden and rapid flow of tide in certain rivers and estuaries which rolls up as wave.

BOREAL 1. Climatic zone having definite winter with snow and short summer that's generally hot and which is characterized by large annual range of temperature; 2. Of, relating to or coming from north.

BORED WELL Well built by drilling hole in ground with auger and installing casing.

BOREHOLE Hole made with drilling equipment to collect samples for analysis or to extract oil or water.

BOREHOLE EXTENSOMETERS Single-point or multi-point instrument designed to measure axial displacement of fixed point or points along its length that can be rod-type or wire-type and are usually grouted into uncased borehole and installed horizontally, vertically or at any angle.

BOREHOLE LOG Practice of making detailed record of geologic formations penetrated by borehole.

BOREHOLE PUMP Narrow shafts drilled vertically into ground used to extract water.

BOREHOLE SAMPLE Material kept in as undisturbed state from borehole drilling.

BORING Rotary drilling.

BORON Chemical element with atomic number of 5, which is metalloid.

BORROW Material excavated from one area to be used as fill material in another area.

BORROW AREA Area from which natural materials (e.g., rock, gravel, soil) used for construction purposes is excavated.

BORROW PIT Specific site within borrow area from which material (e.g., soil, gravel, sand) is excavated for use.

BOSCH NUMBER Measure of diesel smoke.

BOSQUE Dense growth of trees and underbrush.

BOSS 1. Lump-like mass of rock; 2. Cone for opening conduits.

BOTANICAL PESTICIDE Pesticide whose active ingredient is plant-produced chemical (e.g., nicotine, strychnine).

BOTTLE BILL Proposed or enacted legislation which requires returnable deposit on beer or soda containers and provides for retail store or other redemption.

BOTTLENECK 1. To slow traffic; 2. Any narrowing of road.

BOTTOM ASH Part of non-combustible residue of combustion that falls to bottom in furnace or incinerator.

BOTTOM CONTRACTION Decrease in area of overflowing water due to crest of weir contracting nappe.

BOTTOM LAND Area of low-lying ground adjacent to river subject to 1% or greater chance of flooding in any given year.

BOTTOM LAND HARDWOODS Forested freshwater wetlands adjacent to rivers in southeastern United States, especially valuable for wildlife breeding, nesting and habitat.

BOTTOM OUTLET Opening at low level from reservoir generally used for emptying or for scouring sediment and sometimes for irrigation releases.

BOTTOM SAMPLER Apparatus capable of piercing sea bottom and retaining sample of deposit when brought to surface.

BOTTOM SETS Gradually sloping layers of fine grained sediment in seaward end of delta.

BOTTOM SHORE Member nearest wall in series of raking shore supporting wall.

BOTTOM VENTILATION Air motion through media of wastewater filter.

BOTTOM-DUMP SCRAPER Carrying scraper that dumps or ejects its load over cutting edge.

BOTTOM-LAND FOREST Low-lying forested wetland found along streams and rivers, usually on alluvial flood plains.

BOULDER 1. Rock fragment, usually rounded by weathering or abrasion, with average dimension of 12 inches or more that will not pass 12-inch screen; 2. Large mass of stone detached from surrounding land.

BOULDER CLAY Glacial drift that has not been subjected to sorting action of water and therefore contains particles from boulders to clay sizes.

BOULDER-SIZED Particles greater than 12 inches in diameter.

BOUND WATER Water bound to inorganic surface.

BOUNDARY 1. Dividing line or location between two areas; 2. Set of points in closure of set not belonging to interior of that set.

BOUNDARY CONDITION 1. Mathematical statement specifying dependent variable (e.g., hydraulic head of concentration) at boundaries of modeled domain which contain equations of mathematical model; 2. Known or hypothetical conditions at boundary of problem that govern its solution.

BOUNDARY CONDITIONS Definition or statement of conditions or phenomena at boundaries.

BOUNDARY EFFECT Consequence of dissimilarities between model boundary conditions and conditions occurring in prototype at location of model boundaries.

BOUNDARY INTERSECTION Urban street intersection endpoint.

BOUNDARY LAYER 1. Layer of more or less stationary fluid (i.e., water or air) immediately surrounding immersed object in relative motion with fluid; 2. Roughly lowest one or two kilometers of Earth's atmosphere.

BOUNDARY PIEZOMETER Disk piezometer used to quantify pore water pressure in interface between construction and also adjoining soil.

BOUNDARY ROUGHNESS Roughness of bed and banks of stream or river.

BOUNDED WEAK ECHO REGION Radar signature within thunderstorm characterized by local minimum in radar reflectivity at low levels which extends upward into and is surrounded by, higher reflectivity's aloft.

BOUNDING ESTIMATE Estimate of exposure, dose or risk higher than that incurred by person in population with currently highest exposure, dose or risk.

BOW 1. Curved bend in rod, planar surface or in linear formation such as river; 2. Front of vessel.

BOW ECHO Radar echo that is linear but bent outward in bow shape where damaging straight-line winds often occur near "crest" or center of bow echo.

BOWEN RATIO 1. For any moist surface, ratio of heat energy used for sensible heating (conduction and convection) to heat energy used for latent heating (evaporation of water or sublimation of snow); 2. Ranges from about 0.1 for ocean surface to more than 2.0 for deserts where negative values are also possible.

BOWL 1. Round crater (or similar) in ground; 2. Rounded portion of glyph that encloses empty space; 3. Body or pail of scraper; 4. Moldboard or blade of dozer; 5. That portion of sanitary fitting built to include water and having waste outlet; 6. Case that's shut at one end and mates having filter head.

BOWL URINAL Wall-hung, bowl-shaped urinal that can be installed at different heights to suit users.

BOX 1. Cuboid container; 2. Rail car.

BOX CULVERT Reinforced concrete culvert of rectangular cross-section.

BOX DAM Temporary watertight enclosure pumped dry to expose bottom of body of water so that construction can occur.

BOX DRAIN Drain constructed with upright sides and with flat top and bottom.

BOX GIRDER Hollow steel beam with square or rectangular cross-section.

BOX MODEL Computer model used to calculate air pollution concentrations that's based on assumption that pollutants are emitted into box where they are immediately and uniformly dispersed where sides and bottom of box are defined by sidewalls and floor of valley being studied.

BOX SECTION Conduit with rectangular cross-section.

BRACING MEMBER Structure that provides stability and resists lateral loads.

BRACKISH Water containing too much salt to be useful to people but less salt than ocean water.

BRACKISH ICE Ice formed from brackish water.

BRACKISH WATER Water with salinity intermediate between seawater and freshwater (containing from 1,000 to 10,000 mg/l of dissolved solids).

BRAIDED CHANNEL Stream characterized by random interconnected channels divided by islands or bars.

BRAIDED STREAM Stream characterized by interlacing or tangled network of several small branching and reuniting shallow channels that's composed of anabranches.

BRAKE CHECK AREA Location for drivers to inspect their brakes before driving down descent.

BRAKE HORSEPOWER 1. Actual motor input horsepower required to produce hydraulic horsepower from pump (flow and head) taking into account losses incurred within pump due to friction, leakage, etc.; 2. Ratio of hydraulic horsepower to pump efficiency; 3. Horsepower of engine as measured at its drive shaft.

BRAKING TORQUE Force applied at brake wheel to stop motion.

BRANCH 1. Addition to main pipe in piping system; 2. Creek or stream which flows into larger river; 3. Any of parts of something that divides; 4. Woody part of tree, dividing from trunk.

BRANCH DRAIN Drain pipe connecting soil line or plumbing fixtures in building to main line.

BRANCH FITTING Fitting is pipe item used for changing direction, branching or attaching in piping system.

BRANCH INTERVAL Length of soil stack or waste stack which is usually one story high with horizontal branches from one story of building are connected to stack.

BRANCH LINE Piping where fire protection sprinklers are fined and fed by cross main.

BRANCH MANHOLE Sewer or drain manhole that has more than one pipe feeding into it.

BRANCH SEWER Sewer larger in diameter and capacity than lateral sewers that discharge into it.

BRANCH VENT Vent formed by combining two or more vent pipes.

BRASH ICE 1. Accumulation of floating ice made up of fragments not more than 2 meters across; 2. Wreckage of other forms of ice.

BRASS PIPE Pipe composed of 84% to 86% copper and up to 0.06% lead, plus zinc.

BREACH Failed opening (e.g., erosion, gap, rift, hole, rupture) in dam allowing water stored behind dam to flow through in uncontrolled and unplanned manner where controlled breach is constructed opening and uncontrolled breach is unintentional opening which allows uncontrolled discharge from reservoir.

BREACH HYDROGRAPH Flood hydrograph resulting from dam breach.

BREACH OF CONTRACT Unjustifiable failure to perform under terms of contract when performance is due.

BREAK BULK POINT Location where all portions of load are unloaded and distributed.

BREAK-EVEN ANALYSIS Analysis to determine number of units or dollars needed to cover costs.

BREAK-PRESSURE TANK Open pressure-relief container sited on gravity main in place computed to cut back head.

BREAKAWAY Device that separates upon impact.

BREAKAWAY POINT Point on real axis segment of root locus between two real poles where two real closed-loop poles meet and diverge to become complex conjugates.

BREAKDOWN Transition from non-congested to congested area while traveling.

BREAKDOWN FLOW Traffic that transitions from non-congested to congested.

BREAKDOWN PRODUCT 1. Compound derived by chemical, biological or physical action upon pesticide; 2. Natural process that may result in more toxic or less toxic compound and more persistent or less persistent compound.

BREAKDOWN TORQUE Maximum torque in motor without abrupt speed drop.

BREAKER LINE Depth where waves break.

BREAKERS Ocean waves that break, displaying white water that depends on wave steepness and bottom bathymetry.

BREAKING SEAL Breaching air-tightness of water seal in trap of plumbing fitting.

BREAKOUT OF CHLORINE Point where chlorine leaves solution as gas because chlorine feed rate is too high.

BREAKPOINT CHLORINATION Addition of chlorine to water until disinfection demand has been satisfied.

BREAKTHROUGH Crack or break in filter bed that allows passage of floc or particulate matter through filter that will cause increase in filter effluent turbidity.

BREAKTHROUGH CAPACITY Capacity of ion-exchange column at fixed regeneration level.

BREAKUP Time when river whose surface has been frozen from bank to bank for significant portion of its length begins to change to open water flow condition.

BREAKUP DATE Date on which body of water is first observed to be entirely clear of ice and remains clear thereafter.

BREAKUP JAM Ice jam that occurs as result of accumulation of broken ice pieces.

BREAKUP PERIOD Period of disintegration of ice cover.

BREAKWATER 1. Construction in or around harbour designed to break force of sea and to provide shelter for vessels lying inside; 2. On beaches: wooden or concrete barrier, usually perpendicular to shore, intended to prevent movement of sand along coast.

BREAST WHEEL Waterwheel onto which water is led at about axle height and which acts partly by impulse and partly by weight of descending water in buckets.

BREATHING WELL Well where air is alternatively drawn in and blown out.

BREATHING ZONE Area of air where organism inhales.

BRECCIA Conglomerate-like rock made up of angular pieces of volcanic rock usually bound in volcanic ash.

BRECCIATED Rock made up of highly angular, coarse fragments.

BREEDER REACTOR Nuclear reactor that creates fissile material at faster rate than it uses another fissile material as fuel.

BREEDING DENSITY Density of sexually mature organisms in given area during breeding period.

BREEDING POTENTIAL Maximum rate of increase in numbers of individuals of species or population under optimum conditions.

BREEDING RATE 1. Actual rate of increase of new individuals in given population; 2. Breeding potential minus limiting factors.

BREEZE Light wind from 6.5-50 km/hr, force 2-6 on Beaufort scale.

BREEZY 15 to 25 mph winds.

BRIDGE Construction or natural feature that spans divide (e.g., ravine, flood).

BRIDGE APPROACH RAILING Guardrail system attached to bridge rail system.

BRIDGE CAPACITY Design load of bridge structure.

BRIDGE LENGTH Distance measured between end of bridge floors or abutment backwalls.

BRIDGE MAINTENANCE Work performed to maintain bridges to 'as constructed' standards.

BRIDGE OPENING Bridge span raising to allow tall vessels to pass underneath.

BRIDGE OWNER Entity responsible for bridge that's required to perform inspections and maintenance.

BRIDGE RAILING Guard railing on bridge structure.

BRIDGE ROADWAY WIDTH Curb-to-curb bridge deck width.

BRIDGE TRAFFIC Revenue traffic from foreign to connecting carrier.

BRIDGE WATERWAY Area below bridge opening for flow.

BRIDGE, CLASS A Bridge located on federal-aid highway where examination is required every 2 years and in-depth examination every 6 years.

BRIDGE, CLASS B Bridge not located on federal-aid highway used by public where examination is required every 3 years.

BRIDGE, CLASS C Bridge not located on federal-aid highway that can be designated as being private or operating bridge and could be gated or closed to public, if desired.

BRIDGING System of bracing used between floor or other timbers to distribute weight.

BRIGHT BAND 1. Distinct feature observed by radar that denotes freezing level of atmosphere; 2. Enhanced radar echo of snow as it melts to rain.

BRIGHT SURGE ON DISK Bright gaseous surge (stream) emanating from chromosphere.

BRIGHT SURGE ON LIMB Large gaseous surge (stream) that moves outward more than 0.15 solar radius above limb.

BRIGHTNESS 1. Basic visual sensation describing amount of light that appears to emanate from object; 2. Luminance of object.

BRINE 1. Salt water; 2. Water that has quantity of salt, especially sodium chloride, dissolved in it; 3. Water that contains more than 35,000 mg/l of dissolved solids.

BRINE MUD Waste material, often associated with well-drilling or mining, composed of mineral salts or other inorganic compounds.

BRISK 15 to 25 mph winds.

BRISK WIND ADVISORY Small Craft Advisory issued by National Weather Service for ice-covered waters.

BRITISH THERMAL UNIT Unit of energy required to raise 1 lb of water 1 °F.

BRITISH THERMAL UNITS Standard unit for measuring quantity of heat required to raise temperature of 1 pound of water by 1 °F.

BROAD Railway with track gauge wider than 4 ft, 8.5 in. standard-gauge railways.

BROAD IRRIGATION Irrigation with liquid sewage.

BROAD-BASE TERRACE Low wide terrace that permits farm machines to pass over it.

BROAD-CRESTED WEIR Overflow structure on which nappe is supported for appreciable length in direction of flow that causes occurrence of critical flow depth over crest.

BROADBAND Method of signaling where multiple signals share bandwidth of transmission by subdivision of bandwidth into channels based on frequency.

BROADCAST To toss granular material over horizontal surface to obtain thin, uniform layer.

BROADCAST APPLICATION Spreading of pesticides over entire area.

BROADCAST BURNING Prescribed fire ignited in areas with little or no forest canopy present.

BROCKEN SPECTRE Optical phenomenon sometimes occurring at high altitudes when image of observer placed between Sun and cloud is projected on cloud, surrounded by rings of color (glory), as greatly magnified shadow.

BROKEN BACK CULVERT Culvert with one or more breaks in profile slope.

BROKEN BASE Any break in base of rail.

BROKEN LEVEL Layer of atmosphere with 5/8 to 7/8 sky cover (cloud cover).

BROKEN-BACK CULVERT Culvert with one or more breaks in profile slope.

BROMINE Non-metallic chemical element (symbol Br) with atomic number of 35.

BROMOXYNIL Organic compound classified as nitrile herbicide.

BROOK Body of running water smaller than river.

BROOMING Roughing (i.e., cleaning) surface via sweeping action.

BROWN ORES Soils leached of more soluble minerals, but retaining rich amounts of iron oxides and hydroxides.

BROWN PODZOLIC SOIL Freely draining, leached soil profile that has developed acid surface horizons, surface humus and clearly visible enrichment of translocated iron oxide in middle or B-horizon.

BROWN SMOKE Indicates that wood is being heated to its ignition temperature.

BROWN STOCK WASHER Washer and related knotters, vacuum pumps and filtrate used to wash wood pulp after digestion system.

BROWNFIELDS Abandoned, idled or under used industrial and commercial facilities/sites where expansion or redevelopment is complicated by real or perceived environmental contamination.

BROWNIAN MOVEMENT Random movement of microscopic particles suspended in liquid or gas, caused by collisions with molecules of surrounding medium.

BROWNOUT Partial reduction of electrical voltages that often results in lights dimming and motor-driven devices slowing down.

BROWSE To move about while eating parts of plants, especially plants other than pasture, such as shrubs or trees.

BRUSH Wild vegetation, generally larger than grass but smaller than trees.

BRUSH OUT To clear place tree limbs, saplings and debris.

BRUSH RAKE Large apparatus attached to front of tractor to clear land of brush and debris.

BUBBLE 1. Spherically contained volume of air or other gas; 2. System under which existing emissions sources can propose alternate means to comply with set of emissions limitations.

BUBBLE BARRIER Air bubbles discharged from perforated block fixed to base of storage or treatment status across its width or length.

BUBBLE BURSTING Primary mechanism for projection of tiny water droplets in ocean surface.

BUBBLE HIGH Mesoscale area of high pressure, typically associated with cooler air from rainy downdraft area of thunderstorm or complex of thunderstorms.

BUBBLE POINT Temperature (at given pressure) where first bubble of vapor is formed when heating liquid consisting of two or more components.

BUBBLER GAUGE Water stage recording device capable of attaching to LARC for data automation purposes.

BUBBLING FLUIDIZED-BED COMBUSTOR Combustor where majority of bed material remains in fluidized state in principal combustion zone.

BUBBLING SPRING Spring where ascending water currents and/or escaping gases cause water to bubble.

BUCKET 1. Inverted curve at bottom of spillway that deflects water horizontally from steep overflow face to apron; 2. Receptacle fixed to manhole frame to retain grit carried by surface washings; 3. Fitting or attachment to item of mechanical equipment that transports material.

BUCKET AUGER Cylinder or barrel to hold soil, which is forced into barrel by cutting lips.

BUCKET CAPACITY True volume of material that can be carried by equipment's bucket.

BUCKET ELEVATOR Mechanism for hauling flowable bulk materials (most often grain or fertilizer) vertically.

BUCKET LOADER Mobile bucket conveyor for loading loose materials into trucks or railway cars.

BUCKET PUMP 1. Pump comprising of buckets that can be always filled, lifted, emptied and lowered for refilling; 2. Large-capacity reservoir-and-pump for lubricants.

BUCKET RATING Volume that bucket was created to hold.

BUCKET-LADDER DREDGE Mechanical dredge that uses chain of heavy buckets rotating over dredging arm or ladder to excavate and lift material to dredging platform.

BUCKET-WHEEL EXCAVATOR Heavy equipment used in surface mining comprised of rotating wheel with tooth-edged pails used to excavate narrow trenches.

BUDGET Amount of money or resources earmarked for particular institution, activity or time-frame.

BUDGET ESTIMATES Forecasts that are used to plan strategy and budgets.

BUDGET REVIEW COMMITTEE Ad hoc committee of representatives from different areas that coordinates budget activities through formulation phase.

BUDGETARY CONTROL System of management control where actual income and spending are compared with planned income and spending.

BUDGETED COST OF WORK PERFORMED Budgeted cost of work that has actually been performed in carrying out scheduled task during specific time period.

BUDGETED COST OF WORK SCHEDULED Sum of budgets for all work scheduled to be accomplished with given time period.

BUFFER 1. Solution or liquid whose chemical makeup is such that it minimizes changes in pH when acids or bases are added to it; 2. Gap that isolates or separates two things; 3. Anything used to maintain slack or isolate different objects.

BUFFER ACTION Ability of buffer solution to resist changes in pH value on addition of small amount of acid or base.

BUFFER CAPACITY Measure of efficiency of buffer in resisting changes in pH.

BUFFER SOLUTION Solution, containing buffer, that maintains relatively constant pH.

BUFFER STRIP Area of land maintained in permanent vegetation that helps to control air, soil and water quality, along with other environmental problems, dealing primarily on land used in agriculture.

BUFFER STRIPS Strips of grass or other close-growing vegetation that separates waterway (e.g., ditch, stream, creek) from intensive land use area (e.g., subdivision, farm).

BUFFER WIDTH Distance from sidewalk to roadway.

BUFFER ZONE Transitional area between areas of different land use.

BUFFER-SEPARATED LANE Operating lane, in same direction as traffic, that uses 1-ft or wider buffer.

BUFKIT Software tool used by forecasters to examine vertical profile and other aspects of atmosphere.

BUILDER Alkaline phosphate added into detergent to correct its alkalinity or add to its soil-suspending ability.

BUILDING ACTIVITIES Pollutant emitting activities that belong to same industrial grouping, are located on one or more contiguous or adjacent properties and under common control.

BUILDING AUTOMATION SYSTEM System that optimizes start-up and performance of HVAC equipment and alarm systems.

BUILDING CODE Mandatory set of rules on building design, construction, testing, inspection and maintenance to protect public health, safety and welfare.

BUILDING COMMITTEE Source of member power, leadership and problem solving for job.

BUILDING CONNECTION Extension from sewer tap to property or easement line of property to be served.

BUILDING CONTRACTOR Person is responsible for day-to-day oversight of construction site, management of vendors and trades and communication of information to all involved parties throughout course of building project.

BUILDING COOLING LOAD Hourly amount of heat that must be removed from building to maintain indoor comfort (measured in BTU).

BUILDING DEPARTMENT Department of local government in charge of regulation or municipal works.

BUILDING DRAIN Part of lowest piping of drainage system that receives discharge from soil, waste or other drainage pipes inside walls of building (house) and conveys it to building sewer beginning 3 feet outside building wall.

BUILDING ENVELOPE Exterior surface of building's construction (e.g., walls, windows, floors, roof, floor).

BUILDING INSPECTOR Person that ensures construction meets local and national building codes and ordinances, zoning regulations and contract specifications.

BUILDING INSULATION Building materials that form thermal envelope between conditioned building and unconditioned exterior.

BUILDING MAIN Water supply pipe (including fittings) that extends from water main or other source of supply to first distribution branch of building.

BUILDING PERMIT Type of authorization that must be granted by government or other regulatory body before construction of new or existing building can legally occur.

BUILDING REGULATIONS Minimum standards for design, construction and alterations to virtually every building.

BUILDING RELATED ILLNESS Diagnosable illness whose cause and symptoms can be directly attributed to specific pollutant source within building (e.g., Legionnaire's disease, hypersensitivity, pneumonitis).

BUILDING RESTRICTIONS Provisions and specifications in building codes that affect placement, size and appearance of building.

BUILDING SEWER Underground sewer pipe that connects plumbing in your home to public sewer system.

BUILDING STORM DRAIN Drainage system used to get rainwater, surface water and groundwater, linked to building sewer outside building.

BUILDING STORM SEWER 1. Building sewer which conveys only storm water; 2. horizontal piping of storm drainage system which extends from building storm drain, receives its discharge and conveys it to public storm sewer or other point of disposal.

BUILDING SUBDRAIN That portion of building drainage system which does not drain by gravity into building sewer where wastes from subdrain are collected in sump and discharged by pump.

BUILDING TRAP Device that serves to prevent sewer gases from backing up main drain and into home.

BUILT ENVIRONMENT Human- modified environment (e.g., buildings, roads, cities).

BUILT-IN DIRT Substance passed into effluent stream composed of foreign substances incorporated into filter medium.

BULK Unpackaged goods when transported in large volumes.

BULK ASBESTOS Any volume of asbestos fiber of any kind or level or blend of kinds or levels, that's mined or ground with all intent behind getting asbestos.

BULK CONTAINER Big container that can hold free-flowing dry cargo such as cement, grains and ores.

BULK ELECTRICAL CONDUCTIVITY Combined electrical conductivity of all material within doughnut-shaped volume surrounding induction probe.

BULK RICHARDSON NUMBER Non-dimensional number relating vertical stability and vertical shear (i.e., stability divided by shear) where high values indicate unstable and/or weakly-sheared environments and low values indicate weak instability and/or strong vertical shear.

BULK SAMPLE Small portion (e.g., thumbnail size) of suspect asbestos-containing building material collected by asbestos inspector for laboratory analysis to determine asbestos content.

BULK SPREADER Machine for taking cement or alternative substance and distributing it on prepared surface for ground stabilization.

BULKHEAD 1. Pressure-resistant sealed barrier to any fluid in large structure; 2. One-piece unit lowered into guides and seals against frame to close water passage in dam, conduit, spillway, etc.; 3. Wall or partition erected to resist earth, ground or water pressure; 4. Retaining wall along waterfront; 5. Atmosphere holding barrier built for long term management of radon-222 and radon-222 decay product levels in mine atmosphere.

BULKHEAD GATE Gate used either for temporary closure of channel or conduit before dewatering it for inspection or maintenance or for closure against flowing water when head difference is small (e.g., diversion tunnel closure).

BULKING Increase in volume of material due to manipulation (e.g., rock bulks upon being excavated, damp sand bulks if loosely deposited by dumping).

BULKING SLUDGE Sludge that settles poorly because of floc of low density and has sludge volume index (SVI) of ¿ 150 ml/g.

BULKY WASTE Large items of solid waste that size precludes/complicates their handling by normal collection methods.

BULL'S LIVER Inorganic silt of slight plasticity that quakes like jelly from vibration.

BULLDOZER Tractor with attached blade for pushing earth and building debris for coarse preliminary surface grading, demolishing building structures, etc.

BUNKER 1. Coal compartment on tank engine; 2. Space for vessel fuel storage; 3. Hardened shelter.

BUOY Floating object moored to bottom serving as navigation mark.

BUOYANCY 1. Tendency of body to float or to rise when submerged in fluid; 2. Power of fluid to exert upward force on body immersed or partly immersed.

BURDEN Distance between free face and first row of holes or distance between rows of holes parallel to face.

BUREAU OF LAND MANAGEMENT Agency within U.S. Department of Interior, administers 264 million acres of America's public lands, located primarily in 12 Western States that sustains health, diversity and productivity of public lands for use and enjoyment of present and future generations.

BUREAU OF RECLAMATION Federal agency with mission to manage, develop and protect water and related resources in environmentally and economically sound manner in interest of American public.

BURIED CHANNEL Former stream channel which has idled with alluvial or glacial deposits which has been afterwards covered by other substance to degree which makes it almost indiscernible from surface observation.

BURIED RIVER River bed buried beneath stream of basalt or alluvial drift.

BURIED SERVICES standard building services set below-ground.

BURIED-CABLE HEATING SYSTEM Heating system where electrical resistance cable is buried in ground floor concrete slab.

BURN OUT 1. To become extinguished due to lack of fuel; 2. Failure of nuclear reactor heat exchanger surface that can lead to lack of heat-exchange fluids and may, ultimately, result in discharge of radioactive materials.

BURNER Device that burns fuel.

BURNING CONDITIONS Environmental conditions which influence fire in specified fuel organization.

BURNING INDEX Number related to contribution of fire behavior to effort of containing fire.

BURNING OUT Setting fire in control line to consume fuel between edge of forest fire as well as control line.

BURNING RATE 1. Quantity of fuel per unit of time charged to combustion system; 2. Amount of heat released during combustion.

BURST Transient enhancement of solar radio emission, usually associated with active region or flare.

BURST PRESSURE Pressure that pressure vessel like pipe or tube can handle before rupturing.

BURSTING FORCE Tensile force caused by spreading of concentrated force.

BURY BARGE Barge designed to dig trench for underwater pipeline.

BURY LENGTH Depth from your top layer of bottom to underparts of conduit to which hydrant is linked.

BURYING AGENTS Additives that, through physical or chemical means, improve combustibility of substances to which they have been used.

BUS 1. Road vehicle that carries many passengers; 2. Conductor or group of conductors, that serve as common connection for two or more electrical circuits; 3. Three rigid single-phase connectors that interconnect generator and step-up transformer(s) in power plants.

BUS BAY Designated area on side of road for buses to pull out of traffic flow for passenger loading and unloading.

BUS LANE Traffic lane designated for use of buses.

BUS MODE Transportation provided by buses that follow schedules and routes.

BUS PRIORITY SYSTEM Technique to move mass transit vehicles at intersections controlled by traffic signals.

BUS RAPID TRANSIT Form of rapid transit using buses, that have dedicated rights-of-way or laneways, purpose-built bus stations and dedicated traffic signalling.

BUS SHELTER Structure constructed at bus stop, to provide seating and protection from weather for convenience of waiting passengers.

BUS STOP Place where public transport buses stop to allow passengers to board or leave.

BUSBAR Heavy metal conductor used to carry large current.

BUSHING 1. Adapter for joining pipes of different size; 2. Type of bearing, cylindrical lining designed to reduce friction and wear inside hole.

BUSPOOL Services that provided group transportation via buses.

BUST Slang for inaccurate forecast, especially one where significant weather is predicted but does not occur.

BUSWAY Right of way, roadway or lane for exclusive use of buses.

BUTT JOINT In pipe, flat ends that meet but do not overlap.

BUTT-WELD STEEL PIPE PIPE Conduit meant for low pressure steam, water, gas or air service.

BUTTE Isolated hill with steep sides and flat top.

BUTTERFLY GATE Discharge valve that opens just like damper to control gravity flow.

BUTTERFLY VALVE Valve designed for quick closure used for isolating or regulating flow.

BUTTON-DUMP TRUCK Trailer or semitrailer that releases bulk stuff by opening doors in flooring of body.

BUTTRESS DAM Dam consisting of watertight upstream part (e.g., concrete sloping slab) supported at intervals on downstream side by series of buttresses (e.g., walls normal to axis of dam) comprised of reinforced masonry or stonework built against concrete.

BUTTRESS DRAIN Rubble-filled trench system in slope of railway cutting, laid out in herringbone fashion and leading surface water into buttress drains arranged along line of steepest slope.

BUY-BACK CENTER Facility where individuals or groups bring recyclables in return for payment.

BUYER'S RISK Probability that something will go wrong and associated negative consequences.

BY OTHERS Work done by another contractor under subcontract.

BY-PRODUCT Material, other than principal product, generated as consequence of industrial process or as breakdown product in living system.

BY-PRODUCT WASTE Liquid or gaseous substance produced at chemical manufacturing plants or petroleum refineries (except natural gas, distillate oil or residual oil) and combusted in steam generating unit for beat recovery or for disposal, excluding gaseous substances with carbon dioxide levels greater than 50% or carbon monoxide levels greater than 10%.

BYATT Flat lumber supporting decking and paths in trench excavations.

BYE CHANNEL Transport waterways trenched around reservoir designated to handle excess flow from entering streams.

BYLAW Rule made by local authority to regulate its own affairs.

BYPASS 1. Road that passes around something; 2. Intentional diversion of waste streams from any portion of treatment/pretreatment facility; 3. Section of pipe that conducts fluid around some other fixture; 4. Secondary passing of flow, as well as primary flow path.

BYPASS CHANNEL Channel built to carry excess water from stream.

BYPASS FILTER Filter that removes contaminants from fluid by avoiding portion of flow through its filter medium.

BYPASS FLOW Water permitted to be diverted.

BYPASS GATE Gate used to equalize pressure or divert flow to sand trap or waste pit.

BYPASS LANE Unsignalized intersections on two-lane roadways that allow vehicles to pass stopped or turning vehicles.

BYPASS STACK Stack that vents exhaust gases to atmosphere in bypass control apparatus.

BYPASS SWITCH Hardware device that provides fail-safe access port for in-line appliance.

BYPASS VALVE Valve that controls pressure in system by diverting portion of flow.

BYPASS VENT Vent stack parallel to soil or waste stack and connected together at branch intervals.

C

C-FRAME C- or U-shaped frame that connects blade to dozer.

C-HORIZON Layer of soil profile immediately below B horizon and above bedrock, composed of weathered rock little affected by soil-forming processes.

CAB 1. Taxi; 2. Compartment at front of truck or train for driver; 3. Shelter at top of air traffic control or fire tower.

CABLE ENTRY Sealed enclosure where power cable passes from outside to inside building.

CABLE GLAND Connector that holds end of cable in place.

CABLE GRIP Device for propelling vehicle by attaching to wire cable running at constant speed.

CABLE JACKET Cover that mechanically protect insulation and conductor core of cable.

CABLE LADDER Comprises of range of straight lengths as well as different shaped fittings designed to facilitate changing cabling directions or levels easily, without need to modify any components.

CABLE LOCATOR Instrument used for detecting presence and approximate location of buried services in advance of undertaking excavation works.

CABLE PULLING COMPOUND Lubricant used to facilitate pulling of wires through cable duct or conduit.

CABLE PULLOUT UNLOADING METHOD Process where landfill tractor helps in unloading transport trailer by pulling cable network specially set inside transport vehicle for this function.

CABLE SHEATH Protection covering cables.

CABLE TAIL Additional cable left at end of run to ensure that there is sufficient material remaining for connections.

CABLE TRAY Rigid structure to support electric cables; type of raceway, normally having appearance of ladder and open at top to facilitate changes.

CABLE-STAYED BRIDGE Bridge where deck is supported by cables attached directly to pillars.

CABOTAGE LAWS Laws that govern transportation of goods and/or people between two ports in United States.

CADMIUM Metallic chemical element that accumulates in environment with atomic number of 48 that's toxic in tiny amounts.

CAGE Skeleton frame to limit motion of loose piece (e.g., ball valve).

CAGE SCREEN Filtration display in sort of drum, hinged so that it could possibly be lifted or lowered for cleaning.

CAIRN Pile of stones used as marker.

CAISSON Watertight chamber or hallow floating box used in construction work under water.

CALCAREOUS Rock or substance formed of calcium carbonate or magnesium carbonate by biological deposition or inorganic precipitation or containing those minerals in sufficient quantities to effervesce when treated with cold hydrochloric acid.

CALCAREOUS SPRING Spring water comprising significant amount of calcium salts in solution.

CALCINE To heat something without melting in order to drive off water and to decompose carbonates into oxides or to oxidize or reduce it.

CALCINER Device or container where calcination (i.e., moisture and organic matter of phosphate rock is reduced) takes place.

CALCITE Light-colored mineral composed of calcium carbonate that often fills veins in igneous rocks and forms sedimentary rock limestone.

CALCIUM Chemical element, atomic number 20, alkaline earth metal and occurs naturally as carbonate in limestone and as silicate in many rocks.

CALCIUM CARBIDE Material containing 70% to 85% calcium carbide by weight.

CALCIUM HYPOCHLORITE Used in water treatment, dry powder comprising lime and chlorine mixed so that when dissolved in water it discharges active chlorine.

CALCIUM OXIDE Quicklime which becomes calcium hydroxide when water is added.

CALCIUM SULFATE STORAGE PILE RUNOFF Calcium sulfate transports water runoff from or through calcium sulfate pile and precipitation that falls directly on storage pile that may be collected in seepage ditch at base of outer slopes of storage pile.

CALDERA Large circular depression formed by explosion or collapse of volcano.

CALENDAR DAY Timespan of exactly 24 hours that's measured from one midnight to next.

CALENDAR UNIT Smallest unit of time utilized in scheduling job.

CALIBRATE Check or adjust by comparison with standard.

CALIBRATED MODEL Model for which all residuals between calibration targets and corresponding model outputs or statistics computed from residuals, are less than pre-set acceptable values.

CALIBRATED SAND METHOD Weight, in place, of cubic measure of soil.

CALIBRATING GAS Gas of known concentration used to build response curve of analyzer.

CALIBRATION 1. Derivation of set of model parameter values that produces "best" fit to observed data; 2. Process of refining model representation of hydrogeologic framework, hydraulic properties and boundary conditions to achieve desired degree of correspondence between model simulations and observations of flow system.

CALIBRATION OF EQUIPMENT Comparison of measurement values delivered by device under test with those of calibration standard of known accuracy.

CALIBRATION TARGET Measured, observed, calculated or estimated parameters that model must reproduce, at lease approximately, to be considered calibrated.

CALIBRE Diameter or bore of conduit or cylindrical body.

CALICHE 1. Layer of hard clay subsoil or sedimentary rock (e.g., hardpan); 2. Crude form of sodium nitrate used as fertilizer.

CALIFORNIA BEARING RATIO Penetration test for evaluation of mechanical strength of natural ground, subgrades and base courses beneath new carriageway construction.

CALL 1. Visit by ship or boat to port; 2. Decision or judgement.

CALL BOX Box containing telephone for use in emergency.

CALL FOR BIDS Special procedure for generating competing offers from different bidders looking to obtain award of business activity in works, supply or service contracts.

CALL FOR PROPOSALS Invitation to qualified bidders to include proposal to complete project.

CALL FOR TENDERS Invitation to competent contractors and/or suppliers to satisfy what's needed to complete particular project.

CALL LOAN Loan that lender can demand to be repaid at any time.

CALLOUT Annotation that pertains to specific location in body of text or graphic.

CALM Weather condition when no air motion (i.e., wind) is detected.

CALORIE Unit of energy equal to 4.19 Joules.

CALORIFIC VALUE 1. Amount of energy produced by complete combustion of material or fuel; 2. Amount of energy available from item of food when digested.

CALORIFIER Heat exchanger which heats water indirectly by circulating is over heating coil or multiple coils.

CALORIMETER Apparatus for measuring heat generated or absorbed by either chemical reaction, change of phase or some other physical change.

CALVE 1. To separate or break; 2. Become detached.

CAMBER Extra height added to crest of embankment dams to ensure that freeboard will not be diminished by foundation settlement or embankment consolidation.

CAMBIUM MINER Small bug (i.e., larvae of flies) that burrows in tree just under bark.

CAMEL Loaded vessels lashed tightly and then emptied to reduce draft of ship in middle.

CAN Cylindrical container with two ends formed from metal.

CANADIAN GEODETIC VERTICAL DATUM 1928 Geodetic datum derived from general adjustment of Canada's first order level network in 1928.

CANAL Artificial channel, usually open, that conveys water by gravity to farms, municipalities, etc.

CANAL HEADWORKS Beginning of canal.

CANAL LIFT Machine for transporting boats between water at two different elevations.

CANAL LOCK Device used for raising and lowering boats, ships and other watercraft between stretches of water of different levels on river and canal waterways.

CANAL PRISM Shape of canal as seen in cross-section.

CANAL RAPIDS Inclined conduit or chute used to convey water at high velocity to lower elevation.

CANAL SECTION Cross-sectional profile of canal.

CANAL SEEPAGE Water lost to subsurface by seepage through channel bottom or walls.

CANAL SYSTEM System of waterways making up entire irrigation or water transmission system.

CANAL TRIMMER Gear that performs final formation of underside and sides of earthen canal.

CANALIZATION Conversion of river or other waterway to canal.

CANCELLATION CLAUSE Stipulation in agreement that grants (to one or both parties) right to terminate it before its expiration, under specified terms and conditions.

CANCER POTENCY SLOPE FACTOR 1. Indication of chemical's human cancer-causing potential derived using animal studies or epidemiological data on human exposure; 2. Based on extrapolation of high-dose levels over short periods of time to low-dose levels and lifetime exposure period through use of linear model.

CANDELA Luminous intensity, in perpendicular direction, of surface of 1/600,000 m^2 of black body (full radiator) at temperature of freezing platinum (2046 °K) under pressure of 101.325 kPa.

CANDELA PER SQUARE METER Derived unit of luminance.

CANDIDATE METHOD Method of sampling and analyzing surrounding air for pollutant.

CANDIDATE SPECIES Plant or animal species that are candidates for designation as endangered (i.e., in danger of becoming extinct) or threatened (i.e., likely to become endangered), but is undergoing status review by U.S. Fish and Wildlife Service.

CANOPY ANGLE 1. Measure of openness of stream to sunlight; 2. Angle formed by imaginary line from highest structure (e.g., tree, shrub or bluff) on one bank to eye level at mid-channel to highest structure on other bank.

CANOPY-INTERCEPTION Precipitation that falls on and is stored in leaf or trunk of vegetation that refers to either process or volume.

CANT Rate of elevation change between two rails or edges.

CANTILEVER Beam anchored at one end and projecting into space.

CANTILEVERED SIGNAL STRUCTURE Rigidly attached structure to vertical pole used to provide overhead support of signal units.

CANYON LANDFILLING In depression, landfilling method where cover material is obtained within depression.

CANYON WIND Foehn wind channeled through canyon as it descends lee side of mountain barrier.

CAP 1. Layer of clay or other impermeable material, installed over top of closed landfill to prevent entry of rainwater and minimize leachate; 2. Layer of warm air several thousand feet above Earth's surface which suppresses or delays development of thunderstorms; 3. Protective cover or seal; 4. Conduit plug with female threads.

CAP CLOUD Stationary cloud directly above isolated mountain peak, with cloud base below elevation of peak.

CAP ROCK Layer of hard, impervious rock overlying and often sealing in deposit of oil, gas or coal.

CAPABILITY Maximum load that generating unit, generating station or other electrical apparatus can carry under specified conditions for given period of time without exceeding approved limits of temperature and stress.

CAPABLE FAULT Active fault judged able to produce macro-earthquakes.

CAPACITANCE Property of electric circuit or its element that permits it to store charge, defined as ratio of stored charge to potential over that element or circuit.

CAPACITOR Electronic component capable of storing electrical energy in electric field; especially one consisting of two conductors separated by dielectric.

CAPACITOR MOTOR Single-phase induction motor having its primary winding connected to some way to obtain electricity and auxiliary winding connected in series using capacitor to ease starting.

CAPACITOR/CONDENSER Electrical apparatus that stores potential energy in electric field.

CAPACITY 1. Load for which generator, transmission line or system is rated, expressed in kilowatts; 2. Electrical capacitance; 3. Amount of electric power delivered or required; 4. Maximum load that machine, station or system can carry under existing service conditions; 5. Ability to hold, receive or absorb; 6. Maximum amount that can be held.

CAPACITY ASSURANCE PLAN Statewide plan which supports state's ability to manage hazardous waste generated within its boundaries over twenty year period.

CAPACITY CURVE Plot expressing relationship between two aspects of operation.

CAPACITY FACTOR Average power generated, divided by rated peak power.

CAPACITY REDUCER Apparatus that allows adjustment of output signal capacity of air compressor without alteration to any operating state.

CAPILLARITY 1. Degree to which material or object containing minute openings or passages, when immersed in liquid, will draw surface of liquid above hydrostatic level; 2. Phenomenon where water is held in interstices above normal hydrostatic level, due to attraction between water molecules.

CAPILLARY ACTION 1. Movement of water through very small spaces due to molecular forces called capillary forces; 2. Process by which water rises through rock, sediment or soil caused by cohesion between water molecules and adhesion between water and other materials that pulls water upwards.

CAPILLARY ATTRACTION 1. Molecular forces which cause movement of water through very small spaces(e.g., soil particles) regardless of gravity; 2. Adhesive force that holds fluid in capillary or pore space; 3. Function of properties of fluid and surface and dimensions of space where if attraction between fluid and surface is greater than interaction of fluid molecules, fluid will be held in place.

CAPILLARY BREAK Hydrophobic material or space that stops capillary movement.

CAPILLARY CAPACITY Maximum water holding capacity of soil (i.e., upper limit of water available to plants) against gravity.

CAPILLARY FLOW 1. Ability of liquid to flow in narrow spaces without assistance of or even in opposition to, external forces like gravity; 2. Movement of water by capillary action.

CAPILLARY FORCE 1. Molecular forces which cause movement of water through very small spaces, as between soil particles, regardless of gravity; 2. Adhesive force that holds fluid in capillary or pore space; 3. Function of properties of fluid and surface and dimensions of space where if attraction between fluid and surface is greater than interaction of fluid molecules, fluid will be held in place.

CAPILLARY FRINGE Subsurface layer where groundwater seeps up from water table by capillary action to fill pores.

CAPILLARY FRINGE ZONE Zone where groundwater seeps up from water table by capillary action to fill pores.

CAPILLARY GROOVE Space formed between two components to prevent capillary action between them.

CAPILLARY HEAD Potential, expressed in units of water head, that causes water to flow by capillary action.

CAPILLARY INTERSTICE Pore space in sediment small enough for occurrence of appreciable capillary rise.

CAPILLARY LIFE Height that water or alternative liquid rises in capillary tube.

CAPILLARY MIGRATION Movement of water by capillary action.

CAPILLARY MOISTURE Moisture left in soil, along with hygroscopic moisture and water vapor, after gravitational water has drained off.

CAPILLARY MOVEMENT Rise of subsoil water above water table through channels connecting pores in soil.

CAPILLARY OPENING Opening little enough in cross-sectional area to allow state of capillarity.

CAPILLARY POTENTIAL Work required to move unit mass of water from reference plane to any point in soil column.

CAPILLARY PRESSURE Pressure difference across interface between two immiscible fluids arising from capillary forces.

CAPILLARY RISE Height above free water elevation that water will rise by capillary action.

CAPILLARY SPACE Air bubbles that have become embedded.

CAPILLARY STRESS Pore water pressure less than atmospheric value produced by surface tension of pore water acting on meniscus formed in void spaces between soil particles.

CAPILLARY TUBE Narrow tube that conducts small quantity of liquid under pressure.

CAPILLARY WATER 1. Underground water held above water table by capillary attraction; 2. Water subject to influence of capillary action.

CAPILLARY ZONE 1. Zone above water table where water is held by surface tension (i.e., capillary action); 2. Porous material just above water table that may hold water by capillarity in smaller void spaces that's under pressure less than atmospheric.

CAPITAL ASSETS Assets of relatively long-term nature.

CAPITAL BUDGET Budget to evaluate potential investments or expenditures for specific projects or purposes.

CAPITAL COSTS 1. Costs (usually long-term debt) of financing construction and equipment that's usually fixed, one-time expenses which are independent of amount of water produced; 2. All implements, equipment, machinery and inventory used in production of goods and services.

CAPITAL EXPENDITURE Funds spent by company to acquire or upgrade long-term asset.

CAPITAL INVESTMENT 1. General term used to identify any money amount that is to be considered as investment as opposed to annual expense; 2. Can be either interest bearing or non-interest bearing.

CAPPED END Conduit end covered.

CAPPING Region of negative buoyancy below existing level of free convection where energy must be supplied to parcel to maintain its ascent that tends to inhibit development of convection until some physical mechanism can lift parcel to its level of free convection.

CAPPING INVERSION Elevated inversion layer that caps convective boundary layer.

CAPPING PIECE Piece of timber covering heads of series of uprights or other vertical structure.

CAPROCK Geological formation (or group or part of formation) capable of limiting fluid movement out of injection zone.

CAPTIVE Person limited to single transportation mode.

CAPTIVE TRANSIT RIDER Person that does not have vehicle or cannot drive.

CAPTURE Replaced water withdrawn from aquifer.

CAPTURE EFFICIENCY Fraction of organic vapors generated by process that are directed to abatement or recovery device.

CAPTURE SYSTEM Gear that transports particulate matter to air pollution control apparatus.

CAR FLOAT Barge with rail tracks mounted on deck.

CAR HIRE Charge for one transportation line to use another transportation line cars.

CAR SHARING Model where people rent cars for short periods of time.

CAR-SEALED Valve locked in position (i.e., open or closed).

CARBOHYDRATES Sugar, starch or cellulose food source of energy for animal or plant.

CARBON Chemical element with atomic number of 6.

CARBON ABSORBER Add-on control device that uses activated carbon to absorb volatile organic compounds from gas stream.

CARBON ADSORPTION Treatment system that removes contaminants from groundwater or surface water by forcing it through tanks containing activated carbon treated to attract contaminants.

CARBON BLACK Fine black powder of amorphous carbon manufactured by incomplete combustion of natural gas or oil and used as pigment in manufacture of ink and paint.

CARBON CAPTURE AND STORAGE Process of capturing carbon dioxide from emission source, (typically) converting it to supercritical state, transporting it to injection site and injecting it into deep subsurface rock formations for long-term storage.

CARBON CHLOROFORM EXTRACTION Process for evaluating organic pollutants in water.

CARBON CYCLE Physical cycle of carbon through Earth's biosphere, geosphere, hydrosphere and atmosphere that includes such processes as photosynthesis, decomposition, respiration and carbonification.

CARBON DIOXIDE 1. Colorless, odorless, non-poisonous gas that's fourth most abundant constituent of dry air; 2. Product of respiration and fossil fuel combustion used by plants during photosynthesis; 3. Greenhouse gas that traps terrestrial (i.e., infrared) radiation and contributes to potential for global warming.

CARBON DIOXIDE EQUIVALENT 1. Emissions of greenhouse gases are typically expressed in common metric so that their impacts can be directly compared, as some gases are more potent (i.e., have higher global warming potential) than others; 2. International standard practice is to express greenhouse gases in carbon dioxide equivalents.

CARBON DIOXIDE FIRE EXTINGUISHER Extinguisher that emits carbon dioxide, clean, non-contaminating, odorless gas, to smother fire.

CARBON DIOXIDE PLUME Extent underground, in three dimensions, of injected carbon dioxide stream.

CARBON DIOXIDE STREAM Carbon dioxide that has been captured from emission source (e.g., power plant), plus incidental associated substances derived from source materials and capture process and any substances added to stream to enable or improve injection process that does not apply to any carbon dioxide stream that meets definition of hazardous waste under 40 CFR Part 261.

CARBON MONOXIDE Colorless, odorless, poisonous gas produced by incomplete fossil (i.e., carbon or carbon-based) fuel combustion.

CARBON REGENERATION UNIT Enclosed thermal treatment device used to regenerate spent activated carbon.

CARBON TETRACHLORIDE Carcinogenic compound consisting of one carbon atom and four chlorine atoms, once widely used as industrial raw material, as solvent and in production of CFCs.

CARBONACEOUS Of, relating to, rich in or yielding carbon.

CARBONACEOUS STAGE Stage of decomposition that occurs in biological treatment processes when aerobic bacteria, using dissolved oxygen, change carbon compounds to carbon dioxide.

CARBONATE Soil component that reacts with HCl of indeterminate origin (e.g., shell, rock).

CARBONATE ALKALINITY Measure of carbonate and bicarbonate anions in solution.

CARBONATE HARDNESS Measure of water hardness caused by presence of carbonate and bicarbonate anions.

CARBONATE MINERALS Carbonate group minerals.

CARBONATE ROCKS Rocks (e.g., limestone, dolostone) that are composed primarily of minerals (e.g., calcite, dolomite) containing carbonate ion.

CARBONATED SPRING Water where carbon dioxide gas under pressure has been dissolved.

CARBONATION 1. State of having carbon dioxide gas dissolved in liquid; 2. Amount or level of dissolved carbon dioxide remaining in solution.

CARBONATOR Device that carries out carbonation.

CARBONIZATION 1. Act or process of carbonizing; 2. Treatment of fossil fuels by heat in closed container to create coke or gas.

CARBOXYHEMOGLOBIN Hemoglobin where iron is bound to carbon monoxide instead of oxygen.

CARCINOGEN Substance, chemical or physical agent that can cause or aggravate cells to develop cancer.

CARGO Freight carried by ship, aircraft or motor vehicle.

CARGO CRANE Crane designed to transfer cargo between vessel and port.

CARLOAD Contents of automobile for trip.

CARLSON LOAD CELL Instrument for measuring pressure within mass of soil, rock or concrete or at interface between one and other.

CARNIVORE Flesh-eating or predatory organism.

CARNOT EFFICIENCY Theoretical maximum efficiency one can get when heat engine is operating between two temperatures.

CARPOOL Arrangement whereby several people travel together in same car in order to save costs, reduce pollution, etc.

CARRIER 1. Material where test substance is combined for administration to test system; 2. Liquid where active ingredient is dissolved or dispersed.

CARRINGTON LONGITUDE System of fixed longitudes rotating with Sun.

CARRY OVER Quantity of water which continues past inlet.

CARRYING CAPACITY 1. Amount of use recreation area can sustain without loss of quality; 2. Number of individuals of particular species that environment can support; 3. Maximum rate of flow channel, conduit or alternative hydraulic structure can take; 4. Electric load cable or fuse can take without overload.

CARRYING WASTE CONTAINER Receptacle of 35- to 50-gal capability which is taken with collector in back yard carry out service.

CARRYOVER 1. Damaging condition where water droplets are carried out of steam boiler along with dry steam; 2. Liquid or solid particles entrained in vapor evolved by boiling liquid.

CARRYOVER SOIL MOISTURE Moisture stored in soil within root zone during winter, at times when crop is dormant or before crop is planted.

CARTOGRAPHY Art and science of graphically representing features of Earth's surface.

CARTRIDGE FILTER Filter using disposable element.

CARTRIDGE FUSE Electrical fuse used to protect motors and branch circuits where higher amps or volt ratings are required.

CAS REGISTRATION NUMBER Number assigned by Chemical Abstract Service to identify chemical.

CASCADE 1. Waterfall or series of small waterfalls; 2. Stream or sequence of thing or things occurring.

CASCADE AERATOR Inducts air into water flow in order to oxidize iron and reduce dissolved gases.

CASCADING FAILURE Process in system of interconnected parts where failure of one or few parts can trigger failure of other parts.

CASE DRAIN LINE Line convecting fluid from housing to reservoir.

CASE STUDY Brief fact sheet providing risk, cost and performance information on alternative methods and other pollution prevention ideas, compliance initiatives, voluntary efforts, etc.

CASH ALLOWANCE Amount paid out in cash, instead of being reimbursed at later date.

CASH BUDGET Budget or plan of expected cash receipts and disbursements during period.

CASING 1. Pipe lining placed inside drilled hole to prevent collapse of walls of bore hole, to prevent pollutants from entering well and to house pump and pipes; 2. Decorative trim around door or window.

CASING BLOWS Blows, usually of 300 lb hammer fulling 18 in. on to soil sampler casing while making soil boring.

CASING HEAD Simple metal flange welded or screwed onto top of conductor pipe (also known as drive-pipe) or casing and forms part of wellhead system for well.

CASK Thick-walled container (usually lead) used to transport radioactive material.

CAST IRON PIPE Conduit which has had historic use as pressure pipe for transmission of water, gas and sewage and as water drainage pipe made from composite of alloys.

CAST-IN-PLACE CONCRETE Construction where walls and slabs are cast at site in formwork.

CASTELLATUS Cloud that displays at least in its upper part cumuliform protuberances having shape of turrets that give crenellated aspect.

CASUALTY INSURANCE Broadly encompasses insurance not directly concerned with life insurance, health insurance or property insurance.

CATALINA EDDY 1. Coastal eddy forms when upper level large-scale flow off Point Conception interacts with complex topography of Southern California coastline; 2. Counter clockwise circulating low pressure area that forms with its center in vicinity of Catalina Island that predominately occur between April and September with peak in June.

CATALYST Substance that changes speed or yield of chemical reaction without being consumed or chemically changed by chemical reaction.

CATALYTIC COMBUSTION SYSTEM Process that uses catalyst to speed desired oxidation reactions of fuel and so reduce formation of undesired products (i.e., pollutant).

CATALYTIC CONVERTER Emission control device incorporated into exhaust system containing catalyst for converting pollutants into less harmful ones.

CATALYTIC CRACKING Process to convert high-boiling, high-molecular weight hydrocarbon fractions of petroleum crude oils into more valuable gasoline, olefinic gases and other products.

CATALYTIC INCINERATOR Control device that oxidizes volatile organic compounds by using catalyst to promote combustion process.

CATALYTIC REACTION Chemical reaction where amount of catalyst isn't decreased although its existence is required for reaction.

CATARACT 1. Large waterfall; 2. Steep rapids in river; 3. Flood of water.

CATARACT ACTION Excavating backaction of rapid current of water on river bed.

CATASTROPHE Sudden and great disaster causing misfortune, destruction or irreplaceable loss extensive enough to cripple activities in area.

CATASTROPHIC COLLAPSE Unexpected and complete failure of overlying strata due to removal of underlying materials.

CATCH Number of fish captured, regardless if kept or released.

CATCH BASIN Receptacle designed to retain obstructive matter that would not readily pass through sewer.

CATCH FEEDER Irrigating ditch.

CATCHMENT Action of collecting water.

CATCHMENT AREA Area having common outlet (e.g., lake, river) for its surface runoff.

CATCHMENT BASIN 1. Unit watershed; 2. Area where all drainage water passes into common outlet (e.g., stream).

CATCHPIT Receptacle in drainage system where debris is deposited.

CATCHWATER Device which captures surface runoff for use as water supply.

CATEGORICAL National Weather Service precipitation descriptor for 80, 90 or 100% chance of measurable precipitation (0.01 inch).

CATEGORICAL EXCLUSION Class of actions that either individually or cumulatively would not have significant effect on human environment and therefore would not require preparation of environmental assessment or environmental impact statement required by Federal agencies.

CATEGORICAL INDUSTRIAL USER Industrial user subject to national categorical pretreatment standards.

CATEGORICAL PRETREATMENT STANDARD Technology-based effluent limitation for industrial facility discharging into municipal sewer system.

CATEGORICAL STANDARD Standard issued by regulator.

CATENA Series of related soils derived from uniform or similar parent material but exhibiting variations in type due to differences in topography or drainage.

CATENARY Curve described by flexible chain or rope if it is supported at each end and is acted upon only by no other forces than uniform gravitational force due to its own weight.

CATENARY SCAFFOLD Scaffold consisting of platform supported by two essentially horizontal and parallel ropes attached to structural members of building or other structure.

CATERPILLAR GATE Steel gate carried on crawler tracks used to control water flow through spillway.

CATHETOMETER Device used to measure small vertical distances, such as levels of liquid in tubes.

CATHODE Negative pole or electrode of electrolytic cell or system where positive current (i.e., negative terminal) of electricity flows outwards (i.e., electrons flow inwards) and attracts positively charged particles or ions (cations).

CATHODIC PROTECTION Electrical system for prevention of rust, corrosion and pitting of metal surfaces that are in contact with water or soil where low-voltage current is made to flow through liquid (water) or soil in contact with metal, making it cathode of electrochemical cell by concentrating corrosion on auxiliary anodic parts which are deliberately allowed to corrode instead of letting structure corrode.

CATHODIC PROTECTION TESTER Person that tests and evaluates data to determine effectiveness of both galvanic and impressed current systems and gather design data.

CATION Positively charged ion in electrolyte solution attracted to cathode under influence of difference in electrical potential.

CATION EXCHANGE CAPACITY Total capacity of soil to hold exchangeable cations.

CATION-EXCHANGE WATER SOFTENER Process of removing dissolved ionic contaminants in hard water and replacing with sodium ions.

CATIONIC POLYMER Polymer that includes covalently energized subunits that carry net positive charge.

CATIONIC SURFACTANT Surfactant (e.g., positively charged) that shows exceptional disinfectant properties.

CATWALK Elevated walkway.

CAULK Elastic material used to seal joints.

CAULKED JOINT Non-mechanical joint faces sealed with caulk.

CAULKING Sealing material used to seal joints between heterogeneous materials in many kinds of construction and manufacture.

CAULKING CHISEL Tool, with concave edge, used for caulking seams or joints.

CAULKING COMPOUND Soft, plastic, putty-like material that consists of pigment and vehicle.

CAULKING GUN Tool that carries prepackaged cartridges of caulk.

CAULKING HAMMER Hammer used to hit caulking chisel.

CAULKING RECESS Space between bell or hub of one piece of pipe and spigot inserted into it.

CAULKING TOOL Tool employed for driving caulking compound into seams and crevices.

CAUSEWAY Road raised, as to be above water, marshland etc.

CAUSTIC Capable of burning, corroding or destroying organic tissue.

CAUSTIC ALKALINITY Total alkalinity down to carbonate equivalence point (about pH 10.7 in fresh water).

CAUSTIC EMBRITTLEMENT Phenomenon where material of boiler becomes brittle due to accumulation of caustic substances.

CAUSTIC SCRUBBING Procedure for removal of sulfur dioxide and other acid species from gases in different industries.

CAUSTIC SODA Strong alkaline substance used in manufacturing of paint, detergents or drain cleaner.

CAUTION STAGE Stage that, when reached by rising stream, represents level where appropriate officials (e.g., county sheriff, civil defense officials, bypass gate operators) are notified of threat of possible flooding.

CAVE-IN To collapse inward or downward.

CAVING When flowing water action collapses stream banks by undermining.

CAVING SOIL Soil that will fall into uncased excavation.

CAVITATION 1. Formation, in fluid, of vapor bubbles that rapidly collapse ,due to turbulence or temperature that causes pressure in local zones of liquid to fall below vapor pressure; 2. Collapse of pockets or bubbles that drives water with such force that it can cause pitting and is usually accompanied by noise and vibration; 3. Starvation of pumps created by restriction that reduces supply.

CAVITATION DAMAGE Damage caused when partial vacuums formed in liquid by swiftly moving solid body (e.g., propeller) pit and wear away solid surfaces (e.g., metal, concrete) caused by implosion of bubbles of water vapor.

CAVITY Voids greater than diameter of core.

$CBOD_5$ Five-day measure of pollutant parameter carbonaceous biochemical oxygen demand (CBOD).

CDD AND D INSURANCE Comprehensive dishonesty, disappearance and destruction insurance coverage against employee dishonesty, loss of money and securities and theft of materials and equipment.

CEILING 1. Height from ground or sea level to overcast cloud cover; 2. Overhead surface of room; 3. Highest altitude where aircraft can safely maintain flight.

CEILING INSULATION Material, made to resist heat flow, that's installed involving conditioned area and unconditioned attic.

CEILOMETER Device using laser or other light source to determine height of cloud base that uses triangulation to determine height of spot of light projected onto base of cloud by measuring time required for pulse of light to be scattered back from cloud base.

CELERITY Speed with which perturbation to flow propagates through flow domain.

CELL 1. Smallest structural part of living matter capable of functioning as independent unit; 2. convection in form of single updraft, downdraft or updraft/downdraft couplet, typically seen as vertical dome or tower as in towering cumulus cloud; 3. Distinct one-two-or three dimensional model unit representing discrete portion of physical system with uniform properties assigned to it; 4. Compacted solid waste holes where waste is dumped, compacted and covered with layers of dirt on daily basis.

CELL HEIGHT Vertical distance between top and bottom of compacted solid waste enclosed by natural soil or cover material in sanitary landfill.

CELL PAIR Two two separators and reverse-kind membranes.

CELL ROOM Construction holding mercury electrolytic chlor-alkali cells.

CELL VOLUME DETERMINATION Used to estimate biomass of algae in aquatic systems; 2. Number of cells of any organism counted by using microscope and grid or counting cell.

CELL-PAIR RESISTANCE Electrical resistance of one electrodialysis cell pair.

CELL-TYPE INCINERATOR Incinerator with grate areas divided into cells, each of which has its own ash drop, under fire, air control and ash grate.

CELLULAR COFFERDAM Cofferdam comprised of interconnected cells that form watertight wall.

CELLULAR GRAVITY DAM 1. Dam which has outward appearance of gravity dam but is of hollow construction; 2. Dam constructed of concrete and/or masonry on outside but having hollow interior and relying on its weight for stability.

CELLULAR POLYISOCYANURATE INSULATION Insulation produced principally by polymerization of polymeric polyisocyanates, usually in presence of polyhydroxl compounds with addition of catalysts, cell stabilizers and blowing agents.

CELLULAR POLYSTYRENE INSULATION Foam composed principally of polymerized styrene resin processed to form homogeneous rigid mass of cells.

CELLULAR POLYURETHANE INSULATION Insulation composed principally of catalyzed reaction product of polyisocyanurates and polyhydroxi-compounds, processed usually with blowing agent to form rigid foam having predominantly closed cell structure.

CELLULAR-BLOCK MATTRESS Interconnected concrete blocks placed directly on streambank to prevent erosion.

CELLULOSE Complex carbohydrate that forms main constituent of cell wall in most plants and is important in manufacture of numerous products.

CELLULOSE DEBRIS Paper or wood based material laying directly on top of soil vulnerable to wood borers and termite attack.

CELLULOSE FIBER FIBERBOARD Insulation composed principally of cellulose fibers.

CELSIUS 1. Unit of temperature with 0 °C and 100 °C as freezing and boiling points of water, respectively; 2. To convert Fahrenheit temperature to Celsius, multiply it by $5/9$ and then subtract 32; 3. Modern name for centigrade scale.

CEMENT 1. Powdered substance, made with calcined lime and clay, that develops strong adhesive properties when mixed with water; 2. Mineral that binds together fragments in sedimentary rock.

CEMENT LINED PIPE Conduit internally coated with mortar like compound that acts as barrier between carbon steel pipe and process.

CEMENTED SOIL Soil where particles are held together by chemical agent, such as calcium carbonate, such that hand-size sample cannot be crushed into powder or individual soil particles by finger pressure.

CEMENTING When cement slurry is pumped into drilled hole and driven behind casing.

CEMENTITIOUS Densely packed and non-fibrous friable materials.

CENSUS Official count or enumeration of members of population.

CENTER Vertical axis of tropical cyclone, usually defined by location of minimum wind or minimum pressure.

CENTER LINE MARKINGS Yellow lines in middle of road to indicate passing and no passing zones.

CENTER OF BUOYANCY Point where resultant of all buoyant forces on immersed hull are assumed to act.

CENTER OF FLOTATION Center of gravity of water plane of vessel.

CENTER OF GRAVITY Point, near or within body, where its weight can be assumed to act when considering forces on body and its motion under gravity.

CENTER OF MASS Point, near or within body where object's mass can be assumed to be concentrated.

CENTER OF MOMENTS Actual point about which force causes rotation.

CENTER OF PRESSURE Point where total sum of pressure field acts on body, causing force to act through that point.

CENTER PIVOT IRRIGATION Automated sprinkler system involving rotating pipe or boom that supplies water to circular area of agricultural field through sprinkler heads or nozzles.

CENTER-TO-CENTER Distance between center of object and center of another object.

CENTERLINE Line that divides roadway between opposite moving traffic.

CENTERLINE TURNING RADIUS Circular arc formed by turning path of vehicle's outside front tire.

CENTI Prefix symbolizing 10^{-2}.

CENTIGRADE 1. Unit of temperature with 0 °C and 100 °C as freezing and boiling points of water, respectively; 2. To convert Fahrenheit temperature to Celsius, multiply it by 5/9 and then subtract 32.

CENTIGRAM Unit of mass equal to 1/100 of gram.

CENTILITER Unit of volume equal to 1/100 of litre.

CENTIMETER Unit of length equal to 1/100 of meter.

CENTIMETER BURST Solar radio burst in centimeter wavelength range.

CENTISTOKE Unit of kinematic viscosity equal to 1/100 of stoke.

CENTRAL AREA PRICING Pricing of tolls for vehicular traffic during certain days and hours.

CENTRAL BUSINESS DISTRICT Central area of city where there is concentration of certain retail and business activities.

CENTRAL COLLECTION POINT Location were generator of regulated medical waste consolidates wastes originally generated at various locations in his facility.

CENTRAL FLYWAY Important international migration route for many birds.

CENTRAL INCINERATOR Facility that combusts solid waste accumulated from many different sources.

CENTRAL ISLAND Land area surrounded by circular road.

CENTRAL LUBRICATION Where points are lubricated from one place or automatically.

CENTRAL MERIDIAN PASSAGE Passage of Active Region or other feature across longitude meridian that passes through apparent center of solar disk.

CENTRAL NERVOUS Toxicants that deaden central nervous system.

CENTRALIZED TRAFFIC CONTROL Form of railway signalling that consolidates train routing decisions.

CENTRALIZED WASTEWATER TREATMENT SYSTEM Managed system consisting of collection sewers and single treatment plant used to collect and treat wastewater from entire service area.

CENTRATE Water leaving centrifuge after most of solids have been removed.

CENTRIFUGAL COLLECTOR Mechanical system using centrifugal force to remove aerosols from gas stream or to remove water from sludge.

CENTRIFUGAL COMPRESSOR Compressor that achieves pressure rise by adding kinetic energy and/or velocity to continuous flow of fluid through rotor or impeller.

CENTRIFUGAL DEWATERING Removal of water by centrifugal action.

CENTRIFUGAL DRYING Partial drying, by spinning action, that removes excess water, oils or other liquids.

CENTRIFUGAL FAN Mechanical device for moving air or other gases.

CENTRIFUGAL FORCE Effect that tends to move object away from center of circle it is rotating about.

CENTRIFUGAL PUMP Pump that moves water by impeller fixed on rotating shaft that whirls water around, building up enough pressure to force water through discharge outlet and imparting energy to water.

CENTRIFUGAL PUMP CAPACITY Variable that changes directly with pump design, pump speed or impeller diameter.

CENTRIFUGAL PUMP HEAD Variable that varies with square of speed or impeller diameter.

CENTRIFUGAL PUMP POWER Variable that changes with cube of speed or impeller diameter.

CENTRIFUGAL SCREW PUMP Pump designed for handling thick sludge, pumping high viscosity liquids such as crude oils, latex, molasses.

CENTRIFUGAL SEPARATOR Apparatus that uses centrifugal force for separating suspensions consisting of two or more phases of different densities.

CENTRIFUGE Device where mixture of denser and lighter materials (normally dispersed in liquid) is separated by being spun about central axis at high speed.

CENTRIFUGE MOISTURE EQUIVALENT Weight percentage of water (with respect to dry weight) retained by initially saturated 1-cm-thick soil sample after it has been subjected to centrifugal force 1,000 times force of gravity for specific amount of time.

CENTRIFUGE VOLUME Volume of liquid and/or solid separated from liquid exposed to centrifugal force.

CENTRIPETAL DRAINAGE Drainage pattern where rivers drain in towards center of basin (i.e., spokes of wheel).

CENTROID Center of mass.

CERAMIC FILTER Filter dirt, debris and bacteria out of water using inert, inorganic material of small pore size.

CERAMIC-DISC VALVE Valve with two discs that control water flow by gliding back and forth in airtight seal that prevents leaking and ensures super smooth operation.

CERTIFICATE Document serving as evidence.

CERTIFICATE FOR PAYMENT Document verified by authorized professional of construction project that noted work has been completed and is approved for payment by general contractor.

CERTIFICATE OF ANALYSIS Document issued by Quality Assurance that confirms that regulated product meets its product specification.

CERTIFICATE OF COMPLIANCE Document certified by competent authority that supplied good or service meets required specifications.

CERTIFICATE OF INSURANCE Document used to provide information on specific insurance coverage.

CERTIFICATE OF OCCUPANCY Document issued by local government agency or building department certifying building's compliance with applicable building codes and other laws and indicating it to be in condition suitable for occupancy.

CERTIFICATE OF PRACTICAL COMPLETION Document that arks point where Contractor has completed his contractual obligations and can hand over to client.

CERTIFICATE OF TITLE State or municipal-issued document that identifies owner or owners of personal or real property.

CERTIFICATION Professional qualification that certifies person's ability.

CERTIFICATION EXAMINATION Process where professional demonstrates mastery of basic knowledge and skills through written, practical or simulator-based testing.

CERTIFICATION SIGNATURE Signatures of those who co-facilitated risk analysis signifying that methodology, processes and requirements were followed.

CERTIFIED APPLICATOR Person who is responsible for use or supervise use of pesticides.

CERTIFIED WATER RIGHT State-issued document that serves as legal evidence that approved application has been physically developed and water put to beneficial use that establishes: priority date, type of beneficial use and maximum amount of water that can be used.

CERTIFYING AGENCY Individual or agency designated by statute or by other governmental act, to certify conformity with relevant quality standards.

CERTIFYING OFFICIAL Agency official who is responsible for authorizing and granting certification per agency policy.

CESIUM-137 Radioactive isotope of caesium which is formed as one of more common fission products by nuclear fission of uranium-235 and other fissionable isotopes in nuclear reactors and nuclear weapons.

CESSPOOL Underground pit where sewage is held.

CETANE Aliphatic hydrocarbon (hexadecane) used as standard for diesel fuel.

CETANE IMPROVER Chemical which has effect of increasing diesel fuel's Cetane number.

CETANE INDEX Used as substitute for cetane number of diesel fuel calculated based on fuel's density and distillation range.

CETANE NUMBER Performance rating of diesel fuel corresponding to percentage of cetane in mixture with isocetane that would have same performance.

CFC's Fully halogenated paraffin hydrocarbons that contain only carbon, chlorine and fluorine, produced as volatile derivative of methane, ethane and propane.

CGS SYSTEM Variant of metric system based on centimetre as unit of length, gram as unit of mass and second as unit of time.

CHAIN GAUGE Gauge consisting chain, tape or alternative sort of line attached to your weight that's lowered to contact water surface, using its level read from calibrated staff or index.

CHAIN GRATE Type of mechanical stoker for furnace, where grate consists of endless chain that draws solid fuel into furnace as it rotates.

CHAIN OF LOCKS Series of interconnected hydraulic locks on waterway.

CHAIN PIPE VICE Pipes secured by high-tensile steel chain where loops over workpiece and is fastened tightly in place.

CHAIN PUMP Water pump consisting of series of buckets on chain, wheel or drum.

CHAIN REACTION 1. Nuclear reaction where particles produced by fission of one atom trigger fissions of other atoms; 2. Series of events, each one causing or influencing next.

CHAIN-BUCKET DREDGER Ladder-pail dredging machine.

CHALYBEATE SPRING Mineral spring waters containing salts of iron.

CHAMBER Any enclosed space occupying or similar to room.

CHAMFER To bevel or slope edge or corner.

CHANCE 1. National Weather Service precipitation descriptor for 30, 40 or 50% chance of measurable precipitation (0.01 inch); 2. When precipitation is convective in nature, term scattered is used.

CHANGE IN WORK Change to work as explained in contract files.

CHANGE OF AIR One complete change of all atmosphere contained within space or rooms and enclosed spaces.

CHANGE OF STATE Change from one phase to another (e.g., liquid to gas).

CHANGE ORDER Description of changes to be made to work order that describes new work to be done (and any modifications) and price to be paid for this new work.

CHANGEABLE MESSAGE SIGN Traffic sign, often electronic, used on roadways to give information about special events.

CHANGED CONDITIONS Previously unknown and unanticipated situations and circumstances that surface or are revealed after agreement is signed or contract is awarded.

CHANGED USE Major change from use routine approved in relation to registration of pesticide product.

CHANGES TO WORK Changes to work described or stipulated in signed contract and authorized beneath provisions of change order.

CHANNEL 1. Natural or artificial waterway which periodically or continuously contains moving water; 2. Narrow body of water between two land masses; 3. Navigable part of river.

CHANNEL ACCRETION Slow build-up of channel bottom or bank caused by sediment deposit.

CHANNEL AXIS Line joining mid-points of water surface of channel in serial cross-sections.

CHANNEL BARS Lowest prominent geomorphic features higher than channel bed.

CHANNEL BEND U-shaped channel that directs flow from conduit in manhole into main sewer.

CHANNEL COEFFICIENT Roughness (i.e., friction) factor to describe channel character.

CHANNEL CONTRACTION Reduction in channel river width.

CHANNEL DIVERSION Stream channel within hydrologic unit has altered to divert surface water from one location to another.

CHANNEL FLOW ACCRETION Slow increase in stream flow in dry weather caused by influent seepage.

CHANNEL FLOW DEPLETION Slow downstream decline in stream flow on account of seepage in stream to adjoining or underlying groundwater body.

CHANNEL GULLY Slow downstream decline in stream flow on account of seepage in stream to adjoining or underlying groundwater body.

CHANNEL INFLOW Water, that at any instant, is flowing into channel system form surface flow, subsurface flow, base flow and rainfall that has directly fallen onto channel.

CHANNEL INVERT Lowest point in channel.

CHANNEL LEAD Elongated opening in ice cover caused by water current.

CHANNEL LINE Path of most powerful flow of river or stream.

CHANNEL LOSS Reduction of water from channel by capillary action and percolation.

CHANNEL MARGIN DEPOSITS Narrow sand deposits that line channel banks.

CHANNEL MIGRATION Natural process that describes how flowing channel changes with time.

CHANNEL OF APPROACH Channel leading up to control structure (e.g., weir).

CHANNEL PATTERN Plan view of river reach that includes meandering, braiding, sinuosity or straight channels.

CHANNEL PIPE Clay pipe used primarily in manholes collecting branch drains.

CHANNEL PROCESS Channel behavior as it relates to erosion, transportation and sedimentation.

CHANNEL ROUGHNESS Measure of resistivity offered by channel to flow of water.

CHANNEL ROUTING Process of determining progressively timing and shape of flood wave at successive points along river.

CHANNEL SCOUR Erosion by flowing water and sediment on stream channel that results in removal of mud, silt and sand on outside curve of stream bend and bed material of stream channel.

CHANNEL SLOPE Vertical distance divided by horizontal distance along channel centerline.

CHANNEL SPRING Depression spring arising from bank of stream whose channel is below water table.

CHANNEL STABILIZATION Stable channel is neither progressively aggrading nor degrading or changing its cross-sectional area through time where stabilization techniques consist of bank protection and other measures that work to transform unstable channel into stable one.

CHANNEL STORAGE Volume of water at given time in channel, river reach or over flood plain of streams in drainage basin.

CHANNEL TERRACE Contour ridge, built of soil moved from its uphill side, that serves to divert surface water from field.

CHANNEL-PHASE RUNOFF Runoff in open channels.

CHANNELED HIGH WINDS Channeled air through constricted passages producing high winds.

CHANNELIZATION Modification of natural river channel that may include deepening, widening or straightening to permit water to move faster or to drain wet area for farming that's often done for flood control or for improved agricultural drainage or irrigation.

CHANNELIZED INTERSECTION Using islands to direct traffic.

CHANNELIZING LINE MARKINGS Longitudinal line markings located parallel and adjacent to traffic flow.

CHAPARRAL Region of shrubs, typically dry in summer and rainy in winter.

CHAR 1. To burn something to charcoal; 2. To burn slightly or superficially so as to affect color.

CHARACTERISTIC Any one of four categories used in defining hazardous waste: ignitability, corrosivity, reactivity and toxicity.

CHARACTERISTIC SPECIES Species that are special to or especially abundant in particular situation or biotope, which can usually be easily identified.

CHARACTERISTIC SPEED Physical quantity characterizing average speed of motion of particles inside single body or particle system at rest.

CHARACTERISTIC STRESS Stress in presumed production point or limit of proportionality of substance.

CHARACTERISTICS METHOD Numerical method where governing partial differential equations of mathematical model are transformed into characteristic (ordinary differential) equations.

CHARACTERIZATION OF ECOLOGICAL EFFECTS Part of ecological risk assessment that evaluates ability of stressor to cause adverse effects under given circumstances.

CHARACTERIZATION OF EXPOSURE Portion of ecological risk assessment that evaluates interaction of stressor with one or more ecological entities.

CHARCOAL Impure carbon obtained by destructive distillation of wood or other organic matter.

CHARGE 1. Scope of someone's responsibility; 2. Amount of money levied for service.

CHARGE FOR CONDITIONAL WATER SERVICE Amount that consumer who has one source of water supply, but who, for dependability, contracts for second source of supply is charged for readiness to supply up to stated quantity of water from second source.

CHARGE PRESSURE Comprises charge pump, relief valve and two check valves, where charge pump can replenish transmission loop.

CHARGED LINE Hose filled with water under pressure and ready to use.

CHARGING CHUTE Enclosed vertical chute with doors where waste material is dropped down and fed into incinerator.

CHARGING GATE Flat movable cover that closes opening of top-charging furnace.

CHARGING HOPPER Receptacle where materials are placed in enlarged opening that discharge into mixer.

CHARGING PORT Port in hydraulic head of injection pump where fuel passes to fill pumping chamber.

CHARLES' LAW Experimental gas law that describes how gases tend to expand when heated.

CHAROPHYTA Division of freshwater green algae.

CHARPY V-NOTCH TEST High strain-rate test that determines amount of energy absorbed by material when cracking.

CHART OF ACCOUNTS Listing of all accounts used in general ledger of organization.

CHARTER BUS Public transportation bus hired by organization to transport group of people to destination.

CHARTER SERVICE Transportation, fixed rate, contract to have exclusive use of vehicle, vessel or aircraft.

CHASE vertical passage cut into wall for ducts, pipes or wires.

CHECK Land between ridges that confines irrigation water.

CHECK CHAINS Heavy-duty chain and spring assembly made to avoid dump body kickup.

CHECK DAM Small dam/barrier designed to retard flow of water and sediment in channel, used especially to: decrease flow velocity, minimize channel scour and soil erosion and promote deposition of sediment.

CHECK FLOOD FOR BRIDGE SCOUR Flood from storm having flow rate exceeding scour design but not exceeding recurrence interval.

CHECK GATE Gate place in cutoff wall of irrigation ditch.

CHECK IRRIGATION Process of irrigation whereby water flows from one test to another along incline of land.

CHECK STRUCTURE Structure used to regulate upstream water surface and control downstream flow in canal.

CHECK VALVE Device, with hinged disc or flap that opens in direction of normal flow and is forced shut when flows attempt to go in opposite (i.e., backflow) direction of normal flow, that allows fluid or air to pass through it in only one direction.

CHECK-VALVE TUBING PUMP Water sampling tool.

CHECKED SIGNATURE Checked signatures verify that all probability estimates, inputs and outputs and their distributions, were entered correctly into event trees and that any other calculations, figures or tables have been checked.

CHECKERPLATE Metal plates on which non-slip pattern is stamped.

CHECKING Short, transverse cracks that typically do not extend completely through course depth.

CHECKLIST List of tasks to be completed, names to be consulted, conditions to be verified and similar.

CHELATE To form chelate compound by combining metal atom to form ring.

CHELATING AGENTS Any compound that reacts with metal ion to produce chelate.

CHEMAUTOTROPHIC Organism, such as bacterium, that obtains its energy from inorganic reactions using simple compounds, such as ammonia or hydrogen sulphide.

CHEMICAL Of or relating to material or processes not commonly found in nature or in particular product.

CHEMICAL AGENTS Any noxious substances intended for use in military or police operations.

CHEMICAL ANALYSIS Analytical chemistry studies and uses instruments and methods used to separate, identify and quantify matter.

CHEMICAL CASE Grouping of chemically similar pesticide active ingredients (e.g., salts/esters of same chemical) into chemical cases.

CHEMICAL COAGULATION Important unit process in water treatment for removal of turbidity using floc-forming substance.

CHEMICAL COMPOSITION Identities and relative numbers, of elements that make up any particular compound.

CHEMICAL COMPOUND Distinct and pure substance formed by union or two or more elements in definite proportion by weight.

CHEMICAL CONTAMINANT Chemicals toxic to plants and animals in waterways.

CHEMICAL CORROSION Gradual destruction of metal surface due to reaction of surface with substances in its external environment.

CHEMICAL DOSE Amount of chemical agent, expressed in milligrams, taken or absorbed by body.

CHEMICAL ELEMENT 1. Fundamental substance comprising one kind of atom; 2. Simplest form of matter.

CHEMICAL EMERGENCY Situation created by accidental release or spill of hazardous chemicals that poses threat to safety of workers, residents, environment or property.

CHEMICAL ENERGY Net potential energy liberated or absorbed during course of chemical reaction.

CHEMICAL ENGINEERING Branch of engineering that deals with design, construction and operation of industrial process plants and physical, chemical and biological processes to create substances or energy in useful and economic form.

CHEMICAL EQUIVALENT Weight in grams of substance that combines with or displaces one gram of hydrogen.

CHEMICAL FATE STUDIES Studies performed for characterization of physical, chemical and persistence factors that may be used to evaluate transport and transformation processes.

CHEMICAL FEEDER Mechanical device for applying chemicals to liquid.

CHEMICAL FOAM Stable solution of small bubbles containing carbon dioxide with lower density than oil or water and exhibits persistence for covering flat surfaces.

CHEMICAL GAUGING Determining discharges by introducing chemical at known rate into flowing water and determining quantity of chemical in stream at section far enough downstream to ensure thorough mixture of chemical with water.

CHEMICAL MANUFACTURING PLANT Facility engaged in production of chemicals.

CHEMICAL METAL CLEANING WASTE Any wastewater resulting from cleaning of any metal process equipment with chemical compounds, including, but not limited to, boiler tube cleaning.

CHEMICAL NAME Name of chemical compound that shows names of each of its elements or sub-compounds.

CHEMICAL OXYGEN DEMAND 1. Measure of oxygen-consuming capacity of inorganic and organic matter present in wastewater expressed as amount of oxygen consumed in mg/l; 2. Measure of oxygen required to oxidize all compounds, both organic and inorganic, in water and furnishes approximation of amount of organic and reducing material present.

CHEMICAL PRECIPITATION Formation of separable solid substance from solution, either by converting substance into insoluble form or by changing composition of solvent to diminish solubility of substance in it.

CHEMICAL PROTECTIVE CLOTHING Shields or isolates individuals from chemical, physical and biological hazards that may be encountered during hazardous materials operations.

CHEMICAL SLUDGE Sludge created by using chemical to treat wastewater.

CHEMICAL SOLUTION TANK Storage tank where chemicals are placed in solution prior to their introduction to water/wastewater.

CHEMICAL STRESSORS Chemicals released to environment through industrial waste, auto emissions, pesticides and other human activity that can cause illnesses and even death in plants and animals.

CHEMICAL STRUCTURE Arrangement of chemical bonds between atoms in molecule.

CHEMICAL SUBSTANCE Material with specific chemical composition.

CHEMICAL SYMBOL One (1) to three (3) letter international code for chemical element.

CHEMICAL THINNING Any forest thinning where unwanted trees are killed by chemical poisoning.

CHEMICAL TOILET Toilet or chamber pot which treats waste in place with chemical disinfectants and deodorizers, as in airplane lavatories.

CHEMICAL TRANSPORTATION EMERGENCY CENTER Provides information and other assistance to emergency responders.

CHEMICAL TREATMENT Any one of variety of technologies that use chemicals or variety of chemical processes to treat waste.

CHEMICAL WASTE LANDFILL Landfill where protection against danger of harm to health or surroundings from migration of PCBs to land, water or atmosphere is supplied.

CHEMICALLY PURE WATER Water consisting of chemical combination of two volumes of hydrogen with one of oxygen gas, containing absolutely nothing else.

CHEMISTRY MODEL Computer model used in air pollution investigations that simulates chemical and photochemical reactions of pollutants during their transport and diffusion.

CHEMNET Mutual aid network of chemical shippers and contractors that assigns contracted emergency response company to provide technical support if representative of firm whose chemicals are involved in incident is not readily available.

CHEMOSPHERE Portion of lower section of Earth's atmosphere where photochemical processes are prevalent.

CHEMOSTERILANT Chemical that controls pests by preventing reproduction.

CHEMOSYNTHETIC BACTERIA Organisms that use inorganic molecules as source of energy and convert them into organic substances.

CHEMOTROPHIC Movement or growth of organism in response to chemical stimulus.

CHEZY FORMULA Empirical formula that describes mean flow velocity of turbulent open channel flow.

CHIEF FACILITY OPERATOR Responsible for overall supervision and monitoring of performance of solid waste treatment facility and facility personnel.

CHILD RESISTANT PACKAGING 1. Packaging that protects children or adults from injury or illness resulting from accidental contact with or ingestion of residential pesticides that meet or exceed specific toxicity levels required by FIFRA regulations; 2. Term used for protective packaging.

CHILLER Device that generates cold liquid circulated through air-handling unit's cooling coil to cool air supplied to building.

CHILLING EFFECT Lowering of Earth's temperature because of increased particles in air blocking Sun's rays.

CHIMNEY DRAIN Vertical or inclined layer of pervious material in embankment to facilitate and control drainage of embankment fill.

CHINOOK Warm, dry, descending winds that occur on eastern side of Rocky Mountains.

CHINOOK ARCH Foehn cloud formation appearing as bank of altostratus clouds east of Rocky Mountains, heralding approach of chinook that forms in rising portion of standing waves on lee side of mountains.

CHIP CONTROL FILTER Filter intended to prevent only large particles from entering component immediately downstream.

CHIPPING Loosening of shallow rock by light blasting or air hammers.

CHIRONOMID Group of two-winged flying insects who live their larval stage underwater and emerge to fly about as adults.

CHIRP TEST Test that examines basic timing and voltages of both upstream and downstream ports during speed detection protocol.

CHISEL PLOWING Preparing croplands by using special implement that avoids complete inversion of soil as in conventional plowing that can leave protective cover or crops residues on soil surface to help prevent erosion and improve filtration.

CHLOR-ALKALI PROCESS Industrial process for electrolysis of sodium chloride.

CHLORACNE Form of acne caused by exposure to halogenated aromatic hydrocarbons.

CHLORAMINES Any of class of unstable compounds of nitrogen and chlorine.

CHLORDANE Organochlorine insecticide no longer registered for use in U.S.

CHLORIDE Any salt of hydrochloric acid, such as sodium chloride or any binary compound of chlorine and another element or radical.

CHLORIDE INDEX Amount of chlorides in wastewater relative to total amount of chlorides in water supply.

CHLORINATED COPPERAS Generated by chlorinating solution of ferrous sulfate, produces solution of ferrous sulfate and ferric chloride.

CHLORINATED HYDROCARBONS 1. Chemicals containing only chlorine, carbon and hydrogen that linger in environment and accumulate in food chain (e.g., DDT, aldrin, dieldrin, heptachlor, chlordane, lindane, endrin, Mirex, hexachloride, toxaphene); 2. Chlorinated organic compounds including chlorinated solvents (e.g., dichloromethane, trichloromethylene, chloroform).

CHLORINATED ISOCYANURATES Family of chlorine disinfectants used commonly to destroy germs and algae in pools and spas.

CHLORINATED LIME Mix of slaked lime and chlorine used as strong bleaching agent and disinfectant.

CHLORINATED SOLVENT Organic solvent containing chlorine atoms (e.g., methylene chloride, 1,1,1-trichloromethane) with uses including aerosol spray containers, in highway paint and dry cleaning fluids.

CHLORINATION Application of chlorine to drinking water, sewage or industrial waste to disinfect or oxidize undesirable compounds and other biological or chemical results (i.e., aiding coagulation, controlling tastes and odors).

CHLORINATOR Device that adds chlorine, in gas or liquid form, to water or sewage to kill infectious bacteria.

CHLORINE Toxic, green, gaseous chemical element with atomic number of 17.

CHLORINE CONTACT CHAMBER Baffled basin that provides sufficient detention time of chlorine contact with wastewater for disinfection to occur.

CHLORINE DEMAND Quantity of chlorine reduced or converted to inert or less active forms of chlorine by substances in water.

CHLORINE DIOXIDE Reddish-yellow explosive gas used as water disinfectant and to bleach flour and wood pulp.

CHLORINE HYDRATE When chlorine gas comes in touch with water at 49 °F or lower, yellowish ice that forms in chlorinator.

CHLORINE REQUIREMENT Amount of chlorine needed for particular purpose.

CHLORINE ROOM Place where chlorination equipment is housed, with facilities for protecting employees and gear.

CHLORINE-AMMONIA PROCESS Use of chlorine and ammonia to water or of ammonia to water containing chlorine, in such ratios as to supply joined remaining chlorination.

CHLORINE-CONTACT CHAMBER Part of water treatment plant where effluent is disinfected by chlorine.

CHLOROFLUOROCARBONS Class of volatile compounds consisting of carbon, chlorine and fluorine (i.e., freons) that are inert, non-toxic and easily liquefied, with long lifetime, that are not destroyed in lower atmosphere and drift into upper atmosphere destroying stratospheric ozone.

CHLOROGANIC Organic compounds combined with chlorine.

CHLOROPHENOLS Any of several chlorinated derivatives of phenol, some of which are used as antiseptics or disinfectants.

CHLOROPHENOXY Class of herbicides that may be found in domestic water supplies and cause adverse health effects.

CHLOROPHYLL Any of group of green pigments that are found in chloroplasts of plants and in other photosynthetic organisms such as cyanobacteria.

CHLOROPHYTA Unicellular or colonial green algae that are either free-floating, anchored, or present in large aggregations on stagnant water.

CHLOROPLAST Organelle found in cells of green plants and in photosynthetic algae, where photosynthesis takes place.

CHLOROSIS 1. Yellowing of normally green plant parts caused by disease, lack of nutrients, absence of chlorophyll or various air pollutants; 2. Anaemia, due to deficiency of iron, characterized by yellow-green coloration of skin.

CHOKING Fluid flow through constriction area that may cause backwater.

CHOLERA Any of several acute infectious diseases of humans and domestic animals, caused by Vibrio cholera bacterium through ingestion of contaminated water or food.

CHOLINESTERASE Enzyme found in animals that regulates nerve impulses by inhibition of acetylcholine that causes variety of acute symptoms including: nausea, vomiting, blurred vision, stomach cramps and rapid heart rate.

CHOROLOGY 1. Study of spatial distribution of organisms; 2. Science of determining order where places or regions occur or are constructed.

CHROMATOGRAPHY Any of various techniques for qualitative or quantitative separation of components of mixtures of compounds.

CHROMIUM Metallic chemical element with atomic number 24, hard white metal used in stainless steel and other alloys.

CHROMOSPHERE Layer of solar atmosphere above photosphere and beneath transition region and corona.

CHROMOSPHERIC EVENTS Flares that are just chromospheric events without centimetric bursts or ionospheric effects.

CHRONIC Stimulus that lingers or continues for relatively long period of time.

CHRONIC EFFECT 1. Adverse effect on human or animal where symptoms recur frequently or develop slowly over long period of time; 2. Process by which small amounts of toxic substances are taken into body over extended period.

CHRONIC HEALTH EFFECT Possible result of exposure over many years to drinking water contaminant at levels above its maximum contaminant level.

CHRONIC TOXICITY Capacity of substance to cause long-term poisonous health effects in humans, animals, fish and other organisms.

CHRONIC TOXICITY TEST Performed to determine long term toxicity potential of toxicants or other stressors, commonly to aquatic organisms.

CHRONON Unit of time equal to time that photon would take to traverse diameter of electron that's approximately 10^{-24} of second.

CHRONOTHERM Thermostat and clock combination that can activate furnace or air-conditioner system.

CHRYSENE Polycyclic aromatic hydrocarbon.

CHUNK SAMPLE METHOD Determining weight per cubic measure of soil in place.

CHUNK UP Clean up and stack debris after logging area.

CHURN DRILL Large drilling machine that bores large diameter holes in ground.

CHUTE 1. Portion of spillway between gate/crest structure and terminal structure, where open-channel flow conditions will exist; 2. Conduit for conveying free-flowing materials at high velocity to lower elevations; 3. Framework, trough or tube, upon or where objects are made to slide from higher to lower level or where water passes to wheel.

CHUTE GATE Opening in tailgate of dump truck to permit restricted flow during dumping.

CHUTE-FED INCINERATOR Incinerator charged through chute that extends two or more floors above it.

CIENAGA Marshy area where ground is wet due to presence of seepage or springs.

CIENEGA Marshy spring where groundwater bubbles to surface.

CILIATES Any of many protozoa, of phylum Ciliophora, that have many cilia.

CINDER 1. Partially or mostly burnt material that results from incomplete combustion of coal, wood, etc.; 2. Ember; 3. Slag from metal furnace.

CIPOLLETTI WEIR Contracted weir of trapezoidal shape where sides of notch are given slope of 1 horizontal to 4 vertical.

CIRCADIAN Of, relating to or showing rhythmic behaviour with period of 24 hours.

CIRCADIAN RHYTHM Internal body clock that regulates (roughly) 24-hour cycle of biological processes in animals and plants.

CIRCANNUAL Occurring or having cycles of approximately annual periodicity.

CIRCLE 1. Two-dimensional geometric figure, line, consisting of set of all those points in plane that are equally distant from given point (center); 2. Rotary table that supports blade of grader and modulates its angle.

CIRCLE OF INFLUENCE Circular outer edge of depression produced in water table by pumping of water from well.

CIRCLE REVERSE Mechanism that alters angle of grader blade.

CIRCUIT 1. Complete path of electric current, including generating apparatus or other source; 2. Specific segment or section of complete path.

CIRCUIT BREAKER Safety device in electrical circuit that automatically shuts off circuit when it becomes overloaded and can be manually reset.

CIRCUIT MILE For single circuit electric power transmission lines, equal to geographic miles or pole miles.

CIRCUIT PROTECTIVE CONDUCTOR System of conductors joining together all exposed conductive parts and connecting them to main earthing terminal.

CIRCUIT PROTECTOR Device designed to protect electric circuit.

CIRCUIT VENT Vent that serves multiple fixtures on horizontal branch, in single story dwelling.

CIRCULAR INTERSECTION Intersection where traffic passes counterclockwise around center island.

CIRCULAR-TYPE CELLULAR COFFERDAM Construction built of interlocking steel-sheet piling comprising ring-shaped cells joined with joining arcs.

CIRCULATING FLOW Sum of vehicles from different movements passing in front of adjacent upstream splitter island.

CIRCULATING PUMP Centrifugal pump used to circulate gases, liquids or slurries in closed circuit.

CIRCULATION Flow/movement, of fluid (e.g., water, air) in or through given area or volume.

CIRCULATION AREA Area for pedestrian movement.

CIRCULATION PIPE Pipe forming part of primary or secondary circuit of hot-water system.

CIRCULATION TIME-SPACE Difference between total available time and time-space used by pedestrians as they wait to cross roadway.

CIRCULATORY ROADWAY Roundabout requiring vehicles entering to yield.

CIRCUMFERENTIAL FLOW Flow parallel to circumference of storage tank which is circular in plan.

CIRCUMFERENTIAL REINFORCEMENT Reinforcement perpendicular to longitudinal axis of pipe.

CIRCUMNEUTRAL 1. Water with pH between 5.5 and 7.4; 2. PH modifier used in U.S. Fish and Wildlife Service wetland classification system.

CIRCUMPOLAR VORTEX Large-scale cyclonic circulation in middle and upper troposphere centered generally in polar region.

CIRQUE Bowl-like depression carved into mountainside or mountaintop by ice at head of glacier.

CIRRIFORM High altitude ice clouds with very thin wispy appearance.

CIRROCUMULUS Cirriform cloud characterized by thin, white patches, each of which is composed of very small granules or ripples at high altitudes (i.e., 20,000-40,000 feet).

CIRROSTRATUS Cloud of class characterized by composition of ice crystals and often by production of halo phenomena and appearing as whitish and usually somewhat fibrous veil, often covering whole sky and sometimes so thin as to be hardly discernible at high altitudes (i.e., 20,000-40,000 feet).

CIRRUS 1. High-level cloud type characterised by white, delicate filaments or wisps, of white (or mostly white) patches or of narrow bands, found at altitude of above 7,000 metres; 2. Thin tendril-like appendage.

CIRRUS CLOUDS High-level clouds (16,000 feet or higher), composed of ice crystals and appearing in form of white, delicate filaments or white or mostly white patches or narrow bands.

CISTERN Small tank (usually covered) or storage facility used to store water for home or farm.

CITY AND REGIONAL PLANNING Property and community planning that includes regions bigger than one community.

CIVIL ACTION Lawsuit between private parties not criminal in nature.

CIVIL AUTHORITY CLAUSE Insurance policy provision which outlines how loss of business income coverage applies when government entity denies access to covered property.

CIVIL DAWN Time of morning at which Sun is 6° below horizon.

CIVIL DEFENSE AGENCY State and/or local agency responsible for emergency operations, planning, mitigation, preparedness, response and recovery for all hazards.

CIVIL DUSK Time at which Sun is 6° below horizon in evening.

CIVIL EMERGENCY MESSAGE Message issued by National Weather Service in coordination with federal, state or local government to warn general public of non-weather related time-critical emergency which threatens life or property (e.g., nuclear accident, toxic chemical spill).

CIVIL ENGINEER Professional engineer focusing on public works and heavy engineering works.

CIVIL ENGINEERING CONTRACTOR Contractor who focuses on construction.

CIVIL LAW Body of law dealing with private relations between members of community.

CIVIL LAW DOCTRINE OR RULE Rule or law pertaining to disposal of drainage waters, under where owner of higher land has right or easement to dispose of surplus or excess waters from his lands to lower lands, unobstructed by owners thereof.

CIVIL TWILIGHT Period where sun is less than 6° below horizon in morning and evening.

CL Abbreviation on climate outlook maps issued by CPC to indicate areas where equal chances of experiencing below-normal, normal and above-normal conditions are possible.

CLACK VALVE Valve, that when raised from its seat, falls with clacking sound.

CLADISTICS Approach to biological systematics where organisms are grouped based upon synapomorphies only.

CLADOGRAM Branching treelike graphical representation of phylogenetic relationships between organisms showing which taxa have branched from common ancestors.

CLADOPHORA Filamentous green alga.

CLAIM 1. Demand of ownership made for something; 2. Demand compensation or damages through courts.

CLAIMANT 1. Party who initiates lawsuit before court; 2. Person or entity contracted to furnish labor, materials or equipment into contractor or subcontractor to be used in contract.

CLAM SHELL GATE High pressure regulating gate consisting of two curved leaves which open and close over end of conduit used for free discharge into air with minimal cavitation damage.

CLAMSHELL Dredging bucket with hinges like shell of clam.

CLAPOTIS Lapping of water against surface.

CLARIFICATION Clearing action that occurs during wastewater treatment when solids settle out that's often aided by centrifugal action and chemically induced coagulation in wastewater.

CLARIFICATION DRAWING Graphic interpretation of drawings or other contract documents issued by architect as part of addendum, modification, change order or field order.

CLARIFIED SEWAGE 1. Product disposal plant from which effluent appears; 2. Sewage from which suspended matter partially or completely removed.

CLARIFIER Tank where solids settle to bottom and are subsequently removed as sludge.

CLASS Set/category of things having some property/attribute in common and differentiated from others by kind, type or quality.

CLASS A AIRSPACE Generally, airspace from 18,000 feet mean sea level up to and including flight level 600, including airspace overlying waters within 12 nautical miles of coast of 48 contiguous states and Alaska.

CLASS A BRIDGE Bridge located on federal-aid highway where examination is required every 2 years and in-depth examination every 6 years.

CLASS A EXPLOSIVES Possessing detonating (e.g., dynamite) hazard.

CLASS A FIRE Fire involving combustible materials (excluding liquids, gases).

CLASS A WILDFIRE Fire of 0.25 acres or less.

CLASS B AIRSPACE Generally, airspace from surface to 10,000 feet mean sea level surrounding nation's busiest airports in terms of airport operations or passenger enplanements.

CLASS B BRIDGE Bridge not located on federal-aid highway used by public where examination is required every 3 years.

CLASS B EXPLOSIVES Possessing flammable (e.g., propellant explosives) hazard.

CLASS B FIRE Fire involving flammable liquids, gases, oils, paints, fats or greases.

CLASS B WILDFIRE Fire greater than 0.25 acres but less than 10 acres.

CLASS C AIRSPACE Generally, airspace from surface to 4,000 feet above airport elevation above mean sea level surrounding those airports that have operational control tower, are serviced by radar approach control and have certain number of IFR operations or passenger enplanements.

CLASS C BRIDGE Bridge not located on federal-aid highway that can be designated as being private or operating bridge and could be gated or closed to public, if desired.

CLASS C EXPLOSIVES Certain types of manufactured articles containing Class and/or Class B explosives as components, but in restricted quantities.

CLASS C FIRE Fire involving energized electrical equipment.

CLASS C WILDFIRE Fire greater than 10 acres but less than 100 acres.

CLASS D AIRSPACE Generally, airspace from surface to 2,500 feet above airport elevation above mean sea level surrounding those airports that have operational control tower.

CLASS D FIRE Fire involving combustible metal.

CLASS D WILDFIRE Fire greater than 100 acres but less than 300 acres.

CLASS E AIRSPACE Controlled airspace that's not Class A, Class B, Class C or Class D.

CLASS E WILDFIRE Fire greater than 300 acres but less than 1,000 acres.

CLASS F WILDFIRE Fire greater than 1,000 acres but less than 5,000 acres.

CLASS G AIRSPACE Airspace that extends from surface to base of overlying Class E airspace.

CLASS G WILDFIRE Fire of 5,000 acres or more.

CLASS H WELLS Wells that inject fluids that are brought to surface for enhanced recovery of oil or natural gas.

CLASS I Railroad having annual carrier operating revenues of $250 million or more in 1991 dollars.

CLASS I AREA Geographic area designated by Clean Air Act where only small amount or increment of air quality deterioration is permissible.

CLASS I SUBSTANCE One of several groups of chemicals with ozone depletion potential of 0.2 or higher.

CLASS I TWO-LANE HIGHWAYS Two-lane highway where motorists are expected to travel at high speeds.

CLASS I WELLS Technologically sophisticated wells that inject hazardous waste, non-hazardous industrial waste or municipal wastewater into deep, isolated rock formations below lowermost underground source of drinking water.

CLASS II Railroad in United States has operating revenue greater than $20.5 million but less than $277.7 million in 2004 dollars.

CLASS II SUBSTANCE Substance with ozone depletion potential of less than 0.2.

CLASS II TWO-LANE HIGHWAYS Two-lane highways where motorists are not expected to travel at high speeds.

CLASS II WELLS Wells (e.g., salt water disposal, enhanced recovery, hydrocarbon storage) that inject brines and other fluids associated with oil and gas production or storage of hydrocarbons.

CLASS III TWO-LANE HIGHWAYS Two-lane highways that are special cases serving moderately developed areas.

CLASS III WELLS Wells that inject fluids associated with solution mining of minerals (salt solution, in-situ leaching of uranium and sulfur).

CLASS IV WELLS Wells that inject hazardous or radioactive wastes into or above underground source of drinking water and are banned unless authorized under federal or state groundwater remediation project.

CLASS V INJECTION WELL Shallow well used to place variety of fluids at shallow depths below land surface, including domestic onsite wastewater treatment system serving more than 20 people.

CLASS V WELLS Wells not included in Classes I to IV and Class VI that inject non-hazardous fluids into or above underground source of drinking water and are typically shallow, onsite disposal systems; however, this class also includes some deeper injection operations.

CLASS VI WELLS Wells that inject carbon dioxide for purposes of long-term storage known as CO_2 geologic sequestration.

CLASS, PIPE AND FITTINGS Working pressure rating of specific pipe (i.e., cast iron, ductile iron, asbestos cement and some plastic pipe) for use in water distribution systems that includes allowances for surges.

CLASSIFICATION Grouping of subjects or materials according to shared qualities or characteristics.

CLASSIFICATION YARD Railroad yard where incoming freight trains are sorted into various sidings.

CLASSIFIED EXCAVATION Excavation where there are separate prices for common excavation and for rock excavation.

CLASSIFIED PRODUCT Merchandise recorded and bearing label of authorized lab.

CLASSIFIER Device capable of separating materials into defined qualities or characteristics.

CLASSIFIER RAKE Machine for separating coarse and fine particles of granular substances briefly suspended in water.

CLASTIC Rock or sediment composed principally of broken fragments that are derived from preexisting rocks which have been transported from their place of origin (e.g., sandstone).

CLAUS SULFUR RECOVERY PLANT Includes reaction furnace as first processing step and tail gas incinerator as its last.

CLAW BUCKET Pail with positive clamping jaws for garbage pickup, snow removal, debris clean-up, etc.

CLAY Fine-grained soil or fine-grained portion of soil that passes No. 200 United States Standard sieve and can be made to exhibit plasticity (putty-like properties) within range of moisture contents and that exhibits considerable strength when air-dry.

CLAY CONTENT 1. Segment of steel conduit dropped into clay having cord drill; 2. Bit fitted to suction conduit of dredge.

CLAY CUTTER 1. Segment of steel conduit dropped into clay having cord drill; 2. Bit fitted to suction conduit of dredge.

CLAY FILTRATION Refining procedure to adsorb tiny solids.

CLAY LOAM Soil mixture that contains more clay than other types of rock or minerals.

CLAY MINERAL ANALYSIS Separation of clay sized fraction from sample.

CLAY MINERALS Any of group of minerals that occur as minute sheetlike or fibrous crystals in clay.

CLAY PIPE Pipe manufactured from clay and combusted in kiln.

CLAY SIZE Portion of soil finer than 0.002 mm (0.005 mm in some cases).

CLAY SIZE FRACTION Portion of soil finer than 0.002 mm (0.00008 in.).

CLAY SOIL Soil material containing more than 40% clay, less than 45% sand and less than 40% silt.

CLAY SPADE Tool used for trenching or digging clay, gravel, hard-packed earth, etc.

CLAYPAN Compact stratum of partially permeable material rich in clay.

CLEAN 1. Having relatively few impurities; 2. Removal of dirt; 3. Sand or gravel lacking binder.

CLEAN AGGREGATE Sand or gravel, free from clay or silt.

CLEAN AIR STANDARDS Federal law,enforced by Clean Air Act, designed to control air pollution on national level.

CLEAN COAL TECHNOLOGY Any technology not in widespread use prior to Clean Air Act Amendments of 1990 that will achieve significant reductions in pollutants associated with burning of coal.

CLEAN FILL Uncontaminated inert solid material.

CLEAN FUELS Blends or substitutes for gasoline fuels, including compressed natural gas, methanol, ethanol and liquefied petroleum gas.

CLEAN RIVER River that gives no sensible evidence of pollution and may provide drinking water.

CLEAN ROOM Room or environment controlled in such way as to minimize airborne particulate matter.

CLEAN WATER ACT Abbreviation for Federal Water Pollution Control Act of 1948 that establishes basic structure for regulating discharges of pollutants into waters of united states and regulating quality standards for surface waters.

CLEAN WATER STANDARD Enforceable limitation, control, condition, prohibition, standard or ocher requirement.

CLEANABILITY Measure of degree to which something is cleanable and maintain its service life.

CLEANABLE Filter element that can be restored.

CLEANER TECHNOLOGIES SUBSTITUTES ASSESSMENT 1. Document that systematically evaluates relative risk, performance and cost trade-offs of technological alternatives; 2. Serves as repository for all technical data (including methodology and results) developed by pollution prevention or education project.

CLEANING EYE conduit fitting with removable plug which provides access for inspection or cleaning of conduit run.

CLEANING OR ETCHING Act or process to generate desirable surface finish by chemical solution.

CLEANING WATER Process water used to clean surface of product or equipment.

CLEANLINESS LEVEL Standardized method of measuring and describing cleanliness of fluid.

CLEANOUT Opening or drain leading into plumbing system to provide location to insert snake or other implement for cleaning.

CLEANOUT DOOR Opening in side of tank usually at ground level and covered by plate to provide access for removal of sediments from bottom of tank.

CLEANUP Actions taken to deal with release or threat of release of hazardous substance that could affect humans and/or environment.

CLEAR AIR TURBULENCE Sudden severe turbulence occurring in cloudless regions that causes violent buffeting of aircraft.

CLEAR AND GRUB Process of clearing site of all vegetation prior to site work and building construction.

CLEAR CUT Harvesting all trees in one area at one time that can encourage fast rainfall or snowmelt runoff, erosion, sedimentation of streams and lakes and flooding and destroys vital habitat.

CLEAR DISTANCE OF BOLTS Distance between adjacent bolt hole edges.

CLEAR END DISTANCE OF BOLTS Distance between end of member and edges of bolt hole.

CLEAR ICE Thin coating relatively transparent of ice on terrestrial objects, caused by rain that freezes on impact.

CLEAR RUNOUT AREA Total roadside border area available for safe use by errant vehicles.

CLEAR SLOT Local region of clearing skies or reduced cloud cover, with respect to severe thunderstorms, indicating intrusion of drier air that's often seen as bright area with higher cloud bases on west or southwest side of wall cloud.

CLEAR SPAN Unsupported distance between two inside surfaces of span supports.

CLEAR STORAGE DISTANCE Distance available for vehicle storage measured between 6 feet from rail nearest intersection to intersection stop line or normal stopping point on highway.

CLEAR TITLE Full ownership of something, usually land, free of any encumbrances.

CLEAR WELL Reservoir for storing filtered water of sufficient quantity to prevent need to vary filtration rate with variations in demand used to provide chlorine contact time for disinfection.

CLEAR ZONE Roadside border area where vehicle leaving road can travel safely.

CLEAR-AIR TURBULENCE Erratic movement of air masses in absence of any visual cues, especially such movement that causes unexpected turbulence to aircraft.

CLEAR-WATER RESERVOIR Tank or reservoir of filtered water for filter backwashing.

CLEARANCE Procedure used to establish, under tightly controlled discipline, safe environment for maintenance, repair or inspection that includes systematically isolating pertinent equipment from all sources of hazardous energy (e.g., hydraulic, electrical, mechanical, pneumatic, chemical), attaching safety tags or locks to appropriate controls and written statement that documents isolation of equipment.

CLEARANCE INTERVAL Interval for red traffic signal.

CLEARANCE LOST TIME Time between signal phases when intersection is not used by any traffic.

CLEARANCE TIME Time loss at stop other than passenger dwell times.

CLEARCUT Area of forest cleared of trees.

CLEARING 1. Area devoid of vegetation (e.g., trees, shrubs, brush, stumps, exposed roots, down timber, branches, grass, weeds); 2. Removal of all rubbish and all other objectionable material; 3. Open space in fog.

CLEARING AWAY Remove excessive topsoil from site.

CLEARWAY Area projecting from end of runway kept clear of obstructions exceeding height that increases with distance from end of runway.

CLIENT Customer, buyer or receiver of goods or services.

CLIENT AGENCY Public fire service or wildlands management agency, federal or non-federal, which requires and uses National Weather Service fire and forestry meteorological services.

CLIMATE Composite or generally prevailing weather conditions of region, throughout year, averaged over series of years.

CLIMATE CHANGE Non-random change in climate measured over several decades or longer that may be due to natural or human-induced causes.

CLIMATE DIAGNOSTIC CENTER Advance national capabilities to interpret causes of observed climate variations and to apply this knowledge to improve climate models and forecasts and develop new climate products that better serve needs of public and decision-makers.

CLIMATE DIAGNOSTICS BULLETIN Monthly CPC Bulletin reports on previous months' status of ocean-atmosphere climate system and provides various seasonal ENSO-related outlooks.

CLIMATE DIAGNOSTICS CENTER Mission of NOAA's Climate Diagnostics Center is to identify nature and causes for climate variations on time scales ranging from month to centuries.

CLIMATE MODEL Mathematical model for quantitatively describing, simulating and analyzing interactions between atmosphere and underlying surface (e.g., ocean, land, ice).

CLIMATE OUTLOOK Issued by CPC gives probabilities that conditions, averaged over specified period, will be below-normal, normal or above-normal.

CLIMATE PREDICTION CENTER One of several centers under National Centers for Environmental Prediction part of National Weather Service in National Oceanic and Atmospheric Administration that serves public by assessing and forecasting impacts of short-term climate variability, emphasizing enhanced risks of weather-related extreme events, for use in mitigating losses and maximizing economic gains.

CLIMATE SYSTEM System consisting of atmosphere (gases), hydrosphere (water), lithosphere (solid rocky part of Earth) and biosphere (living) that determine Earth's climate.

CLIMATIC CYCLE Slowly varying change of climate about mean that recurs with some regularity but is not necessarily periodic.

CLIMATIC PROVINCE Region of earth's surface characterized by essentially homogeneous climate.

CLIMATIC VARIATION Change in statistical distribution of weather patterns when that change lasts for extended period of time.

CLIMATIC YEAR Continuous 12-month period during which complete annual weather cycle.

CLIMATOLOGICAL OUTLOOK Based upon climatological statistics for region (abbreviated as CL on seasonal outlook maps) that indicates if climate outlook has equal chance of being above normal, normal or below normal.

CLIMATOLOGY Science that deals with phenomena of climates or climatic conditions.

CLIMAX 1. Culmination of ecological development, whereby species are in equilibrium with their environment; 2. Culmination or acme.

CLIMAX FOREST Type of climax community dominated by trees, representing culminating stage of natural succession for its specific locality.

CLIMAX SPECIES Plant species that will remain essentially unchanged in terms of species composition for as long as site remains undisturbed.

CLIMBING LANE Roadway lane that allow slower travel for large vehicles ascending steep grade.

CLIMOMETER Instrument that measures angles of inclination used to measure cloud ceiling heights.

CLOGGING INDICATOR Visual or electrical indication for filter element replacement.

CLONING 1. Obtaining group of genetically identical cells from single cell; 2. Making identical copies of gene.

CLOSE COUPLED PUMP Pump with impeller on same shaft as electric motor that drives pump.

CLOSE NIPPLE Conduit length of nipple is usually specified by overall length with threads.

CLOSE TIMBERING Setting of timber sets and lagging very close together when shaft sinking or tunneling through very loose ground or crushed coal in thick seams.

CLOSED BASIN Basin, without surface outlet for precipitation, draining to some depression or pond where water is lost only by evaporation or percolation.

CLOSED BASIN LAKE FLOODING 1. Flooding that occurs on lakes with either no outlet or relatively small one where water may stay at flood stage for weeks, months or years; 2. Seasonal increases in rainfall cause lake level to rise faster than it can drain.

CLOSED BURNER Sealed-in burner that supplies all air for combustion through burner itself.

CLOSED CANOPY Dense growth of trees where top branches and leaves form ceiling or canopy, that light can barely penetrate to reach forest floor.

CLOSED CENTER VALVE Valve where all ports are closed in center position.

CLOSED CENTRIFUGAL PUMP Centrifugal pump having its impeller built with vanes enclosed within circular disks.

CLOSED COMMUNITY Intentionally limits links with outsiders and outside communities.

CLOSED CONDUIT Pipe used to convey liquid flow.

CLOSED CONTAINER Container sealed using lid or alternative seal that neither liquid nor vapor will escape under normal temperatures.

CLOSED COOLING WATER SYSTEM Cooling water is cooled by chiller or another heat exchanging method in loop.

CLOSED IMPELLER 1. rotor or rotor blade; 2. One that impels, as rotating device used to force fluid in desired direction under pressure.

CLOSED JOINT Deck joint that prevents debris passage.

CLOSED LOOP Hydraulic circuit where pump output is connected to hydraulic pump inlet.

CLOSED LOW Low pressure area with distinct center of cyclonic circulation which can be completely encircled by one or more isobars or height contour lines.

CLOSED SPECIFICATIONS Specifications that require certain trade name products or proprietary processes.

CLOSED STACK Plumbing system not vented.

CLOSED TOP VAN TRUCK Van having sealed top.

CLOSED-CIRCUIT BREATHING APPARATUS Apparatus where wearer's exhalations are reused in closed circuit.

CLOSED-CIRCUIT SURVEILLANCE SYSTEM Network based surveillance system that uses digital picture and audio transmission on local network or on Internet.

CLOSED-CIRCUIT TELEVISION System of video cameras used for surveillance.

CLOSED-END NUT Cap nut for plumbing.

CLOSED-LOOP RECYCLING Reclaiming or reusing wastewater for non-potable purposes in enclosed process.

CLOSED-VENT SYSTEM System not open to atmosphere and composed of piping, connections, and, if necessary, flow-inducing devices that transport gas or vapor from piece or pieces of equipment to control device.

CLOSEOUT Activities taken to ensure adequate completion of project work and also to fulfil administrative conditions.

CLOSET BEND Curved section of length of drain pipe found just below toilet.

CLOSET SPUD Washer that fits inside toilet on base where tank drains into bowl, forming seal here to hold water in tank.

CLOSING DIKE Construction on other side of branch channel of river to prevent or decrease flow going into channel.

CLOSING LINE 1. Line vector that's necessary to complete polygon representing vector sum and thus to render sum zero; 2. Wire rope that closes clamshell or orange-peel bucket and operates as hoisting rope.

CLOSING OF EXCAVATION Backfill placed around foundations and over services to finished ground level.

CLOSTRIDIUM PASTEURIANUM Anaerobic bacterium able to utilize free nitrogen from atmosphere in metabolic processes.

CLOSTRIDIUM PERFRINGENS Spore-forming bacterium common in feces of human and other warmblooded animals that's used experimentally as indicator of past fecal contamination and presence of microorganisms that are resistant to disinfection and environmental stresses.

CLOSURE 1. Procedure landfill operator must follow when landfill reaches its legal capacity for solid ceasing acceptance of solid waste and placing cap on landfill site; 2. Smallest set that both includes given subset and possesses some given property; 3. Device to facilitate temporary and repeatable opening and closing.

CLOSURE FITTING 1. Plug; 2. Cap.

CLOSURE JOINT Cast-in-place concrete to fill segment.

CLOSURE PERIOD period of time beginning with cessation, with respect to waste impoundment of uranium ore processing operations and ending with completion of requirements specified under closure plan.

CLOSURE RAILS Straight or curved rails that are positioned between heel of switch and toe of frog in turnout.

CLOUD Visible aggregate of minute water droplets or ice particles in atmosphere above Earth's surface.

CLOUD AMOUNT Fraction of sky obscured by clouds when observed from particular location.

CLOUD BASE Altitude of base of mass of cloud, measured by ceilometer.

CLOUD BOW Bow, like rainbow, formed in cloud droplets.

CLOUD CEILING Height of cloud base for lowest broken or overcast cloud layer.

CLOUD CHAMBER 1. Lab device where atmosphere could be enlarged adiabatically to create model of cloud; 2. Particle detector used for detecting ionizing radiation.

CLOUD CONDENSATION NUCLEI Small particles in air on which water vapor condenses and forms cloud droplets.

CLOUD GENERATION Impulsive generation of clouds, generally in adiabatic cooling because of rise of atmosphere.

CLOUD LAYER Array of clouds whose bases are at approximately same level.

CLOUD MOVEMENT Direction toward which cloud is moving.

CLOUD SEEDING Act of firing munitions containing silver iodide at passing clouds to make them more likely to produce rain.

CLOUD SHADOW Shadow cast down on surface by clouds.

CLOUD STREETS Rows of cumulus or cumulus-type clouds aligned parallel to low-level flow.

CLOUD TAGS 1. Ragged, detached cloud fragments; 2. Fractus or scud.

CLOUDBURST Intense, torrential downpour of heavy rain that suggests immediate bursting and discharge of entire cloud.

CLOUDY $7/8^{th}$ or more of sky is covered by clouds.

CLOVERLEAF INTERCHANGE Junction of two highways, one crossing over other, having series of exit and entrance ramps, arranged in form of four-leaf clover.

CLUSTER GEAR Set of gears of different sizes mounted as unit on same shaft.

CLUTTER Radar echoes that interfere with observation of desired signals on radar display.

CO-COMPOSTING Controlled aerobic degradation of organics, using more than one feedstock (fecal sludge and organic solid waste).

CO-DISPOSAL Deposition of hazardous wastes with domestic wastes or other similar wastes to fully utilise attenuation process inherent in landfill site.

CO-DOMINANT Describing two or more species that are equally dominant.

CO-FIRE Burning of two fuels in same combustion unit (e.g., coal and natural gas, oil and coal).

CO-FIRED COMBUSTOR Unit combusting hospital waste and/or medical or infectious waste with other fuels or wastes (e.g., coal, municipal solid waste) and subject to enforceable requirement limiting unit to combusting fuel feed stream, 10% or less of weight of which is comprised, in aggregate, of hospital waste and medical or infectious waste as measured on calendar quarter basis.

CO-GENERATION Generation of electricity and other energy jointly, especially utilization of steam left over from electricity generation to produce heat.

CO-GENERATION STEAM GENERATING UNIT Creates electrical (or mechanical) and thermal energy from primary source.

CO-GENERATION SYSTEM System that simultaneously produces power (electrical or mechanical) and cogeneration heat (e.g., to heat water) from same fuel in single thermodynamic process.

CO-GENERATOR Generating facility that produces electricity and another form of useful thermal energy (i.e., heat, steam), used for industrial, commercial, heating or cooling purposes.

COAGULANT Material that causes suspended solids to mass jointly or floc when added to water.

COAGULANT AID Substance added to coagulant to intensify settling of particulate matter by gravity and increase density of floc.

COAGULATION Clumping of particles in wastewater to settle out impurities often induced by chemicals (e.g., lime, alum, iron salts).

COAGULATION BASIN Process where coagulant is rapidly and uniformly dispersed through mass of water in basin.

COAL Black rock formed from prehistoric plant remains, composed largely of carbon and burned as fuel.

COAL CLEANING TECHNOLOGY Pre-combustion process by which coal is physically or chemically treated to remove some of its sulfur so as to reduce sulfur dioxide emissions.

COAL EQUIVALENT Amount of coal that would produce same quantity of heat as given mass of given fuel by combustion.

COAL GAS Poisonous gas (mixture of hydrogen and carbon monoxide) used domestically for cooking and lighting.

COAL GASIFICATION Conversion of coal to gaseous product by one of several available technologies.

COAL REFUSE Legacy of previous coal mining and consists of low-quality coal mixed with rock, shale, slate, clay and other material.

COAL REFUSE DISPOSAL PILE Deposit of coal refuse on or buried in earth and intended as permanent disposal of or long-term storage of such material, but not including coal refuse deposited within mine itself or coal refuse never removed from mine.

COAL TAR Black, viscous tar made by destructive distillation of coal (to make coke and town gas).

COALESCENCE 1. Process by which water droplets in cloud collide and come together to form raindrops; 2. Ability of materials to grow, fuse or bind together.

COALESCENT DEBRIS CONE Debris cone composed of two or more debris cones formed by debris contributed by large streams.

COALESCING SEPARATOR Apparatus that breaks up mixture or emulsion of two non-miscible fluids.

COANDA EFFECT Phenomena where jet flow attaches itself to nearby surface and remains attached even when surface curves away from initial jet direction.

COARSE GRADED AGGREGATE Sample of aggregate comprising higher percentage of large particles.

COARSE GRAINED Grain diameter greater than 0.079 inch for sedimentary rocks or 0.197 inch for igneous or metamorphic rocks.

COARSE GRAINED SAND-SIZED Less than 10% of fine and medium grained sand sizes are present.

COARSE GRAVEL Material passing 75-mm (3-inch) sieve and retained on 19.0-mm (¾-inch) sieve.

COARSE GRAVEL PROTECTION Gravel generally placed in layer upon finished surface to protect finished surface from deterioration or erosion.

COARSE GRAVEL-SIZED Particles greater than 3/4 of inch but less than 3 inches in diameter.

COARSE RACK Stand with bars spaced at distances of 25 mm (1 in.) or more.

COARSE SAND Material passing 4.75-mm sieve (No. 4) and retained on 2.00-mm (No. 10) sieve.

COARSE-AGGREGATE FACTOR Ratio, of amount (mass or solid volume) of coarse aggregate in unit volume of well-proportioned concrete to amount of dry-rodded coarse aggregate compacted into same volume.

COARSE-GRAINED FILTER Filter composed of gravel, rock and slag.

COARSE-GRAINED SOIL Soil with majority of larger grain sizes.

COASTAL Things that are in sea or on land near coast.

COASTAL EFFECTS Climatic differences experienced near shoreline versus weather inland.

COASTAL FLOOD ADVISORY Minor flooding is possible (i.e., over and above normal high tide levels).

COASTAL FLOOD WARNING Flooding that will pose serious threat to life and property is occurring, imminent or highly likely.

COASTAL FLOOD WATCH Flooding with significant impacts is possible.

COASTAL FLOODING Flooding that occurs from storms (e.g., hurricanes, nor′easter, tropical storms) where water is driven onto land from adjacent body of water over and above normal tidal action.

COASTAL MODEL Model of coastal area.

COASTAL PLAIN Flat, low-lying land adjacent to sea coast.

COASTAL WATERS Area from line approximating mean high water along mainland or island as far out as 100 nautical miles including bays, harbors and sounds.

COASTAL WATERS FORECAST Marine forecast for areas (e.g., bays, harbors, sounds) from line approximating mean high water mark (average height of high water over 19-year period) along mainland or near shore islands extending out to as much as 100 nautical miles.

COASTAL ZONE Lands and waters adjacent to coast that exert influence on uses of sea and its ecology or whose uses and ecology are affected by sea.

COASTER GATE Rectangular gate similar in construction and appearance to tractor gate except sealing is accomplished in much same way as ring seal gate.

COASTLINE Shape, outline or boundary of coast.

COATING Thin, protective material applied to outer surface of material.

COBALT-60 Synthetic radioactive isotope of cobalt with half-life of 5.2714 years.

COBBLE 1. Rock fragment, usually rounded or semi-rounded, with average dimension between 3 to 12 inches; 2. Particle of rock that will pass 12 in^2 opening and be retained on 3-inch U.S.A. Standard sieve.

COBBLE-SIZED Particles greater than 3 inches but less than 12 inches in diameter.

COBBLESTONE 1. Rock fragment, usually rounded or semi-rounded, with average dimension between 3 to 12 inches; 2. Particle of rock that will pass 12 in^2 opening and be retained on 3-inch U.S.A. Standard sieve.

COCCI Spherical bacterium.

COCK 1. Valve or tap for controlling flow in plumbing; 2. Upward turn, tilt or angle.

CODE 1. Body of law; 2. Set of rules for converting information into another form or representation.

CODE OF FEDERAL REGULATIONS 1. Document that codifies all rules of executive departments and agencies of Federal Government updated annually; 2. Codification of final rules published daily in Federal Register.

CODE OF PRACTICE Collection of rules issued by professional association that explains how people working in particular profession should behave.

CODE SELECTION Process of choosing appropriate computer code, algorithm or other analysis technique capable of simulating those characteristics of physical system required to fulfill modeling project's objective(s).

COEFFICIENT 1. Constant where algebraic term is multiplied; 2. Number, value or item that serves as measure of some property or characteristic.

COEFFICIENT OF AREA Ratio of area of trickling filter surface receiving wastewater to total area of filter surface.

COEFFICIENT OF COMPRESSIBILITY 1. Slope of one-dimensional compression curve relating void ratio to effective stress; 2. Measure of relative volume change of fluid or solid as response to pressure (or mean stress) change.

COEFFICIENT OF CONSOLIDATION Coefficient that relates change in excess pore fluid pressure with time to excess pore pressure diffusion in soil mass.

COEFFICIENT OF CONTRACTION Ratio between area of jet at vena contracta to area of contracted flow section.

COEFFICIENT OF DETERMINATION Proportion of variance in dependent variable that's predictable from independent variable.

COEFFICIENT OF DISCHARGE Ratio of actual discharge to theoretical discharge (i.e., ratio of mass flow rate at discharge end of nozzle to that of ideal nozzle which expands identical working fluid from same initial conditions to same exit pressures.)

COEFFICIENT OF EARTH PRESSURE Ratio of active earth pressure, normal to plane surface, to corresponding pressure in fluid of same density.

COEFFICIENT OF EARTH PRESSURE AT REST Ratio of earth pressure, normal to plane surface, to corresponding pressure in fluid of same density, when there is no movement.

COEFFICIENT OF EXPANSION Ratio of increase of length, area or volume of body per degree rise in temperature to its length, area or volume, respectively, at some specified temperature with pressure being kept constant.

COEFFICIENT OF FINENESS 1. Measure of size of particles causing turbidity; 2. Ratio of suspended solids to turbidity.

COEFFICIENT OF FRICTION Ratio between magnitude of force of friction which surface produces on object and magnitude of normal force which is produced by surface on object and which is perpendicular to that surface.

COEFFICIENT OF HAZE Measurement of visibility interference in atmosphere.

COEFFICIENT OF HEAT TRANSMISSION Amount of heat transferred through unit area of partition per hour, per degree temperature difference between air on two sides.

COEFFICIENT OF INTERNAL FRICTION 1. Angle of internal friction; 2. Tangent of angle (ϕ).

COEFFICIENT OF KINETIC FRICTION Ratio of force of friction to normal force.

COEFFICIENT OF LINEAR EXPANSION Change in length of specimen one unit long when its temperature is changed by one degree.

COEFFICIENT OF PASSIVE EARTH PRESSURE Ratio between lateral earth pressure and vertical earth pressure.

COEFFICIENT OF PERFORMANCE Ratio of useful heating or cooling provided to work required.

COEFFICIENT OF PERMEABILITY Average velocity of water through total area of soil under hydraulic gradient of 1.

COEFFICIENT OF PERMEABILITY TO WATER Rate of discharge of water under laminar flow conditions through unit cross-sectional area of porous medium under unit hydraulic gradient and standard temperature conditions.

COEFFICIENT OF REGIME Ratio of maximum annual flow at peak to lowest annual flow.

COEFFICIENT OF RUNOFF 1. Dimensionless coefficient relating amount of runoff to amount of precipitation received; 2. Measure of gross storm water in drainage system lost to evaporation, infiltration, exfiltration and runoff.

COEFFICIENT OF SKID RESISTANCE Ratio of load and skid resistance.

COEFFICIENT OF STORAGE Volume of water released from storage per unit surface area of aquifer or aquitard per unit decline in hydraulic head.

COEFFICIENT OF SUBGRADE FRICTION Coefficient of friction between slab and subgrade.

COEFFICIENT OF SUBGRADE REACTION Ratio between soil pressure at any given point of surface of contact and displacement produced by load application at that point.

COEFFICIENT OF THERMAL EXPANSION Fractional change in size of material as it is heated through one degree.

COEFFICIENT OF THERMAL TRANSMISSION Amount of heat transferred through unit area of partition per hour, per degree temperature difference between air on two sides.

COEFFICIENT OF UNIFORMITY Ratio (D60/D10), where D60 and D10 are particle diameters corresponding to 60% and 10% finer on cumulative particle-size distribution curve, respectively.

COEFFICIENT OF VARIATION Ratio of standard deviation to mean (i.e., average).

COEFFICIENT OF VELOCITY Ratio of actual velocity to theoretical velocity.

COEFFICIENT OF VISCOSITY Degree to which fluid resists flow under applied force, expressed as ratio of shearing stress to velocity gradient.

COENOCLINE Gradient of communities (e.g., in transect from summit to base of hill), reflecting changing.

COFFERDAM Temporary structure enclosing all or part of construction area so that construction can proceed in dewatered or have water level controlled where diversion structure diverts stream into pipe, channel or tunnel.

COFFIN Thick-walled container (usually lead) used to transport radioactive material.

COGENERATION Consecutive generation of useful thermal and electric energy from same fuel source.

COHERENT RADAR Radar that utilizes both signal phase and amplitude to determine target characteristics.

COHESION 1. Mutual attraction of soil particles due to molecular and capillary forces in presence of water where attraction is high in clay (especially dry) and low in silt or sand; 2. Various intermolecular forces that hold two solid and liquid particles together; 3. Growing together of normally distinct parts of plant.

COHESIONLESS MATERIALS 1. Soil materials that when unconfined have little or no strength when air-dried and that have little or no cohesion when submerged; 2. Soil that has little tendency to stick together whether wet or dry (e.g., sands, gravels).

COHESIONLESS SOIL 1. Soil materials that when unconfined have little or no strength when air-dried and that have little or no cohesion when submerged; 2. Soil that has little tendency to stick together whether wet or dry (e.g., sands, gravels).

COHESIVE SEDIMENTS Sediments whose resistance to initial movement or erosion is affected mostly by cohesive bonds between particles.

COHESIVE SOIL 1. Predominantly clay and silt soil, fine-grained particles, that sticks together whether wet or dry; 2. Soil that, when unconfined, has considerable strength when air-dried and that has significant cohesion when submerged.

COIL Something wound in form of helix or spiral.

COKE One of basic materials used in blast furnaces for conversion of iron ore into iron or roasting coal in oven.

COKE BURN-OFF Coke removed from surface of fluid catalytic cracking unit catalyst by combustion in catalyst regenerator.

COKE BY-PRODUCT RECOVERY PLANT Facility designed and operated for separation and recovery of coal tar derivatives (by-products) evolved from coal during coking process of coke oven battery.

COKE OVEN Industrial process which converts coal into coke.

COKE-TRAY AERATOR Consist of series of coke-filled trays where water percolates.

COL Dip between mountain peaks in summit-line.

COLD ADVECTION Transport of cold air into region by horizontal winds.

COLD AIR AVALANCHE Downslope flow pulsations that occur at more or less regular intervals as cold air builds up on peak or plateau, reaches critical mass and then cascades down slopes.

COLD AIR DAMMING Phenomenon where low-level cold air mass is trapped topographically.

COLD AIR FUNNEL Funnel cloud or (rarely) small, relatively weak tornado that can develop from small shower or thunderstorm when air aloft is unusually cold.

COLD FRONT Zone separating two air masses, of which cooler, denser mass is advancing and replacing warmer.

COLD JOINT 1. Unplanned joint resulting when concrete surface hardens before next batch is placed against it; 2. Fresh concrete placed on harden concrete.

COLD LOW Any low generally characterized by colder air near its center than around its periphery.

COLD OCCLUSION Frontal zone formed when cold front overtakes warm front and, being colder than air ahead of warm front, slides under warm front, lifting it aloft.

COLD POOL Region of relatively cold air, represented on weather map analysis as relative minimum in temperature surrounded by closed isotherms.

COLD SPRING Non-thermal spring where water has temperature appreciably lower than ambient.

COLD TEMPERATURE CO Standard for automobile emissions of carbon monoxide (CO) emissions to be met at low temperature (i.e., 20 °F).

COLD TROUGH Any trough generally characterized by colder air near its center than its surroundings.

COLD-WATER FISHERY Water or water system which has environment suitable for salmonid fishes.

COLD-WORKED PIPE AND TUBE Those cold forming operations that process unheated pipe and tube products using either water or oil solutions for cooling and lubrication.

COLI-AEROGENES BACTERIA Bacteria mainly inhabiting intestines of man or animals.

COLIFORM 1. Organisms common to intestinal tract of humans and animals; 2. Group of related bacteria whose presence in water indicates fecal pollution and potentially adverse contamination by pathogens.

COLIFORM COUNT Test of water contamination where number of colonies of coliform-bacteria Escherichia coli (E. coli) per 100 milliliter of water is counted.

COLIFORM INDEX Rating of purity of water based on count of fecal bacteria.

COLIFORM-GROUP BACTERIA Bacteria mainly inhabiting intestines of man or animals.

COLIPHAGES Viruses that infect and replicate in coliform bacteria that are indicative of sewage contamination of water and of survival and transport of viruses in environment.

COLLAPSE LOAD Load structure or structural member can bear before failure.

COLLAPSIBLE SOIL Consist of loose, dry, low-density materials that collapse and compact under addition of water or excessive loading.

COLLAR 1. Open end of drill hole; 2. Enlargement of conduit diameter; 3. Neck or line of junction between root of plant and its stem.

COLLAR BOSS Pipe fitting for plumbing stack.

COLLAR CLOUD Generally circular ring of cloud that may be observed on rare occasions surrounding upper part of wall cloud.

COLLARING Starting drill hole.

COLLECTING SYSTEM System of conduits that collects and conveys urban waste water.

COLLECTION Set of items or amount of material procured or gathered together from source.

COLLECTION CENTER Site built to take secondary materials from residences, companies, associations and industrial sites.

COLLECTION EFFICIENCY Fraction of droplets approaching surface that actually deposit on that surface.

COLLECTION FREQUENCY Times refuse collection is offered in certain time period.

COLLECTION LINE Plumbing drain.

COLLECTION METHOD Means where garbage is accumulated from its point of origin.

COLLECTION STOP Stop made to collect solid waste from one or more sites.

COLLECTION SYSTEM Carries wastewater from homes and businesses through underground pipe, to treatment facility.

COLLECTIVE BARGAINING Method of negotiation where employees negotiate as group with their employers.

COLLECTOR 1. Public or private hauler that collects non-hazardous waste and recyclable materials from residential, commercial, institutional and industrial sources; 2. Major sewer which collects sewerage from number of smaller branch sewers.

COLLECTOR ANGLE Optimum angle of solar collector to optimize your system.

COLLECTOR EFFICIENCY Ratio of rate of useful thermal energy leaving collector to useable solar irradiance falling on collector's aperture area.

COLLECTOR HIGHWAY Highways that connect local highways to arterial highways.

COLLECTOR SEWER Lateral sewers that are primarily receive wastewater directly from individual systems or private property facilities.

COLLECTOR SEWERS Pipes used to collect and carry wastewater from individual sources to interceptor sewer that will carry it to treatment facility.

COLLECTOR STREET Road that moves traffic from local to arterial roads.

COLLECTOR SUBSYSTEM Construction required for absorbing solar radiation that converts thermal energy to heat transfer fluid.

COLLECTOR TILT Angle of solar collector assembly.

COLLECTOR-DISTRIBUTOR ROAD Road that moves traffic from local to arterial roads.

COLLOID 1. Particle less than 1 micron in diameter; 2. Stable system of two phases, one of which is dispersed in other in form of very small droplets or particles; 3. Intimate mixture of two substances one of which, is uniformly distributed in finely divided state throughout second substance, called dispersion medium.

COLLOIDAL DISPERSION Heterogeneous system where one substance (e.g., discrete particles, droplets or bubbles) called dispersion phase is dispersed as very fine particles in another substance called dispersion medium.

COLLOIDAL MATTER Particles that can be removed by coagulation, membrane filtration or biochemical action.

COLLOIDAL PARTICLE Items of small size that are floating in solid, liquid or gas.

COLLOIDAL PARTICLES Soil particles that are so small that surface activity has appreciable influence on properties of aggregate.

COLLOIDAL SUSPENSION Method of sediment transport where water turbulence (movement) supports weight of sediment particles, thereby keeping them from settling out or being deposited.

COLLOIDORS Devices that attempt to remove non-settling solids from sewage via surfaces hung vertically to assist in coagulating colloidal matter.

COLLOIDS Very small, finely divided solids (that do not dissolve) that remain dispersed in liquid for long time due to their small size and electrical charge.

COLLUVIAL SOIL Loose, unconsolidated sediments that have been deposited at base of hillslopes by either rainwash, sheetwash, slow continuous downslope creep or variable combination of these processes.

COLLUVIARIUM Opening made at regular intervals in aqueduct for ventilation.

COLLUVIUM General term applied to loose and incoherent deposits, usually at foot of slope and brought there chiefly by gravity.

COLOR CODING System for displaying information by using different colors.

COLOR TEMPERATURE Property of light source: temperature of ideal black body radiator that radiates light of color comparable to that of light source.

COLOR UNIT 1. Produced by 1 mg/l of platinum in form of chloroplatinate ion; 2. Color expressed in units of platinum-cobalt scale.

COLORADO LOW Low pressure storm system that forms in winter in southeastern Colorado or northeastern New Mexico and tracks northeastward across central plains of U.S. over period of several days, producing blizzards and hazardous winter weather.

COLORIMETRY Science of measuring colors and assigning them numeric values.

COLUMNAR ICE Ice consisting of columnar shaped grain where ordinary black ice is usually columnar-grained.

COMBINATION FAUCET Apparatus where flow of water from hot and cold water pipes is controlled and is drawn from common spout.

COMBINATION MEDIUM Filter medium composed of several kinds, levels or arrangements of filter media.

COMBINATION RAILING Traffic railing for bicycle or pedestrian.

COMBINATION WASTE AND VENT Waste piping embodying horizontal wet venting of one or more sinks or floor drains by means of common waste and vent pipe adequately sized to provide free movement of air above flow line of drain.

COMBINATION WELL System of open well connected to other wells or infiltration galleries.

COMBINED AVAILABLE CHLORINE Total chlorine present in water and is still available for disinfection and for oxidation of organic matter.

COMBINED AVAILABLE RESIDUAL CHLORINE Chlorine residual which exists in water in combination with ammonia or organic nitrogen compounds.

COMBINED CARBON Portion of carbon in iron and steel chemically united in form of carbides.

COMBINED CYCLE PLANT Plant which achieves higher efficiency by employing two cycles in tandem (e.g., heat rejected from gas-fired turbine is used to generate steam to operate steam turbine.)

COMBINED DRAIN Pipe that collects sewage and storm water from multiple properties.

COMBINED FOOTING Combination of various footings to support more than one column.

COMBINED HEAT AND POWER Generation of electricity and capture and use of otherwise wasted heat energy byproducts.

COMBINED HYDROGEN Hydrogen chemically united with oxygen.

COMBINED LOAD Sum of two or more different types of load occurring simultaneously.

COMBINED MOISTURE Moisture in coal that cannot be removed by ordinary drying.

COMBINED RESIDUAL CHLORINE Application of chlorine to water to produce residual chlorine.

COMBINED SEAS Combination or interaction of wind waves and swells where separate components are not distinguished.

COMBINED SEWER Sewers that are designed to collect rainwater runoff, domestic sewage and industrial wastewater in same pipe.

COMBINED SEWER OVERFLOW Discharge of untreated sewage and stormwater to stream when capacity of combined storm/sanitary sewer system is exceeded during rainstorms by storm runoff in one piping system.

COMBINED SEWER SYSTEM Wastewater collection system which conveys sanitary wastewaters (domestic, commercial and industrial wastewaters) and storm-water runoff through single pipe to publicly owned treatment works for treatment prior to discharge to surface waters.

COMBINED STRESSES Stresses that acts simultaneously (i.e., axial, shear, flexure, torsion).

COMBINED SYSTEM System of pipework where groundwater and wastewater are combined in single conduit.

COMBINED WASTEWATER Mixture of stormwater or surface water runoff and other wastewater (i.e., domestic or industrial wastewater).

COMBINED WATER Water in molecular combination within clay and glaze materials, which is driven off during water-smoking period of firing.

COMBINED-CYCLE GAS TURBINE Fixed gas turbine that recovers heat from gas turbine exhaust gases to heat water or generate steam.

COMBINED-CYCLE GENERATING STATION Consists of one or more combustion turbines, each with heat recovery steam generator.

COMBINED-CYCLE SYSTEM System where separate source provides exhaust gas to steam generating unit.

COMBUSTIBLE GAS INDICATOR Gas leak detector and confined space monitor.

COMBUSTIBLE WASTE Organic content of solid waste that will burn.

COMBUSTIBLES Material that are capable of burning.

COMBUSTION 1. Burning or rapid oxidation, accompanied by release of energy in form of heat and light; 2. Controlled burning of waste, where heat chemically alters organic compounds, converting into stable inorganics (e.g., carbon dioxide, water); 3. Process where two chemicals are combined to produce heat.

COMBUSTION AIR Air supplied to combustion appliances to be used in combustion of fuels and process of venting combustion gases.

COMBUSTION CHAMBER Actual compartment where waste is burned in incinerator.

COMBUSTION PRODUCT Substance produced during burning or oxidation of material.

COMFORT COOLING TOWERS Cooling towers for heat, ventilation, air conditioning or refrigeration systems.

COMMA CLOUD Synoptic scale cloud pattern with characteristic comma-like shape.

COMMA ECHO Thunderstorm radar echo which has comma-like shape.

COMMAND POST Centralized base of operations established near site of hazardous materials incident.

COMMAND-AND-CONTROL REGULATIONS Specific requirements prescribing how to comply with specific standards defining acceptable levels of pollution.

COMMENCEMENT OF WORK Date when work shall not commence under contract until fully executed, contract has been received by contractor and contractor has been given approval to proceed.

COMMENT PERIOD Time provided for public to review and comment on proposed action or rulemaking after publication in Federal Register.

COMMERCIAL AND INDUSTRIAL FRICTION MATERIAL Asbestos-containing merchandise used in friction material applications (i.e, automotive brakes, clutches).

COMMERCIAL APPLICATOR Any person who uses or supervises use of any pesticide for any purpose or on any property, other than as provided by definition of private pesticide applicator.

COMMERCIAL OPERATION Begins when control of loading of generator is turned over to system dispatcher.

COMMERCIAL RIVER TRIP Trips organized by boating companies that conduct tours for paying passengers or customers.

COMMERCIAL SOLID WASTE All types of solid wastes generated for purposes of trade, business, sport, recreation, education or entertainments, excluding household, agricultural or industrial waste.

COMMERCIAL THINNING Intermediate harvest where merchantable wood removed should cover part or all of cost of harvesting.

COMMERCIAL USER Non-residential user that discharges only segregated domestic wastes or wastes from sanitary conveniences.

COMMERCIAL USER DAY Amount of commercial use within 24-hour period or any portion thereof.

COMMERCIAL VEHICLE Vehicle used for carrying goods or fare-paying passengers.

COMMERCIAL VEHICLE OPERATION Transportation systems for trucks including satellite navigation system, small computer and digital radio in each truck.

COMMERCIAL WASTE Waste emanating from business establishments (i.e., stores, markets, office buildings, restaurants, shopping centers, theaters).

COMMERCIAL WASTE MANAGEMENT FACILITY Treatment, storage, disposal or transfer facility which accepts waste from variety of sources, as compared to private facility which normally manages limited waste stream generated by its own operations.

COMMERCIAL WATER USE Water used for motels, hotels, restaurants, office buildings, other commercial facilities and institutions that may be obtained from public supply or may be self-supplied.

COMMERCIAL WITHDRAWALS Water used for motels, hotels, restaurants, office buildings, other commercial facilities and institutions that may be obtained from public supply or may be self-supplied.

COMMERCIALLY DRY SLUDGE Sludge including maximum 10% moisture by weight (5% if it is to be incorporated into fertilizer merchandise).

COMMINGLED RECYCLABLES Mixed recyclables that are collected together.

COMMINUTED SOLIDS Solids that have been sliced into finer particles.

COMMINUTER Machine that shreds or pulverizes solids to make waste treatment easier.

COMMINUTING SCREEN Apparatus for screening wastewater by cutting and slicing screenings into particles enough fine to pass through screen openings.

COMMINUTION Mechanical shredding, grinding or pulverizing of waste used in both solid waste management and wastewater treatment.

COMMINUTOR Type of machine that grinds up solid materials, especially in waste treatment.

COMMISSIONING Process of assuring that all systems and components of major piece of equipment, process, building or similar are designed, installed and tested according to operational requirements of owner or final client.

COMMITMENTS Promise or agreement to do something in future.

COMMON CARRIER Person or company that undertakes to transport goods for general public or passengers on regular routes at agreed rates.

COMMON EXCAVATION Excavation volumes that are not backfilled with earthwork material.

COMMON EXPOSURE ROUTE Likely way (oral, dermal, respiratory) that pesticide may reach and/or enter organism.

COMMON GREEN TIME Time period when phases at two intersections of interchange are both green.

COMMON MATERIAL All earth materials which do not fall under definition of rock.

COMMON NAME Name where species is known to general public, rather than its taxonomic or scientific name.

COMMON SENSE INITIATIVE Voluntary program to simplify environmental regulation to achieve cleaner, cheaper, smarter results, starting with six major industry sectors.

COMMON SEWER Lateral house sewer shared by more than one residence, directly controlled by public authority.

COMMON TRENCH Trench jointly dug by public service (e.g., electric, telephone) companies.

COMMON VENT Perpendicular vent pipe for multiple drains.

COMMON-FACILITIES PLAN Plans for all common facilities areas and complete narrative description of all proposed activities.

COMMUNITY 1. All members of specified group of species present or inhabiting specific area at specific time; 2. Group of people that see themselves as unit; 3. Species that interact in common area.

COMMUNITY COHESION Ability of community to have unified response when facing problem.

COMMUNITY COMPONENT 1. Any portion of biological community; 2. May pertain to taxonomic group (fish, invertebrates, algae), taxonomic category (phylum, order, family, genus, species), feeding strategy (herbivore, omnivore, carnivore) or organizational level (individual, population, community association) of biological entity within aquatic community.

COMMUNITY RELATIONS Effort to establish two-way communication with public to create understanding of programs and related actions, to ensure public input into decision-making processes related to affected communities and to make certain that Agency is aware of and responsive to public concerns.

COMMUNITY WATER SYSTEM Public water system which serves at least 15 service connections used by year-round residents or regularly serves at least 25 year-round residents.

COMMUTER Person who regularly travels from one place to another.

COMMUTER BUS Bus that carries people across city.

COMMUTER RAIL CAR Rail passenger car obtained by commuter authority for use in commuter rail transportation.

COMMUTER RAIL SERVICE Passenger rail transport service that primarily operates between city and suburbs (i.e., commuter towns).

COMPACT 1. Closely packed; 2. Reduce in volume; 3. Densify.

COMPACT COARSE SAND Substance comprising rough particles which have been compacted.

COMPACT FLUORESCENT LAMP Small fluorescent lamps used as more efficient alternatives to incandescent lighting.

COMPACT LINE SAND Material mostly retained on No. 200 sieve.

COMPACT MATERIAL Small grains or particles of soil with relative compaction of 90%.

COMPACT SECTION Cross-section that can develop plastic moment resistance.

COMPACTED Closely or densely packed material if relative compaction exceeds 90%.

COMPACTED BACKFILL Backfill which has been reduced to bulk by rolling, tamping or soaking.

COMPACTED CUBIC YARD Replaced and compacted fill, equal to 0.765 m^3.

COMPACTED EARTH Closely and firmly united or packed together soil.

COMPACTED EMBANKMENT Embankment which has been reduced in bulk by rolling, tapping or soaking.

COMPACTED FILL Mechanically increasing density of soil, effective at supporting defined load.

COMPACTED VOLUME Volume of material after it has been compressed.

COMPACTING FACTOR Ratio of weight of partially compacted concrete to weight of concrete when fully compacted in same mold.

COMPACTION Increase in density from loose to compacted by reducing voids between particles.

COMPACTION CURVE Curve showing relationship between dry density (dry unit weight) and moisture content of soil for given compactive effort.

COMPACTION EQUIPMENT Equipment used to densify soil.

COMPACTION PIT TRANSFER SYSTEM System where backhoe is used to compact waste before it is pushed into transfer trailer.

COMPACTION RATIO Amount volume of solid waste is reduced compared to its initial volume.

COMPACTION SPECIFICATION Minimum performance standards for material placement preparation and compaction.

COMPACTION TARGETS Density target created by governing agency.

COMPACTION TEST Laboratory compacting procedure whereby soil at known moisture content is placed in specified manner into mold of given dimensions, subjected to compactive effort of controlled magnitude and resulting dry unit weight determined.

COMPACTIVE EFFORT Total energy used to compact specimen.

COMPACTOR Machine that reduces bulk of waste by compaction.

COMPACTOR ATTACHMENT Accessory created (e.g., backhoe) for compaction or driving ability.

COMPACTOR COLLECTION VEHICLE Vehicle with enclosed body containing mechanical devices that convey solid waste into main compartment of body and compress it into smaller volume of greater density.

COMPARATIVE RISK ASSESSMENT Process that generally uses judgment of experts to predict effects and set priorities among wide range of environmental problems.

COMPARTMENT 1. Room or section or chamber; 2. One of parts where area is subdivided.

COMPASS ROSE Graduated circle, usually marked in degrees, indicating directions and printed or inscribed on appropriate medium.

COMPATIBLE 1. Capable of easy interaction; 2. Consistent; congruous.

COMPENSABLE INTEREST Right of property owner to receive just compensation.

COMPENSATING ERROR Error in computation or in recording of accounting data, neutralized by equal and opposite error.

COMPENSATION POINT Amount of light intensity on light curve where rate of photosynthesis exactly matches rate of respiration.

COMPENSATION WATER Water supplied from reservoir to stream in time to offset impacts of abstraction or drought.

COMPETENT 1. Having sufficient skill, knowledge, ability or qualifications; 2. Having jurisdiction or authority over particular issue or question.

COMPETENT BED Rock layer which, during folding, flexes without appreciable flow or internal shear.

COMPETENT PERSON One who is capable of identifying existing and predictable hazards in surroundings or working conditions which are unsanitary, hazardous or dangerous to employees and who has authorization to take prompt corrective measures to eliminate them.

COMPETITION Battle that exists when two or more species have conditions that exceed available supply.

COMPETITIVE BID CONTRACT Contract awarded as outcome of competitive bid.

COMPETITIVE BIDDING/TENDERING Method used to introduce price competition to market where there are only limited number of suppliers.

COMPETITIVE RANGE Proposals that have high chance of being selected for award.

COMPLETE DIVERSION Discharging water into one drainage area that was taken from another drainage area.

COMPLETE TREATMENT Method of treating water that consists of addition of coagulant chemicals, flash mixing, coagulation-flocculation, sedimentation and filtration.

COMPLETE TREE Entire tree from root hairs to leaf hairs.

COMPLETE TREE HARVESTING Harvesting of complete tree, including roots.

COMPLETE WASTE TREATMENT SYSTEM All treatment works involving transport of wastewater from individual homes or buildings to wastewater treatment facility, treatment of wastewater to remove pollutants and disposal (including recycling or reuse) of treated wastewater and residues.

COMPLETED OPERATIONS INSURANCE Covers liability for property damage or injuries to third party once contracted operations cease.

COMPLETED VALUE INSURANCE Coverage, amount of which is derived from contract amount and which is adjusted to final insurable original contractor defaulted.

COMPLETELY MIXED CONDITION Concentration of pollutant does not change across transect of water body.

COMPLETELY-CLOSED DRAIN SYSTEM Individual drain system not open to atmosphere and is equipped and operated with closed vent system and control device.

COMPLETION Conclusion of contract.

COMPLETION DATE Date on which something such as project will be finished.

COMPLEX COVER Combination of vegetation and soil.

COMPLEX GALE/STORM Area for which gale/storm force winds are forecast or are occurring but for which no single center is principal generator of these winds.

COMPLEX TERRAIN 1. Refers to mountainous terrain; 2. Coastal regions and heterogeneous landscapes.

COMPLEXATION Combination of individual atom groups, ions or molecules to create one large ion or molecule.

COMPLEXES Compounds formed by joining of two or more simple salts.

COMPLIANCE Act of meeting all government regulations.

COMPLIANCE COAL Any coal that emits less than 1.2 pounds of sulfur dioxide per million Btu when burned.

COMPLIANCE COATING Coating whose volatile organic compound content does not exceed that allowed by regulation.

COMPLIANCE CYCLE Nine-year calendar year cycle, where public water systems must monitor where each cycle consists of three 3-year compliance periods.

COMPLIANCE MONITORING Collection and evaluation of data, including self-monitoring reports and verification to show whether pollutant concentrations and loads contained in permitted discharges are in compliance with limits and conditions specified in permit.

COMPLIANCE SCHEDULE Negotiated agreement between pollution source and government agency that specifies dates and procedures by which source will reduce emissions and, thereby, comply with regulation.

COMPONENT Smaller, self-contained part of larger entity.

COMPONENT TIDE All primary tides (main lunar, primary solar, N2, K and O) that consist of natural tide.

COMPOSITE Average calculated according to specific criteria.

COMPOSITE ACTION When two different materials are bound together and act together as single unit from structural point of view.

COMPOSITE BEAM Structural beam composed of two or more dissimilar materials joined together to act as unit.

COMPOSITE COLUMN Column with interactive and integral behavior of concrete and structural steel elements.

COMPOSITE CONSTRUCTION Building construction involving dissimilar materials.

COMPOSITE GRADE Highway adjacent grades that have severe impacts on operations.

COMPOSITE HYDROGRAPH 1. Stream discharge hydrograph which includes base flow; 2. Stream discharge hydrograph that corresponds to net rain storm of duration longer than one unit period.

COMPOSITE LINER Geomembrane liner used in municipal solid waste landfills to reduce water pollution.

COMPOSITE POWER VALUE Value of power including both capacity and energy component, usually expressed in mills/kw-hr.

COMPOSITE ROCK-FILL DAM Dam consisting of rock fill on downstream side and earth fill on upstream side.

COMPOSITE SAMPLE 1. Series of water samples taken over given period of time and weighted by flow rate; 2. Collection of individual samples obtained at regular intervals.

COMPOSITE SAMPLE QUALITY Concentration of some parameter obtained and/or tested in composite sample.

COMPOSITE UNIT HYDROGRAPH Sum of individual unit hydrograph discharge values at time values that have been shifted to reflect time of travel from outlets of subareas to outlet of large area.

COMPOST Decayed organic material and animal matter used to fertilize and condition soil.

COMPOSTING Controlled biological decomposition of organic material in presence of air to form humus-like material.

COMPOSTING FACILITIES 1. Offsite facility where organic component of municipal solid waste is decomposed under controlled conditions; 2. Aerobic process where organic materials are ground or shredded and then decomposed to humus in windrow piles or in mechanical digesters, drums or similar enclosures.

COMPOUND Substance made from any combination of elements.

COMPOUND DREDGER Dredger combining suction or suction cutter apparatus with bucket ladder.

COMPOUND FISSURE Progressive fracture originating from horizontal split head that turns up or down or in both directions, in head of rail.

COMPOUND HYDROGRAPH Hydrograph resulting from storms with sequence of rainfall events.

COMPOUND INGREDIENT Ingredient made up of two or more ingredients.

COMPOUND MATERIAL Material made from two or more constituent materials.

COMPOUND METER Meters with two measuring elements and check valve to regulate flow between them.

COMPOUND PIPE Pipe for conveying fluid comprising inner pipe positioned within outer pipe such that walls of inner pipe and outer pipe define gap.

COMPREHENSIVE EAP EXERCISE In depth exercise of EAP that involves interaction of dam owner with state and local emergency management agencies in stressful environment with time constraints.

COMPREHENSIVE ENVIRONMENTAL RESPONSE, COMPENSATION, AND LIABILITY ACT OF 1980 Federal statute that authorized "Superfund," is administered by EPA and provides funding for cleanups and emergency response actions for hazardous substances at worst hazardous waste sites in United States.

COMPREHENSIVE ENVIRONMENTAL RESPONSE, COMPENSATION, AND LIABILITY INFORMATION SYSTEM Federal database that includes all sites which have been nominated for investigation by Superfund program.

COMPREHENSIVE FACILITY REVIEW 1. Review performed on high- or significant-hazard dam every 6 years, which includes field examination and state-of-the-art review of structure's design assumptions, construction practices and integrity under various loading conditions; 2. Detailed examination performed on dams with senior dam engineer.

COMPREHENSIVE FLARE INDEX Indicative of solar flare importance.

COMPREHENSIVE GENERAL LIABILITY Broad policy that protects organization from liability claims related to products coverage, completed operations coverage premise and operations coverage and independent contractors coverage.

COMPREHENSIVE MATERIAL DAMAGE INSURANCE Insurance against loss or damage from fire & perils.

COMPRESSED AIR Air which has been reduced in volume and increased in pressure for use in industrial processes.

COMPRESSED DIAMOND INTERCHANGE Two intersections where ramps that meet minor road are closely spaced.

COMPRESSED GAS Gas which when under pressure is entirely gaseous at -58 °F, including all gases with critical temperature $\leq$ -58 °F.

COMPRESSED NATURAL GAS Alternative fuel for motor vehicles considered one of cleanest because of low hydrocarbon emissions and its vapors are relatively non-ozone producing.

COMPRESSIBILITY 1. Property of soil describing its susceptibility to decrease in volume when subjected to load; 2. Product of pressure and molar volume divided by product of gas constant and absolute temperature; 3. Land property that enables it to deform under load.

COMPRESSIBILITY EFFECT Any of effects that result from changes in flow field about airplane when velocity at some point in field reaches local speed of sound and air ceases to behave as incompressible fluid.

COMPRESSION 1. Reduction in volume of soil mass resulting from increase in effective stress; 2. Force that lends to shorten structural member and density granular substance; 3. Compacting effect of weight of steel-wheel rollers.

COMPRESSION COUPLING Form of shaft-coupling for uniting lengths of driving-shaft end to end, where two ends each have split cone keyed on.

COMPRESSION CURVE Any function relating volume change to effective stress.

COMPRESSION FITTING Coupling used to connect two pipes or pipe to fixture or valve that consists of three parts: compression nut, compression ring and compression seat.

COMPRESSION FORCE Application of force, power, pressure or exertion against object that causes it to become squeezed, squashed or compacted.

COMPRESSION ROLL Drive wheel of steel-wheel roller.

COMPRESSION VALVE Valve opened or closed by raising or lowering horizontal disk by threaded stem.

COMPRESSION-TYPE HYDRANT Operating mechanism at top of hydrant sealed from barrel so that water does not come into contact with mechanism during hydrant use.

COMPRESSIONAL WAVE Longitudinal wave propagated by elastic compression of medium.

COMPRESSIVE STRENGTH Property of material that relates to its ability to resist compression loads.

COMPRESSIVE STRESS Stress on materials that leads to smaller volume or shortening.

COMPRESSIVE TEST Test measuring maximum compressive load before fracturing.

COMPRESSOR Device that produces pressure, such as gas compressor that produces pressurized gas.

COMPROMISE JOINT BAR Special type of rail joint fabricated for connecting 2 rail tracks of different sizes and drillings.

COMPROMISE JOINT RAIL Joint for uniting abutting ends of contiguous rails of different sections or of rails of same section but of different joint drillings.

COMPROMISE RAIL Rail for linking rails having different sections.

COMPUTATION DURATION User-defined time window used in hydrologic modeling.

COMPUTATION INTERVAL User-defined time step used by hydrologic model for performing mathematical computations.

COMPUTATIONAL ENGINE Software language and calculation power of model.

COMPUTER BASE STATION Computer equipment designated to receive data from data collection platforms and translate that data into useable information for decision makers.

COMPUTER CODE Assembly of numerical techniques, bookkeeping and control language that represents model from acceptance of input data and instructions to delivery of output.

COMPUTER-AIDED DESIGN Process of using drawings made by using computer to design machines, buildings, etc.

COMPUTER-AIDED DESIGN AND DRAFTING Technology for design and technical documentation, which replaces manual drafting with automated process.

CONCAVE BANK Outer bank of curved stream, with center of curve toward channel (e.g., undercut slope).

CONCENTRATE AND CONTAIN Waste management technique that relies on wastes being concentrated and contained and kept separate from other environmental media.

CONCENTRATE STREAM Stream where rejected ions and materials are concentrated.

CONCENTRATED FALL Fall concentrated at one point on stream in hydroelectric power development.

CONCENTRATED FLOW PATH Existing hypothetical avenue of concentrated seepage.

CONCENTRATING SOLAR COLLECTOR Solar collector that uses reflective surfaces to concentrate sunlight onto small area, where it is absorbed and converted to heat or, in case of solar photovoltaic devices, into electricity.

CONCENTRATION 1. Relative amount of substance mixed with another substance; 2. Ratio of quantity of any substance present in sample of given volume or given weight compared to volume or weight of sample; 3. Amount of solute in solution measured in suitable units (e.g., parts per million); 4. Procedure of raising dissolved solids per unit volume of solution.

CONCENTRATION FACTOR 1. Amount where element must be increased above its normal crustal abundance in ore to make it commercially extractable; 2. Ratio of average flow of stream during that part of day when pondage or temporarily stored streamflow is being drawn upon to 24-hour average flow.

CONCENTRATION GRADIENT Rate of change in solute concentration per unit distance at given point and in given direction.

CONCENTRATION OF SEDIMENT Dry weight of sediment per unit volume of water-sediment mixture (i.e., mg/l, ppm).

CONCENTRATION OF SOLUTION Number of solute in confirmed level of solvent or solution expressed as weight:weight or weight:volume connection.

CONCENTRATION POLARIZATION Denotes part of polarization of electrolytic cell resulting from changes in electrolyte concentration due to passage of current through electrode and/or solution interface.

CONCENTRATION TANK Detention settling tank where sludge is concentrated by sedimentation or flotation.

CONCENTRATION TIME 1. Time required for storm runoff to flow from hydraulically most remote point (i.e., furthest) of catchment, drainage area or watershed to outlet or point under consideration; 2. Time from end of rainfall excess to inflection point on recession curve.

CONCENTRATION vs. TIME STUDY Plots showing measured concentration of specified compound in solution as function of elapsed time.

CONCENTRATOR 1. Apparatus for separation of dry comminuted ore by exposing it to intermittent puffs of air; 2. Solids contact unit that reduces water content of sludge or slurry.

CONCENTRIC RINGS Quasi-constant intensifying or weakening of intense hurricanes that mark end period of intensification.

CONCEPTUAL MODEL Interpretation of characteristics and dynamics of aquifer system that is based on examination of all available hydrogeological data for modeled area.

CONCEPTUALIZATION ERROR Modeling error where model formulation is based on incorrect or insufficient understanding of modeled system.

CONCORDANT FLOW River system that has same recurrence interval or frequency of occurrence at different locations.

CONCORDANT FLOWS Flows at different points in river system that have same recurrence interval or same frequency of occurrence.

CONCRETE United by coalescence of separate particles or liquid, into one mass or solid (e.g., crushed stone, sand, cement and water).

CONCRETE BARRIER Barrier used as median to protect from temporary hazard.

CONCRETE COVER Minimum distance between surface of embedded reinforcement and outer surface of concrete.

CONCRETE DAM Concrete dam that generally requires sound rock foundation.

CONCRETE LIFT Vertical distance between successive horizontal construction joints.

CONCRETE PIPE Porous pipe used principally for subsoil drainage.

CONCRETE PUMP Machine used for transferring liquid concrete by pumping.

CONCRETE REVETMENT Concrete slabs placed for erosion control, scour protection and soil stabilisation.

CONCRETION Process of aggregating or coalescing into mass.

CONCURRENT 1. Happening at same time; 2. Meeting in one point.

CONCURRENT ENGINEERING Method of designing and developing products, where different stages run simultaneously, rather than consecutively.

CONCURRENT FLOODS Flood flows expected at point on river system below dam at same time flood inflow occurs above dam.

CONCURRENT FLOW HOV LANE Lane in traffic direction for use by high occupancy vehicles.

CONCURRENT FLOW PREFERENTIAL LANE Preferential lane operated in same direction as adjacent mixed flow lanes, separated from adjacent general-purpose freeway lanes by standard lane stripe, painted buffer or barrier.

CONCUSSION Shock or sharp air waves caused by explosion or heavy blow.

CONDEMNATION Process where public entity exercises its powers of eminent domain.

CONDENSATE 1. Water created by cooling steam or water vapor; 2. Liquid product of condensation of gas (i.e. of steam); 3. Any of various condensed quantum states.

CONDENSATE RETURN SYSTEM System that returns heated water condensing within steam piping to boiler and thus saves energy.

CONDENSATION Process where water changes from vapor state (i.e., gas) into liquid or solid state.

CONDENSATION FUNNEL Funnel-shaped cloud associated with rotation and consisting of condensed water droplets.

CONDENSATION LEVEL Height where parcel of air becomes saturated when it is lifted dry-adiabatically.

CONDENSATION NUCLEUS Small particles on which water vapor condenses.

CONDENSATION SAMPLING Gas-sampling technique where gas is passed through tubes immersed in refrigerant, thus trapping and condensing various fractions of gas mixture.

CONDENSATION TRAIL 1. Contrail; 2. Visible line of condensation flowing from edge of airplane wing or similar structure.

CONDENSER 1. Device designed to condense gas into liquid, either as part of still, steam engine, refrigerator or similar machine; 2. lens (or combination of lenses) designed to gather light and focus it onto specimen or part of mechanism.

CONDITIONAL INSTABILITY State of layer of unsaturated air when its lapse rate of temperature is less than dry-adiabatic lapse rate but greater than moist-adiabatic lapse rate.

CONDITIONAL REGISTRATION Permits registration "conditional" upon submission of additional data.

CONDITIONAL WATER SUPPLY Water supply agreement "conditional" upon something.

CONDITIONALLY EXEMPT GENERATORS Persons or enterprises which produce less than 220 pounds of hazardous waste per month.

CONDITIONALLY UNSTABLE AIR Atmospheric condition that exists when environmental lapse rate is less than dry adiabatic lapse rate but greater than moist adiabatic lapse rate.

CONDITIONS OF CONTRACT Include 'general conditions' that are common to all types of contracts, as well as 'special conditions' that are peculiar to specific contract.

CONDUCTANCE Measure of degree that object conducts electricity, equal to reciprocal of its resistance.

CONDUCTION Flow of heat in response to temperature gradient within object or between objects that are in physical contact.

CONDUCTIVITY Measure of ability of material to conduct electrical charge.

CONDUCTIVITY BRIDGE Modified Kelvin bridge for measuring very low resistances.

CONDUCTOR 1. Person who takes tickets on public transportation; 2. Something that can transmit electricity, heat, light or sound.

CONDUCTOR PIPE Large diameter pipe set into ground to provide initial stable structural foundation for borehole or oil well.

CONDUIT 1. Pipe or channel for conveying water; 2. Duct or tube where electrical cables may be pulled; 3. Covered portion of spillway between gate or crest structure and terminal structure; 4. Portion of outlet works between intake structure and gate chamber and/or control structure.

CONE Surface of revolution formed by rotating segment of line around another line that intersects first line.

CONE OF DEPRESSION 1. Depression, roughly conical in shape, produced in water table or other piezometric surface, by pumping of water from well at given rate; 2. Volume of cone will vary with rate of withdrawal of water.

CONE OF INFLUENCE 1. Depression, roughly conical in shape, produced in water table or other piezometric surface, by pumping of water from well at given rate.

CONE PENTEROMETER TESTING Direct push system used to measure lithology based on soil penetration resistance where sensors in tip of cone of DP rod measure tip resistance and side-wall friction, transmitting electrical signals to digital processing equipment on ground surface.

CONE VALVE Valve where moving stopper is conical and opened by unscrewing stopper from seat and turning it 90°.

CONE-SHEET Type of high-level igneous intrusion found in partly eroded central volcanic complexes.

CONFIDENCE INTERVAL Range defined such that there is fixed probability that value of measured parameter will fall within range.

CONFIDENCE LEVEL Probability that measured quantity will fall within given confidence interval.

CONFIDENCE LIMITS For statistical sample pair of values that delimit interval for which there is certain probability that true value of some deduced variable lies between those values.

CONFIDENTIAL BUSINESS INFORMATION Material that contains trade secrets or commercial or financial information that has been claimed as confidential by its source (e.g., pesticide, new chemical formulation registrant).

CONFIDENTIAL STATEMENT OF FORMULA List of ingredients in new pesticide or chemical formulation.

CONFINED AQUIFER 1. Aquifer bounded above and below by confining beds (e.g., impermeable material) where hydraulic head is above top of aquifer; 2. Aquifer under pressure so that when aquifer is penetrated by well, water will rise to level associated with pressure.

CONFINED DISPOSAL FACILITY Engineered area for containment of dredged material.

CONFINED GROUNDWATER Groundwater under pressure significantly greater than atmospheric, with its upper limit bottom of bed with hydraulic conductivity distinctly lower than that of material where confined water occurs.

CONFINED SPACE 1. Space that has limited or restricted means for entry and exit; 2. Space large enough to and so configured that person can bodily enter and perform assigned work; 3. Space not designed for continuous occupancy.

CONFINED-WATER WELL Well that taps confined aquifer.

CONFINEMENT Concrete with closely-spaced transverse reinforcement to restrain concrete perpendicular to applied stress.

CONFINING BED Hydrogeologic unit of less permeable material (e.g., layer of rock, unconsolidated sediments) that retards movement of water in and out of aquifer and possesses very low hydraulic conductivity.

CONFINING LAYER Body of impermeable or low permeability that bounds aquifer.

CONFINING STRATUM Confining layer right above or below one bearing water.

CONFINING UNIT Hydrogeologic unit of less permeable material bounding one or more aquifers.

CONFINING ZONE Geological formation (or group or part of formation) capable of limiting fluid movement out of injection zone.

CONFIRMATION Process where model of specific study area is built and tested to prove that model design and implementation is adequate and no major phenomenon has been overlooked.

CONFLICT Traffic movements at intersection.

CONFLICT MONITOR Device that automatically puts signal into flash mode until issue is resolved.

CONFLICTING FLOW RATE Number of conflicting vehicles at unsignalized intersection observed divided by observation time.

CONFLICTING MOVEMENTS Traffic occupying same space and time.

CONFLUENCE 1. Point where two rivers, streams, etc. join and become one; 2. Pattern of wind flow where air flows inward toward axis oriented parallel to general direction of flow.

CONFLUENT 1. Wind that converges, especially when viewed on weather chart; 2. Flowing together; 3. Blending into one; 4. Describing cells in culture that merge to form mass; 5. Triangle exactly same size as another triangle.

CONFLUENT GROWTH Continuous bacterial growth covering all or part of filtration area of membrane filter where bacteria colonies are not discrete.

CONFLUENT STREAM Stream that merges with another.

CONFORMITY MARK Label, postage, label or alternative mark put on item by its own producer to show its certification status.

CONGENER 1. Plant or animal of same taxonomic genus as another; 2. Any of group of structurally-related compounds.

CONGESTION 1. Excess of traffic; 2. Hindrance or blockage of passage resulting in overfilling or overcrowding.

CONGESTION PRICING Demand management strategy fee paid by users to enter restricted area.

CONGLOMERATE Sedimentary rock composed of rounded gravel (pebbles, cobbles and boulders) cemented together, usually found with sandstone, larger than 2 millimeters in diameter.

CONGRESSIONAL AUTHORIZATION Act by Congress of United States which authorizes use of public funds to carry out prescribed action.

CONGRESSIONAL ORGANIC ACT OF 1890 Act that assigned responsibility of river and floor forecasting for benefit of general welfare of Nation's people and economy to Weather Bureau and subsequently National Weather Service.

CONGRESSIONALLY AUTHORIZED RECLAMATION PROJECT Congressionally approved Bureau of Reclamation project that has been authorized for specific purposes.

CONICAL BURNER Hollow, cone-shape combustion chamber that has exhaust vent at its point and door at its base where waste materials are charged and air is delivered to burning solid waste inside cone.

CONIFERS Cone-bearing trees or shrubs, mostly evergreens (i.e., pine, cedar, spruce.)

CONING 1. Pattern of wildfire plume dispersion in neutral atmosphere, where plume attains form of cone with its vertex at top of stack; 2. Conduct of plume when it enlarges to form nearly conical shape; 3. Growth of cone shaped depression in top or top layer of fluid that constant volume will be removed in reduced amount.

CONJUGATE DEPTH Depth upstream and downstream of hydraulic jump when momentum fluxes are equal for given discharge.

CONJUGATE DEPTHS Depth upstream and downstream of hydraulic jump whose momentum fluxes are equal for given discharge (i.e., volume flux).

CONJUGATE POINTS Two points on Earth's surface, at opposite ends of geomagnetic field line.

CONJUNCTIVE USE Coordinated use of surface water and groundwater resources.

CONNATE WATER Water trapped in pores of sedimentary rocks as they were deposited.

CONNECTED PIPING All underground piping including valves, elbows, joints, flanges and flexible connectors attached to tank system where regulated substances flow.

CONNECTION BAND Coupling that fits over adjoining conduit stops that holds conduit together either by friction or mechanical bond.

CONNECTIVITY Ability of transportation network to make connection between two or more points in network.

CONNECTOR 1. Device for connecting together two wires, cables or conduits, allowing electricity or fluid to flow but also allowing easy disconnection and reconnection when necessary; 2. Highway or freeway road which connects to another highway or freeway.

CONNECTOR FITTING Section of fitting or component connected to piece of conduit by being inserted into that conduit.

CONSECUTIVE DIGESTION Digestion by thermophile, followed promptly by digestion and concentration under mesophilic conditions.

CONSENT DECREE Legal document, approved by judge, that formalizes agreement reached between agency and potentially responsible parties.

CONSENT OF SURETY Written consent of surety on performance bond and/or labor and material payment bond to contract changes (e.g., change orders, reduction in contractor's retainage).

CONSEQUENCE ASSESSMENT Study of describing and quantifying relationship between risk and adverse outcome.

CONSEQUENCES 1. Potential loss of life or property damage downstream of dam caused by floodwaters released at dam or by waters released by partial or complete failure of dam that includes effects of landslides upstream of dam on property located around reservoir; 2. Potential number of lives that would be lost from dam failure and uncontrolled reservoir release (considers load, failure modes and estimated population distribution warning times); 3. Interface between two successive placements or pours of concrete (i.e., construction joint) where bond and not permanent separation, is intended.

CONSEQUENT Following as result, inference or natural effect (e.g., rivers path determined by natural slope and landscape).

CONSEQUENTIAL DAMAGES Damages that can be proven to have occurred because of failure of one party to meet contractual obligation.

CONSEQUENTIAL LOSS Indirect loss resulting from insured's inability to use business property or equipment.

CONSERVANCY SYSTEM Type of waste disposal and/or refuse where wastes are collected separately and disposed off.

CONSERVATION 1. Increasing efficiency of energy use, water use, production or distribution; preserving and renewing human and natural resources; 2. Lack of change in measurable property of isolated physical system (e.g., conservation of energy, mass, momentum).

CONSERVATION EASEMENT Easement restricting landowner to land uses that that are compatible with long-term conservation and environmental values.

CONSERVATION OF LAND Policy tool to mitigate urban sprawl, protect natural areas and habitat and provide ecosystem services.

CONSERVATION OF MOMENTUM PRINCIPLE Two objects (in isolated system) that collide have same summed momentum before and after.

CONSERVATION STORAGE Storage of water for later release for usual purposes (e.g., municipal water supply, power, irrigation) in contrast with storage capacity used for flood control.

CONSERVATIVE POLLUTANT Pollutants that do not readily degrade in environment and which are mitigated primarily by natural stream dilution after entering receiving bodies of waters.

CONSIGNEE Receives waste for recycling or disposal.

CONSIGNOR Person or firm that delivers consignment to carrier for transporting it to named consignee.

CONSIST Train or vehicle composition.

CONSISTENCY Relative ease with which soil can be deformed.

CONSISTENCY LIMITS Boundaries (determined by laboratory tests) of moisture content in soil between liquid state and plastic state (known as liquid limit), between plastic state and semisolid state (known as plastic limit) and between semisolid state and solid state (known as shrinkage limit).

CONSISTENT ESTIMATOR Rule for computing estimates that states as sample size gets large, estimate gets closer to true value of parameter.

CONSOCIATION 1. Subdivision of association having one dominant species of plant; 2. Intimate union.

CONSOLE 1. Instrument with displays and input device used to monitor and control electronic system; 2. Ornamental member jutting out of wall to carry superincumbent weight.

CONSOLIDATED ICE COVER Ice cover formed by packing and freezing together of floes, brash ice and other forms of floating ice.

CONSOLIDATION Gradual compression of cohesive soils of low permeability naturally due to imposition of weight that drives water and air out of voids in soil.

CONSOLIDATION CURVE Any function relating volume change to time.

CONSOLIDATION GROUTING Strengthening area of ground by injecting grout.

CONSOLIDATION SEDIMENTATION Procedure where substances become solid and coherent under action of gravity.

CONSOLIDATION SETTLEMENT Reduction in volume due to expulsion of water under increased load.

CONSOLIDATION TEST Performed to determine magnitude and rate of volume decrease that laterally confined soil specimen undergoes when subjected to different vertical pressures.

CONSOLIDOMETER Apparatus used for one-dimensional consolidation/compression testing.

CONSTANCY Unscientific notion that species of organisms do not change over time.

CONSTANT AMPLITUDE FATIGUE THRESHOLD Nominal stress below range that would cause fatigue failure after infinite repetitions.

CONSTANT ANGLE ARCH DAM Arch dam where angle subtended by any horizontal section is constant throughout whole height of dam.

CONSTANT PRESSURE CHART 1. Weather map representing conditions on surface of equal atmospheric pressure; 2. Isobaric chart.

CONSTANT RADIUS ARCH DAM Arch dam where every horizontal segment or slice of dam has approximately same radius of curvature.

CONSTANT SPRING Variation in discharge from maximum to minimum of spring does not exceed one third of its average.

CONSTANT WARNING TIME DETECTION Way of detecting rail traffic that provides uniform warning time for traffic approach that are not changing speeds after detection.

CONSTANT-HEAD BOUNDARY Boundary in discretized groundwater flow model domain (node) where hydraulic head remains same over time period considered.

CONSTANT-HEAD NODE Location in discretized groundwater flow model domain (node) where hydraulic head remains same over time period considered.

CONSTITUENT Chemical or biological substance in water, sediment or biota that can be measured by analytical method.

CONSTITUENT OF CONCERN Specific chemicals that are identified for evaluation in site assessment process.

CONSTRICTION 1. Compression; 2. Obstacle that confines flow.

CONSTRUCTION 1. Manner where something is built; 2. Geometric figure of arcs and line segments that's drawable with straightedge and compass.

CONSTRUCTION AND DEMOLITION WASTE Waste building materials, dredging materials, tree stumps and rubble resulting from construction, remodeling, repair and demolition of homes, commercial buildings and other structures and pavements that may contain lead, asbestos or other hazardous substances.

CONSTRUCTION AND DEMOLITION WASTES Building material waste and debris caused by construction, repair, remodelling and demolition.

CONSTRUCTION BAN If area's planning requirements for correcting non-attainment is not approved, construction or modification of any major stationary source of pollutant is not permitted for which area is in non-attainment.

CONSTRUCTION BUDGET Amount of money allotted for specific building or remodeling project.

CONSTRUCTION CAPITAL Major, non-recurring expenditure in construction.

CONSTRUCTION CONTRACT Legal binding agreement, for both owner and builder, that executed job will receive specific amount of compensation or how compensation will be distributed.

CONSTRUCTION CONTRACTOR Individual or corporation appointed by clients to carry out construction works.

CONSTRUCTION COST Expense incurred by contractor for labor, material, equipment, financing, services, utilities, etc., plus overheads and contractor's profit.

CONSTRUCTION DEFECT Deficiency in design or construction of building or structure resulting from failure to design or construct in reasonably workmanlike manner, and/or in accordance with buyer's reasonable expectation.

CONSTRUCTION DOCUMENTS PHASE Preparation of drawings and specifications establishing requirements for construction of project.

CONSTRUCTION ESTIMATOR Person who calculates price of time, labor and materials for specific job and then puts together bid package.

CONSTRUCTION JOINT Joints placed to facilitate construction, reduce initial shrinkage stresses and cracks, allow time for later installation or allow for subsequent placing of other materials.

CONSTRUCTION LOAD Load that structure is subjected to during construction.

CONSTRUCTION LOAN Loan to contractor adequate to cover cost of construction.

CONSTRUCTION MANAGEMENT Professional service that uses specialized, project management techniques to oversee planning, design and construction of project, from its beginning to its end.

CONSTRUCTION MANAGEMENT FIRM Company that performs responsibilities of construction supervisor and/or staff.

CONSTRUCTION MANAGER Person that plans, coordinates, budgets and supervises construction projects from start to finish.

CONSTRUCTION PHASE Period during which contractor takes control of construction site to carry out pathworks.

CONSTRUCTION PROGRAM Plan that sets out construction operation sequence.

CONSTRUCTION SCHEDULE Shows individual tasks and phases of project, their associated durations, their sequencing requirements and their dependencies upon one another for timely completion of projects.

CONSTRUCTION STAGE PLANNING Commences with acceptance of tender and extends till completion of contract.

CONSTRUCTION START Beginning of excavation for footings or foundation.

CONSTRUCTION TIME Time from commencement of site work to handing over finished job.

CONSTRUCTION WARRANTY Guarantee for construction given by seller or building contractor of new house or building.

CONSTRUCTION WORK Carrying out of any building, civil engineering or engineering construction work.

CONSTRUCTIVE Created to give legal effect to something for equitable reasons.

CONSULTANT Person whose occupation is to be consulted for their expertise, advice or help in area or specialty.

CONSULTING ENGINEER Expert in planning, design and construction of both public and private infrastructures.

CONSUMER SURPLUS 1. Value of commodity, good or opportunity above cost to consumer; 2. Measured using willingness to pay, as specified in federal guidelines for water resources planning.

CONSUMER WASTE Substances that were used and discarded by client and/or consumer.

CONSUMPTION RESIDUES Wastes that result from final consumption of goods or services.

CONSUMPTIVE USE Water removed from available supplies without return to water resource system.

CONSUMPTIVE WASTE Water returning to atmosphere without benefiting humanity.

CONSUMPTIVE WATER USE 1. Use which lessens amount of water available for another use; 2. Quantity of water not used for manufacturing, plant or animal tissue.

CONTACT AERATOR Tank where sewage settled on bed of stone, cement-asbestos or other surfaces is treated by aeration with compressed air.

CONTACT BED Bed filled with coke or other coarse material used to purifying sewage left for contact to promote bacterial action and then filtered off.

CONTACT FILTER Filter that provides desired chemical and turbidity removal.

CONTACT GROUTING Filling, with cement grout, any voids existing at contact of two zones of different materials usually carried out at low pressure.

CONTACT LOAD Sediment particles that roll or slide along in almost continuous contact with streambed.

CONTACT PESTICIDE 1. Chemical that kills pests when it touches them, instead of by ingestion; 2. Soil that contains minute skeletons of certain algae that scratch and dehydrate waxy-coated insects.

CONTACT RECREATION Recreational activities (i.e., swimming, kayaking) where contact with water is prolonged or intimate and where there is likelihood of ingesting water.

CONTACT STABILIZATION Modification of conventional activated sludge process where on or two aeration tanks are used to separate re-aeration of return sludge for some time before it is permitted to be mixed with primary effluent requiring treatment.

CONTACT TANK Tank where contact between chemicals and other materials reacts with contaminants in liquid.

CONTAINER Item where objects, materials or data can be stored or transported.

CONTAINER FREIGHT STATION Facility where freight shipments are consolidated, deconsolidated and/or staged between transport legs.

CONTAINMENT LEVEE Dike or embankment to contain stream flow.

CONTAMINANT 1. Any physical, chemical, biological or radiological substance or matter that has adverse effect on air, water or soil; 2. Impurity; 3. Material added by humans or natural activities that may, in sufficient concentrations, render environment unacceptable for biota.

CONTAMINANT FATE Chemical changes and reactions that change chemical nature of contaminant, effectively removing contaminant from subsurface hydrologic system.

CONTAMINANT TRANSPORT MODEL Model describing movement of contaminants in environment.

CONTAMINANT TRANSPORT VELOCITY Rate where contamination moves through aquifer.

CONTAMINATE To make something dangerous or toxic by introducing impurities or foreign matter.

CONTAMINATED LAND Land with hazardous substances in or on it that are reasonably likely to have significant adverse effects on environment (including human health).

CONTAMINATED NON-PROCESS WASTEWATER Water, including precipitation runoff that, during manufacturing or processing, comes into incidental contact with any raw material, intermediate product, finished product, by-product or waste product by means of precipitation runoff, accidental spills, accidental leaks or discharges from safety showers and related personal safety equipment.

CONTAMINATED RUNOFF Overflow that contacts any raw material, intermediate product, finished product, by product or waste product.

CONTAMINATION 1. Degradation of quality compared to original or natural conditions due to human activity; 2. Introduction into water, air and soil of microorganisms, chemicals, toxic substances, wastes or wastewater in concentration that makes medium unfit for its next intended use.

CONTAMINATION LEVEL Measure of pollution.

CONTAMINATION SOURCE INVENTORY Inventory of contaminant sources within delineated State Water-Protection Areas.

CONTEMPLATED CHANGE NOTICE Form issued to construction contractor requesting its quotation for proposed change in scope of work that would be included in its contract.

CONTENTS Volume of water in reservoir.

CONTENTS HAZARD CLASSIFICATION Categorization of possible risk of construction contents as ordinary, high or low.

CONTIGUOUS 1. Actual contact with; 2. Also, near or adjacent to.

CONTIGUOUS ZONE Band of water extending farther from outer edge of territorial sea to up to 27.6 miles from baseline, where state can exert limited control for purpose of preventing or punishing "infringement of its customs, fiscal, immigration or sanitary laws and regulations within its territory or territorial sea.

CONTINENTAL AIR MASS Dry air mass originating over large land area.

CONTINENTAL CLIMATE Relatively dry climate with very hot summers and very cold winters (e.g., central parts of Asia and North America).

CONTINENTAL DRIFT Slow movement of continents explained by plate tectonics.

CONTINENTAL RISE Wide, gentle incline found between continental slope and abyssal plain.

CONTINENTAL SHELF Zone bordering continent and extending to ocean depth from which there is steep descent toward greater ocean depth.

CONTINENTAL SLOPE Steep, narrow fringe separating coastal zone from deep ocean.

CONTINGENCIES Used in appraisal and feasibility estimates to estimate overruns on quantities, changed site conditions, change orders, etc.

CONTINGENCY Amount of money which party to contract has to pay to other party if he or she does not fulfill contract according to specification.

CONTINGENCY FUND Reserve of money set aside to cover possible unforeseen future expenses.

CONTINGENCY PLAN 1. Document setting out organized, planned and coordinated course of action to be followed in case of fire, explosion or other accident that releases toxic chemicals, hazardous waste or radioactive materials that threaten human health or environment; 2. Alternative plan to be put into operation if needed, especially in case of emergencies or if primary plan fails.

CONTINGENCY PLANNING Plan devised for outcome other than in expected.

CONTINGENCY RESERVE Money to provide cushion against financial surprises.

CONTINGENT AGREEMENT Any agreement under which rights or obligations of party are subject to happening of stated contingency.

CONTINGENT VALUATION Survey method asking for maximum values that users would pay for access to particular activity.

CONTINGENT VALUATION EXPENDITURE Sum individuals would pay to avoid unwanted effects.

CONTINUING ENVIRONMENTAL PROGRAMS Pollution control programs that will not be finished by definable time period.

CONTINUITY Structural continuity where two or more simple spans are connected end-to-end.

CONTINUITY EQUATION States that rate where mass enters system is equal to rate where mass leaves system plus accumulation of mass within system.

CONTINUOUS DISCHARGE Routine release to environment that occurs without interruption, except for infrequent shutdowns for maintenance, process changes, etc.

CONTINUOUS DISPOSAL System of tailings disposal and management where tailings are dewatered by mechanical processes after generation.

CONTINUOUS EMISSION MONITORING SYSTEM System of gas analyzers, gas sampling system, temperature, flow and opacity monitors that are integrated with data acquisition system to demonstrate environmental regulatory compliance of various industrial sources of air pollutants.

CONTINUOUS FLOW PUMP Displacement pump that doesn't change or reverse direction of flow.

CONTINUOUS FOREST INVENTORY Series of uniformly spaced permanent sample plots that are measured periodically to quantify forest conditions and changes.

CONTINUOUS FURNACE Furnace operated on uninterrupted cycle.

CONTINUOUS HORSEPOWER 1. Power recommended by manufacturer for satisfactory operation under specified continuous-duty conditions; 2. Amount of horsepower engine is capable of developing for more than 24 hours.

CONTINUOUS INTERSTICE Interstice in granular substance which is linked with other interstices.

CONTINUOUS IRRIGATION System whereby each irrigator gets allotted amount of water throughout season.

CONTINUOUS MECHANICAL VENTILATION Ventilation (i.e., mechanical) system that operates non-stop.

CONTINUOUS MODEL Model that tracks periods between precipitation events, as well as events themselves.

CONTINUOUS OPERATIONS Method where business maintains continuity of output to internal systems and customers through uninterrupted delivery of critical services or functions.

CONTINUOUS PROCESS Materials, either dry bulk or fluids that are being processed are continuously in motion, undergoing chemical reactions or subject to mechanical or heat treatment.

CONTINUOUS RECORDER Recorder that saves instantaneous data values at high frequencies.

CONTINUOUS SAMPLE Flow of water, waste or other material from particular place in plant to location where samples are collected for testing.

CONTINUOUS SAMPLING Method that records all of activity.

CONTINUOUS SLUDGE-REMOVAL TANK Sedimentation tank equipped for constant removal of sludge in restricted speed.

CONTINUOUS STREAM Stream that does not have interruptions in space (i.e., its course).

CONTINUOUS VENT Vertical vent that's continuation of drain, soil pipe or waste pipe to which vent connects.

CONTINUOUS WASTE Several fixtures linked to exactly same trap.

CONTINUOUS WASTE AND VENT Single pipe, upper portion of which is vent, while lower portion functions as waste drain.

CONTINUOUS WELDED RAIL Rails welded together by utilising flash butt to form single, continuous rail.

CONTINUOUS-FEED INCINERATOR Incinerator where solid caste is charged nearly always to keep constant speed of burning.

CONTINUOUS-FLOW INTERCHANGE Diamond interchange where vehicles cross before attempting to turn across opposing traffic direction.

CONTINUOUS-FLOW IRRIGATION System of irrigation water delivery where each irrigator receives their allotted quantity of water at continuous rate.

CONTINUOUS-FLOW TANK Tank that serves both as storage tank and as settling tank, where liquid flows at regular rate.

CONTINUOUS-RECORD STATION Site where data are collected with sufficient frequency to define daily mean values and variations within day.

CONTINUOUSLY REGENERATING TRAP Apparatus that removes particulates in diesel engine exhaust using nitrogen dioxide to oxidize particulates.

CONTINUOUSLY REINFORCED CONCRETE PAVEMENT
Concrete pavement that doesn't require transverse contraction joints.

CONTINUUM STORM General term for solar noise lasting for hours and sometimes days.

CONTOUR 1. Line of constant elevation; 2. Line on map or chart delineating points that have same altitude or other plotted quantity.

CONTOUR BASIN Basin formed by edges or levees that follows constant ground elevation.

CONTOUR CHECK Regions within irrigated section of land divided by edges which are sited along contours.

CONTOUR DITCH Irrigation ditch laid out approximately on contour.

CONTOUR FARMING Conservation-based method of farming where all farming operations (e.g., tillage, planting) are performed perpendicular to (rather than up and down) slope.

CONTOUR FLOODING Method of irrigation by flooding from contour ditches.

CONTOUR FLYING Technique of flying at constant altitude of less than 75 feet above ground.

CONTOUR GAUGE Gauge that traces contours of surface when pushed against.

CONTOUR GRADIENT Imaginary line on surface of earth having constant inclination with horizontal (i.e., slope).

CONTOUR INTERVAL Difference in elevation represented by space between adjacent contour lines.

CONTOUR IRRIGATION Describes how farmers can use contour-furrow irrigation as means of preventing soil erosion and maintaining even watering of crops.

CONTOUR LINE Line on map through points of equal elevation, often height above sea level.

CONTOUR MAP Topographic map that portrays relief or elevation differences by use of lines (contours) indicating equal elevation.

CONTOUR PLOWING Soil tilling method that follows shape of land to discourage erosion.

CONTOUR SCALING Crust forming over top of sandstones and limestones that follows contour of surface in place of bedding planes of rocks.

CONTOUR STRIP FARMING Contour farming where row crops are planted in strips, between alternating strips of close-growing, erosion resistant forage (i.e., grass, grain, hay) crops.

CONTRACT Agreement between two or more parties, to perform specific job or work order, often temporary or of fixed duration and usually governed by written agreement, that law will enforce.

CONTRACT ADMINISTRATION Management of contracts made with customers, vendors, partners or employees.

CONTRACT BOND Bond that guarantees contract is fulfilled.

CONTRACT CARRIER Trucking or shipping company that serves specific customers and may refuse to convey passengers or freight for payment.

CONTRACT CHANGE Contract alteration that should be consented and acknowledged in writing by all parties represented in initial contract.

CONTRACT CLOSEOUT Occurs when contract has met all terms of contract and all administrative actions have been completed, all disputes settled and final payment has been made.

CONTRACT DATE Date on which contract is awarded.

CONTRACT DOCUMENTS Written documents that define roles, responsibilities and work under contract that are legally-binding to both parties.

CONTRACT DOCUMENTS STAGE Preparing final detailed drawings and specifications.

CONTRACT ITEM Procurement item of work where price is provided in contract.

CONTRACT LABS Laboratories under contract that analyze samples taken from waste, soil, air and water or carry out research projects.

CONTRACT LOGGING Person doing all or part of logging.

CONTRACT OVERRUN Amount where original contract price, as adjusted by change orders, differs from total cost of project at completion.

CONTRACT PAYMENT BOND Legal contract, type of bond, that guarantees certain employees, subcontractors and suppliers are protected against non-payment.

CONTRACT PERFORMANCE BOND Bonds that ensure satisfactory completion of project by contractor.

CONTRACT PERIOD Number of working days or calendar days, from specified commencement date to specified completion data, as provided for in contract.

CONTRACT PRICE Price for goods or services to be received in contract.

CONTRACT RATE Repayment or water service rate set forth in contract to be paid by district to United States.

CONTRACT RENT Monthly rent agreed to or contracted for, regardless of any furnishings, utilities, fees, meals or services that may be included.

CONTRACT TIME Maximum time allowed in contract for completion of all work contained in contract documents.

CONTRACT UNDERRUN Difference between contract price and closing expense.

CONTRACTED WEIR Weir shorter that width of channel whose flow it measures, so that effects from crest and sides is negligible on flow.

CONTRACTED-OPENING DISCHARGE MEASUREMENT Determination by indirect measurement of peak discharge following flood through observation of high-water marks and channel and bridge geometry at bridge constriction.

CONTRACTING AGENCY Department, agency, corporation, establishment, partnership, individual or government body that enters into contracts.

CONTRACTING OFFICER Individual entrusted by organization with authority to enter into, administer, renew or terminate contract s, along with related determinations and findings.

CONTRACTING OUT To agree by contract to pay someone outside organization to perform task.

CONTRACTING REACH Channel reach where flow is accelerating.

CONTRACTION 1. Reduction in cross-sectional area of conduit along its longitudinal axis; 2. Extent where cross-sectional area of jet, nappe or stream is decreased after passing orifice, weir or notch; 3. Shortening of muscle when it is used.

CONTRACTION JOINT 1. Joints placed in concrete to provide for volumetric shrinkage of monolithic unit or movement between monolithic units; 2. Joints that have no bond between concrete surfaces forming joint.

CONTRACTION METHOD OF FLOW MEASUREMENT Indirect measurement of peak discharge using highwater mark and geometry at constriction.

CONTRACTION SCOUR Scour that occurs when water accelerates as it flows through opening smaller than upstream.

CONTRACTOR Person or company hired to perform specific tasks.

CONTRACTOR DEFAULT Contractor failure to perform provisions of contract.

CONTRACTOR'S AFFIDAVIT Notarized statement by contractor must issue before contractor can sue for lack of payment.

CONTRACTOR'S EQUIPMENT FLOATER Form of inland marine insurance, often on all-risk basis, covering various kinds of equipment.

CONTRACTOR'S LIABILITY INSURANCE Covers costs of lawsuits by all third-party claims.

CONTRACTOR'S OPTION Stipulation in contract that gives contractor right to select certain materials or methods from approved list.

CONTRACTUAL LIABILITY Liability that one party assumes on behalf of another via contract.

CONTRAFLOW LANE Lane where traffic flows in opposite direction of lanes surrounding.

CONTRARIES 1. In opposite direction; 2. Substances that have detrimental effect upon reprocessing or recycling.

CONTRIBUTING AREA Area in drainage basin that contributes water to streamflow or recharge to aquifer.

CONTROL 1. Channel section (i.e., constriction) where bed and bank conditions give quantifiable indicator of flow; 2. Security mechanism, policy or procedure that can counter system attack, reduce risks and resolve vulnerabilities.

CONTROL AREA Part of power system or combination of systems, that common electrical generation allocation scheme is applied.

CONTROL CENTER Point of collection and management of all data from field.

CONTROL CHARTS Tool used to determine whether manufacturing or business process is in state of statistical control or not.

CONTROL CONDITION Traffic signs and controls that regulate intersections.

CONTROL CONSOLE Electronic device used to control multiple items at once.

CONTROL DELAY Delayed caused by vehicles reducing speed prior to intersection.

CONTROL DEVICE Measure of protection that restricts user access to devices.

CONTROL FACTOR Ratio of minimum compressive strength to average compressive strength.

CONTROL FLOAT Float switch that turns on and off in response to water level.

CONTROL FLUME Open conduit or artificial channel with constricted section to measure flow.

CONTROL HOUSE Structure on stream or canal used to regulate flow or stage of stream or to prevent intrusion of saltwater.

CONTROL JOINT 1. Joints placed in concrete to provide for control of initial shrinkage stresses and cracks of monolithic units; 2. Unbonded joints to provide weak areas for cracking; 3. Joints that will transfer moment but not shear unless keyed.

CONTROL LIMITS 1. Horizontal lines drawn on statistical process control chart, usually at distance of 3 standard deviations of plotted statistic from statistic's mean; 2. Limitations regarding personal vulnerability to dangerous materials.

CONTROL LINE Building of natural barriers to forest fire.

CONTROL LOOP Group of components working together as system to achieve and maintain desired value of system variable by manipulating value of another variable.

CONTROL OF ACCESS Condition that limits types of vehicles on roadway and road design that limits access points for vehicles.

CONTROL POINTS Small monuments securely embedded in surface of dam monitored to detect horizontal and vertical movement of dam.

CONTROL REACH Cross-section where discharge is uniquely related to depth of flow or to energy head.

CONTROL ROD Any of number of steel tubes, containing boron or other neutron absorber, that are inserted into core of nuclear reactor in order to control its rate of reaction.

CONTROL ROOM Room serving as center of monitoring building, controlling operation, etc.

CONTROL SECTION Section of channel where for given discharge, depth of flow is known.

CONTROL STOP Apparatus installed in supply piping to modulate or shut off flow of water into flush valve.

CONTROL STRATEGY Planned set of controls, derived from current product and process. understanding, that assures process.

CONTROL STRIP Short pavement section compacted to desired density and used as compaction standard for particular job.

CONTROL STRUCTURE Structure on stream or canal used to regulate flow or stage of stream or to prevent intrusion of saltwater.

CONTROL SUBSTANCE Drug or chemical whose manufacture, possession or use is regulated by government.

CONTROL SYSTEM Manages, commands, directs or regulates behavior of other devices or systems using control loops.

CONTROL TECHNIQUE GUIDELINES Documents designed to assist state and local authorities to achieve and maintain standards for certain sources through reasonably available control technologies.

CONTROL VALVE Apparatus used to control fluid flow by varying size of flow passage as directed by signal from controller.

CONTROL WORKS Equipment that regulates flow.

CONTROLLED Refers to device that controls traffic signal and intersection.

CONTROLLED AIRSPACE Term for different classifications of airspace (Class A, B, C, D and E) with defined dimensions where air traffic control services are provided.

CONTROLLED AREA Any area to which access is managed by or for DOE to protect individuals from exposure to radiation and/or radioactive material.

CONTROLLED BURNING Any fire intentionally ignited to meet specific land management objectives, such as to reduce flammable fuels, restore ecosystem health, recycle nutrients or prepare area for new trees or vegetation.

CONTROLLED FILL Known material either from site or specially imported material.

CONTROLLED LOW STRENGTH MATERIAL Mixture of pozzolan, Portland cement, water, coarse aggregate and occasionally soil, typically used for pipe bedding and backfill also known as flowable fill or cement-slurry backfill.

CONTROLLED PROCESS Process under intentional control by individual.

CONTROLLED PRODUCT Product or substance listed in one of six hazard classes in Controlled Products Regulations part IV, defined for Workplace Hazardous Materials Information System.

CONTROLLED REACTION Chemical reaction under temperature and pressure conditions maintained within safe limits to produce desired product or process.

CONTROLLED STORAGE Total storage portion that can be controlled in reservoir.

CONTROLLED SURFACE MINE DRAINAGE Surface mine drainage pumped or siphoned from active mining area.

CONTROLLED TIPPING Solid waste is disposed of into dug pit and is regularly covered with soil to avoid attracting disease vectors such as flies and rodents.

CONTROLLED WASTE Waste subject to legislative control in either its handling or its disposal.

CONTROLLED-AIR INCINERATOR Uses less than required quantity of air for complete combustion in primary chamber that by starving combustion process, volatile part of waste is gasified and combustible gas produced is mixed with air, ignited by gas or oil burner and completely burned in secondary combustion chamber or after burner chamber.

CONTROLLER ASSEMBLY Electrical mechanism for controlling traffic signal operation.

CONTROLLER UNIT Electrical component responsible for display of traffic signals.

CONURBATION Continuous aggregation of built-up urban communities created as result of urban sprawl.

CONVECTION 1. Transport of heat and moisture by movement of fluid; 2. Vertical transport of heat and moisture in atmosphere.

CONVECTION CURRENT 1. Thermally produced vertical air flow; 2. Current in fluid that results from thermal convection.

CONVECTIVE AVAILABLE POTENTIAL ENERGY Measure of amount of energy available for convection that's directly related to maximum potential vertical speed within updraft and thus, higher values indicate greater potential for severe weather.

CONVECTIVE CLOUDS Vertically developed family of clouds are cumulus and cumulonimbus with height of bases ranging from as low as 1,000 feet to bit more than 10,000 feet.

CONVECTIVE COLUMN Above heat source, warm air that rises.

CONVECTIVE CONDENSATION LEVEL Level in atmosphere that air parcel, if heated from below, will rise dry adiabatically, without becoming colder than its environment just before parcel becomes saturated.

CONVECTIVE EQUILIBRIUM State of atmosphere where vertical temperature profile is determined by balance between radiative and convective fluxes (i.e., no buoyancy force).

CONVECTIVE INHIBITION 1. Measure of amount of energy needed in order to initiate convection; 2. Numerical measure of strength of "capping," typically used to assess thunderstorm potential.

CONVECTIVE INSTABILITY Dry mid-level air that advects over very warm and moist air in lower troposphere.

CONVECTIVE OUTLOOK Forecast containing area(s) of expected thunderstorm occurrence and expected severity over contiguous United States, issued several times daily.

CONVECTIVE OVERDEVELOPMENT Convection that covers sky with clouds, thereby cutting off sunshine that produces convection.

CONVECTIVE PRECIPITATION Moisture-laden air that first expands and rises and then cools and condenses will produce moisture.

CONVECTIVE TEMPERATURE Approximate temperature that air near ground must warm to in order for surface-based convection to develop, based on analysis of sounding.

CONVENTIONAL DIAMOND INTERCHANGE Diamond interchange used when right-of-way is constrained.

CONVENTIONAL FILTRATION Method of treating water that consists of filtration.

CONVENTIONAL FILTRATION TREATMENT Series of processes including coagulation, flocculation, sedimentation and filtration resulting in substantial particulate removal.

CONVENTIONAL MINE Underground excavation or pit for creation of minerals.

CONVENTIONAL POLLUTANTS Pollutants typical of municipal sewage (e.g., organic waste, sediment, acid, BOD, TSS, fecal coliform bacteria, viruses, nutrients, oil, grease) for which municipal secondary treatment plants are typically designed.

CONVENTIONAL ROAD Road that permits direct access to homes and businesses.

CONVENTIONAL SEPTIC SYSTEM Wastewater treatment system consisting of septic tank and typical trench or bed subsurface wastewater infiltration system.

CONVENTIONAL SITE ASSESSMENT Assessment where most of sample analysis and interpretation of data is completed off-site.

CONVENTIONAL SYSTEMS Systems that have been traditionally used to collect municipal wastewater in gravity sewers and convey it to central primary or secondary treatment plant prior to discharge to surface waters.

CONVENTIONAL TILLAGE Traditional method of farming where soil is prepared for planting by completely inverting it with moldboard plow.

CONVENTIONAL TILLING Tillage operations considered standard for specific location and crop and that tend to bury crop residues.

CONVENTIONAL TREATMENT Wastewater treatment involving preliminary, sedimentation, flotation, trickling filter, activated sludge stages and chlorination.

CONVERGENCE 1. State of tending to unique solution; 2. Act of moving toward union, uniformity or limiting value; 3. Evolution of similar structures or traits in unrelated species in similar environments; convergent evolution; 4. Compression of fluid or gas into smaller volume.

CONVERGING TUBE Tube for which diameter decreases, usually at uniform rate, along its longitudinal axis from end where liquid enters.

CONVERSION 1. Chemical reaction wherein substrate is transformed into product; 2. Transformation of wastes; 3. Transition section in conduit connecting sections, each having different hydraulic elements.

CONVERSION FACTOR Manufacturing cost of product relative to cost of raw materials.

CONVEX BANK Where center of curvature in stream bank is away from channel.

CONVEYANCE Measure of flow carrying capacity of channel section.

CONVEYANCE LOSS Loss of water from pipes, channels, conduits and ditches due to leakage, seepage, evaporation and transpiration.

CONVEYANCE SYSTEM EFFICIENCY Ratio of volume of water delivered to users in proportion to volume of water introduced into conveyance system.

CONVEYING SYSTEM Mechanical handling equipment that moves materials from one location to another.

CONVEYOR Device that transports material by belts, cables or chains.

CONVEYOR APRON Consist of two strands of roller chain separated by overlapping plates.

CONVEYOR BELT Continuous flexible band moved by series of rollers to transport objects or material from one place to another, especially within factory or mine.

CONVEYOR DECLINE Mechanical transport from higher elevation to lower elevation.

CONVEYOR FEEDER Short conveyor designed to transport materials to another conveyor.

CONVEYOR SCREW Revolving auger flights that move bulk materials along tube or trough.

CONVEYOR SWING ANGLE Angles to left and right that machine can swing from centerline.

COOLANT Medium, usually fluid, used to draw heat from object.

COOLING COIL Coiled tube containing refrigerant used as heat exchanger in refrigerators, air conditioning systems and other cooling device.

COOLING DEGREE DAYS Form of Degree Day used to estimate energy requirements for air conditioning or refrigeration, typically calculated as how much warmer mean temperature at location is than 65 $^{\circ}$F on given day.

COOLING ELECTRICITY USE Amount of electricity used to meet building cooling load.

COOLING POND Man-made body of water primarily formed for purpose of supplying cooling water to nearby power plant or industrial facility such as petroleum refinery, pulp and paper mill, chemical plant, steel mill or smelter.

COOLING SPRAY For cooling system, water spray directed into hot gas or air.

COOLING SYSTEM Apparatus employed to keep temperature of structure or device from exceeding limits imposed by needs of safety and efficiency.

COOLING TOWER 1. Structure that helps remove heat from water used as coolant (e.g., in electric power generating plants); 2. Device that dissipates heat from water-cooled systems by spraying water through streams of rapidly moving air.

COOLING TOWER BLOWDOWN WATER Water released from cooling tower to maintain proper water mineral concentration.

COOLING TOWER MAKE-UP WATER Water added to cooling tower to replace water lost to evaporation or blowdown.

COOLING TOWER PRECIPITATION Precipitation occurring in vicinity of cooling towers due to air in hot gas plume cooling.

COOLING WATER 1. Water used for cooling purposes, (e.g., condensers, nuclear reactors); 2. Clean wastewater released from heat transfer system.

COOPERATIVE AGREEMENT Formal document that states obligations to one or more other parties providing authority to issue funding, transfers money, property, services or anything of value.

COOPERATIVE ELECTRIC UTILITY Electric utility legally established to be owned by and operated for benefit of those using its service.

COOPERATIVE OBSERVER Individual (or institution) who takes precipitation and temperature observations-and in some cases other observations (e.g., river stage, soil temperature, evaporation-at or near their home or place of business).

COOPERATIVE STATION REPORT Weather Service form documenting station management, exposure, topography, driving instructions, payment information, hydrometeorologic equipment and observing information.

COOPERATIVELY-OWNED ELECTRIC UTILITY Customer-owned electric utility that was created to transmit and distribute power in rural areas.

COORDINATED OPERATION Operation of two or more interconnected electrical systems to achieve greater reliability and economy.

COORDINATED UNIVERSAL TIME 1. Under international agreement, local time at prime meridian that passes through Greenwich, England; 2. Known as "Z time," "Zulu Time" or prior to 1972, Greenwich Mean Time (GMT).

COORDINATION Practice where two or more interconnected electric power systems augment reliability of bulk electric power supply by establishing planning and operating standards by exchanging pertinent information regarding additions, retirements and modifications, to bulk electric power supply system and by joint review of these changes to assure that they meet predetermined standards.

COPPER Reddish-brown, malleable, ductile metallic element with high electrical and thermal conductivity and atomic number 29.

COPPER NAPHTHENATE 1. Copper salt of naphthenic acid; 2. Wood preservative.

COPPER PIPE STRAPS Copper straps used to fasten copper conduit to construction.

COPPER SULFATE 1. Forms blue crystals and is used in agriculture, dyeing, electroplating and many other applications where reducing agents reduce it to red cuprous sulphate.

COPPER TYPE K PIPE Corrosion-resistant, copper conduit ideal for residential water supply lines identified by green stripe.

COPPER TYPE L PIPE Corrosion-resistant, copper conduit ideal for residential water supply lines identified blue stripe.

COPPER TYPE M PIPE Corrosion-resistant, copper conduit ideal for residential water supply lines identified red stripe.

COPPER-CHROME ARSENATE Wood preservative that has been used for timber treatment mix of chromium, copper and arsenic formulated as oxides or salts and is recognizable for greenish tint it imparts to timber.

COPPERAS Green, crystalline salt of divalent iron and sulfuric acid used as pigment and in water treatment.

CORAL Hard substance made of limestone skeletons of marine polyps.

CORAL REEF Ridge of limestone, composed chiefly of coral skeletons, coral sands and solid limestone resulting from organic secretion of calcium carbonate that generally occur along continents and islands.

CORDUROY Road made of logs laid crosswise on ground or on other logs.

CORE 1. Cylindrical piece of underground formation cut and raised by rotary drill with hollow bit; 2. Cutoff wall of clay inside earth dam or embankment; 3. Bony process which forms central axis of horns in many animals; 3. Inner part of nuclear reactor where nuclear reaction takes place; 4. Piece of soft iron, inside windings of electromagnet, that channels magnetic field; 5. Central part of protein structure consisting in mostly hydrophobic amino acids.

CORE CITY Planning concept calling for buildings to be developed very densely in center of city.

CORE DEPTH Distance between concrete slab top and bottom reinforcement.

CORE DRILL Rotary drill, usually diamond drill, equipped with hollow bit and core lifter for examination.

CORE DRILLING Drilling that uses hollow, cylindrical drill to make holes through surface.

CORE PROGRAM COOPERATIVE AGREEMENT Assistance agreement whereby government agency supports states or tribal governments with funds to help defray cost of non-item-specific administrative and training activities.

CORE PUNCH Penetration by vehicle into heavy precipitation core of thunderstorm.

CORE SAMPLE Sample of rock, soil or other material obtained by driving hollow tube into undisturbed medium and withdrawing it with its contained sample.

CORE WALL Wall of substantial thickness built of impervious material, usually of concrete or asphaltic concrete, in body of embankment dam to prevent seepage.

CORE WIDTH Width across face of core between side rails.

CORING Production of core by means of drilling.

CORIOLIS FORCE Fictitious force used to account for apparent deflection (caused by rotation of Earth) of body in motion with respect to Earth, as seen by observer on Earth.

CORN SNOW ICE Rotten granular ice.

CORNER BREAKS Concrete slab crack at intersect of longitudinal and transverse joints.

CORNER EFFECTS Small-scale convergence effect that can be quite severe that occurs around steep islands and headlands.

CORONA 1. Outermost layer of solar atmosphere characterized by low densities and high temperatures; 2. White or colored circle or set of concentric circles of light of small radius seen around luminous body (e.g., Sun, Moon).

CORONAL HOLE Extended region of corona, exceptionally low in density and associated with unipolar photospheric regions.

CORONAL RAIN Material condensing in corona and appearing to rain down into chromosphere as observed at solar limb above strong sunspots.

CORONAL TRANSIENTS General term for short-time-scale changes in corona, but principally used to describe outward-moving plasma clouds.

CORPORATION COCK VALVE Valve connected to water main that may be excavated, turned off and disconnect water service pipe.

CORRASION Erosion by abrasion caused by wind-blown or water-borne sand, stream-borne or glacier-borne stones or collisions between stones under influence of ocean waves.

CORRECTED NOISE LEVEL Index of corrected sound emitted from industrial site.

CORRECTIVE ACTION Use of Director-approved methods to assure that wells within area of review do not serve as conduits for movement of fluids into underground sources of drinking water.

CORRELATED SHEAR Output of mesocyclone detection algorithm indicating 3-dimensional shear region (i.e., vertically correlated) that's not symmetrical.

CORRELATION 1. Relation between two or more variables; 2. Measure of similarity between variables of functions; 3. Statistical test that sets numerical value to amount of interdependence.

CORRELATIVE ESTIMATE Amount determined by correlation that represents similar value of amount for any particular period.

CORRIDOR Narrow strip of land reserved for location of transmission lines, pipelines and service roads.

CORROSION Dissolution and wearing away (i.e., erosion) of metal caused by chemical (i.e., rusting, acids) reaction between water and pipes, chemicals touching metal surface, contact between two metals, electrochemical action or electrolytic reaction.

CORROSION CONTROL Measures that are implemented in various fields to control corrosion.

CORROSION EMBRITTLEMENT Embrittlement caused in certain alloys by exposure to corrosive environment.

CORROSION EXPERT Person who is expert and qualified to engage in practice of corrosion control.

CORROSION FATIGUE Mechanical degradation of material under joint action of corrosion and cyclic loading.

CORROSION RESISTANT Resistance of material against adverse element reaction that can corrode material.

CORROSION TESTING Determines resistance of materials to corrosion under certain environmental conditions, including temperature, humidity and salt water.

CORROSIVE 1. Chemical agent that reacts with surface of material causing it to deteriorate or wear away; 2. Chemical that destroys or irreversibly alters living tissue by direct chemical action at site of contact.

CORROSIVE SUBSTANCE Substances that can damage tissues when they come in contact with them.

CORROSIVE WEAR Relating to any form of wear dependent on chemical or corrosive process.

CORRUGATED RAIL Wear on rail surface in wave pattern.

CORRUGATION Alternate ridges and grooves in surface of pavement.

COSMIC RAY Extremely energetic (relativistic) charged particle.

COSMIC RAYS Energetic particle originating outside our solar system.

COST ACCOUNTING Recording of all costs incurred in business in way that can be used to improve its management.

COST ALLOCATION Procedures used to allocate joint project costs to specific purposes in multipurpose project.

COST ANALYSIS Detailed outline of potential risks and gains of projected venture.

COST APPROACH Real estate valuation method that surmises that price buyer should pay for piece of property should equal cost to build equivalent building.

COST BENEFIT ANALYSIS Methodology to help assess case for project by estimating net cost or benefit of project.

COST BREAKDOWN Systematic process of identifying individual elements that comprise total cost of good, service or package.

COST BUDGETING Tool to estimate costs or necessary efforts for projects, work packages or activities in project management.

COST CENTER Division or project of organization to which costs can be specifically allocated.

COST CHECK Regular approximation review to offer up to date evaluation of price with regards to improvement of design or building.

COST CLASS Classification that groups related cost categories.

COST CODE Method of grouping individual costs by codes based on their nature or function.

COST CONTROL Accounting procedure designed to identify and reduce business expenses.

COST EFFECTIVENESS Economic efficiency, obtaining best method for least amount of money.

COST ENGINEERING Process where engineer will budget, plan and monitor projects.

COST ESTIMATE Approximation of cost of program, project or operation.

COST ESTIMATING Approximating cost of program, project or operation.

COST FLOW Process of allocating costs between inventory and cost of sales.

COST FUNCTION Any mathematical relationship between activity and its cost.

COST OF CAPITAL Opportunity cost of making specific investment.

COST OF OVERRUN Prices beyond initially expected, budgeted or contracted.

COST OF QUALITY Methodology that allows organization to determine extent to which its resources are used for activities that prevent poor quality, that appraise quality of organization's products or services and that result from internal and external failures.

COST OF WORK Sum of all expenses incurred by contractor in performance of contract.

COST PER MILE Total cost involved in running vehicle/equipment item for one mile.

COST PER TON PER MINUTE Unit utilized in price comparisons between transport and direct-haul operations.

COST PERFORMANCE INDEX Measures value of work completed compared to actual cost spent on project.

COST PLAN Total cost limit of budget.

COST PLUS FIXED FEE Cost-reimbursement contract that provides for payment to contractor of negotiated fee fixed at inception of contract.

COST PLUS INCENTIVE FEE Cost-reimbursement contract that provides for initially negotiated fee to be adjusted later by formula based on relationship of total allowable costs to total target costs.

COST RECOVERY Legal process by which potentially responsible parties can be required to reimburse for money spent during any actions.

COST REVIEW Planned, systematic reassessments of estimated final cost of scope of work in cost class.

COST SHARING Publicly financed program where society, as beneficiary, shares part of cost with those who must actually install controls.

COST TYPES Four different cost types for any scope of work: authorized, expended, contractually committed or estimated.

COST VALUE Analysis of unit or material costs versus services required to function.

COST VARIANCE Difference between cost's actual amount and its budgeted or planned amount.

COST-BENEFIT ANALYSIS Quantitative evaluation of costs which would be incurred versus overall benefits to society of proposed action.

COST-EFFECTIVE Returning benefit that justifies initial investment.

COST-EFFECTIVE ALTERNATIVE Alternative control or corrective method identified after analysis as being best available in terms of reliability, performance and cost.

COST-PLUS Contract where contractor is paid for all of its allowed expenses, plus additional payment to allow for profit.

COST/BENEFIT RATIO Ratio of present value of investment decision or project with its initial cost.

COSTING Estimation of cost of process or product.

COULEES 1. Small streams or dry streambeds; 2. Deep gulch or ravine usually dry in summer.

COULOMB Unit of electric charge conveyed in 1 second by current of 1 ampere.

COULOMMETRIC TITRATION Concentration of titrant is equivalent to generating current and volume of titrant is equivalent to generating time.

COUNTER-FLOW LANE Lane operating in opposing direction of traffic flow.

COUNTERBALANCE PRESSURE VALVE Pressure control valve that keeps back pressure to prevent load from falling.

COUNTERBALANCE VALVE Valves designed to create back pressure at return line of actuator to prevent losing control over load.

COUNTERDRAIN Drain at foot of embankment of canal or watercourse, for carrying off water that may soak through.

COUNTERFLOW Flow in opposite direction; or flow of two fluids in opposite directions.

COUNTERMEASURE Any action taken to counteract or correct another.

COUNTRY ROCK Stone native to certain area.

COUNTY WARNING AND FORECAST AREA Group of counties for which National Weather Service Forecast Office is responsible for issuing warnings and weather forecasts.

COUNTY WARNING AREA Group of counties for which National Weather Service Forecast Office is responsible for issuing warnings.

COUPLED MODEL Refers to numerical model which simulates two systems and which takes into account effects of each component on other.

COUPLED NODES Modes of acoustic oscillation that affect one another.

COUPLER Device that connects locomotives or cars.

COUPLET Adjacent maxima of radial velocities of opposite signs.

COUPLING Conduit fitting that joins two things together.

COURSE 1. Path that something or someone moves along; 2. To run or flow (e.g., oil through engine); 3. To run through or over.

COVER 1. Landfill lid; 2. Set of sets, whose union contains given set; 3. Buyer's purchase on open market of goods similar or identical to goods contracted for after seller has breached contract of sale by failure to deliver goods contracted for; 4. Insurance contract coverage by insurance contract.

COVER CROP Crop that provides temporary protection for delicate seedlings and/or provides cover canopy for seasonal soil protection and improvement between normal crop production periods.

COVER MATERIAL Soil or another substance used to cover compacted solid waste in sanitary landfill.

COVERAGE 1. Extent of insurance protection; 2. Measured amount that something or someone is covered; 3. Average number of reads representing given nucleotide in reconstructed sequence.

COVERED SLUDGE DRYING BED Enclosure, covered with glass, that subjects sludge to radiant heating but shields it from precipitation.

COYOTE HOLES Horizontal tunnels where explosives are packed for blasting high rock face.

CPVC PIPE Conduit, made from chlorinated polyvinyl chloride, that can handle temperatures up to 200 $^{\circ}$F.

CRACK Separation formed in ice cover of floe that does not divide it into two or more pieces.

CRACK MONITORS Measure movements transverse and along joint or crack.

CRACKING 1. Formation of cracks on surface; 2. Thermal decomposition of substance (e.g., crude petroleum to produce gasoline).

CRACKING MOMENT Moment corresponding to tensile bending stress where failure will occur in concrete.

CRADLE 1. Machine on rockers, used in washing out auriferous earth; 2. Concrete bed in underside of trench that supports conduit and partly encloses it; 3. Ribbing for vaulted ceilings and arches intended to be covered with plaster.

CRADLE-TO-GRAVE SYSTEM Procedure where hazardous materials are identified and followed as they are produced, treated, transported and disposed of by series of permanent, linkable, descriptive documents (e.g., manifests).

CRAMMING Plugging conduit to stop flow prior to repair.

CRANE Mechanical lifting machine or device.

CRASH CUSHION Device to reduce damage to vehicles, motorists and structures resulting from collision.

CRASH TEST Test involving controlled crashing of vehicle.

CRASHWORTHY Capable of withstanding crash.

CRATER Large circular depression formed by impact, explosion or volcano collapse.

CRATON Part of Earth's crust that has survived splitting and merging of continents.

CRAWL SPEED Slowest speed vehicle will travel.

CRAWLING Propulsion by forcing rotation of movable tracks.

CRAZING Network pattern of fine cracks in concrete that do not penetrate much below surface.

CREEK 1. Relatively small stream of water that serves as natural drainage course for drainage basin of nominal (i.e., small size); 2. Small inlet or bay, narrower and extending farther into land than cove; 3. Recess in shore of sea or of river; 4. Inner part of port used as dock for small boats.

CREEP 1. Slow movement of rock debris or soil usually imperceptible except to observations of long duration; 2. Time-dependent strain or deformation with time; 3. Gradual flow or deformation of material under stress; 4. Displacement of particles under pressure during compaction.

CREOSOTE 1. Pale yellow oily liquid, containing phenols and similar compounds, obtained by destructive distillation of wood tar; 2. Brown liquid obtained from coal tar used as wood preservative.

CREPUSCULAR RAYS Alternating bands of light and dark (rays and shadows) seen at Earth's surface when Sun shines through clouds.

CRESCENT PUMP Internal gear pump where number of teeth on inner gear differs from outer gear by more than one.

CREST 1. Top surface of dam, dike, spillway or weir, to which water must rise before passing over structure excluding any parapet wall, railing, curb, etc.; 2. Crown of roadway or level of walkway which crosses dam; 3. Highest stage or level of flood wave as it passes point; 4. Highest point in wave; 5. Summit of hill or mountain ridge.

CREST CONTROL Device on spillway dam crest that controls flow by lifting or lowering crest.

CREST GATE Gate on crest of spillway to control discharge or reservoir water level.

CREST GAUGE Gauge used to obtain record of flood crests at sites where recording gauges are installed.

CREST LENGTH Measured length of dam along axis, centerline crest (e.g., top) or roadway surface on crest.

CREST STRUCTURE Portion of spillway between inlet channel and chute, tunnel or conduit, which does not contain gates.

CREST WIDTH Thickness or width of dam at level of crest (top) of dam (excluding corbels or parapets) where term thickness is used for gravity and arch dams and width is used for other dams.

CRIB 1. Bin for drying or storing grain; 2. Small room or covered structure, especially one of rough construction, used for storage or penning animals; 3. Confined space; 4. Hovel, roughly constructed building best suited to shelter of animals but used for human habitation; 5. Boxy structure traditionally built of heavy wooden timbers, to support existing structure from below, as with mineshaft or building being raised off its foundation in preparation for being moved.

CRIB DAM Gravity dam built up of boxes, cribs, crossed timbers or gabions, filled with earth or rock.

CRIB HEIGHT Difference in elevation between top and bottom of base log.

CRIB WALL Gravity retaining wall constructed of stacked concrete or timber members that are filled with free draining material.

CRIB WEIR Low diversion weir of log cribs or wire-mesh gabions filled with rock.

CRIBWORK Building (structural) crib.

CRITERIA 1. Numeric values and narrative standards that represent contaminant concentrations that are not to be exceeded in receiving environmental media (surface water, groundwater, sediment) to protect beneficial uses; 2. Standard or test where individual things or people may be compared and judged.

CRITERIA MAXIMUM CONCENTRATION EPA national water quality criteria recommendation for highest instream concentration of toxicant or effluent that organisms can be exposed for brief period of time without causing acute effect.

CRITERIA POLLUTANTS 1970 amendments to Clean Air Act required EPA to set National Ambient Air Quality Standards for certain pollutants known to be hazardous to human health and welfare where EPA identified and set standards for six pollutants: ozone, carbon monoxide, total suspended particulates, sulfur dioxide, lead and nitrogen oxide.

CRITERION Standard rule or test on which judgment or decision can be based.

CRITICAL ACTIVITIES OR WORK ITEMS Those activities or work items that have positive float.

CRITICAL DAMPING Minimum amount of damping that prevents free oscillatory vibration.

CRITICAL DENSITY 1. Density of substance at its thermodynamic critical point; 2. Density where capacity occurs for vehicular facility.

CRITICAL DEPTH 1. Depth of flow where discharge is maximum for given specific energy; 2. Depth of flow where given discharge occurs with minimum specific energy; 3. Depth of flow where Froude number equals one.

CRITICAL DISCHARGE 1. Maximum discharge for given specific energy; 2. Discharge that will occur with minimum specific energy.

CRITICAL EFFECT First adverse effect or its known precursor, that occurs as dose rate increases.

CRITICAL FLOW 1. Flow where Froude number is equal to one; 2. Flow at critical depth; 3. In reference to Reynolds' critical velocities, point where flow changes from streamline or non-turbulent to turbulent flow.

CRITICAL GRADIENT 1. Minimum channel slope that will produce critical flow; 2. Gradient where flotation of particles begins.

CRITICAL GROUP Large group of related organisms that, although variations exist between them, cannot be divided into smaller groups of equivalent taxonomic rank to parent group.

CRITICAL HABITAT Areas designated as critical for survival and recovery of threatened or endangered species.

CRITICAL HEADWAY Headway needed for vehicle to enter traffic stream.

CRITICAL HEIGHT Maximum height where vertical or sloped bank of soil will stand unsupported under given set of conditions.

CRITICAL HYDRAULIC GRADIENT Hydraulic gradient where effective stress of soil becomes negligible (i.e., quicksand).

CRITICAL IMPACT ANGLE Impact angle of vehicle that will most likely cause failure.

CRITICAL IMPACT POINT Location on vehicle where crash has greatest potential to cause failure.

CRITICAL LANE GROUPS Lane groups with highest volume-to-capacity ratios.

CRITICAL LIFE STAGE Period of time in organism's lifespan where it is most susceptible to adverse effects caused by exposure to toxicants, usually during early development (i.e., egg, embryo, larvae).

CRITICAL MASS Amount of fissile material needed to support self-sustaining nuclear chain reaction.

CRITICAL MOISTURE Moisture of soil samples achieved by rise in moisture content as result of deformation from steady load.

CRITICAL PATH Sequence of dependent steps that determine minimum time needed to carry out operation.

CRITICAL PATH METHOD Algorithm for process planning that identifies critical and non-critical tasks, preventing time-frame problems and process bottlenecks.

CRITICAL PHASE Longest signal cycle in sum of signal phases.

CRITICAL POINT Temperature and pressure where vapor density of gas and liquid phases of fluid are equal, where point there is no difference between gas and liquid.

CRITICAL PRESSURE Pressure of gas or vapor in its critical state.

CRITICAL RAINFALL PROBABILITY Probability that actual precipitation during rainfall event will cause river or stream to rise above flood stage and exceed flash flood guidance value.

CRITICAL SEGMENT Roadway segment highly probable to break down first if traffic, roadway and conditions don't change.

CRITICAL SHEAR STRESS Shear stress required to move soil particle.

CRITICAL SILENCER Absorptive silencer recommended for those conditions where ambient noise level in environment is very low.

CRITICAL SLOPE 1. Maximum angle with horizontal at which sloped bank of soil or rock of given height will stand unsupported; 2. Slope that will sustain given discharge at uniform critical depth in given channel.

CRITICAL SPECIES Species for which there is evidence that loss of species from site is likely to cause unacceptable impact on commercially or recreationally important species, threatened or endangered species, abundances of variety of other species or structure or function of community.

CRITICAL SPEED 1. Speed where capacity occurs along roadway segment; 2. Angular velocity that excites natural frequency of rotating object.

CRITICAL VELOCITY 1. Velocity where flow of liquid changes from laminar to turbulent; 2. Mean velocity when discharge is critical; 3. Velocity with which fluid can flow through given conduit where flow is streamlined without becoming turbulent; 4. Speed that falling object reaches when gravity and air resistance equalize on object.

CRITICAL VOLUME-TO-CAPACITY RATIO Proportion of intersection capacity available for vehicles in critical lane groups.

CRITICAL-DEPTH DISCHARGE MEASUREMENT Determination (by indirect measurement) of peak discharge following flood based on field survey of high-water marks and channel and control-section geometry.

CRITICALITY 1. Being extremely important; 2. Point where nuclear chain reaction becomes self-sustaining.

CRITICALITY ASSESSMENT Assessment that identifies key assets and infrastructure.

CROCHET Sudden deviation in sunlit geomagnetic field (i.e., H component) associated with large solar flare X-ray emission.

CROP CONSUMPTIVE USE Amount of water transpired during plant growth plus what evaporated from soil surface and foliage in crop area.

CROP IRRIGATION REQUIREMENT Quantity of water, exclusive of effective precipitation, needed for crop production.

CROP MOISTURE INDEX Index to assess short-term crop water conditions and needs across major crop-producing regions.

CROP ROTATION Pattern of changing crops grown in specific field from year to year in order to control pests and maintain soil fertility.

CROP SUBSIDY Price support paid by government to farmers.

CROP WATER REQUIREMENT Crop consumptive use plus water required to provide leaching requirements.

CROPLAND Land able to be plowed or tilled, capable of growing crops.

CROPPING PATTERN Acreage distribution of different crops in any one year in given farm area (e.g., county, water agency, farm).

CROSS Place or thing that intersect and lead off in four directions.

CROSS BRACE Cross-beam connections that adds rigidity to construction.

CROSS BRACING System utilized to reinforce building structures where diagonal supports intersect.

CROSS BRIDGING Diagonal bracing between adjacent floor joists, placed near center of joist span to prevent joists from twisting.

CROSS CONNECTION Connection made between two or more distinct things.

CROSS CONTAMINATION Movement of underground contaminants from one level or area to another due to invasive subsurface activities.

CROSS DRAINAGE Structure that carries discharge from natural stream across canal that intercepts stream.

CROSS FLOW Pedestrian flow perpendicular to another pedestrian flow.

CROSS-CONNECTION Actual or potential connection between drinking water system and unapproved water supply or other source of contamination.

CROSS-DITCH Shallow ridges from which water drains to prevent erosion.

CROSS-MEDIA Pollution conversion type.

CROSS-SECTION 1. Depicts shape of channel where stream flows, measured by surveying stream bed elevation across stream on line perpendicular to flow; 2. Elevation view of dam formed by passing plane through dam perpendicular to axis.

CROSS-SECTIONAL AREA Area of stream, channel or waterway perpendicular to direction of flow.

CROSS-SLOPE Transverse slope of pavement surfaces to provide drainage gradient so water will runoff surface.

CROSS-VALLEY WIND SYSTEM Thermally driven wind that blows during daytime across axis of valley toward heated sidewall.

CROSSCURRENT Turbulent stretch of water caused by multiple currents and directions.

CROSSFLOW FILTRATION Filtration where majority of feed flow travels tangentially across surface of filter.

CROSSING 1. Place where river passes over shallow shoal; 2. Relatively short and shallow length of river between bends.

CROSSING TIME Distance curb-to-curb divided by pedestrian speed.

CROSSOVER 1. Connection of two distinct piping systems which contain potable water; 2. Means where crossing is made.

CROSSTALK Undesirable signals from neighbouring transmission circuit.

CROSSWALK Place where pedestrians have right-of-way to cross street.

CROSSWALK LINES White line painted on roadway that signals crosswalk.

CROSSWALK OCCUPANCY TIME Time when crosswalk is completely or partially occupied by pedestrians.

CROSSWIND COMPONENT Winds that are not parallel, with or against line of travel.

CROSSWIND LEG Short climbing flight path at right angles to departure end of runway.

CROSSWIND RUNWAY Runway that allows aircraft to land when wind crosses runway.

CROWDING FORCE Force generated by crowding arm of loader bucket toward machine.

CROWN 1. Highest point of interior of circular conduit, pipe or tunnel (i.e., soffit); 2. Point in arch dam which generally corresponds with where height of dam is maximum; 3. Highest part of hill; 4. Raised center of road; 5. Top of levee or embankment; 6. Part of plant where root and stem meet; 7. Top of tree; 8. Where direction of flow changes from up to downward in trap; 9. Maximum point of interior of drain or sewer.

CROWN FIRE Fire where flames travel from tree to tree at level of tree's crown or top.

CROWN GATE Head gate (i.e., lock of canal).

CROWN WEIR Highest point on internal bottom surface of crown of plumbing trap.

CROWNED DRAINAGE Where road or highway is lifted, along centerline, above its longitudinal margin to drain water.

CROWNING Movement of fire from understory into crown of forest canopy.

CRUDE OIL Unrefined oil found underground and consists mainly of hydrocarbons.

CRUDE SEWAGE Fluid discharge from medical, domestic and industrial sanitary appliances.

CRUDE WASTEWATER Fluid discharge of domestic sewage, industrial waste and stormwater runoff before it receives any treatment.

CRUMB RUBBER Ground rubber fragments size of sand or silt used in rubber or plastic products or processed further into reclaimed rubber or asphalt products.

CRUSH CAPACITY Passenger capacity of vehicle with absolute minimum space.

CRUSHED GRAVEL Gravel which has been produced by machine.

CRUSHED ROCK Rock which has been reduced in size by machine.

CRUSHER Machine that reduces rocks to smaller and more uniform sizes.

CRUSHER SCREENINGS Crushed stone or gravel with gradation of 0 to 0.25 in.

CRUST 1. Outermost layer of lithosphere of Earth; 2. More solid, dense or hard layer on surface or boundary.

CRYOGENIC SCRAP RECOVERY Form of scrap recovery which uses liquid nitrogen.

CRYOGENIC SYSTEM System that produces and monitors effects of very low temperatures.

CRYOLOGY 1. Science of physical aspects of snow, ice, hail and sleet and other forms of water produced by temperatures below 0 °C; 2. Science of refrigeration.

CRYOTURBATION Any of several forms of disturbances within soils and subsoils as result of freeze-thaw processes in periglacial conditions.

CRYPTOCRYSTALLINE Whose crystalline nature only becomes apparent at high levels of magnification.

CRYPTOSPORIDIUM Microorganism commonly found in lakes and rivers that are highly resistant to disinfection that can cause gastrointestinal illness.

CRYSTAL Solid composed of array of atoms or molecules possessing long-range order and arranged in pattern which is periodic in three dimensions.

CRYSTALLINE 1. Of, relating to or composed of crystals; 2. Having regular three-dimensional molecular structure.

CRYSTALLINE ROCKS Rocks (e.g., igneous, metamorphic) consisting wholly of crystals or fragments of crystals.

CUBATURE Process of determining solid or cubic contents of body.

CUBIC CENTIMETER Unit of volume equal to 1 cm^3.

CUBIC CONTENT Volume of solid.

CUBIC FEET PER MINUTE 1. Measure of volume of substance flowing within fixed period of time; 2. With regard to indoor air, refers to amount of air, in ft^3, exchanged with outdoor air in minute's time.

CUBIC FEET PER SECOND 1. Unit expressing flow rate or of discharge in ft^3) passing reference point in 1 second (e.g., 7.48 gal/s, 448.8 gal/min (gpm), 1.98 acre-ft/day or 0.02832 m^3/s).

CUBIC FEET PER SECOND PER SQUARE MILE Average number of ft^3/s flowing from each mi^2 of area drained by stream, assuming that runoff is distributed uniformly in time and area.

CUBIC FEET PER SECOND-DAY Volume represented by flow of 1 ft^3/s for 24 hours (e.g., 86,000 ft^3, 1.983471 acre-feet, 646,317 gallons).

CUBIC FOOT Unit of volume equal to 1 ft^3.

CUBIC FOOT PER HOUR Unit of flow rate equal to 1 ft^3/hr.

CUBIC FOOT PER MINUTE Unit of flow rate equal to 1 ft^3/min.

CUBIC FOOT PER SECOND Unit of flow rate equal to 1 ft^3/s.

CUBIC FOOT PER SECOND PER SQUARE MILE Average number of ft^3/s of water flowing from each mi^2 of area drained, assuming runoff is distributed uniformly in time and area.

CUBIC FOOT PER SECOND-DAY 1. Volume represented by flow of 1 ft^3/s for 24 hours; 2. Approximately 86,400 ft^3, 1.98347 acre-feet, 646,317 gallons or 2,446.6 m^3.

CUBIC INCH Unit of volume equal to 1 in^3.

CUBIC METER Unit of volume equal to 1 m^3.

CUBIC METER PER MOLE Unit of molar volume equal to 1 $m^3/mole$.

CUBIC METER PER SECOND Unit of volume flow rate equal to 1 m^3/s.

CUBIC SCALE Approximation of cubic measure volume of wood fiber to improve usage and depletion of forest inventory.

CUBIC YARD Unit of volume equal to 1 yd^3.

CUBIC YARD BANK MEASUREMENT Unit of excavation in cut or natural bed.

CUBIC YARD COMPACTED MEASUREMENT Unit of excavation in fill/embankment after compaction.

CUBIC YARD LOOSE MEASUREMENT Unit of excavation in machine, stockpile or in uncompacted fill/embankment.

CUBIC YARD STRUCK MEASUREMENT Unit of capacity of machine measured by striking off ends and sides of container by straight edge.

CUBING To multiply number or quantity by itself three times.

CUL-DE-SAC Circular area at end of dead end street to allow cars to turn around.

CULLET Crushed glass melted for reuse.

CULTURAL EUTROPHICATION Increasing rate at which water bodies "die" by pollution from human activities.

CULTURAL PATRIMONY Ancestral heritage and entitlement.

CULTURAL RESOURCES Any building, site, district, structure or object significant in history, architecture, archeology, culture or science.

CULTURE Process of growing bacterial or other biological entity in artificial medium.

CULTURES AND STOCKS 1. Infectious agents and associated biologicals including cultures from medical and pathological laboratories; 2. Cultures and stocks of infectious agents from research and industrial laboratories; 3. Waste from production of biologicals; 4. Discarded live and attenuated vaccines; 5. Culture dishes and devices used to transfer, inoculate and mix cultures.

CULVERT Conduit for free passage of surface drainage water under road, structure, highway, railroad, canal or other embankment.

CULVERT FLOW DISCHARGE MEASUREMENT Field survey of channel and culvert geometry, high water marks, of peak discharge immediately after flooding.

CUMULATIVE BATCHING Measuring more than one ingredient of batch in same container by bringing batcher scale into balance at successive total weights as each ingredient is accumulated in container.

CUMULATIVE ECOLOGICAL RISK ASSESSMENT Consideration of total ecological risk from multiple stressors to given ecozone.

CUMULATIVE EXPOSURE Sum of exposures of organism to pollutant over period of time.

CUMULATIVE RUNOFF Overall volume of runoff of defined interval.

CUMULATIVE TOXICITY Continuous exposure to toxin.

CUMULATIVE WORKING LEVEL MONTHS Sum of lifetime exposure to radon working levels expressed in total working level months.

CUMULIFORM Descriptive of all clouds with vertical development in form of rising mounds, domes or towers.

CUMULIFORM ANVIL Thunderstorm anvil with visual characteristics resembling cumulus-type clouds that arises from rapid spreading of thunderstorm updraft.

CUMULOCIRRUS Small cumulus cloud at high altitude having whiteness and delicacy of cirrus.

CUMULONIMBUS Cloud, with tall structure and flat base, often associated with thunderstorms.

CUMULONIMBUS CLOUD Thunderhead, characterized by strong vertical development in form of mountains or huge towers topped at least partially by smooth, flat, often fibrous anvil.

CUMULOSE SOILS Soils derived from remains of vegetation accumulated in shallow water basins or in poorly drained areas on uplands in mountainous and subarctic regions (e.g., muck and peat).

CUMULOSTRATUS Cumulus cloud with base that extends horizontally as stratus cloud.

CUMULUS Large white puffy cloud that develops through convection.

CUMULUS BUILDUPS Clouds which develop vertically due to unstable air.

CUMULUS CLOUDS Detached clouds, generally dense and with sharp outlines, showing vertical development in form of domes, mounds or towers.

CUMULUS CONGESTUS Large cumulus cloud with great vertical development, usually with cauliflower-like appearance, but lacking characteristic anvil of towering cumulus.

CUNETTE Trench dug in moat to allow for drainage or as extra obstacle for attackers.

CUNETTE CHANNEL Narrow bottom channel that carries low flow.

CUNIT Unit of volume equal to 100 ft^3.

CUPRICHLORAMINE Mixture of copper sulfate, ammonia and chlorine used as algicide.

CURB COCK Water service shutoff valve located in water service pipe near curb and between water main and building.

CURB EXTENSION Extension of sidewalk out into vehicle lane.

CURB RAMP Ramp sloped from top surface of curb to surface of street.

CURB STOP Water service shutoff valve located in water service pipe near curb and between water main and building usually operated by wrench or valve key and is used to start or stop flows in water service line to building.

CURBSIDE COLLECTION Method of collecting recyclable materials and waste at homes, community districts or businesses.

CURIE Unit of radioactivity equal to 3.7×10^{10} disintegration's per second.

CURRENT 1. Movement of water in certain direction; 2. Time rate of flow of electric charge.

CURRENT DENSITY Amount of electric current flowing per unit cross-sectional area of material.

CURRENT DIAGRAM Diagram in diagrammatic expansion of current-current amplitude.

CURRENT DIFFERENCE Difference between time of slack water (or minimum current) or strength of current in any locality and time of corresponding phase of tidal current at reference station.

CURRENT EFFICIENCY Efficiency with which charge (electrons) is transferred in system facilitating electrochemical reaction.

CURRENT FINISH DATE Date where activity is currently expected to be completed.

CURRENT METER 1. Device used to measure speed of flowing water (i.e., velocity) or current in river; 2. Measures current of electrical circuit.

CURRENT POLE Calibrated pole used in measuring surface water current speed and direction.

CURRENT START DATE Deliberate and calculated estimate of specific point in time where activity and or schedule activities is expected to be initiated.

CURRENT TABLES Provide predicted times and heights of high and low waters associated with vertical movement of tide.

CURRENT YEAR DOLLARS Dollar values of given year that include effects of inflation or escalation for that year or reflect price levels expected to prevail during year at issue.

CURTAIN 1. Perpendicular obstacle to flow of water; 2. Flat area of wall which connects two bastions or towers.

CURTAIN DRAIN Drain constructed at upper end of area to be drained, to intercept surface or groundwater flowing toward protected area from higher ground and carry it away from area.

CURTAIN GROUTING Process of pressure grouting deep holes under dam or in abutment to form watertight barrier and effectively seal seams, fissures, fault zones or fill cavities in foundation or abutment.

CURTAIN WALL Cutoff wall that extends deep below overflow masonry dam on pervious bases to prevent underflow.

CURVATURE FACTOR Power decrease due to one or more bends in conduit.

CURVATURE FRICTION Friction from bends when tendon moves against duct.

CURVE 1. Bend without angles; 2. Graph of continuous map from one-dimensional space.

CURVED BUTTRESS DAM Buttress dam that is curved in plan.

CURVED GRAVITY DAM Gravity dam curved in plan.

CUSEC Unit of volume equal to 1 ft^3/s of flow in water.

CUSHION SEAL JOINT Reinforced neoprene pad rigidly attached to each side of joint.

CUSHION UNDERFRAME Railway car designed so that forces acting on coupler and sill prevent shocks and impact stresses.

CUT To lower existing grade (i.e., surface level) or make notch or channel by cutting or digging.

CUT BANK Outside bank of stream continually undergoing erosion.

CUT-AND-COVER Method of construction for shallow tunnels where trench is excavated and roofed over with overhead support system strong enough to carry load of what is to be built above tunnel.

CUT-AND-FILL process of construction whereby amount of material from cuts roughly matches amount of fill needed to make nearby embankments.

CUTBANK Outer edge of bend in stream or river, where water cuts into bank leaving opposite side flat.

CUTICULAR TRANSPIRATION Transpiration of gases or vapor directly through external membranes.

CUTIE-PIE Instrument used to measure radiation levels.

CUTOFF 1. Road, path of channel that provides shorter or quicker path; 2. Device that stops flow of current; 3. Impervious construction that prevents water from passing through foundation material (e.g., dam).

CUTOFF DEPTH Depth below excavation reached by cofferdam walls or sheet piling.

CUTOFF HEIGHT Specified distance of surface high record from chord representing 25 ft on longitudinal scale.

CUTOFF LOW Closed upper-level low which has become completely displaced from basic westerly current and moves independently of that current.

CUTOFF RATIO Ratio of length of cutoff to stream's original length.

CUTOFF TRENCH Excavation in foundation of dam that's later to be filled with impervious material to limit seepage beneath dam.

CUTOFF WALL Wall of impervious material (e.g., concrete, asphaltic concrete, timber, steel sheet piling, impervious grout curtain) constructed in foundation and abutments to reduce seepage beneath and adjacent to dam.

CUTS Excavation through hill or mountain for purposes of building road through it.

CUTTERHEAD Revolving blades fitted to head of hydraulic dredge suction line.

CUTTING 1. Excavating; 2. Lowering grade.

CUTTING SCREEN Apparatus to grind, shred or macerate stuff removed from wastewater.

CUTTINGS Spoils left by conventional drilling with hollow stem auger or rotary drilling equipment.

CUTWATER Wedge of bridge pier, that resists flow of water and ice.

CYANIDE Salt of hydrocyanic acid (i.e., water soluble poison).

CYANOPHYTA Group of phytoplankton and periphyton organisms (i.e., algae) with blue pigment in addition to green pigment called chlorophyll that cause nuisance water-quality conditions in lakes, streams and slow-flowing rivers.

CYCLE 1. Regularly recurring succession of events (e.g., seasons); 2. Complete rotation of anything.

CYCLE FAILURE Interrupted traffic condition where queued vehicles are unable to depart intersection.

CYCLE LENGTH Time to complete single sequence.

CYCLE OF FLUCTUATION Time taken for rise and fall of water table.

CYCLE TIME Total elapsed time to complete operation or set of operations.

CYCLE-CONTROL JOINT Slab joint that allows cycling of integral bridges and attached approach slabs.

CYCLIC DEPLETION Withdrawal of water from source in excess.

CYCLIC MOBILITY Phenomenon where cohesionless soil loses shear strength during earthquake ground vibrations and acquires degree of mobility sufficient to permit intermittent movement up to several feet, as contrasted to liquefaction where continuous movements of several hundred feet are possible.

CYCLIC RECOVERY Rise in elevation of watertable due to additions of water over secular period.

CYCLIC STORM Thunderstorm that undergoes cycles of intensification and weakening (i.e., pulses) while maintaining its individuality.

CYCLIC STRESS Distribution of forces that repeatedly change over time.

CYCLING Power plant operation to meet intermediated portion of load (9 to 14 hours/day).

CYCLOGENESIS Formation or intensification of cyclone or low-pressure storm system.

CYCLONE Area of low pressure where winds rotate counterclockwise in Northern Hemisphere and clockwise in Southern Hemisphere.

CYCLONE COLLECTOR Device that uses centrifugal force to remove large particles from polluted air.

CYCLONIC CIRCULATION Circulation (or rotation) that is in same sense as Earth's rotation (i.e., counterclockwise in Northern Hemisphere).

CYCLONIC FLOW Wind that swirls counter-clockwise in northern hemisphere or clockwise in southern hemisphere.

CYCLONIC PRECIPITATION Precipitation resulting from leading edge of warm, moist air mass (warm front) meeting cool and dry air mass (cold front).

CYCLOPEAN DAM Gravity dam where mass masonry consists primarily of large one-man or derrick stone embedded in concrete.

CYCLOSTROPHIC FORCE Force acts radially outward from center of rotation (low pressure area).

CYLINDER 1. Space where piston travels inside reciprocating engine or pump; 2. Solid figure bounded by cylinder and two parallel planes intersecting cylinder.

CYLINDER GATE 1. Gate that resembles large barrel, reinforced to withstand external pressure, with no top or bottom;

2. Outlet gate with vertical hollow cylinder being raised and lowered to expose openings in outer wall where water enters central area of gate.

CYLINDRICAL IMPELLER Round impeller for fluid pumps.

CYPRESS DOME Small, isolated, circular, depressional, forested wetlands, where cypress predominates, that have convex silhouettes when viewed from distance.

D

D DOWEL Steel reinforcing bars that provide connection between concrete slabs.

D REGION Daytime layer of Earth's ionosphere approximately 50 to 90 km in altitude.

D-FACTOR Directional distribution factor that's proportion of traffic traveling in peak direction during selected hour.

D-LOAD Test load where cracks, no more than 0.01 inch, are generated in three-edge bearing test.

D-LOAD ULTIMATE Maximum three edge-bearing test load pipe can carry.

D-RING Metal ring used to tie-down items and shaped like letter 'D' used primarily as lashing point.

D16 Particle diameter at 16 percent finer on distribution curve of size versus mass.

D50 Particle diameter at 50 percent finer on distribution curve of size versus mass.

D85 Particle diameter at 85 percent finer on distribution curve of size versus mass.

DAILY AVERAGE EFFLUENT LIMITATION Highest allowable average of daily discharges over calendar month.

DAILY CAPACITY Quantity of waste capable to be processed or landfilled daily in facility.

DAILY COVER Layer of compressed soil or earth which is laid on top of day's deposition of waste on operational landfill site.

DAILY DISCHARGE Discharge of pollutant measured during any 24-hour period that reasonably represents calendar day for purposes of sampling.

DAILY FLOOD PEAK Maximum mean daily discharge occurring in stream during given flood event.

DAILY MAXIMUM LIMIT Maximum allowable discharge of pollutant during calendar day.

DAILY MAXIMUM LIMITATION Peak value that should not be exceeded by any measurement.

DAILY MEAN SUSPENDED-SEDIMENT CONCENTRATION Time-weighted mean concentration of suspended sediment passing stream cross-section during 24-hour day.

DAILY RECORD STATION Site where data are collected with sufficient frequency to develop record of one or more data values per day.

DAILY SERVICE VOLUME Maximum total daily volume that can be sustained in both directions in given segment without violating loss of service criteria.

DAILY TYPE OF TIDE Tides where only one high and one low water occur in day.

DAILY VEHICLE MILES TRAVELED Number of vehicles that have traveled over route distance during course of day.

DAM 1. Barrier built to impound or divert water for purpose of storage or control of water; 2. Firebrick wall or stone, which forms front of hearth of blast furnace.

DAM FAILURE Catastrophic type of failure characterized by sudden, rapid and uncontrolled release of impounded water or likelihood of such uncontrolled release.

DAM FAILURE HYDROGRAPH Flood hydrograph resulting from dam breach prepared for specific location downstream from dam.

DAM FOUNDATION Excavated surface or undisturbed material upon which dam is placed.

DAM SAFETY Art and science of ensuring integrity and viability of dams such that they do not present unacceptable risks to public, property and environment.

DAM SAFETY DEFICIENCY Physical condition capable of causing sudden uncontrollable release of reservoir water by partial or complete failure of dam, appurtenant structure or facility.

DAM SAFETY ISSUE Dam safety related issues and concerns that if not adequately addressed could/would lead to failure or malfunction resulting in uncontrolled release of stored water that would place downstream population potentially at risk or compromise agency's ability to detect developing adverse dam performance and prudently respond to that performance.

DAM TENDER Person responsible for daily or routine operation and maintenance activities of dam and its appurtenant structures.

DAM TOP THICKNESS Thickness or width of dam at level of crest (top) of dam (excluding corbels or parapets) where term thickness is used for gravity and arch dams and width is used for other dams.

DAM-BREAK ANALYSIS Calculation of possible failures in dam due to flooding events.

DAMAGE FUNCTION Used to represent relation between maximal damage and damage factor.

DAMP Being in state between dry and wet.

DAMPING Reduction in oscillation magnitude by dissipation of energy.

DAMPING RATIO 1. Ratio of actual damping to critical damping, critical damping being minimum amount of damping which prevents free oscillatory vibration; 2. Maximum water elevation, including flood surcharge, that dam is designed to withstand.

DANGEROUS AIR CONTAMINATION Atmosphere presenting threat of causing death, injury, acute illness or disablement due to presence of flammable or explosive, toxic or otherwise injurious or incapacitating substances.

DARCIAN VELOCITY Rate of discharge of groundwater per unit area of porous medium measured at right angle to direction of groundwater flow.

DARCY'S FORMULA Formula for calculating loss of head in conduits as result of friction.

DARK MODE Signal indication missing at typically signalized area.

DARK SURGE ON DISK Dark gaseous ejections visible in H-alpha.

DART LEADER Faint, negatively charged channel that travels more or less directly and continuously from cloud to ground.

DART VALVE Drain for well bailer that opens automatically when rested on ground.

DATA Information, especially in scientific or computational context or with implication that it is organized.

DATA CALL-IN Part of Office of Pesticide Programs (OPP) process of developing key required test data, especially on long-term, chronic effects of existing pesticides, in advance of scheduled Registration Standard reviews.

DATA COLLECTION PLATFORM Electronic equipment device that gathers digital and analog data from monitoring devices and remotely relays that data to central computer for interpretation and archiving.

DATA COMPARABILITY Characteristics that allow information from many sources to be of definable or equivalent quality so that this information can be used to address program objectives not necessarily related to those for which data were collected.

DATA GAP Lack of any valid study or studies.

DATA LOGGER Microprocessor-based data acquisition system designed specifically to acquire, process and store data.

DATA POINT Location on river/stream for which observed data is input to hydrologic forecast procedures or included in public hydrologic products.

DATA QUALITY OBJECTIVES 1. Characteristics or goals that are determined by monitoring or interpretive program to be essential to usefulness of data; 2. Qualitative and quantitative statements of overall level of uncertainty that decision-maker will accept in results or decisions based on environmental data.

DATALOGGER Instrument that records data as function of time.

DATAWEB Electronic presentation of Bureau of Reclamation's Project Data book and contains historical, statistical and technical information on projects of Bureau of Reclamation.

DATE OF COMMENCEMENT OF WORK Date that work is authorized to begin on site.

DATE OF SUBSTANTIAL COMPLETION Date in progress of work when work or designated portion thereof is sufficiently complete in accordance with contract documents so owner can occupy or utilize work for its intended use.

DATEOMETER Small calendar disk attached to motors and equipment to indicate year where last maintenance service was performed.

DATUM 1. Surface or point relative to which measurements of height and/or horizontal position are reported; 2. Vertical datum is horizontal surface used as zero point for measurements of gauge height, stage or elevation; 3. Horizontal datum is reference for positions given in terms of latitude-longitude, State Plane coordinates or Universal Transverse Mercator coordinates; 4. Piece of information; 5. Fixed starting point of scale or operation.

DATUM PLANE Horizontal plane that ground elevations or water surface elevations are referenced.

DAUGHTER PRODUCT Remaining nuclide left over from radioactive decay.

DAWN Time of morning at which Sun is 6° below horizon.

DAY Unit of time equal to 24 hours, 1440 minutes or 86,400 seconds.

DAY LABOR Work done where worker is hired and paid one day at time, with no promise that more work will be available in future.

DAY LENGTH Duration of period from sunrise to sunset.

DAY TANK 1. Deaerating tank; 2. Tank used to store distillate or diesel fuel for boilers, emergency generators or other stationary engine driven or oil fired equipment.

DAY-DEGREES Accumulation of temperatures above certain threshold during given period.

DAYMARK Mark (e.g., river markers) designed to help navigators to find their way during daylight hours.

DBZ Non-dimensional "unit" of radar reflectivity which represents logarithmic power ratio (in decibels) with respect to radar reflectivity factor.

DDT Synthetic organic compound (i.e, dichlorodiphenyl-trichloroethane) used as insecticide that has half-life of 15 years and can collect in fatty tissues of certain animals.

DE-INKING Industrial process of removing printing ink from paper fibers of recycled paper to make deinked pulp.

DEACTIVATION Mechanical or chemical removal of dissolved oxygen from water to reduce its corroding power.

DEAD CAPACITY Reservoir capacity from which stored water cannot be evacuated by gravity.

DEAD END 1. End of water main which is not connected to other parts of distribution system; 2. Street or conduit that goes nowhere or is blocked on one end.

DEAD END OF LINE End of pipeline that will not lead back to oil storage tank.

DEAD LOAD Weight of structure itself, including weight of fixtures or equipment permanently attached to it.

DEAD PLATE GRATE Stationary grate where no air passes.

DEAD STORAGE Volume (e.g., storage) that lies below lowest outlet invert and cannot readily be withdrawn from reservoir by gravity.

DEAD WATER Non-circulating water.

DEAD WEIGHT RELIEF VALVE Valve where unrelieved weight of stopper is force that will keep valve shut.

DEAD WELL Oil well no longer used for production.

DEAD-END MAIN Water main over fifty feet long and not being fed from both ends at time of installation.

DEAD-END STREET Street with single one way in or out.

DEADHEAD 1. Train or truck moved between cities with no passengers or freight, in order to make it available for service; 2. Employee of transportation company (e.g., pilot) traveling as passenger for logistical reasons.

DEADMAN 1. Buried object that acts as anchor; 2. Anchor used to keep pipework from separating.

DEADMAN CONTROL Foot pedal or brake valve kept in depressed position while locomotive is operating that when released, initiates brake.

DEADMEN Anchors drilled or cemented into ground to provide additional reactive mass for DP sampling rigs.

DEADWEIGHT TONNAGE Carrying capacity of vessel.

DEAERATION Process of removing dissolved gasses from liquid.

DEATH Cessation of life and all associated processes.

DEBRIS 1. Scattered fragments; 2. Loose material.

DEBRIS BASIN Capture sediment, gravel, boulders and vegetative debris that are washed out of canyons during storms.

DEBRIS CLOUD Rotating "cloud" of dust or debris, near or on ground, often appearing beneath condensation funnel and surrounding base of tornado.

DEBRIS CONE Debris deposited in conical shape with surface slope greater than 10-degrees (perpendicular to contours), usually transported by small streams or snow avalanches.

DEBRIS DAM Fixed dam across stream channel for retention of sand, gravel, driftwood or other debris.

DEBRIS FAN Sloping mass of boulders, cobbles, gravel, sand, silt and clay formed by debris flows at mouth of tributary.

DEBRIS FLOW Flash flood consisting of mixture of rocks and sediment containing less than 40% water, by volume that forms debris fan.

DEBRIS JAM Debris obstructing movement of fluid.

DECA Prefix symbolizing 10^1.

DECADAL Occurring over 10-year period.

DECALITER Unit of measurement equal to 1O liters.

DECAMETER Unit of measurement equal to 10 meters.

DECANTATION Act of separating supernatant liquid from solid sediment.

DECANTATION TEST Evaluation to determines amount of material finer than No. 200 sieve.

DECANTING Process where liquid is removed.

DECAY Breaking down or rotting of organic matter through action of bacteria, fungi or other organisms.

DECAY HEAT Heat released as result of radioactive decay.

DECAY PRODUCTS Degraded radioactive materials, often referred to as "daughters" or "progeny."

DECELERATION LANE Lane that enables vehicle making exit turn from roadway to slow to safe speed before ramp ahead.

DECENT, SAFE, AND SANITARY DWELLING Dwelling which meets local housing and occupancy codes.

DECHLORINATION Removal of chlorine from substance.

DECHLORLNATION Removal of excessive rates of chlorine from potable water supply by inclusion of sulfur dioxide.

DECI Prefix symbolizing 10^{-1}.

DECIBEL 1. Unit (1/10 bel) for expressing relative intensity of sounds on scale from zero for average least perceptible sound to about 130 for average level at which sound causes pain to humans.

DECIDUOUS Trees, bush or shrub that sheds leaves annually.

DECILITER Unit of measurement equal to 1/10 of litre.

DECIMETER Unit of measurement equal to 1/10 of meter.

DECISION CRITERIA Set of formal criteria used to determine if event is threatening and what action to implement.

DECISION MAKERS Designated personnel responsible for interpreting data that has been collected and determining if event is threatening enough to issue alert or warning.

DECISION MAKING Second of five Early Warning System components consisting of processes and facilities necessary to translate incoming data about threatening event into decisions to alert or warn population at risk.

DECK 1. Any raised flat surface that can be walked on; 2. Top of hollow reinforced concrete dam.

DECK DRAIN Flat strainer fitted flush with surface that caps drain outlet.

DECK GIRDER Board or composite that serves as supporting structure for perpendicular deck joists.

DECK JOINT Structural difference between deck element and another element of construction.

DECK PLATE Metal plate meant to offer secure ground.

DECK SCREEN Two or more displays set above one another to classify exactly same run of material.

DECK SLAB Surface structure supporting vehicles or pedestrians to cross over.

DECK TRUSS Roadway with load-bearing superstructure composed of truss.

DECKING Separating charges of explosives by inert material which prevents passing of concussion and placing primer in each charge.

DECLINATION Latitude that Sun is directly over at given time that's 23° N at summer solstice, 23° S at winter solstice and 0° (over equator) at spring and autumn equinoxes.

DECODER Device that decodes scrambled electronic signal.

DECOMMISSIONING To take out of service or to render unusable.

DECOMPOSED Saprolitic rock essentially reduced to soil with relic rock texture that can be molded or crumbled by hand.

DECOMPOSERS Any organism that feeds off decomposing organic material, especially bacterium or fungi.

DECOMPOSITION 1. Breakdown of matter by bacteria and fungi, changing chemical makeup and physical appearance of materials; 2. Subdividing failure mode into discrete elements or sequential events so that failure mode could be better understood and probabilities can be more reasonably estimated for each step in event tree.

DECOMPRESSION PRESSURE VALVE Pressure control valve that controls speed where controlled energy of compressed fluid is discharged.

DECONCENTRATOR Device for removing suspended or dissolved material from feedwater.

DECONTAMINATION Removal of harmful substances (e.g., noxious chemicals, harmful bacteria) or radioactive material from exposed individuals, rooms and furnishings in buildings or exterior environment.

DECONTAMINATION AREA Enclosed space adjoining and connected to controlled area for decontamination of workers, materials and gear.

DEDICATED FUNDS Funds available for expenditure.

DEDICATION 1. Deliberate or negligent surrender of all rights to property; 2. Ceremony marking official completion or opening.

DEDUCTIBLE CLAUSE Clause in insurance policy stipulating that insured will be liable for specified initial amount of each loss, injury, etc. and that insurance company will be liable for any additional costs up to insured amount.

DEDUCTION Sum that can be removed from calculations.

DEED Legal contract showing bond in form of document.

DEEP DRAFT WATERWAY Navigation channels provided for movement of self-propelled vessels with drafts greater than 15 feet.

DEEP FOUNDATION Foundation that transfers building loads past weak soil, farther down to supporting soil or rock.

DEEP MANHOLE Inspection chamber for sewers linked to surface via entrance shaft.

DEEP PERCOLATION Movement of water by gravity downward through soil profile, beyond root zone, to regions of deeper groundwater aquifers.

DEEP PERCOLATION LOSS Water that percolates downward through soil beyond reach of plant roots.

DEEP SEA Deeper part of sea or ocean where no light penetrates.

DEEP SEA DEPOSIT Deposit from deeper parts of ocean, especially those beyond edge of continental shelf, typically below below 600 feet (182 m, 100 fathoms) or more.

DEEP SEAL TRAP Plumbing trap, with seal of 4 in. (100 mm) or more, larger than traditional trap and can handle much greater flow.

DEEP SEEPAGE Infiltration which reaches water table.

DEEP SHAFT SEWAGE TREATMENT Sewage treatment where screened raw sewage starts under and flows up rotating shaft with compressed air.

DEEP WATER 1. Water deeper than half length of surface waves; 2. Waters suitable for deep-draft ships.

DEEP WELL Well created to access groundwater in underground aquifers where pumping head is too great to permit use of suction pump.

DEEP WELL PUMP Multi-stage centrifugal pump for lifting water from deep, small-diameter wells.

DEEP-WATER HABITATS Permanently flooded lands that lie below deep-water boundary of wetlands.

DEEP-WATER WAVE Surface wave length of which is less than twice depth of water.

DEEP-WELL INJECTION Deposition of raw or treated, filtered hazardous waste by pumping it into deep wells, where it is contained in pores of permeable subsurface rock.

DEEP-WELL TURBINE PUMP Multi-stage borehole pumps whose pump shaft extends up to ground level, where it is coupled to dry installed motor or gearbox.

DEEPENING Decrease in central pressure of surface low pressure system (i.e., storm is intensifying).

DEFAULT Failing to compete.

DEFAULT VALUE Preset values when data is missing.

DEFECT Fault or malfunction.

DEFECT IN TITLE Any recorded instrument that would prevent grantor from giving clear title.

DEFECTIVE WORK Work in progress or finished goods which fail to pass inspection because they do not conform to standards laid down.

DEFECTS LIABILITY PERIOD Set period of time after construction project has been completed during which contractor has right to return to site to remedy defects.

DEFERRED CHARGE Long-term prepaid expense treated as asset on balance sheet and is carried forward until it is actually used.

DEFERRED CREDIT Credit received not initially reported as income.

DEFERRED LIABILITY Obligation for which settlement is not required until later period.

DEFERRED MAINTENANCE practice of postponing maintenance activities such as repairs on both real property and personal property in order to save costs, meet budget funding levels or realign available budget monies.

DEFERRIZATION Treatment of water to remove soluble iron salts.

DEFICIENCY 1. Inadequacy or incompleteness; 2. Insufficiency, especially of something essential to health; 3. Amount where number of double points on curve is short of maximum for curves of same degree.

DEFICIENCY VERIFICATION ANALYSIS Analysis that verifies existence or non-existence of Safety of Dams (SOD) deficiencies at dam.

DEFINING procedure that orders carbon particles according to size in activated-carbon filter.

DEFLAGRATION To burn with sudden and startling combustion.

DEFLATION 1. Removal of loose surface material by force of wind erosion (e.g., blowouts); 2. Decrease in general price level.

DEFLECT Decrease in vertical diameter of flexible pipe.

DEFLECTION Upstream or downstream movement of dam or dike.

DEFLOCCULATING AGENT Material added to suspension to prevent settling.

DEFLUORIDATION Removal of excess fluoride in drinking water to prevent staining of teeth.

DEFOAMANT Common additive that retards formation of stable foam on surface of oil in sump, gearbox or reservoir.

DEFOLIANT Herbicide that removes leaves from trees and growing plants.

DEFOLIANTS Substances used to deprived leaves from plants.

DEFOLIATOR 1. Insect that ruins leaves; 2. Substance that causes plants to drop their leaves.

DEFORESTATION Process of destroying forest and replacing it with something else.

DEFORMATION Change in shape or size.

DEFORMATION ZONE Change in shape of fluid mass by variations in wind, specifically by stretching and/or shearing.

DEFORMED ICE Ice which has been squeezed together and forced upwards and downwards in places.

DEGASIFICATION 1. Water treatment that removes dissolved gases from water; 2. Removal of gas from something, especially from liquid.

DEGENERATIVE DISEASE Result of continuous process based on degenerative cell changes, affecting tissues or organs, which will increasingly deteriorate over time.

DEGRADATION 1. Geologic process where stream beds, sandbars, floodplains and bottoms of other water bodies are lowered in elevation by removal of material from boundary, by action of wind and water; 2. Dislocation of matter by biological activity.

DEGRADATION CONSTANT Term used to address decay of contaminant concentration due to factors other than dispersion.

DEGRADATION PRODUCTS Compounds resulting from transformation of organic substance through chemical, photochemical, and/or biochemical reactions.

DEGRADED Condition of quality of water that has been made unfit for some specified purpose.

DEGREASING Removal of grease from something.

DEGREE ANGLE Measurement of plane angle, defined so that full rotation is 360 degrees.

DEGREE CELSIUS Unit of temperature where freezing point is $0\,^{\circ}C$ and boiling point is $100\,^{\circ}C$.

DEGREE DAY Measure that gauges amount of heating or cooling needed for building using $65\,^{\circ}F$ as baseline.

DEGREE FAHRENHEIT 1. Unit of temperature where freezing point is $32\,^{\circ}F$ and boiling point is $212\,^{\circ}F$.

DEGREE OF ARC Unit of angle ($180^{\circ} = \pi$ x radian) equal to 60 minutes of arc or 3600 seconds of arc.

DEGREE OF COMPACTION Measure of compactness of solid soil particles in given volume.

DEGREE OF CONSOLIDATION Ratio, expressed as percentage, of amount of consolidation at given time within soil mass, to total amount of consolidation obtainable under given stress condition.

DEGREE OF CURVE Central angle to ends of arc or chord of agreed length.

DEGREE OF DENSITY Calculation of particle compaction.

DEGREE OF PURIFICATION Measure of completeness of removal or destruction of special impurities or substances or organic components.

DEGREE OF SATURATION 1. Ratio of liquid to total volume of voids in porous material; 2. Measure of how much road or intersection is experiencing demand compared to its total capacity.

DEGREE OF SLOPE Angle above or below horizontal of slope.

DEGREE OF TREATMENT Measure of removal and/or reduction of special impurities or compound or organic components.

DEGREE-HOUR Number of degrees Fahrenheit where hourly average indoor temperature is below or above standard temperature.

DEHUMIDIFIER Device for removing moisture content from air.

DEHUMIDIFY Reduce moisture in body of air.

DEHYDRATION Act or process of removing water from something.

DEIONIZE Remove ions from solution using ion-exchange process.

DELAMINATION 1. Blending of body of water which is now stratified; 2. Separation of layers of laminar composite material; 3. Formation of gastrula by splitting of blastoderm.

DELAY Device used to obtain detonation of charges at separate times.

DELAY PATTERN Order of firing charges obtained by arranging delays to fire separate holes or series of holes at different times.

DELAY TIME Time during which some action is awaited.

DELAYED DENSITY-DEPENDENT Scenario where mortality among host population is determined by population density of host in consecutive generations.

DELAYED PASSING MANEUVER Average bicyclist inability to make immediate passing maneuver.

DELEGATED STATE State (e.g., other governmental entity) that has received authority to administer regulatory program in lieu of federal counterpart.

DELINEATOR Perambulator that records distances and delineates profile (e.g., road).

DELINEATORS Reflective materials (e.g., roadway embedded reflectors).

DELIQUESCE 1. To melt and disappear; 2. To become liquid by absorbing water from atmosphere.

DELIST Use of petition process to have facility's toxic designation rescinded.

DELIVERED PRODUCT Green power fed into same electric transmission and distribution system that serves end user.

DELIVERED WATER Water delivered to user that may be used for either inside uses or outside watering, does not include precipitation or distribution losses and may apply to metered or unmetered deliveries.

DELIVERY Act of conveying something.

DELIVERY BOX Enclosure for measurement.

DELIVERY/RELEASE 1. Amount of water delivered to point of use and amount released after use; 2. Difference between amount delivered and amount released is consumptive use.

DELTA 1. Alluvial deposit, often in shape of Greek letter "delta", which is formed where stream drops its debris load on entering body of quieter water; 2. Landform at mouth of river where it empties into body of water; 3. Angle subtended at center of circular arc.

DEMAND 1. Maximum water use under specified condition; 2. Rate at which electric energy is used whether at given instant or averaged over any designated period of time; 3. Measure of maximum load over short period of time; 4. Desire to purchase goods and services.

DEMAND FACTOR Fractional amount of some quantity being used relative to maximum amount that could be used by same system.

DEMAND FLOW RATE Vehicle count during analysis period that's converted to hourly rate.

DEMAND HORSEPOWER Number of horsepower needed to propel specified weight at specified velocity.

DEMAND SCHEDULING Method of irrigation scheduling whereby water is delivered to users as needed and which may vary in flow rate, frequency and duration.

DEMAND STARVATION Amount of green time when there is no queue present to be discharged and there are no arrivals due to signalization.

DEMAND SYSTEM Breathing apparatus self-contained and supplies air on demand.

DEMAND VOLUME Arriving number of vehicles to utilize facility.

DEMAND-RESPONSIVE Form of transportation where vehicles alter their routes based on transport demand.

DEMAND-SIDE WASTE MANAGEMENT Prices whereby consumers use purchasing decisions to communicate to product manufacturers that they prefer environmentally sound products packaged with least amount of waste, made from recycled or recyclable materials and containing no hazardous substances.

DEMAND-TO-CAPACITY RATIO Ratio of roadway demand (e.g., vehicles) to supply (e.g., capacity).

DEMANGANIZATION Act of removing compounds of manganese from water.

DEMERSAL Fish eggs or organisms that hatch and/or live on bottom of body of water.

DEMINERALIZATION 1. Chemical engineering treatment process where minerals are removed from liquid; 2. Loss of minerals, especially those of calcium, from bone.

DEMOGRAPHIC TRANSITION Process that represents transition from high birth and death rates to low birth and death rates as country develops from pre-industrial to industrialized economic system.

DEMOGRAPHICS Relating to statistical study of human populations.

DEMOGRAPHY Study of statistics (e.g., births, deaths, income, diseases of community).

DEMOLITION Process of demolishing or destroying buildings or other structures.

DEMURRAGE Compensation paid for detention of freight vehicle during delayed loading or unloading.

DENATURE 1. To subject to conditions altering its original form or state; 2. To take away natural characteristic or inherent property.

DENDRITES Thin branch-like growth of ice on water surface.

DENDRITIC Channel pattern of streams with tributaries that branch to form tree-like pattern.

DENDRITIC CRYSTALS Crystal that develops with typical multi-branching tree-like form.

DENDRITIC DRAINAGE Branching drainage pattern.

DENDROCHRONOLOGY Science that uses spacing between annual growth rings of trees to date their exact year of formation.

DENITRIFICATION Biological reduction of nitrate by denitrifying bacteria in soil that usually resulting in escape of nitrogen gas to air.

DENSE FIRE SPRINKLER SYSTEM Dry-pipe sprinkler system where all sprinklers are open and speed-of-temperature-rise, temperature sensitivity or pressure-sensitive devices activate water supply control valve.

DENSE FOG ADVISORY 1. Issued when fog reduces visibility to 1/8 mile or less over widespread area; 2. Marine advisory for widespread or localized fog reducing visibilities to regionally or locally defined limitations not to exceed 1 nautical mile.

DENSE NON-AQUEOUS PHASE LIQUID Non-aqueous phase liquids (e.g., chlorinated hydrocarbon solvents, petroleum fractions) with specific gravity greater than 1.0 that sink through water column until they reach confining layer.

DENSE SMOKE ADVISORY Advisory for widespread or localized smoke reducing visibilities to regionally or locally defined limitations not to exceed 1 nautical mile.

DENSE-GRADED AGGREGATE Aggregate containing very small particles that reduce void spaces.

DENSITY 1. Mass per unit volume; 2. Measure of how heavy specific volume of solid, liquid or gas is in comparison to water; 3. Ratio of one quantity to another quantity; 4. Weight of substance per unit of volume of substance.

DENSITY ALTITUDE Altitude relative to standard atmospheric conditions where air density would be equal to indicated air density at observation location.

DENSITY CURRENT Flow of water maintained by gravity through large body of water (e.g., reservoir, lake) and retaining its unmixed identity because of difference in density.

DENSITY FLOW Movement of water resulting from flow of different density, salinity or temperature.

DENSITY OF SNOW Ratio, expressed as percentage, of volume which given quantity of snow would occupy if it were reduced to water versus volume of snow.

DENSITY SEPARATION Division of material into fractions of changing density by flotation in atmosphere or liquid.

DENSITY STRATIFICATION Layers of water in body of water with layer of cold dense water at bottom and, at temperatures above freezing, warmer layer floating on top.

DENSITY-INDEPENDENT Scenario where size of population whose effect is not dependent on number of individuals in population.

DENTATE 1. One of series of upright obstructions designed to dissipate energy as in case of stilling basin or drop structure; 2. Having teeth or tooth-like projections.

DENTATE SILL One of series of upright obstructions (e.g. concrete block) designed to dissipate energy as in case of channel, stilling basin or drop structure.

DENTATED SILL Notched sill that reduces force of flowing water and resulting scour.

DENUDATION Laying bare by washing away of overlying or action of running water.

DEOXYGENATION Removal of oxygen atoms or molecules from material.

DEOXYGENATION CONSTANT Speed of biochemical oxidation of organic matter in water under aerobic conditions.

DEPARTURE HEADWAY 1. Average time for vehicles departing queue; 2. Average time of unqueued vehicles passing stop line.

DEPENDABLE CAPACITY Capacity that can be relied upon to carry system load for specified time interval and period, provide assumed reserve, and/or meet firm power obligations.

DEPENDABLE YIELD Maximum annual supply of water development available on demand.

DEPLETION 1. To permanently remove water from system, stream, river or basin resulting from consumptive use; 2. Progressive withdrawal at rate greater than that of replenishment.

DEPLETION CURVE 1. Hydrograph portion where water depletion from storage-stream channels, surface soil and groundwater that can be drawn for base flow, direct runoff or total flow; 2. Part of hydrograph extending from point of termination of Recession Curve to subsequent rise or alternation of inflow due to additional water becoming available for stream flow.

DEPLETION HYDROGRAPH hydrograph of discharge from water bodies depicting drainage from groundwater, swamps and lakes during long rainless periods of no recharge.

DEPOSIT 1. Sediment or rock not native to its present location or is different from surrounding material; 2. Sum of money or other asset given as initial payment, to show good faith or to reserve something for purchase.

DEPOSITION 1. Material settling out of water onto streambed; 2. Mechanical or chemical processes where sediments accumulate in temporary resting place; 3. Formation of snow or frost directly from water vapor; 4. Act of depositing material by natural processes; 5. Production of thin film of material onto existing surface.

DEPOSITION PROCESSES Removal of particulate matter from air stream by deposit.

DEPOSITION SEDIMENTATION Process where sediments are released, settled from or dropped from erosional system.

DEPOSITORIAL CLAY Clayey sediment carried by running water from its place of origin.

DEPOT Storage facility.

DEPRECIATION Measurement of decline in value of assets.

DEPRECIATION FACTOR Percent of adjusted tax basis that's depreciated within depreciation period.

DEPRECIATION RATE Percent rate where asset is depreciated in any one of methods for computing depreciation.

DEPRESSANT Agent that inhibits flotation of mineral or minerals.

DEPRESSED SEWER Sewer that runs full under gravity flow at pressure above atmosphere in sewer.

DEPRESSION 1. Region of low atmospheric pressure usually accompanied by low clouds and precipitation; 2. Area lower in topography than its surroundings.

DEPRESSION HEAD Difference between water table level and water level in well.

DEPRESSION SPRING Spring where earth's surface is coincident with water table.

DEPRESSION STORAGE 1. Volume of water contained (stored) in elevation concavity; 2. Small low points in undulating terrain that can store precipitation that otherwise would become runoff.

DEPRESSURIZATION Condition that occurs when air pressure inside structure is lower that air pressure outdoors.

DEPTH FILTER Variety of filters that use porous filtration medium to retain particles throughout medium.

DEPTH FILTRATION Filtration that utilizes thickness of filter media.

DEPTH MICROMETERS Measure movements transverse and along joint or crack.

DEPTH OF CUTOFF Vertical distance that cutoff penetrates into dam foundation.

DEPTH OF FLOTATION Depth to which body sinks below its plane of flotation.

DEPTH OF FLOW Vertical distance from bed of stream to water surface.

DEPTH OF RUNOFF Total runoff from drainage basin, divided by its area.

DEPTH OF SCOUR Vertical distance sediment is eroded by scour.

DEPURATION Removal of impurities, especially from bodily fluids.

DEPURATION PHASE Part of bioconcentration evaluation following uptake period during which organisms have been in running water to which no evaluation compound is added.

DEPURATION RATE CONSTANT Growth correction and use as indicator of bioaccumulation potential.

DERAIL Device placed on railway tracks causing train to come off tracks.

DERATE To lower rated capability of any rated equipment or material.

DERECHO Widespread and usually fast-moving windstorm associated with convection.

DERELICT LAND Land that has become damaged by industrial or other development and beyond beneficial use without treatment.

DERIVED PRODUCTS Processed base data on Doppler radar.

DERIVED WAVE Tide created by tidal changes in interconnected waterways.

DERIVED WORKING LIMIT Limit to radiation intensity and dose to which workers may be exposed during working hours.

DERMAL ABSORPTION/PENETRATION Process by which chemical penetrates skin and enters body as internal dose.

DERMAL CORROSION Production of irreversible damage of skin following application of test substance for up to 4 hours.

DERMAL EXPOSURE Exposure to chemical by entry though skin.

DERMAL IRRIGATION Inflammatory altercation of skin after application of substance.

DERMAL TOXICITY Ability of pesticide or toxic chemical to poison people or animals by contact with skin.

DERRICK Non-mobile tower equipped with hoist (e.g., crane).

DESALINATION 1. Removing salts from ocean or brackish water by using various technologies to provide freshwater; 2. Removal of salts from soil by artificial means (e.g., leaching).

DESALINIZATION Removal of dissolved salts from water by natural means (leaching) or by specific water treatment processes.

DESCALING Removal of scale (i.e., rust) from surface of metal.

DESCRIPTIVE MODEL Describes how things actually work with its environment.

DESERT Barren area of land or desolate terrain.

DESERT PAVEMENT Desert surface covered with closely packed rock fragments.

DESERT VARNISH Dark coating found on exposed rock surfaces in arid environments.

DESERTIFICATION Tendency toward more prominent desert conditions in region when soil erosion is so severe that plants and animals can no longer exist.

DESICCANT 1. To dry up; 2. Remove moisture from substance; 3. Chemical drying agent (i.e., calcium oxide, silica gel) used as drying agent because of its high affinity to water.

DESICCATION 1. Process of drying out; 2. Process used to thoroughly dry air; 3. To remove virtually all moisture from air; 4. Process where area experiences entire loss of its stored water as result of withdraw, decrease of precipitation or escalation in evaporation.

DESIGN Plan for structure and functions of artifact, building or system.

DESIGN ALTERNATIVES Identification of options that all meet requirements and standards during all design phases.

DESIGN ANALYSIS Studying characteristics (e.g., form, structure, construction, features) of design of thing.

DESIGN BASIS EARTHQUAKE Earthquake which structure is required to safely withstand with repairable damage.

DESIGN CAPACITY 1. Theoretical maximum output of system in given period under ideal conditions; 2. Maximum volume that something is designed for.

DESIGN CHANGE Modification conducted to product that can happen at any stage in product development process.

DESIGN CONTINGENCY Estimate included in cost plan for minor changes during development of design and documents for items that may change or be added.

DESIGN CRITERIA Hypothetical flood used in sizing of dam and associated structures to prevent dam failure by overtopping, especially for spillway, outlet works, channel culvert and inlet sizing.

DESIGN DEVELOPMENT STAGE Period that commences once final concept design, along with estimated budgets based on this, are approved.

DESIGN DISCHARGE Discharge estimated for bridge design.

DESIGN DISPLACEMENT Method of design to limit maximum lateral displacement from earthquake.

DESIGN DOCUMENTS Records prepared by designer.

DESIGN ENGINEER person who works on products and systems that involve adapting and using complex scientific and mathematical techniques.

DESIGN ENGINEERING FIRM In charge of design, strategies and specifications to successfully finish layout of job.

DESIGN FACTOR 1. Application (i.e., not actual calculation); 2. Minimum safety factor.

DESIGN FLOOD Value of instantaneous peak discharge adopted for design of particular project or any of its structures.

DESIGN FLOOD FOR BRIDGE SCOUR Method of determining bridge foundation design to resist damage from estimated scour.

DESIGN FLOOD FOR WATERWAY OPENING Probability of exceedance for peak discharge, volume, stage or wave crest elevation selected for highway or bridge design.

DESIGN FLOOD FREQUENCY Method of design using frequency of flood occurrence through probability distribution.

DESIGN FLOW Used for steady-state waste load allocation modeling.

DESIGN FORCE Force applicable to particular structural component design.

DESIGN HIGHWATER ELEVATION Calculated water surface elevation for hydraulic structure design.

DESIGN HOUR Activity levels during hour to determine street dimensions.

DESIGN HOUR FACTOR Percentage of Annual Average Daily Traffic (AADT) occurring during peak hour.

DESIGN LIFE Period of time during which item is expected by its designers to work within its specified parameters.

DESIGN LOAD Total load on structural system for most severe combination of loads and forces which it is designed to sustain.

DESIGN PHASE When one or more designs are developed for project.

DESIGN RESPONSE SPECTRA Smooth, broad-banded spectra appropriate for specifying level of seismic design force or displacement, for earthquake-resistant design purposes.

DESIGN RULE Regulation that requires producers to design products that conform to standards.

DESIGN SPAN Distance between centers of supporting beam components.

DESIGN SPECTRUM Potential stress ranges behaving in most unfavorable way due to worst possible load states.

DESIGN SPEED Speed to determine geometric features of road during design phase.

DESIGN STAGE Process of taking on and developing approved concept design.

DESIGN STORM 1. Precipitation pattern defined for use in design of hydrologic system; 2. Storm characteristics that hydraulic structure is designed to withstand.

DESIGN STRENGTH 1. Load-bearing capacity of member computed on basis of allowable stresses which are assumed in design; 2. Assumed values for strength of concrete and yield stress of steel on which theoretical ultimate strength of section is computed.

DESIGN SUMMARY Document that summarizes designers' development of design that results in specifications.

DESIGN TEAM Members that brings their skills and abilities toward resolution of design issue.

DESIGN VALUE Value derived from representative value for use in design verification procedure.

DESIGN VALUE NUMBER Means of classifying or rating.

DESIGN VEHICLE Least maneuverable vehicle that routinely uses street or facility utilized during design phase.

DESIGN VOLUME Expected traffic to utilize roadway.

DESIGN WATER DEPTH Depth of water utilized in design process.

DESIGN WATER LEVEL Maximum water elevation, including flood surcharge, that dam is designed to be able to withstand.

DESIGN WAVE Wave condition with specified return period used in design of coastal designs.

DESIGN WEIGHT Maximum weight permitted, with no risk of early wear or failure.

DESIGN WIND Most severe wind reasonably possible.

DESIGN WIND PRESSURE Wind pressure used in design of wind loads for buildings or other structures.

DESIGN WORKING PRESSURE Normal operating pressure to which component is designed to operate continuously with safe margin below point where material will yield or burst.

DESIGN-BID-BUILD Project delivery method where agency contracts with separate entity for project design and construction.

DESIGN-BUILD Method of project delivery where team works under single contract to design and construct project.

DESIGN-BUILDER Individual or legal entity that reviews bid proposal and then develops design and cost proposal.

DESIGNATED BICYCLE ROUTE Dedicated bikeway system with signage.

DESIGNATED FACILITY Facility that has entered into agreement to render health services, hazardous waste treatment, storage or disposal.

DESIGNATED FLOODWAY Channel of water course and those portions of adjoining flood plain required to provide for passage of selected flood with small increase in flood stage above that of natural conditions.

DESIGNATED FREQUENCY FLOOD Probability that flood will be equaled or exceeded in any one year where 100-year flood has 0.01 annual exceedance probability (i.e., 1 chance in 100 that this flood flow level will be equaled or exceeded in any given year).

DESIGNATED POLLUTANT Air pollutant that's neither criteria nor hazardous pollutant but for which new source performance standards exist.

DESIGNATED REPRESENTATIVE Person who is named for purpose of caring for, escorting or receiving monetary payments on behalf of dependent.

DESIGNATED USES Those uses specified in water quality standards for each water body or segment, whether or not they are being attained.

DESIGNER 1. Person who designs something; 2. Person who designs things as profession.

DESIGNER BUGS Microbes developed through biotechnology that can degrade specific toxic chemicals at their source in toxic waste dumps or in groundwater.

DESIGNERS' OPERATING CRITERIA Detailed operating criteria which stress designer's intended use and operation of equipment and structures in interest of safe, proper and efficient use of facilities.

DESILTING BASIN Space or structure constructed just below diversion structure of canal to remove bed, sand and silt loads.

DESILTING WORKS Basins constructed just below diversion structure of canal to remove bed, sand and silt loads.

DESIRE LINE Path that pedestrians or vehicles take informally rather than taking sidewalk or set route.

DESLUDGE Removal of accumulated sludge.

DESMIDIACEAE One of four families of Charophyte green algae in order Desmidiales.

DESORPTION 1. Release or removal of adsorbed material from surface of solid adsorbent; 2. All processes by which chemicals move from solid phase and concentrate them in liquid phase (groundwater).

DESTINATION FACILITY Facility where waste is shipped for treatment and destruction, recycling, incineration, and/or disposal.

DESTRATIFICATION Vertical mixing within lake or reservoir to totally or partially eliminate separate layers of temperature, plant or animal life.

DESTROYED MEDICAL WASTE Regulated medical waste that has been ruined, torn apart or mutilated through thermal treatment, melting, shredding, grinding, tearing or breaking, so that it is no longer generally recognized as medical waste, but has not yet been treated (excludes compacted regulated medical waste).

DESTRUCTION AND REMOVAL EFFICIENCY Percentage that represents number of molecules of compound removed or destroyed in incinerator relative to number of molecules entering system (e.g., DRE equal to 99.99% means that 9,999 molecules are destroyed for every 10,000 that enter.)

DESTRUCTION FACILITY Facility that destroys regulated medical waste.

DESTRUCTION OR ADVERSE MODIFICATION Direct or indirect alteration that appreciably diminishes value of critical habitat for conservation of listed species.

DESTRUCTIVE DISTILLATION Heating of material (e.g., coal or wood) in inert atmosphere, at high temperature such as to cause decomposition.

DESTRUCTIVE TESTING Tests carried out to specimen's failure, in order to understand specimen's performance or material behaviour under different loads.

DESULFURIZATION Removal of sulfur from fossil fuels to reduce pollution.

DETACHABLE CONTAINER REAR LOADER Detachable container system with roll-out containers.

DETACHABLE CONTAINER SYSTEM Includes container body and fastener arrangement.

DETAIL DRAWING Drawing, at relatively large scale, of part of building, machine, etc., with dimensions or other information for use in construction.

DETAILED FRACTURE Break near surface of rail head.

DETAILER One who gives details of something.

DETAILING 1. Drawing preparation; 2. Ornamentation on something that has been constructed.

DETECT To determine presence of compound.

DETECTABLE LEAK RATE Smallest leak (e.g., from storage tank), expressed in terms of gallons- or liters-per-hour, that test can reliably discern with certain probability of detection or false alarm.

DETECTABLE WARNING Surface pattern that permits vision impaired people to detect street crossings and drop-offs.

DETECTION First of five Early Warning System components consisting of processes and equipment necessary to collect information about threatening event and response of dam and reservoir and relay that information to decision makers.

DETECTION CRITERION Predetermined rule to ascertain whether tank is leaking or not.

DETECTION LAG Time true anomaly is detected minus time anomaly actually happened.

DETECTION ZONE Intersection where vehicles are detected by signal controllers.

DETECTOR Device capable of registering specific substance or physical phenomenon and that optionally sounds alarm or triggers warning.

DETECTOR CHECK VALVE Swing check valve that protects against backflow by either back-pressure or back-siphonage from cross-connection between potable water systems and substances in 'low hazard' conditions in fire sprinkler systems.

DETENTION BASIN Storage structures (e.g., small unregulated reservoir) built upstream from populated areas that delay conveyance of water downstream to prevent runoff and/or debris flows from causing property damage and loss of life that can be periodically excavated to maintain their storage capacity.

DETENTION DAM Dam built to store streamflow or surface runoff and to control release of such stored water.

DETENTION PERIOD Time required for volume of water or wastewater to pass through tank at given rate of flow.

DETENTION RESERVOIR Ungated reservoir for temporary storage of flood water.

DETENTION STORAGE Volume of water, other than depression storage, existing on land surface as flowing water which has not yet reached channel.

DETENTION TANK Flow-control structure used to contain stormwater and wastewater for limited period of time.

DETENTION TIME Theoretical (i.e., calculated) time required for given amount of water or wastewater to pass through tank at given rate of flow.

DETERGENT Synthetic (i.e., non-soap) washing agent that helps to remove dirt and oil.

DETERMINISTIC METHODOLOGY Method where chance of occurrence of variable involved is ignored and method or model used is considered to follow definite law of certainty and not probability.

DETERMINISTIC MODEL Mathematical model where behavior of every variable is completely determined by governing equations and initial states of variables.

DETONATING CORD Round, flexible textile or plastic cord with center core of high explosive.

DETONATION Practically instantaneous decomposition or combustion of unstable compound, with tremendous increase in volume.

DETONATOR Device to start explosion(e.g., fuse or cap).

DETOUR Diversion from original route.

DETRITAL SEDIMENTS 1. Solid fragments of inorganic or organic material that come from weathering of rock and are carried and deposited by wind, water or ice; 2. Material that settles to bottom of liquid.

DETRITUS 1. Loose material that results directly from disintegration; 2. Heavier mineral debris moved by natural watercourses, usually in bed-load form; 3. Sand, grit and other coarse material removed by differential sedimentation in relatively short period of detention.

DETRITUS SLIDE Slow movement of surface soil downhill.

DETRITUS TANK Chamber for removing large heavy suspended matter from sewage.

DEUTERIUM OXIDE Form of water that contains larger than normal amount of hydrogen isotope deuterium, rather than common hydrogen-1 isotope that makes up most of hydrogen in normal water.

DEUTRIC Alteration of igneous rock by activity of volatiles based on magma throughout later periods of consolidation.

DEVELOPED LENGTH Length of conduit measured along center line of conduit and fittings.

DEVELOPED WATER 1. Water controlled and managed for variety of uses; 2. Artificially induced flow of stream; 3. Groundwater unnaturally brought to surface.

DEVELOPING GALE/STORM Indicate that gale/storm force winds are not now occurring but are expected before end of forecast period.

DEVELOPMENT 1. Process of growth or directed change; 2. Work done to well to increase flow; 3. process where mature multicellular organism or part of organism is produced by addition of new cells.

DEVELOPMENT DOCUMENT Report prepared during development of effluent limitation guideline by EPA that provides data and methodology used to develop limitations guidelines and categorical pretreatment standards for industrial category.

DEVELOPMENT EFFECTS Adverse effects (e.g., altered growth, structural abnormality, functional deficiency, death) observed in developing organism.

DEVELOPMENT LENGTH Reinforcement bar required to be embedded or projected into column to establish desired bond strength.

DEVELOPMENT-INDUCED DISPLACEMENT Population movement, forcing of communities and individuals out of their homes for purposes of economic development.

DEVELOPMENTAL TOXICITY Any structural or functional alteration that interferes with homeostasis, normal growth, differentiation, development or behavior and is caused by environmental insult.

DEVIATOR STRESS Difference between major and minor principal stresses in triaxial test.

DEW Moisture in air that condenses on solid surfaces when air is saturated with water vapor and temperatures fall below dew point temperature.

DEW POINT Temperature to which air with given quantity of water vapor must be cooled to cause condensation of vapor in air.

DEW POINT DEPRESSION Difference in degrees between air temperature and dew point.

DEW POINT FRONT Narrow zone (mesoscale feature) of extremely sharp moisture gradient and little temperature gradient that separates moist air from dry air.

DEW POINT TEMPERATURE Temperature at which dew begins to form or vapor begins to condense into liquid.

DEWATER 1. Removing water by pumping, drainage or evaporation; 2. Removal and control of groundwater from pores or other open spaces in soil or rock formations to extent that allows construction activities to proceed through use of deep wells and wellpoints; 3. Remove or separate portion of water in sludge or slurry to dry sludge so it can be handled and disposed of; 4. Remove or drain water from tank or trench.

DEWATERABLE Capable of having water removed.

DEWATERING 1. Any of various techniques for removal of water, either from solid or from structure; 2. Lowering of groundwater table.

DEWATERING PUMP Pump that removes water.

DEWPOINT DEPRESSION Difference between temperature and dew point temperature at certain height in atmosphere.

DEWPOINT HYGROMETER Device for determining dewpoint by computing relative humidity.

DIABATIC Process which occurs with addition or loss of heat.

DIAGNOSTIC MODEL Computer model used to calculate air pollution concentrations.

DIAGNOSTIC TOOL Any artifact which, because of form, shape or function, provides chronological or manufacturing information.

DIAGNOSTICS Process of determining state of or capability of component to perform its function(s).

DIAGONAL CRACKING Inclined crack at surface.

DIAGONAL CURB RAMP Curb ramp placed diagonally at intersection.

DIAGRAM Plan, drawing, sketch or outline to show how something works or show relationships between parts of whole.

DIAL GAUGE Any device for comparing physical property of two objects or object with standard.

DIAL GAUGES Measure movements transverse and along joint or crack.

DIAL INDICATOR 1. Any device for comparing physical property of two objects or object with standard; 2. Apparatus used to discover when axis of two rotating shafts are in line.

DIALYSATE Depleting fluid of salt by electrodialysis.

DIALYSIS 1. Method of separating molecules or particles of different sizes by differential diffusion through semipermeable membrane; 2. Utilization of this method for removal of waste products from blood in case of kidney failure.

DIAMETER Any straight line between two points on circumference of circle that passes through center of circle.

DIAMOND DRILL Rotary drill that makes holes in hard nonwood surfaces (i.e., ceramic, porcelain, glass, stone, marble, granite) that get their extra cutting power from little pieces of embedded diamond powder.

DIAMOND DUST 1. Fall of non-branched (snow crystals are branched) ice crystals in form of needles, columns or plates; 2. Ground-level cloud composed of tiny ice crystals.

DIAMOND INTERCHANGE Interchange for freeway crossing minor road.

DIAPHRAGM 1. Sheet of thin zone or facing, made of flexible impervious material (e.g., asphaltic concrete, plastic concrete, steel, wood, copper, plastic); 2. Flexible membrane separating two chambers and fixed around its periphery that distends into one or other chamber as difference in pressure in chambers varies; 3. Thin, semi-rigid membrane which vibrates to produce sound; 4. Floor slab, metal wall panel, roof panel having sufficiently large in-plane shear stiffness and sufficient strength to transmit horizontal forces to resisting systems.

DIAPHRAGM PACKING Packing between stiff members in relative movement.

DIAPHRAGM PRESSURE TANK Tank (closed) divided into compartments by diaphragm.

DIAPHRAGM PUMP Positive displacement pump that moves water by reciprocating motion of diaphragm in chamber having inlet and outlet check valves.

DIAPHRAGM WALL Sheet, thin zone or facing made of relatively impervious material (e.g., concrete, steel, wood, plastic).

DIAPHRAGM-TYPE CELLULAR COFFERDAM Cellular cofferdam consists of circular arcs at inner and outer sides, which are connected by straight diaphragm walls.

DIAPHRAGM-TYPE CONTROLLER Management apparatus that relies upon positive or negative pressure causing motion of diaphragm.

DIAPHRAGM-TYPE EARTHFILL Embankment dam that is constructed mostly of pervious material and having diaphragm of impermeable material which forms water barrier.

DIATOM Any of group of minute unicellular algae having siliceous covering of great delicacy.

DIATOMACEOUS EARTH Chalk-like material (fossilized diatoms) used to filter out solid waste in wastewater treatment plants that's active ingredient in some powdered pesticides.

DIATOMITE Chalk-like material (fossilized diatoms) used to filter out solid waste in wastewater treatment plants that's active ingredient in some powdered pesticides.

DIATOMS Single-celled (unicellular), colonial or filamentous algae with siliceous cell walls constructed of two overlapping parts.

DIAZINON Insecticide.

DIBENZOFURANS Group of organic compounds, some of which are toxic.

DICAMBA Herbicide 3,6-Dichloro-2-methoxybenzoic acid used to control weeds in cereal crops.

DICHLOROCLIPHENYLTRI-CHLOROETHANE Organic pesticide.

DICOFOL Pesticide used on citrus fruits.

DIE-FORMED RING Packing band to get rid of emptiness in stuffing box.

DIEL 1. Is of or pertaining to 24-hour period of time; 2. Regular daily cycle.

DIELDRIN 1. Organochlorine insecticide no longer registered for use in United States; 2. Degradation product of insecticide aldrin; 3. Cyclodiene insecticide that's persistent organic pollutant.

DIELECTRIC Electrically insulating or non-conducting material considered for its electric susceptibility.

DIELECTRIC STRENGTH Maximum electric field that pure material can withstand under ideal conditions without breaking down.

DIELECTRIC UNION Used to join pipes made from dissimilar metals to avoid electrolysis that leads to corrosion and pipe failure and is made from galvanized steel and lead-free brass.

DIESEL 1 Standard diesel fuel more volatile and flows more easily than Diesel 2.

DIESEL 2 Standard diesel fuel less volatile and flows less easily than Diesel 1.

DIESEL ENGINE Internal combustion engine which operates using diesel fuel.

DIESEL FUEL Mixture of hydrocarbons (boiling points in range of 150 to 380 $^{\circ}$C) that are obtained from petroleum.

DIESEL INDEX Denote ignition properties of diesel fuel calculated based on fuel's density and distillation range.

DIESELING Condition that can occur in engines, whereby engine keeps running after being turned off, due to engine kicking back and drawing fuel through carburetor, into engine and igniting it without spark.

DIETARY LC_{50} Statistically derived estimate of concentration of test substance in diet that would cause 50% mortality to test population under specified conditions.

DIFFERENCE IN ELEVATION Perpendicular measurement between two surfaces.

DIFFERENTIAL DIAPHRAGM Diaphragm that uses flexible disc to separates interior of gauge from process.

DIFFERENTIAL DRY-PIPE PRESSURE VALVE Dry-pipe sprinkler valve that has large clapper on air side that bears directly on smaller water side clapper.

DIFFERENTIAL GAUGE Gauge that measures difference between pressures at two points.

DIFFERENTIAL HEAD Condition where water pressure on upstream and downstream sides of object differ.

DIFFERENTIAL INTAKE Intake formed by hollow, low diversion dam with longitudinal slot for intercepting component of stream flow and conducting it to pipe line at one or both ends of dam.

DIFFERENTIAL MOTION Cloud motion that appears to differ relative to other nearby cloud elements.

DIFFERENTIAL PRESSURE Visual indicator, designed to measure and illustrate difference between two pressure points within process system.

DIFFERENTIAL PRESSURE VALVE Valve controlled (i.e., opening) with rising differential pressure.

DIFFERENTIAL PRICING Changing pricing for customers due to factors (e.g., time of day).

DIFFERENTIAL SETTLEMENT Unequal settling of building's foundation that can result in damage to structure.

DIFFERENTIAL SURGE TANK Surge tank that has vertical riser about same diameter as pipe where flow of water into main tank is limited by capacity of openings in riser or opening around base of riser.

DIFFERING SITE CONDITIONS When contractor encounters unknown physical conditions that differ materially from ordinarily encountered at work site.

DIFFLUENCE Flowing off on all sides.

DIFFRACTED WAVE Wave whose front has been changed in direction by obstacle or other non-homogeneity in medium, other than by reflection or refraction.

DIFFRACTION Breaking up of electromagnetic wave as it passes geometric structure (e.g., slit), followed by reconstruction of wave by interference.

DIFFRACTION ANALYSIS Utilizes diffraction patterns of laser beam passed through any object to quickly measure geometrical dimensions of particle.

DIFFUSE FIELD Acoustic field where sound waves reach point from all directions (i.e., sound pressure is same everywhere).

DIFFUSE ICE Poorly defined ice edge limiting area of dispersed ice usually on leeward side of area of floating ice.

DIFFUSED AIR Type of aeration that forces oxygen into sewage by pumping air through perforated pipes inside holding tank.

DIFFUSED SURFACE WATER Water on surface of land because of rain, melting snow or floods.

DIFFUSED-AIR AERATION Takes air, compresses it and then discharges air below water surface of aerator through some type of air diffusion device.

DIFFUSER 1. Aerating device consisting of membrane with fine pores, where air is blown to generate small bubbles; 2. Mechanical device designed to control characteristics of fluid at entrance to thermodynamic open system; 3. Any device that or spreads out or scatters.

DIFFUSER PLATE Porous plate that sits above water surface that protects sand surface, so that when water is added, it trickles down into filter.

DIFFUSER TUBE Porous tube used for aeration of liquid or for dispersion of gases or vapors.

DIFFUSING WELL Well used to add water to aquifer.

DIFFUSION 1. Movement of suspended, dissolved particles or molecules from more concentrated to less concentrated area; 2. Dissipation of energy associated with flood wave that results in attenuation of flood wave; 3. Scattering of something; 4. Intermingling of molecules of fluid due to random thermal agitation.

DIFFUSION COEFFICIENT Constant of proportionality which relates mass flux of solute to solute concentration gradient.

DIFFUSION VANE Fixed or removable casting in pump between impeller and casing with liquid passages designed to convert velocity head to pressure head.

DIFLUENCE Pattern of wind flow where air moves outward (i.e., "fan-out" pattern) away from central axis oriented parallel to general direction of flow.

DIGESTED SLUDGE Sludge or thickened mixture of sewage solids with water that has been decomposed by anaerobic bacteria.

DIGESTER In wastewater treatment, closed tank unit where bacterial action is induced and accelerated in order to break down organic matter and establish proper carbon to nitrogen ratio.

DIGESTER COILS System of hot water or steam pipes installed in digestion tank to heat contents.

DIGESTER SYSTEM Each continuous digester or each batch digester used for cooking of wood in white liquor and associated flash tank(s), blow tank(s), chip steamer(s) and condenser(s).

DIGESTION Biochemical decomposition of organic matter, assisted by microorganisms, resulting in partial gasification, liquefaction and mineralization of pollutants.

DIGESTION CHAMBER Lower part of Imhoff tank where sludge is decomposed by anaerobic bacteria.

DIGESTION TANK First tank of septic tank system where organic material is processed.

DIGIT 1. Whole numbers from 0 to 9; 2. 1/12 apparent diameter of sun or moon; 3. Narrow extremity of human hand or foot (i.e., finger, thumb or toe).

DIGITAL Property of representing values as discrete, usually binary, numbers rather than continuous spectrum.

DIGITAL ENCODER Combination circuit that generates specific code at its outputs in response to one or more active inputs.

DIGITAL INDICATION SYSTEM System that tracks operation by using digital display calibrators.

DIGITAL PROCESSING Computers perform wide variety of signal processing operations.

DIGITAL READOUT Numeric display.

DIGITIZATION Representation of continuous process, field or surface by numerical (digital) values.

DIGITIZE To convert data from map or graphical form to digital form for use by computer programs.

DIGITIZER Device that converts analog information into digital format.

DIKE 1. Low wall or embankment that can act as barrier to prevent spill from spreading or restrain river or stream from flooding; 2. To dig ditch; 3. To raise earthwork; 4. Slim stone formation that cuts over construction of surrounding stone.

DILATANCY 1. Expansion of cohesionless soils when subject to shear deformation; 2. Viscosity, of some substances, increases with shear or pressure.

DILUENT Any liquid or solid material used to lessen concentration or carry active ingredient.

DILUTE AND DISPERSE Waste disposal that relies on dilution and dispersion of source material.

DILUTION Reduction of concentration of substance by additional of solvent.

DILUTION FACTOR Ratio of parts of solute to total number of parts.

DILUTION GAUGING Adding of known amount of concentration to fluid and determining resultant dilution of solution downstream.

DILUTION RATIO Relationship between volume of water in stream and volume of incoming water.

DILUTION WATER Water added to maintain or dilute system.

DILUVIUM 1. Inundation or flood; 2. Deposit of sand, gravel, etc. made by oceanic flooding.

DIMENSION Measure of spatial extent in particular direction.

DIMENSION LINE Line with end limiters that reveal space between points.

DIMENSIONAL ANALYSIS Study of dimensions of physical quantities used to obtain information about large complex systems.

DIMENSIONAL STABILITY Degree that material maintains its original dimensions when subjected to outside forces.

DIMENSIONAL TOLERANCES Designed-in clearance or interference between two parts.

DIMENSIONLESS NUMBER Physically meaningful ratio of parameters that's dimensionless.

DIMENSIONLESS UNIT HYDROGRAPH Unit graph that compares unit hydrographs for hydrologic evaluation of small watersheds and for hydrologic design of soil and water resource management practices.

DIMICTIC Lakes and reservoirs that freeze over and normally go through two stratifications and two mixing cycles year.

DIMINISHING PIPE Pipe of diminishing diameter which acts as reducer.

DIMINUTION Gradual reduction of size, intensity, extent or importance.

DIMPLE SPRING Spring that goes below watertable and occurs in small depressions.

DINOCAP Fungicide used primarily by apple growers to control summer diseases.

DINOSEB Herbicide of dinitrophenol family also used as fungicide and insecticide.

DIOXIN Any of family of compounds known chemically as dibenzo-p-dioxins.

DIOXINS Broad range of toxic or carcinogenic halogenated polycyclic compounds that occur as byproducts of herbicides.

DIP 1. Angle between planar feature (e.g., sedimentary bed, fault) and horizontal plane; 2. Slope of layers of soil or rock; 3. Lower section of road or geological feature; 4. Lower curve of plumbing trap or depth of water in trap; 5. 20° to 45°.

DIP STICK Stick or rod used to measure depth of liquid.

DIPMETER 1. Instrument designed for determining direction and angle of dip of geological formations; 2. Device to record depth below ground level of surface of water in borehole or piezometric tube.

DIPPER DREDGE Power shovel mounted on barge for marine use.

DIPPING 20° to 45°.

DIQUAT Contact herbicide that produces desiccation and defoliation.

DIRECT ACCESS Arrangement where customers can purchase electricity directly from any supplier in competitive market, using transmission and distribution lines of electric utilities to transport electricity.

DIRECT CHARGES Prices which can be recognized especially having merchandise, service or action.

DIRECT CIRCULATION State where cold air sinks and warm air rises that generates wind.

DIRECT COMPENSATION Payment that employee receives from their employer.

DIRECT CURRENT Electrical current flowing in one direction only and essentially free from pulsation.

DIRECT DIGITAL CONTROLS Application of microprocessor technology to building environmental controls.

DIRECT DISCHARGE Wastewater discharged directly into stream or other receiving body.

DIRECT DISCHARGER Municipal or industrial facility which introduces pollution through defined conveyance or system (e.g., outlet pipes, point source).

DIRECT EFFECTS Direct impact of one on another when not mediated or transmitted through third.

DIRECT ENERGY One of two elements of solar energy.

DIRECT ENERGY CONVERSION Converts charged particle's kinetic energy into voltage.

DIRECT EXPENSES Expense incurred that varies directly with changes in volume of cost object.

DIRECT FILTRATION Method of treating water which consists of addition of coagulant chemicals, flash mixing, coagulation, minimal flocculation and filtration.

DIRECT FIRE PRESSURE Pressure required in pipes of water distribution system to provide adequate fire streams when fire hose is connected directly to hydrant, without use of fire engine pumping.

DIRECT FLOOD DAMAGE Damage done to property, structures, goods, etc., by flood as measured by cost of replacement and repairs.

DIRECT HIT Close approach of tropical cyclone to particular location.

DIRECT IN-LINE MECHANICAL LINKAGE Control rods directly linked to valve sections.

DIRECT IRRIGATION Use of wastewater effluent by spraying on land.

DIRECT LABOR Labor employed by authority commissioning work, not by contractor.

DIRECT MATERIAL Materials and supplies that are consumed during manufacture of product and which are directly identified with that product.

DIRECT MATERIALS Materials and supplies that are consumed during manufacture of product and which are directly identified with that product.

DIRECT OXIDATION Direct combination of substances with oxidants (e.g., sparging, air diffusion, use of chemical agents).

DIRECT PHOTOLYSIS Transformations of organic compounds that are consequence of that compound's own light absorption.

DIRECT PUSH Technology used for performing subsurface investigations by driving, pushing, and/or vibrating small-diameter hollow steel rods into ground.

DIRECT RUNOFF 1. Water that flows over ground surface or through ground directly into streams, rivers or lakes; 2. Runoff entering stream channels promptly after rainfall or snow melt that forms major part of flood hydrograph and excludes base flow.

DIRECT SHEAR TEST Shear test where soil or rock under applied normal load is stressed to failure by moving one section of specimen container (shear box) relative to other section.

DIRECT SOLAR GAIN 1. Using sun's heat without interference; 2. When heat collection, storage and distribution all occur within same space.

DIRECT SOLAR RADIATION Component of solar radiation received by Earth's surface only from direction of Sun's disk (i.e., it has not been reflected, refracted or scattered).

DIRECT-ACTING PUMP Displacement reciprocating pump where steam or power piston is connected to pump piston by means of rod, without crank motion or flywheel.

DIRECT-DUMP TRANSFER SYSTEM Unloading of solid waste directly from collection vehicle into open-top transfer trailer or container.

DIRECT-FEED INCINERATOR Incinerator that accepts solid waste directly into its combustion chamber.

DIRECTION Ten-minute average wind direction measurements in degrees clockwise from true North.

DIRECTION LIMIT SWITCH Mechanical switches that contains features of safety switch with forced actuation.

DIRECTION OF IRRIGATION Flow direction of irrigation water.

DIRECTION SWITCH Device that determines direction of rotation of electric motor.

DIRECTIONAL CONTROL VALVE Valve that allow fluid flow into different paths from one or more sources.

DIRECTIONAL DESIGN HOUR VOLUME Volume in predominant one-way direction of travel during design hour.

DIRECTIONAL DISTRIBUTION Traffic split during peak design time.

DIRECTIONAL FLOW RATE Traffic flow rate in one direction of transportation facility.

DIRECTIONAL INTERCHANGE Infrastructure of two highways when high volume traffic transfers between two legs of interchange.

DIRECTIONAL SEGMENT Two-lane highway length in one travel direction.

DIRECTIONAL SHEAR Component of wind shear that is due to change in wind direction with height (e.g., southeasterly winds at surface and southwesterly winds aloft).

DIRECTIONAL SPLIT Directional distribution of hourly volume on highway.

DIRECTOR 1. Person responsible for permitting, implementation and compliance; 2. Authorized representative.

DIRT CAPACITY Amount of dirt that particular filter will hold.

DISAPPEARING SOLAR FILAMENT Sudden (timescale of minutes to hours) disappearance of solar filament (prominence).

DISASTER Event that demands crisis response beyond scope of any single line agency or service, presents threat and requires resources beyond what are available locally.

DISC 1. To harrow farmland; 2. Thin, flat, circular plate or similar object.

DISCHARGE 1. Flow of liquid, gas or other substance; 2. Rated output of pump.

DISCHARGE ALLOWANCE Amount of pollutant that plant is permitted to discharge.

DISCHARGE AREA Cross-sectional area at point of flow.

DISCHARGE CAPACITY Maximum rate of flow capable of passing.

DISCHARGE COEFFICIENT Ratio of actual discharge to theoretical discharge.

DISCHARGE CONSENT Permission for discharge of stormwater, wastewater or other contaminants into water or onto or into land.

DISCHARGE CONVEYOR Conveyor that carries and deposits material away from equipment.

DISCHARGE CURVE Curve that expresses relation between discharge of stream or open conduit at given location and stage or elevation of liquid surface at or near that location.

DISCHARGE HEAD 1. Maximum height reached by pipe after pump; 2. Pressure at discharge of pump.

DISCHARGE HOSE Conduit connecting discharge outlet of pump and delivery point.

DISCHARGE HYDROGRAPH Statistical plot of discharge or flow.

DISCHARGE LENGTH SCALE Square root of cross-sectional area of any discharge outlet.

DISCHARGE MEASUREMENT Determination of quantity of water flowing per unit of time.

DISCHARGE MONITORING REPORT Form used (including any subsequent additions, revisions or modifications) to report self-monitoring results by NPDES permittees.

DISCHARGE OF FILL MATERIAL Addition of fill material into public waters or lands.

DISCHARGE PIPE Conduit joining discharge outlet of pump to delivery point.

DISCHARGE POINT Point within disposal site where dredged or fill material is released.

DISCHARGE RATING CURVE Curve that expresses relation between discharge of stream or open conduit at given location and stage or elevation of liquid surface at or near that location.

DISCHARGE STANDARDS Regulatory standards for wastewater discharge based on performance of treatment and control technologies.

DISCHARGE TABLE 1. Table showing relation between gauge height and discharge of stream or conduit at given gaging station; 2. Table showing relation between two mutually dependent quantities or variable over given range of magnitude.

DISCHARGE VALVE Conduit (control) valve for reducing and increasing flow.

DISCONNECTING SWITCH Electrical switch used in electrical circuits to isolate section which has electricity from section with no electricity.

DISCONNECTING TRAP Device to trap, remove or separate deleterious, hazardous or undesirable matter.

DISCONTINUE Particles were present within unit above but are no longer present.

DISCONTINUOUS EASEMENT Easement or servitude (i.e., right of way, right to draw water) that requires for its exercise or enjoyment action by party claiming it.

DISCONTINUOUS INTERSTICE Small open space or interstice not connected with another interstice.

DISCOUNT RATE Interest rate used to discount future cash flows of financial instrument.

DISCRETIONARY FUNDS Money spent on non-necessary things.

DISCRETIZATION Process of subdividing continuous model and/or time domain into discrete segments or cells.

DISCRETIZATION ERROR Error introduced by discrete representation of continuous variable.

DISDROMETER Equipment that measures and records size distribution of raindrops.

DISEASE VECTOR Any agent that carries and transmits infectious pathogen into another living organism.

DISINCENTIVE Something that discourages particular behaviour.

DISINFECTANT Any oxidant added to water in any part of treatment/distribution process intended to kill/inactivate pathogenic microorganisms.

DISINFECTANT BY-PRODUCT Compound formed by reaction (i.e., byproduct) of disinfectant (e.g., chlorine) with organic material in water supply.

DISINFECTANT CONTACT TIME Time disinfection product needs to work effectively.

DISINFECTANT TIME Time it takes water to move from point of disinfectant application (or previous point of residual disinfectant measurement) to point before or at point where residual disinfectant is measured.

DISINFECTION Treatment with disinfectant materials in order to destroy harmful microorganisms.

DISINTEGRATION Wearing away or falling to pieces of rocks or strata, produced by atmospheric action, frost, ice, etc.

DISK Visible surface of Sun (or any heavenly body) projected against sky.

DISK METER Meter for measuring fluids that has disk that controls alternate filling and emptying of measuring chambers.

DISK SCREEN Rotating disks mounted at even intervals on multiple, parallel, vertical shafts to remove debris.

DISMEMBERED STREAM Tributary left as independent stream after lower part of drainage system was submerged by invasion body of water.

DISPATCHER Person who controls movements of vehicles.

DISPERSANT 1. Chemical agent used to break up concentrations of organic material (e.g., spilled oil); 2. Substance used to prevent settling or clumping of particles suspended in liquid.

DISPERSE To scatter in different directions.

DISPERSING AGENT Agent used to assist in separating individual fine soil particles and to prevent them from flocculating when in suspension.

DISPERSION 1. Process by which some of water molecules and solute molecules travel more rapidly than average linear velocity and some travel more slowly; 2. Spreading of solute in direction of groundwater flow (longitudinal dispersion) or direction perpendicular to groundwater flow (transverse dispersion); 3. Distortion of shape of seismic-wave train because of variation of velocity with frequency; 4. Process of separating radiation into various wavelengths; 5. Separation of visible light by refraction or diffraction; 6. Degree of scatter of data.

DISPERSION COEFFICIENT 1. Measure of spreading of flowing substance due to nature of porous medium, with its interconnected channels distributed at random in all directions; 2. Sum of coefficients of mechanical dispersion and molecular diffusion in porous medium.

DISPERSION INDEX Ratio used to measure dispersion of probability distribution.

DISPERSION TECHNIQUE Any technique which attempts to affect the. concentration of pollutant in ambient air.

DISPERSIVE CLAY Clay that's highly erosive, subject to severe damage and/or failure that deflocculates in presence of water.

DISPERSIVITY Scale dependent property of aquifer that determines degree that dissolved constituent will spread in flowing groundwater comprised of three directional components (longitudinal, transverse and vertical).

DISPLACED THRESHOLD Runway threshold located at location other than beginning or end.

DISPLACEMENT 1. Relative movement of any two sides of fault; 2. Quantity displaced by floating body (e.g., boat), where weight of displaced liquid being equal to that of displacing body; 3. Amount of fluid that will pass via pump, motor or cylinder within revolution or stroke; 4. Process of extracting soluble substances from organic material and like, whereby quantity of saturated solvent is displaced or removed, for another quantity of solvent.

DISPLACEMENT METER Meter where all of fluid passes in almost completely isolated quantities and are counted and indicated in terms of volume or weight units by register.

DISPLACEMENT PUMP Reciprocating type of pump where piston draws water into closed chamber and then expels it under pressure.

DISPLACEMENT SAVINGS Saving realized by displacing purchases of natural gas or electricity from local utility by using landfill gas for power and heat.

DISPLACEMENT TIME Average time transferring liquid is in tank or channel.

DISPLACEMENT VELOCITY Speed of displacement of settling tank contents by inflowing liquids.

DISPOSABLE ELEMENT Replaced by end of its own service life.

DISPOSABLE FILTER Filters usually made of paper-like material and cannot be reused.

DISPOSABLE WASTE CONTAINER Bags designed for keeping solid waste.

DISPOSABLES Consumer products, other items and packaging used once or few times and discarded.

DISPOSAL 1. Final placement, destruction or getting rid of something; 2. Surplus or banned pesticides or other chemicals; 3. Polluted soils; 4. Any drums containing hazardous materials from removal actions or accidental releases; 5. Arrangement, categorization or classification of things.

DISPOSAL AREA Place where solid waste may be dumped.

DISPOSAL BY DILUTION Process whereby raw or treated sewage is discharged into water bodies such as rivers.

DISPOSAL FACILITIES Repositories for solid waste, including landfills and combustors intended for permanent containment or destruction of waste materials, excluding transfer stations and composting facilities.

DISPOSAL FACILITY Any site where hazardous waste is intentionally placed and where waste will remain after closure.

DISPOSAL LEVY Tax imposed to accelerate waste avoidance, reuse and recycling.

DISPOSAL SITE 1. Waters where special disposal actions are allowed and consisting of bottom surface area and any overlying volume of water; 2. Interim or ultimately approved and exact geographical area where ocean or land dumping of wastes is allowed under conditions established in licenses.

DISPOSAL SITE DESIGNATION STUDY Collection, analysis and interpretation of all available pertinent data and information on proposed disposal site prior to use.

DISPOSAL WELL Well where waste fluids can be injected for safe disposal.

DISPUTE RESOLUTION Process of resolving disputes between parties.

DISSECTED Cut by erosion into valleys, hills and upland plains.

DISSIPATER Apparatus that reduces erosive force of running water.

DISSOCIATION Process where compound body breaks up into simpler constituents by way of action of heat on gaseous or volatile substances.

DISSOLVE To enter into solution.

DISSOLVED Material in representative water sample that passes through 0.45-micrometer membrane filter.

DISSOLVED AIR Mixture of dispersed air in fluid.

DISSOLVED CONSTITUENT Operationally defined as constituent that passes through 0.45-micrometer filter.

DISSOLVED OXYGEN Amount of free oxygen (oxygen gas) dissolved in water that's indicator of water quality.

DISSOLVED SOLIDS Disintegrated organic and inorganic material dissolved in water that's indicator of salinity or hardness.

DISSOLVED WATER Mixture of water and fluid that's dispersed.

DISSOLVED-OXYGEN SAG CURVE Graph along path of stream or channel of uptake of dissolved oxygen related to biochemical oxidation of organic matter, plus re-oxygenation from consumed atmospheric oxygen and biological photosynthesis.

DISSOLVED-SOLIDS CONCENTRATION Quantity of dissolved material in sample of water.

DISTANCE Length or interval of time.

DISTANCE GAUGE Instrument used to quantify persistent measurement.

DISTANCE MEASURING EQUIPMENT Electronic system that measures distance between aircraft and ground station by timing propagation delay of signals.

DISTILLATION 1. Act of purifying liquids through boiling, so that steam or gaseous vapors condense to pure liquid; 2. Act of falling in drops or act of pouring out in drops; 3. Separation of volatile parts by heating and condensing.

DISTILLED WATER Water formed by condensation.

DISTORTED MODEL Hydraulic model where horizontal and vertical scales are different.

DISTORTION Conscious departure from scaling law often necessitated by complex set of prototype and laboratory conditions.

DISTORTION-INDUCED FATIGUE Cracks that develop at or near connections between elements due to secondary stresses.

DISTRESS State where soil cave-in is about to happen.

DISTRIBUTARIES Diverging streams which do not return to main stream, but discharge into another stream or ocean.

DISTRIBUTARY STREAM 1. Smaller stream that conducts flow away from larger, main channel, especially on deltas; 2. Opposite of tributary stream.

DISTRIBUTED SAMPLE Sample (e.g., soil) that's been completely blended and is thus not representative of its own initial features.

DISTRIBUTION 1. Way where something is shared out among group or spread over area; 2. Delivery of electricity to retail customer's home or business through low voltage distribution lines; 3. Conduit where services are delivered to many locations inside building.

DISTRIBUTION BOX Chamber that lies between septic tank and drainfield and serves to evenly distribute wastewater evenly amongst drainfield lines.

DISTRIBUTION COEFFICIENT Quantity of solute, chemical or radionuclide absorbed by solid per unit weight of solid divided by quantity dissolved in water per unit volume of water.

DISTRIBUTION EFFICIENCY Measure of uniformity of irrigation water distribution over field.

DISTRIBUTION GRAPH Unit hydrograph of direct runoff modified to show proportions of volume of runoff that occurs during successive equal units of time.

DISTRIBUTION HYDROGRAPH Unit hydrograph of direct runoff modified to show proportions of volume of runoff that occurs during successive equal units of time.

DISTRIBUTION LINE Low voltage electric power line usually 69 kilovolts or less.

DISTRIBUTION LOSS Loss of water from pipes, channels, conduits and ditches due to leakage, seepage, evaporation or evapotranspiration that's not available for further use.

DISTRIBUTION PIPE Conduit that delivers potable water to fixtures.

DISTRIBUTION RATIO Ratio of maximum potential 24-hour use of wastewater to your high-speed filter to particular average daily use.

DISTRIBUTION RESERVOIR Storage reservoirs that store treated water for supplying water during emergencies and also to help in absorbing hourly fluctuations in normal water demand.

DISTRIBUTION SYSTEM 1. Network of pumps and pipes leading from treatment plant to customers' plumbing systems; 2. System of ditches or conduits and their appurtenances, which conveys irrigation water from main canal to farm units; 3. Portion of electric system dedicated to delivering electric energy to end user.

DISTRIBUTION TILE Tiles in agricultural that receive effluent from septic tank.

DISTRIBUTOR Fixed or movable apparatus used to apply liquid in steady speed; 2. Device that distributes electric current.

DISTRICT 1. Entity that has contract for delivery of irrigation water; 2. Administrative division of area.

DISTRICT HEATING System for distributing heat generated in centralized location for residential and commercial heating requirements.

DISTRICT MAP Border map of region broken up into districts.

DISTURBANCE Any event or series of events that disrupt ecosystem, community or population structure and alters physical environment.

DISTURBANCE TIME Time of mixing.

DISTURBED SAMPLE Natural structure of soil gets modified or destroyed during sampling operation.

DISUSE OF LAND Land no longer being used or in use.

DITCH 1. Long, narrow excavation; 2. Constructed open channel for conducting water.

DITCH CHECK Small barrier constructed across swale or drainage ditch to reduce velocity of flowing water thereby allowing sediment to settle.

DITCH LINING Covering interior surface of ditch with material to reduce and/or prevent erosion, exfiltration or infiltration.

DITCH OXIDATION Large holding tank with oval shape built on surface of ground, lined with impermeable lining that allows wastewater to be exposure to air for diffusion of oxygen.

DITCHER Machine that digs ditches and piles dirt to side.

DIURNAL Related to actions which are completed in course of calendar day (i.e., daily) and which typically recur every calendar day.

DIURNAL CYCLES Variations in meteorological parameters (e.g., temperature, relative humidity) over course of day which result from rotation of Earth about its axis and resultant change in incoming and outgoing radiation.

DIURNAL FLUCTUATIONS Variations (e.g., watertable, stream flow) occurring within 24-h period and related to rotation of earth.

DIURNAL FORCE Force (via tides) having period of one lunar day.

DIURNAL INEQUALITY Difference in height between successive high (or low) tides.

DIURNAL TEMPERATURE RANGE Temperature difference between minimum at night (low) and maximum during day (high).

DIURNAL TIDE When only one high tide and one low tide occur per day.

DIURNAL WAVE Air pressure daily variation at single location.

DIURON Algicide and herbicide of arylurea class that inhibits photosynthesis.

DIVE CULVERT Bent conduit having unequal legs, employed for drawing off water when summit of bend is higher than supply and discharge end (longer leg) is lower than supply.

DIVER Someone who works underwater.

DIVERGE 1. To separate; 2. To tend into different direction.

DIVERGENCE 1. Expansion or spreading out of vector field; 2. Degree to which two or more things spread apart; 3. Speed of increase of unit volume of fluid which, in atmosphere, is roughly proportional to perpendicular speed.

DIVERGING Spreading apart (e.g., air, water, traffic) into different streams.

DIVERGING DIAMOND INTERCHANGE Diamond interchange where two directions of traffic on non-freeway road cross to opposite side on both sides of bridge at freeway.

DIVERGING TUBE Increasing diameter conduit along its direction of flow.

DIVERSION 1. Taking of water from stream or other body of water into canal, pipe or other conduit for human use (e.g., water supply); 2. Channel constructed across slope for purpose of intercepting surface runoff and divert it to some convenient discharge point.

DIVERSION AREA Part of adjoining region past standard groundwater or watershed divide that may contain water.

DIVERSION CANAL Waterway used to divert water from its natural course.

DIVERSION CAPACITY Flow which can be passed through canal headwork at dam under normal head.

DIVERSION CHAMBER Chamber or box that contains device for diverting or drawing off all or part of flow or for discharging portions of total flow to various outlets.

DIVERSION CHANNEL Waterway used to divert water from its natural course.

DIVERSION CUT Reservoir channel that joins incoming flow to discharge channel below dam that used to redirect flooding.

DIVERSION DAM Dam built to divert water from waterway or stream into different watercourse.

DIVERSION DITCH Canal to divert water away.

DIVERSION DUTY OF WATER Irrigation water diverted at intake of canal system usually expressed in depth or irrigable area under system.

DIVERSION GATE Gate which may be closed to divert water from main conduit or canal to lateral or some other channel.

DIVERSION INLET Conduit or tunnel upstream from intake structure.

DIVERSION MANHOLE Manhole that can redirect incoming flow to different conduit.

DIVERSION RATE Percentage of waste materials diverted from traditional disposal (e.g., landfilling, incineration) to be recycled, composted or re-used.

DIVERSION TERRACE Terrace constructed to divert water from terrace system or to divert water from its natural drainage channel.

DIVERSION TUNNEL Waterway used to divert water from its natural course.

DIVERSION VALVE Valve at junction of pipe tee used to change flow from one branch to another.

DIVERSION WORKS Structures and/or facilities necessary that conduct flows to and through alternative routes.

DIVERSITY Distribution and abundance of different kinds of plant and animal species (biological taxa) and communities in specified location.

DIVERSITY FACTOR Sum of individual demands divided by maximum demand of whole system.

DIVERSITY INDEX Numerical expression of evenness of distribution of aquatic organisms.

DIVERT To direct flow away from its natural course.

DIVERTER VALVE Valve that diverts flow from one source to multiple destinations.

DIVERTING WEIR Weir over which excess water, stormwater or wastewater is allowed to flow.

DIVIDE 1. Ridge or high area of land that separates one drainage basin from another; 2. High ground that forms boundary of watershed; 3. To split or divide into two or more parts.

DIVIDED HIGHWAY Road where two directions of traffic are separated by entrance barrier or strip of land.

DIVIDING STREAMLINE In blocked flow region upwind of mountain barrier, streamline that separates blocked flow region near ground from streamlines above which go over barrier.

DIVIDING STREAMLINE HEIGHT Height above ground of dividing streamline, as measured far upwind of mountain barrier.

DIVINING ROD Rod used for dowsing, technique of divination used to locate subterranean sources of water, metal, other mineral resources.

DIVISION BOX Structure for dividing flow into several streams.

DIVISIONAL ISLAND Elongated structures in center of roads that act as slip roads to bypass traffic.

DNA Deoxyribonucleic acid, self-replicating material present in nearly all living organisms as main constituent of chromosomes.

DNA HYBRIDIZATION Use of segment of DNA, called DNA probe, to identify its complementary DNA used to detect specific genes.

DO-NOTHING ALTERNATIVE Status-quo is maintained.

DOBSON UNIT Unit used to measure abundance of ozone in atmosphere that's equivalent to 2.69 x 10^{16} molecules of ozone per centimeter squared.

DOCK 1. Structure attached to shore for loading and unloading vessels; 2. Place of arrival and departure of train in railway station.

DOCKAGE Charge levied for docking vessel.

DOCTOR BLADE Blade used for scraping off excess from rotating screen.

DOGS Wedges that force gate to seat tightly.

DOLDRUMS Regions on either side of equator where air pressure is low and winds are light.

DOLOMITE Sedimentary rock consisting chiefly of magnesium carbonate.

DOLPHIN Mooring post on wharf or beach.

DOMAIN Geographical area over which simulation is performed.

DOME Structure curved both in lengthwise and crosswise direction.

DOMESTIC APPLICATION Pesticide application in and around houses, office buildings, motels and other living or working areas.

DOMESTIC CONSUMPTION Quantity or quantity per capita, of water consumed in municipality or district for domestic uses or purposes during given period.

DOMESTIC FILTER Apparatus designed for purification of water from various supply systems and open water reservoirs, lakes and rivers.

DOMESTIC GARBAGE Residential waste that may contain paper, metal, glass, food and/or vegetation.

DOMESTIC METER Device for measurement and billing residential services.

DOMESTIC WASTE Solid waste, composed of garbage and rubbish, which normally originates in private home or apartment house.

DOMESTIC WASTEWATER Wastewater originating from residential activities such as restroom usage, washing, bathing, food preparation and laundry.

DOMESTIC WATER USE Water for household purposes (e.g., watering of livestock, drinking, food preparation, bathing, washing clothes and dishes, flushing toilets, watering lawns and gardens).

DOMESTIC WITHDRAWALS Water for household purposes (e.g., watering of livestock, drinking, food preparation, bathing, washing clothes and dishes, flushing toilets, watering lawns and gardens).

DOMINANCE FREQUENCY Percentage of units where certain species has largest quantity.

DOMINANT DISCHARGE Particular magnitude of flow that is sometimes referred to as "channel forming" discharge.

DOMINANT PLANT Plant species controlling environment.

DOPPLER RADAR Radar that can measure radial velocity, instantaneous component of motion parallel to radar beam (i.e., toward or away from radar antenna).

DOPPLER SHIFT Change in frequency or wavelength of wave for observer who is moving relative to wave source.

DORMANCY State or characteristic of being quiet or inactive restfulness.

DORTMUND TANK Deep tank, with conical bottom to which liquid sewage is supplied by pipe reaching down nearly to bottom where resulting upward flow assists sedimentation of sludge.

DOSAGE Dose per time unit (e.g., mg/day).

DOSE 1. Actual quantity of chemical administered to organism or that it is exposed; 2. Amount of substance available for interaction with metabolic processes after crossing outer boundary of organism and reaches specific tissue (e.g., liver).

DOSE EQUIVALENT Unit that quantifies biological effectiveness of absorbed dose of ionizing radiation.

DOSE RATE 1. Dose per time unit (e.g., mg/day); 2. Quantity of radiation absorbed per unit time.

DOSE RESPONSE Shifts in toxicological responses of individual (e.g., alterations in severity) or populations (e.g., alterations in incidence) that are related to changes in dose of any given substance.

DOSE RESPONSE CURVE Graphical representation of relationship between dose of stressor and biological response thereto.

DOSE STANDARD Regulatory standard that takes regulated facility to restrict its emissions to degree required to make sure that no person receives effective dose equivalent greater in relation to given amount.

DOSE-RESPONSE ASSESSMENT 1. Estimating potency of chemical; 2. Process of determining relationship between dose of stressor and specific biological response; 3. Evaluating quantitative relationship between dose and toxicological responses.

DOSE-RESPONSE RELATIONSHIP Quantitative relationship between amount of exposure to substance and extent of toxic injury or disease produced.

DOSIMETER Instrument to measure dosage.

DOSING APPARATUS 1. Device for regulating application of measured chemical quantities; 2. Device for regulating flow of wastewater or water to filters or equipment.

DOSING CHAMBER Tank where effluent is fed and held until specific volume is collected to be removed.

DOSING RATIO Maximum speed of application of wastewater to given filter area divided by average speed on that area.

DOSING SIPHON Mechanism installed in septic tank that converts small continuous flows into intermittent release of effluent.

DOSING TANK Tank where effluent is fed and held until specific volume is collected to be removed.

DOT REPORTABLE QUANTITY Quantity of substance specified in U.S. Department of Transportation regulation that triggers labeling, packaging and other requirements related to shipping such substances.

DOUBLE BLOCK AND BLEED Hydraulic manifold that combines two or more valves into one component.

DOUBLE CHECK-VALVE Backflow prevention device designed to protect water supplies from contamination.

DOUBLE COAGULATION Use of coagulant at two different stages in treatment process.

DOUBLE CONNECTOR Short conduit with extended parallel screw thread at every end.

DOUBLE CURVATURE ARCH DAM Arch dam that is curved vertically as well as horizontally, with undercutting of heel and in most instances, downstream overhang near crest.

DOUBLE FILTRATION Separation of solids from fluids by interposing two filters where only fluid can pass.

DOUBLE HUB Conduit having bell on both ends.

DOUBLE LOCK Lock with two parallel chambers.

DOUBLE OFFSET Where one conduit moves around second conduit to continue its original orientation.

DOUBLE TRACK Section of railway line or tramline consisting of two parallel tracks, one being used in one direction and other in opposite direction.

DOUBLE WASH/RINSE Method for decontaminating (e.g., remove PCBs) non-porous surfaces.

DOUBLE-ACTING GATE Gate on dump truck that's hinged top and bottom for dumping choice.

DOUBLE-ACTING PUMP Reciprocating pump where hydraulic fluid performs on both sides of piston.

DOUBLE-MASS CURVE Plot on arithmetic cross-section paper of cumulated values of one variable against cumulated values of another or against computed values of same variable for concurrent period of time.

DOUBLE-POLE SWITCH Switch that controls two circuits concurrently.

DOUBLE-SEAL MANHOLE COVER Cast-iron plate with two fitting into cast-iron frame.

DOUBLE-STRENGTH PIPE Specification standard for large, vitrified clay pipe deeply trenched.

DOUBLE-SUCTION IMPELLER Impeller where water enters on both sides.

DOUBLE-SUCTION PUMP Centrifugal pump whose impeller pair is arranged back to back on pump shaft.

DOUBLE-THROW SWITCH Switch that connects input terminal to one of two output terminals.

DOUBLE-WALL COFFERDAM Cofferdam consisting of two parallel sets of sheet piles driven into ground around excavation.

DOWN TIME 1. Amount of time lost due to forces beyond one's control; 2. Period of time set aside for rest and relaxation.

DOWN-VALLEY WIND Thermally driven wind directed down valley's axis, usually occurring during nighttime that's part of along-valley wind system.

DOWNBURST Strong downdraft current of air from cumulonimbus cloud, often associated with intense thunderstorms.

DOWNDRAFT 1. Low-pressure zone created on lee side when wind blows over promontory; 3. Small-scale column of air that rapidly sinks toward ground (e.g., downward current), usually accompanied by precipitation as in shower or thunderstorm.

DOWNGRADIENT 1. Direction that groundwater flows (i.e., "downstream" for surface water); 2. Position down along gradient from starting position.

DOWNSLOPE FLOW Thermally driven wind directed down mountain slope and usually occurring at night that's part of along-slope wind system.

DOWNSPOUT Vertical pipe or conduit that carries rainwater from scupper, guttering of building to lower roof level, drain, ground or storm water runoff system.

DOWNSTREAM In same direction as stream or other flow or toward direction where flow is moving.

DOWNSTREAM FACE Inclined surface of dam away from reservoir.

DOWNSTREAM PROCESSORS Industries dependent on crop production (e.g., canneries, food processors).

DOWNSTREAM SLOPE Slope or face of dam away from reservoir water.

DOWNTIME Amount of time lost due to forces beyond one's control.

DOWNWARD VERTICAL VELOCITY Sinking air.

DOWNWASH Deflection of air downward relative to object that causes deflection.

DOWNWELLING RADIATION Component of radiation directed toward Earth's surface from Sun or atmosphere, opposite of upwelling radiation.

DOWSING Practice of seeking water or other substances with aid of forked stick or similar pointing device.

DP HOLE Hole in ground made with DP equipment.

DPD Dominant wave period (seconds) is period with maximum wave energy.

DPD METHOD Method of measuring chlorine residual in water using diethyl-phenylene-diamine.

DRAFT 1. Act of drawing or removing water; 2. Action or act of pulling something along (e.g., water pump) or back; 3. Depth of water needed to float ship.

DRAFT DEPTH Depth measured perpendicularly from water surface to bottom of boat, ship, etc.

DRAFT PERMIT 1. Document prepared indicating tentative decision to issue, deny, modify, revoke and reissue, terminate or reissue permit; 2. Preliminary permit drafted and published that's subject to public review and comment before final action on application.

DRAFT SPECIFICATIONS Specifications that create overall quality and standards to be preserved.

DRAFT TUBE Conduit that connects runner exit to tail race where water is being finally discharged from Francis turbine.

DRAFT-TUBE LOSS Decrease in energy caused by eddies and surface friction.

DRAG Resistance of static surface to something moving through or across it.

DRAG COEFFICIENT Dimensionless quantity used to quantify drag or resistance of object in fluid environment, such as air or water.

DRAG SCARIFICATION Towing one or more rows of anchor chains, shark fin barrels, tractor pads, alone or in various combinations, to break up and possibly spread slash and to loosen forest floor and topsoil or expose mineral soil.

DRAG SCRAPER Towed scraper to carry, grade and spread all types of material.

DRAG SHOVEL Shovel with hinge-and-stick mounted bucket that digs while being pulled inward.

DRAIN 1. Conduit allowing liquid to flow out of otherwise contained volume; 2. Access point or conduit for rainwater that drains directly downstream in drainage basin without going through sewers or water treatment in order to prevent or belay floods.

DRAIN CHUTE Tapered drain fitting attached to inlet or outlet of manhole to facilitate rodding through.

DRAIN COCK Tap or valve (i.e., faucet) used to draw off water.

DRAIN PIPE Conduit for carrying fluid (e.g., rainwater, flood water, wastewater) being drained.

DRAIN PLUG Stopper that seals drain.

DRAIN RODS Plumbing equipment that consists of stiff, but flexible, rods used to attempt to unblock drains and sewers.

DRAIN TILE Drainage system that collects excess water, from soil below surface and delivers it to another location.

DRAIN-BACK SYSTEM Solar water-heating system tank that allows all water from solar collectors and related piping connections to drain into tank reservoir, protecting system from both freezing and overheating.

DRAIN-DOWN SYSTEM Solar water-heating system that consists of open loop (water heated by collector goes to hot water faucet) system where collectors are filled with domestic water under house pressure when there is no danger of freezing.

DRAINABLE SLUDGE Sludge that can be dewatered readily by gravity.

DRAINAGE 1. Process of removing surface or subsurface water from soil or area; 2. Technique to improve productivity of some agricultural land by removing excess water from soil where surface drainage is accomplished with open ditches and subsurface drainage uses porous conduits (drain tile) buried beneath soil surface.

DRAINAGE AREA 1. Land area from which water drains toward common watercourse (e.g., natural basin, stream, river, reservoir); 2. Topographically defined area of land drained by river/stream or system of connecting rivers/streams such that all outflow is discharged through specific delivery point (i.e., single outlet).

DRAINAGE AREA OF STREAM Specific location, area upstream from location, measured in horizontal plane, that has common outlet at site for its surface runoff from precipitation that normally drains by gravity into stream.

DRAINAGE BASIN 1. Watershed; 2. Land area where precipitation runs off into streams, rivers, lakes and reservoirs; 3. Area of land that drains water, overland run-off, including tributaries and impoundments, sediment and dissolved materials to common outlet at some point along stream channel; 4. Part of surface of earth that contains drainage system with common outlet for its surface runoff.

DRAINAGE BLANKET Layer of pervious material placed directly over foundation material to facilitate drainage of foundation and/or embankment.

DRAINAGE CANAL Artificial canal built to drain water from area having no natural outlet for precipitation accumulation.

DRAINAGE CHANNEL Channel sloped downstream and to ditch to: impede, reduce or redirect flow.

DRAINAGE CURTAIN Line of vertical wells or boreholes to facilitate drainage of foundation and abutments and to reduce water pressure.

DRAINAGE DENSITY Relative density of natural drainage channels in given area usually expressed in terms of miles (length) of natural drainage or stream channel per mi^2 of drainage area and obtained by dividing total length of stream channels in area in miles by area in mi^2.

DRAINAGE DISTRICT Local bodies formed for purpose of draining, ditching and improving land for agricultural and sanitary purposes.

DRAINAGE DITCH Small to moderate trench created to drain water from low-lying areas or to channel water from more distant source.

DRAINAGE DIVIDE 1. Rim of drainage basin; 2. Boundary line, along topographic ridge or along subsurface formation, separating two adjacent drainage basins.

DRAINAGE EASEMENT Easement for purpose of orderly flow of water.

DRAINAGE EQUILIBRIUM Condition existing in area where quality of water reaching water table from all sources.

DRAINAGE FACILITY Engineered system designed to control flooding by transporting stormwater from urban areas to body of water.

DRAINAGE FEATURE Patterns formed by streams and rivers or roadside items that provide drainage.

DRAINAGE FILL 1. Lightweight concrete which is placed on roofs or floors to promote drainage; 2. Base course of granular material placed between sub-grade and floor slab to retard capillary rise of moisture.

DRAINAGE GALLERY Gallery in masonry dam parallel to top of dam, to intercept seepage from upstream face and conduct it away from downstream face, to reduce uplift forces in dam foundation and in body of dam.

DRAINAGE HEAD Furthest or highest elevation in drainage area.

DRAINAGE LAYER Layer of pervious material in earthfill dam to relieve pore pressures or to facilitate drainage of fill.

DRAINAGE MORPHOMETRY Drainage pattern study.

DRAINAGE PATTERN Governed by topography of land and shaped by streams, rivers and lakes in particular drainage basin.

DRAINAGE PIPING All or any part of drain pipe of plumbing system.

DRAINAGE RIGHT-OF-WAY Landowner may not divert or redirect natural occurring waterway from his or her property causing damages to another property.

DRAINAGE RIGHTS Legal principles regarding ability of property owner to divert natural flow of water onto another property.

DRAINAGE SWALE Wide, shallow ditch that carries away runoff water.

DRAINAGE SYSTEM Collection of surface and/or subsurface drains, together with structures and pumps used to remove surface or groundwater.

DRAINAGE SYSTEMS Pattern formed by streams, rivers and lakes in particular drainage basin.

DRAINAGE TERRACE Graded terrace with relatively deep channel.

DRAINAGE TUNNEL Tunnel excavated under some region requiring drainage.

DRAINAGE WATER Any water, usually precipitation or irrigation water, that enters, interacts with and then leaves landscape.

DRAINAGE WELL Well designed for purpose of removing gravity-drained water from saturated, porous medium.

DRAINAGE WELLS 1. Vertical wells downstream of or in downstream shell of embankment dam to collect and control seepage through and under dam where line of such wells forms "drainage curtain"; 2. Well drilled to carry excess water off agricultural fields.

DRAINED SHEAR TEST Test of triaxial compression of cohesive soil consolidated under normal load.

DRAINER Valley or basin from which air drains continuously during nighttime rather than becoming trapped or pooled.

DRAINS 1. Vertical wells downstream of or in downstream shell of embankment dam to collect and control seepage through and under dam where line of such wells forms "drainage curtain"; 2. Well drilled to carry excess water off agricultural fields.

DRAW 1. To drain by emptying; 2. Little watercourse; 3. Natural depression, swale, small valley or gully.

DRAW-DOOR WEIR Weir consisting of row of piles, frames or piers.

DRAW-OFF PIPE Conduit that takes water from plumbing system to tap or other points of use.

DRAW-OFF TAP Faucet with nozzle bent downward.

DRAWBAR HORSEPOWER Horsepower available to pull at drawbar in rear of locomotive or tractor.

DRAWDOWN 1. Lowering of surface elevation of body of water, water surface of well, water table, piezometric surface adjacent to well, tank or reservoir resulting from withdrawal of water therefrom; 2. Process of depleting reservoir or groundwater storage; 3. Difference between water level and lower water level in reservoir within particular time.

DRAWDOWN CURVE Plot of decline of water table or piezometric level versus distance from pumping well or versus time at given distance from pumping well, resulting from continuous pumping from well discharging at known rate.

DREDGE 1. To dig under water; 2. Machine that digs under water; 3. Any instrument used to gather or take; 4. Very fine mineral matter held in suspension in water.

DREDGED BULKHEAD Bulkhead constructed in natural ground by driving sheet piles and then soil is removed from its front by dredging.

DREDGED LEVEL Elevation where ground has been drudged.

DREDGED MATERIAL Sediment excavated or removed from bottoms of navigable waters.

DREDGING Scooping or suction of underwater material.

DREDGING WELL Opening through dredging vessel that bucket ladders work.

DRIFT 1. Food organisms, including algae, plankton and even larval fish, dislodged and moved by river current; 2. Small, nearly horizontal tunnel; 3. Collection of loose earth and rocks or boulders caused by retreat of continental glaciers, wind, ice or water; 4. Distance where current flows in given time; 5. Difference between measured and fixed values.

DRIFT BARRIER Obstacle built to reduce downward deposits.

DRIFTING ICE Pieces of floating ice moving under action of wind and/or currents.

DRIFTING SNOW Uneven distribution of snowfall/snow depth caused by strong surface winds.

DRILL 1. Tool used to remove material so as to create hole; 2. Light furrow or channel made to put seed into, when sowing.

DRILL COLLAR Heavy, thick-walled tube used between drill pipe and bit in drill stem.

DRILL EXERCISE Activity designed to evaluate single emergency response function.

DRILL FEED Rate that drill bit is inserted into object in order to bore hole.

DRILL JUMBO Rock drilling machine usually used in underground mining.

DRILL PIN Pin in lock which enters key stem.

DRILL PIPE Thin-walled, piping used on drilling rigs, that's hollow to allow drilling fluid to be pumped down hole through bit and back up annulus.

DRILL POSITIONER Tool that manages perpendicular drilling.

DRILL ROD Form of bar stock machined to produce tools (e.g., drill bits, taps, dowel pins, hammers, reamers).

DRILL STEEL Hollow steel that links drill to bit.

DRILL STRING Column or string, of drill pipe that transmits drilling fluid and torque to drill bit.

DRILL-BLASTER Individual responsible for stripping, blasting and removal of rock for crushing and other demands.

DRILLABILITY Relative ease that hole can be drilled.

DRILLED SHAFT Deep foundation system that uses large diameter cylinder constructed of concrete and reinforcing steel.

DRILLER'S STROKE Height of fall of drop-hammer weight to drive casing or soil sampling tool.

DRILLING AND PRODUCTION FACILITY Drilling equipment and gear utilized in production of subterranean resources.

DRILLING CORE System that drills and collects core (cylindrical) sample in state it is found (i.e., in-situ).

DRILLING FLUID Fluid used to lubricate bit and convey drill cuttings to surface with rotary drilling equipment.

DRILLING LINE Wire rope to hoist or lower drill pipe or casing.

DRILLING LOG Comprehensive day-to-day record of drilling stone passed.

DRILLING MUD Fluid used to aid drilling of boreholes into earth.

DRILLING OIL Fluid component of mud for rotary drilling.

DRILLING RIG Equipment for drilling holes in stone or earth.

DRILLINGS Cuttings generated via drilling.

DRINKING WATER Water intended to be ingested by humans.

DRINKING WATER EQUIVALENT LEVEL Protective level of exposure related to potentially non-carcinogenic effects of chemicals that are also known to cause cancer.

DRINKING WATER STANDARDS Standards that regulate drinking water to protect public health.

DRINKING WATER STATE REVOLVING FUND Provides capitalization grants to states to develop drinking water revolving loan funds to help finance system infrastructure improvements, assure source-water protection, enhance operation and management of drinking-water systems and otherwise promote local water-system compliance and protection of public health.

DRINKING WATER SUPPLY Water available and safe to drink or to use for food preparation by public.

DRINKING-WATER STANDARD OR GUIDELINE Threshold concentration for constituent or compound in public drinking-water supply, designed to protect human health.

DRIP IRRIGATION Irrigation system where water is applied directly to root zone of plants by means of applicators (e.g., orifices, emitters, porous tubing, perforated pipe and so forth) operated under low pressure.

DRIVE CASING Heavy duty steel casing driven along with sampling tool in cased DP systems that keeps hole open between sampling runs and is not removed until last sample has been collected.

DRIVE POINT PROFILER Exposed groundwater DP (direct-push) system used to collect multiple depth-discrete groundwater samples.

DRIVER POPULATION FACTOR Parameter for mix of driver types in traffic stream by trip purpose.

DRIVEWAY Short private road that leads to house or garage.

DRIVING RAIN INDEX Quantity that gives measure of amount of wind-driven rainfall on vertical surface.

DRIZZLE 1. Precipitation consisting of numerous minute droplets of water less than 0.5 mm in diameter; 2. Shed slowly in minute drops or particles.

DROP 1. Structure in open conduit or canal installed for purpose of dropping water to lower level and dissipating its energy; 2. Small mass of liquid just large enough to hold its own weight via surface tension; 3. Upstream and downstream difference in water surface elevations.

DROP CHANNEL Short, steep channel between inlet and outlet of drop manhole.

DROP ELBOW Elbow having lugs for attaching it to wall or joist.

DROP INLET Below-ground, square box, precast concrete structure with grate opening set at ground level to allow water to enter into structure.

DROP L L-shaped pipe fitting that can be directly attached to structure frame.

DROP MANHOLE Shaft where sewage is dropped from higher to lower level.

DROP PIPE Suction conduit below deep well pump.

DROP STRUCTURE Structure that conveys water to lower elevation and dissipates excess energy resulting from drop.

DROP TEE Tee-shaped pipe fitting that can be attached to structure framework.

DROP-DOWN CURVE Longitudinal profile of water surface in channel that depth of water has been diminished by sudden drop in invert.

DROP-OFF Recyclable materials collection method where individuals bring them to designated collection site.

DROP-SIZE DISTRIBUTION Distribution of rain drops or cloud droplets of specified sizes.

DROPLET Very small mass of liquid just large enough to hold its own weight via surface tension.

DROPPED LANE Through lane that becomes mandatory turn or exit lane on roadway.

DROPPING HEAD State where inflow to reservoir or storage facility is less than outflow such that water level reduces.

DROUGHT 1. Prolonged period of less-than-normal precipitation such that lack of water causes serious hydrologic imbalance; 2. Deficiency of moisture that results in adverse impacts on people, animals or vegetation over sizeable area.

DROUGHT ASSESSMENTS At end of each month, CPC issued long-term seasonal drought assessment.

DROUGHT INDEX 1. Computed value that is related to some of cumulative effects of prolonged and abnormal moisture deficiency; 2. Index of hydrological drought corresponding to levels below mean in streams, lakes and reservoirs.

DROWNED VALLEY Valley that, having been flooded by sea, now exists as bay or estuary.

DROWNED WEIR When water level on downstream side of weir is above top surface of weir.

DRUM 1. Barrel or large cylindrical container for liquid transport and storage; 2. Rotating cylindrical member that transmits compaction forces to surface materials.

DRUM GATE 1. Movable crest gate in form of sector of cylinder hinged at centerline; 2. Type of spillway gate consisting of long hollow drum that may be held in its raised position by water pressure in flotation chamber beneath drum.

DRUM SCREEN Rotating cylindrical screen used to trap solid material then scraped off above waterline from sewage.

DRUM TRAP Prevent sewer gases from coming in to building via large water vessel.

DRY ADIABAT Line of constant potential temperature on thermodynamic chart.

DRY ADIABATIC LAPSE RATE Rate at which temperature of parcel of dry air decreases as parcel is lifted in atmosphere equal to 5.5 °F per 1,000 feet (9.8 °C per km).

DRY CONNECTION Water or sewer main connection completed when main is dry.

DRY CRACK Crack visible at surface but not going right through ice cover and therefore it is dry.

DRY DENSITY Dry mass of soil per unit volume.

DRY DEPOSITION Deposition from gases and dust particles.

DRY FEEDER Equipment that provide accurate dosing of bulk material required in chemical and process engineering.

DRY FLOOD PROOFING Sealed against floodwaters where all areas below flood protection level are made watertight.

DRY FLUE GAS DESULFURIZATION Sulfur dioxide control system positioned downstream of steam generating unit.

DRY ICE Carbon dioxide, frozen in solid state, that sublimes at temperatures above -109.3 $^{\circ}$F.

DRY LINE Boundary separating moist and dry air masses.

DRY LINE BULGE Bulge in dry line, representing area where dry air is advancing most strongly at lower levels.

DRY LINE STORM Any thunderstorm that develops on or near dry line.

DRY MASS Mass of residue present after drying in oven at 105 $^{\circ}$C, until mass remains unchanged that represents total organic matter, ash and sediment, in sample.

DRY MICROBURST Microburst with little or no precipitation reaching ground that's most common in semi-arid regions.

DRY PUNCH Surge of drier air that's normally synoptic-scale or mesoscale process.

DRY SAMPLING Procedure of sampling soil by augering hole.

DRY SLOT Zone of dry (and relatively cloud-free) air which wraps east- or northeastward into southern and eastern parts of synoptic scale or mesoscale low pressure system.

DRY SOIL Soil that has been dried at until it reaches constant weight.

DRY SOLIDS Basis for measuring dry sludge weight.

DRY SPELL Drawn-out period where weather has been dry, for abnormally long time; shorter and not as severe as drought.

DRY SUSPENDED SOLIDS Weight of suspended matter in sample after drying.

DRY THUNDERSTORM High-based thunderstorm when lightning is observed, but little if any precipitation reaches ground.

DRY UNIT WEIGHT Weight of solid particles per unit of total volume.

DRY VENT Vent that does not transport wastewater.

DRY WEATHER FLOW 1. Streamflow which results from precipitation that infiltrates into soil and eventually moves through soil to stream channel; 2. Wastewater flow in sewer system during periods of dry weather with minimum infiltration.

DRY WEIGHT 1. Weight after all moisture is removed; 2. Weight of animal tissue after it has been dried in oven at 65 $^{\circ}$C until constant weight is achieved that represents total organic and inorganic matter in tissue.

DRY WELL 1. Deep hole, covered and usually lined or filled with rocks that failed to penetrate aquifer; 2. Underground structure that disposes of unwanted water (e.g.,surface runoff, stormwater and gray water) that holds water until it soaks into ground.

DRY-ADIABATIC Adiabatic process in hypothetical atmosphere where no condensation (i.e., moisture) of its water vapor occurs and no liquid water is present.

DRY-BULB TEMPERATURE Temperature of air measured by thermometer freely exposed to air, but shielded from radiation and moisture.

DRY-DENSITY/MOISTURE RELATIONSHIP Relationship between moisture content and dry density of soil.

DRY-PIPE FIRE SPRINKLER SYSTEM System where conduits are filled with pressurized air or nitrogen, rather than water.

DRY-PIPE VALVE Valve in system where conduits are filled with pressurized air or nitrogen, rather than water.

DRY-SEAT PIPE THREAD Pipe threads where sealing is function of root and crest interference.

DST INDEX Geomagnetic index describing variations in equatorial ring current.

DUAL ENTRY Accounting method where each transaction is recorded as both credit and debit.

DUAL POROSITY ELEMENT Element that contains two media of different porosity in parallel.

DUAL POROSITY FILTER Filter consisting of dual layers with high porosity layer on top of low porosity layer.

DUAL SYSTEM Two-conduit plumbing system.

DUAL VENT Vent pipe that serves two fixtures and connects at junction of trap arms.

DUAL-ARROW SIGNAL SECTION Signal designed to include both yellow and green arrow.

DUAL-FLUSH CISTERN Variation of flush toilet that uses two buttons or handles to flush different amounts of water.

DUAL-MODE VEHICLE Vehicle that can run on conventional road or automated guideway.

DUAL-PHASE EXTRACTION Active withdrawal of both liquid and gas phases from well usually involving use of vacuum pump.

DUCKFOOT BEND Pipe bend at bottom of shaft column or rising main fitted with horizontal base sufficiently strong for weight of rising main to rest upon it.

DUCT 1. Pipe, tube or canal which carries gas or liquid from one place to another; 2. Enclosure or channel for electrical cable runs.

DUCTILE Capable of being pulled or stretched into thin wire by mechanical force without breaking.

DUCTILE FAILURE Long-term failure developed failure in malleable materials by extensive plastic deformation or necking.

DUCTILE IRON PIPE Pipe made of ductile cast iron commonly used for potable water transmission and distribution.

DUCTILITY Property of material that allows it to be formed into thin sections without breaking.

DUCTILITY FACTOR Ratio of total deformation at maximum load to elastic-limit deformation.

DUFF Decaying vegetable matter on forest floor.

DUMMY ACTIVITY Simulated activity of sorts, one that's of zero duration and is created for sole purpose of demonstrating specific relationship and path of action on arrow diagramming method.

DUMP Site used to dispose of solid waste without environmental controls.

DUMP ANGLE Degree of slope dump body attains at top of its lift.

DUMP BODY TRUCK Storage section of dump truck.

DUMP TIME Time taken to empty mechanical equipment.

DUMPED Method of compacting soil by dumping soil into place with no compactive effort.

DUMPER Small one-man diesel-powered vehicle often used to carry loads and material around.

DUMPING Disposal of something no longer needed.

DUMPING WASTE CONTAINER Metal container that can be raised above body of truck to discharge its contents.

DUMPSTER Large trash receptacle designed to be hoisted up by garbage truck in order to be emptied.

DUNE Ridge or hill of sand piled up by wind.

DUNE STABILIZATION Means used to prevent erosion including vegetation and fencing.

DUNES Bed forms with triangular profile that advance downstream due to net deposition of particles on steep downstream slope.

DUPLEX Made up of two parts.

DUPLEX PUMP Direct acting reciprocating steam pump that have two steam and two water cylinders.

DUPLICATE Second aliquot or sample treated same as original sample in order to determine precision of analytical method.

DUPLICATION OF PLANT Setup of back-up components for vital mechanical services.

DURABILITY Permanence by virtue of power to resist stress or force.

DURABILITY FACTOR Measure of change (with time) in property of material as result of exposure to influence which has potential of causing deterioration.

DURATION Amount of time or particular time interval.

DURATION COMPRESSION Shortening of project's completion period, without reducing its scope, usually resulting in higher total cost.

DURATION CURVE 1. Cumulative frequency curve that shows percent of time during which specified units of items (e.g., discharge, head, power) were equaled or exceeded in given period; 2. Integral of frequency diagram.

DURATION OF CONGESTION Number of hours in 24-hour period where traffic is congested.

DURATION OF ICE COVER Time from freeze-up to breakup of ice cover.

DURATION OF STRONG GROUND MOTION "Bracketed duration" or time interval between first and last acceleration peaks that are equal to or greater than 0.05g.

DURATION OF SUNSHINE Amount of time sunlight detected at given point.

DUSK Time at which Sun is 6° below horizon in evening.

DUST 1. Fine, dry particles of matter found in air and covering surface of objects, typically consisting of soil lifted up by wind, pollen, hair, etc.; 2. Totally disconnected set of points with fractal structure.

DUST BURDEN Weight of dust suspended in unit volume.

DUST COLLECTOR System used to enhance quality of air released from industrial and commercial processes by collecting dust and other impurities from air or gas.

DUST DEVIL Small, rapidly rotating wind made visible by dust, dirt or debris it picks up.

DUST LOADING Quantity of dust in gas stream, usually expressed as grains of dust per cubic foot of air.

DUST PALLIATIVE Agents applied to various road surfaces to prevent dust clouds and their environmental impact.

DUST WHIRL Rotating column of air rendered visible by dust.

DUST-HANDLING EQUIPMENT Gear that manages particulate matter.

DUSTFALL JAR Open container used to collect large particles from air for measurement and analysis.

DUTCH MATTRESS Woven mat of brushwood and reeds used to protect from scour.

DUTY OF CARE Legal obligation imposed on individual requiring that they adhere to standard of reasonable care while performing any acts that could foreseeably harm others.

DUTY OF WATER Quantity of water required to satisfy irrigation water requirements in given area.

DWELL TIME Period of time that system or element of system remains in given state.

DWELL TIME VARIABILITY Time interval distribution between bus doors opening and closing along routes.

DWELLING Habitation (i.e., place or house where person lives).

DYKE 1. Long, narrow hollow dug from ground to conduct water; 2. Embankment that prevents overflow of water from natural watercourse.

DYNAMIC COMPACTION Method of compacting soil by dropping heavy weight onto loose soil.

DYNAMIC COMPRESSOR Constant pressure that uses rotating vane or screw to move and/or compress gas.

DYNAMIC DISCHARGE HEAD Static discharge head plus any friction losses from hose, pipes, couplings and fittings.

DYNAMIC ENVELOPE Vertical white ground markings, extending on both sides of railroad track.

DYNAMIC EQUILIBRIUM Condition achieved when average sand load transported by flowing water is in balance with sand load being supplied by tributaries.

DYNAMIC EXIT GATE OPERATING MODE Operation of exit gate based on minimum distance of vehicle track clearance.

DYNAMIC FORCE Process that energy from action increases or decreases.

DYNAMIC FORCE APPLIED Maximum static force multiplied by dynamic load factor.

DYNAMIC HEAD 1. Total equivalent height that fluid is to be pumped plus friction losses in conduit; 2. Total of static and velocity heads of flowing fluid at measurement point.

DYNAMIC HYDROPLANING High-speed phenomenon where thin film of water lifts wheels off runway.

DYNAMIC ICE Pressure due to moving ice cover or drifting ice.

DYNAMIC LIFTING Forced uplifting of air from various atmospheric processes, (e.g., weather fronts, cyclones).

DYNAMIC LOAD Any load not static (e.g., wind load, moving live load).

DYNAMIC LOAD ALLOWANCE In addition to static loads, measure of dynamic effects exerted by moving vehicles on bridges.

DYNAMIC LOADING Load that's time varying and can be accelerating or decelerating (e.g., live, wind, earthquake, snow).

DYNAMIC MODEL Mathematical model of flow in open channel that solves complete unsteady flow equations (e.g., St. Venant equations for one-dimensional problems).

DYNAMIC PENETRATION TEST Test that uses moving parts to calculate loading capacity and compaction properties of your soil.

DYNAMIC PRESSURE 1. Vertical distance from reference point (e.g., pump centerline) to hydraulic grade line of operating pump; 2. Kinetic energy per unit volume of fluid particle.

DYNAMIC PRICING Flexible toll pricing strategy based on congestion levels.

DYNAMIC ROUTING Hydraulic flow routing based on solution of St. Venant Equation(s) to compute changes of discharge and stage with respect to time at various locations along stream.

DYNAMIC SEAL Seals for when there is motion between surfaces.

DYNAMIC STRENGTH Resistance to loads applied abruptly.

DYNAMIC SUCTION HEAD Static suction head plus any suction line friction losses.

DYNAMIC VEHICLE TIRE FORCE Time-varying force applied perpendicul by tire to road surface by moving vehicle.

DYNAMIC WAVE ROUTING MODEL Computerized hydraulic routing program whose algorithms incorporate complete one-dimensional equations of unsteady flow.

DYNAMICS Any forces that produce motion or effect change.

DYNAMOMETER Device used to place load on engine and measure its performance.

DYNE Unit of force equal to force that produces acceleration of 1 cm/s^2 on mass of one gram.

DYSPHOTIC ZONE Region of Photic zone that lies below compensation level and where light penetration is such that oxygen production by photosynthesis is exceeded by oxygen consumption by respiration.

DYSTROPHIC Affecting with wasting of body tissues, of genetic origin or due to inadequate or defective nutrition.

DYSTROPHIC LAKES Acidic, shallow bodies of water that contain much humus, many plants and/or other organic matter but few fish.

E

E REGION Daytime layer of Earth's ionosphere roughly between altitudes of 85 and 140 km.

E-19, REPORT ON RIVER GAUGE STATION Report to be completed every 5-10 years providing complete history of river station and all gauges that have been used for public forecasts since establishment of station that should be updated anytime significant change occurs at forecast point.

E-19A, ABRIDGED REPORT ON RIVER GAUGE STATION
Abridged version of E-19, E-19a was used to update E-19 as additional information or changes occur at station during intervening five year period.

E-3, FLOOD STAGE REPORT Form that Service Hydrologist/Hydrology Focal Point completes to document dates where forecast points are above flood stage, as well as crest dates and stages.

E-5, MONTHLY REPORT OF RIVER AND FLOOD CONDITIONS
Monthly narrative report covering flooding which occurred over past month.

E-7, FLOOD DAMAGE REPORT Report to be completed anytime there is reported flood damage or loss of life as direct result of flooding.

E. COLI 1. Bacteria present in intestine and feces of warm-blooded animals; 2. Member species of fecal coliform group of indicator bacteria where concentrations are expressed as number of colonies per 100 ml of sample.

EAR Projection on pipe fitting that can be attached to wall.

EAR INSERT DEVICE Hearing protective device designed to be inserted into ear canal and to be held in place principally by virtue of its fit.

EAR MUFF Hearing protective apparatus designed to cover person's ears for hearing protection or for warmth.

EAR PROTECTORS Protects against hearing damage from excessive sound.

EARLIEST EVENT OCCURRENCE TIME In Critical Path Method, earliest point in time that all activities that precede event will be completed.

EARLIEST FINISH In Critical Path Method, earliest start time increased with its duration estimate.

EARLIEST START In Critical Path Method, first time where activity may begin.

EARLY FINISH DATE In Critical Path Method, earliest possible point where as yet to be completed elements of schedule activity may potentially be completed.

EARLY LIFE STAGE TOXICITY TEST Chronic toxicity test using sensitive early life stages like embryos or larvae to predict effects of toxicants on organisms.

EARLY START DATE In Critical Path Method, specific point that represents earliest possible point where as yet to be completed segments of specific schedule activity or project in general can begin.

EARLY WARNING SYSTEM Designed system that will ensure timely recognition of threatening event and provide reliable and timely warning and evacuation of population at risk from dangerous flooding associated with large operational releases or dam failure that must address five components: detection, decision making, notification, warning and evacuation.

EARLY WARNING SYSTEM FEASIBILITY Study of feasibility of installing some level of Early Warning System to meet target times for alerts and warnings.

EARNED VALUE Approach where you monitor project plan, actual work and work completed value to see if project is on track and shows how much of budget and time should have been spent, considering work completed so far.

EARTH 1. Any general rock-based material; 2. Planet, third out from Sun.

EARTH BERM Mound, narrow ledge or shelf of dirt.

EARTH DAM Embankment dam where more than 50% of total volume is formed of compacted earth material, where seepage through dam is controlled by designed use of upstream blankets and/or internal cores constructed using compacted soil of very low permeability.

EARTH DRILL Auger-type drill with bucket.

EARTH LINING Compacted layer of earth on surface of canal or other excavation.

EARTH PRESSURE Pressure or force exerted by soil on any boundary.

EARTH PRESSURE CELL Device designed to provide estimate of normal stress in soil.

EARTH RESERVOIR Water storage facility formed using excavated material as embankments.

EARTH RESOURCES TECHNOLOGY SATELLITE Satellites that recognize and track natural resources of planet.

EARTH TANK Livestock Drinking facility formed by excavating and creating earthen dam.

EARTH'S SHADOW Shadow of Earth that falls onto surface of Moon that's facing night side of Earth.

EARTHFILL DAM Embankment dam where more than 50% of total volume is formed of compacted earth material generally smaller than 3-inch size where seepage through dam is controlled by designed use of upstream blankets and/or internal cores constructed using compacted soil of very low permeability.

EARTHQUAKE Sudden motion or trembling in Earth caused by abrupt release of accumulated stress along fault or volcanic activity.

EARTHQUAKE LOADING Application of earthquake-generated agitation (i.e., load, pressure)to structure.

EARTHQUAKE, MAXIMUM CREDIBLE Earthquake(s) associated with specific seismotectonic structures, source areas or provinces that would cause most severe vibratory ground motion or foundation dislocation capable of being produced at site under currently known tectonic framework.

EARTHQUAKE, MAXIMUM DESIGN Postulated seismic event, specified in terms of specific bedrock motion parameters at given site, that is used to evaluate seismic resistance of manmade structures or other features at site.

EARTHQUAKE, OPERATING BASIS Earthquakes for which structure is designed to resist and remain operational.

EARTHQUAKE, SAFETY EVALUATION Earthquake, expressed in terms of magnitude and closest distance from dam site or in terms of characteristics of time history of free-field ground motions, for which safety of dam and critical structures associated with dam are to be evaluated.

EARTHQUAKE, SYNTHETIC Earthquake time history records developed from mathematical models that use white noise, filtered white noise and stationary and non-stationary filtered white noise or theoretical seismic source models of failure in fault zone.

EARTHWORK 1. Any one or combination of operations involved in altering or movement of earth; 2. Any structure made from earth.

EASEMENT Right, by person or public authority, to access or cross someones land for specific purpose.

EASTERLIES Any winds with components from east.

EASTERLY WAVE Low level disturbance of tropical origins that can develop into tropical cyclones.

EBB 1. Receding movement of tide; 2. Gradual decline.

EBB CHANNEL Tidal channel where ebb currents are stronger than flood currents.

EBB CURRENT Movement of tidal current away from coast or down estuary.

EBB TIDE Period between high tide and next low tide where sea is receding.

EBBING AND FLOWING SPRING Spring that exhibits tidal characteristics with predictable regularity.

EBULLITION Act of boiling.

ECCENTRIC FITTING Any fitting that center line is offset from that of run of pipe.

ECCENTRICITY Dimensionless quantity describing elliptical shape of planet's orbit.

ECHO 1. Energy back scattered from target (e.g., precipitation, clouds) and received by and displayed on radar screen; 2. Reflected sound heard again by its initial observer.

ECHO LOCATION Biological sonar used by several kinds of animals where animals emit calls out to environment and listen to echoes that return from various objects near them.

ECHO TOPS Height above ground of center of radar beam using tilt or scan, that contains highest elevation where reflectivity's greater than 18 dBZ can be detected.

ECLIPSE Darkening of heavenly body when another heavenly body completely/partly blocks off its light.

ECLIPTIC 1. Great circle drawn on terrestrial globe; 2. Apparent annual path of Sun in sky.

ECO-MANAGEMENT AND -AUDIT SCHEME Premium management instrument for companies and other organisations to evaluate, report and improve their environmental performance.

ECOCATASTROPHE Major destructive upset in balance of nature especially when caused by action of humans.

ECOCLINE Gradation from one ecosystem to another when there is no sharp boundary between two.

ECOLOGICAL BALANCE Equilibrium between and harmonious coexistence of, organisms and their environment.

ECOLOGICAL EFFECTS STUDIES Studies performed for development of information on non-human toxicity and potential ecological impact of test substances and their degradation and activation products.

ECOLOGICAL EFFICIENCY Efficiency with which energy is transferred from one trophic level to next.

ECOLOGICAL ENTITY Term referring to species, group of species, ecosystem function or characteristic or specific habitat or biome.

ECOLOGICAL EXPOSURE Exposure of non-human organism to stressor.

ECOLOGICAL FACTOR Any factor, abiotic or biotic, that influences living organisms.

ECOLOGICAL IMPACT Effect that man-caused or natural activity has on living organisms and their non-living (abiotic) environment.

ECOLOGICAL INDICATOR 1. Characteristic of ecosystem related to or derived from, measure of biotic or abiotic variable, that can provide quantitative information on ecological structure and function that can contribute to measure of integrity and sustainability; 2. Plant or animal species, communities or special habitats with narrow range of ecological tolerance.

ECOLOGICAL INDICATORS Communicate information about ecosystems and impact human activity has on ecosystems to groups such as public or government policy makers.

ECOLOGICAL INTEGRITY Living system exhibits integrity if, when subjected to disturbance, it sustains and organizes self-correcting ability to recover toward biomass end-state that's normal for that system.

ECOLOGICAL NICHE Describes how organism or population responds to distribution of resources and competitors and how it in turn alters those same factors.

ECOLOGICAL PYRAMID Diagram that shows relative amounts of energy or matter or numbers of organisms within each trophic level in food chain or food web.

ECOLOGICAL RISK ASSESSMENT Application of formal framework, analytical process or model to estimate effects of human actions(s) on natural resource and to interpret significance of those effects in light of uncertainties identified in each component of assessment process.

ECOLOGICAL STUDIES Studies of biological communities and habitat characteristics in NAWQA Study Units to evaluate effects of physical and chemical characteristics of water and hydrologic conditions on aquatic biota and to determine how biological and habitat characteristics differ among environmental settings.

ECOLOGICAL/ENVIRONMENTAL SUSTAINABILITY Maintenance of ecosystem components and functions for future generations.

ECOLOGY Branch of biological science which deals with relationships between living organisms and their environments.

ECONOMIC ANALYSIS Procedure that includes both tangible and intangible factors to evaluate various alternatives.

ECONOMIC CONSERVATION Direction of natural resources to be able to keep up routine production of commodity in maximal degree possible.

ECONOMIC DEPRECIATION Measure of decrease in value of asset over time.

ECONOMIC DEVELOPMENT Process where nation improves economic, political and social well-being of its people.

ECONOMIC EFFICIENCY Implies economic state where every resource is optimally allocated to serve each individual or entity in best way while minimizing waste and inefficiency.

ECONOMIC GROUNDWATER YIELD Maximum rate where water can be withdrawn artificially from aquifer without depleting supply or altering chemical character of water.

ECONOMIC LIFE Expected period of time during which asset is useful to average owner.

ECONOMIC OBSOLESCENCE When value of property decreases due to external factors in neighborhood or immediate area.

ECONOMIC POISONS Chemicals used to control pests and to defoliate cash crops (e.g., cotton).

ECONOMIC RATIO State when two substances or substances acting jointly are each performing at their maximum capacity.

ECONOMICS Study of resource allocation, distribution and consumption; of capital and investment; and of management of factors of production.

ECONOMIZER 1. Heat exchange device in boiler that improves efficiency and saves fuel; 2. Person who avoids waste.

ECOREGION Area of similar climate, landform, soil, potential natural vegetation, hydrology or other ecologically relevant variables.

ECOSPHERE 1. "Bio-bubble" that contains life on Earth, in surface waters and in air; 2. Portion of atmosphere where it is possible to breathe without technological assistance.

ECOSYSTEM 1. Complex system composed of community of people, animals, plants and bacteria as well as chemical and physical environments; 2. Interacting system of biological community and its non-living environmental surroundings.

ECOSYSTEM STRUCTURE Attributes related to instantaneous physical state of ecosystem (e.g., species population density, species richness or evenness, standing crop biomass).

ECOTONE 1. Habitat created by juxtaposition of distinctly different habitats; 2. Edge habitat; 3. Ecological zone or boundary where two or more ecosystems meet.

ECSTATIC Global and coincident change in sea level.

EDAPHIC FACTORS Abiotic factor relating to physical or chemical composition of soil found in particular area.

EDDY Circular current of air or water moving against main current.

EDDY CURRENT 1. Fluid with flow directions different from general flow; 2. Induced electric current formed within body of conductor when it is exposed to time varying magnetic field.

EDDY FLOW Motion of fluid having local velocities and pressures that fluctuate randomly.

EDDY LOSS Loss of energy in flow caused by eddy currents (e.g., velocity changes).

EDGE CRACKING Formation of crescent-shaped cracks near surface edge.

EDGE LINE Lines that mark border of road.

EDGE LINE MARKINGS Solid lines at edge of travel lane for undivided roads.

EDUCTOR Jet pump that utilizes kinetic energy of one liquid to cause flow of another.

EFFECTIVE AREA Cross-sectional area of process media or available surface area of absorbent contacted by gas or liquid.

EFFECTIVE AREA OF ORIFICE Area of orifice multiplied by its coefficient of discharge.

EFFECTIVE AVAILABLE TIME-SPACE Number of turning vehicles during walk and pedestrian clear intervals.

EFFECTIVE CONCENTRATION Point estimate of toxicant concentration that would cause observable adverse effect (e.g., death, immobilization, serious incapacitation) in given percentage of test organisms.

EFFECTIVE CORROSION INHIBITOR RESIDUAL Concentration of corrosion inhibitor sufficient to form protective coating on interior walls of pipe, reducing its corrosion.

EFFECTIVE DAMPING Measure of energy dissipation from system.

EFFECTIVE DATE Date upon which something is considered to take effect, which may be past, present or future date.

EFFECTIVE DEPTH Distance from compression fiber for beam to centroid of tension reinforcement (e.g., steel).

EFFECTIVE DIAMETER Particle diameter corresponding to 10% finer on accumulative gradation curve.

EFFECTIVE DOSE EQUIVALENT Sum of products of dose equivalent to organ or tissue and weighting factors applicable to each of body organs or tissues that are irradiated.

EFFECTIVE DURATION 1. Time when rainfall rate is greater than infiltration rate; 2. Calculation used to approximate duration of callable bond.

EFFECTIVE FORCE Force transmitted through soil or rock mass by intergranular pressures.

EFFECTIVE GRAIN SIZE 1. Diameter of particles in assumed rock or soil that would transmit water at same rate as rock or soil under consideration and composed of spherical particles of equal size and arranged in specific manner; 2. Single particle diameter that best depicts bed material properties.

EFFECTIVE GREEN TIME Traffic signal cycle length minus effective red signal time.

EFFECTIVE HEAD Reduced value for hydraulic head with losses accounted for.

EFFECTIVE HEIGHT OF EMISSION Height above ground level where plume of waste gas is estimated to become approximately horizontal.

EFFECTIVE LEAKAGE AREA Area of special nozzle-shaped hole that would leak same amount of air as building does at pressure of 4 Pascals.

EFFECTIVE OPENING Cross-sectional area of opening that water is released from water supply conduit.

EFFECTIVE PEAK GROUND ACCELERATION Acceleration most closely related to structural response and to damage potential of earthquake.

EFFECTIVE POROSITY 1. Ratio of volume of soil or rock mass that can be drained of fluid by gravity to total volume of mass.

EFFECTIVE PRECIPITABLE WATER That part of precipitable water which, in theory, can actually fall as precipitation.

EFFECTIVE PRECIPITATION 1. Part of precipitation that produces runoff; 2. Amount of precipitation actually added and stored in soil; 3. Part of precipitation falling on irrigated area effective in meeting consumptive use requirements.

EFFECTIVE PRECIPITATION, EXCESS Precipitation in excess of infiltration capacity, evaporation, transpiration and other losses.

EFFECTIVE PRESSURE Quantity relating to operation of reciprocating engines and is valuable measure of engine's capacity to do work independent of engine displacement.

EFFECTIVE PRESTRESS Stress remaining in concrete from prestressing after removal of prestress.

EFFECTIVE RAINFALL 1. Component of storm hyetograph which is neither retained on land surface nor which infiltrates into soil; 2. Produces overland flow that results in direct runoff hydrograph from sub-area of catchment.

EFFECTIVE RANGE Portion of instruments design range where it has acceptable accuracy.

EFFECTIVE RED TIME Traffic signal cycle length minus effective green signal time.

EFFECTIVE SOIL SIZE Grain size larger than 10% by weight of soil particles in grading curve of soil.

EFFECTIVE STACK HEIGHT Sum of actual physical height of top of stack, plus any plume rise due to buoyancy or initial momentum of rising effluent, minus any downwash such as stack downwash, building downwash or terrain downwash.

EFFECTIVE STORAGE Useable contents of storage container.

EFFECTIVE TERRESTRIAL RADIATION Difference between upwelling infrared or terrestrial radiation emitted from Earth and downwelling infrared radiation from atmosphere.

EFFECTIVE THREAD Length of thread which makes seal on conventional machined pipe thread.

EFFECTIVE TOPOGRAPHY Topography as seen by approaching flow, which may include not only actual terrain but also cold air masses trapped within or adjacent to actual topography.

EFFECTIVE VELOCITY 1. Velocity of groundwater percolating through water-bearing material; 2. Volume of groundwater passing through unit cross-sectional area in unit of time divided by effective porosity.

EFFECTIVE VELOCITY PRESSURE Pressure exerted by wind effects assuming both Importance Factor and Drag Coefficient equal unity.

EFFECTIVE WALKWAY WIDTH Portion of pedestrian walkway free of obstructions.

EFFERVESCENCE Escape of gas from solution in liquid.

EFFICACY Ability to produce desired effect under ideal testing conditions.

EFFICIENCY State or quality of being efficient.

EFFICIENCY FACTOR Ratio of some measure of performance to expected value.

EFFICIENT ESTIMATOR Optimal estimator of quantity of interest.

EFFLUENT 1. Wastewater, treated or untreated, that flows out of treatment plant, sewer or industrial outfall; 2. Outflow from particular source (i.e., stream that flows from lake, liquid waste that flows from factory or sewage-treatment plant).

EFFLUENT CHARGE Fee or tax imposed on generator of effluent discharged to public collection system or to public resource.

EFFLUENT DATA Information necessary to determine identity, amount, frequency, concentration, temperature or other characteristics (to extent related to water quality) of any pollutant.

EFFLUENT GUIDELINES Technical documents which set effluent limitations for given industries and pollutants.

EFFLUENT LIMITATION Maximum amount of quantities, rates and concentrations specific substance or characteristic that can be present in effluent discharge without violating water quality standards in receiving waters.

EFFLUENT LIMITATIONS GUIDELINES Regulation that establishes national technology-based effluent requirements for specific industrial category.

EFFLUENT SEEPAGE Diffuse discharge of groundwater to ground surface.

EFFLUENT STANDARD Maximum amount of quantities, rates and concentrations specific substance or characteristic that can be present in effluent discharge without violating water quality standards in receiving waters.

EFFLUENT STREAM Any watercourse where all or portion of water volume came from Phreatic zone or zone of saturation by way of groundwater flow or baseflow.

EFFLUENT TROUGH Open channel where effluent flows.

EFFLUENT WEIR Dam at outflow end of watercourse.

EFFLUX TUBE Tube where fluid is discharged.

EFFLUX VELOCITY Average flow rate of material emitted into atmosphere from source such as smokestack.

EGG-SHAPED SEWER Sewer conduit with egg shaped cross-section.

EIGHTY-FIFTH (85TH) PERCENTILE SPEED Speed value that 85% of vehicles are not exceeding.

EJECTOR 1. Device used to disperse chemical solution into water being treated; 2. Jet jump for lifting water or withdrawing air from space.

EJECTOR WELL POINT Water is pumped and vacuum that promotes drying of rock is created by ejector mounted within well point.

EL NIÑO Warming of ocean current along coasts of Peru and Ecuador generally associated with dramatic changes in weather patterns of region where major El Niño event generally occurs every 3 to 7 years and is associated with changes in weather patterns worldwide.

ELAPSED TIME Time or difference between beginning time and ending time.

ELASTIC COMPRESSION Movement of soil particles as result of load.

ELASTIC LIMIT Level of stress where solid undergoes greater change in strain than predicted by Hooke's law.

ELASTIC MOVEMENT Motion under load which is recoverable when bed is removed.

ELASTIC RESTRAINT SYSTEM Collection of structural elements providing restraint of seismically isolated structure for non-seismic lateral loads.

ELASTIC STRAIN Deformation per unit of length produced by force acting on object that returns to its original shape and size when deforming force is removed.

ELASTICITY Property by virtue of which material deformed under load can regain its original dimensions when unloaded.

ELBOW 1. Pipe fitting having two openings which causes run of pipe to change direction 90 °; 2. Any turn or bend like that of elbow, in fitting, wall, building, coastline, etc.

ELECTRIC FIELD Region of space around charged particle or between two voltages, which exerts force on charged objects in its vicinity.

ELECTRIC HEATING SYSTEM System where electrical energy is converted to heat.

ELECTRIC HEATING SYSTEM OFF-PEAK System where electric heat source is its main component with supplemental heating source for times of peak electrical demand.

ELECTRIC LOCOMOTIVE Locomotive powered by electricity from overhead lines, third rail or on-board energy storage.

ELECTRIC PLANT Facility containing prime movers, electric generators and auxiliary equipment for converting mechanical, chemical, and/or fission energy into electric energy.

ELECTRIC POWER INDUSTRY Public, private and cooperative electric utility systems of United States taken as whole.

ELECTRIC POWER SYSTEM Physically connected electric generating, transmission and distribution facilities operated as unit under one control.

ELECTRIC PUMP Machine for raising or transferring fluids via electricity.

ELECTRIC SET Gear that produces electricity.

ELECTRIC SUBMERSIBLE PUMP Pump that can operate fully immersed in fluid.

ELECTRIC UTILITY Corporation, person, agency, authority or other legal entity or instrumentality that owns and/or operates facilities for generation, transmission, distribution or sale of electric energy primarily for use by public.

ELECTRIC WELL LOG Record of certain electrical characteristics (such as resistivity and conductivity) of formations traversed by borehole.

ELECTRIC-FUSION-WELDED PIPE Conduits formed by hot or cold rolling plate and welding seam.

ELECTRIC-RESISTANCE-WELDED PIPE Conduit manufactured by cold-forming sheet of steel into cylindrical shape.

ELECTRICAL CONDUCTIVITY Measure of salt content of water.

ELECTRICAL DEMAND Energy requirement placed upon utility's generation at given instant or averaged over any designated period of time usually expressed in kilowatts.

ELECTRIFICATION Adaptation (e.g., home, city, industry, railroad) for electric power.

ELECTRIFIED TRACK Railway system that supplies electric power to railway trains.

ELECTROCHEMICAL CORROSION Occurs when electrons from atoms at surface of metal are transferred to suitable electron acceptor or depolarizer.

ELECTROCHEMICAL GAUGING Method for measuring flow of water based on nearly linear relationship between concentration and electrical conductivity of salt solution added to water as tracer.

ELECTROCHEMICAL SOIL STABILIZATION Stiffening of specific clays generated by passing of direct electric current utilizing anode of polyvalent metal.

ELECTRODE Terminal where electric current passes between metallic and non-metallic parts of electric circuit.

ELECTRODIALYSIS Process that uses electrical current applied to permeable membranes to remove minerals from water often used to desalinize salty or brackish water.

ELECTRODYNAMIC SEPARATOR Apparatus used for separation non-ferrous metals from cohesionless dry mixtures.

ELECTROHYDRAULIC SERVO VALVE Electrically operated valve that controls how hydraulic fluid is sent to actuator.

ELECTROLYSIS Chemical change produced by passing electric current through conducting solution or molten salt.

ELECTROLYSIS/ELECTROLYTIC PITTING Migration (i.e., pitting) of metal particles between two dissimilar metals.

ELECTROLYTE 1. Substance that, in solution or when molten, ionizes and conducts electricity; 2. Any of various ions (such as sodium or chloride) that regulate electric charge on cells and flow of water across their membranes.

ELECTROLYTIC CHLORINE Chlorine generated by electrolytic dissociation of hydrochloric acid or certainly one of its salts.

ELECTROMAGNETIC GEOPHYSICAL METHODS Ways to measure subsurface conductivity via low-frequency electromagnetic induction.

ELECTROMAGNETIC UNIT Prefix denoting practical quantification of theoretical electromagnetic unit.

ELECTROMAGNETISM Unified fundamental force that combines aspects of electricity and magnetism and is one of four fundamental forces.

ELECTROMETER 1. Device used to detect and measure static electricity; 2. Precision voltmeter that draws almost no current from circuit.

ELECTROMETRIC TITRATION Titration where end point is ascertained from observation.

ELECTROMOTIVE FORCE Electrical pressure available to cause flow of current when electrical circuit is closed.

ELECTRON Subatomic particle having negative charge and orbiting nucleus.

ELECTRON CAPTURE DETECTOR Item of laboratory equipment used coupled to gas chromatograph for detection and quantification of very minute amounts halogenated organic compounds.

ELECTRONIC FILTER Circuits that perform signal processing functions.

ELECTRONIC SCREENING System that uses transponder for trucks to bypass weigh and inspection stations.

ELECTRONIC TOLL COLLECTION Automated toll collection system that determines whether vehicle enrolled in program and electronically debits accounts without requiring vehicles to stop.

ELECTROOSMOSIS Migration of ions in solvent through capillaries of membrane under influence of applied electric field.

ELECTROPLATING WASTEWATER Surface plating operations where metal is dipped in electroplating solution of various types of metals and then rinsed.

ELECTROSTATIC FIELD Electric field produced by static electric charges.

ELECTROSTATIC PRECIPITATOR Device that removes particles using electric forces from gas stream (smoke) after combustion occurs.

ELECTROSTATIC SEPARATOR Device for separating particles by mass in low energy charged beam.

ELEMENT 1. One of basic conditions of atmosphere (e.g., wind, visibility, runway visual range, weather, obscurations, sky condition, temperature and dew point, pressure); 2. Any one of simplest chemical substances that cannot be decomposed in chemical reaction or by any chemical means and made up of atoms all having same number of protons.

ELEMENTARY NEUTRALIZATION UNIT Apparatus employed for neutralizing wastes that are hazardous only because they exhibit corrosivity characteristic or because they are listed solely for characteristic of corrosivity.

ELEVATED GRAVITY TANK Water storage tank located above ground level where water is distributed via gravity.

ELEVATED STORAGE SYSTEM Elevated storage tank to handle specific fire flow requirement.

ELEVATED TANK Elevated structure supporting water tank constructed at height sufficient to pressurize water supply system for distribution of potable water and to provide emergency storage for fire protection.

ELEVATED TEMPERATURE INVERSION Layer of atmosphere where air temperature increases with height.

ELEVATION 1. Height of point above plane of reference (e.g., benchmark); 2. Distance of celestial object above horizon.

ELEVATION HEAD Part of hydraulic head that is attributable to elevation of measuring point above given datum.

ELEVATION LOSS Pressure loss caused by raising water through conduit to higher elevation.

ELIGIBLE COSTS Construction costs for wastewater treatment works upon which EPA grants are based.

ELLIPTICAL PIPE Conduit with slightly greater capacity than round conduit.

ELONGATE Increase of vertical diameter of flexible pipe.

ELUTRIATION Process of separating lighter particles from heavier ones by means of upward directed stream of gas or liquid.

ELUTRIATOR Device for separating lighter particles from heavier ones by means of upward directed stream of gas or liquid.

ELUVIATION 1. Sideways or downward movement of dissolved or suspended material within soil caused by rainfall; 2. Creation of geological deposits (eluvial deposits) by in situ weathering or weathering plus gravitational movement or accumulation.

ELUVIUM Residual deposits of soil, dust and rock particles produced by action of wind.

EMAGRAM One of four thermodynamic diagrams used to display temperature lapse rate and moisture content profiles in atmosphere.

EMAP DATA Environmental monitoring data collected under auspices of Environmental Monitoring and Assessment Program.

EMBANKMENT 1. Earth structure top that's higher than adjoining surface; 2. Fill material, usually earth or rock, placed with sloping sides and with length greater than its height.

EMBANKMENT DAM Any dam constructed of excavated natural materials (e.g., earthfill dam, rockfill dam, tailings dam, hydraulic fill dam, rolled fill dam).

EMBANKMENT FOUNDATION Ground surface where embankment is placed.

EMBANKMENT WALL Retaining wall from top of which supported earth normally rises at slope.

EMBAYMENT Topographical feature that used to be bay.

EMBEDDED CULVERT Culvert with lower invert located below streambed.

EMBEDDEDNESS Degree that gravel-sized and larger particles are surrounded or enclosed by finer-sized particles.

EMBEDMENT Embedded structure (e.g., steel structure) embedded in concrete that transmits external loads to it.

EMBEDMENT LENGTH Distance rebar is inserted into concrete beyond critical section.

EMERGENCY 1. Condition of serious nature which develops unexpectedly and endangers structural integrity of dam or endangers downstream property and human life and requires immediate action; 2. Event that demands crisis response beyond scope of any single line agency or service and that presents threat to community or larger area; 3. Event that can be controlled within scope of local capabilities.

EMERGENCY ACTION PLAN Predetermined plan of action to be taken to reduce potential for property damage and loss of life in area affected by dam break, excessive spillway release or large flood.

EMERGENCY ACTION PLAN EXERCISE Activity designed to promote emergency preparedness, test or evaluate EAP's, procedures or facilities, train personnel in emergency management duties and demonstrate operational capability.

EMERGENCY AND HAZARDOUS CHEMICAL INVENTORY Annual report by facilities having one or more extremely hazardous substances or hazardous chemicals above certain weight limits.

EMERGENCY BROADCAST SYSTEM Federally established network of commercial radio stations that voluntarily provide official emergency instructions or directions to public during emergency.

EMERGENCY CLASSIFICATION LEVELS Phased system where dam operating organizations classify dam safety emergency incidents into response levels according to how severe they are at time of observation and as to time of occurrence.

EMERGENCY ESCAPE RAMP Traffic ramp that enables vehicles having braking problems to stop safely.

EMERGENCY EVACUATION ZONE I Emergency evacuation zone immediately below dam and located on both sides of river or stream.

EMERGENCY EVACUATION ZONE II Second emergency evacuation zone, beyond emergency evacuation zone I and also located on both sides of river or stream.

EMERGENCY EVACUATION ZONE III Outermost emergency evacuation zone, extending beyond emergency evacuation zone II and also located on both sides of river or stream.

EMERGENCY EVACUATION ZONES Geographical areas delineated in inundation areas downstream from dam (hazard generator) that define potential area of impact and allow prioritizing evacuation activities based on proximity of populations at risk to hazard in terms of distance and flood wave travel times.

EMERGENCY EXEMPTION Provision that can grant temporary exemption to state or another federal agency to allow use of pesticide product not registered for that particular use.

EMERGENCY EXERCISE Activity designed to promote emergency preparedness, evaluate emergency operations, policies, plans, procedures and facilities, train personnel in emergency management and response duties and demonstrate operational capability.

EMERGENCY GATE 1. First gate in series of flow controls, remaining open while downstream gates or valves are operating; 2. Gate, usually located between emergency and regulating gates, used in closed position to permit servicing of downstream regulating gate(s) or valve(s) or downstream conduit when normal means of water control is not available.

EMERGENCY LOCATOR TRANSMITTER Emergency location radio beacon that's used on aircraft.

EMERGENCY LOCK Air lock built to hold and allow rapid passing of complete shift of workers.

EMERGENCY MANAGEMENT System by which mitigation, preparedness, response and recovery activities are undertaken to save lives and protect property from all hazards.

EMERGENCY MANAGEMENT AGENCY Any state or local agency responsible for emergency operations, planning, mitigation, preparedness, response and recovery for all hazards.

EMERGENCY ON-SCENE COORDINATOR One employee (at least) either on premises or on call with responsibility for coordinating all emergency response measures.

EMERGENCY OPERATION CENTER Center responsible for carrying out emergency management and ensuring operation continuity.

EMERGENCY OPERATIONS CENTER Location of facility where responsible officials gather during emergency to direct and coordinate emergency operations, to communicate with other jurisdictions and with field emergency forces and to formulate protective action decisions and recommendations during emergency.

EMERGENCY OPERATIONS PLAN Plan, usually developed in accord with guidance contained in Guide for Development of State and Local Emergency Operations Plans, Civil Preparedness Guide 1-8 and other similar guides.

EMERGENCY POWER Independent source of electrical power that supports important electrical systems on loss of normal power supply.

EMERGENCY PREPAREDNESS PLAN Predecessor to Emergency Action Plan (EAP).

EMERGENCY PROGRAM MANAGER Individual responsible on day-to-day basis for jurisdiction's effort to develop capability for coordinated response to and recovery from effects of emergencies and large-scale disasters.

EMERGENCY REMOVAL ACTION 1. Steps taken to remove contaminated materials that pose imminent threats to local residents (e.g., removal of leaking drums, excavation of explosive waste); 2. State record of such removals.

EMERGENCY RESERVE FUND Money reserved or required by contract to be reserved by operating entity for use in emergency situations involving facilities under entity's jurisdiction.

EMERGENCY RESPONSE VALUES Concentrations of chemicals, published by various groups, defining acceptable levels for short-term exposures in emergencies.

EMERGENCY SERVICES Services provided in order to minimize impact of flood already happening.

EMERGENCY SHUTDOWN Safely shut down operating system.

EMERGENCY SIGNAL Signal used in times of emergency for obtaining help.

EMERGENCY SPILLWAY Any secondary spillway designed to be operated infrequently, possibly in anticipation of some degree of structural damage or erosion to spillway that would occur during operation.

EMERGENCY STOP SWITCH Switch is used as safety measure to stop hazards.

EMERGENCY SUSPENSION Suspension of pesticide product registration due to imminent hazard.

EMERGENCY-VEHICLE HYBRID BEACON Beacon that assigns right-of-way to emergency vehicles.

EMERGENCY-VEHICLE TRAFFIC CONTROL SIGNAL Traffic control signal that assigns right-of-way to emergency vehicle.

EMERGENT PLANTS Erect, rooted, herbaceous plants that may be temporarily or permanently flooded at base but do not tolerate prolonged inundation of entire plant.

EMERGING ENVIRONMENTAL PROBLEMS Problems that may be new and/or are becoming known because of better monitoring and use of indicators.

EMERGING FLUX REGION Area on Sun where new magnetic flux is erupting.

EMINENT DOMAIN Right of government over private property within its jurisdiction.

EMISSION 1. Pollution discharged into atmosphere from smokestacks, other vents and surface areas of commercial or industrial facilities; 2. Something that is emitted or sent out (i.e., residential chimneys, motor vehicle, locomotive or aircraft exhausts).

EMISSION CAP Limit designed to prevent projected growth in emissions from existing and future stationary sources from eroding any mandated reductions.

EMISSION CONTROL DEVICE Device for purpose of activating, modulating, delaying or deactivating emissions.

EMISSION DATA Data related to emission parameters and facility identification.

EMISSION FACTOR Amount of pollutant discharged per unit of time or volume.

EMISSION INVENTORY Listing, by source, of amount of air pollutants discharged into atmosphere of community used to establish emission standards.

EMISSION MEASUREMENT SYSTEM Emissions testing device small and light enough to be transported.

EMISSION SOURCE Pollutant origination point or area.

EMISSION STANDARD Maximum amount of air polluting discharge legally allowed from single source, mobile or stationary.

EMISSIONS TRADING Creation of surplus emission reductions at certain stacks, vents or similar emissions sources and use of this surplus to meet or redefine pollution requirements applicable to other emissions sources.

EMISSIVITY Ability of surface to emit radiant energy compared to that of black body at same temperature and with same area.

EMPIRICAL MODEL Representation of real system by mathematical description based on experimental or observed data rather than on general physical laws.

EMPLOYER Person, firm or other entity which pays for or hires services of another person.

EMPLOYER'S LIABILITY INSURANCE Protects employers from financial loss if worker has job-related injury or illness not covered by workers' compensation.

EMTO Prefix symbolizing 10^{-15}.

EMULSIFIER Chemical that aids in suspending one liquid in another that's usually organic chemical in aqueous solution.

EMULSIOD Colloid that's easily dispersed.

EMULSION Stable suspension of small droplets of one liquid in another with which it is immiscible.

ENCAPSULATION Treatment of asbestos-containing material with liquid that covers surface with protective coating or embeds fibers in adhesive matrix to prevent their release into air.

ENCASE To enclose.

ENCLOSED 1. Fenced in or surrounded; 2. Held within container.

ENCLOSED PROCESS Process designed and operated so that there is no intentional release of chemical.

ENCLOSED STORAGE AREA Any area covered by roof under that metallic minerals are stored prior to further processing or loading.

ENCLOSED TRICKLING FILTER Trickling filter surrounded or closed off on all sides.

ENCLOSURE 1. Structure that surrounds or closes off on all sides; 2. Putting airtight, impermeable, permanent barrier around asbestos-containing materials to prevent release of asbestos fibers into air.

ENCROACHMENT 1. Seawater that penetrates freshwater aquifer; 2. Intrusion upon another's possessions or rights.

ENCRUSTATION Crust or hard coating of anything upon or within body, as deposit of lime, sediment, etc., from water on inner surface of steam boiler.

ENCUMBRANCES Any interest, right, burden or liability attached to title of land (i.e., lien, mortgage).

END CONTRACTION When length of weir is less than width of channel and nappe contracts either one or both sides of weir.

END DISTANCE OF BOLTS Distance along line of force between center of hole and component end.

END EVENT Final event in Critical Path Method.

END MANHOLE Manhole located at upstream end of sewer and having no inlet pipe.

END MORAINE Ridge of sediment piled at front edge of glacier.

END POINT 1. Stage in titration where change in color of indicator indicates that no more titrant should be added; 2. defined occurrence during observation period of experiment or study.

END SECTION Culvert flared end attachment to reduce erosion and enhance hydraulic efficiency.

END TREATMENT Extruder for barrier designed for frontal vehicle impacts.

END USE Application or function for which something is designed or for which it is ultimately used.

END USER Consumer of products for purpose of recycling that excludes products for re-use or combustion for energy recovery.

END-OF-ROADWAY MARKER Device that alerts end of roadway.

END-OF-THE-PIPE Technologies (e.g., scrubbers on smokestacks, catalytic converters on automobile tailpipes) that reduce emissions of pollutants after they have formed.

END-RESULT Specifications requiring contractor to take product responsibility.

END-RESULT SPECIFICATION Specification that requires contractor to take entire responsibility for supplying each item of construction.

END-USE PRODUCT Pesticide formulation for field or other end use that excludes products used to formulate other pesticide products.

ENDANGERED OR THREATENED SPECIES Any species listed in Endangered Species Act, where endangered species are plants and animals that have become so rare they are in danger of becoming extinct and threatened species are plants and animals that are likely to become endangered within foreseeable future.

ENDANGERED SPECIES Species (i.e., animals, birds, fish, plants, other living organisms) in imminent danger of extinction throughout all or significant portion of its range.

ENDANGERED SPECIES ACT Act that provides framework for protection of endangered and threatened species.

ENDANGERMENT Construction, operation, maintenance, conversion, plugging or abandonment of injection well or performance of other injection activities, by owner or operator in manner that allows movement of fluid containing any contaminant into USDW, if presence of that contaminant may cause violation of any primary drinking water regulations or may adversely affect health of persons.

ENDOCRINE SYSTEM Collection of ductless glands in animals that secrete hormones, which influence growth, gender and sexual maturity.

ENDOGENOUS Produced, originating or growing from within.

ENDOTHERMIC 1. Of chemical reaction that absorbs heat energy from its surroundings; 2. Of animal whose body temperature is regulated by internal factors.

ENDRIN Pesticide toxic to freshwater and marine aquatic life that produces adverse health effects in domestic water supplies.

ENDURANCE LIMIT STRENGTH Stress level below where material does not fail after infinite cycles.

ENERGY Force or action of doing work measured in terms of work it is capable of doing.

ENERGY ANALYSIS Analysis informs policy and investment decisions that lead to more resilient, reliable and efficient energy systems.

ENERGY AUDIT Assess how much energy you consume and evaluate what measures you can take to be more energy efficient.

ENERGY AVERAGE LEVEL Quantity calculated by taking ten times common logarithm of arithmetic average of antilogs of one-tenth of each of levels being averaged.

ENERGY BUDGET Balance sheet of energy income against expenditure.

ENERGY CONSERVATION Effort made to reduce consumption of energy by using less of energy service.

ENERGY DISSIPATER Device constructed in waterway to reduce kinetic energy of fast flowing water to prevent erosion of stream channel.

ENERGY DISSIPATION 1. Loss of kinetic energy of moving water due to internal turbulence, bottom friction, large rocks, debris or other obstacles that impede flow; 2. Process where energy is transformed from initial to final form.

ENERGY EFFICIENCY Goal to reduce amount of energy required to provide products and services.

ENERGY FLOW Flow of energy through food chain.

ENERGY GRADE LINE 1. Total energy in flow of section with reference to datum line is sum of elevation of pipe centerline, piezometric height (i.e., pressure head) and velocity head; 2. Total energy slopes (drops) in direction of flow (slope of line is energy gradient) except where energy is added by mechanical devices.

ENERGY GRADIENT Line representing elevation of total energy head of flow.

ENERGY HEAD Elevation of hydraulic grade line at any section of waterway plus velocity head of mean velocity of water in that section.

ENERGY HELICITY INDEX Index that incorporates vertical shear and instability designed for purpose of forecasting supercell thunderstorms.

ENERGY LINE 1. Total energy in flow (i.e., head of flow) of section with reference to datum line (i.e., sum of elevation of pipe centerline, pressure head and velocity head); 2. Line connecting levels of energy heads of surface water equal to hydraulic grade line plus velocity head.

ENERGY MANAGEMENT SYSTEM Control system capable of monitoring environmental and system loads and adjusting HVAC operations accordingly in order to conserve energy while maintaining comfort.

ENERGY RECOVERY Obtaining energy from waste through variety of processes (e.g., combustion).

ENERGY RECOVERY PROCESS Process that recovers energy content of non-recyclable waste materials directly by conversion into heat, electricity or fuel.

ENERGY SOURCE Primary source that provides power converted to electricity through chemical, mechanical or other means where sources include coal, petroleum and petroleum products, gas, water, uranium, wind, sunlight, geothermal, etc.

ENERGY STORAGE DEVICE Apparatus used for storing electric energy when needed and releasing it when required.

ENERGY-GRADIENT LINE Total energy at various plotted points along axis of flow and joined by line.

ENFORCEABLE REQUIREMENTS Conditions or limitations in permits issued that, if violated, could result in issuance of compliance order or initiation of civil or criminal action under federal or applicable state laws.

ENFORCEMENT Federal, state or local legal actions to obtain compliance with laws, rules, regulations or agreements and/or obtain penalties or criminal sanctions for violations.

ENFORCEMENT DECISION DOCUMENT Document that provides explanation to public of EPA's selection of cleanup alternative at enforcement sites on National Priorities List.

ENGINE Mechanical device which converts energy into useful motion or physical effects.

ENGINE BURN FRACTURE Progressive fracture originating in spots where driving wheels have slipped on rail head surface.

ENGINE RATING Engine certified to deliver power output.

ENGINE SAFETY CONTROL Apparatus that protects against devastating damage by shutting engine down.

ENGINE SIDESCREEN Rugged screen that fits on engine housing of vehicle used at sanitary landfill to keep paper and other objects from accumulating and impairing engine performance.

ENGINEER Person who is qualified or professionally engaged in designing, building or maintaining.

ENGINEER'S LEVEL Telescope that is attached to spirit-tube level, all revolving around vertical axis and is mounted on tripod.

ENGINEERED CONTROLS Method of managing environmental and health risks by placing barrier between contamination and rest of site, thus limiting exposure pathways.

ENGINEERING GEOLOGIST Application of geology to engineering study for purpose of assuring that geological factors regarding location, design, construction, operation and maintenance of engineering works are recognized and accounted for.

ENGINEERING JUDGMENT Decision based on available data.

ENGINEERING STUDY Comprehensive analysis and evaluation of available pertinent information and application of appropriate principles, provisions and practices.

ENGLACIAL Occurring or located within glacier.

ENGLACIAL LOAD Debris, originally accumulated on surface of glacier, which has been buried in interior of glacier by subsequent falls of snow.

ENGLACIAL STREAM Meltwater river that flows inside glacier or ice sheet.

ENHANCED INSPECTION AND MAINTENANCE Improved automobile inspection and maintenance program aimed at reducing automobile emissions that contains, at minimum, more vehicle types and model years, tighter inspection and better management practices.

ENHANCED V Pattern seen on satellite infrared photographs of thunderstorms, where thunderstorm anvil exhibits V-shaped region of colder cloud tops extending downwind from thunderstorm core.

ENHANCED WORDING Strong wording or emphasis used in zone forecast issued by National Weather Service Forecast Office used in tornado and severe thunderstorm watches when potential for strong/violent tornadoes or unusually widespread damaging straight-line winds, is high (e.g., "some thunderstorms may be severe").

ENHANCEMENT Improvement of facility beyond its originally designed purpose or condition.

ENHANCEMENT FLOW Improved flows that result in better stream conditions for aquatic, terrestrial and other resources.

ENLARGEMENT LOSS Energy loss by friction in flowing fluid when it moves into cross-sectional area of sudden enlargement.

ENRICHMENT Addition of nutrients (e.g., nitrogen, phosphorus, carbon compounds) from sewage effluent or agricultural runoff to surface water, greatly increases growth potential for algae and other aquatic plants.

ENSEMBLE Collection of numerical model results that show slightly different possible outcomes.

ENSEMBLE HYDROLOGIC FORECASTING Process whereby continuous hydrologic model is successively executed several times for same forecast period by use of varied data input scenarios or perturbation of key variable state for each model run.

ENSEMBLES Reference to set of computer models run under concept of Ensemble Forecasting (multiple predictions from ensemble of models with slightly different initial conditions used as input and/or slightly different versions of models).

ENSIGN VALVE Earliest type of Bureau of Reclamation needle valve, operated by water pressure from forebay and mounted on upstream side of dam.

ENSO DIAGNOSTIC DISCUSSION Addresses current oceanic and atmospheric conditions in Pacific and seasonal climate outlook for following one to three seasons.

ENTERIC Of, relating to, within or by way of intestines.

ENTERIC BACTERIA Bacteria of intestines.

ENTEROBACTER AEROGENES Any of group of rod-shaped bacteria of family Enterobacteriaceae.

ENTEROCOCCI Any of group of streptococci bacteria, of genus Enterococcus, that inhabit human gastrointestinal tract and have great resistance to antibiotics.

ENTEROCOCCUS BACTERIA Commonly found in feces of humans and other warmblooded animals where presence in water is indication of fecal pollution and possible presence of enteric pathogens (e.g., Streptococcus feacalis, Streptococcus "Bacteria.")

ENTRAIN To trap bubbles in water either mechanically through turbulence or chemically through reaction.

ENTRAINED AIR Creation of microscopic air bubbles in concrete.

ENTRAINMENT 1. Process by which aquatic organisms, suspended in water, are pulled through pump or other device; 2. Carrying away of bed material produced by erosive action of moving water; 3. Any of several processes where solid or liquid is put into motion by fluid.

ENTRAINMENT SEPARATOR Equipment that removes entrapped droplets for vapor stream in desalting procedure.

ENTRAINMENT ZONE Shallow region at top of convective boundary layer where fluid is entrained into growing boundary layer from overlying fluid by collapse of rising convective plumes or bubbles.

ENTRANCE GATE Gate used at grade crossings to prevent lane usage.

ENTRANCE HEAD Head required to cause fluid flow into conduit.

ENTRANCE LOSS Energy loss by friction in flowing fluid when it moves into cross-sectional area of sudden contraction.

ENTRANCE REGION Region upstream from wind speed maximum in jet stream (jet max), where air is approaching (entering) region of maximum winds and therefore is accelerating that results in vertical circulation that creates divergence in upper-level winds in right half of entrance region (as would be viewed looking along direction of flow) and results in upward motion of air in right rear quadrant (or right entrance region) of jet max.

ENTRANCE WELL Well or opening at surface of ground, constructed to receive surface water directed to sewer.

ENTRAPPED AIR Air pockets in concrete due to improper mixing, consolidating and placement.

ENTRENCHED STREAM River that flows in narrow trench cut into level land.

ENTROPY Amount of energy not available for work during certain process.

ENTRY FLOW Traffic entering roundabout.

ENVIRONMENT 1. All biological, chemical, social and physical factors that organisms are exposed; 2. Sum of all external conditions (surroundings) affecting life, growth, development and survival of organism.

ENVIRONMENT CANADA Canadian Federal Government department responsible for issuing weather forecasts and weather warnings in Canada.

ENVIRONMENT QUALITY STANDARDS Threshold concentrations of contaminants in order to protect both environment and human health from adverse effects.

ENVIRONMENTAL AGENCY Government agency in charge of formulation and management of public policy on issues concerned with and changing natural ecosystem.

ENVIRONMENTAL ANALYSIS Systematic process for consideration of environment factors in land management actions.

ENVIRONMENTAL ASSESSMENT Environmental analysis prepared to determine whether federal action would significantly affect environment and thus require more detailed environmental impact statement where if not, finding of no significant impact is written and if so, environmental impact statement is written.

ENVIRONMENTAL AUDIT Independent assessment of current status of party's compliance with applicable environmental requirements or of party's environmental compliance policies, practices and controls.

ENVIRONMENTAL CONDITIONS Environment condition (e.g., soil, surface water, air) of hazardous levels that exceeds standard or threshold.

ENVIRONMENTAL DAMAGE ASSESSMENT Procedures where monetary value may be set on environmental damage.

ENVIRONMENTAL DESIGN Process of addressing surrounding environmental parameters when devising plans, programs, policies, buildings or products.

ENVIRONMENTAL ENGINEERING Engineers that use principles of engineering, soil science, biology and chemistry to develop solutions to environmental problems.

ENVIRONMENTAL EQUITY/JUSTICE Equal protection from environmental hazards for individuals, groups or communities regardless of race, ethnicity or economic status.

ENVIRONMENTAL EXPOSURE 1. Human exposure to pollutants originating from facility emissions; 2. Threshold levels are not necessarily surpassed, but low-level chronic pollutant exposure is one of most common forms of environmental exposure.

ENVIRONMENTAL FATE Destiny of chemical or biological pollutant after release into environment.

ENVIRONMENTAL FATE DATA Data that characterize pesticide's fate in ecosystem, considering factors that foster its degradation (light, water, microbes), pathways and resultant products.

ENVIRONMENTAL FORECASTING Estimating intensity, nature and timing of external forces that may effect performance of firm, disrupt its plans or force change in its strategies.

ENVIRONMENTAL FRAMEWORK Natural and human-related features of land and hydrologic system (e.g., geology, land use, habitat) that provide unifying framework for making comparative assessments of factors that govern water-quality conditions within and among Study Units.

ENVIRONMENTAL HALF-LIFE Time taken for half amount of material to disappear.

ENVIRONMENTAL IMPACT ASSESSMENT Process of examining anticipated environmental effects of proposed project.

ENVIRONMENTAL IMPACT STATEMENT Document required of federal agencies by National Environmental Policy Act (NEPA) for major projects or legislative proposals significantly affecting environment used to evaluate range of alternatives when solving problem would have significant effect on human environment that describes positive and negative effects of undertaking and cites alternative actions.

ENVIRONMENTAL INDICATOR Measurement, statistic or value that provides proximate gauge or managerially and scientifically useful evidence of effects of environmental management programs or of state or condition of environment.

ENVIRONMENTAL INFORMATION DOCUMENT Written evaluation describing environmental impacts of planned activity.

ENVIRONMENTAL JUSTICE Fair treatment of people of all races, cultures, incomes and educational levels with respect to development and enforcement of environmental laws, regulations and policies.

ENVIRONMENTAL LAPSE RATE Rate of decrease of air temperature with height, usually measured with radiosonde.

ENVIRONMENTAL LIEN Charge, security or encumbrance on property's title to secure payment of cost or debt arising from response actions, cleanup or other remediation of hazardous substances or petroleum products.

ENVIRONMENTAL MANAGEMENT SYSTEMS Set of processes and practices that enable organization to reduce its environmental impacts and increase its operating efficiency.

ENVIRONMENTAL MEDIUM Major environmental category that surrounds or contacts humans, animals, plants and other organisms (e.g., surface water, groundwater, soil, air) and where chemicals or pollutants move.

ENVIRONMENTAL MONITORING FOR PUBLIC ACCESS AND COMMUNITY TRACKING Joint federal (EPA, NOAA and USGS) program to provide timely and effective communication of environmental data and information through improved and updated technology solutions that support timely environmental monitoring reporting, interpreting and use of information for benefit of public.

ENVIRONMENTAL NOISE Summary of noise pollution from outside, caused by transport, industrial and recreational activities.

ENVIRONMENTAL POLICY FORMULATION Measures taken by firm wanting to boost its environmental performance.

ENVIRONMENTAL PROTECTION Practice of protecting natural environment by individuals, organizations and governments.

ENVIRONMENTAL PROTECTION AGENCY Environmental Protection Agency's mission is to protect human health and to safeguard natural environment.

ENVIRONMENTAL REPORT Publication prepared for real estate holding that identifies potential or existing environmental contamination liabilities.

ENVIRONMENTAL RESPONSE TEAM Federal (EPA) experts who provide around-the-clock technical assistance to EPA regional offices and states during all types of hazardous waste site emergencies and spills of hazardous substances.

ENVIRONMENTAL REVIEW Process of reviewing project and its potential environmental impacts to determine whether it meets federal, state and local environmental standards.

ENVIRONMENTAL SAMPLE Water sample collected from aquifer or stream for purpose of chemical, physical or biological characterization of sampled resource.

ENVIRONMENTAL SETTING Land area characterized by unique combination of natural and human-related factors (e.g., row-crop cultivation, glacial-till soils).

ENVIRONMENTAL SITE ASSESSMENT Process of determining whether contamination is present on parcel of real property.

ENVIRONMENTAL STUDIES Multidisciplinary academic field which systematically studies human interaction with environment in interests of solving complex problems.

ENVIRONMENTAL SUSTAINABILITY Long-term maintenance of ecosystem components and functions for future generations.

ENVIRONMENTAL TEMPERATURE Temperature where inanimate body of same shape and size as given organism will come to equilibrium with its surroundings when placed at same point in space as organism.

ENVIRONMENTAL TEMPERATURE SOUNDING 1. Instantaneous or near-instantaneous sounding of temperature as function of height; 2. Sounding or vertical profile is usually obtained by balloon-borne instrument, but can also be measured using remote sensing equipment.

ENVIRONMENTAL TOBACCO SMOKE Mixture of smoke from burning end of cigarette, pipe or cigar and smoke exhaled by smoker.

ENVIRONMENTAL TRANSFORMATION PRODUCT Chemical substance caused by activity of environmental procedures on parent compound that alters molecular identity of parent compound.

ENVIRONMENTAL/ECOLOGICAL RISK Potential for adverse effects on living organisms associated with pollution of environment by effluents, emissions, wastes or accidental chemical releases, energy use or depletion of natural resources.

ENVIRONMENTALLY TRANSFORMED Chemical substance whose chemical structure changes as result of action of environmental processes on it.

ENVIRONMENTALLY-RELATED MEASUREMENTS Data collection or investigation involving assessment of chemical, physical or biological factors in environment that affect human health or quality of life.

ENZYME Globular protein that catalyses biological chemical reaction.

EOLIAN 1. Materials carried, deposited, produced or eroded by wind; 2. Windblown.

EPHEMERAL CREEK Stream that flows only in direct response to precipitation and whose channel is at all times above water table.

EPHEMERAL STREAM Creek that flows only in direct response to precipitation and whose channel is at all times above water table.

EPICENTER 1. Focal point directly above origin; 2. Point on Earth's surface located vertically above origin where first rupture and first seismic disturbance motion occur.

EPIDEMIOLOGY 1. Investigation of factors contributing to disease or toxic effects in studies of general population; 2. Study of distribution of disease or other health-related states and events in human populations, as related to age, sex, occupation, ethnicity and economic status in order to identify and alleviate health problems and promote better health.

EPIFAUNA Animals that live on benthos.

EPIGENE Of or pertaining to region just below surface of earth.

EPILIMNION Upper layer of water in thermally stratified lake or reservoir.

EPIPEDON Horizon that forms near soil surface.

EPIPELAGIC Of, pertaining to or inhabiting water from surface of sea down to 200 m.

EPISODE, POLLUTION Air pollution incident in given area caused by concentration of atmospheric pollutants under meteorological conditions that may result in significant increase in illnesses or deaths that may also describe water pollution events or hazardous material spills.

EPOCH 1. Unit of geological time; 2. Distinctive period in history.

EPT RICHNESS INDEX Total number of distinct taxa within insect orders Ephemeroptera (mayflies), Plecoptera (stoneflies) and Trichoptera (caddisflies), that are composed primarily of species considered to be relatively intolerant to environmental alterations that usually decreases with pollution.

EQUAL-FALLING PARTICLES Particles possessing equal terminal velocities and specific gravities.

EQUAL-WIDTH INCREMENT SAMPLE Composite sample of water collected across section of stream with equal spacing between verticals and equal transit rates within each vertical that yields representative sample of stream conditions.

EQUALIZER Culvert that enables standing water to rest in common level in regards to bank.

EQUALIZING BASIN Holding basin where variations in flow and composition of liquid are averaged.

EQUALIZING BED Material laid beneath pipeline in trench to provide uniform support for pipeline.

EQUALIZING RESERVOIR Regulates flow of water below intermittently operated hydropower station or peaking power plant.

EQUILIBRIUM 1. Balanced; 2. State at which radioactivity of consecutive elements within radioactive series is neither increasing nor decreasing.

EQUILIBRIUM DEPTH Minimum water depth for condition of no sediment transport.

EQUILIBRIUM DRAWDOWN Ultimate, constant drawdown for steady rate of pumped discharge.

EQUILIBRIUM LEVEL On sounding, level above level of free convection at which temperature of rising air parcel again equals temperature of environment.

EQUILIBRIUM LOAD Amount of sediment that system can carry for given discharge without overall accumulation (deposit) or scour (degradation).

EQUILIBRIUM SURFACE DISCHARGE Steady rate of surface discharge which results from long-continued, steady rate of net rainfall, with discharge rate equal to net rainfall rate.

EQUILIBRIUM TIME Time when flow conditions become substantially equal to those corresponding to equilibrium discharge or equilibrium drawdown.

EQUINOX Time when Sun crosses Earth's equator, making night and day of approximately equal length all over Earth and occurring about March 21 (spring or vernal equinox) and September 22 (autumnal equinox).

EQUIPMENT One or more assemblies capable of performing complete function.

EQUIPOTENTIAL LINE 1. Line connecting points of equal hydraulic head (potential); 2. Line, in field of flow, such that total head is same for all points on line and therefore direction of flow is perpendicular to line at all points where set of such lines provides contour map of potentiometric surface.

EQUIVALENCY Any body of procedures and techniques of sample collection and/or analysis for parameter of interest that has been demonstrated in specific cases to produce results not statistically different to those obtained from reference method.

EQUIVALENT 1. Similar or identical in value; 2. Having equal ability to combine.

EQUIVALENT DIAMETER Diameter of hypothetical sphere composed of material having same specific gravity as that of actual soil particle and of such size that it will settle in given liquid at same terminal velocity as actual soil particle.

EQUIVALENT DIRECT RADIATION Heat load on steam heating system in building and is expressed in so many thousands of square feet.

EQUIVALENT FLUID Hypothetical fluid having unit weight such that it will produce pressure against lateral support presumed to be equivalent to that produced by actual soil.

EQUIVALENT FLUID PRESSURE Simplification of backfill earth pressure against retaining structures.

EQUIVALENT METHOD Any method of sampling and analyzing for air pollution which has been demonstrated to be, under specific conditions, acceptable alternative to normally used reference methods.

EQUIVALENT PER MILLION Equivalent weight of solute per million unit weights of solution.

EQUIVALENT PIPES Imaginary pipes which are used to determine head loss and flow of discharge considering that flow of discharge and head loss in actual piping system is same as that of equivalent pipe.

EQUIVALENT POTENTIAL TEMPERATURE Temperature parcel at specific pressure level and temperature would have if it were raised to 0 mb, condensing all moisture from parcel and then lowered to 1,000 mb.

EQUIVALENT POTENTIAL TEMPERATURE RIDGE Axis of relatively high values of equivalent potential temperature.

EQUIVALENT SINGLE-AXLE LOAD Numerical factor to establish damage relationship for comparing effects of axle's carrying different loads.

EQUIVALENT SIZE Diameter of hypothetical sphere composed of material having same specific gravity as that of actual soil particle and of such size that it will settle in given liquid at same terminal velocity as actual soil particle.

ERA Unit of time, smaller than eons and greater than periods.

ERADICATION 1. Act of plucking up by roots; 2. Utter destruction.

ERGONOMIC DESIGN Science of equipment design intended to maximize productivity by reducing operator fatigue and discomfort.

ERGONOMICS Science of design of equipment, especially so as to reduce operator fatigue, discomfort and injury.

ERODE To wear away or remove land surface by wind, water or other agents.

EROSION 1. Progressive wearing away of land by wind, water or other natural agents; 2. Gradual destruction or diminution of something.

EROSION CONTROL Practice of preventing or controlling wind or water erosion in agriculture, land development, coastal areas, river banks and construction.

EROSION CONTROL MATTING Stabilization matting to reinforce slopes and support plant growth for erosion control.

EROSIVE VELOCITY Rate of flow of stream that, when exceeded, will result in erosion or its banks and/or bed.

ERRATIC Boulder transported from one location to another, usually by glacier and left behind when ice melted.

ERROR Difference between measured or calculated value and true one.

ERUPTIVE Solar activity levels with at least one radio event (10 cm) and several chromospheric events per day (Class C Flares).

ERUPTIVE PROMINENCE ON LIMB Solar prominence that becomes activated and is seen to ascend from Sun.

ESCALATED BASE PRICE Base price estimate adjusted for inflation up to time of tender.

ESCALATION Increase or rise, especially one to counteract perceived discrepancy.

ESCALATION CLAUSE Clause in agreement that guarantees change in agreement price once particular factor beyond control of either party affecting value has been determined.

ESCALATOR CLAUSE Clause in lease or contract that guarantees change in agreement price once particular factor beyond control of either party affecting value has been determined.

ESCAPE Leakage or outflow, as of steam or liquid or electric current through defective insulation.

ESCAPE CLAUSE Any clause, term or condition in contract that allows party to that contract to avoid having to perform contract or cancel it.

ESCAPE ROUTE Planned means of escape from dangerous situation.

ESCAPE STAIR Specially developed stair for rapid evacuation.

ESCAPEMENT 1. Cliff or steep slope that separates two levels or gently sloping areas; 2. Cliff or steep slope edging higher land.

ESCARPMENT Steep face or edge of ridge.

ESCHERICHIA COLI 1. Bacteria present in intestine and feces of warmblooded animals; 2. Member species of fecal coliform group of indicator bacteria.

ESKER Long, narrow, sinuous ridge created by deposits from stream running beneath glacier.

ESSENTIAL ELEMENT 1. Structural or geologic feature or equipment item whose failure under particular loading condition or set of circumstances being considered would create dam safety deficiency; 2. Equipment item or procedure required for safe operation of dam or reservoir.

ESSENTIAL MINERALS Primary mineral that's igneous rock presence is essential to defining root name of that rock.

ESTABLISHED TREATMENT TECHNOLOGIES Technologies for which cost and performance data are readily available.

ESTHETICS Emotional judgment about what is beautiful or pleasing.

ESTIMATE 1. Document or verbal notification, specifying how much job will probably cost; 2. To calculate roughly.

ESTIMATE AT COMPLETION Forecasted cost of project as project progresses.

ESTIMATED COST Approximation of cost of program, project or operation.

ESTIMATED DESIGN LOAD Sum of useful heat transfer, heat transfer to or from connected piping and heat transfer that occurs in any auxiliary apparatus connected to HVAC system.

ESTIMATED ENVIRONMENTAL CONCENTRATION Estimated pesticide concentration in ecosystem.

ESTIMATED MAXIMUM LOAD Calculated maximum heat transfer that system may be called upon to provide in HVAC system.

ESTIMATED VALUE OF CONCENTRATION Reported when analyte is detected and all criteria for positive result are met.

ESTIMATES Documents or verbal notifications, specifying how much job will probably cost.

ESTIMATING Document or verbal notification, specifying how much job will probably cost based material in-hand and material required to complete job.

ESTIMATOR Person who computes amounts of materials, labor, costs and prices to complete job.

ESTIMATOR'S CONTINGENCY Approximate cost that covers identifiable items that may have been overlooked, anticipated growth in quantities, lack of confidence in input data or risk associated with cost item.

ESTUARIAL STORAGE Storage of water for domestic water supplies in estuaries when suitable inland sites for reservoirs have been exhausted.

ESTUARINE Formed in estuary by alluvial deposition.

ESTUARINE HABITAT Tidal habitats and adjacent tidal wetlands that are usually semi-enclosed by land but have open, partly obstructed or sporadic access to open ocean and where ocean water is at least occasionally diluted by freshwater runoff from land.

ESTUARINE WETLANDS 1. Tidal wetlands in low-wave-energy environments where salinity of water is greater than 0.5 parts per 1,000 and is variable owing to evaporation and mixing of seawater and freshwater; 2. Tidal wetlands of coastal rivers and embayments, salty tidal marshes, mangrove swamps and tidal flats.

ESTUARY Area of mixing of fresh water and salt water where river flows into ocean and where tidal effects are evident.

ESTURINE WATERS Deepwater tidal habitats and tidal wetlands that are usually enclosed by land but have access to ocean and are at least occasionally diluted by freshwater runoff from land (e.g., bays, mouths of rivers, salt marshes, lagoons).

ESTURINE ZONE Area near coastline that consists of estuaries and coastal saltwater wetlands.

ETA MODEL Now referred to as North American Meso is one of operational numerical forecast models run at NCEP where Eta is run four times daily, with forecast output out to 84 hours.

ETHANOL Simple aliphatic alcohol, with boiling point of 173.12 $^{\circ}$F, formally derived from ethane by replacing one hydrogen atom with hydroxyl group that's alternative automotive fuel derived from grain and corn usually blended with gasoline to form gasohol.

ETHNOBOTANY Scientific study of relationships between people and plants.

ETHOGRAM Catalogue of typical behaviour patterns of species.

ETHYLENE DIBROMIDE Chemical used as agricultural fumigant and in certain industrial processes that's extremely toxic and found to be carcinogen in laboratory animals.

EUGLENOIDS Group of algae that usually are free-swimming and rarely creeping that have ability to grow either photosynthetically in light or heterotrophically in dark.

EUGLENOPHYTA Group of algae that usually are free-swimming and rarely creeping that have ability to grow either photosynthetically in light or heterotrophically in dark.

EUPHOTIC ZONE Upper layer of ocean or lake that receives enough light for photosynthesis to occur.

EURYHALINE Able to tolerate various saltwater concentrations.

EURYTHERMIC Organisms able to tolerate wide range of temperature.

EURYTOPIC Describing plant or animal found in wide range of environments.

EUTROPHIC Nutrient enrichment of body of water that contains more organic matter than existing biological oxidization processes can consume.

EUTROPHIC LAKE Body of water that has lost much of its dissolved oxygen, due to excessive nutrients that causes algae to grow, that causes normal aquatic life to die off.

EUTROPHIC LAKES Shallow, murky bodies of water with concentrations of plant nutrients causing excessive production of algae.

EUTROPHICATION Process in water where more organic matter is produced than existing biological oxidization processes can consume.

EVACUATION Fifth of five Early Warning System components consisting of plans, personnel, equipment and facilities needed to move population at risk to safety that involves taking protective actions to leave area of risk until hazard has passed and area is safe for return.

EVACUATION WARNING Public warning message that local officials would issue following declaration of Response Level

III by personnel of dam operating organization intended to notify population at risk to evacuate flood inundation areas.

EVALUATION REPORT Final evaluation report is written document that describes how you monitored and evaluated your program.

EVAPORABLE WATER Water present in set cement paste in capillaries or held by surface forces, measured as that removable by drying under specified conditions.

EVAPORATION 1. Process where water is changed from liquid or solid state into gaseous (i.e., vapor) state; 2. Vaporization and sublimation that takes place at temperature below boiling point; 3. Water vapor losses from water surfaces, sprinkler irrigation and other related factors into atmosphere.

EVAPORATION AREA Body of water and adjoining wetland surface area that could lose moisture by evaporation.

EVAPORATION DEMAND Maximum potential evaporation generally determined using evaporation pan.

EVAPORATION DISCHARGE Gaseous discharge to atmosphere from vegetation and soil.

EVAPORATION OPPORTUNITY Ratio of rate of evaporation from land or water surface in contact with atmosphere, to evaporability under existing atmospheric conditions.

EVAPORATION PAN Pan (US National Weather Service class pan is 4 feet in diameter, 10 inches deep, initially filled with 8 inches of water, set up on timber grillage so that top rim is about 16 inches from ground) used to hold water during observations for determination of quantity of evaporation at given location.

EVAPORATION POND Artificial ponds with very large surface areas that are designed to efficiently evaporate water by sunlight and exposure to ambient temperatures.

EVAPORATION PONDS Areas where sewage sludge is dumped and dried.

EVAPORATION RATE Rate at which chemical or material changes into vapor.

EVAPORATION RETARDANT Long-chain organic material that retards evaporation when spread on water film.

EVAPORATION, TOTAL Sum of water lost from given land area during any specific time by transpiration from vegetation, building of plant tissue, by evaporation from water surfaces, moist soil, snow and interception.

EVAPORATION-MIXING FOG Fog that forms when evaporation of water raises dew point of adjacent air.

EVAPORATIVE COOLER Device that cools air through evaporation of water.

EVAPORATIVE EQUILIBRIUM In enclosed area, process when escape and return of vapor above liquid reaches equilibrium where vapor becomes saturated and no further change in vapor pressure, density or liquid temperature occurs.

EVAPORATIVITY Potential rate of evaporation.

EVAPORATOR Equipment used to evaporate solvent from solution.

EVAPORITE DEPOSITS Minerals formed by evaporation of water containing salts that are common in arid climates.

EVAPORITE MINERALS Deposits of minerals formed by evaporation of water containing salts that are common in arid climates.

EVAPOTRANSPIRATION 1. Sum of evaporation, sublimation and transpiration; 2. Collective term that includes water lost (discharged to atmosphere) through evaporation from soil and surface-water bodies and by plant transpiration in specific time period.

EVENT Point in time where certain conditions have been fulfilled and something noteworthy happens.

EVENT MODELS Models that simulate process for singular event.

EVENT-BASED MODEL Model that simulates some hydrologic response to precipitation event.

EVENT-ON-NODE Network diagramming technique where events are represented by boxes (or nodes) connected by arrows to show sequence where events are to be completed.

EVENT-ORIENTED SYSTEMS Systems that present information in relation to occasions.

EVERGREEN 1. Plants, especially trees, that do not shed their leaves seasonally; 2. Clause which causes automatic renewal of contract unless action is taken.

EVOLUTION Process of accumulating change.

EX PARTE Proceeding where one of involved parties is not present.

EXA Prefix symbolizing 10^{18}.

EXACT CHANGE LANE Toll lane that requires correct fare.

EXAMINATION REPORT Report that documents condition of facility during examination, operation and maintenance activities accomplished since last examination and recommendations necessary for continued safe and efficient operation of facility.

EXCAVATION 1. Action or process of digging or removing earth; 2. Cavity formed by cutting, digging or scooping.

EXCEEDANCE Violation of pollutant levels permitted by environmental protection standards.

EXCEEDANCE PROBABILITY Probability of event above stated magnitude occurring during any one year.

EXCESS ACTIVATED SLUDGE Wasted from return sludge line.

EXCESS AIR Oxygen that isn't consumed during combustion that absorbs otherwise usable heat and transports it away.

EXCESS EFFECTIVE PRECIPITATION Precipitation in excess of infiltration capacity, evaporation, transpiration and other losses.

EXCESS EMISSIONS Discharges of regulated air pollutant released into environment at rate above permitted levels for facility.

EXCESS LAND Irrigable land, other than exempt land, owned by any landowner in excess of maximum ownership entitlement under applicable provision of Reclamation law.

EXCESS PORE PRESSURE Excess pressure over hydrostatic pressure.

EXCESS PRESSURE PUMP Pump for installation on wet pipe sprinkler systems as means of establishing excess pressure for suppression of false alarms due to water surges (i.e., water hammer).

EXCESS RAIN Effective rainfall in excess of infiltration capacity.

EXCESS RAINFALL Rainfall neither retained on land surface nor infiltrated into soil.

EXCESS SLUDGE Sludge removed from treatment process to keep ratio of biomass to food supplied in wastewater in balance.

EXCESS WAIT TIME Difference between scheduled and actual wait times.

EXCESSIVE HEAT Combination of high temperatures (significantly above normal) and high humidities.

EXCESSIVE HEAT OUTLOOK Combination of temperature and humidity over certain number of days designed to provide indication of areas of country where people and animals may need to take precautions against heat during May to November.

EXCESSIVE HEAT WARNING Issued within 12 hours of onset of following criteria: heat index of at least 105 °F for more than 3 hours/day for 2 consecutive days or heat index more than 115 °F for any period of time.

EXCESSIVE HEAT WATCH Issued when heat indices in excess of 105 °F (41 °C) during day combined with nighttime low temperatures of 80 °F (27 °C) or higher are forecast to occur for two consecutive days.

EXCESSIVE I/I Quantities of infiltration and inflow that can be economically eliminated from sewer system.

EXCESSIVE PRECIPITATION Precipitation (e.g., rain) of unusually high rate.

EXCESSIVE RAINFALL Rainfall where rate of fall is greater than certain adopted limits, chosen with regard to normal precipitation (excluding snow) of given place or area.

EXCHANGE CAPACITY Capacity to exchange ions as measured by quantity of exchangeable ions in soil or rock.

EXCITATION Act of exciting, awakening or putting in motion.

EXCLUSION One of several situations that permits deletion of one or more of items required.

EXCLUSIONARY ORDINANCE Zoning that excludes classes of persons or businesses from particular neighborhood or area.

EXCLUSIVE BUS LANE Lane restricted to buses.

EXCLUSIVE FLOOD CONTROL CAPACITY Reservoir capacity assigned to sole purpose of regulating flood inflows to reduce flood damage downstream.

EXCLUSIVE FLOOD CONTROL STORAGE CAPACITY Space in reservoir reserved for sole purpose of regulating flood inflows to abate flood damage.

EXCLUSIVE LANE Designated lane that specified vehicles must use.

EXCLUSIVE OFF-STREET BICYCLE PATHS Bicycle paths, separated from motor vehicle traffic, that does not allow pedestrians.

EXCLUSIVE TRANSIT FACILITY Transportation element for transit vehicle use only.

EXCLUSIVE TURN LANE Turn lane designated for turning vehicles.

EXECUTION Formal process where contract is made valid and put into binding effect.

EXEMPT LAND Irrigation land in district that acreage limitation and pricing provisions do not apply.

EXEMPT SOLVENT Specific organic compounds not subject to requirements of regulation because they are deemed to be of negligible photochemical reactivity.

EXEMPTED AQUIFER Underground bodies of water defined in Underground Injection Control program as aquifers that are potential sources of drinking water though not being used as such and thus exempted from regulations barring underground injection activities.

EXEMPTION Federal or state permission for water system not to meet certain drinking water standard involving MCL, treatment technique or both, if system cannot comply due to compelling economic or other factors or because system was in operation before requirement or MCL was instituted and exemption will not create public health risk.

EXERCISE Simulated emergency condition carried out for purpose of testing and evaluating readiness of community or organization to handle particular type of emergency.

EXFILTRATION 1. Loss of water from drainage system as result of percolation or absorption into surrounding soil; 2. Filtering out (i.e., movement of substance through barrier).

EXHAUST DUCT Low-pressure pneumatic conveyors to convey dust, particles, shavings, fumes or chemical hazardous components from air.

EXHAUST EMISSIONS Substances that come out of exhaust system into atmosphere.

EXHAUST FAN Mechanical ventilation device that withdraws air via suction.

EXHAUST GAS ANALYZER Instrument for measurement of carbon monoxide among other gases in exhaust.

EXHAUST GAS RECIRCULATION Nitrogen oxide emissions reduction technique used in petrol or gasoline and diesel engines.

EXHAUST LINE Passage open to atmosphere.

EXHAUST OPENING Exhaust grille or any type of opening where air is exhausted from space.

EXHAUST SYSTEM Subsystem of engine that takes products of combustion chamber through catalytic converter and silencer to exhaust pipe.

EXISTENCE VALUE Value people place on simply knowing area or feature continues to exist in particular condition.

EXISTING GROUND Earth's surface as it is prior to any work.

EXISTING LAND-USE MAP Map indicating use to which lands in any specified area are put at time of preparing map.

EXISTING USES Uses actually attained in water body on or after November 28, 1975, whether or not they are included in water quality standards.

EXIT Passage or gate from someplace to another.

EXIT CHANNEL 1. Channel downstream from terminal structure that conveys releases back to "natural" stream or river; 2. Outlet channel.

EXIT CONFERENCE Discussion following facility review examination involving examination team members and interested representatives of water uses, project and region.

EXIT DOOR Last door on exit route that opens out while building is inhabited.

EXIT FLOW Traffic exiting roundabout.

EXIT GATE Automatic door raised to leave.

EXIT GATE CLEARANCE TIME Time of exit gate arm delay descent after entrance gate arm descends.

EXIT GRADIENT Potential drop between two adjacent equipotential lines divided by distance between.

EXIT LEVEL Level of exit stairway in building where exterior exit door or exit passageway leads to exterior.

EXIT LOSS Reduction of head occurring when flow of air or water is voided from confining construction and presumes lesser rate.

EXIT REGION Region downstream from wind speed maximum in jet stream (jet max), where air is moving away from region of maximum winds and therefore is decelerating that results in divergence in upper-level winds in left half of exit region (as would be viewed looking along direction of flow) and results in upward motion of air in left front quadrant (or left exit region) of jet max.

EXIT STORY Story where exterior door provides direct access at about exact same amount into public thoroughfare or to approved open space.

EXOSPHERE Upper most layer of Earth's atmosphere, where atmospheric gases can escape into outer space.

EXOTHERMIC 1. That releases energy in form of heat; 2. That releases heat during its formation and absorbs it during its decomposition; 3. Animal whose body temperature is regulated by external factors (i.e., cold-blooded.)

EXOTIC SPECIES Species (plants or animals) not indigenous (non-native) to region.

EXPANDING PLUG Flexible drain stopper designed for positive shut-off and zero leakage.

EXPANDING REACH Reach where flow is decelerating due to channel cross-section change.

EXPANSION Increase in volume of soil mass.

EXPANSION BEND Loop in pipe that provides flexibility for thermal expansion.

EXPANSION CHAMBER Tuned exhaust system used to enhance its power output by improving its volumetric efficiency.

EXPANSION COUPLING Yielding joint or coupling for so uniting parts of machine or structure that expansion, as by heat, is prevented from causing injurious strains.

EXPANSION FIT Condition of optimum clearance between parts where cold inner member is placed inside warmer outer member to allow temperature to equalize.

EXPANSION FLOW Flow in conduits where increased cross-sectional area yields decline in speed and increase in pressure.

EXPANSION JOINT 1. Separation between adjoining parts of concrete structure that is provided to allow small relative movements (e.g., those caused by temperature changes) to occur independently where flexible filler is provided in joint, reinforcement does not pass through joint; 2. Joint that permits pipe to move as result of expansion.

EXPANSION JOINT FILLER Flexible substance used to fill openings.

EXPANSION PIPE Hot water system vent that permits escape of steam or excess pressure.

EXPANSION TANK Small tank used to protect closed water heating systems and domestic hot water systems from excessive pressure.

EXPANSION VALVE Component in refrigeration and air conditioning systems that controls amount of refrigerant released into evaporator thereby controlling superheat.

EXPANSIVE SOILS Soils that expand when water is added and shrink when they dry out.

EXPENDING BEACH Beach shaped such that energy of waves is absorbed.

EXPENDITURES Act of paying out for charges.

EXPERIMENTAL ANIMALS Animals used in experiments that seek to control variables that affect behavior or biological system.

EXPERIMENTAL PLOT Place of earth laid out to discover results of particular approach to treatment or state, frequently split into subplots.

EXPERIMENTAL PRODUCT Product in final stages of testing and evaluation.

EXPERIMENTAL START DATE First date test substance is applied to test system.

EXPERIMENTAL TECHNOLOGY Technology that has not yet been proven possible.

EXPERIMENTAL USE PERMIT Obtained by manufacturers for testing new pesticides or uses thereof whenever they conduct experimental field studies to support registration on 10 acres or more of land or one acre or more of water.

EXPERIMENTAL WELLS Class V wells that are used to test new technologies.

EXPLICIT SCHEME Scheme where governing equations of numerical model are arranged to update dependent variables in terms of previously known values only.

EXPLODED VIEW Drawing with principal parts separated from each other, showing relationship/order of assembly of various parts.

EXPLOIT Excavate in such manner as to utilize material in particular vein or layer and waste or avoid surrounding material.

EXPLORATION Process of penetrating or ranging over for purposes of discovery.

EXPLORATORY DRILLING Drilling for purpose of determining physical properties and boundaries of reservoir.

EXPLORATORY PROGRAM Chain of activities or evaluations carried out to show advice.

EXPLOSIMETER Device used for detecting explosive gases.

EXPLOSION VENTING Explosion protection that provides planned pathway for expanding gases to escape.

EXPLOSION WAVE Shock wave generated by explosion.

EXPLOSION-PROOF APPARATUS Electric equipment enclosed in case which is effective at withstanding explosion.

EXPLOSIVE Chemical that causes sudden, almost instantaneous release of pressure, gas and heat when subjected to sudden shock, pressure or high temperatures.

EXPLOSIVE DEEPENING Decrease in minimum sea-level pressure of tropical cyclone of 2.5 mb/hr for at least 12 hours or 5 mb/hr for at least six hours.

EXPLOSIVE DUST Mixture of particulate matter and air in such proportions that's liable to catch fire or explode upon ignition when dispersed in air.

EXPLOSIVE LIMITS 1. Amounts of gas or vapor in air that form explosive mixtures; 2. Expressed as lower and upper limits and give range of vapor concentrations in air that will explode if ignition source is present.

EXPLOSIVES FACTOR Term expressing amount of explosives used to break quantity of rock.

EXPONENTIAL GROWTH Growth in value of quantity, where rate of growth is proportional to instantaneous value of quantity.

EXPORT 1. To sell (goods) to foreign country; 2. To carry away.

EXPORTS Municipal solid waste and recyclables transported outside state or locality where they originated.

EXPOSURE Location of building or structure in relation to wind or sun.

EXPOSURE ASSESSMENT Identifying pathways by which toxicants may reach individuals, estimating how much of chemical individual is likely to be exposed to and estimating number likely to be exposed.

EXPOSURE CONCENTRATION Concentration of chemical or other pollutant representing health threat in given environment.

EXPOSURE DOSE EFFECT RELATIONSHIP Effect of exposure to some known pollutant on measured period of time.

EXPOSURE HAZARD Chance that particular building and its contents may sustain loss or damage from fire in neighboring property.

EXPOSURE INDICATOR Characteristic of environment measured to provide evidence of occurrence or magnitude of response indicator's exposure to chemical or biological stress.

EXPOSURE LEVEL Concentration of chemical at absorptive surfaces of organism.

EXPOSURE LIMITS Limits that protect workers against adverse effects of exposure to substances.

EXPOSURE PATHWAY Path from sources of pollutants via, soil, water or food to man and other species or settings.

EXPOSURE PERIOD Time of continuous contact between agent and target.

EXPOSURE ROUTE Way chemical or pollutant enters organism after contact (i.e., ingestion, inhalation, dermal absorption).

EXPOSURE-RESPONSE RELATIONSHIP Relationship between exposure level and incidence of adverse effects.

EXPRESS BUS Bus that operates on faster schedule by limiting stops and taking quicker routes.

EXPRESS SERVICE Service that operates more quickly than ordinary service.

EXPRESSED WARRANTY Warranty created by overt words or actions of seller.

EXPRESSION OF SLOPE Rate of rise or fall with respect to horizontal distance (i.e., gradient).

EXPRESSWAY Divided highway where intersections and direct access to adjacent properties have been eliminated.

EXTENDED AERATION PROCESS Modification of activated sludge process which provides biological treatment for removal of biodegradable organic wastes under aerobic conditions.

EXTENDED FORECAST DISCUSSION Issued once day and is primarily intended to provide insight into guidance forecasts for 3- to 5-day forecast period.

EXTENSION AGREEMENT Extension agreement is written agreement between all parties to extend terms or contract milestone dates.

EXTENSION OF TIME Allow construction period to be extended where there is delay that's not contractor's fault.

EXTENSIONS Document used to extend term of contract beyond its original end date.

EXTENSIVE GARDEN 1. Gardens that have thinner soil depths and require less management and less structural support than intensive gardens; 2. Gardens that do not require artificial irrigation where plants are low-maintenance, hardy species that do not have demanding habitat requirements.

EXTENT OF CONGESTION Number of people or vehicles affected by geographic distribution of congestion.

EXTERIOR Outside part, parts or surface of something.

EXTERNAL AUDIT Examination conducted by independent accountant.

EXTERNAL COMBUSTION ENGINE Heat engine where working fluid, contained internally, is heated by combustion in external source, through engine wall or heat exchanger.

EXTERNAL FLOATING ROOF Storage tank commonly used to store large quantities of petroleum products such as crude oil or condensate.

EXTERNAL LOAD CRUSHING STRENGTH TEST Test performed to determine if pipe has sufficient strength to withstand external crushing load.

EXTERNAL SEALING BAND Flexible wrap applied outside of pipe, box, section or manhole joint to prevent fluid or solid transmission.

EXTERNAL TENDON Tendon outside of structural flange or web.

EXTERNAL THREAD Screw thread on outside of cone or cylinder.

EXTERNAL WORKS Any works carried out to external structure or building project.

EXTERNAL-EXTERNAL TRIP Trip with origin and destination outside study area but travels through study area.

EXTERNAL-INTERNAL TRIP Trip with either origin or destination inside study area.

EXTINCTION COEFFICIENT Measure of water clarity.

EXTINGUISHER Device that can be used to put out fire.

EXTINGUISHING MEDIA Chemical mixture made to fight specific forms of fire (e.g., oil, grease, electrical, organic combustibles).

EXTIRPATED SPECIES Species which has become extinct in given area.

EXTRA Contract work beyond what is due.

EXTRA FIRE HAZARD Presence of flammable hazards.

EXTRA WORK Something done in addition to requirements of contract.

EXTRA WORK ORDER Change in described contract or order following extra work beyond.

EXTRA-BOARD Worker available to cover for regular employee.

EXTRACT FAN Fan built to take out and exhaust air.

EXTRACTABLE ORGANIC HALIDES 1. Organic compounds that contain halogen atoms (e.g., chlorine); 2. Semi-volatile and extractable by ethyl acetate from air-dried streambed sediment where concentration is reported as micrograms of chlorine per gram of dry weight of streambed sediment.

EXTRACTION Separation process consisting in separation of substance from matrix.

EXTRACTION PROCEDURE Determining toxicity by procedure which simulates leaching where if certain concentration of toxic substance can be leached from waste, that waste is considered hazardous.

EXTRACTION SITE Area from which dredged or fill material proposed for discharge will be taken.

EXTRACTION WELL Discharge well used to remove groundwater or air.

EXTRACTOR COLUMN Apparatus used to take out solute from saturated solutions.

EXTRADOS Curved upstream surface of horizontal arch elements in arch dam.

EXTRAORDINARY STORM Storm that, due to prolongation or severity, creates weather conditions that cause significant amounts of damage.

EXTRAPOLATION Estimation of unknown values by extending or projecting from known values.

EXTRATERRESTRIAL RADIATION Theoretically-calculated radiation flux from Sun at top of atmosphere, before losses by atmospheric absorption.

EXTRATROPICAL Term used in advisories and tropical summaries to indicate that cyclone has lost its "tropical" characteristics.

EXTRATROPICAL CYCLONE Cyclone in middle and high latitudes often being 2,000 kilometers in diameter and usually containing cold front that extends toward equator for hundreds of kilometers.

EXTRATROPICAL LOW Low pressure center which refers to migratory frontal cyclone of middle and higher latitudes.

EXTREME EVENT LIMIT STATES Limit states that represent loads or events that are economically prohibitive to design to.

EXTREME ULTRAVIOLET Portion of electromagnetic spectrum from approximately 100 to 1,000 angstroms.

EXTREME WIND WARNING Informs public of need to take immediate shelter in interior portion of well-built structure due to onset of extreme tropical cyclone winds.

EXTREMELY HAZARDOUS SUBSTANCE Hundreds of hazardous chemicals identified as toxic to provide focus for state and local emergency planning activities.

EXTREMELY LOW FREQUENCY Portion of radio frequency spectrum from 30 to 3,000 hertz.

EYE 1. Relatively calm center in hurricane that's more than one half surrounded by wall cloud; 2. Organ through which animals see.

EYE OF STORM Region of calm weather right in middle of storm.

EYE WALL Organized band of cumuliform clouds that immediately surrounds center (eye) of hurricane where fiercest winds and most intense rainfall typically occur near eye wall.

EYEBAR Straight bar with hole at each end to allow it to be fixed to other components.

EYEBROW POP Rapid upward movement of Juliet's eyebrows during expression of surprise or skepticism.

F

F CORONA Of white-light corona (i.e., corona seen by eye at total solar eclipse), that portion that is caused by sunlight scattered or reflected by solid particles (dust) in inter-planetary space.

F REGION Upper layer of ionosphere, approximately 120 to 1500 km in altitude, subdivided into F1 and F2 regions where F2 region is most dense and peaks at altitudes between 200 and 600 km and F1 region is smaller peak in electron density, which forms at lower altitudes in daytime.

FABRIC FILTER Cloth device that catches dust particles from industrial emissions.

FABRIC MATTRESS Hollow mattress having cells filled with concrete mixture.

FABRICATE Construct/manufacture something.

FABRICATION Act of framing, constructing or manufacturing.

FABRICATOR Person who fabricates or manufactures something.

FACE 1. Exposed surface of dam materials (e.g., earth, rockfill, concrete), upstream and downstream; 2. External surface of structure, vertical surface of rock exposed by blasting or excavating; 3. Cutting end of drill hole.

FACE OF CURB Roadway surface side of curb that's vertical or sloping.

FACE WALING Waling throughout end of trench.

FACE-TO-FACE DIMENSION Distance from face of inlet port to face of outlet port.

FACEPIECE Part of various forms of mask or helmet that covers nose and/or mouth and eyes.

FACILITIES Structures associated with irrigation projects, municipal and industrial water systems, power generation facilities, including all storage, conveyance, distribution and drainage systems.

FACILITIES PLANS Plans and studies related to construction of treatment works necessary to comply with Clean Water Act or RCRA.

FACILITY Railroad, highway, airport, walkway or other public work used for transportation.

FACILITY EMERGENCY COORDINATOR Representative of facility covered by environmental law (e.g., chemical plant) who participates in emergency reporting process with Local Emergency Planning Committee.

FACING 1. With reference to wall or concrete dam, coating of different material, masonry or brick, for architectural or protection purposes (e.g., stonework facing, brickwork facing); 2. With reference to embankment dam, impervious coating or face on upstream slope of dam.

FACT SHEET 1. Document summarizes principal facts and significant factual, legal, methodological and policy questions considered in preparing draft permit and tells how public may comment; 2. Document that must be prepared for all draft individual permits for NPDES major dischargers, NPDES general permits, NPDES permits that contain variances, NPDES permits that contain sewage sludge land application plans and several other classes of permittees.

FACTOR One of elements, circumstances or influences which contribute to produce result.

FACTOR OF SAFETY Ratio of ultimate strength of material to allowable or working stress.

FACTORED LOAD Load multiplied by certain factor designated by codes of practice to determine strength of structural members.

FACTORY FINISHED Finished in factory prior to installation.

FACULA Bright region of photosphere seen in white light seldom visible except near solar limb.

FACULTATIVE Able to perform particular life function or to live generally, in more than one way.

FACULTATIVE BACTERIA Bacteria that can live under aerobic or anaerobic conditions.

FACULTATIVE POND Waste stabilization pond used for biological treatment of industrial and domestic wastewater where top layer is aerobic and bottom layer is anaerobic.

FAHRENHEIT 1. Unit of temperature with 32 °F and 212 °F as freezing and boiling points of water, respectively; 2. To convert Celsius temperature to Fahrenheit, multiply it by $9/5$ and then add 32.

FAIL-SAFE That does not cause undue damage in event of failure.

FAILURE 1. State or condition of not meeting desirable or intended objective; 2. Incident resulting in uncontrolled release of water from dam more severe and hazardous than breach and destroyed and made useless (i.e., ceases to function as dam).

FAILURE ANALYSIS Process of collecting and analyzing data to determine cause of failure, often with goal of determining corrective actions or liability.

FAILURE MODE Physically plausible process for dam failure resulting from existing inadequacy or defect related to natural foundation condition, dam or appurtenant structures design, construction, materials incorporated, operations and maintenance or aging process, which can lead to uncontrolled release of reservoir.

FAILURE POTENTIAL ASSESSMENT Judgment of potential for failure of essential element within expected life of project where five terms are used to describe assessment: negligible, low, moderate, high and urgent.

FAILURE RATE Frequency where engineered system or component fails.

FAIR Usually used at night to describe less than 3/8 opaque clouds, no precipitation, no extremes of visibility, temperature or winds that describes generally pleasant weather conditions.

FALL 1. Rapid change in elevation; 2. Amount of slope given to item to allow for adequate flow without ponding.

FALL LINE 1. Imaginary line marking boundary between ancient, resistant crystalline rocks of Piedmont province of Appalachian Mountains and younger, softer sediments of Atlantic Coastal Plain province in Eastern United States; 2. Along rivers, this line commonly is reflected by waterfalls; 3. Skiing term, indicating line of steepest descent of slope.

FALL VELOCITY Falling or settling rate of particle in given medium.

FALL WIND Strong, cold, downslope wind.

FALL-STREAK HOLE Large gap, usually circular or elliptical, that can appear in cirrocumulus or altocumulus clouds.

FALLING HEAD PERMEAMETER Apparatus used to measure flow rate associated with passage of water through granular solid under action of diminishing pressure head.

FALLING LIMB Portion of hydrograph where runoff is decreasing.

FALLING OBJECT PROTECTION STRUCTURE Shields equipment operator from danger above.

FALLING PRESSURE Indicates mass of atmosphere above point is decreasing, that creates convergence of atmosphere with falling pressure in center.

FALLING TIDE Movement of tidal current away from shore or down tidal river or estuary.

FALLOUT Event of small airborne particles falling to ground in significant quantities as result of major activity (i.e., volcano eruption, sandstorm, nuclear explosion).

FALLOW Cropland, tilled or untilled, allowed to lie idle during whole or greater part of growing season.

FALLSTREAK Streaks or wisps of precipitation falling from cloud but evaporating before reaching ground.

FALSE BOTTOM Internal partition in container, enabling concealment of objects from cursory examination of item's contents.

FALSE SET Temporary support for forepoles used in tunneling soft ground.

FALSEWORK Temporary framework used in building of bridges and arched structures to hold items in place until structure is able to support itself.

FAN Electrical or mechanical device for moving air.

FANNING 1. Pattern of plume dispersion in stable atmosphere, where plume fans out in horizontal and meanders about at fixed height; 2. Very small fragment of tea leaf.

FAR-SIDE STOP Transit stop immediately after intersection.

FARAD Unit of electric capacitance of capacitor where 1 coulomb of charge causes potential difference of 1 volt.

FARE Money paid for transport.

FARE COLLECTION SYSTEM Components that automate fare payment of public transportation network.

FAREBOX Box where passengers drop their bus fare.

FARM DRAIN Artificial watercourse on farm that contains no natural portions from its confluence with river or stream to its headwaters.

FARM DUTY WATER Volume of water seasonally delivered for irrigation.

FARM POND Pool of water formed by dam or pit, to supply water for livestock, recreation. and wildlife and to control gully erosion.

FARM WATER LOSS Water delivered to farm that is not made available to crop to be irrigated.

FASCINE Cylindrical bundle of small sticks of wood, bound together.

FAT Specialized animal tissue with high oil content, used for long-term storage of energy.

FATAL ACCIDENT RATE Indicator of accidents used to classify dangerousness that's measure of individual risk expressed as estimated number of fatalities per 100 million hours (approximately 1,000 employee working lifetimes) of this same activity.

FATAL FLAW 1. Any problem, lack or conflict (real or perceived) that will destroy solution or process; 2. Negative effect that cannot be offset by any degree of benefits from other factors.

FATALITY RATE Multiplication factor based on estimated severity of flood, potential warning times and judgment related to understanding of flood severity.

FATHOM Unit of length equal to 6 feet.

FATHOMETER Depth finder that uses sound waves to determine depth of water.

FATIGUE Material failure (e.g., cracking) caused by stress on material.

FATIGUE CRACKING Series of interconnected, sealed pavement distress cracks in road surface.

FATIGUE DESIGN LIFE Number of years or cycles that design is expected to resist fatigue failure.

FATIGUE FAILURE Tendency of material to fracture by means of progressive brittle cracking under repeated alternating or cyclic stresses of intensity considerably below normal strength.

FATIGUE LIFE Number of cycles of fluctuating stress and strain of specified nature that material will sustain before failure occurs.

FATIGUE LIMIT Highest stress that material can withstand for infinite number of cycles without breaking.

FATIGUE LIMIT STATE Maximum stress material can withstand cyclic stress without breaking.

FATIGUE RESISTANCE Skill of material to withstand stresses.

FATIGUE STRENGTH Highest stress that material can withstand for given number of cycles without breaking.

FATIGUE TEST Process of progressive localized permanent structural change occurring in material subjected to conditions that produce fluctuating stresses and strains.

FATIGUED Weakening of material caused by repeatedly applied loads.

FAUCET Exposed plumbing fitting (e.g., tap, spigot).

FAUCET EAR Projection on faucet that serves as means of attachment.

FAULT 1. Fracture or fracture zone in Earth along which there has been displacement of two sides relative to one another and that is parallel to fracture; 2. Break in rocks along which movement has occurred; 3. Shear with significant continuity that can be correlated between observation points.

FAULT BRECCIA Rock consisting of angular clasts that was formed by tectonic forces.

FAULT PLANE Surface formed between two rock blocks that slip one with other during earthquake.

FAULT SPRING Spring sourced by aquifer encountering vertical fracture in Earth's crust.

FAULT, ACTIVE Fault which, because of its present tectonic setting, can undergo movement from time to time in immediate geologic future.

FAULT, CAPABLE 1. Active fault judged capable of producing macroearthquakes; 2. Structural relationship to capable fault such that movement on one fault could be reasonably expected to cause movement on other; 3. Established patterns of microseismicity that define fault, with historic macroseismicity that can reasonably be associated with fault.

FAULT-BLOCK Uplifted section of rock bounded on both sides by faults.

FAULTING Movement which produces relative displacement along fracture in rock.

FAUNA All animals associated with given habitat, country, area or period.

FDA ACTION LEVEL Regulatory level recommended by U.S. Environmental Protection Agency (EPA) for enforcement by Food and Drug Administration (FDA) when pesticide residues occur in food commodities for reasons other than direct application of pesticide.

FEASIBILITY Determination that something can be done.

FEASIBILITY ESTIMATE Estimate used for determining economic feasibility of project, probable sequence and cost for construction of project and as guide in choice between alternative locations or plans.

FEASIBILITY PHASE Precedes design phase and where project concept is examined from different perspectives to assess advisability of proceeding with it.

FEASIBILITY STUDY 1. Analysis of practicability of proposal that's required in some planning processes to examine situation and determine if workable solution can be developed and implemented usually recommends selection of cost-effective alternative; 2. Small-scale investigation of problem to ascertain whether proposed research approach is likely to provide useful data.

FEASIBILITY STUDY/REPORT Documented output of feasibility study that summarizes results to review proposed solution and investigate project alternatives for purpose of identifying if project is really feasible, cost-effective and profitable.

FEASIBLE That can be done in practice.

FEATHER To blend edge of new material smoothly into old surface.

FEATURED Of high public interest and demand.

FECAL Bacteria (i.e., Streptococcus) found in intestinal tracts of warmblooded animals other than humans.

FECAL BACTERIA 1. Bacteria found in intestinal tracts of mammals; 2. Microscopic single-celled organisms (primarily fecal coliforms and fecal streptococci) found in wastes of warm-blooded animals where presence in water is used to assess sanitary quality of water for body-contact recreation or for consumption and possible presence of pathogenic (disease producing) organisms.

FECAL COLIFORM 1. Bacteria found in intestinal tracts of mammals; 2. Microscopic single-celled organisms (primarily fecal coliforms and fecal streptococci) found in wastes of warm-blooded animals where presence in water is used to assess sanitary quality of water for body-contact recreation or for consumption and possible presence of pathogenic (disease producing) organisms.

FECAL STREPTOCOCCAL BACTERIA Present in intestines of warmblooded animals and are ubiquitous in environment that's characterized as gram-positive, cocci bacteria that are capable of growth in brain-heart infusion broth.

FECES Digested waste material discharged from bowels.

FEDERAL EMERGENCY MANAGEMENT AGENCY Federal agency responsible for enforcing legislation for disaster and emergency planning and response that administers National Flood Insurance Program.

FEDERAL ENERGY REGULATORY COMMISSION Primary responsibility (established in 1977, replacing Federal Power Commission) of ensuring Nation's consumers adequate energy supplies at just and reasonable rates and providing regulatory incentives for increased productivity, efficiency and competition.

FEDERAL FACILITY Any buildings, installations, structures, land, public works, equipment, aircraft, vessels and other vehicles and property, owned by or constructed or manufactured for purpose of leasing to, federal government.

FEDERAL IMPLEMENTATION PLAN Plan to achieve attainment of air quality standards, used when state is unable to develop adequate plan.

FEDERAL LANDS HIGHWAY Financial resources and transportation engineering assistance for public roads that service transportation needs of Federal and Indian lands.

FEDERAL MOTOR VEHICLE CONTROL PROGRAM All federal actions aimed at controlling pollution from motor vehicles by such efforts as establishing and enforcing tailpipe and evaporative emission standards for new vehicles, testing methods development and guidance to states operating inspection and maintenance programs.

FEDERAL ORGANIZATIONS Agencies, departments or their components of Federal Government.

FEDERAL SNOW SAMPLER Snow sampler consisting of five or more sections of sampling tubes, one which has steel cutter on end and combined snowpack measuring depth is 150 inches.

FEDERAL WATER POLLUTION CONTROL ACT OF 1948 Establishes basic structure for regulating discharges of pollutants into waters of United States and regulating quality standards for surface waters.

FEDERAL-AID HIGHWAY Financial assistance for construction, maintenance and operation of highways.

FEDERAL-AID SYSTEM Authorized funds by Congress to assist States in construction, reconstruction and improvement of highways and bridges.

FEED PIPE Pipe for supplying boiler.

FEED PUMP Pump that supplies water to boiler to produce steam.

FEEDBACK 1. Critical assessment of process or activity or of their results; 2. To generate high-frequency sound by allowing speaker to cause vibration of sound generator of musical instrument connected by amplifier to speaker.

FEEDER Tributary stream.

FEEDER BANDS 1. Lines or bands of low-level clouds that move (feed) into updraft region of thunderstorm, usually from east through south (i.e., parallel to inflow); 2. Spiral-shaped bands of convection surrounding and moving toward, center of tropical cyclone.

FEEDER SERVICE Vehicles that transfer cargo to/from larger vessels.

FEEDLOT Confined area for controlled feeding of animals for market.

FEEDSTOCK Any bulk raw material constituting principal input for industrial process.

FEEDWATER Water added to boiler to replace that lost during operation.

FEEDWATER HEATERS Power plant component used to preheat water delivered to steam generating boiler.

FEEDWATER SYSTEM System used to optimize thermal efficiency of steam cycle.

FEEDWATER TREATMENT Water treatment used to control alkalinity, prevent scaling, correct pH and to control conductivity.

FELDSPAR Group of light-colored minerals often found as crystals in intrusive igneous rocks that's most common rock-forming mineral.

FEMA Federal agency responsible for enforcing legislation for disaster and emergency planning and response that administers National Flood Insurance Program.

FEMALE Component having end that encloses similar component.

FEMALE COUPLING Internally threaded conduit.

FEMALE THREAD Helical groove in cylindrical hole.

FEMTO Prefix symbolizing 10^{-15}.

FEN Peat-accumulating wetland that generally receives water from surface runoff and (or) seepage from mineral soils in addition to direct precipitation.

FENDER 1. Panel of car that encloses wheel area; 2. Structure that protects bridge structure; 3. Shield over bicycle wheel that protects rider.

FERMENTATION Any of many anaerobic biochemical reactions where enzyme catalyses conversion of one substance into another.

FERREL CELL Middle latitude cell marked by sinking motion near 30° and rising motion near 60° latitude.

FERRIC SULFATE Soluble iron salt.

FERROUS METALS Magnetic metals derived from iron or steel.

FERROUS SULFATE Soluble iron salt.

FERRYBOAT Vessel used to ferry passengers, vehicles or goods across open water.

FERTILIZER Any of large number of natural or synthetic materials, including manure and nitrogen, phosphorus and potassium compounds, spread on or worked into soil to increase its fertility.

FESCUE Hardy grass that's member of genus Festuca.

FESCUE GRASS Shade tolerant grass.

FETCH 1. Straight line distance across body of water subject to wind forces; 2. Straight line space between rim of dam as well as farthest reservoir shoreline; 3. Length over which wind can create waves; 4. Effective distance which waves have traversed in open water, from their point of origin to point where they break.

FEW National Weather Service convective precipitation descriptor for 10% chance of measurable precipitation (0.01 inch) that's used interchangeably with isolated.

FEW CLOUDS Official sky cover classification for aviation weather observations that's descriptive of sky cover of 1/8 to 2/8.

FIBER Thread/filament from which textile, vegetable tissue or mineral substance is formed.

FIBER-REINFORCED COMPOSITE Composite building material that consists of fibers, matrix and modifiers.

FIBERGLASS Composite material made from fine fibres of spun glass held together with resin.

FIBRIL Linear pattern in H-alpha chromosphere of Sun, as seen through H-alpha filter, occurring near strong sunspots and plage or in filament channels.

FIBROSIS Formation of (excess) fibrous connective tissue in organ.

FIDELITY Degree to which system accurately reproduces input.

FIELD Wide, open space usually used to grow crops or to hold farm animals; 2. Space throughout which force works.

FIELD CAPACITY 1. Amount of water held in soil against pull of gravity; 2. Depth of water retained in soil after ample irrigation or heavy rain when rate of downward movement has substantially decreased.

FIELD CARRYING CAPACITY Amount of soil moisture or water content held in soil after excess water has drained away and rate of downward movement has decreased.

FIELD CHARACTERIZATION Review of historical, on-and off-site, as well as surface and subsurface data and collection of new data to meet project objectives.

FIELD CHECK Onsite review of materials, work and workers.

FIELD DITCH Permanent irrigation ditch constructed in or with earth materials, to convey water from source of supply to field or fields in irrigation system.

FIELD DUTY WATER Area of land that can be irrigated with unit volume of irrigation water.

FIELD ENGINEER Engineer who works at project site.

FIELD GROUNDWATER VELOCITY True velocity of groundwater percolating through water-bearing substance.

FIELD MOISTURE Water held in soil above water table.

FIELD MOISTURE DEFICIENCY Amount of water required to restore moisture in soil to field moisture capacity.

FIELD ORDER Written instructions for making minor change in contract not involving adjustment to contract amount or any change in its completion time.

FIELD PERMEABILITY COEFFICIENT Rate of flow of water through unit cross-sectional area under unit hydraulic gradient at prevailing temperature.

FIELD SUPERINTENDENT Run day-to-day operations on construction site and control short-term schedule.

FIELD TEST Test of something conducted in conditions under which it was designed to actually operate, especially out in real world instead of in laboratory or workshop.

FIELD TESTING Test in environment where it will be used.

FIELD TILE Short lengths of clay pipe that are installed as subsurface drains.

FIELD TREATMENT Application of wood preservative performed onsite.

FIELD TRIAL Practical test conducted in field.

FIELD WASTE 1. Wastes generated as result of field activities; 2. Part of applied irrigation water that isn't consumed or evaporated as it passes over irrigated land.

FIELDWORK Work done out in real world rather than in controlled conditions.

FIFRA CANCELLATION Refers to Section 6 (b) of Federal Insecticide, Fungicide and Rodenticide Act which authorizes cancellation of pesticide registration if unreasonable adverse effects to environment and public health develop when product is used according to widespread and commonly recognized practice or if its labeling or other material required to be submitted does not comply with FIFRA provisions.

FIFRA PESTICIDE INGREDIENT Ingredient of pesticide that must be registered with EPA under Federal Insecticide, Fungicide and Rodenticide Act.

FIFTY-YEAR FLOOD Flood that has 2% chance of occurring in any given year.

FILAMENT 1. Long, sheet-like prominences nearly vertical to solar surface; 2. Mass of gas suspended over photosphere by magnetic fields and seen as dark lines threaded over solar disk.

FILAMENT CHANNEL Broad pattern of fibrils in chromosphere, marking where filament may soon form or where filament recently disappeared.

FILAMENTOUS Single algae cells that form long visible chains, threads or filaments.

FILE TRANSFER PROTOCOL Method of data transfer.

FILL 1. Manmade deposits of fill, natural soils, process of depositing or rock products and waste materials designed and installed in such manner as to provide drainage, yet prevent movement of soil particles due to flowing water; 2. Earth or broken rock structure or embankment; 3. Soil or loose rock used to raise grade that has no value except as bulk (i.e., height where substance will be set).

FILL CAP FILTER Reservoir tank opening filter.

FILL DAM Any dam constructed of excavated natural materials (e.g., earthfill dam, rockfill dam, tailings dam, hydraulic fill dam, rolled fill dam) or of industrial wastes.

FILL FACE Location where solid waste is unloaded and compacted at landfill.

FILL MATERIAL Any material used for primary purpose of replacing aquatic area with dry land or changing bottom elevation of water body.

FILL SLOPE Embankment face that raises desired road profile above ground.

FILLER GATE Small, sliding, sluice gates, placed at convenient point on main gate.

FILLING 1. Depositing dirt, mud or other materials into aquatic areas to create more dry land, usually for agricultural or commercial development purposes, often with ruinous ecological consequences; 2. Opposite of deepening; 3. General increase in central pressure of low pressure system.

FILM BADGE Piece of photographic film worn on torso to act as dosimeter and record wearer's radiation exposure.

FILM FLOW Gravity flow of continuous liquid film down solid tube having one free surface.

FILM PRESSURE Difference between surface tension of pure liquid and surface tension of liquid with substance adsorbed on it.

FILTER 1. Device which separates suspended, dissolved or particulate matter from fluid, solution or other substance; 2. One or more layers of granular material which is incorporated in embankment dam and is graded (either naturally or by selection) to allow seepage through or within layers while preventing migration of material from adjacent zones.

FILTER AID Group of inert materials that can be used in filtration pretreatment to aid filtering.

FILTER BAG Remove particles from gas or water systems and chemical-processing equipment.

FILTER BED Layer of gravel or other material where liquid is passed in order to remove impurities.

FILTER BLANKET Pervious materials that provide drainage, yet prevent soil movement.

FILTER BLOCK Filter designed for trickling filter of sewage treatment plant.

FILTER BOTTOM Filter is designed to go over lower intake.

FILTER BOX Filter that protects pump from debris.

FILTER CAKE 1. Deposit of mud on walls of drill hole; 2. Dense mass of substances that gradually accumulates on filter.

FILTER CLOGGING When particles fill voids of filter that results in deterioration and reduced capacity.

FILTER CLOTH Cloth that will separate fine particles from medium.

FILTER CRIB Water treatment works intake consisting of wooden crib in excavation in bed of stream.

FILTER CROP Crops planted close together, across slope, to retard runoff and causes deposit of soil.

FILTER EFFICIENCY Measure how efficiently filter removes particles from flow as particles pass through filter.

FILTER ELEMENT Apparatus that performs actual filtration and/or dewatering.

FILTER FABRIC Geotextiles that allows water to pass while preventing soil movement.

FILTER FLOODING Process of cutting back outflow from trickling filter to less than inflow, to increase level in filter above media.

FILTER GALLERY Alongside of stream, underground gallery or tunnel that collects water that filters through earth.

FILTER HOUSING Enclosure that covers filter element and directs flow.

FILTER LOADING Weight of biochemical oxygen demand in liquid.

FILTER MEDIUM Portion of filtration system that provides separation of liquid and solids.

FILTER OPERATING CONSOLE Filtration plant equipment where controls and gauges are mounted.

FILTER PACK Coarse sand or fine gravel placed between borehole wall and screen.

FILTER PERFORMANCE Variables that describe functions and characteristics of filter.

FILTER PLANT Facility that employs variety of techniques designed to deliver clear, safe potable water.

FILTER PONDING Flows that are not able to flow through certain filters, due to clogging, that create ponds.

FILTER POOLING Formation of ponds on surface of trickling filter, due to biofilm growth and media degregation.

FILTER PRESS Apparatus that separates liquids and solids using pressure filtration.

FILTER PRESSURE DIFFERENTIAL Differential pressure across filter element.

FILTER RATE Amount of flow filtered over given period of time.

FILTER RATED FLOW Maximum flow rate of fluid that filter was created to process.

FILTER RATING Rated capability of filter to capture particles.

FILTER RUN Time between between backwashing of filter media or changing of filter medium.

FILTER STRAINER Apparatus employed to capture much larger contaminants in flow.

FILTER STRIPS Strips of grass or other close-growing vegetation that separates waterway (ditch, stream, creek) from intensive land use area (e.g., subdivision, farm) used for removing sediment, organic matter and other pollutants from runoff and wastewater.

FILTER UNDERDRAINS System for gathering water that offer low head loss and equal distribution of backwash air or water.

FILTER UNLOADING Release of contaminant that was initially captured by filter medium.

FILTER ZONE 1. One or more layers of granular material that is incorporated in embankment dam and is graded (either naturally or by selection) to allow seepage through or within layers while preventing migration of material from adjacent zones; 2. Layer or combination of layers of pervious materials designed and installed in such manner as to provide drainage, yet prevent movement of soil particles due to flowing water.

FILTER-PRESSED SLUDGE Sludge that has been dewatered in filter press.

FILTERABLE SOLIDS Solids kept on membrane.

FILTERED Pertains to constituents in water sample passed through filter of specified pore diameter, most commonly 0.45 micrometer or less for inorganic analytes and 0.7 micrometer for organic analytes.

FILTERED WASTEWATER First filtering stage of wastewater that uses mechanical filtering to remove suspended solids.

FILTERED, RECOVERABLE Amount of given constituent in solution after part of representative water suspended sediment sample that has passed through filter has been extracted.

FILTRATE Liquid that has been passed through filter and has been separated from filtride.

FILTRATION 1. Mechanical separation of liquid from undissolved particles floating in it; 2. Treatment process, under control of qualified operators, for removing solid (particulate) matter from water by means of porous media (e.g., sand, manmade filter) that's often used to remove particles that contain pathogens.

FILTRATION FILTER RATING Proportion of quantity of particles equal to and greater than given size in influent fluid to number of particles equal to and greater than same size in effluent fluid.

FILTRATION PLANT Facility that works to filter and purify by removing chemicals, hazardous materials and toxic matters from source.

FILTRATION RATE Rate of application of flow to filter.

FILTRATION RATIO Ratio of number of particles greater than given size in influent fluid to number of particles greater than same size in effluent fluid.

FILTRATION SPRING Springs with small flow rates where source water has filtered through permeable earth.

FIN DRAIN Narrow trench back filled with single size aggregate to aid passage of groundwater and/or surface water flow along drain system.

FINAL ACCEPTANCE Acceptance of work as completed, including any deficiencies known to exist.

FINAL ACUTE VALUE Estimate of concentration of toxicant corresponding to cumulative probability of 0.05 in acute toxicity values for all genera for which acceptable acute tests have been conducted on toxicant.

FINAL APPROACH Last part of landing before touchdown, involving descending flight path directly in direction of runway.

FINAL CLOSURE Occurs when all hazardous waste management units at facility are closed according to closure regulations.

FINAL COMPLETION Satisfaction or waiver of all of conditions for completion of project.

FINAL COVER Multi-layered system of various materials which are primarily used to reduce amount of storm water that will enter landfill after closing.

FINAL CURE Time needed to get substance or mixture to react and achieve 100% of its rated physical properties.

FINAL DESIGN When all key issues have been broadly considered in detailed design.

FINAL EFFLUENT After pollutants have been removed from effluent.

FINAL FILTER Final stage in multi-stage filter system.

FINAL INSPECTION Last review of product, including requests from customers, before payment.

FINAL LAYOUT Drawing showing completed layout of connected parts.

FINAL PAYMENT Payment that brings balance to zero or completely satisfies obligation.

FINAL SEDIMENTATION Separates activated solids (i.e., sludge) from water in settling tank.

FINAL SEDIMENTATION BASIN Where effluent from waste effluent oxidizing procedure is passed in tank to eliminate settleable solids.

FINAL SETTLING BASIN Where effluent from waste effluent oxidizing procedure is passed in basin to eliminate settleable solids.

FINANCIAL ANALYSIS Procedure that considers only tangible factors when evaluating various alternatives.

FINANCIAL ASSURANCE FOR CLOSURE Documentation or proof that owner or operator of facility (e.g., landfill, waste repository) capable of paying projected costs of closing facility and monitoring it afterwards.

FINDING OF NO SIGNIFICANT IMPACT Document prepared by federal agency showing why proposed action would not have significant impact on environment and thus would not require preparation of Environmental Impact Statement.

FINE GRADING Cut or fill material to match as close as possible to finish elevations.

FINE GRAINED Grain diameter between 0.004 inch and 0.016 inch for sedimentary rocks or 0.039 inch for igneous or metamorphic rocks.

FINE GRAINED SAND-SIZED Less than 10% of medium and coarse grained sand sizes are present.

FINE GRAINED SOIL Soil with majority of smaller grain sizes.

FINE GRAVEL Stone consisting of particles with diameter of 0.08 to 0.25 in.

FINE GRAVEL-SIZED Particles greater than No. 4 sieve but less than 3/4 of inch in diameter.

FINE RACK Grid of bars having clear spaces between bars of 1 in. (25 mm) or less.

FINE SAND Sand composed of grains ranging from 0.008 to 0.002 inches in diameter.

FINE SCREEN Screen openings typically ranging from 0.06 to 0.25 in.

FINE SEDIMENT LOAD Portion of suspended load of stream composed of particles not big enough to be deposited in altering parts stream bed.

FINELY GRADED AGGREGATE Aggregate of small particles added to fill cracks, minor depressions and holes.

FINENESS 1. Ratio of length to width; 2. Ratio, in precious metal, of primary metal to any additives or impurities.

FINENESS MODULUS Empirical figure obtained by adding total percentage of sample of aggregate retained on each of specified series of sieves and dividing sum by 100.

FINES 1. Portion of soil finer that No. 200 U.S. Standard sieve; 2. Clay or silt particles in soil.

FINGER DRAINS Series of parallel drains of narrow width (instead of continuous drainage blanket) draining to downstream toe of embankment dam.

FINISH COAT Final coating applied.

FINISH DATE Calendar date that goal, objective or project is to be completed.

FINISH GRADE Final elevation and contour of ground after cutting or filling and conforming to proposed design.

FINISH GRADING Addition or removal of soil within two inches of existing grade to allow for proper drainage.

FINISH-TO-FINISH RELATIONSHIP Logical relationship where successor activity cannot finish until predecessor activity has finished.

FINISH-TO-START RELATIONSHIP Logical relationship where successor activity cannot start until predecessor activity has finished.

FINISHED GRADE Elevation or surface of earth after all earthwork has been completed (also finish grade) required by specifications.

FINISHED SIZE Dimensions of item in its final form.

FINISHED WATER Water that has passed through all processes in water treatment plant and is ready to be delivered to consumers.

FINISHER Craftsman who completes something.

FINISHING 1. Last treatment or coating of surface; 2. Third stage in conduits including setup of fixtures as well as other open parts.

FINITE DIFFERENCE METHOD Numerical model that uses mathematical technique to obtain approximate solution to governing partial differential equation (in space and time) on regular spaced mesh-or grid-points.

FINITE ELEMENT METHOD Method of solving governing equations of numerical model by dividing spatial domain into elements in each of which solution of governing equations is approximated by some continuous function on irregular spaced mesh-or grid-points.

FINITE STRIP METHOD Variant of finite element method used for bridges, structure and beam design and structural analysis.

FIRE Chemical reaction involving bonding of oxygen with carbon or other fuel, with production of heat and presence of flame or smouldering.

FIRE ALARM Device that warns people in building of possible fire or other condition requiring evacuation.

FIRE ALARM INDICATOR Device that reveals place of fire inside secure region.

FIRE ALGAE Free-swimming unicells characterized by red pigment spot.

FIRE AREA Aggregate floor area enclosed and bounded by fire walls, fire barriers, exterior walls or fire-resistance-rated horizontal assemblies of building.

FIRE BARRIER Fire-resistant structure used to prevent spread of fire for prescribed period of time.

FIRE BOOSTER High pressure water pumps designed to increase fire fighting capacity of building by boosting pressure in hydrant service when mains is not enough or when tank fed.

FIRE BOSS Person in charge of safety of mine (e.g., excavation).

FIRE BREAK Fire resistant fitting or construction.

FIRE BROOM Broom used for beating out flames in tall grass, sweeping or scraping duff, initial attack, controlling small "break-overs" and mop-up.

FIRE CAMP Campsite for firefighters and support personnel.

FIRE CANOPY Horizontal, fire-resistive construction designed to prevent flames from window from igniting contents of floors above.

FIRE CISTERN Supply tank for firefighting.

FIRE CLIMAX Climax community for which fire is dominating control factor.

FIRE COMPARTMENT Spread of fire can be restricted by sub-dividing buildings into number of discrete compartments.

FIRE CONTROL Prevention and monitoring of forest fires and grass fires.

FIRE CONTROL DAMPER Apparatus designed to close air duct in event of fire.

FIRE CONTROL EQUIPMENT Instruments used in forest fire control.

FIRE CONTROL IMPROVEMENTS Structures primarily used for control of forest fires.

FIRE CONTROL PLANNING Design of facilities and processes to shield woods and similar land areas from fire.

FIRE CUT Diagonal cut along end of joist or beam where it enters masonry wall.

FIRE DAMAGE Physical damage to property as result of burning.

FIRE DAMPER Damper that closes off air duct automatically in event of fire so as to restrict passage of fire and smoke.

FIRE DANGER Description of combination of constant and variable factors that affect initiation, spread and ease of control of wildfire.

FIRE DANGER RATING System that allows fire managers to estimate fire danger for given area.

FIRE DANGER RATING AREA Geographic area of relatively homogeneous climate, fuels and topography, tens of thousands of acres in size, where fire danger can be assumed to be uniform.

FIRE DECK Parting line of block and head of engine where reliable temperatures and maximum pressures of combustion are created.

FIRE DEMAND 1. Amount of water required to extinguish fire; 2. Water required for fire fighting in given area.

FIRE DEMAND RATE Rate of flow desired in selected pressure for fire suppression.

FIRE DEPARTMENT CONNECTION Component in system where fire department connects hose lines from pumper truck.

FIRE DETECTOR Apparatus for detecting one or more of products or phenomena resulting from fire, such as smoke, heat, infrared and/or ultraviolet light radiation or gas.

FIRE DEVIL Whirling column of fire found in large fiery areas such as wildfires.

FIRE DISTRIBUTION SYSTEM System of conduits or mains to supply water for fire fighting.

FIRE DISTRICT Entity responsible for prevention and extinguishment of fires and regulation of fire hazards within geographic boundaries.

FIRE EDGE Boundary of fire at given moment.

FIRE ENDURANCE Measure of elapsed time during which material or assembly continues to exhibit fire resistance under specified conditions of test and performance.

FIRE ENGINE Vehicle used by firefighters to pump water to fight fire.

FIRE ENGINEERING Application of science principles to protect people, property and their environments from harmful and destructive effects of fire and smoke.

FIRE ESCAPE Device used for escape from structure during fire.

FIRE EXTINGUISHER Mobile apparatus that can be used to put out fire.

FIRE FIGHTING SHAFT Shaft that provides fire and rescue service with safe area from which to undertake firefighting operations.

FIRE FLOW Quantity of water available for fire-protection purposes in excess of that required for other purposes.

FIRE FLOW TEST Performed to determine maximum flow rate water distribution system can deliver to make sure that adequate water supply is available.

FIRE GRADING Designation based on contents and operations conducted in building or structure or on flame-spread rating of interior contents.

FIRE GUARD Mesh screen placed in front of fire to prevent sparks escaping.

FIRE HAZARD Something that could easily catch or cause fire.

FIRE HEADQUARTERS Location for operations against fire.

FIRE HOSE REEL Hose reel located strategically in building to ensure proper coverage of water to combat fire.

FIRE HYDRANT Device used by firefighters to obtain water from main supply, consisting of barrel, connections for hose and valve assembly.

FIRE INSPECTION Examination of location to determine risk of fire breaking out and equipment and procedures to be used in event of fire.

FIRE LIABILITY Legal obligation for taking reasonable measures to prevent fires.

FIRE LIMITS Boundaries drawn around center of city where buildings had to be constructed of brick or stone rather than wood.

FIRE LINE Gap in vegetation or other combustible material that acts as barrier to slow or stop progress of fire.

FIRE LOAD Quantity of combustible matter in given area or quantity of heat that can be generated by its combustion.

FIRE MAIN Conduit conveying water to extinguish fire.

FIRE MODELLING Simulation developed to provide detail of fire and gas flow.

FIRE PACK Backpack designed for use by firefighters that includes tools and supplies.

FIRE PERFORMANCE CHARACTERISTIC Response of product, material or assembly of such products and/or materials to prescribed source of heat or flame under controlled fire conditions.

FIRE POINT Lowest temperature where vapor of that fuel will continue to burn for at least 5 seconds after ignition by open flame.

FIRE PRESSURE Pressure needed in water main to transfer water long distances or deliver at high pressure.

FIRE PREVENTION Educating public to take precautions to prevent potentially harmful fires and be educated about surviving them.

FIRE PREVENTION CODE OR ORDINANCE Law enacted by political jurisdiction to enforce fire prevention and safety regulations.

FIRE PROGRESS MAP Map maintained on large fire to show at given times location of fire, deployment of suppression forces and progress of suppression.

FIRE PROPAGATION INDEX Propensity of material to support fire propagation beyond ignition zone in terms of chemical heat release rate.

FIRE PROTECTION Study and practice of mitigating unwanted effects of potentially destructive fires.

FIRE PROTECTION DISTRICT Entity created to provide fire prevention and/or protection.

FIRE PROTECTION ENGINEER Accredited professional who focuses on fire detection, suppression, mitigation, human behavior and maintaining tenable environment for evacuation.

FIRE PROTECTION RATING Time that materials and assemblies used as opening protection have withstood fire exposure.

FIRE PUMP Water pump hat provides pressure to combat fire.

FIRE PUMPER Vehicle (i.e., fire truck) designed primarily for firefighting operations.

FIRE RATING Grading of resistance to fire.

FIRE RESISTANCE Measure of its ability to withstand effects of fire.

FIRE RESISTANT RATING Duration for which passive fire protection system can withstand standard fire resistance test.

FIRE RESISTING FINISH Coating used for fire protection.

FIRE RESISTING FLOOR/WALL Flooring or wall that divides building into separate compartments to reduce spread of fire.

FIRE RETARDANT Any of several substances used to delay or prevent combustion when used as coating or component of combustible material.

FIRE RETARDANT COATING Coating applied to materials for fire protection.

FIRE RETARDANT RATING Evaluation of building materials and assemblies for fire resistance and protection.

FIRE RISK Possibility to cause fire.

FIRE SAFETY OFFICER Responsible for ensuring and maintaining fire safety for company or facility.

FIRE SCAR Damaged area of wood in tree or shrub, caused by fire.

FIRE SEASON Moves around country, waxing and waning, in particular area when fires are most prone to happen.

FIRE SEPARATION Construction that separates or subdivides building to prevent spread of fire.

FIRE SERVICE Service in fire department or other firefighting organization.

FIRE SETTING Method of traditional mining where fires were set against rock face and doused with liquid, to cause stone fracture by thermal shock.

FIRE SPRINKLER SYSTEM Active fire protection method providing adequate pressure and flow rate onto which fire sprinklers are connected.

FIRE STREAM Stream of extinguishing agent (e.g., water) after it leaves fire conduit and nozzle until it reaches desired point.

FIRE SUPPLY Amount of water necessary for fire fighting.

FIRE SUPPRESSION Systems used to extinguish or prevent spread of fire in building.

FIRE SUPPRESSION ORGANIZATION Management structure having specific responsibilities in fire suppression.

FIRE SWITCH Control switch for elevators that enables fire fighters to assume direct control.

FIRE TEST EXPOSURE SEVERITY Measure of degree of fire vulnerability.

FIRE TOOL CACHE Supply of fire tools and equipment assembled in planned quantities or standard units at strategic point for exclusive use in wildland operations.

FIRE TOWER Watchtower to search for fires in forest.

FIRE TRAIL Rural road built specifically for purpose of fire management purposes.

FIRE TRIANGLE Three elements necessary and sufficient to sustain fire: fuel, oxidizer and heat.

FIRE TRUCK Vehicle used by firefighters, fitted with firefighting gear, to gain access to upper stories of building.

FIRE VENT Vent cover that opens automatically when heat exceeds specified temperature.

FIRE VENTING Means of preventing spread of flames and smoke in event of fire.

FIRE WARDEN Individual employed to prevent or extinguish fires.

FIRE WATCH Act of watching for occurrence of fires.

FIRE WEATHER Forecast issued to inform area firefighting and land management agencies of conditions that are ideal for wildland fire combustion and rapid spread.

FIRE WEATHER STATION Sensor to quantify weather components significant to forest fire prevention and management.

FIRE WHIRLWIND Vortex of wind formed by fire.

FIRE WIND Thermally driven wind blowing radially inward toward fire, produced by horizontal temperature differences between heated air above fire and surrounding cooler free atmosphere.

FIRE-RATED SYSTEM Construction materials and layouts analyzed and rated according to fire safety standards.

FIRE-RESISTANT COATING Coating to prevent failure due to fires.

FIRE-RESISTANT FLUID Lubricants that are difficult to ignite and do not propagate flame from ignition source.

FIRE-RESISTANT GREASE Grease comprised of fire-resistant base oil and graphitic carbon or oleophilic graphite.

FIRE-RETARDANT-TREATED Combustion retardant of construction material (e.g., plywood).

FIRE-SERVICE CONNECTION Water connection for sole purpose of fire-fighting.

FIRE-SERVICE DETECTOR CHECK METER Device for use on fire service connections consisting of weighted check with disk meter in bypass.

FIRE-SERVICE METER Meter designed to measure both domestic and fire service water usage through single water line.

FIREBRAND 1. Any source of heat, natural or man-made, capable of igniting wildland fuels; 2. Flaming or glowing fuel particles that can be carried naturally by wind, convection currents or gravity into unburned fuels; 3. Torch or other burning stick with flame at one end.

FIREBREAK Area cleared of all flammable material to prevent fire from spreading across it.

FIREGROUND Region where fire-fighting operations are carried on.

FIREMAN'S PANEL Control unit that has specialized buttons and indicators for firemen.

FIREPROOF Resistant to damage from fire.

FIREPROOFING Fire-resistant coating or substance.

FIRESTORM Fire whose intensity is greatly increased by in rushing winds.

FIRM ENERGY Non-interruptible energy guaranteed by supplier to be available at all times, except for uncontrollable circumstances.

FIRM FIXED PRICE CONTRACT Contract that provides for price that's not subject to any adjustment on basis of contractor's cost experience in performing contract.

FIRM POWER Non-interruptible power guaranteed by supplier to be available at all times.

FIRM PRICE CONTRACT Contract (i.e., fixed price contract) that provides for price that's not subject to any adjustment on basis of contractor's cost experience in performing contract.

FIRM YIELD Maximum quantity of water that can be guaranteed with some specified degree of confidence during specific critical period.

FIRN Old snow on top of glaciers, granular and compact but not yet converted into ice that's transitional stage between snow and ice.

FIRST DRAW Water that comes out when tap is first opened.

FIRST FOOD USE Use of pesticide on food or in manner that otherwise would be expected to result in residues in food, if no permanent tolerance, exemption from requirement of tolerance or food additive regulation for residues of pesticide on any food has been established for pesticide.

FIRST LAW OF THERMODYNAMICS Law of physics that states that heat absorbed by system either raises internal energy of system or does work on environment.

FIRST RESPONDERS Individuals who are likely to witness or discover hazardous substance release and who have been trained to initiate emergency response sequence by notifying proper authorities of release.

FIRST STAGE BIOCHEMICAL OXYGEN DEMAND That part associated with biochemical oxidation of carbonaceous (as distinct from nitrogenous) material.

FIRST TRACK Right-of-way distance occupied by tracks or lanes.

FIRST-DRAW SAMPLE One-liter sample of tap water that has stood motionless in plumbing pipes for at least 6-hours and is collected without flushing tap.

FIRST-ORDER REACTION Reaction that proceeds at rate that depends linearly on only one reactant concentration.

FIRST-STAGE BIOCHEMICAL OXYGEN DEMAND Oxygen demand related to biochemical oxidation of carbonaceous substance.

FIRTH Narrow inlet of sea.

FISCHER & PORTER PUNCHED TAPE RECORDER GAUGE Precipitation gauge which converts weight into code disk position.

FISH AND WILDLIFE Any non-domesticated member of animal kingdom that includes, without limitation, any mammal, fish, bird, amphibian, reptile, mollusk, crustacean, arthropod or other invertebrate and that includes any part, product, egg or offspring thereof or dead body or parts thereof.

FISH COMMUNITY All members of specified group of fish present in specific area at specific time in common area.

FISH LADDER Inclined trough which carries water from above to below dam so that fish can easily swim upstream.

FISH SCREEN Obstacle designed to prevent fish from swimming or being drawn into structure.

FISH TOTAL LENGTH Straight-line distance from anterior point of fish specimens snout, with mouth closed, to posterior end of caudal (tail) fin, with lobes of caudal fin squeezed together.

FISH WEIR Type of fish ladder.

FISH & WILDLIFE SERVICE Principal federal agency responsible for conserving, protecting and enhancing fish, wildlife and plants and their habitats for continuing benefit of American people that's agency within U.S. Department of Interior.

FISHERY Aquatic region where certain species of fish lives.

FISHING Operation of recovering object left or dropped in drill hole.

FISHWAY Inclined trough which carries water from above to below dam so that fish can easily swim upstream.

FISSILE 1. Easily split along grain; 2. Capable of undergoing nuclear fission.

FISSION 1. Process where bacterium splits to form two daughter cells; 2. Process of splitting nucleus of atom into smaller particles.

FISSURE Crack, groove, deep furrow or opening in rock.

FISSURED SOIL Soil material that has tendency to break with little resistance.

FIT CLASSIFICATIONS Comparative tightness of assembly based on how mating parts interact.

FIT UP To prepare.

FITMENT Proper positioning and orientation of thing for it to serve its designed purpose.

FITTING 1. Any functional accessory; 2. Electrical accessory intended primarily to perform mechanical rather than electrical function; 3. Part of piping system that joins lengths of pipe (excluding valves).

FITTING GAIN Space in fitting for conduit length.

FITTING-UP BOLT Bolt used to hold members together temporarily while they are being connected permanently.

FIVE HUNDRED MB 1. Pressure surface (geopotential height) in troposphere equivalent to about 18,000 feet above sea level; 2. Level of atmosphere at which half mass of atmosphere lies above and half below, as measured in pressure units.

FIVE-HUNDRED-YEAR FLOOD Flood with 0.2% chance of being exceeded every year.

FIVE-YEAR, SIX-HOUR PRECIPITATION EVENT Maximum 6-hour precipitation event with probable recurrence interval of once in 5 years.

FIX Determination of location.

FIX SAMPLE Adding chemicals in field that prevent water quality indicators of interest in sample from changing before laboratory measurements are made.

FIXATION 1. Act of uniting chemically with solid substance or in solid form; 2. State of resistance to evaporation or volatilization by heat.

FIXATIVE Substance that fixes, protects or preserves.

FIXED 1. Chemically stable; 2. Not changing.

FIXED AMOUNT-FREQUENCY SCHEDULING Method of irrigation scheduling that involves water delivery at fixed rate or fixed volume and at constant intervals.

FIXED ASSET Assets that are purchased for long-term use rather than for sale.

FIXED AXLE GATE Gate having wheels or rollers mounted on end posts of gate, particularly suited for high head situations.

FIXED BEARING Bearing that prevents differential longitudinal translation to abutting elements.

FIXED BED AND BANKS Bed and banks of channel unable to translate due to water forces.

FIXED BED MODEL Model where bed and side materials are non-erodible and deposition does not occur.

FIXED BRIDGE Bridge with permanent vertical and horizontal alignment.

FIXED BUDGET Budget that does not change or flex when sales or some other activity increases or decreases.

FIXED CAPITAL Capital invested in fixed assets.

FIXED CAPITAL COST Capital cost that occurs at start of project regardless of size or architecture.

FIXED CARBON Solid combustible residue that remains after coal particle is heated and volatile matter is expelled.

FIXED CHARGE Unavoidable expense.

FIXED CONE VALVE Cylinder gate mounted with axis horizontal.

FIXED COST Cost of business that do not vary with output or sales.

FIXED COSTS Cost of business which does not vary with output or sales.

FIXED DISPLACEMENT HYDRAULIC SYSTEM Hydraulic system that generates continuous flow rate.

FIXED DISTRIBUTOR Distributor consisting of perforated pipes, notched troughs, sloping boards or sprinkler nozzles that discharges fluid at constant rate.

FIXED EARTH SUPPORT Deep piling such that soil under line of excavation provides required restraint against deformations and rotations.

FIXED EXPENSES Costs that do not fluctuate with changes in production level or sales volume.

FIXED FEE Set amount paid for work or service.

FIXED FILM DENITRIFICATION Anaerobic treatment process that uses attached growth.

FIXED GRATE Stationary grate.

FIXED GROUNDWATER Water assumed to be permanently attached to fine-grained particles.

FIXED GUIDEWAY Public transportation facility that uses separate right-of-way for exclusive public transportation use.

FIXED LADDER Vertical ladder mounted permanently to structure.

FIXED LIABILITIES Debts (e.g., bonds, mortgages or loans) that are payable over term exceeding one year.

FIXED LIGHT Light that emits constant beams.

FIXED NOZZLE Nozzle with fixed position.

FIXED OBSTRUCTION Pathway obstructed by non-movable object.

FIXED PACKER Adjunct of refuse container system that compacts refuse at site of generation into detachable container.

FIXED PRICE INCENTIVE FEE CONTRACT Contract where buyer or purchaser pays seller or provider fixed amount.

FIXED ROLLER GATE Gate having wheels or rollers mounted on end posts of gate, particularly suited for high head situations.

FIXED ROOF Roof permanently affixed to structure.

FIXED ROUTE Form of public transit bus or ferry service without any regular deviation from schedules or course.

FIXED SIGNAL Fixed location signal indicating condition affecting movement of train or engine.

FIXED SITES Monitoring sites in NAWQA Study Units at which most comprehensive suites of data are collected.

FIXED SOLIDS Residue that remain after firing sample in furnace.

FIXED TIME Times which can be relatively persistent in cycle.

FIXED WHEEL GATE Gate having wheels or rollers mounted on end posts of gate, particularly suited for high head situations.

FIXED-BAR GRILLE Grille with bars that can't be moved.

FIXED-LOCATION MONITORING Sampling of environmental or ambient medium for pollutant concentration at one location continuously or repeatedly.

FIXED-PIN HINGE Pin for light, domestic door hinges that cannot be removed.

FIXED-PLATFORM TRUCK Truck designed to utilize load platform.

FIXED-SPRAY NOZZLE Steady-output nozzle.

FIXED-STATION MONITORING Repeated long-term sampling or measurement of parameters at representative points for purpose of determining environmental quality characteristics and trends.

FIXING Something to aid attachment during construction.

FIXTURE 1. Something fixed in place (e.g., permanent appliance); 2. Lighting unit.

FIXTURE BRANCH Any conduit which connects several plumbing fixtures.

FIXTURE DRAIN Section of drainage conduit between trap serving fixture and another section conduit.

FIXTURE SUPPLY PIPE Water supply conduit connecting fixtures to fixture branch conduit.

FIXTURE UNIT Design factor equal to 1 ft^3 (7.5 gal) of water drained in 1-1/4 pipe over one minute.

FLAGELLATA Cell or organism with one or more whip-like appendages called flagella.

FLAGGER Person who uses flag or signal to control traffic.

FLAGMAN Person who carries flag for signalling.

FLAIL 1. Hammer hinged to axle so that's can be used to break or crush material; 2. Tool used for threshing.

FLAIL MILL Rotating gear which uses chains to open or distribute solid waste and shred or crush substances.

FLAME ARRESTER Device which allows gas to pass through it but stops flame in order to prevent larger fire or explosion.

FLAME IONIZATION DETECTOR Laboratory analytical method used as screening technique for semi-volatile organic compounds that are extractable from water in methylene chloride.

FLAME PHOTOMETRY Quantitative analysis of metals by passing solution of their salts into flame and measuring intensity of specific spectral lines.

FLAMEPROOF ENCLOSURE Enclosure for electrical apparatus that will withstand any explosion of prescribed flammable gas.

FLAMMABLE Any material that ignites easily and will burn rapidly.

FLAMMABLE LIQUID Combustible liquid that can easily catch fire.

FLANGE 1. Ridge that prevents sliding motion; 2. Rib or rim for strength or for attachments; 3. Rim or collar attached to one end of pipe to give support.

FLANGE UNIT Housing designed to enable mounting with bolts.

FLANGED JOINT Joint consisting of two companion flanges.

FLANGED PIPE Conduit provided with flanges at both ends.

FLANKING Erosion at end of structure due to flow.

FLANKING INLET Inlet for drainage at low point of road in sag vertical curve.

FLANKING LINE Line of cumulus or towering cumulus clouds connected to and extending outward from most active part of supercell, normally on southwest side.

FLAP 1. Anything broad and flexible that hangs loose; 2. Hinged leaf of table.

FLAP GATE 1. Gate hinged along one edge, usually either top or bottom edge; 2. Device that allows water to flow in one direction only.

FLAP VALVE Check valve put on end of conduit to allow flow out, but prevent anything from flowing in.

FLAPPER ACTION VALVE Valve where output signal control pressure is modulated with pivoted flapper.

FLARE 1. Control device that burns hazardous materials to prevent their release into environment that may operate continuously or intermittently, usually on top of stack; 2. Sudden eruption of energy on solar disk lasting minutes to hours, from which radiation and particles are emitted; 3. Widening of object with otherwise roughly constant width; 4. Source of brightly burning light or intense heat used to attract attention.

FLARE ANGLE Angle between wall and structure centerline.

FLARE OUT Part of aircraft landing sequence where nose is raised while engine idled.

FLARED APPROACH Right turning lane that shared to permit vehicles to turn right with other vehicles in lane.

FLARED FITTING Seals and holding fitting with preformed flare at terminus.

FLARED INTERSECTION Pavement widened at intersection area by one or more traffic lanes.

FLARED JOINT Joint between two conduits by flaring end of one to pass over other.

FLARELESS FITTING Seal and grip fitting that doesn't use flare.

FLARING INLET Converging entrance to conduit used to facilitate water entry.

FLASH Sudden, brief illumination of conductive channel associated with lightning, which may contain multiple strokes with their associated stepped leaders, dart leaders and return strokes.

FLASH COAT Thin layer of shotcrete for sealing or bonding.

FLASH DRYER Apparatus designed to remove free moisture for instant drying of materials.

FLASH EVAPORATOR Separation equipment used in chemical industry that allow for partial or total component separation.

FLASH FLOOD 1. Flood which follows within few hours (usually less than 6 hours) of heavy or excessive rainfall, dam or levee failure or sudden release of water impounded by ice jam; 2. Flood of short duration with relatively high peak rate of flow and local impact into normally dry area.

FLASH FLOOD GUIDANCE Forecast guidance produced by River Forecast Centers containing rainfall threshold values which must be exceeded in order to produce flash flood.

FLASH FLOOD STATEMENT Statement by National Weather Service which provides follow-up information on flash flood watches and warnings.

FLASH FLOOD TABLE Pre-computed forecast crest stage values for small streams for variety of antecedent moisture conditions and rain amounts.

FLASH FLOOD WARNING Issued to inform public, emergency management and other cooperating agencies that flash flooding is in progress, imminent or highly likely.

FLASH FLOOD WATCH Issued to indicate current or developing hydrologic conditions that are favorable for flash flooding in and close to watch area, but occurrence is neither certain or imminent.

FLASH MIXER Apparatus used to agitated mixture quickly after screening out debris and coagulation processes.

FLASH MULTIPLICITY Number of return strokes in lightning flash.

FLASH POINT Lowest temperature at which evaporation of substance produces sufficient vapor to form ignitable mixture with air.

FLASH RANGE Temperature range over which plant operates.

FLASHBOARD Board placed temporarily upon milldam to raise water above its usual level.

FLASHBOARD CHECK GATE Gate that opens to ease pressure on flashboard.

FLASHBOARDS Temporary structural barriers, consisting of either timber, concrete or steel, anchored to crest of spillway to raise retention water level (e.g., increase reservoir storage) but which may be quickly removed in event of flood by tripping device or by deliberately designed failure of flashboard or its supports.

FLASHER 1. Device that switches light on and off; 2. Indicator or turn signal.

FLASHING Blinking of light source.

FLASHING MODE Traffic signal repetitively turned on and off.

FLASHING RING Ring placed around pipe to secure to surface.

FLASHING-LIGHT SIGNALS Two red lights, flashing alternately, mounted in horizontal line.

FLASHY STREAM Stream that rapidly collects flows and produces flood peaks rapidly after precipitation event.

FLAT Having no variations in height.

FLAT RATE Charge that's same in all cases.

FLAT SLOPE For specific discharge, conduit sloped less in relation to critical value.

FLAT-CRESTED WEIR Measuring weir that's crest is horizontal and length is much larger than compared with height of water passing over it.

FLAT-PLATE SOLAR COLLECTOR Insulated metal box with cover and dark-colored absorber plate where solar radiation is absorbed by absorber plate and transferred to fluid that circulates through collector in tubes.

FLATBED Open freight vehicle with no sides that's designed to carry heavy and/or out-sized loads.

FLATCAR Railroad freight car without sides or roof.

FLATIRONS Triangular-shaped landforms along mountain ranges formed by erosion of steeply inclined rock layers or hogbacks.

FLEET ANGLE Angle between position of rope or cable at extreme end wrap on drum and line drawn perpendicular to axis of drum used to indicate how effective or efficient rope or cable is for raising load.

FLEETING AREA Area, within defined boundaries, where individual barges are moored and/or assembled for towing.

FLEXIBILITY Capable of being flexed or bent without breaking.

FLEXIBILITY FACTOR Ratio of flexibility of bend to that of straight pipe having same length and cross-section.

FLEXIBLE ARMORED REVETMENT Revetment using interlocking, precast concrete blocks that can bend to shape of land.

FLEXIBLE BUDGET Budget that adjusts or flexes for changes in volume of activity.

FLEXIBLE JOINT Assembly designed to safely absorb temperature expansion or contraction of construction materials, to absorb vibration, to hold together or allow movement.

FLEXIBLE MEMBRANE Semi-permeable material or material that acts as barrier to prevent transmission of certain substances, yet allows for passage of others.

FLEXIBLE PAVEMENT Surface made of mixture of aggregates and bitumen that's compacted on bed of granular layer.

FLEXIBLE PIPE 1. Pipe designed to transmit backfill load to soil at sides of pipe; 2. Conduit that can bend before breaking.

FLEXURAL MODULUS OF STRENGTH Ratio of stress to strain in flexural deformation.

FLEXURAL RIGIDITY Resistance offered by structure while undergoing bending.

FLEXURAL STRENGTH Stress in material just before it yields in flexure test.

FLIGHT 1. Planks used to gather and transfer settled sludge or floating scum; 2. Shaped material forming thread of screw.

FLIGHT LEVEL Internationally agreed standard altitudes that are based on atmospheric pressure rather than on true height.

FLIGHT PLAN Document filed by pilot prior to departure, containing essential information about flight.

FLIGHT SERVICE STATION Air traffic facility that provides services and information to aircraft pilots.

FLIGHT SEWER Sewer with step series that reduces velocity on steep grade.

FLIGHT VISIBILITY Average forward horizontal distance, from cockpit to aircraft in flight, where prominent unlighted objects may be seen and identified.

FLING Near field long period pulse from strong ground motion resulting in unidirectional ground heave after rupture.

FLIP BUCKET Energy dissipater located at downstream end of spillway and shaped so that water flowing at high velocity is deflected upwards in trajectory away from foundation of spillway.

FLOAT 1. Buoyant device used to support something in water or another liquid; 2. Trowel used for finishing concrete surfaces or smoothing plaster; 3. Funds committed to be paid but not yet paid.

FLOAT ARM Stick joining float sphere to inlet valve in toilet.

FLOAT BALL Sphere that automatically regulates level of liquid in tank.

FLOAT GAUGE Apparatus used to measure level within storage tank.

FLOAT GAUGING Measuring velocity of neutral buoyancy object and multiplying by average cross-sectional area to determine discharge.

FLOAT RECORDING PRECIPITATION GAUGE Rain gauge where rise of float within instrument with increasing rainfall is recorded.

FLOAT RUN Length timed for float to flow downstream.

FLOAT SWITCH Sensor switch used to detect level of liquid within tank.

FLOAT TUBE Conduit intake held below water surface by float.

FLOAT VALVE Mechanism, using lever and floating ball, for filling water tanks.

FLOATABLE DAYS Number of days during recreation season on which it is safe to allow floating activities on recreation facilities.

FLOATABLE FLOWS River flows which make rafting and other floating recreation possible.

FLOATING BOOM Temporary floating barrier used to contain spill.

FLOATING COVER Gas tight cover reducing harmful emissions.

FLOATING PAN Evaporation pan where evaporation is measured from water in pan floating in larger body of water.

FLOATING PIPELINE System of floats for conduits and electrical wires.

FLOATING ROOF Storage tank commonly used to store large quantities of petroleum products.

FLOC Clumps of soft or fluffy solids suspended in liquid sewage by biological or chemical action.

FLOCCULATION 1. Process by which clumps of solids in water or sewage aggregate through biological or chemical action so they can be separated from water or sewage; 2. Step in water filtration where alum is added to cause particles to clump together.

FLOCCULATION AGENT Process of destabilizing colloidal particles by neutralizing their surface charges.

FLOCCULATION LIMIT Soil water content when in state of deflocculated sediment.

FLOCCULATION RATIO Void ratio of soil in state of deflocculated sediment.

FLOCCULATION TANK Tank where floc is propagated.

FLOCCULATOR Machine that induces formation of floc.

FLOE Accumulation of frazil flocs (e.g., "pan") or single piece of broken ice.

FLOE ICE Flat mass of ice floating in body of water.

FLOOD 1. Temporary rise in water levels resulting in inundation of areas not normally covered by water; 2. Overflow of water from lake or other body of water due to excessive rainfall or other input of water; 3. Flowing in of tide, opposed to ebb.

FLOOD ATTENUATION Weakening or reduction in force or intensity of flood.

FLOOD BASALT Result of volcanic eruption(s) that coats large stretches of land or ocean floor with basalt lava.

FLOOD BASE ELEVATION Highest recorded flood elevation for area.

FLOOD BASIN System to control flooding by using submerged land to temporarily store flood waters.

FLOOD BENEFITS Determined value of flood control with regards to damages prevented and public safety.

FLOOD BOUNDARY Line drawn or outer edge of colored (inundation) area on inundation map to show limit of flooding.

FLOOD CATEGORIES Terms defined for each forecast point which describe or categorize severity of flood impacts in corresponding river/stream reach where each flood category is bounded by upper and lower stage where flood categories used in National Weather Service are: Minor Flooding - minimal or no property damage, but possibly some public threat, Moderate Flooding - some inundation of structures and roads near stream with some evacuations of people and/or transfer of property to higher elevations, Major Flooding - extensive inundation of structures and roads with significant evacuations of people and/or transfer of property to higher elevations, Record Flooding - flooding which equals or exceeds highest stage or discharge at given site during period of record keeping.

FLOOD CHANNEL Enlarged stream channel carrying water during floods.

FLOOD CONTROL Methods used to reduce or prevent detrimental effects of flood waters.

FLOOD CONTROL POOL Reservoir volume above active conservation capacity and joint use capacity reserved for flood runoff and then evacuated as soon as possible to keep that volume in readiness for next flood.

FLOOD CONTROL STORAGE Storage of water in reservoirs to abate flood damage.

FLOOD CREST Maximum height of flood wave as it passes certain location.

FLOOD CURRENT Movement of tidal current toward shore or up tidal river or estuary.

FLOOD DAMAGE REPORT (E-7) Report to be completed anytime there is reported flood damage or loss of life as direct result of flooding.

FLOOD ENVELOPE CURVE Mathematical expression that provides upper and lower bounds of observed maximum instantaneous peak discharges for hydrologic region.

FLOOD EVENT 1. Rise in streamflow to crest and its subsequent recession caused by precipitation, snow melt, dam failure or reservoir releases; 2. Event having some percent probability of occurring in any given year.

FLOOD FLOW Stream discharge during flood.

FLOOD FREQUENCY Probability that flood will occur in given year.

FLOOD FREQUENCY CURVE Graph with recurrence intervals of floods, showing number of times per year on average (plotted as abscissa) that floods of magnitude (indicated by ordinate) are equaled or exceeded.

FLOOD GATE Gate to control flood release from reservoir.

FLOOD HAZARD Hazards that occur due to contact with water that occur because of flooding.

FLOOD HYDROGRAPH Graph showing, for given point on stream, discharge, height or other characteristic of flood with respect to time.

FLOOD IRRIGATION Irrigation system where field is ponded with water to particular depth.

FLOOD LEVEL Highest level that flowing water will reach to cause sufficient inundation.

FLOOD LEVEL RIM Top edge where water can overflow from plumbing fixture or device.

FLOOD LOSS REDUCTION MEASURES Four basic strategies for reducing flood losses: prevention, property protection, emergency services and structural projects.

FLOOD OF RECORD Highest observed river stage/discharge at given location during period of record keeping.

FLOOD PEAK Maximum stage or mean discharge occurring in stream during given flood event.

FLOOD PLAIN Alluvial plain that may or may not experience occasional or periodic flooding.

FLOOD PLANE Position occupied by water surface of stream during particular flood.

FLOOD POOL Reservoir volume above active conservation capacity and joint use capacity reserved for flood runoff and then evacuated as soon as possible to keep that volume in readiness for next flood.

FLOOD POOL INDEX Ratio of flood control pool depth to depth below pool, multiplied by percent of time reservoir water surface will be within flood control pool.

FLOOD POTENTIAL OUTLOOK Issued to alert public of potentially heavy rainfall that could send area rivers and streams into flood or aggravate existing flood.

FLOOD PREVENTION Measures that are taken in order to keep flood problems from getting worse.

FLOOD PROBABILITY Probability distribution that describes flood will occur within given time period.

FLOOD PROBLEMS Problems and damages that occur during flood as result of human development and actions.

FLOOD PROFILE Graph of water surface elevation (ordinate) of flood plotted against distance downstream (abscissa).

FLOOD PROTECTION WORKS Structures constructed to protect public and infrastructure from flood damage.

FLOOD RESERVOIR Ponding area of water for surface runoff temporary storage.

FLOOD ROUTING Process of determining progressively timing, shape and amplitude of flood wave (i.e., hydrograph) as it moves past dam or downstream to successive points along river.

FLOOD SEVERITY UNDERSTANDING Understanding as to what degree flooding might affect downstream population.

FLOOD SOURCE AREA Drainage basin section where conditions favor flood condition development.

FLOOD SPREADING Flood management and water harvesting that increases groundwater recharge, makes soil more fertile and increases nutrients in soil.

FLOOD STAGE Established elevation where rise in water surface level creates hazard to lives, property or commerce.

FLOOD STAGE REPORT (E-3) Form that Service Hydrologist/Hydrology Focal Point completes to document dates where forecast points are above flood stage, as well as crest dates and stages.

FLOOD STATEMENT Statement issued to inform public of flooding along major streams where there is not serious threat to life or property.

FLOOD STORAGE Retention of water or delay of runoff either by planned operation, as in reservoir or by temporary filling of overflow areas, as in progression of flood wave through natural stream channel.

FLOOD SURCHARGE Storage volume between top of active storage and design water level.

FLOOD WARNING Statement issued to inform public of flooding along larger streams where there is serious threat to life or property.

FLOOD WATCH Statement issued to inform public and cooperating agencies that current and developing hydrometeorological conditions are such that there is threat of flooding, but occurrence is neither certain nor imminent.

FLOOD WAVE Rise in water by precipitation, snow melt, runoff, dam failure or reservoir releases.

FLOOD ZONE Area that borders stream and is subject to equal frequency floods.

FLOOD, 100-YEAR Flood level with 1% chance of being equaled or exceeded in any given year.

FLOOD, SAFETY EVALUATION Largest flood for which safety of dam and appurtenant structure is to be evaluated.

FLOOD-CONTROL STORAGE Storage of water in reservoirs to abate flood damage.

FLOOD-CONTROL WORKS Methods of flood control including planting vegetation to retain extra water, terracing hillsides to slow flow downhill and construction of floodways.

FLOOD-FREQUENCY CURVE Plot of flood discharge versus recurrence interval.

FLOOD-RELIEF CHANNEL Channel that bypasses flood water.

FLOODED FIXTURE Where liquid starts to overflow top or rim of structure.

FLOODED ICE Ice which has been flooded by melt water or river water and is heavily loaded by water and wet snow.

FLOODGATE Adjustable gate or valve used to control flow of water through sluice.

FLOODING Overflowing of water.

FLOODPLAIN Area of low-lying ground adjacent to river subject to 1% or greater chance of flooding in any given year.

FLOODPLAIN INFORMATION STUDIES Reports usually prepared by U.S. Army Corps of Engineers following survey of flood-impacted community.

FLOODPLAIN REGULATIONS Regulations that protect life, health and property from flooding.

FLOODPROOFING Process of making resistant to flooding.

FLOODWALL Long, narrow embankment usually built to protect land from flooding.

FLOODWAY Part of floodplain that facilitates passage of flood water discharge.

FLOOR DRAIN Plumbing fixture installed in floor to remove water.

FLOOR FLANGE Fitting on conduit to cover hole in floor where conduit passes.

FLOOR HOLE Opening measuring less than 12 in. (2.5 cm) in any floor, roof or platform where materials but not persons may fall (e.g., belt hold, pipe opening, slot opening).

FLOOR OPENING Opening measuring less than 12 in. (2.5 cm) in any floor, roof or platform where materials but not persons may fall (e.g., belt hold, pipe opening, slot opening).

FLOOR OUTLET Electrical outlet placed flush with floor surface.

FLOOR PLATE Plate set in or forming part of floor.

FLOOR SWEEP Capture of heavier-than-air gases that collect at floor level.

FLOORBEAM Alternate name for crossbeam.

FLORA All plant life (and bacteria) associated with given habitat, country or period.

FLOTATION 1. State of floating; 2. Ability to stay on surface of soft ground or snow; 3. Process of separating minerals by agitating mixture with water and detergents.

FLOURY SOIL Soil that's powder when dry and like clay when wet.

FLOW 1. Volume of water that passes given point within given period of time; 2. Qualitative reference of air parcel(s) with respect to its direction of movement.

FLOW AUGMENTATION Release of water stored in reservoir or other impoundment to increase natural flow of stream.

FLOW BOG Peat-bog with surface that fluctuates with every increase or diminution of water.

FLOW CELL Sensor that when immersed in fluid measures continuously.

FLOW CHANNEL Portion of flow net bounded by two adjacent flow lines.

FLOW CHART Diagram that represents algorithm, workflow or process.

FLOW COEFFICIENT Variable for efficiency of fluid flow via orifice.

FLOW CONTROL GATE Barrier that assists with flow in channel.

FLOW CONTROL VALVE Valve that regulates flow or pressure of fluid.

FLOW CURVE 1. Plot of stress vs. flow rate obtained from theological measurements; 2. Points obtained from standard liquid limit test and plotted on graph representing moisture content as ordinate on arithmetic scale and number of blows as abscissa on logarithmic scale.

FLOW DEMAND Flow necessary to meet requirements.

FLOW DIAGRAM Diagram representing flow or set of dynamic relationships in system.

FLOW DIVIDER VALVE Pressure compensated flow regulator that divides flow.

FLOW DURATION CURVE Cumulative frequency curve that shows percentage of time that specified discharges are equaled or exceeded at given location.

FLOW DURATION PERCENTILES Values on scale of 100 that indicate percentage of time for which flow is exceeded.

FLOW EQUALIZATION Process that minimizes variability of water and wastewater.

FLOW FAILURE Failure where soil mass moves over relatively long distances in fluidlike manner.

FLOW GRADIENT Process that follows path of steepest descent in landscape.

FLOW INDEX 1. Slope of water content versus percussion cup or cone penetration method.

FLOW INDICATOR Apparatus that signals whether flow exists in waterway.

FLOW LINE 1. Idealized path followed by particles of water; 2. Path that particle of water follows in its course of seepage under laminar flow conditions; 3. Pipe run that connects flowing well to storage tank.

FLOW MEASUREMENT DEVICES Instruments (e.g., flowmeters, weirs, calibrated bucket and stopwatch) that measure leakage quantities.

FLOW METER Measures, monitors and records rate of flow, pressure or discharge of liquids or gasses.

FLOW NET Graphical representation of flow lines and equipotential (piezometric) lines used in study of seepage phenomena.

FLOW PATH Subsurface course water molecule or solute would follow in given groundwater velocity field.

FLOW PRESSURE Pressure where valve is wide open and flowing full.

FLOW RATE Rate where fluid passes specified cross-section per unit time.

FLOW RATIO Ratio of traffic actual flow rate to saturation flow rate at intersection.

FLOW REGULATOR Maintains predefined flow rate near-constantly and mostly independently from prevailing line pressure.

FLOW RESISTANCE Measure of substances ability to resist flow.

FLOW SEPARATION Process by which separation eddy forms on windward or leeward sides of bluff objects or steeply rising hillsides.

FLOW SLIDE Failure of sloped bank of soil where movement of soil mass does not take place along well-defined surface of sliding.

FLOW SPLITTING Splitting of stable airflow around mountain barrier, with branches going around left and right edges of barrier, often at accelerated speeds.

FLOW STRUCTURE Structure of igneous rocks where stream lines or flow lines of magma are revealed.

FLOW SWITCH Device that monitors flow of air, steam or liquid.

FLOW WELL Well that naturally discharges water at surface.

FLOW-LINE AQUEDUCT Aqueduct where hydraulic gradient is lower than its crown that prevents aqueduct from flowing full.

FLOW-NOZZLE METER Differential pressure flow meter used for pipe flow measurement.

FLOW-PROPORTIONAL COMPOSITE SAMPLE Volume of sample changes in proportion to flow.

FLOWABLE Pesticide and other formulations where active ingredients are finely ground insoluble solids suspended in liquid.

FLOWAGE 1. Water that floods onto adjacent land; 2. Flowing or overflowing of liquid.

FLOWAGE EASEMENT 1. Right or easement to overflow, submerge or flood certain lands; 2. Right to prohibit building on certain floodways.

FLOWING ARTESIAN WELL Well drilled into confined aquifer with enough hydraulic pressure for water to flow to surface without pumping.

FLOWING WELL Well drilled into confined aquifer with enough hydraulic pressure for water to flow to and rises above surface without pumping (e.g., Artesian well).

FLOWMETER Gauge indicating velocity of wastewater moving through treatment plant or of any liquid moving through various industrial processes.

FLOWNET Graphical representation of two-dimensional steady-state groundwater flow.

FLOWPATH Underground route for groundwater movement, extending from recharge (intake) zone to discharge (output) zone (e.g., shallow stream).

FLOWPATH STUDY Network of clustered wells located along flowpath extending from recharge zone to discharge zone.

FLOWSHEET Schematic representation of how different stages in process are interconnected.

FLOWTHROUGH CHAMBER Top section of sedimentation chamber.

FLOWTHROUGH PROCESS TANK Tank that forms integral part of production process where there is steady, variable, recurring or intermittent flow of materials during operation of process.

FLOWTHROUGH TEST Toxicity evaluation where water is renewed constantly in test chambers.

FLOWTHROUGH TIME Time needed to get volume of liquid to pass through.

FLUCTUATING FLOWS Water released from dam that varies in volume with time.

FLUCTUATING ZONE Area of sandbar or vegetation zone within range of fluctuating flow.

FLUCTUATION Variation in water level, up or down, as result of project operation.

FLUE Pipe or duct that carries gaseous combustion products away from point of combustion.

FLUE GAS Gaseous combustion product (e.g., smoke) that comes out of chimney, stove or furnace.

FLUE GAS ANALYSIS Tests used both for efficiency and emissions purposes.

FLUE GAS DESULFURIZATION Technology that employs sorbent, usually lime or limestone, to remove sulfur dioxide from gases produced by burning fossil fuels.

FLUE GAS LOSS Sensible heat energy in flue gases carried away due to flue gas temperature.

FLUE GAS RECIRCULATION Nitrogen oxide control technique that significantly reduces emissions by recirculating flue gases from boiler exhaust duct into main combustion chamber.

FLUE GAS SCRUBBER Equipment for removing particulates from products of combustion from industrial gas output.

FLUE GAS WASHER Equipment for removing particulates from products of combustion from industrial gas output.

FLUE LINING Conduit installed inside of chimney, intended to contain combustion products, direct them to outside atmosphere and protect chimney walls from heat and corrosion.

FLUE PIPE Pipe where exhaust gases and other byproducts of combustion are expelled.

FLUE TERMINAL Device to improve draught of chimney.

FLUE TILE Tile used to line chimney flue.

FLUENCE Time integrated flux.

FLUFF Exceptionally combustible remnants from auto recycling.

FLUID Matter which flows (e.g., gas or liquid).

FLUID ANALYSIS Process where contamination amount of fluid might be discovered.

FLUID FRICTION Force that resists movement of solid object through fluid (i.e., liquid, gas).

FLUID MECHANICS Branch of mechanics dealing with gases and liquids, either at rest or in motion.

FLUID SAMPLE Fluid obtained for analyzing fluid properties and/or amount of contamination.

FLUIDIC Of or pertaining to devices that depends on pressures and flows of fluid for operation.

FLUIDICS Branch of engineering and technology concerned with construction of devices that use flow and pressure of fluid in circuits analogous to electronic ones.

FLUIDITY State of being fluid rather than viscous.

FLUIDIZED Mass of solid particles made to flow like liquid by injection of water or gas.

FLUIDIZED BED Industrial unit process where quantity of solid particulate substance is caused to float in stream of fluid in order to create conditions where many chemical, combustion and drying reactions take place efficiently.

FLUIDIZED BED INCINERATOR Incinerator that uses bed of hot sand or other granular material to transfer heat directly to waste.

FLUIDIZED-BED COMBUSTION Fuel particles suspended in heated, fluidity bed of ash and other particulate materials where jets of air are blown to provide oxygen required for combustion and/or gasification.

FLUME 1. Shaped, open-channel flow sections that force flow to accelerate for purposes of measurement; 2. Artificial channel, often elevated above ground used to carry fast flowing water.

FLUORESCENCE Emission of light by material when stimulated by absorption of radiation or of subatomic particle.

FLUORESCING DYES Non-protein molecules that absorb light and re-emit it at longer wavelength.

FLUORIDATION Addition of chemical to increase concentration of fluoride ions in drinking water to reduce incidence of tooth decay.

FLUORIDE 1. Any salt of hydrofluoric acid; 2. Binary compound of fluorine and another element or radical.

FLUORIDES 1. Gaseous, solid or dissolved compounds containing fluorine that result from industrial processes; 2. Any salts of hydrofluoric acid.

FLUORINE Chemical element with atomic number of 9.

FLUOROCARBONS Any of number of organic compounds analogous to hydrocarbons where one or more hydrogen atoms are replaced by fluorine.

FLURRIES Intermittent light snowfall of short duration (generally light snow showers) with no measurable accumulation (trace category).

FLUSH 1. To force large amounts of water through system to clean out piping or tubing and storage or process tanks; 2. Group of birds that have suddenly started up from undergrowth, trees, etc.

FLUSH BALL Part of flush valve that seals water inside toilet tank or allows passage of water into toilet bowl during flush cycle.

FLUSH BUSHING Conduit fitting that reduces diameter of female-threaded conduit fitting.

FLUSH HYDRANT Hydrant used for flushing water line of silt, rust, debris or stagnant water.

FLUSH PIPE Conduit joining cistern to some sanitary fitting.

FLUSH TANK Tank holding water or sewage for periodically flushing.

FLUSH VALVE Part inside tank of toilet that moves water into bowl.

FLUSHED Rushing or overspreading flow to clean.

FLUSHING Method used to clean water distribution lines by passing large amount of water through system.

FLUSHING CHAMBER Tank that holds water for flushing.

FLUSHING CISTERN Cistern provided with valve or device for controlling discharge of stored water into toilet or urinal.

FLUSHING MANHOLE Manhole built with opening gate to ensure sewage flows could possibly be collected and after that released to flush downstream segment of conduit.

FLUSHING MECHANISM Apparatus that disposes of human urine and feces by using water to flush.

FLUSHOMETER Device that flushes urinals automatically at regular intervals.

FLUVIAL Pertaining to, growing or living in streams or ponds.

FLUVIAL DEPOSIT Sedimentary deposit consisting of material transported by suspension or laid down by river or stream.

FLUVIAL EROSION Detachment of material from stream bed and side banks.

FLUVIAL GEOMORPHOLOGY Study of form and function of streams and rivers.

FLUVIAL INDEX Incline of flow curve of standard liquid-limit test of soil, plotted with water content as ordinate on arithmetic scale and number of flows as abscissa on logarithmic scale.

FLUVIAL SEDIMENT Particles derived from rocks or biological materials which are transported by, suspended in or deposited by streams.

FLUVIAL SOIL Soil placed by river or stream.

FLUVIOGLACIAL Describing stream formed from glacial meltwater.

FLUX 1. Flowing or flow; 2. Substance used to help metals fuse together; 3. Rate of transfer of energy (or another physical quantity) through given surface.

FLUX STANDARD Regulatory standard that restricts quantity of radon that will emanate per m^2 of controlled substance per second, averaged over one source.

FLY ASH Non-combustible residual particles, expelled by flue gas, that are by-product of coal-fired power plants.

FLYWAY Specific air route taken by birds during migration.

FMIN Lowest radiowave frequency that can be reflected from ionosphere.

FOAM Mass of small bubbles formed on surface of liquid by agitation or fermentation that's used for fire suppression to cool fire prevent fuel contact with oxygen that suppresses combustion.

FOAM INLET Connection,by fire fighters, for foam-making equipment.

FOAM SEPARATION Chemical process where foam is used to separate components of solution when they differ in surface activity.

FOAM-IN-PLACE INSULATION Material used for homes that hardens on site.

FOEHN Warm, dry wind on lee side of mountain range, warmth and dryness of air being due to adiabatic compression as air descends mountain slopes.

FOEHN PAUSE Temporary cessation of foehn at ground due to formation or intrusion of cold air layer which lifts foehn off ground.

FOES Maximum ordinary mode radiowave frequency capable of reflection from sporadic E region of ionosphere.

FOF2 Maximum ordinary mode radiowave frequency capable of reflection from F2 region of ionosphere.

FOG Water droplets suspended in air at Earth's surface.

FOG DRIP Water dripping to ground during fog.

FOG SEAL Specially formulated asphalt emulsion (i.e., thin liquid oil) applied to existing asphalt pavement surface.

FOGBOW Rainbow that has white band that appears in fog and is fringed with red on outside and blue on inside.

FOGGING Applying pesticide by rapidly heating liquid chemical so that it forms very fine droplets that resemble smoke or fog used to destroy mosquitoes, black flies and similar pests.

FOLD To bend so that it comes in contact with itself.

FOLIATION 1. Process of forming into leaf or leaves; 2. Property of being divided into plates or layers, due to cleavage structure of one of constituents.

FOLLOW-UP HEADWAY Time difference between two vehicles entering major street under continuous queuing condition.

FOLLOWER Sleeve on conduit die that aligns die together with conduit.

FOLLOWER DENSITY Number of followers in directional traffic stream over unit length.

FOOD CHAIN Sequence of organisms, each of which uses next, lower member of sequence as food source.

FOOD PROCESSING WASTE Food residues produced during agricultural and industrial operations.

FOOD WASTE Uneaten food and food preparation wastes from residences and commercial establishments (e.g., grocery stores, restaurants, produce stands, institutional cafeterias, kitchens, employee lunchrooms).

FOOD WEB Feeding relationships by which energy and nutrients are transferred from one species to another.

FOOD-CHAIN CROPS Crops grown for human consumption and animal feed for animals whose products are consumed by humans.

FOOD/MICROORGANISM RATIO Food to microorganism ratio that measures food provided to bacteria in aeration tank.

FOOT 1. Unit of length equal to 12 inches; 2. One of number of projections on cylindrical drum of tamping roller.

FOOT PER HOUR Unit of velocity equal to number of feet transited in one hour.

FOOT PER MINUTE Unit of velocity equal to number of feet transited per in one minute.

FOOT PER SECOND Unit of velocity equal to number of feet transited per in one second.

FOOT VALVE Valve of steam engine that opens to air pump from condenser.

FOOT-POUND Unit of energy equal to work done when mass of one pound, accelerated at rate of 1 ft/s^2, has transited one foot.

FOOTBATH Small basin or bath designed for washing feet.

FOOTBRAKE VALVE Valve that controls air pressure delivered to or discharged by brake chambers.

FOOTING 1. Sill under foundation; 2. Ground, in relation to its load bearing and friction qualities; 3. Portion of foundation of structure that transmits loads directly to soil.

FOOTPRINT Space required by piece of equipment.

FOOTWALL 1. Section of rock that extends below diagonal fault line; 2. Under wall of enclosed vein.

FORAGE Vegetation used for animal consumption.

FORAGE FISH Small fish that produce prolifically and are consumed by predators.

FORB Weed or broad-leafed plant.

FORCE Physical quantity that denotes ability to push, pull, twist or accelerate body.

FORCE ACCOUNT Payment method for construction work when existing agreement on cost exists.

FORCE CUP Plunger.

FORCE MAIN Pipelines that convey wastewater under pressure from discharge side of pump or pneumatic ejector to discharge point.

FORCE PER SQUARE INCH Unit of pressure equal to 1 lb/in^2.

FORCE PUMP Pump used to move water or other liquid under pressure.

FORCED CHANNELING Channeling of upper winds along valley's axis when upper winds are diverted by underlying topography.

FORCED DRAFT FAN Type of fan supplying pressurized air to system.

FORCED OUTAGE Unscheduled shut down of generating unit or other facility for emergency or other unforeseen reasons.

FORD 1. Place where road crosses stream under water; 2. Location where stream is shallow and bottom has good footing.

FOREBAY 1. Water behind (upstream) of dam or hydroelectric plant intake structure; 2. Impoundment immediately upstream from larger body of water.

FOREBAY AREA Natural groundwater basin functioning as recharge reservoir to artesian basin.

FORECAST Statement of prediction.

FORECAST CREST Highest elevation of river level or stage, expected during specified storm event.

FORECAST GUIDANCE Computer-generated forecast materials used to assist preparation of forecast, (e.g., numerical forecast models).

FORECAST ISSUANCE STAGE Stage which, when reached by rising stream, represents level where RFCs need to begin issuing forecasts for non-routine (flood-only) forecast point.

FORECAST PERIODS Official definitions for National Weather Service products: Today (sunrise to sunset), This afternoon (noon till 6 p.m.), This evening (6 p.m. till sunset), Tonight (sunset till sunrise), Tomorrow (sunrise to sunset of following day).

FORECAST POINT Location that represents area (i.e., reach of river), where forecast warning services are made available to public.

FORECAST TO COMPLETE Approximation of value that remains to be invested inside price category as of given date.

FORECAST VALID FOR Period of time forecast is in effect beginning at given day, date and time and ending at given day, date and time.

FOREIGN WATER Water that originated in different drainage basin.

FOREPOLE Plank driven ahead of tunnel face to support roof or wall during excavation.

FORESHOCK Small earth tremor which precedes mainshock in earthquake sequence.

FORESHORE Part of shore between ordinary high-and low-watermarks and generally crossed by tide each day.

FORESIGHT Sighting on point of unknown elevation from instrument of known elevation.

FOREST Dense collection of trees covering relatively large area.

FOREST ECOLOGY Scientific study of interrelated patterns, processes, flora, fauna and ecosystems in forests.

FOREST ECONOMICS Branch of forestry that provides essential information that helps forest policy decision making at both sectoral level and at forest level.

FOREST INFLUENCE Effects on climate, water and land caused by forest development.

FOREST MANAGEMENT Branch of forestry concerned with overall administrative, economic, legal and social aspects, as well as scientific and technical aspects.

FOREST MANAGEMENT CYCLE Periods that happen in managing of forests.

FOREST MANAGEMENT PLAN Specific statement of objectives for forests, followed by series of activities that will take place in order to meet those objectives.

FOREST PRACTICE Establish standards for timber harvesting, pre-commercial thinning, road construction, fertilization, forest chemical application and other forest practices applications.

FOREST PROTECTION Preservation or improvement of national forest lands.

FOREST RENEWAL Process where new tree seedlings become established after forest trees have been harvested or have died from fire, insects or disease.

FOREST RESIDUALS Limbs and branches after trees are harvested for lumber and pulp.

FOREST SERVICE Agency within U.S. Department of Agriculture with mission to sustain health, diversity and productivity of Nation's forests and grasslands to meet needs of present and future generations.

FOREST TYPE Categorization of forest into three major types: Tropical, Temperate or Boreal (i.e., Taiga).

FOREST TYPE LABEL Map symbol that codes information about forest type.

FOREST TYPE LINE Map or photo line indicating type of forest.

FORESTRY Science and practice of planting and growing trees in forests.

FORM LOSS OF HEAD Reduction of head attributable to change in shape of waterway.

FORMALDEHYDE Colorless, pungent and irritating gas used chiefly as disinfectant and preservative and in synthesizing other compounds like resins.

FORMATION Any sedimentary, igneous or metamorphic material represented as unit in geology that's generally called rock, but not necessarily meeting definition of rock.

FORMATION FLUID Naturally occurring liquids and gases contained in geologic formations.

FORMULATION 1. Substances comprising all active and inert ingredients in pesticide; 2. Act, process or result of mixing, combining, diluting, formulating or reducing.

FORMULATION PROCESS First phase performed by Early Warning System design team, which includes Early Warning System reliability, local capabilities assessment and conceptual level designs for Early Warning System.

FORMWORK Temporary mould, made from planks, where concrete is poured.

FORWARD FLANK DOWNDRAFT Main region of downdraft in forward or leading, part of supercell, where most of heavy precipitation is.

FOSSIL FUEL Fuel derived from ancient organic remains (e.g., peat, coal, crude oil, natural gas).

FOSSILIFEROUS Greater than 40% fossils.

FOSSORIAL INSECTS Insects that live in soil.

FOUL AIR Air that has noticeable, abnormal, physiological effect on humans.

FOUL AIR DUCT Suction line in tunnel ventilation system.

FOUL WATER Mix of surface and waste water.

FOULED To clog, obstruct, entangle or make dirty.

FOULING Adhesion of foreign material onto surface.

FOUNDATION 1. Lower part of structure that transmits loads directly to soil; 2. Excavated surface upon which dam is placed; 3. Portion of valley floor that underlies and supports dam structure.

FOUNDATION DRAIN Exterior foundation drainage system placed outside foundation wall near wall footing and covered with gravel piped for drainage.

FOUNDATION DRAINS Tile or pipe for collecting seepage within foundation.

FOUNDATION MATERIAL Surface of upper part of earth mass carrying load of structure.

FOUNDATION TRENCH Trench built at and into foundation of dam and filled with clay or other impermeable substances to prevent water from seeping beneath dam.

FOUNTAIN 1. Natural source of water; 2. spring; 3. Artificial water feature consisting of one or more streams of water originating from statue or structure.

FOUNTAIN AERATOR Apparatus that adds oxygen to fountain water.

FOUNTAIN FLOW Occurs when material at flow front is pushed forward and "fountains" to channel wall.

FOUNTAIN HEAD Source of water head in saturated, confined aquifer.

FOUNTAIN JET Ornamental structure where one or more jets of water are pumped into air.

FOUNTAINHEAD Upper end of confined-aquifer conduit where it intersects land surface.

FOUR ZERO ONE (401) CERTIFICATION Requirement of Section 401(a) of Clean Water Act that all federally issued permits be certified by state where discharge occurs.

FOUR-WAY VALVE Valve that controls direction of fluid flow in hydraulic circuit that controls direction of movement of work cylinder or rotation of fluid motor.

FOURNEYRON WHEEL Water wheel where water enters from below.

FOURWAY VALVE Valve that may be placed to direct flow to all or any of four different settings.

FRACTOCUMULUS Cumulus cloud presenting ragged, shredded appearance, as if torn.

FRACTURE 1. Any break or rupture formed in ice cover or floe due to deformation; 2. Break in rock formation due to structural stresses (e.g., faults, shears, joints, planes of fracture cleavage) where no movement has occurred.

FRACTURE SPRING Spring where waters flows from rock joints or other fractures.

FRACTURE TOUGHNESS Method of expressing material's resistance to crack propagation.

FRACTURE ZONE 1. Area that has great number of fractures; 2. Linear oceanic faulting feature resulting from action of offset mid-ocean ridge axis segments.

FRACTURE-CRITICAL MEMBER Member in tension where failure would probably cause bridge portion or entire structure to collapse.

FRACTURING Deformation process whereby ice is permanently deformed and fracture occurs.

FRACTUS Ragged, detached cloud fragments (i.e., scud).

FRAME To construct by fitting or uniting together various parts.

FRAME ACTION Continuity between two components oriented for stiffer behavior.

FRANGIBLE Able to be broken.

FRAZIL Fine spicules, plates or discoids of ice suspended in water.

FRAZIL ICE Loose, randomly oriented ice formed by accumulation of ice crystals in water too turbulent to freeze solid.

FRAZIL SLUSH Agglomerate of loosely packed frazil which floats or accumulates under ice cover.

FREAK WAVE Wave of much greater height and steepness than other waves in prevailing sea or swell system.

FREE ACCELERATION TEST Process for measuring exhaust emissions from vehicles.

FREE AIR Air not under restraint (i.e., normal atmospheric air).

FREE AREA Sum of total area openings (e.g., conduit) for ventilation system.

FREE ATMOSPHERE Part of atmosphere that lies above frictional influence of Earth's surface.

FREE AVAILABLE CHLORINE Amount of chlorine available in water.

FREE AVAILABLE RESIDUAL CHLORINE Portion of total available residual chlorine composed of dissolved chlorine gas, hypochlorous acid, and/or hypochlorite ion remaining in water after chlorination.

FREE CHLORINE Both hypochlorous acid and hypochlorite ion or bleach commonly added to water systems for disinfection.

FREE CONVECTION Convection of contents of fluid by means of currents induced by buoyancy force which is dependent on acceleration acting upon density changes in fluid which are induced by heat.

FREE FIELD Situation where no sound reflections occur.

FREE FLOW Continuous flow from storage or supply unit.

FREE GROUNDWATER Unconfined groundwater whose upper boundary is free water table.

FREE LIQUID Liquid that easily separates from solid portion of waste.

FREE MOISTURE Moisture that can be removed by ordinary drying.

FREE OXYGEN Molecular oxygen available for respiration by organisms.

FREE PRODUCT Petroleum hydrocarbon in liquid free or non-aqueous phase.

FREE RADICAL Molecule, ion or atom that has one or more unpaired electrons and is generally highly reactive.

FREE RESIDUAL CHLORINATION Application of chlorine or chlorine compounds to water or wastewater to produce free available chlorine residual directly or through destruction of ammonia or certain organic nitrogenous compounds.

FREE SURFACE Surface of fluid subject to zero parallel shear stress.

FREE SURFACE ENERGY Energy associated with intermolecular forces at interface between two media.

FREE WATER Water not bound to inorganic surface and can flow freely.

FREE WAVE Wave separated from its source.

FREE WEIR Submerged weir.

FREE-FLOW SPEED Average speed motorist would travel if no congestion or obstructions.

FREEBOARD 1. Distance between water level and top of something that contains or restrains it; 2. Vertical distance from sand surface to underside of trough in sand filter.

FREEWAY Road designed for high-speed motor vehicle operations.

FREEWAY AUXILIARY LANE Freeway lane to connect on- and off-ramps.

FREEWAY DIVERGE SEGMENT Freeway segment where single traffic stream divides into more traffic streams.

FREEWAY FACILITY Extended length of freeway composed of continuously connected basic freeway, weaving, merging and diverging segments.

FREEWAY FACILITY CAPACITY Capacity of critical segment of freeway facility.

FREEWAY MERGE SEGMENT Where two or more traffic streams of freeway combine to single traffic stream.

FREEWAY SECTION Segment of freeway where there are no interruptions to traffic flow.

FREEWAY WEAVING SEGMENT Segments where two or more traffic streams cross paths along freeway minus traffic control devices.

FREEZE When surface air temperature is expected to be 32 °F or below over widespread area for climatologically significant period of time.

FREEZE WARNING Issued during growing season when surface temperatures are expected to drop below freezing over large area for extended period of time, regardless whether or not frost develops.

FREEZE-THAW CYCLE Water moving from states of frozen to thawed to frozen, etc.

FREEZE-THAW DAMAGE Damage to concrete caused by extreme temperature variations as noted by random pattern cracking that's accelerated by presence of water and commonly more severe on south-facing side of structures.

FREEZE-UP Blocking of pipes due to freezing.

FREEZELESS WATER FAUCET Faucet that releases water on exterior but has its valve on interior face of wall.

FREEZEOUT Deeply frozen over for long periods of time.

FREEZEUP DATE Date on which water body was first observed to be completely frozen over.

FREEZING DRIZZLE Drizzle that falls as liquid but freezes into glaze or rime upon contact with cold ground or surface structures.

FREEZING DRIZZLE ADVISORY Issued when freezing rain or freezing drizzle is forecast but significant accumulation is not expected.

FREEZING FOG Fog droplets of which freeze upon contact with exposed objects and form coating of rime and/or glaze.

FREEZING INDEX Number of Fahrenheit degree-days (above and below 32 °F) between highest and lowest points on cumulative degree-days time curve for one freezing season.

FREEZING LEVEL Altitude where air temperature first drops below freezing.

FREEZING NUCLEI Particles in atmosphere that water freezes to.

FREEZING PROCESS Phase transition where liquid turns into solid when its temperature is lowered below its freezing point.

FREEZING RAIN Rain that falls as liquid but freezes into glaze upon contact with ground.

FREEZING RAIN ADVISORY Issued when freezing rain or freezing drizzle is forecast but significant accumulation is not expected.

FREEZING SPRAY Accumulation of freezing water droplets on vessel caused by some appropriate combination of cold water, wind, cold air temperature and vessel movement.

FREEZING SPRAY ADVISORY Advisory for accumulation of freezing water droplets on vessel at rate of less than 2 cm/hr caused by some appropriate combination of cold water, wind, cold air temperature and vessel movement.

FREEZING TIME Time taken to lower temperature of product from its initial temperature to given temperature.

FREEZUP JAM Ice jam formed as frazil ice accumulates and thickens.

FREIGHT CAR Railway wagon for carrying goods.

FRENCH DRAIN Covered ditch containing layer of stone, pervious material or perforated pipe that redirects water away from area.

FREON Any of several non-flammable refrigerants based on halogenated hydrocarbon.

FREQUENCY Number of repetitions of periodic process in certain time period.

FREQUENCY CURVE Curve that expresses relation between frequency distribution plot, with magnitude of variables as abscissas and number of occurrences of each magnitude in given period as ordinates.

FREQUENCY DEMAND SCHEDULING Method of irrigation scheduling similar to demand scheduling, but typically involves fixed duration of delivery.

FREQUENCY DISTRIBUTION Mathematical function showing number of instances where variable takes each of its possible values.

FREQUENCY RESPONSE Quantitative measure of output spectrum of system or device in response to stimulus.

FRESH AIR FILTER Filter that removes dust and particles from outside air.

FRESH AIR INLET Line of conduit that opens into outside air.

FRESH AIR RATE Percent of clean air supplied to space or equipment.

FRESH SLUDGE Sludge before it is stabilized.

FRESH WASTEWATER Wastewater that dissolved oxygen can be quantified.

FRESHET 1. Annual spring rise of streams in cold climates as result of snow melt; 2. Flood caused by rain or melting snow.

FRESHWATER Water that contains less than 1,000 mg/l of dissolved solids.

FRESHWATER CHRONIC CRITERIA Highest concentration of contaminant that freshwater aquatic organisms can be exposed to for extended period of time (4 days) without adverse effects.

FRIABLE 1. Capable of being crumbled, pulverized or reduced to powder by hand pressure; 2. Descriptive of rock or mineral that crumbles naturally or is easily broken, pulverized or reduced to powder.

FRIABLE ASBESTOS Any material containing more than one% asbestos and that can be crumbled or reduced to powder by hand pressure.

FRICTION 1. Resistance to motion when one body is sliding or tending to slide over another; 2. Mechanical resistive force of one object on another object's relative movement when in contact with first object.

FRICTION FACTOR Variable used to describe energy loss in pipe based on velocity of fluid and resistance due to friction.

FRICTION HEAD Decrease in total head caused by friction.

FRICTION HORSEPOWER Power lost especially in internal-combustion engine through friction between parts of machine itself.

FRICTION LAYER Layer within atmosphere between earth's surface and 1 km above surface where friction affects wind speed and wind direction.

FRICTION LOSS Loss of pressure due to fluid's viscosity near surface of conduit.

FRICTION NUMBER 1. Friction at slip speed; 2. Numerical, dimensionless value of friction given by specific measuring device.

FRICTION PILE Pile that utilizes frictional resistance force between pile surface and adjacent soil.

FRICTION SLOPE Change in friction head versus length of conduit.

FRICTION WEIR Not submerged weir.

FRICTIONAL SOIL Any free-running type of soil (e.g., sand, gravel) where strength depends on friction between particles.

FRINGE WATER Water of capillary fringe.

FROG 1. Part of railway switch or turnout where running-rails cross; 2. Depression in upper face of pressed or handmade clay brick; 3. Small tailless amphibian of order Anura that typically hops.

FRONT 1. Boundary or transition zone between two air masses of different density and thus (usually) of different temperature; 2. Foremost side of something or end that faces direction it normally moves.

FRONT FOOT Foot measured along front of piece of property.

FRONT LOADING REFUSE TRUCK Refuse trucks provided with fluid cylinder operated lift arms and forks.

FRONT LOT LINE Boundary line of lot that abuts public street.

FRONT OF LEVEE Side of levee alongside river.

FRONT-MOUNT HOIST Hydraulic cylinder mounted vertically at front of dump truck.

FRONTAGE ROAD Smaller road alongside main road (e.g., expressway) for access to adjacent properties.

FRONTAL ANALYSIS Form of chromatography where pure sample flows through column and each component breaks through at different time depending on its affinity for adsorbent.

FRONTAL FLOW Portion of flow that passes over upstream side of grate.

FRONTAL INVERSION Temperature inversion that develops aloft when warm air overruns cold air behind front.

FRONTAL PRECIPITATION Precipitation that results when leading edge of warm, moist air mass interacts with cool, dry air mass.

FRONTAL SLOPE Steepness of weather frontal surface that depends on latitude of front, wind speed and temperature difference between air masses.

FRONTAL SURFACE Perpendicular projected surface according to trajectory of fluid.

FRONTOGENESIS 1. Initial formation of front or frontal zone; 2. Increase in horizontal gradient of air mass property, principally density and development of accompanying features of wind field that typify front.

FROST Formation of thin ice crystals on ground or other surfaces in form of scales, needles, feathers or fans.

FROST ACTION Weathering process caused by freezing and thawing cycles of water in surface pores, cracks and other openings.

FROST ADVISORY Issued during growing season when widespread frost formation is expected over extensive area where surface temperatures are usually in mid-30s Fahrenheit.

FROST BOIL Upwelling that occur from swelling and subsequent disintegration caused by frost action.

FROST BOTTOM Water meter with bottom that will freeze and pop off before line freezes.

FROST BOX Insulated box to protect water conduits or devices.

FROST CRACK Vertical cracks in trees caused by alternating freezing and thawing temperatures.

FROST HEAVE Raising of surface due to accumulation of ice in underlying soil or rock.

FROST HOLLOW Low-lying area where frosts occurs frequently.

FROST LINE Greatest depth that ground may be expected to freeze.

FROST POINT Dew point below freezing.

FROST WEDGING Occurs as result of expansion of water when it is converted to ice and cracks, filled with water, are forced further apart when it freezes.

FROSTBITE Human tissue damage caused by exposure to intense cold.

FROSTLINE Maximum depth where soil is frozen.

FROSTPROOF CLOSET Toilet without integral trap that has its trap and control valve for water supply installed below frost line.

FROTH Bulk of tiny bubbles (i.e., foam).

FROTH FLOTATION Process for selectively separating hydrophobic materials from hydrophilic, by first grinding ore and mixing it with water and then adding surfactant.

FROUDE NUMBER Ratio of inertial forces to gravitational forces in flow (when ratio less than one, flow is tranquil; when greater than one, flow is rapid and when equal to one, flow is critical).

FROUDE NUMBER MODEL Model designed to emphasize similarity of gravitational and inertial forces (i.e., Froude Number), while other forces (e.g., viscous (i.e., Reynolds Number)) may not be reproduced as accurately.

FROZEN State that something freezes to ice form.

FROZEN DEW When liquid dew changes into tiny beads of ice.

FROZEN FOG Fog consisting of fine ice crystals suspended in air.

FROZEN SOIL Ground, including rock or soil, at or below freezing point of 0 °C (32 °F) of water.

FRY Life stage of fish between egg and fingerling stages.

FUEL Substance consumed to provide energy through combustion, chemical or nuclear reaction.

FUEL CELL Device where intrinsic chemical free energy of fuel and oxidant is catalytically converted to direct current energy.

FUEL CYCLE Progression of fuel through series of differing stages.

FUEL ECONOMY STANDARD Corporate Average Fuel Economy Standard effective in 1978 that enhanced national fuel conservation effort imposing miles-per-gallon floor for motor vehicles.

FUEL EFFICIENCY Proportion of energy released by fuel combustion converted into useful energy.

FUEL ELEMENT Structured group of fuel rods for nuclear reactors.

FUEL EVAPORATIVE EMISSION Vaporized fuel discharged in engine fuel system.

FUEL OIL Heaviest commercial fuel that can be obtained from crude oil.

FUEL REPLACEMENT ENERGY Electric energy generated at hydroelectric plant as substitute for energy which would have been generated by thermal electric plant.

FUEL SWITCHING 1. Pre-combustion process whereby low-sulfur coal is used in place of higher sulfur coal in power plant to reduce sulfur dioxide emissions; 2. Illegally using leaded gasoline in motor vehicle designed to use only unleaded.

FUEL TRANSFER PUMP Pumps for transferring fuel between tanks.

FUEL-AIR RATIO Ratio of air to fuel supply flow rates where ideal ratio burns all fuel with no excess air.

FUGITIVE DUST Dust not emitted from definable point sources (e.g., industrial smokestacks where sources include open fields, roadways, storage piles).

FUGITIVE EMISSIONS Emissions not caught by capture system.

FUGITIVE SOURCE Emissions not released through stack, vent, duct, pipes or other confined air stream.

FUGITIVE VOLATILE ORGANIC COMPOUNDS Volatile organic compounds that are emitted from applicators and flash-off areas that are not emitted in oven.

FUGITIVE WATER Pumping, flow, release, escape or leakage of any water for purposes of water supply, transport, storage or disposal.

FUJITA SCALE Scale of tornado intensity where wind speeds are inferred from analysis of wind damage.

FUJIWHARA EFFECT Binary interaction where tropical cyclones within certain distance (300-750 nautical miles depending on sizes of cyclones) of each other begin to rotate about common midpoint.

FULCRUM Pivot for lever.

FULL ACTUATED OPERATION All signal phases are actuated with change and clearance intervals at predetermined duration.

FULL CONTROL OF ACCESS To give preference to thorough traffic by providing access connections to selected roads.

FULL GATE Maximum gate position of turbine for particular head.

FULL HYDRAULIC CAPACITY Designed capacity of pipe or conduit.

FULL IRRIGATION SERVICE LAND Irrigable land now receiving or to receive, its sole and generally adequate water supply through facilities which have been or are to be constructed by, rehabilitated by or replaced by Bureau of Reclamation.

FULL LOAD SPEED Speed where rated full-load torque is delivered at rated power output.

FULL LOAD TORQUE Torque required to produce rated power of electrical motor at full-load speed.

FULL POOL 1. Volume of water in reservoir at normal water surface; 2. Reservoir level that would be attained when reservoir is fully utilized for all project purposes, including flood control.

FULL STOP Vehicle slowing to less than 5 mi/hr.

FULL-ACTUATED OPERATION Operation of traffic control signal where all signal phases function on actuation.

FULL-CELL PROCESS Method of impregnating wood with preservatives that involves sealing wood in pressure chamber and applying vacuum in order to remove air and moisture from wood cells.

FULL-PHYSICS NUMERICAL MODEL Computer model used to calculate air pollution concentrations that uses full set of equations describing thermodynamic and dynamic state of atmosphere and can be used to simulate atmospheric phenomena.

FULL-SCALE EXERCISE Activity where emergency preparedness officials respond in coordinated manner to timed, simulated incident but includes mobilization of field personnel and resources and actual movement of emergency workers, equipment and resources required to demonstrate coordination and response capability.

FULL-TIDE COFFERDAM Cofferdam that's height is sufficient to exclude water at all tidal periods.

FULLER'S EARTH Fine clay used in fulling cloth.

FULLY ACTUATED CONTROL Intersections where all phases of traffic signal require movement detection.

FUMAROLE Opening in ground that emits steam and gases due to volcanic activity.

FUME Tiny particles trapped in vapor in gas stream.

FUME SCRUBBER Wet scrubber where acid fumes of contaminated vapor are re-solubilized when passed through liquid rich environment to remove and/or clean.

FUME SUPPRESSION SYSTEM Equipment comprising any system used to inhibit generation of emissions.

FUMIGANT Substance or mixture of substances that produces gas, vapor, fume or smoke intended to destroy insects, bacteria or rodents.

FUMIGATION Pattern of plume dispersion produced when convective boundary layer grows upward into plume trapped in stable layer and elevated plume is suddenly brought downward to ground, producing high surface concentrations.

FUNCTIONAL ANNEX Emergency operations plan element that describes jurisdiction's plan for functioning in that component area of activity during emergencies.

FUNCTIONAL CLASSIFICATION Process where roads, streets and highways are grouped into classes according to service they provide.

FUNCTIONAL EXERCISE Activity where participants respond in coordinated manner to timed, simulated incident that parallels real operational event as close as possible.

FUNCTIONAL SPACE Set of functions between two fixed sets.

FUNCTIONAL UNBUNDLING Electric service is traditionally provided on bundled basis, meaning that generation, transmission and distribution services are provided as single package.

FUNCTIONALLY EQUIVALENT COMPONENT Component that performs same function or measurement and which meets and/or exceeds performance specifications of another component.

FUNDAMENTAL FREQUENCY Lowest frequency of periodic waveform.

FUNDAMENTAL PARTICLES Any of subatomic particles that does not consist of other, smaller particles (e.g., gauge bosons, leptons and quarks).

FUNDAMENTALLY DIFFERENT FACTORS Those components of petitioner's facility that are determined to be so unlike those components considered by EPA during effluent limitation guideline and pretreatment standards rulemaking that facility is worthy of variance from effluent limitations guidelines or categorical pretreatment standards.

FUNGI Eukaryotic single-celled or multi-nucleate organisms that live by decomposing and absorbing organic material where they grow.

FUNGICIDE Pesticides which are used to control, deter or destroy fungi.

FUNGISTAT Chemical that keeps fungi from growing.

FUNGUS Mold, mildew, yeast, mushroom and puffball, organism lacking in chlorophyll (i.e., are not photosynthetic) and which are usually non-mobile, filamentous and multicellular.

FUNNEL CLOUD Condensation funnel extending from base of towering cumulus, associated with rotating column of air not in contact with ground (i.e., different from tornado).

FUNNELLING Process whereby wind is forced to flow through narrow opening between adjacent land areas, resulting in increased wind speed.

FURNACE Device that provides heat.

FURNITURE ZONE Section of sidewalk between curb and clear path where street furniture and amenities are provided.

FURRED Pipe lined with hard lime or other salts deposited from water.

FURROW 1. Natural or man-made narrow depression in earth's surface; 2. Narrow trenchlike plowed depression in earth surface to keep surface water away from slopes of cuts.

FURROW IRRIGATION Surface irrigation whereby water is applied at upper (higher) end of field and flows in trenches to lower end.

FUSE 1. Thin core of black powder surrounded by wrappings, which when lit at one end, will burn to other at fixed speed; 2. String-like core of PETN, high explosive, contained within waterproof reinforced sheath.

FUSE PLATE Plate used to connect cut in wide flange beam.

FUSE PLUG SPILLWAY Form of auxiliary spillway consisting of low embankment designed to be overtopped and washed away during exceptionally large flood.

FUSED SWITCH Switch in series with fuse in single housing.

FUSION Nuclear reaction where nuclei combine to form more massive nuclei with concomitant release of energy.

FUSION DRILL Mining and cutting tool that burns out blasthole.

FUSION REACTOR Thermonuclear reactor that relies on controlled fusion of light elements into heavier elements, with attendant release of enormous amounts of energy.

FUTURE LIABILITY Refers to potentially responsible parties' obligations to pay for additional response activities beyond those specified in Record of Decision or Consent Decree.

FUTURE WITHOUT What would occur if no action were taken.

G

GABION Wire basket, filled with stones used to stabilize banks of water course and to enhance habitat.

GABION DAM Crib dam when built with gabions.

GAGE 1. Device for indicating magnitude or position of thing in specific units, when such magnitude or position undergoes change (e.g., elevation of water surface, velocity of flowing water, discharge, pressure of water, amount or intensity of precipitation, depth of snowfall); 2. Thickness of wire or sheet metal.

GAGING STATION Particular site on watercourse (e.g., stream, canal, lake, reservoir) where systematic observations of hydrologic data are measured.

GAIN Factor where signal is multiplied.

GAINING STREAM Stream or reach that receives water from zone of saturation.

GALE Extratropical low or area of sustained surface winds of 34 knots (39 mph) to 47 knots (54 mph).

GALE WARNING Warning of sustained surface winds or frequent gusts, in range of 34 knots (39 mph) to 47 knots (54 mph) inclusive, either predicted or occurring and not directly associated with tropical cyclone.

GALE WATCH Watch for increased risk of gale force wind event for sustained surface winds or frequent gusts, of 34 knots (39 mph) to 47 knots (54 mph), but its occurrence, location, and/or timing is still uncertain.

GALL Growth found on surface of plants caused by burrowing of insect larvae into living tissues.

GALLERY 1. Subterranean construction that gathers and carries water and/or wastewater; 2. Passageway within body of dam, its foundation or abutments used for inspection, foundation grouting, and/or drainage.

GALLON Unit of volume equal to 231 in^3, 128 fluid ounces or 8.34 pounds of water.

GALLON PER DAY Unit of volume, equal to 231 in^3, 128 fluid ounces or 8.34 pounds of water per day.

GALLON PER MINUTE Unit of volume, equal to 231 in^3, 128 fluid ounces or 8.34 pounds of water per minute.

GALLON PER SECOND Unit of volume, equal to 231 in^3, 128 fluid ounces or 8.34 pounds of water per second.

GALVANIZE To coat metal (e.g., iron, steel) with thin layer of metal (e.g., zinc).

GALVANIZED STEEL PIPE Steel conduit galvanized to withstand corrosion.

GAME FISH Species like trout, salmon or bass, caught for sport.

GAMETES Eggs or sperm.

GAMMA 1. Unit of magnetic field intensity equal to 1×10^{-5} Gauss; 2. Equal to 1 nanotelsa.

GAMMA RAY Type of electromagnetic radiation with very short wavelength and high energy level.

GAMMA RAYS Very high frequency electromagnetic radiation emitted as consequence of radioactivity.

GANG OF WELLS Group of wells, connected by pipe manifold, from which water is drawn by single pump.

GANTRY Overhead structure that supports machines or operating parts, (e.g., gantry crane).

GANTRY CRANE Fixed or traveling, bent-supported crane for handling heavy equipment.

GAP 1. Opening in anything made by breaking or parting; 2. Space or time between two vehicles; 3. Opening allowing passage or entrance; 4. Mountain or hill pass.

GAP ACCEPTANCE Process where driver of vehicle completes lane change.

GAP OUT Actuated operation for given phase where phase terminates due to lack of vehicle calls within specific period of time.

GAP WINDS Strong winds channeled through gaps in Pacific coastal ranges, blowing out into Pacific Ocean or into waterways of Inside Passage.

GAP-GRADED HOT MIX ASPHALT Hot asphalt mix with non-uniform distribution of aggregate sizes.

GARBAGE Animal and vegetable waste resulting from handling, storage, sale, preparation, cooking and serving of foods.

GARBAGE DISPOSAL Electric device in kitchen drain that gets rid of food waste by shredding it to tiny bits that can be washed away.

GAS Matter in state intermediate between liquid and plasma that can be contained only if it is fully surrounded by solid.

GAS CAP Natural gas, above oil pool, that provides primary drive mechanism (i.e., expansion) for oil recovery.

GAS CHROMATOGRAPH Instrument that identifies molecular composition and concentrations of various chemicals in water and soil samples.

GAS CHROMATOGRAPHY Laboratory analytical method used as screening technique for semi-volatile organic compounds that are extractable from water in methylene chloride.

GAS COCK VALVE Gas line shutoff valve.

GAS DOME Collects gas from tank and forwards collected gas to storage tank.

GAS LAWS Thermodynamic laws pertaining to perfect gases, including Boyle's law, Charles' law, Dalton's law and equation of state.

GAS THERMOMETER Measures temperature by variation in volume or pressure of gas.

GAS TRAP Drain or sewer pipe section (e.g., S- or P-trap) for trapping gaseous fraction of mixtures.

GAS VENT Passageway (e.g., pipe) to allow escape of gas.

GAS-COOLED FAST-BREEDER REACTOR Atomic reactor that uses gas for coolant.

GAS-LIFT FLOW AREA When subsurface water is under pressure from gas.

GAS-SOLUBILITY FACTOR Volume of gas liberated from volume of oil.

GAS:OIL RATIO Ratio of volume of gas to oil at standard conditions.

GASAHOL Mixture of gasoline and ethanol derived from fermented agricultural products containing at least 9% ethanol where emissions contain less carbon monoxide than those from gasoline.

GASEOUS SUPERSATURATION Condition of higher levels of dissolved gases in water due to entrainment, pressure increases or heating.

GASIFICATION Conversion of solid material (e.g., coal into gas) for use as fuel.

GASOLINE VOLATILITY Property of gasoline whereby it evaporates into vapor.

GATE Movable barrier or device that controls flow in conduit, pipe, canal, ditch or tunnel without obstructing any portion of waterway when in fully open position.

GATE CHAMBER 1. Room from which gate or valve can be operated or sometimes where gate is located; 2. Concrete portion of outlet works containing gates between upstream and downstream conduits and/or tunnels.

GATE DAM Barrier built across river, comprising series of gates which when fully open allow flood to pass without appreciably increasing flood level upstream of barrage.

GATE HANGER Device used to maintain set gate opening.

GATE HOUSE Structure located at entrance to large property, especially dwelling.

GATE LIFT Apparatus on back of vehicle that operates by mechanically raising or lowering.

GATE STEM Cylinder rod connected to gate slide.

GATE STRUCTURE Portion of spillway between inlet channel and chute, tunnel or conduit, which contains gates (e.g., radial gates).

GATE VALVE Valve with circular-shaped closing element (i.e., utilizes disc moving at right angle to flow) to regulate rate of flow.

GATE YARD Volume of landfill solid waste computed at landfill entry as % of truck volume.

GATE-TYPE HYDRANT Hydrant having one main valve.

GATED PIPE Portable pipe with small gates installed along one side for distributing irrigation water to corrugations or furrows.

GATED SPILLWAY Overflow section of dam restricted by use of gates that can be operated to control releases from reservoir to ensure safety of dam.

GATED WYE Hose appliance where two or more hoselines brought into one hose line.

GATING DEVICE Device designed to allow penetration of vehicle when impacted.

GAUGE 1. Device for indicating magnitude or position of thing in specific units, when such magnitude or position undergoes change; 2. Act or operation of registering or measuring magnitude or position of thing when these characteristics are undergoing change; 3. Operation, including both field and office work, of measuring discharge of stream of water in waterway; 4. Thickness of wire or sheet metal.

GAUGE BOX Tide gauge that uses float technology.

GAUGE COMPOUND Type of pressure gauge.

GAUGE CORRELATION Stream or conduit stage-discharge relationship of gauge height versus discharge.

GAUGE DATUM Horizontal surface used as zero point for measurement of stage or gauge height.

GAUGE FLUID LEVEL Apparatus that displays fluid amount.

GAUGE GLASS Glass tube for water gauge.

GAUGE HEIGHT Water-surface elevation referred to some arbitrary gauge datum (zero point).

GAUGE MANOMETER Devices where liquid columns are used to measure difference in pressure between two points or point and atmosphere.

GAUGE POINT Apparatus that measures water level using sharp fixed point.

GAUGE POUNDS PER SQUARE INCH Pressure relative to atmospheric pressure 0 at sea level.

GAUGE PRESSURE 1. Absolute pressure minus atmospheric pressure; 2. Pressure within closed container as measured with gauge.

GAUGE VACUUM Gauge that measures pressures less than atmospheric.

GAUGE VALUES Values that are recorded, transmitted, and/or computed from gaging station.

GAUGE WATER Instrument indicating level of water.

GAUGE ZERO Elevation of zero stage.

GAUGING Measurement of flow per unit of time at specific point.

GAUGING PIG Apparatus used to prove pipe roundness and indicate obstructions or internal damage.

GAUGING STATION Specific location on stream where systematic observations of hydrologic data are obtained through mechanical or electrical means.

GAUSS Unit of magnetic induction.

GAUSSIAN DISTRIBUTION Distribution of many random variables as symmetrical bell-shaped graph (i.e. normal distribution).

GAUSSIAN PLUME MODEL Computer model used to calculate air pollution concentrations.

GAUSSIAN PUFF MODEL Computer model used to calculate air pollution concentrations.

GDR On buoy report, direction, in degrees clockwise from true North, of GSP, reported at last hourly 10-minute segment.

GEAR Wheel with grooves (i.e., teeth) on outer circumference to interlock and convey motion from another gear.

GEAR CASE Enclosure to protect gear train and shafts.

GEAR PUMP Hydraulic pump that uses meshing of gears to pump fluid by displacement.

GEAR RATIO Ratio of speeds of rotation of initial and final gears in gear train.

GEAR REDUCTION UNIT Gear construction where rpm of input shaft is greater than output shaft.

GEAR SEGMENT Piece for receiving or communicating reciprocating motion from or to cogwheel.

GEAR STEP Percentage of increase over lower gear.

GEAR TRAIN System of interconnected gears.

GEARBOX Enclosed gear train.

GEIGER COUNTER Instrument used for detecting and measuring ionizing radiation.

GEIGER THRESHOLD Lowest applied voltage for which Geiger counter will operate.

GEL Semi-solid to almost solid colloid of solid and liquid.

GEL COAT Material applied to provide high-quality finish.

GEL PORE Emptiness after cement gel is dried off by heating.

GEL SPACE RATIO Ratio of volume of hydrated solids in pores to total volume of pores available before hydration.

GEL STRENGTH Shear stress measured at low shear rate after mud has set quiescently for period of time.

GEL TIME Time it takes for mixed resin to harden.

GEL WATER Gel that contains excessive amount of water.

GELIFLUCTION Freeze-thaw action upon waterlogging topsoils which induces downslope movement.

GELLING Transform liquid to semisolid.

GENE Segment of DNA or RNA from cell's or organism's genome.

GENE EXCHANGE Genetic combinations resulting from exchange of genes.

GENE FLOW Transfer of genes from one population to another.

GENE FREQUENCY Fraction of gene copies that are of particular allele in defined population.

GENECOLOGY Branch of ecology dealing with genetic variation in different environments.

GENERAL AVIATION Section of aviation industry that excludes both military and commercial passenger transport.

GENERAL BENEFIT Overall benefit from specified public works to society at large.

GENERAL CIRCULATION 1. Worldwide system of winds or ocean currents; 2. Totality of large-scale organized motion for entire global atmosphere.

GENERAL CIRCULATION MODELS Computer simulations that reproduce Earth's weather patterns to predict change in weather and climate.

GENERAL CONTRACT Contract where one prime contractor is responsible for entire project.

GENERAL CONTRACTOR Contractor responsible for managing project and hiring subcontractors.

GENERAL DRAWING Drawing showing plan, dimensions, elevations, cross-sections and other basic details of project.

GENERAL ENVIRONMENT Factors and conditions (e.g., atmospheric, terrestrial) that generally affect everyone.

GENERAL ESTIMATE METHOD Simple estimating method based on 1 ft^2 cost.

GENERAL FOREMAN Worker responsible for general management, organisational and operational control of construction site.

GENERAL PERMIT Permit applicable to class or category of dischargers.

GENERAL REPORTING FACILITY Facility having one or more hazardous chemicals above 10,000 pound threshold for planning quantities.

GENERAL SCOUR Scour at bridge caused by decrease in channel width that decreases flow area and increases velocity.

GENERAL WIND Winds produced by synoptic-scale pressure systems on which smaller-scale or local convective winds are superimposed.

GENERAL-PURPOSE BUCKET Bucket for general loading, carrying, dumping and performing clean-up work.

GENERAL-PURPOSE LANES Traffic lanes available for public use without restrictions or tolls.

GENERAL-USE SWITCH Switch intended for use in general distribution and branch circuits.

GENERALLY RECOGNIZED AS SAFE Designation by FDA that chemical or substance (including certain pesticides) added to food is considered safe by experts and so is exempted from usual food additive tolerance requirements.

GENERATE To produce as result of chemical or physical process.

GENERATED CONTAMINANT Contamination created by internal processes.

GENERATION CURVE Population density in certain phase of growth plotted on graph.

GENERATION DISPATCH AND CONTROL Aggregating and dispatching (sending off to some location) generation from various generating facilities, providing backup and reliability services.

GENERATOR 1. Machine that converts mechanical energy into electrical energy; 2. Facility or mobile source that emits pollutants into air or releases hazardous waste into water or soil; 3. Any person, by site, whose act or process produces regulated medical waste or whose act first causes such waste to become subject to regulation.

GENERATOR NAMEPLATE CAPACITY Full-load continuous rating of generator, prime mover or other electric power production equipment under specific conditions as designated by manufacturer.

GENERATOR SET Combination of electrical generator and engine to form single piece of equipment that produces electrical power.

GENETIC Relating to genetics or genes.

GENETIC DRIFT Overall shift of allele distribution in isolated population, due to random fluctuations in frequencies of individual alleles of genes.

GENETIC ENGINEERING Process of inserting new genetic information into existing cells in order to modify specific organism for purpose of changing one of its characteristics.

GENETIC POLLUTION Uncontrolled flow of genes into wild populations.

GENETICS Branch of biology that deals with transmission and variation of inherited characteristics (i.e., chromosomes, DNA).

GENOME Complete genetic information of organism.

GENOTOXIC Pertaining to agents known to damage DNA.

GEO 1. Inlet, gully or cleft in face of cliff; 2. Prefix for earth.

GEOBOTANICAL ANOMALY Local concentration of one or more elements in ecological assemblage or plants that may indicate presence of ore deposit or concentration of hydrocarbons.

GEOBOTANY Branch of biogeography concerned with geographic distribution of plant species.

GEOCENTRIC Having or symbolizing Earth at center.

GEOCHEMICAL ANOMALY Concentration of one or more elements in soil, rock, water or vegetation.

GEOCHEMICAL DISTRIBUTION Explanation of relative and absolute amounts of chemical elements.

GEOCLINE Earth's crust section subjected to downward gradient during geological time span.

GEODESIC Shortest line between two points on specific surface.

GEODESY Measurement and representation of Earth, its gravitational field and geodynamic phenomena in three-dimensional, time-varying space.

GEODETIC Relating to geodesy.

GEOGNOSY Geological study of Earth's structure and composition.

GEOGRAPHER Specialist in geography.

GEOGRAPHIC INFORMATION SYSTEM Computer system designed for storing, manipulating, combining, analyzing and displaying data in geographic context.

GEOGRAPHY Study of physical properties of Earth.

GEOHYDROLOGIC Interpretation of geology from hydrologic characteristics of strata or rock.

GEOHYDROLOGY Geological study of character, source and mode of groundwater that's branch of hydrology relating to subsurface or subterranean waters.

GEOLOGIC EROSION Natural erosion, over long periods of time, caused by geological processes.

GEOLOGIC SEQUESTRATION Process of injecting carbon dioxide, which has been compressed into liquid state, into deep subsurface.

GEOLOGIC SEQUESTRATION PROJECT 1. Injection well or wells used to emplace carbon dioxide stream beneath lowermost formation containing USDW; 2. Wells used for geologic sequestration of carbon dioxide that have been granted waiver of injection depth requirements.

GEOLOGICAL CONTROL Local rock formation or clay layer that limits (within engineering time frame) vertical and/or lateral movement of stream at particular point.

GEOLOGICAL ENGINEER Professional engineer who designs structures and solves problems related to earth and its resources.

GEOLOGICAL FORMATION Layer of rock made up of certain type of rock or combination of types.

GEOLOGICAL LOG Detailed description of all underground features (depth, thickness, type of formation) discovered during drilling of well.

GEOLOGICAL MAP Map made to show geological features.

GEOLOGICAL REPORT Reports that present, analyse and summarise geological field data for industry and research purposes.

GEOLOGICAL SURVEY Systematic investigation of geology underground for purpose of creating map or model.

GEOLOGICAL TIME Interval of time relating to geologic history of Earth.

GEOLOGIST Person skilled at geology.

GEOLOGY Science that deals with physical history of Earth, rocks of which it is comprised and physical changes which Earth has undergone or is undergoing.

GEOMAGNETIC Of or pertaining to magnetism of Earth.

GEOMAGNETIC ELEMENTS Components of geomagnetic field at surface of Earth.

GEOMAGNETIC FIELD Magnetic field observed in and around Earth.

GEOMAGNETIC STORM Worldwide disturbance of Earth's magnetic field, distinct from regular diurnal variations.

GEOMEMBRANE Impermeable membrane used to line body of water (e.g., canals).

GEOMETRIC DELAY Delay experienced by vehicle negotiating intersection without other vehicles.

GEOMETRIC DESIGN Highway engineering design that deals with positioning of physical elements of roadway according to standards and constraints.

GEOMETRIC FEATURE Features of objects constructed by geometric elements (e.g., points, lines, curves or surfaces).

GEOMETRIC SIMILARITY Similarity in form or shape.

GEOMORPHIC Pertaining to form or general configuration of Earth or of its surface features.

GEOMORPHIC CHANNEL UNITS Fluvial geomorphic descriptors of channel shape and stream velocity.

GEOMORPHOLOGY Geological study of configuration, characteristics, origin and evolution of land forms and Earth features.

GEOPHONE Microphone that detects and records seismic vibrations.

GEOPHYSICAL EVENTS Flares (Importance two or larger) with Centimetric Outbursts (maximum of flux higher than Quiet Sun flux, duration longer 10 minutes) and/or strong SID.

GEOPHYSICAL EXPLORATION Applied branch of geophysics that uses physical methods to measure physical properties of surface and subsurface.

GEOPHYSICS Study of physical characteristics and properties of Earth including: geodesy, seismology, meteorology, oceanography, atmospheric electricity, terrestrial magnetism and tidal phenomena.

GEOPOLITICS Study of effects of geography on international politics.

GEOPOTENTIAL HEIGHT Height above sea level of pressure level.

GEOSERE Series of ecological climax communities that follow each other in geologic time and change according to changing conditions.

GEOSPHERE Solid body of Earth.

GEOSTATIONARY ORBITING ENVIRONMENTAL SATELLITE Satellites orbiting at 22,370 miles above Equator with same rotational velocity as Earth.

GEOSTATIONARY ORBITING ENVIRONMENTAL SATELLITE DATA COLLECTION SYSTEM Satellite-based system collects variety of environmental data from locations in Western Hemisphere.

GEOSTATIONARY SATELLITE Satellite that rotates at same rate as Earth, remaining over same spot above equator.

GEOSTROPHIC WIND Wind affected by Coriolis force, blows parallel to isobars and whose strength is related to pressure gradient (i.e., spacing of isobars).

GEOSYNCHRONOUS Term applied to any equatorial satellite with orbital velocity equal to rotational velocity of Earth.

GEOSYNCLINE Large, linear depression in Earth's crust where sediment accumulates.

GEOSYNTHETIC Synthetic product used to stabilize terrain.

GEOTECHNICAL ENGINEER Engineer concerned with behavior of earth materials.

GEOTECHNICAL PROCESSES Processes that alter properties of soil.

GEOTECHNOLOGY Of or relating to soil and bedrock mechanics.

GEOTEXTILE Any strong, permeable fabric used in earthworks.

GEOTEXTILES Any fabric or textile (natural or synthetic) used as engineering material in conjunction with soil, foundations or rock.

GEOTHERMAL Relating to Earth's internal heat.

GEOTHERMAL DIRECT HEAT RETURN FLOW WELLS Class V wells that dispose of spent geothermal fluids following extraction of heat used directly (without conversion to electricity or passage through heat exchanger) to heat homes, swimming pools, etc.

GEOTHERMAL ELECTRIC POWER WELLS Class V wells that dispose of spent geothermal fluids following extraction of heat for production of electricity.

GEOTHERMAL ENERGY Heat under ground used to heat water and make steam to turn generator turbines and make electricity.

GEOTHERMAL GRADIENT Rate of increasing temperature with respect to increasing depth.

GEOTHERMAL/GROUND SOURCE HEAT PUMP Underground coils to transfer heat from ground to inside of building.

GEOTROPIC Of or pertaining to gravity.

GERMICIDAL TREATMENT Treatment against bacterial, yeast and fungal infections, while also preventing reinfestation.

GERMICIDE Any compound that kills disease-causing microorganisms.

GEYSER Geothermal feature of Earth where there is opening in surface that contains superheated water that periodically erupts in shower of water and steam.

GIANT 1. Tall species of particular animal or plant; 2. Big-diameter jet nozzle for hydraulic excavating.

GIANT SALVINIA Aquatic fern prohibited in United States by federal law that's invasive, rapidly growing plant that covers surfaces of lakes and streams forming floating mats that shade and crowd out important native plants.

GIARDIA LAMBLIA Microorganism (protozoan) in feces of humans and animals frequently found in rivers and lakes, which, if not treated properly, can cause severe gastrointestinal ailments.

GIARDIASIS Disease that results from infection by protozoan parasite Giardia Intestinalis.

GIGA Prefix symbolizing 10^9.

GIGAWATT Unit of power equal to 1 billion watts.

GIGAWATT-HOUR One billion watt-hours of electrical energy.

GIRDER Beam used as main horizontal support in building or structure.

GIS Computer system designed for storing, manipulating, combining, analyzing and displaying data in geographic context.

GLACIAL Of or relating to presence and activities of ice or glaciers.

GLACIAL DEPOSIT Landform created by large rock or stones deposited when glacier withdraws.

GLACIAL DRIFT All material of glacial origin transported by glacier and deposited by ice or meltwater from glacier.

GLACIAL LAKE Lake that derives its water or much of its water, from melting of glacial ice and occupies basin produced by glacial erosion.

GLACIAL MORAINE Mass of loose rock, soil and earth deposited by edge of glacier.

GLACIAL OUTWASH Stratified detritus (chiefly sand and gravel) "washed out" from glacier by meltwater streams and deposited in front of or beyond end moraine or margin of active glacier.

GLACIAL STREAM Stream where flow is from melting glacier.

GLACIAL STRIATION Scratches cut into bedrock by glacial abrasion.

GLACIAL STRIATIONS Lines carved into rock by overriding ice, showing direction of glacial movement.

GLACIAL TILL Material deposited by glaciation, usually composed of wide range of particle sizes, that has not been subjected to sorting action of water.

GLACIATION Transformation of cloud particles from water drops to ice crystals.

GLACIER Huge mass of ice that moves very slowly downslope or outward due to gravity.

GLACIER BURST Abrupt release of water impounded by glacier.

GLACIER DAMMED LAKE Lake formed when glacier flows across mouth of adjoining valley and forms ice dam.

GLACIER SNOW Snow compacted that becomes part of glacier.

GLACIER WIND Shallow downslope wind above surface of glacier, caused by temperature difference between air in contact with glacier and free air at same altitude.

GLACIOFLUVIAL Pertaining to water flowing on, in or against glaciers or ice sheets.

GLAND 1. Compressible sleeve made of copper or brass; 2. Secretory structure on surface of organ.

GLAND JOINT Conduit joint that permits movement resulting from thermal expansion or contraction.

GLAND NUT Nut with one conical face that applies pressure.

GLAND RING Ring that puts uniform pressure on packing.

GLARE SCREEN Clear panel or filter placed over screen to prevent glare.

GLASS CONTAINERS Containers like bottles and jars for drinks, food, cosmetics and other products.

GLAZE Ice formed by freezing precipitation covering ground or exposed objects.

GLAZED FROST Ice coating that occurs when freezing rain or drizzle hits surface.

GLAZED WARE Stoneware conduit and fitting.

GLEY Hydric soil that's sticky, greenish-blue-grey in color and low in oxygen.

GLIDE SLOPE Imaginary line extending from end of runway approach up to landing aircraft.

GLOBAL COMMONS Earth's unowned natural resources.

GLOBAL FORECAST SYSTEM One of operational forecast models run at NCEP.

GLOBAL MARITIME DISTRESS AND SAFETY SYSTEM Intended to provide more effective and efficient emergency and safety communications and disseminate Maritime Safety Information to all ships on world's oceans regardless of location or atmospheric conditions.

GLOBAL POSITIONING SYSTEMS Space-based radio positioning systems that provide 24-hour, three-dimensional position, velocity and time information to suitably equipped users anywhere on or near surface of Earth.

GLOBAL SAFETY FACTOR Factor of safety resulting from combining partial safety factors.

GLOBAL TEMPERATURE CHANGE Net result of four primary factors including greenhouse effect, changes in incoming solar radiation, altered patterns of ocean circulations and changes in continental position, topography and/or vegetation.

GLOBAL WARMING Increase in near surface temperature of Earth.

GLOBAL WARMING POTENTIAL Ratio of warming caused by substance to warming caused by similar mass of carbon dioxide.

GLOBE VALVE Valve consisting of movable disk-type element and stationary ring seat in spherical body.

GLOETZL CELL Instrument for measuring pressure within mass of soil, rock or concrete or at interface between one and other.

GLORY Optical effect characterized by concentric rings of color (red outermost and violet innermost) surrounding shadow of observer's head when shadow is cast onto cloud deck below observer's elevation.

GLORY-HOLE SPILLWAY Giant funnel spillway.

GLOVEBAG Polyethylene or polyvinyl chloride bag-like enclosure affixed around asbestos-containing source (most often thermal system insulation) permitting material to be removed while minimizing release of airborne fibers to surrounding atmosphere.

GLUCOSE Simple sugar that's important energy source in living organisms and is component of many carbohydrates.

GMN Minute of hour that GSP occurred, reported at last hourly 10-minute segment.

GNEISS Metamorphic rock that displays distinct banding of light and dark mineral layers.

GO-DEVIL 1. Weight dropped into bore or oil well to explode cartridge previously lowered; 2. Device driven through pipe by pressure to clear away obstructions; 3. Rough sled or dray used for dragging logs, hauling stone, etc.

GOGGLE VALVE Valve to positively isolate downstream flow.

GONDOLA 1. Hanging platform or car for transporting people or cargo; 2. Open railway car with low sides, used to carry heavy freight; 3. Small long, narrow boat.

GOOSENECK Portion of water service connection between distribution system water main and meter that's sometimes called pigtail.

GORE Triangular piece of land where roads meet.

GORE AREA Triangular area between lanes of roadway and ramp.

GOVERNING INSTRUMENTS Legal documents or other instruments under which registrant was organized or created or under which it will operate.

GOVERNOR SWITCH Switch that removes power from motor or actuates control circuit to reduce speed.

GPS Space-based radio positioning systems that provide 24-hour, three-dimensional position, velocity and time information to suitably equipped users anywhere on or near surface of Earth.

GRAB SAMPLE Single sample collected at particular time and place that represents composition of water, air or soil only at that time and place.

GRABEN Elongated block of Earth's crust, bounded by faults, that has dropped relative to surrounding area.

GRADATION 1. Proportion of material of each particle size (i.e., soil, sediment, sedimentary rock); 2. Frequency distribution (i.e., variation) of various sizes; 3. Sequence of gradual, successive stages (e.g., geological process, stream bed evolution).

GRADATION CURVE Graph showing grain size versus accumulated percent of material finer than that grain size.

GRADE 1. Elevation of surface or surface slope; 2. Inclination (slope) of pipeline, conduit, stream channel; 3. Elevation of invert of bottom of pipeline, canal, culvert or conduit; 4. Natural ground surface in reference to horizontal plane; 5. Harsh scraping and/or cutting.

GRADE AND SLOPE CONTROL System that controls machine for transverse slope and longitudinal grade.

GRADE AND SLOPE EQUIPMENT Devices that both level ground and change inclination.

GRADE ASSISTANCE Support given to machine when descending elevation to resist gravitational force on its weight.

GRADE BREAK Location of specified slope change.

GRADE CONTROL STRUCTURE Structure built across drainageway that prevents gully erosion.

GRADE CORRECTION Correction applied to slope over distance.

GRADE CROSSING At-grade crossing between railroad line and ordinary road, with tracks and road at same level.

GRADE CROSSING PROTECTION SYSTEM Railroad crossing warning device that predicts arrival time of trains to grade crossing.

GRADE LINE Longitudinal reference line or slope.

GRADE NO. 1 FUEL OIL Light distillate fuel with lowest boiling point.

GRADE NO. 2 FUEL OIL Home heating fuel with boiling point greater than No. 1.

GRADE NO.s 4-6 FUEL OIL Fuel oils to ease pumping and burning that need preheating.

GRADE PERCENTAGE Slope of grade referenced to tangent of angle of that surface to horizontal.

GRADE PIN Rod, used with string, that indicates required grade to be filled.

GRADE RING Ring placed on top of manhole to raise manhole cover flush with surface.

GRADE SECTION Part of cross-section graded to given elevations.

GRADE SEPARATION Structure for crossing two or more surface transport axes at different heights.

GRADE STAKE Stake indicating amount of cut or fill required to bring ground to specified level or to mark predetermined elevation.

GRADE TUNNEL Tunnel with top elevation above hydraulic gradient.

GRADE-CONTROL STRUCTURE Structure built to stabilize and reduce channel bed erosion.

GRADED AGGREGATE Aggregate made up from particles of distinct, given sizes.

GRADED BEDDING Sediment deposit where grain size within bed decreases upwards.

GRADED FILTER Filter that consists of several layers of unconsolidated material with coarsest layer on bottom.

GRADED RIVER River that's slope and channel have developed such that it neither degrades nor aggrades its bed.

GRADED SLOPE Downstream gradient of stream considered graded.

GRADED STANDARD SAND Ottawa sand that has been subjected to accurate grading between No. 30 and No. 100 sieves.

GRADED STREAM Streams that receive and carry away equal amounts of sediment.

GRADER Machine with centrally located blade that can be angled to cast to either side, with independent hoist control on each side.

GRADIENT 1. General slope or rate of change in vertical elevation per unit of horizontal distance of water surface of flowing stream; 2. Slope along specific route (i.e., road surface, channel, pipe); 3. Rate of change with respect to distance of variable quantity in direction of maximum change; 4. Gradual change in color.

GRADIENT HIGH WINDS Winds that usually cover large area and are due to synoptic-scale, extra-tropical low pressure systems.

GRADIENT OF SLOPE Change in rise over change in fall.

GRADIENT SPEED Theoretical wind speed at 600 m (12,000 ft) that can be calculated from isobars.

GRADIENTER Surveyor's instrument for determining grades.

GRADING Work of ensuring level base or slope for construction work.

GRADING BUCKET Bucket on machine used for finish work in construction.

GRADING CURVE Graphical rendering of percentages of different particle sizes.

GRADING LIMITS Range of allowable particle sizes, according to sieve analysis evaluation, for general construction and highway work.

GRADING PLAN Prepared as part of permitting process, plan that shows existing and proposed site topography plus limits of grading and disturbance.

GRADIOMETER Any of several instruments used to measure gradient of physical property.

GRADUAL COMMENCEMENT Commencement of geomagnetic storm that has no well-defined onset.

GRADUAL CONTRACTION Constant decrease of cross-sectional area.

GRADUATED ACTING Device that gives throttling control.

GRADUATED FARE Fare proportional to distance traveled or passenger ride time.

GRAIN LOADING Rate at which particles are emitted from pollution source.

GRAIN SHAPE FACTOR Particle shape factor of perfect sphere is 1.0 and can be as low as 0.1 for very irregular shapes.

GRAIN SIZE Diameter of particle determined by sieve or sedimentation methods comprising particular soil that's used to characterize size of particle.

GRAIN SIZE CLASSIFICATION Soil classified according to its grain size.

GRAIN SIZE DISTRIBUTION Graph showing grain size versus accumulated percent of material finer than that grain size.

GRAM Unit of mass equal to 1/1,000 kilogram.

GRAM PER CUBIC CENTIMETER Unit of density equal to 1 gram/cm^3.

GRAM REACTION Reaction obtained when bacteria in laboratory are subjected to certain staining procedure called Gram's stain.

GRANITE Light-colored, coarse-grained intrusive igneous rock with quartz and feldspar as dominant minerals and typically peppered with mica and hornblende.

GRANITIC ROCK Light-colored, coarse-grained intrusive igneous rock with quartz and feldspar as dominant minerals and typically peppered with mica and hornblende.

GRANODIORITE Coarse-grained intrusive igneous rock with less quartz and more feldspar than true granite and typically darker.

GRANULAR ACTIVATED CARBON TREATMENT Filtering system often used in small water systems and individual homes to remove organics.

GRANULAR DUST Finely powdered stone or rock dust.

GRANULAR FILL Broken or crushed stone, gravel, reclaimed miscellaneous aggregate or mixture thereof.

GRANULAR SOIL Gravel, sand or silt with little or no clay content.

GRANULATION Cellular structure of photosphere visible at high spatial resolution.

GRANULOMAS Inflammatory nodules found in many diseases, consisting of histiocytes (macrophages) attempting to wall off substances they perceive as foreign but are unable to eliminate.

GRAPHITE Planes of carbon atoms arranged in hexagonal arrays with planes stacked loosely.

GRAPPLE 1. Clamshell-type bucket having three or more jaws; 2. Tool with claws or hooks which is used to catch or hold something.

GRAPPLE DREDGE Dredger that operates with clamshell, orange-peel or other bucket.

GRAPPLE LEG Primary legs of grapple.

GRAPPLER Tool consisting of several hooks for grasping and holding.

GRASS Any plant of family Poaceae, characterized by leaves that arise from nodes in stem and leaf bases that wrap around stem.

GRASSCYCLING Source reduction activities where grass clippings are left on lawn after mowing.

GRASSED BUFFERS Strips of grass or other close-growing vegetation that separates waterway (ditch, stream, creek) from intensive land use area (e.g., subdivision, farm).

GRASSED WATERWAY Natural or constructed watercourse or outlet shaped or graded and established in suitable vegetation for disposal of runoff water without erosion.

GRASSLAND Area dominated by grass or grasslike vegetation.

GRATE Horizontal metal grill where water, ash or small objects can fall, while larger objects cannot.

GRATE SIFTINGS Materials that fall from solid waste fuel bed through grate openings.

GRATED WASTE Fitting with flush grating to prevent foreign matter entering waste system.

GRATING Barrier that has parallel or crossed bars blocking passage but admitting air.

GRAUPEL 1. Snow pellets; 2. Small hail.

GRAVEL Loose rounded fragments of rock that will pass 3-inch sieve and be retained on No. 4 U.S. Standard sieve (3/16 inch).

GRAVEL BLANKET Thin layer of gravel spread over area either of natural ground, excavated surface or embankment.

GRAVEL FILL loose aggregation of rock fragments layer to fill excavation.

GRAVEL FRACTION Fraction of particles between 2.4 and 0.078 in. (60 and 2 mm).

GRAVEL GUARD Screen that prevents gravel from entering.

GRAVEL PACK Downhole filter designed to prevent production of unwanted formation sand.

GRAVEL PUMP Centrifugal pump designed for continuous handling of most difficult higher abrasive slurry, containing big solids.

GRAVEL SURFACING Layer of gravel spread over area intended for vehicular or personnel traffic.

GRAVEL-WALL WELL Well that uses larger slot sizes in well screen section.

GRAVELFILL Gravel used to fill holes or spaces.

GRAVIMETRIC Of or pertaining to measurement by weight (i.e., local gravitational field).

GRAVITATIONAL CONVECTION Convection of contents of fluid by means of currents induced in fluid by buoyancy forces which are dependent on gravity acting upon density differences within fluid.

GRAVITATIONAL POTENTIAL Potential energy that unit mass would have at any point.

GRAVITATIONAL WATER Water that moves through soil due to force of gravity.

GRAVITY Resultant force on Earth's surface, of attraction by Earth's masses and centrifugal pseudo-force caused by Earth's rotation.

GRAVITY CIRCULATION Plumbing circuit that uses temperature differential to move water around piping system.

GRAVITY DAM Concrete structure proportioned so that its own weight provides major resistance to forces exerted on it.

GRAVITY DRAINAGE Water drainage by gravity.

GRAVITY FEED Supply system making use of gravity to maintain flow of material.

GRAVITY FILTER Filter used in water purification to collect solid products.

GRAVITY FLOW Movement of water that goes downslope under effect of gravity on sloping bottom.

GRAVITY GROUNDWATER Water withdrawn in saturation zone by force of gravity.

GRAVITY IRRIGATION Irrigation method that applies irrigation water to fields by letting it flow from higher level supply canal through ditches or furrows to fields at lower level.

GRAVITY MAIN Conduit unpressurized and relies on gravity to move water through main.

GRAVITY PIPE Conduit not designed for water under pressure.

GRAVITY SEPARATION METHODS Uses specific gravity for treatment of mineral particles.

GRAVITY SPRING Spring that water releases in response to direct action of gravity.

GRAVITY SUPPLY Water distribution system that relies on gravity to supply water instead of pressure.

GRAVITY SYSTEM Water supply system that relies entirely on force of gravity to create pressure and cause water to flow through system.

GRAVITY TANK water storage tank where water is stored at atmospheric pressure and distributed by gravity.

GRAVITY WATER Water supplied by gravity flow.

GRAVITY WAVE Wave created by action of gravity on density variations in stratified atmosphere (e.g., lee waves, mountains waves).

GRAY Unit of absorbed dose of ionizing radiation equal to 1 Joule/kg.

GRAY WATER Water not clean enough to be potable but not contaminated with fecal matter or other source of dangerous bacteria or dangerous or noxious materials.

GRAYBODY Hypothetical "body" that absorbs some constant fraction of all electromagnetic radiation incident upon it.

GREASE 1. Oily or fatty matter; 2. Lubricate.

GREASE REMOVAL TANK Tank that encourages flotation of oil and grease.

GREASE SKIMMER Apparatus that removes grease or scum in wastewater treatment tanks.

GREASE TRAP Plumbing device designed to intercept most greases and solids before they enter wastewater disposal system.

GREAT CIRCLE TRACK Shortest distance between two points on sphere and when viewed on 2-dimensional map track will appear curved (e.g., swell waves).

GREAT LAKES FREEZE-UP/BREAK-UP OUTLOOK National Weather Service product to keep mariners informed of projected freeze-up date or break-up date of ice on Great Lakes.

GREAT LAKES MARINE FORECAST National Weather Service coded summary appended to each of Great Lakes Open Lakes forecasts.

GREAT LAKES STORM SUMMARY National Weather Service forecast product providing updated information whenever storm warning is in effect on any of Great Lakes.

GREAT LAKES WEATHER BROADCAST National Weather Service product containing observation summary prepared to provide Great Lakes mariners with listing of weather observations along or on Lakes.

GREEN ALGAE Unicellular or colonial algae with chlorophyll pigments similar to those in terrestrial green plants.

GREEN BELT Area of agricultural land around urban area protected from large-scale housing.

GREEN INFRASTRUCTURE Array of products, technologies and practices that use natural systems or engineered systems that mimic natural processes, to enhance overall environmental quality and provide utility services.

GREEN LINE One of strongest (and first-recognized) visible coronal lines.

GREEN MANURE Crop ploughed into soil to improve its fertility.

GREEN ROOF Planted over existing roof structures and consist of waterproof, root-safe membrane covered by drainage system, lightweight growing medium and plants.

GREEN SLUDGE Sewage sludge where small bacterial decomposition has taken place.

GREEN TAGS Represent technology and environmental attributes of electricity generated from renewable sources.

GREEN TIME Duration of green signal for intersection movement.

GREEN TIME RATIO Ratio of green time to signal cycle length.

GREEN VITRIOL Common name for copperas, iron vitriol or ferrous sulfate.

GREENHOUSE EFFECT 1. Warming of Earth's atmosphere attributed to buildup of carbon dioxide or other gases; 2. Atmospheric heating caused by solar radiation being readily transmitted inward through Earth's atmosphere but longwave radiation less readily transmitted outward, due to absorption by certain gases in atmosphere.

GREENHOUSE GAS Gas (e.g., carbon dioxide, methane) that contributes to potential climate change that absorbs terrestrial radiation and contributes to greenhouse effect.

GREENHOUSE GASES Gases that contributes to greenhouse effect when released into atmosphere.

GREENSAND Greenish sandstone containing glauconite.

GREY WATER Water not clean enough to be potable but not contaminated with fecal matter or other source of dangerous bacteria or dangerous or noxious materials.

GRID 1. System of interconnected power lines and generators for delivery of electricity; 2. Rectangular array of squares or rectangles of equal size.

GRID OPERATOR Entity that oversees delivery of electricity over grid to customer, ensuring reliability and safety.

GRID ROLLER Soil compaction roller that brings about increase in soil density.

GRID SYSTEM WATER MAINS Interconnecting system of water mains in crisscross or rectangular pattern.

GRIDIRON Water distribution system where all conduits are joined with other conduits at road junctions.

GRINDER PUMP Mechanical device that shreds solids and raises sewage to higher elevation through pressure sewers.

GRIP Channel cut through grass verge.

GRIT Collection of hard, small materials.

GRIT CATCHER Baskets to remove debris and grit from sewers.

GRIT CHAMBER Long, narrow tanks that are designed to slow down flow so that solids will settle out of water.

GRIT CHANNEL Enlargement in collection line where grit can easily settle out of flow.

GRIT COLLECTOR Device placed in grit chamber to carry deposited grit to point of collection for ultimate disposal.

GRIT COMPARTMENT Portion of grit chamber where grit is collected and stored before removal.

GRIT REMOVAL Process used to remove sand, silt and grit from water.

GRIT WASHER Apparatus for washing organic matter from grit.

GRIZZLIE 1. Coarse screen used to remove oversize pieces from earth or blasted rock; 2. Gate or closure on chute.

GRIZZLY 1. Coarse screen used to remove oversize pieces from earth or blasted rock; 2. Gate or closure on chute.

GROIN 1. Contact between upstream or downstream face of dam and abutments; 2. Structure constructed perpendicular to shoreline (i.e., across beach) to trap sand.

GROOVED COUPLING Clamp-kind gasketed conduit coupling.

GROSS ALPHA PARTICLE ACTIVITY Total measured alpha particle activity in water sample.

GROSS ALPHA/BETA PARTICLE ACTIVITY Total radioactivity due to alpha or beta particle emissions as inferred from measurements on dry sample.

GROSS AVAILABLE HEAD Pressure difference between two points in water stream.

GROSS CROP VALUE Sum of annual receipts from sale of crops produced.

GROSS DUTY OF WATER Relation between quantity of water diverted from source of supply and total area of land irrigated by canal system.

GROSS ERROR Errors so large that cannot be attributed to either systematic or random errors.

GROSS GENERATION Total amount of electrical energy produced by generating station or stations, measured at generator terminals.

GROSS HORSEPOWER Power output of engine running with no external engine-driven accessories.

GROSS POWER-GENERATION POTENTIAL Installed power generation capacity that landfill gas can support.

GROSS REGISTERED TONNAGE Vessel's total internal volume.

GROSS VEHICLE WEIGHT Combined weight of vehicle and all passengers and cargo components.

GROSS WEIGHT 1. Total weight of vehicle and its freight, fuel, passengers, etc.; 2. Total weight of product and its packaging.

GROUND AIR Gas that opens to surface in interstices of aeration zone.

GROUND BLIZZARD WARNING When blizzard conditions are solely caused by blowing and drifting snow.

GROUND CLUTTER Pattern of radar echoes from fixed ground targets (e.g., buildings, hills) near radar.

GROUND COVER 1. Plants grown to keep soil from eroding; 2. Something that overlays area of land.

GROUND DUCT Channel in ground that contains utility pipes, cables and services.

GROUND EFFECT Aerodynamic effects occurring due to vehicle's body or appendages moving in close proximity to ground.

GROUND FOG Fog produced over land by cooling of lower atmosphere as it comes in contact with ground.

GROUND GARBAGE Shredded garbage.

GROUND HEAT FLUX Rate of transfer of heat from ground to Earth's surface that's component of surface energy budget.

GROUND ICE All types of ice contained in freezing and frozen ground.

GROUND KEY VALVE Tapering valve that can be opened or closed with 90° turn of stopper.

GROUND MORAINE Irregular blanket of till deposited under glacier.

GROUND MOTION General term including all aspects of ground motion, namely particle acceleration, velocity or displacement, from earthquake or other energy source.

GROUND MOTION PARAMETERS Numerical values representing vibratory ground motion, (e.g., particle acceleration, velocity, displacement, frequency content, predominant period, spectral intensity, duration).

GROUND RECEIVE SITES Satellite dish and associated computer which receives signals from GOES satellite, decodes information and transmits it to another site for further processing.

GROUND SILLS Submerged walls that protect stream bed.

GROUND SOURCE HEAT PUMP Electrically powered systems that tap stored energy of Earth.

GROUND STROKE Current that propagates along ground from point where direct stroke of lightning hits ground.

GROUND WIRE Wire that provides conducting path to earth that's independent of normal current-carrying path.

GROUND-FAULT CIRCUIT INTERRUPTER Electrical device designed to protect people (not equipment) from electrical shock.

GROUND-LEVEL CONCENTRATION Concentration in air of pollutant to which human being is normally exposed (i.e., between ground and height of 6 feet above).

GROUND-LEVEL STORAGE Shallow tank for water distribution where outlet at or below ground-level.

GROUND-PENETRATING RADAR Geophysical method that uses high frequency electromagnetic waves to obtain subsurface information.

GROUNDED ICE Ice that has run aground or is contact with ground underneath it.

GROUNDWATER Water that exists beneath earth's surface in underground streams, wells and aquifers that's part of subsurface water in saturated zone; 2. Water stored underground in rock crevices and in pores of geologic materials that make up Earth's crust; 3. Phreatic water.

GROUNDWATER ARTERY Confined subsurface zone of permeable material saturated, under artesian pressure and surrounded by less-permeable material.

GROUNDWATER BASIN Groundwater system that has defined boundaries and may include more than one aquifer of permeable materials, which are capable of furnishing significant water supply.

GROUNDWATER CASCADE Near-vertical flow of groundwater over groundwater barrier.

GROUNDWATER DAM Structure that obstructs groundwater flow through aquifer.

GROUNDWATER DECREMENT Groundwater gradually extracted from subsurface reservoir.

GROUNDWATER DEPLETION CURVE Recession curve where slope of curve represents depletion rate (i.e., runoff) of groundwater.

GROUNDWATER DISCHARGE 1. Movement of groundwater from subsurface to surface; 2. Water released from zone of saturation; 2. 3. Groundwater entering near coastal waters which has been contaminated by landfill leachate, deep well injection of hazardous wastes, septic tanks, etc.

GROUNDWATER DISCHARGE AREA Area that groundwater is discharged (e.g., wells, springs, percolation).

GROUNDWATER DISINFECTION RULE 1996 amendment of Safe Drinking Water Act that requires disinfection as for all public water systems, including surface waters and groundwater systems.

GROUNDWATER DIVIDE Boundary between two groundwater basins represented by high point in water table or piezometric surface.

GROUNDWATER DRAIN Drain that carries away groundwater.

GROUNDWATER FLOW 1. Rate of water flow in aquifer or soil; 2. Movement of water in zone of saturation; 2. Streamflow that results from precipitation that infiltrates into soil and eventually moves through soil to stream channel (i.e., baseflow, dry-weather flow).

GROUNDWATER FLOW MODEL Application of mathematical model to represent regional or site-specific groundwater flow system.

GROUNDWATER FLOW SYSTEM Underground pathway by which groundwater moves from areas of recharge to areas of discharge.

GROUNDWATER HYDROLOGY Branch of hydrology that specializes in groundwater, its occurrence and movements, its replenishment and depletion, properties of rocks that control groundwater movement and storage and methods of investigation and utilization of groundwater.

GROUNDWATER INCREMENT 1. Water from all sources moving down to groundwater; 2. Recharge.

GROUNDWATER INFILTRATION Process where groundwater moves into underground water and sewer systems.

GROUNDWATER INVENTORY Approximation of water kept under earth's surface within defined area.

GROUNDWATER IRRIGATION Water irrigation by subterranean sources.

GROUNDWATER LEVEL Level, at particular site, below which subsoil and rock masses of earth are fully saturated with water.

GROUNDWATER LOWERING Removing water from ground at rate faster than it can recharge.

GROUNDWATER MINING Pumping of groundwater for irrigation or other uses, at rates faster than rate at which groundwater is being recharged.

GROUNDWATER MODELING CODE Computer code used in groundwater modeling to represent non-unique, simplified mathematical description of physical framework, geometry, active processes and boundary conditions present in reference subsurface hydrologic system.

GROUNDWATER OUTFLOW That part of discharge from drainage basin that occurs through groundwater.

GROUNDWATER OVERDRAFT Pumping of groundwater for irrigation or other uses, at rates faster than rate at which groundwater is being recharged (i.e., in excess of safe yield).

GROUNDWATER PRESSURE HEAD Hydrostatic pressure exerted by fluid at equilibrium at given point within fluid.

GROUNDWATER PROVINCE Place defined with general likeness in incidence of groundwater.

GROUNDWATER RECESSION Recession curve of streamflow.

GROUNDWATER RECHARGE Process where water moves downward from surface water (i.e., flow from precipitation, infiltration from streams and other sources of water) to groundwater storage.

GROUNDWATER RESERVOIR Confined aquifer between two impervious zones or confining units.

GROUNDWATER RUNOFF That part of runoff that has passed into ground, has become groundwater and has been discharged into stream channel as spring or seepage water.

GROUNDWATER STORAGE Significant amounts of water stored in ground.

GROUNDWATER TABLE Below ground surface where ground is wholly saturated with water and where water pressure is equal to atmospheric pressure.

GROUNDWATER UNDER DIRECT INFLUENCE OF SURFACE WATER Any water beneath surface of ground with significant occurrence of insects or other microorganisms, algae or large-diameter pathogens or with significant and relatively rapid shifts in water characteristics (e.g., turbidity, temperature, conductivity, pH) which closely correlate to climatological or surface water conditions.

GROUNDWATER, CONFINED Groundwater under pressure significantly greater than atmospheric, with its upper limit bottom of bed with hydraulic conductivity distinctly lower than that of material where confined water occurs.

GROUNDWATER-STORAGE CURVE Curve that shows quantity of groundwater available for runoff at given rates of groundwater flow.

GROUP 1 ORDINARY FIRE HAZARD Quantity of combustibles is moderate, combustibility of contents is low, storage heights are 8 feet or less and moderate rates of heat release expected.

GROUP 2 ORDINARY FIRE HAZARD Quantity of combustibles is moderate to high, combustibility of contents is moderate to high, storage heights do not exceed 12 feet and moderate to high rates of heat release expected.

GROUP 3 ORDINARY FIRE HAZARD Combustibility of contents is high with fires of high rates of heat release are expected.

GROUP I STORM WATER DISCHARGE Storm water point source subject to effluent limitations guidelines, new source performance standards or toxic pollutant effluent standards or located at industrial plant or in plant associated areas.

GROUP VELOCITY Speed at which particular wave front or swell train advances.

GROUP VENT Branch vent that serves two or more traps.

GROUT Fluid mixture of cement and water or sand, cement and water injected into soil, rock, concrete or other construction material used to seal joints and cracks in rock foundation, to seal openings or to lower permeability and/or provide additional structural strength.

GROUT BLANKET Area of foundation systematically grouted to uniform shallow depth.

GROUT CAP Concrete filled trench or pad encompassing all grout lines constructed to impede surface leakage and to provide anchorage for grout connections.

GROUT CURTAIN Barrier produced by injecting grout into vertical zone, usually narrow (horizontally) and in foundation to reduce seepage under or around dam.

GROUT CUTOFF Barrier produced by injecting grout into vertical zone, usually narrow (horizontally) and in foundation to reduce seepage under or around dam.

GROUTING Filling cracks and crevices with cement mixture.

GROWING DEGREE DAY Number of degrees that average temperature is above baseline value.

GROWING SEASON Frost-free period of year (i.e., time between last killing frost of spring and first killing frost of autumn) during which climate is such that crops can be produced.

GROWLER Similar to bergy bit, but smaller, extending less than 1 meter above sea surface and occupying area of 20 m^2 or less.

GROWTH Increase in size, number, value or strength.

GROWTH FACTOR Multiplier to current demands to estimate future demands.

GROWTH RATE Number of generations (i.e., bacteria) per unit time.

GRUBBING Removal of stumps, roots and vegetable matter from ground surface after clearing and prior to excavation.

GUANO Dung from sea bird or bat.

GUARANTEED MAXIMUM COST Total expense established in contract.

GUARANTEED MAXIMUM PRICE Cost-type contract where contractor is paid for actual costs incurred plus fixed fee subject to ceiling price.

GUARANTY Undertaking to answer for payment of debt, performance of contract or duty.

GUARD Part of machine that blocks access to dangerous parts.

GUARD BOARD Board positioned on border of scaffolding to stop items from falling.

GUARD GATE First gate in series of flow controls, remaining open while downstream gates or valves are operating.

GUARD LOCK Lock situated between enclosed basin or canal and tide water of harbor or river so that craft can pass either way at all times of tide.

GUARDED Protected, covered or shielded.

GUARDRAIL Rail set alongside dangerous place in order to improve safety.

GUIDE BANK Embankments with stone pitching in slopes facing water to guide river.

GUIDE SIGN Signs that provide directional and mileage information to specific destinations.

GUIDE SPECIFICATION Minimum requirements to be used as basis for evaluation.

GUIDE VANE Fixed grooves found in turbines that help direct water or gas around bends at maximum efficiency.

GUIDE WALL Parallel walls constructed along side of wall as guide to clamshell used for excavation of diaphragm wall trenches.

GUIDEWAY Track where component in machine or automated transit vehicle is guided.

GULF OPEN LAKES FORECAST National Weather Service marine forecast product for U.S. waters within Great Lake not including waters covered by existing Nearshore Waters Forecast.

GULF STREAM Warm water current extending from Gulf of Mexico and Florida up U.S. east coast then east northeast to Iceland and Norway.

GULLY Trench, ravine or narrow channel that was worn by water flow.

GULLYING Small-scale stream erosion.

GUMBO Fine, silty soil that when wet becomes very thick and heavy.

GUSSET PLATE Plate for connecting beams and girders to columns.

GUST Rapid fluctuation of wind speed.

GUST EFFECT FACTOR Ratio of maximum wind response to mean wind response of structure.

GUST FRONT Leading edge of gusty surface winds from thunderstorm downdrafts.

GUSTINESS Air turbulence close to ground caused by obstructions that prevent direct flow.

GUSTNADO Small, whirlwind that forms as eddy in thunderstorm outflows.

GUT Narrow passage of water.

GUTTATION Removal of liquid water through stomata from healthy leaf or plant stem as result of root pressure.

GUTTER 1. Channel in surface intended for drainage of water; 2. Duct or channel beneath eaves of building to carry rain water.

GUTTER TOOL Scoop tool tailored to fit inside gutters for cleaning.

GUYOT Flat-topped submarine mountain.

GYPSUM Mineral consisting of hydrated calcium sulphate.

GYRATORY CRUSHER Machine for crushing raw material.

GYRES Oceanic current systems of planetary scale driven by global wind system.

H

H-ALPHA Absorption line of neutral hydrogen that falls in red part of visible spectrum and is convenient for solar observations.

H-COMPONENT OF GEOMAGNETIC FIELD Components of geomagnetic field at surface of Earth.

H5 500 millibar level height near 18,000 ft in standard atmosphere.

H8 850 millibar level height near 5,000 ft in standard atmosphere.

HABIT Tendency or practice of animal or plant.

HABITAT Place where population (e.g., human, animal, plant, microorganism) lives and its surroundings, both living and non-living.

HABITAT CAPABILITY Estimated carrying capacity of area to support wildlife, fish or sensitive plant population.

HABITAT INDICATOR Physical, chemical or biological attribute measured to characterize conditions necessary to support organism, population, community or ecosystem in absence of stressors (e.g., pollutants).

HABITAT LOSS Destruction, fragmentation or degradation of habitat.

HABITAT MANAGEMENT Management to conserve, protect and restore habitat areas for plant and animal species.

HABITAT QUALITY INDEX Qualitative description of in-stream habitat and riparian conditions surrounding reach sampled.

HABITUATION Process of becoming accustomed to internal or external stimulus.

HAGEN-POISEUILLE LAW Physical law that gives pressure drop in incompressible and Newtonian fluid in laminar flow flowing through pipe of constant cross-section.

HAGUE LINE North Atlantic boundary between U.S. and Canada fishing waters as determined by World Court in Hague, Netherlands.

HAIL Showery precipitation in form of irregular pellets or balls of ice more than 5 mm in diameter, falling from cumulonimbus cloud.

HAIL CONTAMINATION Limitation in NEXRAD rainfall estimates whereby abnormally high reflectivity's associated with hail are converted to rainfall rates and rainfall accumulations.

HAIL INDEX Indication of whether thunderstorm structure of each storm identified is conducive to production of hail.

HAIL SIZE Diameter of hailstones.

HAIL SPIKE Area of reflectivity extending away from radar immediately behind thunderstorm with extremely large hail.

HAINES INDEX Index composed of stability term and moisture term that indicates potential for large fire growth.

HALF-LIFE 1. Time required for any specified property to decrease by half; 2. Time taken for radioactivity of specified isotope to half its original value.

HALF-ROUND PIPE Conduit cut in half along its length.

HALF-TIDE LEVEL Plane midway between mean high and mean low water.

HALF-VALUE LAYER Thickness of any given material where 50% of incident energy has been attenuated.

HALOCLINE Strong, vertical salinity gradient.

HALOGEN 1. Type of incandescent lamp with higher energy-efficiency that standard ones; 2. Any element (i.e., fluorine, chlorine, bromine, iodine and astatine) that form salt by direct union with metal.

HALOGENATED FLUOROCARBONS Human-made compounds that include: chlorofluorocarbons, halons, methyl chloroform, carbon tetrachloride, methyl bromide and hydrochlorofluorocarbons.

HALOGENATION Reaction of halogen with something.

HALOMORPHIC SOIL Soil that contains large amounts of soluble salts.

HALON 1. Hydrocarbon where one or more hydrogen atoms have been replaced by halogens; 2. Bromine-containing compounds with long atmospheric lifetimes whose breakdown in stratosphere causes depletion of ozone.

HALOPHYTE Any plant that tolerates environment having high salt content.

HALOPHYTIC 1. Plants that thrive in soils that contain salty or alkaline (e.g., sodium) soil; 2. Salt-loving.

HAMADA High, desert landscape from which most of sand has been removed by wind erosion.

HAMMER DRILL Rotary drill with hammering action.

HAMMER MILL High-speed machine that uses hammers and cutters to crush, grind, chip or shred solid waste.

HAMMERMILL Machine for shredding or crushing aggregate material.

HAMMERMILL CRUSHER Stone crusher that uses hammers or flails.

HAND BORING Drilling of holes by hand for site investigations or for exploration of shallow mineral deposits.

HAND WHEEL Wheel or disk used in machinery as convenient form of circular crank.

HANDICAPPED PERSON Individual with condition that restricts their ability to function physically, mentally or socially.

HANDLING TIGHT Sufficiently tight so that coupling cannot be removed except by use of wrench.

HANGER Support for conduit.

HANGING DAM Mass of ice composed mainly of frazil or broken ice deposited underneath ice cover in region of low flow velocity.

HANGING VALLEY Valley cut across by deeper valley or cliff.

HARBOR Sheltered expanse of water, adjacent to land, that ships may dock or anchor.

HARD CONVERSION Change in measurement units.

HARD HAT Helmet used on construction sites to protect head from falling objects.

HARD SUCTION Suction hose used in drafting operations (e.g., vacuum to draw water from portable water tank, pool).

HARD WATER Water high in mineral content.

HARDENING Process of reducing surface vulnerability by making surface harder.

HARDNESS 1. Water-quality indication of concentration of alkaline salts in water, mainly calcium and magnesium; 2. Property of water that causes formation of insoluble residue when water is used with soap and scale in vessels where water has been allowed to evaporate; 3. Physical-chemical characteristic that commonly is recognized by increased quantity of soap required to produce lather.

HARDPAN Hard, impervious layer that limits downward movement of water and roots; 2. Dry lake.

HARDWOOD Wood from any dicotyledonous tree, without regard to its hardness.

HARMONIC Component frequency of signal of wave that's integer multiple of fundamental frequency.

HARMONIC MEAN FLOW Number of daily flow measurements divided by sum of reciprocals of flows (i.e., reciprocal of mean of reciprocals).

HARROW Agricultural tool that loosens and works ground surface.

HARVEST Number of fish that are caught and kept.

HASS Inner curve of pipe bend.

HATCHERY Place for hatching fish eggs.

HAUL DISTANCE Distance measured along center line or most direct practical route between center of mass of excavation and center of mass of fill as finally placed.

HAUL TIME Travelling time from excavation to dump area.

HAULER Garbage collection company that offers complete refuse removal service.

HAUNCH Squat vertical support structure.

HAUNCHES 1. Outside areas between springline and bottom of pipe; 2. Sides of lower third of circumference of pipe.

HAUNCHING 1. Concrete support at sides of drain or sewer conduit above bedding; 2. Placing bedding material around haunch of conduit.

HAZARD 1. Something that creates potential for adverse consequences (e.g., loss of life, property damage, potential for radiation, chemical or other pollutant to cause human illness or injury); 2. Inherent toxicity of compound.

HAZARD ASSESSMENT Evaluating effects of stressor or determining margin of safety for organism by comparing concentration which causes toxic effects with estimate of exposure to organism.

HAZARD CATEGORY Division of criteria within hazard classes.

HAZARD CLASS Group of materials that share common major hazardous property (e.g., radioactivity, flammability).

HAZARD CLASSIFICATION Rating on potential consequences of failure.

HAZARD COMMUNICATION STANDARD OSHA regulation that requires chemical manufacturers, suppliers and importers to assess hazards of chemicals that they make, supply or import and to inform employers, customers and workers of these hazards through MSDS information.

HAZARD EVALUATION Component of risk evaluation that involves gathering and evaluating data on types of health injuries or diseases that may be produced by chemical and on conditions of exposure under which such health effects are produced.

HAZARD IDENTIFICATION Determining if chemical or microbe can cause adverse health effects in humans and what those effects might be.

HAZARD INSURANCE Risks covered by homeowners insurance policy.

HAZARD POTENTIAL Possible adverse incremental consequences that result from failure or misoperation.

HAZARD POTENTIAL CLASSIFICATION System that categorizes according to degree of adverse incremental consequences of failure or misoperation.

HAZARD QUOTIENT Ratio of estimated site-specific exposure to single chemical from site over specified period to estimated daily exposure level, at which no adverse health effects are likely to occur.

HAZARD RATIO Term used to compare animal's daily dietary intake of pesticide to its (Lethal Dose) LD_{50} value.

HAZARD REDUCTION Treatment of risk that reduces and/or removes it.

HAZARDOUS AIR POLLUTANTS Air pollutants (e.g., asbestos, beryllium, mercury, benzene, coke oven emissions, radionuclides, vinyl chloride) that are not covered by ambient air quality standards but which, as defined in Clean Air Act, may present threat of adverse human health effects or adverse environmental effects.

HAZARDOUS ATMOSPHERE When atmosphere has too much or too little oxygen, contains flammable, combustible or explosive agents or contains contaminants that could pose immediate threat to life or interfere with person's ability to escape unaided from confined space.

HAZARDOUS CHEMICAL Any hazardous material requiring MSDS under OSHA's Hazard Communication Standard.

HAZARDOUS MATERIAL Substance, natural or man-made, intrinsically dangerous or otherwise poses safety hazard.

HAZARDOUS MATERIALS Materials that pose potential for grave, immediate, future and genetic injury and illness when handled without proper equipment and precautions.

HAZARDOUS MATERIALS RESPONSE TEAM Team of specially trained personnel who respond to hazardous materials incident.

HAZARDOUS POLLUTANT Pollutants that are known or suspected to cause cancer or other serious health effects.

HAZARDOUS RANKING SYSTEM Principal screening tool to evaluate risks to public health and environment associated with abandoned or uncontrolled hazardous waste sites.

HAZARDOUS SEAS WARNING Warning for wave heights and/or wave steepness values meeting or exceeding locally defined warning criteria.

HAZARDOUS SEAS WATCH Watch for increased risk of hazardous seas warning event to meet Hazardous Seas Warning criteria but its occurrence, location, and/or timing is still uncertain.

HAZARDOUS SUBSTANCE Any material that poses threat to human health and/or environment (i.e., substances that are toxic, corrosive, ignitable, explosive, chemically reactive).

HAZARDOUS WASTE By-products of society that can pose substantial or potential hazard to human health or environment when improperly managed.

HAZARDOUS WASTE INCINERATION Converts waste into ash, flue gas and heat.

HAZARDOUS WASTE INDICATORS 1. Indicators that may signal presence of hazardous materials include stressed vegetation or unusual lack of vegetation; 2. Dead or sick domestic stock, wildlife or birds; 3. Fish kills or otherwise unexplained stream sterility or diminished species and numbers of flora and fauna; 4. Unusual coloration or discoloration of land surface and acrid or other chemical odors.

HAZARDOUS WASTE LANDFILL Excavated or engineered site where hazardous waste is deposited and covered.

HAZARDOUS WASTE MANAGEMENT FACILITY Facilities that receive hazardous wastes for treatment, storage or disposal.

HAZARDOUS WASTE MANAGEMENT UNIT Containers, tanks, drip pads, buildings, incinerators, furnaces, landfills, impoundments, waste piles and land treatment units that protect human health and environment from risks posed by hazardous waste.

HAZARDOUS WASTE MINIMIZATION Reducing amount of toxicity or waste produced by facility via source reduction or environmentally sound recycling.

HAZARDOUS WASTE SITES Any spill, authorized or unauthorized dumping, abandoned or inactive waste disposal sites containing or suspected of containing hazardous materials.

HAZARDOUS WEATHER OUTLOOK Narrative statement produced by National Weather Service to provide information regarding potential of significant weather expected during next 1 to 5 days.

HAZARDS ANALYSIS Procedures used to identify potential sources of release of hazardous materials from fixed facilities or transportation accidents or to compare hazards to determine which present greater or lesser risks to community.

HAZARDS ASSESSMENT Provides emergency managers, planners, forecasters and public advance notice of potential hazards related to climate, weather and hydrological events.

HAZARDS IDENTIFICATION Providing information on which facilities have extremely hazardous substances, what those chemicals are, how much there is at each facility, how chemicals are stored and whether they are used at high temperatures.

HAZE Aggregation in atmosphere of very fine, widely dispersed, solid or liquid particles or both, giving air opalescent appearance that subdues colors.

HAZEN NUMBER Unit of measurement for water color.

HCFC'S Hydrochlorofluorocarbon that are second generation of fluorine based refrigerants that replaced chlorofluorocarbons (CFC's).

HDRAIN Hourly Digital Rainfall Product of WSR-88D.

HEAD 1. Differential of pressure or vertical distance that would statically result from velocity of moving fluid; 2. Height of water above specified point; 3. Vertical distance between two points in fluid (e.g., pool height and tailwater height); 4. Deposits near top of geological succession; 5. Back-pressure against pump.

HEAD BAY Space immediately above lock in water canal.

HEAD BOX Delivers pulp into its paper machine.

HEAD DEPENDENT BOUNDARY Linear combination of head and flux at boundary.

HEAD END Forward portion of train.

HEAD FLUME Channel formed to stop scouring by moving water.

HEAD GATE 1. Gate upstream of lock or canal; 2. Floodgate that controls flow of water.

HEAD INCREASER Apparatus that reduces tailwater pressure and boosts discharge.

HEAD INJURY CRITERIA Measure of probability of head injury arising from impact.

HEAD LOSS Energy lost (i.e., decrease in total head) from flowing fluid due to friction, transitions, bends, etc.

HEAD PRESSURE 1. Internal energy of fluid due to pressure exerted on its container; 2. Pressure pump must supply to lift water to higher elevation.

HEAD RACE 1. Channel which directs water to water wheel; 2. Forebay.

HEAD TANK Water storage tank elevated to supply pressure.

HEAD, STATIC 1. Difference in elevation between pumping source and point of delivery; 2. Vertical distance between two points in fluid; 3. Static flow pressure in pipe, channel or duct.

HEAD, VELOCITY Vertical distance that would statically result from velocity of moving fluid; 2. Pressure needed to increase fluid flow speed.

HEAD-GATE DUTY OF WATER Water requirement or duty measured at point of diversion.

HEADCUTTING Channel bed degradation (i.e. change in elevation).

HEADER 1. Large pipe to which series of smaller pipes are connected; 2. Raised tank that supplies water at constant pressure.

HEADING 1. In tunnel, digging face and its work area; 2. Where waterway or pipeline redirects water; 3. Material for heads of casks, barrels, etc.

HEADRACE Water (i.e., impoundment) immediately upstream from dam or hydroelectric plant intake structure.

HEADSPACE Vapor mixture trapped above solid or liquid in sealed vessel.

HEADWALL 1. Upstream wall; 2. Retaining wall at outlet of drain or culvert; 3. Highest cliff of glacial cirque.

HEADWARD EROSION Erosion which occurs in upstream end of valley of stream, causing it to lengthen its course in such direction.

HEADWATER 1. Source and upper part of stream, reservoir, dam or powerhouse; 2. Small streams that come together to form river.

HEADWATER ADVISORY PROGRAM Program that uses Antecedent Precipitation Index method of estimating runoff, unit hydrograph theory and stage-discharge ratings to produce hydrologic forecasts for headwater basins.

HEADWATER ADVISORY TABLE Pre-computed matrix of values allows forecaster to ascertain anticipated crest or rise on small river or stream for variety of rainfall events and soil moisture conditions.

HEADWATER BASIN Basin at headwaters of river.

HEADWATER CONTROL Regulation to control flooding from impounding area.

HEADWATER DEPTH 1. Depth of water above culvert inlet bottom; 2. Amount of energy available to convey water through culvert; 3. Water depth upstream from structure or contraction.

HEADWATERS 1. Source of river; 2. Set of streams that feed into river's beginning.

HEADWAY 1. Interval of time or distance between fronts of two vehicles moving in succession in same direction; 2. Clearance beneath object.

HEADWORKS Any structure at head or diversion point of waterway.

HEALTH ADVISORY Non-regulatory levels of contaminants in drinking water that may be used as guidance in absence of regulatory limits.

HEALTH ADVISORY LEVEL Non-regulatory health-based reference level of chemical traces (usually in ppm) in drinking water at which there are no adverse health risks when ingested over various periods of time.

HEALTH AND SAFETY PLAN Primary management document for health and safety on site.

HEALTH AND SAFETY STUDY Any study of any effect of chemical substance or mixture on health or environment or on both.

HEALTH ASSESSMENT Evaluation of available data on existing or potential risks to human health posed by Superfund site.

HEALTH HAZARD Something that could potentially endanger person's health.

HEALTH RISK ASSESSMENT Survey that collects relevant information about health status and health risk factors of individual or population.

HEAP Soil carried above sides of body or bucket.

HEAPED BUCKET Machine bucket that scoops more material than bucket volume itself.

HEARING 1. Proceeding where discussions are heard; 2. To perceive sounds through ear.

HEARING LOSS Inability to hear sound.

HEARING PROTECTIVE DEVICE Apparatus to protect from high noise exposure.

HEARTWOOD Wood nearer heart of stem or branch, different in color from sapwood.

HEAT 1. To cause increase in temperature of object or space; 2. Thermal energy.

HEAT ACTUATING DEVICE Apparatus used to activate fire sprinklers, alarms or appliances due to expansion of heated air.

HEAT ADVISORY Issued within 12 hours of onset of following conditions: heat index of at least 105 °F but less than 115 °F for less than 3 hours/day or nighttime lows above 80 °F for 2 consecutive days.

HEAT BALANCE State of thermal equilibrium where heat increase is equivalent to heat reduction.

HEAT BRIDGE Area or component of object that has higher thermal conductivity than surrounding materials.

HEAT BUDGET Balance between incoming heat absorbed by earth and outgoing heat escaping in form of radiation.

HEAT CAPACITY Amount of thermal energy required to raise temperature of system by one temperature unit without any change of phase.

HEAT CONDUCTOR Substance that conducts heat.

HEAT CONTENT Sum of internal energy of system plus product of its volume and pressure.

HEAT DETECTOR Device that detects heat and sounds alarm.

HEAT ENDURANCE Ability to tolerate high ambient temperatures.

HEAT ENGINE Any device which converts heat energy into mechanical work, by moving heat from source to sink.

HEAT EXCHANGER Any device that transfers heat from one fluid to another via separating wall.

HEAT EXCHANGER TANK Tank that encourages mixing and heat transfer.

HEAT EXHAUSTION Mild form of heat stroke, characterized by faintness, dizziness and heavy sweating.

HEAT FLUX Amount of heat transferred per unit area per unit time from or to surface.

HEAT INDEX Accurate measure of how hot it really feels when relative humidity is added to actual air temperature.

HEAT INSULATION Material that reduces heat transfer.

HEAT ISLAND EFFECT Elevated temperatures over urban area caused by structural and pavement heat fluxes and pollutant emissions.

HEAT LIGHTNING Lightning that occurs at distance such that thunder is no longer audible.

HEAT LOAD Amount of heat energy that would need to be added to space to maintain temperature in acceptable range.

HEAT LOSS Measure of total transfer of heat from inside to outside of building.

HEAT MIRROR Film suspended inside double insulating glass to improve insulating properties.

HEAT OF COMBUSTION Total energy released as heat when substance undergoes complete combustion with oxygen under standard conditions.

HEAT OF FORMATION Enthalpy change associated with formation of one mole of compound from its elements in their standard states.

HEAT OF FUSION Amount of heat absorbed by unit mass of material, at its melting point, that causes phase transition from solid to liquid.

HEAT OF HYDRATION Heat generated when water and portland cement react.

HEAT OF SOLUTION Heat evolved or absorbed when substance dissolves.

HEAT PIPE Device that can quickly transfer heat from one point to another.

HEAT PUMP Electric device with both heating and cooling capabilities that extracts heat from one medium at lower (heat source) temperature and transfers it to another at higher temperature (heat sink), thereby cooling first and warming second (e.g., takes thermal energy from groundwater and transfers it to space being heated or removes heat from building and transfers it to groundwater.)

HEAT PUMP/AIR CONDITIONING RETURN FLOW WELLS Class V wells that re-inject groundwater that has been passed through heat exchanger to heat or cool buildings.

HEAT RECOVERY Device that recaptures heat that would be lost to venting.

HEAT RECOVERY SYSTEM Process by which exhaust air preheats supply air when outdoor air is cooler than inside air or if outside air is warmer than inside air, then exhaust air will cool supply air.

HEAT RELEASE RATE Rate that combustion releases energy.

HEAT RESISTANCE Resists deterioration when heated.

HEAT STROKE Condition resulting from excessive exposure to intense heat, characterized by high fever, collapse and sometimes convulsions or coma.

HEAT TRANSFER Transmission of thermal energy via conduction, convection or radiation.

HEAT TRANSFER LIQUID Liquid that transports thermal energy.

HEAT TRANSMISSION Transfer of heat from one place to another with time.

HEAT TRANSMISSION COEFFICIENT Coefficient utilized to calculate heat transmission by conduction, convection or radiation through materials or structures.

HEAT TREATMENT Processes used to alter physical or chemical properties of material.

HEAT WAVE Period of abnormally and uncomfortably hot and unusually humid weather.

HEAT-ABSORBING GLASS Glass that reflects only small percentage of radiation.

HEATER Device that produces and radiates heat.

HEATING AND VENTILATING ENGINEER Professional that's qualified to design building's heating and ventilation system.

HEATING BATTERY Heat exchanger that warms air ducted across hot water conduits.

HEATING CAPACITY Amount of heat that HVAC equipment can provide.

HEATING COIL Element that converts electrical energy into heat that radiates out in all directions.

HEATING DEGREE DAYS Form of degree day used to estimate energy requirements for heating.

HEATING ELEMENT Converts energy into heat through resistance coil.

HEATING LOAD Quantity of heat energy that would need to be added to space to maintain temperature in acceptable range.

HEATING MEDIUM Any solid or fluid used to convey heat from source to substance or space.

HEATING PLANT Facility, composed of equipment, used for heating enclosed spaces.

HEATING RATE Combination of time and temperature.

HEATING SEASON Part of year that heating equipment is required.

HEATING SYSTEM Mechanism for maintaining temperatures at acceptable level.

HEATING VALUE Amount of heat released during combustion of specified amount of it.

HEATING, VENTILATING AND AIR-CONDITIONING Mechanical systems comprised of heating, air conditioning and ventilation systems for residential or commercial buildings.

HEAVE Upward movement of land surfaces or structures due to vertical faulting of rock or subsurface expansion of soil or rock resulting from phenomena (e.g., moisture absorption, removal of overburden, driving of piles, frost action, loading of adjacent area).

HEAVY FREEZING SPRAY Accumulation of freezing water droplets on vessel at rate of 2 cm/hr or greater caused by some appropriate combination of cold water, wind, cold air temperature and vessel movement.

HEAVY FREEZING SPRAY WARNING Warning for accumulation of freezing water droplets on vessel at rate of 2 cm/hr or greater caused by some appropriate combination of cold water, wind, cold air temperature and vessel movement.

HEAVY FREEZING SPRAY WATCH Watch for increased risk of heavy freezing spray event to meet Heavy Freezing Spray Warning criteria but its occurrence, location, and/or timing is still uncertain.

HEAVY LIQUIDS Substance with high density and low relative viscosity.

HEAVY METAL Any metal that has specific gravity greater than about 5.

HEAVY METALS Metal of relatively high density or of high relative atomic weight.

HEAVY MINERALS Minerals with specific gravity greater than or equal to 2.87.

HEAVY RAIL TRANSIT Transit rail system for heavy and high-speed trains.

HEAVY SLUDGE Sludge with low moisture content.

HEAVY SNOW 1. Snowfall accumulating to 4 inches or more in depth in 12 hours or less; 2. Snowfall accumulating to 6 inches or more in depth in 24 hours or less.

HEAVY SNOW WARNING Issued by National Weather Service when snowfall of 6 inches or more in 12 hours or 8 inches or more in 24 hours is imminent or occurring.

HEAVY SPECIALIZED CARRIER Company equipped and authorized to transport equipment and goods that require special trucking equipment.

HEAVY SURF ADVISORY Advisory issued by National Weather Service for fast moving deep water waves which can result in big breaking waves in shallow water.

HEAVY VEHICLE Vehicle that uses more than four wheels to operate.

HEAVY WATER Water containing deuterium instead of normal hydrogen.

HEAVY-DUTY BUCKET Excavator bucket designed for harsh conditions.

HEAVY-DUTY VEHICLE Bus or truck designed for heavy work.

HECTARE Unit of area equal to 2.47 acres.

HECTO Prefix symbolizing 10^2.

HECTOPASCAL Unit of pressure equal to millibar.

HEIGHT Height above sea level of pressure level.

HEIGHT OF DAM Vertical distance between lowest point on crest of dam and lowest point in original streambed.

HEIGHT, ABOVE GROUND Maximum height from natural ground surface to top of dam.

HEIGHT, HYDRAULIC Vertical difference between maximum design water level and lowest point in original streambed.

HEIGHT, STRUCTURAL Vertical distance between lowest point of excavated foundation to top of dam.

HELD WATER Water above watertable.

HELICAL Spiral.

HELICAL GEAR Gears where leading edges of teeth are set at angle to axis of rotation.

HELICITY Property of moving fluid which represents potential for helical flow (i.e., flow which follows pattern of corkscrew) to evolve.

HELIOTHERMOMETER Instrument for determining intensity of solar radiation.

HELIPORT Facility designed to let helicopters take off and land.

HELIUM Colorless and inert gas with atomic number of 2.

HELPER ENGINE Locomotive used to assist in train movement.

HEMOGLOBIN Iron-containing substance in red blood cells that transports oxygen from lungs to rest of body.

HEMOLYSIS Destruction of red blood cells and subsequent release of hemoglobin, at normal end of cell's life.

HENRY Unit of inductance equal to electromotive force of 1 volt in closed circuit with uniform rate of change of current of 1 ampere/second.

HEPA FILTER Air filter capable of trapping minimum of 99.97% of submicron airborne particulate contaminants down to particle size of 0.3 micron.

HEPATITIS Inflammation of liver.

HEPTACHLOR Insecticide banned on some food products.

HERBACEOUS 1. With characteristics of herb; 2. Plant with no persistent woody stem above ground.

HERBICIDE Chemical pesticide designed to control or destroy plants, weeds or grasses.

HERBICIDES Substance used to kill plants.

HERBIVORE Animal that feeds on plants.

HERPETOFAUNA Reptiles and amphibians.

HERRINGBONE DRAIN Drain filled with stones laid in herringbone pattern.

HERRINGBONE GEAR Side-to-side combination of two helical gears of opposite hands.

HERTZ 1. Number of complete electromagnetic cycles or waves in one second of electrical or electronic circuit; 2. Unit of frequency equal to cycle/s.

HETEROGENEITY Characteristic of medium where material properties vary from point to point everywhere.

HETEROTROPHIC ORGANISM Organism that utilizes organic compound to obtain carbon that's essential for growth and development.

HETEROTROPHIC ORGANISMS Species that are dependent on organic matter for food.

HEURISTIC MODEL Representation of real system by mathematical description based on reasoned, but unproven, argument.

HI-CUBE Boxcar developed to take advantage of available railway clearances to haul more product.

HICKEY Tool for making smooth, semicircular bends in conduit.

HIGH 1. Region of high pressure; 2. Anticyclone.

HIGH ALTITUDE Elevation over 8,000 feet above sea level.

HIGH CLOUDS Clouds that have bases between 16,500 and 45,000 feet in mid-latitudes.

HIGH DAM Dam over 300 feet high.

HIGH DUMP BUCKET Bucket that increases lift and reach for easier loading.

HIGH DUMPER Dump truck with high discharge heights.

HIGH DUTY OF WATER Irrigation requirement for fixed quantity of water to serve large area of land.

HIGH END DOSE ESTIMATE Estimate of dose level received anyone in defined population greater than 90th percentile of all individuals in that population, but less than exposure at highest percentile in that population.

HIGH END EXPOSURE ESTIMATE Estimate of exposure level received anyone in defined population greater than 90th

percentile of all individuals in that population, but less than exposure at highest percentile in that population.

HIGH ENERGY EVENT Flares (class two or more) with outstanding centimetric bursts and SID.

HIGH FREQUENCY Portion of radio frequency spectrum between 3 and 30 MHz.

HIGH HAZARD Hazard classification where more than 6 lives would be in jeopardy and excessive economic loss (urban area including extensive community, industry, agriculture or outstanding natural resources) would occur as direct result of failure.

HIGH INTENSITY DISCHARGE Generic term for mercury vapor, metal halide and high pressure sodium lamps and fixtures.

HIGH PLATFORM Train platform built at higher elevation for loading.

HIGH RAIL Outer or elevated rail of curved track.

HIGH RISK OF SEVERE THUNDERSTORMS Severe weather is expected to affect more than 10% of area.

HIGH SEAS FORECAST Marine forecasts for major oceans of world.

HIGH SURF Large waves breaking on or near shore resulting from swells spawned by distant storm.

HIGH SURF ADVISORY Issued when breaking wave action poses threat to life and property within surf zone.

HIGH SURF WARNING Issued when breaking wave action results in especially heightened threat to life and property within surf zone.

HIGH TIDE Maximum height reached by each rising tide.

HIGH WATER ELEVATION Water surface elevation when natural water body is at or near capacity.

HIGH WATER MARK Maximum rise of water over land.

HIGH WIND 1. Sustained wind speeds of 40 mph or greater lasting for 1 hour or longer; 2. Winds of 58 mph or greater for any duration.

HIGH WIND ADVISORY Issued by National Weather Service when high wind speeds may pose hazard.

HIGH WIND WARNING Issued by National Weather Service when high wind speeds may pose hazard or is life threatening.

HIGH WIND WATCH Issued by National Weather Service when there is potential of high wind speeds developing that may pose hazard or is life threatening.

HIGH-CALCIUM LIME Quicklime produced from limestone containing over 98% calcium carbonate.

HIGH-CAPACITY FILTER High-speed filter with more pleats when compared to standard filter.

HIGH-DENSITY POLYETHYLENE Material used to make plastic bottles and other products that produces toxic fumes when burned.

HIGH-EFFICIENCY PARTICULATE AIR FILTER Filter that removes 99.97% of particles that have size greater-than or equal to 0.3 micrometers.

HIGH-LEVEL INVERSION Cooling of subsiding air, in anticyclone, that forms 1,000 feet (300 m) or more above surface.

HIGH-LEVEL LIGHTING Lighting that provides that background light.

HIGH-LEVEL NUCLEAR WASTE FACILITY Plant designed to handle disposal of used nuclear fuel, high-level radioactive waste and plutonium waste.

HIGH-LEVEL RADIOACTIVE WASTE Waste generated in core fuel of nuclear reactor, found at nuclear reactors or by nuclear fuel reprocessing.

HIGH-LINE CONDUIT Elevated pipe that transports water from reservoir.

HIGH-LINE JUMPERS Pipes or hoses connected to fire hydrants and laid on top of ground to provide emergency water service for isolated portion of distribution system.

HIGH-OCCUPANCY TOLL LANE Traffic lane that's available to high-occupancy vehicles without charge.

HIGH-OCCUPANCY VEHICLE Motor vehicle carrying passengers over certain threshold.

HIGH-OCCUPANCY VEHICLE LANE Traffic lane for exclusive use of vehicles with driver more than one passenger.

HIGH-PRECIPITATION STORM Thunderstorm where heavy precipitation (often including hail) falls on trailing side of mesocyclone.

HIGH-PRECIPITATION SUPERCELL Thunderstorm where heavy precipitation (often including hail) falls on trailing side of mesocyclone.

HIGH-PRESSURE FIRE SYSTEM Individual high-pressure water supply network for direct hydrant to fire hose connections.

HIGH-PRESSURE GATE Gate consisting of rectangular leaf encased in body and bonnet and equipped with hydraulic hoist for moving gate leaf.

HIGH-PRESSURE STEAM HEATING SYSTEM Heating system that uses steam at pressures above 15 psig.

HIGH-PRESSURE WATER HEATING SYSTEM Boiler to radiator heating system where water above 350 °F is used to conduct heat.

HIGH-RATE DIGESTION Digester designed for treating highly biodegradable solid wastes and mixed waste streams consisting of solid and liquid fraction.

HIGH-RATE FILTER Filter where media size decreases (i.e., top material is coarse with finer material farther down).

HIGH-RISK COMMUNITY Community located within vicinity of numerous sites of facilities or other potential sources of environmental exposure/health hazards which may result in high levels of exposure to contaminants or pollutants.

HIGH-SPEED COMPACTOR Soil- and refuse-compaction roller equipped with four octagonal wheels.

HIGH-SPEED STREAM Feature of solar wind having velocities that are about double average solar wind values.

HIGH-TEMPERATURE GAS-COOLED REACTOR Nuclear reactor that uses graphite-moderated with once-through uranium fuel cycle.

HIGH-TEST HYPOCHLORITE Powdered lime and chlorine mixture that consists of calcium hypochlorite having chlorine content greater than 65%.

HIGH-TO-LOW-DOSE EXTRAPOLATION Process of prediction of low exposure risk to humans and animals from measured high-exposure-high-risk data involving laboratory animals.

HIGH-VELOCITY AIR FILTER Air filter with lowest pressure drop.

HIGH-VOLUME SAMPLING Method for assessing vapor concentrations and distributions in subsurface.

HIGH-WATER LUNITIDAL INTERVAL Average of all high water intervals for all phases of Moon.

HIGH-WATER MARK Mark showing highest level reached by body of water.

HIGHEST DOSE TESTED Highest dose of chemical or substance tested in study.

HIGHLY FRACTURED Spacing 0.3 ft to 1 ft.

HIGHLY WEATHERED Entire section is discolored greater than 50%.

HIGHWALL Unexcavated face of exposed overburden and coal in surface mine.

HIGHWAY Any public road for vehicular travel.

HIGHWAY ADVISORY RADIO Radio stations that broadcast highway and traffic information.

HIGHWAY CAPACITY Road ability to accommodate traffic volume.

HIGHWAY TRAFFIC SIGNAL Devices placed at roadway to guide, warn and regulate traffic flow.

HIGHWAY TRUCK Vehicle that transports loads without surpassing legal highway limitations.

HIGHWAY-RAIL GRADE CROSSING Intersection where roadway crosses railroad tracks at same elevation.

HILSENHOFF'S BIOTIC INDEX Indicator of organic pollution that uses tolerance values to weight taxa abundances that usually increases with pollution.

HINDERED SOLIDS SEPARATION Solids settling in thickening rather than in clarifying mode.

HINDRANCE Something that causes problems to bicyclist.

HINGE 1. Jointed or flexible device that allows pivoting; 2. Median of upper or lower half of probability distribution; 3. One of four cardinal points (i.e., east, west, north or south).

HINGE CRACK Crack caused by significant changes in water level.

HINTERLAND 1. Land immediately adjacent to and inland from coast; 2. Remote or undeveloped area.

HISTOGRAM Graphical display of numerical data in form of upright bars, with area of each bar representing frequency.

HISTOGRAPH Line graph formed by joining midpoints of histogram column tops.

HISTORIC FLOWS Collection of recorded flow data for stream during period of time where steam gauges were in operation.

HISTORIC SITE Place where important event in history took place and is protected by law as memorial to historical event and/or person.

HOAR FROST Deposit of interlocking crystals formed by direct sublimation on objects, usually those of small diameter freely exposed to air, (e.g., tree branches, plants, wires, poles).

HOARFROST Dew-drops that have undergone deposition and frozen into ice crystals to form white deposit on exposed surface, when air is cold and moist.

HODOGRAPH Polar coordinate graph which shows vertical wind profile of lowest 7,000 meters of atmosphere.

HOE Shovel that digs by pulling broom-and-stick-mounted bucket toward itself.

HOGBACK Ridge formed by erosion of resistant, steeply inclined sedimentary layers.

HOIST Mechanism by which bucket or blade is lifted or process of lifting it.

HOLDING AREA WAITING TIME Time, on average, pedestrian waits before crossing intersection.

HOLDING POND Pond or reservoir, usually made of earth, built to store polluted runoff.

HOLDING TIME Maximum amount of time sample may be stored before analysis.

HOLDING VALVE Pressure control valve to stop loss of fluid.

HOLDOVER STORAGE Part of storage, at end of drawdown, in reservoir to be used in dry part of year.

HOLLOW DAM Dam supported at intervals on downstream side by series of buttresses.

HOLLOW GRAVITY DAM Dam which has outward appearance of gravity dam that's constructed of concrete and/or masonry on outside but having hollow interior and relying on its weight for stability.

HOLLOW JET VALVE 1. Valve having closing member that moves upstream to shut off flow; 2. Device for regulating high-pressure outlets.

HOLLOW STEM AUGER DRILLING Conventional drilling method that uses augurs to penetrate soil.

HOME RANGE Area where animal or group of animals regularly travels in search of food or mates.

HOME SIGNAL Railway signal controlling entry to next section of line.

HOME-BASED TRIP When home of trip maker is either origin or destination.

HOMEOWNER WATER SYSTEM Any water system which supplies piped water to single residence.

HOMOGENEITY Characteristic of medium where material properties are identical everywhere.

HOMOGENEOUS EARTHFILL DAM Embankment dam construction throughout of more or less uniform earth materials, except for possible inclusion of internal drains or blanket drains.

HOMOLOGOUS FLARES Solar flares that occur repetitively in same active region, with essentially same position and with common pattern of development.

HOOD Metal covering that leads to vent to suck away smoke or fumes.

HOOD CAPTURE EFFICIENCY Ratio of emissions captured by hood and directed into control or disposal device, expressed as percent of all emissions.

HOODOO Tall, thin spire of rock that protrudes from bottom of arid basins and badlands.

HOOK ECHO Radar reflectivity pattern characterized by hook-shaped extension of thunderstorm echo, usually in right-rear part of storm (relative to its direction of motion).

HOPPER Storage bin or funnel loaded from top and discharges through door or chute in bottom.

HOPPER CAR Railroad car used to transport bulk materials in hopper.

HOPPER DREDGE Vessel with pump system that sucks up mixture of sand or soil and water and discharges it in hopper of vessel.

HORIZON 1. Distant line along which Earth and sky appear to meet; 2. Specific layer of soil or strata.

HORIZONTAL BRANCH Drain branch extending laterally from soil, waste stack or building drain.

HORIZONTAL PIPE Any conduit or fitting installed in horizontal position or which makes angle of less than 45° with horizontal.

HORIZONTAL PUMP Centrifugal pumps with horizontal shaft axis.

HORIZONTAL SPLIT HEAD Horizontal progressive defect originating inside of rail head and progressing horizontally in all directions.

HORIZONTAL WELL Well dug at angle of at least 80° to vertical bore.

HORIZONTAL-FLOW TANK Tank with length equal to at least twice width and flow moves in horizontal direction.

HORNBLENDE Black blade-like mineral common in igneous and metamorphic rocks.

HORNFELS Fine-grained, gray-green metamorphic rock produced by "baking" of sedimentary rocks by igneous intrusion where sedimentary features may still be preserved.

HORSEHEAD Temporary support for forepoles used in tunneling soft ground.

HORSEPOWER Measurement of work/power where 1 HP is equal to 550 lb-ft/s or 76 m-kg/s or 746 watts.

HORSEPOWER HOUR Work performed or energy consumed by working at rate of one horsepower for 1-hour.

HORSESHOE CONDUIT Conduit with domed cross-section above invert.

HORST Uplifted section of rock bounded on both sides by faults.

HOSE COUPLING Connector on end of hose to connect it with another hose, tap or apparatus.

HOSE THREAD Standard screw thread equal to 12 threads/in on 3/4 inch pipe.

HOST 1. Organism, typically bacterium, into which gene from another organism is transplanted; 2. Animal infected or parasitized by another organism.

HOT BOX Overheated axle box, bearing and bearing enclosure.

HOT MIX ASPHALT Combination of stone, sand or gravel bound together by asphalt cement.

HOT SPRING Natural spring producing water whose temperature is greater than about 20° C (68° F).

HOT WATER HEATING SYSTEM Heating system where water is heated by boiler and circulated throughout building.

HOUR Unit of time equal to 60 minutes or 3600 seconds.

HOURLY PRECIPITATION DATA Contains data on nearly 3,000 hourly precipitation stations.

HOUSE CISTERN Holds rainwater for domestic purposes.

HOUSE CONNECTION Conduit that takes potable water into home or wastewater to sewer.

HOUSE LINE Private standpipe hoseline.

HOUSE SEWER Conduit extending from outside foundation to public sewer.

HOUSE SUBDRAIN Lowest part of building drainage system where drainage pipes meet and convey discharge to sewer.

HOUSE TRAP Creates seal with water that prevents sewer gas from passing from drain pipes back into building.

HOUSEHOLD DETERGENT Surfactant in laundry and household cleaners.

HOUSEHOLD FILTER Filter linked to water supply in residence.

HOUSEHOLD HAZARDOUS WASTE Hazardous products used and disposed of by residential as opposed to industrial consumers.

HOUSEHOLD WASTE Solid waste, composed of garbage and rubbish, which normally originates in private home or apartment house.

HOV LANE Restricted traffic lane for exclusive use of vehicles with driver and at least one passenger.

HOWELL BUNGER VALVE Cylinder gate mounted with axis horizontal.

HUB 1. Central part of wheel; 2. Enlarged end of bell and spigot cast-iron pipe.

HUB END Conduit end connection.

HUB HEIGHT Wind turbine rotor's height above ground.

HUBLESS PIPE Simple-end conduit without that cannot adjoin by itself.

HUMAN ENVIRONMENT Natural and physical environment and relationship of people with that environment.

HUMAN EQUIVALENT DOSE Dose that when administered to humans produces effect equal to that produced by dose in animals.

HUMAN EXPOSURE EVALUATION Describing nature and size of population exposed to substance and magnitude and duration of their exposure.

HUMAN HEALTH ADVISORY Guidance provided, in absence of regulatory limits, to describe acceptable contaminant levels in drinking water or edible fish.

HUMAN HEALTH RISK Likelihood that given exposure or series of exposures may have damaged or will damage health of individuals.

HUMAN SUBJECT Living individual about whom investigator conducting research obtains data through intervention or interaction with individual or identifiable private information.

HUMIC ACID Complex mixture of organic acids produced by decomposition of vegetable matter.

HUMID HEAT Heat capacity of humid air expressed per unit mass of dry air in mixture.

HUMIDITY Measure of water vapor content of air.

HUMIDITY RECOVERY Change in relative humidity over given period of time.

HUMIFICATION 1. Process of humifying; 2. Conversion into humus.

HUMMOCK Hillock of broken ice which has been forced upward by pressure.

HUMMOCKED ICE Ice piled haphazardly one piece over another to form uneven surface.

HUMP Mound of earth.

HUMP YARD Classification or marshalling yard that has hump where trains are shunted and then run on their own by gravity.

HUMUS 1. Organic portion of soil remaining after prolonged microbial decomposition; 2. Dark fluffy material composed chiefly of decayed vegetation or animal matter.

HUMUS SLUDGE Sloughed particles of biomass from trickling filter media that are removed from water being treated in secondary clarifiers.

HUNDREDWEIGHT Unit of mass equal to 100 pounds.

HUNTING Fluctuating around central value without stabilizing.

HURRICANE Tropical cyclone in Atlantic, Caribbean Sea, Gulf of Mexico or eastern Pacific, which maximum 1-minute sustained surface wind is 64 knots (74 mph) or greater.

HURRICANE FORCE WIND WARNING Warning for sustained winds or frequent gusts, of 64 knots (74 mph) or greater, either predicted or occurring and not directly associated with tropical cyclone.

HURRICANE FORCE WIND WATCH Watch for increased risk of hurricane force wind event for sustained surface winds or frequent gusts, of 64 knots (74 mph) or greater, but its occurrence, location, and/or timing is still uncertain.

HURRICANE LOCAL STATEMENT Public release prepared by local National Weather Service offices in or near threatened area giving specific details for its county/parish warning area on weather conditions, evacuation decisions made by local officials and other precautions necessary to protect life and property.

HURRICANE SEASON Part of year having relatively high incidence of tropical cyclones where in Atlantic, Caribbean and Gulf of Mexico and central North Pacific, hurricane season is period from June through November and in eastern Pacific, May 15 through November 30.

HURRICANE WARNING Announcement that hurricane conditions (sustained winds of 74 mph or higher) are expected somewhere within specified coastal area.

HURRICANE WATCH Announcement that hurricane conditions (sustained winds of 74 mph or higher) are possible within specified coastal area.

HUSTLER Vehicle used to maneuver material at shipside.

HYBRID BEACON Traffic control device to increase motorists' awareness of pedestrians at crosswalks.

HYBRID MODEL Model combining at least two modeling techniques (e.g., physical and numerical) in closely coupled fashion.

HYBRID SOLAR SYSTEM Generate power as common grid-tie solar systems do but use batteries to store energy.

HYDER FLARE Filament-associated two-ribbon flare, often occurring in spotless regions.

HYDRANT Outlet from main often consisting of upright pipe with valve attached from which fluid can be tapped.

HYDRANT BARREL Top segment of hydrant that prevents freezing by covering operating mechanism.

HYDRANT BONNET Removable top cap on hydrant.

HYDRANT CAP Removable top portion of hydrant.

HYDRANT DRAIN Drain for removing water from hydrant barrel.

HYDRANT PITOMETER Apparatus for ascertaining velocity of water from hydrant.

HYDRANT SIZE Net diameter of main valve seat of hydrant.

HYDRANT WRENCH Tool used to remove fire hydrant caps and open or close valve of hydrant.

HYDRATE Solid compound containing or linked to water molecules.

HYDRATED LIME Calcium hydroxide or slake lime that's highly caustic and can burn skin and eyes.

HYDRATION 1. Incorporation of water molecules into complex with those of another compound; 2. Chemical reaction where substance (i.e., cement) combines with water, giving off heat to form crystalline structure in its setting and hardening.

HYDRATION WATER Water that joins with salts when they crystallize.

HYDRAUCONE Draft tube where emerging water impinges on plate.

HYDRAULIC 1. Powered by water; 2. Having to do with water in motion.

HYDRAULIC ACCUMULATOR pressure storage reservoir where non-compressible hydraulic fluid is held under pressure applied by external source.

HYDRAULIC ACTUATOR Apparatus responsible for moving and controlling mechanism or system using hydraulic pressure.

HYDRAULIC BORE Standing wave that progresses upstream.

HYDRAULIC CONDUCTIVITY Rate of groundwater flow per unit of time through cross-section of unit area in unit hydraulic gradient.

HYDRAULIC CONTROL VALVE Apparatus where dimensions of operating channel vary as result of action of hydraulic fluid passing through it.

HYDRAULIC DEPTH Hydraulic depth is ratio of cross-sectional area to top width at any given elevation.

HYDRAULIC DREDGE Floating pump that excavates and pumps material through pipeline to different location using centrifugal pump.

HYDRAULIC EFFICIENCY 1. Ability of hydraulic structure or element to conduct water with minimum energy loss; 2. Efficiency of pump or turbine to impart energy to or extract energy from water.

HYDRAULIC ELEMENTS Hydraulic properties of fluid flowing in channel.

HYDRAULIC ENGINEERING Study concerned with flow and conveyance of fluids.

HYDRAULIC EXCAVATION Utilization of water under pressure to remove material.

HYDRAULIC FILL Fill material transported and deposited using water.

HYDRAULIC FILL DAM Dam constructed of materials, often dredged, that are conveyed and placed by suspension in flowing water.

HYDRAULIC FILL STRUCTURE Dam or impoundment made of hydraulic fill.

HYDRAULIC FLOW Atmospheric flow similar in character to flow of water over obstacle.

HYDRAULIC FLOW NET Procedure of computing pressures and velocities in curving stream.

HYDRAULIC FLUIDS Liquids made of many kinds of chemicals, oil or water.

HYDRAULIC FRACTURING Creation of fractures within reservoir that contains oil or natural gas to increase flow and maximize production.

HYDRAULIC FRICTION Flow resistance exerted on perimeter or contact surface between stream and its containing conduit.

HYDRAULIC FRICTION COEFFICIENT Actual to theoretical discharge ratio in frictionless conduit or open channel without turbulence.

HYDRAULIC GRADE LINE Line whose plotted ordinate position represents sum of pressure head plus elevation head for various positions along given fluid flow path, (e.g., along pipeline, groundwater streamline) that's equal to water surface in open channel and lies below energy grade line by amount equal to velocity head at section.

HYDRAULIC GRADIENT 1. Change in total hydraulic head per unit distance of flow at given point and in direction of flow; 2. Direction of groundwater flow due to changes in depth of water table.

HYDRAULIC HEAD Measurement of liquid pressure above geodetic datum.

HYDRAULIC HEIGHT Vertical difference between maximum design water level and lowest point in original streambed.

HYDRAULIC HORSEPOWER Energy imparted to fluid flowing through pump or compressor.

HYDRAULIC JUMP 1. Constant discharge that experiences change in flow depth, from below critical depth to above critical depth, where velocity changes from supercritical to subcritical, that dissipates energy; 2. Steady disturbance in lee of mountain, where airflow passing over mountain suddenly changes from region of low depth and high velocity to region of high depth and low velocity.

HYDRAULIC LIFT TANK Tank holding hydraulic fluid for closed-loop mechanical system that uses compressed air or hydraulic fluid to operate lifts, elevators and other similar devices.

HYDRAULIC LOAD Amount of flow wastewater treatment plant, facility or system can process over fixed period.

HYDRAULIC LOADING Flows to wastewater treatment plant or treatment process.

HYDRAULIC LOSS Loss in fluid power due to flow friction within system.

HYDRAULIC MEAN DEPTH 1. Ratio of flow area to wetted perimeter; 2. Right cross-sectional area of stream of water divided by length of that part of its periphery in contact with its containing conduit.

HYDRAULIC MODEL Physical scale model of river, sewer or storm water system used for engineering studies.

HYDRAULIC MOLE Micro-tunneling machine worked by hydraulics.

HYDRAULIC PERFORMANCE CURVE Plot of head versus discharge.

HYDRAULIC PERMEABILITY Flow of water through unit cross-sectional area of soil normal to direction of flow when hydraulic gradient is unity.

HYDRAULIC POWER power transmitted by controlled circulation of pressurized fluid.

HYDRAULIC POWER UNIT Consists of four main components: prime mover, pump, valve and reservoir.

HYDRAULIC PROFILE 1. Vertical section of aquifer's potentiometric surface; 2. Graphical representation of hydraulic grade line through treatment plant.

HYDRAULIC PROPERTIES Properties of solid and rock that govern entrance of water and capacity to hold, transmit and deliver water (e.g., porosity, effective porosity, specific retention, permeability, direction of maximum and minimum permeability).

HYDRAULIC PUMP Converts mechanical power into hydraulic energy that generates flow with enough power to overcome pressure induced by load at pump outlet.

HYDRAULIC RADIUS 1. Ratio of flow area to wetted perimeter; 2. Right cross-sectional area of stream of water divided by length of that part of its periphery in contact with its containing conduit; 3. For porous material, ratio of porosity to specific surface area.

HYDRAULIC RAM Water pump where flow of water is intermittently halted by valve so that flow is forced upward through open pipe (i.e., reservoir).

HYDRAULIC REACH Segment of river or stream between major changes in hydraulic nature.

HYDRAULIC RELIEF VALVE Device used to prevent excessive hydraulic pressure.

HYDRAULIC RESERVOIR Container for holding fluid required to supply system.

HYDRAULIC ROUGHNESS Measure of frictional resistance water experiences when passing over surface.

HYDRAULIC SIMILITUDE Indication of flow system relationship between model and prototype.

HYDRAULIC SLOPE Incline of hydraulic gradient.

HYDRAULIC SLUICING Process of moving materials by water.

HYDRAULIC STRUCTURE Structure submerged or partially submerged in any body of water that disrupts flow.

HYDRAULIC STRUCTURES Structure, in body of water, that disrupts natural flow of water.

HYDRAULIC SYSTEM System where fluid is used to move energy from motor to actuator.

HYDRAULIC TEST Piping system test where all conduits are sealed and filled with water to design pressure.

HYDRAULIC TESTER Device that can measure fluid flow, pressure and temperature.

HYDRAULIC TRANSIENT Pressure fluctuations caused by sudden increase or decrease in flow velocity, usually associated with rapid closure or opening of valve in pipeline.

HYDRAULIC TURBINE Converts potential energy of fluid flow to mechanical energy.

HYDRAULIC VALVE Valve controlled through hydraulic energy.

HYDRAULICKING Mining using pressurized water jets to move rock or sediment.

HYDRAULICS Having to do with mechanical properties of water in motion and application of these properties in engineering.

HYDRAZINE Reagent used to scavenge oxygen from boiler water.

HYDRIC 1. Characterized by or thriving in, abundance of moisture; 2. Of or relating to hydrogen.

HYDRIC SOIL Soil wet long enough to periodically produce anaerobic conditions, thereby influencing growth of plants.

HYDRIDE Compound of hydrogen with more electropositive element.

HYDRO Related to water.

HYDROCARBON Compound consisting only of carbon and hydrogen atoms.

HYDROCARBONS Chemical compounds that consist entirely of carbon and hydrogen.

HYDROCHLOROFLUOROCARBONS Compounds containing carbon, hydrogen, chlorine and fluorine that have potential to destroy stratospheric ozone.

HYDRODYNAMICS Scientific study of fluids in motion.

HYDROELECTRIC Generates electricity by converting energy of moving water or of steam escaping under high pressure.

HYDROELECTRIC PLANT Electric power plant that produces electricity from power of rushing water turning turbine-generators.

HYDROELECTRIC POWER Electrical energy produced by flowing water.

HYDROELECTRIC POWER WATER USE Use of water in generation of electricity at plants where turbine generators are driven by falling water.

HYDROELECTRIC UNIT Electric generator and turbine combination that is driven by water flow, thereby converting mechanical energy to electric energy.

HYDROELECTRICITY Electricity produced from energy of moving water.

HYDROFLUORIC ACID Solution of hydrogen fluoride in water that's weak but very corrosive acid, used in production of fluorine compounds and in metallurgy.

HYDROFLUOROCARBONS Man-made compounds containing hydrogen, fluorine and carbon that do not have potential to destroy stratospheric ozone.

HYDROGEN Lightest chemical element with atomic number of 1.

HYDROGEN CHLORIDE Inorganic compound that forms hydrochloric acid when dissolved in water.

HYDROGEN CYANIDE Colorless, very poisonous, volatile liquid, used in production of dyes, plastics and fumigants.

HYDROGEN FLUORIDE Colorless fuming liquid or gas that dissolves in water to form hydrofluoric acid.

HYDROGEN ION CONCENTRATION Concentration of hydrogen ions in solution expressed usually in moles/liter or in pH value.

HYDROGEN PEROXIDE Colorless liquid, soluble in water, used as mild antiseptic, bleaching agent, oxidizing agent and chemical reagent.

HYDROGEN SULFIDE Gas emitted during organic decomposition that's by-product of oil refining and burning.

HYDROGENATION Chemical reaction of hydrogen with another substance.

HYDROGENESIS Chemical reaction that involves loss of water molecule from reacting molecule.

HYDROGEOCHEMISTRY Chemistry of groundwater and surface water.

HYDROGEOLOGIC CONDITIONS Conditions stemming from interaction of groundwater and surrounding soil and rock.

HYDROGEOLOGIC CYCLE Natural process of recycling water from atmosphere down to (and through) Earth and back to atmosphere again.

HYDROGEOLOGIST Person who studies and works with groundwater.

HYDROGEOLOGY Geology of groundwater, with particular emphasis on movement of water and chemical, physical and biological properties.

HYDROGRAPH Graphical representation of water property (i.e., stage, discharge, velocity, available power, rate of runoff) as function of time at particular point on watercourse.

HYDROGRAPH SEPARATION Process where storm hydrograph is separated into baseflow components and surface runoff components.

HYDROGRAPH, BREACH Flood hydrograph resulting from dam breach.

HYDROGRAPH, DAM FAILURE Flood hydrograph resulting from dam breach.

HYDROGRAPH, FLOOD Flood hydrograph resulting from dam breach prepared for specific location downstream from dam.

HYDROGRAPHER Person who measures and describes physical features of Earth's surface waters.

HYDROGRAPHIC Of or pertaining to hydrography.

HYDROGRAPHIC DATUM Level used for referencing depths of water and heights of predicted tides or water level observations.

HYDROGRAPHIC SURVEY 1. Science of measurement and description of water features; 2. Instrumental survey to measure and determine characteristics of streams and other bodies of water within area (e.g., position and elevation of high water marks).

HYDROGRAPHY Scientific study of physical aspects of all waters on Earth's surface.

HYDROISOBATHS Contours of similar depth of subsoil watertable below ground surface.

HYDROISOPLETH MAP Map revealing fluctuation of watertable with respect to time and space.

HYDROKINETICS Branch of physics concerned with movement of fluids.

HYDROLOGIC BUDGET Accounting of inflow, outflow, storage of hydrologic unit.

HYDROLOGIC CYCLE 1. Movement or exchange of water between atmosphere and Earth; 2. Natural recycling process powered by Sun that causes water to evaporate into atmosphere, condense and return to Earth as precipitation.

HYDROLOGIC EQUATION Equation balancing hydrologic budget, water inventory equation which expresses basic principle that during given time interval total inflow to area must equal total outflow plus net change in storage.

HYDROLOGIC INDEX STATIONS Continuous-record gaging stations that have been selected as representative of streamflow patterns for their respective regions.

HYDROLOGIC INVENTORY Procedure of getting and assessing data as well as other information regarding hydrologic scenarios and occasions.

HYDROLOGIC MODEL Conceptual or physically-based procedure for numerically simulating process or processes which occur in watershed.

HYDROLOGIC REGIME Characteristic behavior and total quantity of water involved in drainage basin.

HYDROLOGIC SEQUENCE Series of soil sections taken from differentiated parent material that demonstrates increasing lack of drainage downslope.

HYDROLOGIC SERVICE AREA Geographical area assigned to Weather Service Forecast Office's/Weather Forecast Office's that embraces one or more rivers.

HYDROLOGIC SERVICES General term referring to operations, products, verbal communications and related forms of support provided by National Weather Service for Nation's streams, reservoirs and other areas affected by surface water.

HYDROLOGIC SOIL GROUP Soils classified into groups based on runoff potential.

HYDROLOGIC UNIT Geographical area representing part or all of surface drainage basin or distinct hydrologic feature (e.g., reservoir, lake) as delineated by U.S. Geological Survey on State Hydrologic Unit Maps where each hydrologic unit is assigned number.

HYDROLOGIC UNIT CODE Number used to identify geographic area representing part or all of surface drainage basin or distinct hydrologic feature.

HYDROLOGICAL CYCLE Continuous process where water is circulated throughout Earth and its atmosphere.

HYDROLOGICAL DATA STATION Place where data as well as other hydrological info is obtained and saved.

HYDROLOGISTS Individuals who study applied science of hydrology and solve hydrologic problems.

HYDROLOGY Science that treats properties, occurrence, circulation and distribution of waters of Earth and interaction with land surface and underlying soils and rocks.

HYDROLYSIS Decomposition of organic compounds by interaction with water.

HYDROLYTIC TANK Wastewater sedimentation tank where biochemical processes are encouraged.

HYDROMECHANICS Branch of mechanics dealing with gases and liquids, either at rest or in motion.

HYDROMET System of data collection platforms that gathers hydrometeorological data and transmits that data via GOES Satellite to computer downlink.

HYDROMETEOR Particle of condensed water (e.g., liquid, snow, ice, graupel, hail) in atmosphere.

HYDROMETEOROLOGICAL REPORT Series of hydrometeorological reports published by National Weather Service addressing meteorological issues related mainly to developing estimates of probable maximum precipitation used in determination of probable maximum flood for design of water control structures.

HYDROMETEOROLOGICAL TECHNICIANS Individuals whose duties include data collection, quality control, gauge network maintenance, as well as gathering and disseminating of data and products at National Weather Service Forecast Offices.

HYDROMETEOROLOGISTS Individuals who have combined knowledge in fields of both meteorology and hydrology which enables them to study and solve hydrologic problems where meteorology is factor.

HYDROMETEOROLOGY Interdisciplinary science involving study and analysis of interrelationships between atmospheric and land phases of water as it moves through hydrologic cycle.

HYDROMETER Device for measuring specific gravity of fluids.

HYDROMETRIC PENDULUM Apparatus consisting of hollow ball of ivory or metal suspended by treated from center of graduated quadrant and held in stream to measure velocity of water by inclination given to thread.

HYDROMETRY Branch of hydrostatics dealing with measurement of specific gravity using hydrometers.

HYDROMORPHIC SOIL Soil with characteristics caused by excess water.

HYDRONIC Ventilation system using heated or cooled water pumped through building.

HYDROPHILIC Having tendency to mix with, dissolve in or be wetted by water.

HYDROPHYTE Any plant growing in water or on substrate that's at least periodically deficient in oxygen as result of excessive water content.

HYDROPLANING To skim surface of water body while moving at high speed.

HYDROPNEUMATIC Water system, usually small, where water pump is automatically controlled by pressure in compressed air tank.

HYDROPNEUMATIC TANK SYSTEM System that maintains water system pressure at constant level.

HYDROPNEUMATICS Of, relating to or operating by means of both water and air or other gas.

HYDROPONICS Cultivation of plants in nutrient solution rather than in soil.

HYDROSEEDING Method of seeding that uses slurry of seed in water.

HYDROSPHERE Region that includes all Earth's liquid water, frozen water, floating ice, frozen upper layer of soil and small amounts of water vapor in Earth's atmosphere.

HYDROSTATIC Pressures, forces or equilibrium resulting from fluid at rest.

HYDROSTATIC EXCESS PRESSURE Equal to total pressure minus effective pressure plus neutral pressure.

HYDROSTATIC HEAD Measure of pressure at given point in liquid related to vertical height of column of liquid that would produce same pressure.

HYDROSTATIC JOINT Spigot-and-socket joint in water main.

HYDROSTATIC LEVEL Level to which very top of column of water would rise, if free to achieve this, from artesian aquifer basin or from conduit under pressure.

HYDROSTATIC PORE PRESSURES Pore water pressures or groundwater pressures applied under conditions of no flow where magnitude of pore pressures grows linearly with depth below soil surface.

HYDROSTATIC PRESS Large ram, surface of which is acted on by liquid in contact with small ram.

HYDROSTATIC PRESSURE Pressure exerted by water at any given point in body of water at rest.

HYDROSTATIC PRESSURE RATIO Ratio between pressure on vertical plane due to soil and that that would exist at same point in liquid of same density as soil.

HYDROSTATIC SLUDGE REMOVAL Discharge when surface of water in clarifier is higher than surface of water in sludge well or hopper.

HYDROSTATIC STRENGTH Ability of body to withstand hydrostatic stress.

HYDROSTATIC TEST Pressure test of tanks, conduits, etc. for strength and leaks.

HYDROSTATIC UPLIFT Upward pressure against foundation transmitted by groundwater.

HYDROSULFITE Any salt or ester of dithionous acid.

HYDROTESTING Pressure test of tanks, conduits, etc. for strength and leaks.

HYDROTHERMAL Of or relating to emanations of hot water that are rich in minerals or to rocks formed from them.

HYDROUNIT Electric generator and turbine combination that is driven by water flow, thereby converting mechanical energy to electric energy.

HYDROXIDE ALKALINITY Alkalinity due to hydroxyl ions.

HYETAL COEFFICIENT For any month at given station, ratio of monthly normal precipitation to one-twelfth of annual normal precipitation.

HYETAL REGIONS Region where amount and seasonal variation of rainfall are of given type.

HYETOGRAPH Graphical representation (e.g., bar graph) of rainfall intensity versus time.

HYETOGRAPHY Science that deals with geographical distribution of rain.

HYETOLOGY Scientific study of precipitation.

HYGROGRAPH Any of several forms of automated hygrometer that record humidity.

HYGROMETER Instrument which measures humidity of air.

HYGROMETRIC CONTRACTION Contraction of substances as they consume (exude) moisture.

HYGROMETRIC EXPANSION Expansion and contraction of materials as they absorb or exude moisture.

HYGROMETRY Measurement of moisture content of gases, but especially of humidity of atmosphere.

HYGROSCOPIC Absorbing or attracting moisture from air.

HYGROSCOPIC COEFFICIENT Moisture weight percentage of soil after equilibration with water vapor saturated atmosphere at particular temperature.

HYGROSCOPIC MOISTURE Moisture held in soil in equilibrium with atmospheric water vapor.

HYGROSCOPIC SALTS Salts that react with moisture.

HYGROSCOPIC WATER Water absorbed from atmosphere and held very tightly by soil particles, so that it is unavailable to plants in amounts sufficient for them to survive.

HYGROSCOPICITY State of readily taking up and retaining water.

HYGROSTAT Device that maintains constant level of humidity.

HYPERBARIC Of, relating to or utilizing greater than normal pressure.

HYPERCRITICAL FLOW Supercritical flow.

HYPERSENSITIVITY DISEASES Diseases characterized by allergic responses to pollutants and/or most clearly associated with indoor air quality are asthma, rhinitis and pneumonic hypersensitivity.

HYPERTROPHIC State where body of water is grossly enriched with plant nutrients.

HYPOCENTER 1. Point or focus within Earth that is center of earthquake and origin of its elastic waves (i.e., slip responsible for earthquake originates); 2. Location within Earth where sudden release of energy is initiated.

HYPOCHLORINATION Application of hypochlorite compounds to water or wastewater for purpose of disinfection.

HYPOCHLORINATORS Devices used to dispense chlorine solutions made from hypochlorites.

HYPOCHLORITE Any salt of hypochlorous acid (i.e., used as household bleach).

HYPOCHLORITE OF LIME Mix of slaked lime and chlorine where calcium oxychloride is active component.

HYPOID GEAR Spiral bevel gear whose axis does not intersect with axis of meshing gear.

HYPOLIMNETIC Pertaining to lower, colder portion of lake or reservoir that is separated from upper, warmer portion (i.e., epilimnion) by thermocline.

HYPOLIMNION Lower layer of water, in thermally stratified lake, with essentially uniform colder temperatures.

HYPORHEIC ZONE 1. Groundwater habitats created by movement of river water from active channel to areas to side and beneath active channel; 2. Uniquely adapted organisms that can provide food for fish live in groundwater habitat.

HYPOTHERMIA Rapid, progressive mental and physical collapse that accompanies lowering of body temperature.

HYPOXIA WATERS Waters with dissolved oxygen concentrations of less than 2 ppm, level generally accepted as minimum required for most marine life to survive and reproduce.

HYPOXIC WATERS Waters with dissolved oxygen concentrations of less than 2 ppm, level generally accepted as minimum required for most marine life to survive and reproduce.

HYPSOGRAPHY Elevation measurement system based on sea level datum.

HYSTERESIS Property of system such that output is not strict function of input.

I

I/I ANALYSIS Analysis demonstrating possible excessive or non-excessive infiltration and/or inflow to sewer system.

ICE Water in frozen (solid) form.

ICE AGE Time of long-term cooler Earth surface and atmospheric temperatures that resulted in widespread glaciation.

ICE ANVIL Cumulonimbus cloud formed of tiny ice crystals with anvil-shaped top.

ICE APRON Thin mass of snow and ice adhering to mountainside.

ICE BOOM Floating structure designed to retain ice.

ICE BRIDGE Continuous ice cover of limited size extending from shore to shore like bridge.

ICE CRYSTALS Barely visible crystalline form of ice that has shape of needles, columns or plates.

ICE DAM Mass of ice composed mainly of frazil or broken ice deposited underneath ice cover in region of low flow velocity.

ICE EVAPORATION LEVEL Height where ice crystals entering dry air will change directly into water vapor by sublimation.

ICE FOG Type of fog formed by direct freezing of supercooled water droplets with little growth directly from vapor.

ICE GORGE Gorge or opening left in jam after it has broken.

ICE JAM Stationary accumulation that restricts or blocks streamflow.

ICE MODEL Model where formation of ice, ice conditions or ice forces are simulated.

ICE NUCLEUS Any particle that serves as nucleus in formation of ice crystals in atmosphere.

ICE PACK Large floating mass of ice.

ICE PELLETS Pellets of ice composed of frozen or mostly frozen raindrops or refrozen partially melted snowflakes.

ICE PRESSURE Pressure ridge develops in ice cover as result of stress regime established within plane of ice.

ICE PUSH 1. Compression of ice cover particularly at front of moving section of ice cover; 2. Pressure exerted by sheet of floating ice as it expands.

ICE RUN Flow of ice in river.

ICE SHEET Huge mass of ice, formed on land by compaction and recrystallization of old snow, that moves very slowly downslope or outward due to gravity.

ICE SHOVE On-shore ice push caused by wind and currents, changes in temperature, etc.

ICE STORM Occasions when damaging accumulations of ice are expected during freezing rain situations.

ICE STORM WARNING Issued by National Weather Service when freezing rain produces significant and possibly damaging accumulation of ice.

ICE TWITCH Downstream movement of small section of ice cover.

ICE WEDGE Vertical wedge of subterranean ice.

ICEBERG Piece of glacier which has broken off and is floating in sea.

ICELANDIC LOW Semi-permanent, subpolar area of low pressure in North Atlantic Ocean.

ICHTHYOLOGY Scientific study of fish.

ICING Coating of ice on solid object.

IDEAL GAS LAW Equation of state of ideal gas where one mole of ideal gas has volume of 22.4143 liters at temperature of 0 °C along with pressure of 1 atmosphere.

IDEAL GAS LAWS Thermodynamic laws applying to perfect gases.

IDENTIFICATION CODE Unique code assigned to each generator, transporter and treatment, storage or disposal facility by regulating agencies to facilitate identification and tracking of chemicals or hazardous waste.

IDENTIFICATION INDEX Series of numbers/letters that indicate blends and grades.

IDLE TIME Non-productive time of employees or machines due to work stoppage from any cause.

IDLER GEAR System used to transmit rotation of main shaft of motor to another rotating device.

IGNEOUS Rock that forms from solidification of molten lava or magma.

IGNEOUS ROCKS Rocks that have solidified from molten or partly molten material.

IGNIMBRITE Deposit left by pyroclastic flow from volcano, consisting of ash, pumice lapilli and lithic fragments.

IGNITABLE Capable of burning or causing fire.

IGNITION SWITCH Switch that activates main electrical systems for vehicle.

ILLUVIATION Accumulation of suspended material and soluble compounds leached from overlying stratum.

IMHOFF CONE Clear, cone-shaped container, marked with graduations, used to measure volume of settleable solids in specific volume of water.

IMHOFF TANK Chamber suitable for reception and processing of sewage that consists of V-shaped settling compartment above tapering sludge anaerobic digestion chamber with gas vents.

IMMEDIATE BIOCHEMICAL OXYGEN DEMAND 1. First volume of oxygen taken up by wastewater promptly upon being introduced to water containing dissolved oxygen. 2. obvious BOD for quarter-hour at 68 °F as ascertained in conventional lab process.

IMMEDIATE STAGE BIOCHEMICAL OXYGEN DEMAND First volume of oxygen employed by polluted liquid promptly upon being inserted into water containing dissolved oxygen.

IMMEDIATELY DANGEROUS TO LIFE AND HEALTH Maximum level that healthy individual can be exposed to chemical for 30 minutes and escape without suffering irreversible health effects or impairing symptoms.

IMMEDIATELY DANGEROUS TO LIFE OR HEALTH Exposure to airborne contaminants "likely to cause death or immediate or delayed permanent adverse health effects or prevent escape from such environment."

IMMERSE To put under surface of liquid.

IMMERSED TUBE Underwater tunneling used principally for underwater crossings.

IMMINENT HAZARD One that would likely result in unreasonable adverse effects on humans or environment or risk unreasonable hazard to endangered species during time required for pesticide registration cancellation proceeding.

IMMINENT THREAT High probability that exposure is occurring.

IMMISCIBILITY Inability of two or more substances or liquids to readily dissolve into one another (e.g., soil and water).

IMMISSION Act of sending or letting in.

IMMOBILIZATION Process where organic soil matter is decomposed and absorbed by microorganisms and thus not available to plants.

IMMOBILIZE To hold by strong chemical attraction.

IMPACT 1. Estimated loss associated with risk; 2. Force or energy of collision of two objects; 3. Change in chemical, physical or biological quality or condition of water body caused by external sources.

IMPACT ANGLE Acute angle formed by plane tangent to ground surface and tangent to trajectory.

IMPACT LOAD Load applied or acted on structure with short duration.

IMPACT LOSS Head lost as result of impact of particles of water.

IMPAIRED Condition of quality of water that has been adversely affected for specific use by contamination or pollution.

IMPAIRMENT Detrimental effect on biological integrity of water body caused by impact that prevents attainment of designated use.

IMPEDANCE Measure of opposition to flow.

IMPELLER Rotating part of centrifugal pump, compressor or other machine designed to move (i.e., pump or lift) fluid by rotation.

IMPELLER EYE Center of impeller where fluid enters.

IMPELLER PUMP Pump using mechanical agency to provide continuous power to move liquids.

IMPERIAL GALLON Measure of fluid gallon equal to approximately 1.2 US gallons.

IMPERMEABILITY Incapacity of rock to transmit fluid.

IMPERMEABILITY FACTOR Storm water flow depends upon imperviousness of surface over which rainfall takes place.

IMPERMEABLE Material that does not permit passage of liquids.

IMPERMEABLE BOUNDARY Conceptual representation of natural feature (e.g., fault, depositional contact) that places boundary of significantly less-permeable material laterally adjacent to aquifer.

IMPERMEABLE LAYER Layer of solid material (e.g., rock, clay) that does not allow water to pass through.

IMPERMEABLE ROCK Rock that's impermeable and will not allow fluid to pass through it.

IMPERMEABLE STRATA Not able to be broken or allow fluids to pass through.

IMPERVIOUS 1. Incapable of being penetrated; 2. Soil where spacing of particles permits extremely slow passage of water.

IMPERVIOUS BED Bed or stratum that will not allow fluid to pass through it.

IMPERVIOUS CORE Zone of low permeability material in embankment dam sometimes referred to as central core, inclined core, puddle clay core, rolled clay core or impervious zone.

IMPERVIOUS MATERIAL Relatively waterproof.

IMPERVIOUS TILE Tile that absorbs 0% to 0.5% of water.

IMPERVIOUS ZONE Zone (e.g., central core, inclined core, puddle clay core, rolled clay core) of low permeability material in embankment dam.

IMPERVIOUSNESS COEFFICIENT Ratio of impervious to total area.

IMPINGE To collide or strike.

IMPINGEMENT Impact of fluid flow upon or against surface.

IMPINGEMENT FILTER Filter that induces solution to change direction and particles to adhere to filter medium.

IMPLEMENTATION Process of moving idea from concept to reality.

IMPLEMENTATION MONITORING Documents whether or not management practices were applied as designed.

IMPLEMENTATION PROCESS Procurement, installation and implementation of all components of Early Warning System that includes inspecting and exercising system.

IMPLICIT SCHEME Scheme where governing equations of numerical model are arranged to obtain solutions for dependent variables simultaneously at all grid points corresponding to any one time.

IMPLIED WARRANTY Not written but exists under law.

IMPORT Water piped or channeled into area.

IMPORTANCE FACTOR Multiplier that increases or decreases base design loads.

IMPORTS Municipal solid waste and recyclables that have been transported to state or locality for processing or final disposition (but that did not originate in that state or locality).

IMPOUNDING DAM Dam where tailings are collected and settled.

IMPOUNDING RESERVOIR Basin constructed in valley of stream or river for purpose of holding stream flow.

IMPOUNDMENT Body of water created by barrier (e.g., dam, dike, floodgate).

IMPROVED DISCHARGE Volume, composition and location of discharge following construction of outfall improvements or of treatment system improvements to meet merit levels or discharge characteristics or implementation of program to improve operation and maintenance of existing treatment system or to eliminate or control introduction of pollutants into treatment works.

IMPROVED LAND Land that has certain utilities and services available to it.

IMPROVED SINKHOLE Naturally occurring karst depression or other natural crevice found in volcanic terrain and other geologic settings which have been modified for purpose of directing and emplacing fluids into subsurface.

IMPROVED VENTURI FLUME 1. Device used to measure flow of water in open channel; 2. Calibrated device, based on principle of critical flow, consisting of broad and flat converging section, narrow downward sloping throat and diverging upward sloping section developed to measure flow of water in open channels.

IMPROVEMENT Structural measures for betterment, modernization or enhancement of existing facility or system to improve social, economic and environmental benefits of project.

IMPULSE 1. Integral of force over time; 2. General term for any large-scale or mesoscale disturbance capable of producing upward motion (lift) in middle or upper parts of atmosphere.

IMPULSE LINE Small bore pipe used to connect point in pipe where process variable is to be measured by instrument.

IMPULSE PUMP Device for forcing running water to higher level by using kinetic energy of flow.

IMPULSE-REACTION TURBINE Turbine with impulse and reaction forces.

IMPULSIVE NOISE Noise that includes unwanted sharp sounds.

IMPURITY Component or additive that renders something else impure.

IN SITU 1. Original location; 2. Unmoved unexcavated; 3. Remaining at site or in subsurface.

IN VACUUM SERVICE Gear operating with internal pressure 5 kPa below ambient pressure.

IN VHAP SERVICE Piece of equipment either contains or contacts fluid (liquid or gas) that's at least 10% by weight volatile hazardous air pollutant.

IN VITRO Testing or action outside organism (e.g., inside test tube or culture dish).

IN VIVO Testing or action inside organism.

IN VOC SERVICE Piece of equipment contains or contacts process fluid that's at least 10% VOC by weight.

IN-CLOUD LIGHTNING Lightning that takes place within cloud.

IN-LINE FILTRATION Pre-treatment method where chemicals are mixed by flowing water.

IN-PLACE SHELTERING Protective action that involves taking cover in upper levels of building able to withstand high flood levels.

IN-PLANE BENDING Plate that bends in its own plane.

IN-PROCESS CONTROL TECHNOLOGY Conservation of substances and water throughout creation operations to decrease quantity of wastewater to be released.

IN-ROADWAY LIGHTS Signals to warn approaching condition on or adjacent to roadway that might not be readily apparent or require reduced speed and/or stop.

IN-SHIFT MOVING TIME Part of non-mechanical delay time during which machine is moving or being transported.

IN-SHIFT REPAIR TIME Part of mechanical delay time when machine is actually undergoing repair plus time during which machine is waiting to be repaired or for repair parts, mechanics or facilities.

IN-SHIFT SERVICE TIME Part of mechanical delay time when machine is actually undergoing service plus time machine is waiting for service parts, mechanics or repair facilities.

IN-SITU FLUSHING Introduction of large volumes of water, at times supplemented with cleaning compounds, into soil, waste or groundwater to flush hazardous contaminants from site.

IN-SITU FOSSIL FUEL RECOVERY WELLS Class V wells that are used to recover lignite, coal, tar sands and oil shale by injecting water, air, oxygen, solvents, combustibles or explosives into underground or oil shale beds to free fossil fuels so they can be extracted.

IN-SITU LEACH METHODS Processes including purposeful introduction of appropriate leaching solutions into ore body to dissolve precious minerals.

IN-SITU OXIDATION Technology that oxidizes contaminants dissolved in groundwater, converting them into insoluble compounds.

IN-SITU SAMPLING SYSTEMS Non-extractive samplers or in-line samplers.

IN-SITU STRIPPING Treatment system that removes or "strips" volatile organic compounds from contaminated ground or surface water by forcing airstream through water and causing compounds to evaporate.

IN-SITU VITRIFICATION Technology that treats contaminated soil in place at extremely high temperatures, at or more than 3,000 °F.

INACTIVE CAPACITY Reservoir capacity exclusive of and above dead capacity from which stored water is normally not available because of operating agreements or physical restrictions.

INACTIVE FACILITY Facility that no longer receives solid waste.

INACTIVE LAYER Depth of material beneath active layer.

INACTIVE STORAGE Storage volume of reservoir between crest of invert of lowest outlet and minimum operating level.

INACTIVE STORAGE CAPACITY Portion of capacity below which reservoir is not normally drawn and that is provided for sedimentation, recreation, fish and wildlife, aesthetic reasons or for creation of minimum controlled operational or power head in compliance with operating agreements or restrictions.

INACTIVE WASTE DISPOSAL SITE Disposal site or portion of it where additional waste material has not been deposited within past year.

INBOUND Trip into main area.

INCH Unit of length equal to 1/12 of foot.

INCH-DEGREES Product of inches of rainfall multiplied by temperature in degrees above freezing (Fahrenheit scale), used as measure of snow melting capacity of rainfall.

INCH-FOOT CHARGE Fire line connection charge that's separate from domestic water charge based on cost of service for providing water to fire line.

INCHES OF MERCURY Unit of barometric pressure equal to pressure exerted by column of mercury 1 inch.

INCHES OF RUNOFF Volume of water from runoff of given depth over entire drainage.

INCIDENT Event that causes or may cause interruption (e.g., traffic accident).

INCIDENT COMMAND POST Facility located at safe distance from emergency site, where incident commander, key staff and technical representatives can make decisions and deploy emergency manpower and equipment.

INCIDENT COMMAND SYSTEM Combination of facilities, equipment, personnel, procedures and communications operating within common organizational structure to effectively respond to emergency or disaster.

INCIDENT COMMANDER Person in charge of on-scene coordination of response to incident, usually senior officer in fire department.

INCIDENT DELAY Delay that's function of both duration and traffic demand.

INCIDENT RADIATION Incoming radiation (i.e., radiation that strikes surface).

INCINERATION Destruction by controlled burning at high temperatures.

INCINERATION AT SEA Disposal of waste by burning at sea on specially-designed incinerator ships.

INCINERATOR Furnace for burning waste or refuse under controlled conditions.

INCIPIENT EROSION Start of erosion process.

INCIPIENT MOTION Flow condition at which given size bed particle just begins to move.

INCISED CHANNEL Channel where stream bed has dropped and is disconnected from its floodplain.

INCISED RIVER River that erodes its channel by process of degradation.

INCISED STREAM Stream that has cut its channel into bed of valley through degradation.

INCLINED STAFF GAUGE Gauge placed on slope of stream bank and graduated so that scale reads directly in vertical depth.

INCLINOMETER Instrument that measures offset from vertical.

INCLUDED GAS Gases in isolated interstices in either zones of saturation or aeration.

INCOMPATIBLE FLUIDS Mixed fluids that could have negative impact on system.

INCOMPATIBLE WASTE Waste unsuitable for mixing with another waste or material because it may react to form hazard.

INCORPORATED INTO SOIL Injection of solid waste into soil.

INCREASER Conduit fitting where one end is larger than other.

INCREMENTAL LOSS OF LIFE Measurement for assessing damage after dam fails that's difference between projected loss of life had dam not been built and projected loss of life with dam in place.

INCRUSTANTS Solids that form on interior of conduits due to water hardness.

INCUBATION To develop life within.

INDEMNIFICATION Legal requirement that EPA pay certain end-users, dealers and distributors for cost of stock on hand at time pesticide registration is suspended.

INDEPENDENT ASSURANCE Verification by another party of product and/or reliability of test results.

INDEPENDENT POWER PRODUCER Non-utility owned power producer that's not regulated utility, government agency or qualifying facility under Public Utility Regulatory Policies Act of 1978.

INDEX Number representing property or ratio (e.g., coefficient).

INDEX NUMBER Number reflecting value or quantity compared with standard or base value.

INDEX OF BIOTIC INTEGRITY Aggregated number or index, based on several attributes or metrics of fish community that provides assessment of biological conditions.

INDEX OF LIQUIDITY Water content of sample minus water content in plastic limit, divided by index of plasticity.

INDEX OF PLASTICITY Difference between water content of clay at its liquid and plastic limits, showing range of water contents over which clay is plastic.

INDEX OF WETNESS Ratio of precipitation for given year over mean annual precipitation.

INDEX ORGANISMS Organisms that signal increased likelihood of pathogen originating from same source as index organism and thus serve predictive function.

INDEX PERIOD Sampling period during which selection is based on temporal behavior of indicator and practical considerations for sampling.

INDEX PRECIPITATION Index that can be used to adjust for bias in regional precipitation, often quantified as expected annual precipitation.

INDIAN COUNTRY All land within limits of any Indian reservation under jurisdiction of United States government, notwithstanding issuance of any patent, and, including rights-of-way running through reservation including all Indian allotments, Indian titles that have not been extinguished, including rights-of-ways running through same.

INDIAN RESERVATION All land within limits of any Indian reservation under jurisdiction of United States Government, notwithstanding issuance of any patent and including rights-of-way running through reservation.

INDIAN SUMMER Unseasonably warm period near middle of autumn, usually following substantial period of cool weather.

INDICATED HORSEPOWER Power produced in reciprocating engine by working of cylinders.

INDICATED NUMBER Amount obtained by taking reciprocal of very best potential dilution (lowest potential sample of bacterial density).

INDICATING FLOOR STAND Operating device valve that suggests degree of valve opening.

INDICATOR 1. Organism, species or community which indicates certain environmental conditions; 2. Substance that shows visible change, usually of color, at desired point in chemical reaction; 3. Device that indicates result of measurement (e.g., pressure gauge or moveable scale).

INDICATOR BYPASS Apparatus that shows alternative flow path is used.

INDICATOR DIFFERENTIAL PRESSURE Apparatus for signaling pressure difference between two points.

INDICATOR DISPLAY Apparatus that shows state, value or level of something.

INDICATOR GAUGE Instrument used to monitor operation and/or condition of system.

INDICATOR LIGHT Light used to monitor operation and/or condition of system.

INDICATOR METER Instrument designed to display values.

INDICATOR PANEL System panel designed to communicate with lights and/or signals.

INDICATOR POST Posts used to actuate and indicate open or shut status of buried gate valves.

INDICATOR SITES Stream sampling sites located at outlets of drainage basins with relatively homogeneous land use and physiographic conditions where most indicator-site basins have drainage areas ranging from 20 to 200 mi^2.

INDICATOR SPECIES Organism whose presence, absence or abundance reflects specific environmental condition.

INDICATOR VALVE Valve that suggests open or closed position.

INDIGENOUS Native to given area.

INDIGENOUS COMMUNITY BALANCED Biotic community typically characterized by diversity, capacity to sustain itself through cyclic seasonal changes and presence of necessary food chain species.

INDIGENOUS SPECIES Species that originally inhabited particular geographic area.

INDIRECT BANK PROTECTION Works built in-front of river banks.

INDIRECT CHARGE Charge that's part of total charge, but not directly related to it (e.g., overhead).

INDIRECT CHARGES Costs not associated with labor, materials, commissions and supplies.

INDIRECT DISCHARGE Introduction of pollutants from non-domestic source into publicly owned waste-treatment system.

INDIRECT DRAIN Pipe that does not connect directly with building's drainage system.

INDIRECT EFFECTS Impact of one organism or species on another mediated or transmitted by third.

INDIRECT FLOOD DAMAGE Expenditures (other than repair) made as result of flood (e.g., relief and rescue work, removing silt and debris).

INDIRECT HIT Locations that do not experience direct impact from tropical cyclone, but do experience hurricane force winds (either sustained or gusts) or tides of at least 4 feet above normal.

INDIRECT MATERIAL Materials that are used in production process, but they are not directly traceable to cost object.

INDIRECT SOLAR Energy that isn't directly powered by sun.

INDIRECT SOLAR GAIN Control solar radiation reaching area adjacent but not part of living space.

INDIRECT SOURCE Any facility or building, property, road or parking area that attracts motor vehicle traffic and, indirectly, causes pollution.

INDIRECT WASTE PIPING Waste pipe that does. not connect directly with drainage system.

INDIVIDUAL DRAIN SYSTEM All process drains connected to first common downstream junction box.

INDIVIDUAL SYSTEM Individual, independently owned alternate wastewater treatment works.

INDIVIDUAL VENT Vent that connects to fixture drain of trap or trapped fixture being vented.

INDOLE Organic compound found in coal tar and produced in gut by bacterial decomposition of tryptophan.

INDOOR AIR Breathable air inside habitable structure or conveyance.

INDOOR AIR POLLUTION Chemical, physical or biological contaminants in indoor air.

INDOOR CLIMATE Temperature, humidity, lighting, air flow and noise levels in habitable structure or conveyance.

INDUCED DRAFT FAN System is under negative pressure where fan is at exit end of path of flow.

INDUCED RECHARGE Water entering into aquifer from stream or body of water as result of lowering water table or potentiometric head in aquifer.

INDUCED SIPHONAGE Siphonage of water from fixture trap.

INDUCED TRAFFIC Predicted congestion benefits of new road are eroded by new traffic.

INDUCED-DRAFT WATER-COOLING TOWER Water-cooling tower having one or more fans.

INDURATED Cemented, hardened or rocklike condition.

INDURATED BED Channel bed that's been hardened.

INDURATED SOIL Soil cemented into hard mass that will not soften on wetting.

INDURATION Geological process of hardening mass or formation.

INDUSTRIAL CONSUMPTION Quantity of water consumed in municipality or district for mechanical, trade and manufacturing purposes, in given period, generally one day.

INDUSTRIAL COOLING TOWER Tower used to dispose of unwanted heat from chiller.

INDUSTRIAL DISCHARGE Waste produced by industrial activity which includes any material rendered useless during manufacturing process.

INDUSTRIAL INCINERATOR System that burns industrial waste.

INDUSTRIAL OCCUPANCY Any building, portion of building or group of buildings used for manufacture, assembly, service, mixing or packaging.

INDUSTRIAL OCCUPANCY HIGH HAZARD Industrial occupancy including adequate amounts of highly combustible and flammable or volatile substances that make up unique fire hazard.

INDUSTRIAL OCCUPANCY LOW HAZARD Industrial occupancy where combustible content isn't more than 4.5 kg (10 lb) or 105 500 kJ (100,000 Btu) per sq foot of floor space.

INDUSTRIAL OCCUPANCY MEDIUM HAZARD Industrial occupancy where combustible content is more than 4.5 kg (10 lb) or 105 500 kJ (100,000 Btu) per sq foot of floor area and not classified as high hazard industrial occupancy.

INDUSTRIAL PARK Area zoned and planned for purpose of industrial development.

INDUSTRIAL POLLUTION PREVENTION Combination of industrial source reduction and toxic chemical use substitution.

INDUSTRIAL PROCESS WASTE Residues produced during manufacturing operations.

INDUSTRIAL SILENCER Muffler for use on exhaust of internal combustion engines where fire hazards exist and minimum silencing is required.

INDUSTRIAL SLUDGE Semi-liquid residue or slurry remaining from treatment of industrial water and wastewater.

INDUSTRIAL SOLID WASTE Solid waste generated by manufacturing or industrial processes that isn't hazardous waste.

INDUSTRIAL SOURCE REDUCTION Practices that reduce amount of any hazardous substance, pollutant or contaminant entering any waste stream or otherwise released into environment that also reduces threat to public health and environment associated with such releases.

INDUSTRIAL USER Non-governmental, non-residential user of publicly owned treatment works that discharges more than equivalent of 25,000 gal/day of sanitary wastes.

INDUSTRIAL WASTE Unwanted materials from industrial operation.

INDUSTRIAL WASTE DAM Embankment dam, usually built in stages, to create storage for disposal of waste products from industrial process.

INDUSTRIAL WATER USE Water used for industrial purposes (e.g., fabrication, processing, washing, cooling) and includes such industries as steel, chemical and allied products, paper and allied products, mining and petroleum refining.

INDUSTRIAL WITHDRAWALS Water used for industrial purposes (e.g., fabrication, processing, washing, cooling) and includes such industries as steel, chemical and allied products, paper and allied products, mining and petroleum refining.

INEFFECTIVE FLOW When high ground or some other obstruction (e.g., levee prevents water from flowing into subsection), area up to that point is ineffective for conveying flow and is not used for hydraulic computations until water surface exceeds top elevation of obstruction.

INELASTIC Deformed material by load that cannot regain its original dimensions when unloaded.

INERT Not readily reacting with other elements or compounds.

INERT ATMOSPHERE Non-reactive gas atmosphere.

INERT GAS Gas which does not undergo chemical reactions.

INERT INGREDIENT Pesticide components (e.g., solvents, carriers, dispersants, surfactants) that are not active against target pests.

INERTIAL SEPARATOR Device that uses centrifugal force to separate particles by allowing materials to bypass inlet and get ejected.

INFECTIOUS AGENT Any organism (e.g., pathogenic virus, parasite, bacterium) that's capable of invading body tissues, multiplying and causing disease.

INFECTIOUS WASTE Hazardous waste capable of causing infections in humans (e.g., contaminated animal waste, human blood and blood products, isolation waste, pathological waste, needles, scalpels, broken medical instruments).

INFILTRATION 1. Downward movement of water from atmosphere, through ground surface into soil or porous rock; 2. Gradual flow of liquid into substance (e.g., soil) through pores of small openings; 3. Penetration of water from soil into sewer or other pipes through defective joints, connections or manhole walls; 4. Technique of applying large volumes of waste water to land to penetrate surface and percolate through underlying soil.

INFILTRATION CAPACITY Maximum rate at which water can enter soil at particular point under given set of conditions.

INFILTRATION CAPACITY CURVE Graph showing time-variation of infiltration rate which would occur if supply were continually in excess of infiltration capacity.

INFILTRATION COEFFICIENT Ratio of infiltration rate to precipitation or sprinkler irrigation rates.

INFILTRATION DITCH Linear ditch that collects rainwater to manage stormwater runoff, prevent flooding/downstream erosion and improve water quality by allowing water to quickly seep into ground.

INFILTRATION DIVERSION Moving stream water using perforated pipes placed underneath stream bed.

INFILTRATION GALLERY Buried conduits for intercepting, collecting and transferring groundwater.

INFILTRATION INDEX Average constant value of infiltration from rainfall and runoff records.

INFILTRATION RATE 1. Rate of water entry into soil; 2. Rate at which groundwater enters infiltration ditch or gallery, drain, sewer or other underground conduit.

INFILTRATION WATER Water soaked into soil from ground level.

INFILTRATION/INFLOW Excess water that flows into sewer pipes from groundwater and stormwater.

INFLATABLE GASKET Inflatable seals that works by introduction of inflation pressure into cavity to seal it.

INFLECTION POINT 1. Point on hydrograph separating falling limb from recession curve; 2. Any point on hydrograph where curve changes concavity.

INFLOW 1. Water that flows into body of water; 2. Entry of extraneous rain water into sewer system from sources other than infiltration (e.g., basement drains, manholes, storm drains, street washing).

INFLOW BANDS Bands of low clouds, arranged parallel to low-level winds and moving into or toward thunderstorm.

INFLOW DESIGN FLOOD 1. Flood flow above which incremental increase in downstream water surface elevation due to failure of dam or other water impounding structure is no longer considered to present unacceptable threat to downstream life or property; 2. Flood hydrograph used in design of dam and its appurtenant works particularly for sizing spillway and outlet works and for determining maximum storage, height of dam and freeboard requirements.

INFLOW JETS Local jets of air near ground flowing inward toward base of tornado.

INFLOW NOTCH Radar signature characterized by indentation in reflectivity pattern on inflow side of storm.

INFLOW STINGER Beaver tail cloud with stinger-like shape.

INFLOWING LOAD CURVE 1. Curve that expresses relation between discharge of stream or open conduit at given location and stage or elevation of liquid surface at given gaging station; 2. Graph showing relation between water height (ordinate) and volume of water flowing in channel (abscissa) per unit of time.

INFLUENCE AREA Surface area surrounding well or number of wells where watertable or piezometric surface is lowered by withdraw of water.

INFLUENCE BASIN Basin surrounding well or number of wells where watertable or piezometric surface is lowered by withdraw of water.

INFLUENT 1. Water, wastewater or other untreated liquid flowing into reservoir, stream, basin or treatment plant; 2. Organism having important effect on plant or animal community.

INFLUENT SEEPAGE Movement of gravity water in unsaturated zone, from ground surface to water table.

INFLUENT STREAM Any watercourse where all or portion of surface water flows back into ground.

INFLUENT WATER Water partially flowing into reservoir, basin, treatment process or treatment plant.

INFLUENT WEIR Weir at inflow end of treatment basin or tank.

INFORMATION COLLECTION REQUEST Description of information to be gathered in connection with rules, proposed rules, surveys and guidance documents that contain information-gathering requirements.

INFRARED RAYS Wavelengths longer than 780 nanometers.

INFRARED SATELLITE IMAGERY Satellite imagery senses surface and cloud top temperatures by measuring wavelength of electromagnetic radiation emitted from these objects.

INFRARED SPECTROPHOTOMETRY Interaction of infrared radiation with matter.

INFRASOUND Sound waves having frequency below human audible range (below 20 Hz).

INGESTED MATERIAL HALF-LIFE Time required for half amount of ingested substance to be discharged from your body naturally.

INGROUND TANK Device or portion of device, situated to any degree within ground.

INHALABLE DIAMETER Aerodynamic diameter of particle considered to be inhalable for organism, it is used to refer to particles that are capable of being inhaled and may be deposited anywhere within respiratory tract from trachea to alveoli.

INHALABLE PARTICLES All dust capable of entering human respiratory tract.

INHALATION LC_{50} Amount of material, given all at once, which causes death of 50% (one half) of group of test animals.

INHERENT ASH Part of ash content of coal that's structurally part of coal itself and cannot be separated from it by any mechanical means.

INHERENTLY LOW EMISSION VEHICLE Vehicle that meets exhaust emission standards and produces minimum or no evaporative emissions.

INHIBITION 1. Process of stopping or retarding chemical reaction; 2. Decline in growth rate of test organism.

INHIBITION CONCENTRATION Point estimate of toxicant concentration that would cause given percent reduction in non-lethal biological measurement of test organisms (e.g., reproduction, growth).

INHIBITORY SUBSTANCE Substances elaborated by specific strains of bacteria that are lethal against other strains of same or related species.

INHIBITORY TOXICITY Level of toxicity that reduces rate of reproduction of organism.

INITIAL ABSTRACTION All losses prior to surface runoff (e.g., interception, infiltration, evaporation and surface depression storage).

INITIAL ABSTRACTION RETENTION Total rainfall volume that does not cause runoff.

INITIAL CONDITIONS 1. Conditions prevailing prior to event; 2. Specified values for dependent variable (e.g., water levels, velocities) at beginning of model simulation (i.e., first estimate of variables model is trying to solve).

INITIAL CRUSHER Any crusher where non-metallic minerals can be fed without prior crushing in plant.

INITIAL DETENTION Volume of water on ground, either in depressions or in transit, at time active runoff begins.

INITIAL LOSS Rainfall that precedes start of surface runoff.

INITIAL MOISTURE DEFICIENCY Quantity, usually expressed in depth of water in inches upon unit area, by which actual water content of given soil zone (usually root zone) in such area is less than field capacity of such zone at beginning of rainy season.

INITIAL PHASE Of geomagnetic storm, that period when there may be increase of middle-latitude horizontal intensity.

INITIAL QUEUE Demand at beginning of period.

INITIAL QUEUE DELAY Delay to all vehicles in analysis period due to any initial queue at start of analysis period.

INITIAL RESPONDERS Individuals who are likely to witness or discover hazardous substance release and who have been trained to initiate emergency response sequence by notifying proper authorities of release.

INITIAL SEED EXAMINATION First formal onsite examination.

INITIAL WATER DEFICIENCY Quantity, usually expressed in depth of water in inches upon unit area, by which actual water content of given soil zone (usually root zone) in such area is less than field capacity of such zone at beginning of rainy season.

INJECTATE Fluids injected into underground injection well.

INJECTION Subsurface discharge of fluids through well.

INJECTION INTERVAL Geological formation targeted to receive injected fluids.

INJECTION PUMP Device that pumps fuel into cylinders of engine.

INJECTION WELL Well into which fluids are injected for purposes (e.g., waste disposal, improving recovery of crude oil, solution mining).

INJECTION ZONE Geological formation, group of formations or part of formation that receives fluids through well.

INLAND 1. Interior of land; 2. Away from coast.

INLAND CARRIER Transportation line that hauls export or import traffic between ocean ports and inland points.

INLAND FRESHWATER WETLANDS Swamps, marshes and bogs found inland beyond coastal saltwater wetlands.

INLAND WATERS Permanent water bodies inland from coastal zone and areas whose properties and use are dominated by permanent, seasonal or intermittent occurrence of flooded conditions.

INLET 1. Body of water let into coast (e.g., bay, cove, fjord, estuary); 2. Connection to closed drain. 3. Passage that leads into cavity.

INLET CHANNEL Concrete lined portion of spillway between approach channel and gate or crest structure.

INLET CONTROL Amount of water culvert barrel can carry is limited by culvert entrance.

INLET STRUCTURE Concrete lined portion of spillway between approach channel and gate or crest structure.

INLET TIME Time required for rain in falling on most remote point of tributary area to flow across ground surface to inlet of sewer.

INLET WELL Opening that allows water to be drained into sewer.

INNER LINER Continuous layer of material placed inside tank or container that protects construction materials of tank or container from contained waste or reagents used to treat waste.

INNINGS Acreage recovered from ocean.

INNOVATIVE TECHNOLOGIES New or inventive methods to treat effectively hazardous waste and reduce risks to human health and environment.

INNOVATIVE TREATMENT TECHNOLOGIES Technologies whose routine use is inhibited by lack of data on performance and cost.

INOCULATE To graft by inserting buds.

INOCULUM 1. Bacteria or fungi injected into compost to start biological action; 2. Medium containing organisms, usually bacteria or virus, introduced into cultures or living organisms.

INORGANIC 1. Containing no carbon; 2. Matter other than plant or animal; 3. Substances that are of mineral origin.

INORGANIC CHEMICALS Chemical substances of mineral origin, not of basically carbon structure.

INORGANIC CONTAMINANTS Mineral-based compounds (e.g., metals, nitrates and asbestos).

INORGANIC MATTER Matter which is not derived from living organisms and contains no organically produced carbon.

INORGANIC SOIL Soil with less than 20% organic matter in upper 16 inches.

INORGANIC SOLID DEBRIS Non-friable inorganic solids contaminated with hazardous wastes that are incapable of passing through 0.374 inch standard sieve; and that require cutting or crushing and grinding in mechanical sizing equipment prior to stabilization.

INPUTS 1. To enter data; 2. Something fed into process.

INSANITARY 1. Of or pertaining to lack of sanitation; 2. Unsanitary, dirty, unhealthy.

INSCRIBED CIRCLE Circle within polygon tangent to each side.

INSECTICIDE Pesticide compound specifically used to kill or prevent growth of insects.

INSECTICIDES Substance used to kill insects.

INSERT VALVE Valve (e.g., shutoff, check) inserted into conduit while under pressure.

INSOLATION 1. Incoming solar radiation that reaches unit area over period of time; 2. Solar heating; 3. Sunshine.

INSPECTION Act of examining something.

INSPECTION AND MAINTENANCE 1. Activities to ensure that vehicles' emission controls work properly; 2. Applies to wastewater treatment plants and other anti-pollution facilities and processes.

INSPECTION CERTIFICATE Document certifying that merchandise was in good condition at time of inspection.

INSPECTION CHAMBER Clean-out generally installed at property line of building.

INSPECTION COVER Section which can be taken out to permit review of internals.

INSPECTION DOOR Opening that may be cleared to allow person to move through.

INSPECTION FITTING Conduit fitting with removable plug.

INSPECTION LIST Critical items to check.

INSPECTOR Person employed to inspect something.

INSTABILITY Tendency for air parcels to accelerate when they are displaced from their original position.

INSTALL To connect, set up or prepare something for use.

INSTALLATION INSPECTOR Individual who has demonstrated knowledge required to properly inspect installation of something.

INSTALLED CAPACITY Total of capacities shown on nameplates of generating units in power plant.

INSTANTANEOUS ACCELERATION Acceleration of object at specific moment of time.

INSTANTANEOUS DISCHARGE Volume of water that passes point at particular instant of time.

INSTANTANEOUS EFFLUENT LIMITATION Maximum allowable concentration in wastewater discharge at any time, as measured in grab sample.

INSTANTANEOUS MAXIMUM LIMIT Maximum allowable concentration of pollutant determined from analysis of any discrete or composite sample collected, independent of flow rate and duration of sampling event.

INSTANTANEOUS SPEED Speed of object at certain instant of time.

INSTANTANEOUS UNIT HYDROGRAPH Theoretical, ideal, unit hydrograph that has infinitesimal duration.

INSTITUTIONAL SOLID WASTE Solid waste from government offices, schools, hospitals, recreational facilities, etc.

INSTITUTIONAL WASTE Waste generated at institutions (e.g., schools, libraries, hospitals, prisons).

INSTITUTIONALIZED POPULATIONS People in schools, hospitals, nursing homes, prisons, federal buildings or other facilities that require special care or consideration during emergencies by virtue of their dependency on others for appropriate protection.

INSTREAM FLOW REQUIREMENTS Amount of water flowing through defined stream channel needed to sustain instream values (e.g., flows designated for fish and wildlife).

INSTREAM USE Use of water that does not require withdrawal or diversion from its natural watercourse (e.g., navigation, recreation, hydro-electric power generation, water quality improvement, fish propagation, recreation).

INSTRUCTIONS TO BIDDERS Bidding documents directions describing how to submit bids.

INSTRUMENT APPROACH PROCEDURE Predetermined maneuvers that allow pilots to align aircraft with runway and fly towards it without being able to see outside windows.

INSTRUMENT FLIGHT RULES General weather conditions pilots can expect at surface and applies to weather situations at airport during which pilot must use instruments to assist takeoff and landing.

INSTRUMENT LANDING SYSTEM Collection of radio navigation aids used by pilots for approach guidance to specific airport runway.

INSTRUMENT PANEL Flat, vertical surface onto which controls, instruments and displays are mounted.

INSTRUMENT RUNWAY Runway with air navigation aids to permit aircraft landing under restricted visibility conditions.

INSTRUMENT SHELTER Boxlike structure designed to protect temperature measuring instruments from exposure to direct sunshine, precipitation and condensation, while at same time providing adequate ventilation.

INSTRUMENTATION 1. Any device used to monitor performance of structure during its construction and throughout its useful life; 2. Arrangement of devices installed into or near dams that provide for measurements that can be used to evaluate structural behavior and performance parameters of structure.

INSULATED 1. Stream or reach of stream that neither contributes water to zone of saturation nor receives water from it (i.e., separated from zones of saturation by impermeable bed); 2. Material that prevents the loss of temperature or intrusion of sound.

INSULATED FLANGE Flanges equipped with plastic pieces that insulate and prevent flow of electric current.

INSULATED JOINT Rail joint that prevent electrical current from flowing between ends of two adjoining rails.

INSULATED STREAM Stream or stream reach hydrologically separated from saturated zone of underlying formation by aquitard.

INTACT Spacing greater than 6 ft.

INTAKE Place where water, air or other fluid is taken into pipe or conduit.

INTAKE AREA Surface area where water is initially absorbed before reaching aquifer or groundwater basin.

INTAKE HEADING Where channel enters control works.

INTAKE HOSE Hose that connects water to pump.

INTAKE MANIFOLD Part of engine that supplies fuel or air mixture to cylinder.

INTAKE PIPE 1. Conduit carrying water by gravity from source to intake well; 2. Conduit at head of system where water enters.

INTAKE SCREEN Screen designed to prevent foreign objects from entering.

INTAKE STRUCTURE 1. Concrete portion of outlet works, including trash racks and/or fish screens, upstream from tunnel or conduit portions; 2. Entrance to outlet works.

INTAKE WORKS Structure for obtaining water from supply source for purposes of hydroelectric power engineering, water supply or irrigation.

INTANGIBLE ASSET Any valuable property of business not physical by nature, including intellectual property, customer lists and goodwill.

INTANGIBLE FACTORS Factors that affect decision, but that cannot be expressed in monetary terms.

INTANGIBLE FLOOD DAMAGE Estimates of damage done by disruption of business, danger to health, shock and loss of life and in general all costs not directly measurable which require large element of judgment for estimating.

INTEGRAL BRIDGE Bridge without expansion joints.

INTEGRAL VISTA View perceived from within mandatory area of specific landmark or panorama located outside such boundary.

INTEGRATED CURBSIDE COLLECTION Residential service combining collection of recyclables and refuse concurrently.

INTEGRATED DRAINAGE Drainage developed during geomorphic maturity in arid region, characterized by coalescence of drainage basins as result of headward erosion in lower basins or spilling over from upper basins.

INTEGRATED EARLY WARNING SYSTEM Flood warning system tied into adjacent systems so that multiple users can access basin data and monitor event.

INTEGRATED EXPOSURE ASSESSMENT Cumulative summation of magnitude of exposure to toxic chemical in all media.

INTEGRATED FLOW CURVE Mass diagram with discharge as ordinate and time as abscissa.

INTEGRATED PEST MANAGEMENT Mixture of chemical and other, non-pesticide, methods to control pests.

INTEGRATED POLLUTION CONTROL Development and optimization of methods that limit spread of pollutants.

INTEGRATED RESOURCE PLANNING Public planning process and framework within which costs and benefits of both demand and supply side resources are evaluated to develop least total cost mix of utility resource options.

INTEGRATED RISK INFORMATION SYSTEM EPA's electronic data base containing latest descriptive and quantitative regulatory information on chemical constituents.

INTEGRATED SOLID WASTE MANAGEMENT Strategic approach to sustainable management of solid wastes.

INTEGRATED WASTE MANAGEMENT Using variety of practices to handle municipal solid waste (e.g., source reduction, recycling, incineration, landfilling).

INTEGRATOR Device or meter that continuously measures and calculates total flow.

INTEGRATOR OR MIXED-USE SITE Stream sampling site located at outlet of drainage basin that contains multiple environmental settings.

INTELLIGENT TRANSPORTATION SYSTEM Advanced applications that aim to provide innovative services relating to different modes of transport and traffic management.

INTENSELY FRACTURED Spacing less than 0.3 ft.

INTENSIFIER 1. Chemical agent used to intensify lights or shadows of photograph; 2. Device that converts low pressure fluid into higher-pressure fluid power.

INTENSITY Measurable amount of property.

INTENSITY OF CONGESTION Amount of transport system congestion that identifies different congestion levels versus total.

INTENSITY SCALE Arbitrary scale to describe degree of shaking at particular place.

INTENSIVE FIXED SITES Basic Fixed Sites with increased sampling frequency during selected seasonal periods and analysis of dissolved pesticides for 1 year.

INTERBASIN TRANSFER Physical transfer of water from one watershed to another.

INTERBEDDED Occurring between beds of rock.

INTERCEPTING CHANNEL Channel excavated at top of earth cuts or foot of slopes to intercept surface flow.

INTERCEPTING DITCH Ditch constructed to prevent surface water from flowing in contact with toe of embankment or causeway or down slope of cut.

INTERCEPTING DRAIN Drain constructed at upper end of area to be drained, to intercept surface or groundwater flowing toward protected area from higher ground and carry it away from area.

INTERCEPTING SEWER Large sewer that receives flow from number of sewers and conducts wastewater to treatment plant.

INTERCEPTION 1. Capture of precipitation above ground surface (e.g., by vegetation, by buildings) and returned to atmosphere through evaporation; 2. Procedure of redirecting wastewater from primary sewer right into conduit leading straight into treatment facility.

INTERCEPTION STORAGE REQUIREMENTS Water caught by plants, before rainfall reaches ground, at onset of rainstorm.

INTERCEPTOR Device to trap, remove or separate deleterious, hazardous or undesirable matter (e.g., oil, grease, gasoline, sand, sediment) from normal waste conveyed through it.

INTERCEPTOR SEWER Sewer network component that receives flow from trunk sewer lines and storm water runoff and directs it to wastewater treatment plant.

INTERCEPTOR SEWERS Large sewer lines that, in combined system, control flow of sewage to treatment plant.

INTERCEPTOR TRENCH Deep depression on edge of storage unit that contains leakage.

INTERCHANGE Highway junction where traffic may change from one road to another without crossing stream of traffic.

INTERCHANGE DENSITY Density of freeway determined over 6 mile segment of freeway (3 mile upstream and 3 mile downstream).

INTERCITY TRANSPORTATION Transportation system between cities.

INTERCONNECTED SYSTEM System consisting of two or more individual power systems normally operating with connecting tie lines.

INTERCONNECTION 1. Connection (i.e., physical, logical) between multiple things; 2. Connecting renewable-energy systems to power grid.

INTERCONNECTOR Structure that allows energy to flow between networks.

INTERFACE 1. Boundary between two substances (i.e., liquid and solid, liquid and gas) or two phases; 2. Contact zone between two fluids of different chemical or physical makeup.

INTERFACIAL TENSION Strength of film separating two immiscible fluids (e.g., oil and water) usually measured in dynes/cm or millidynes/cm.

INTERFERENCE Effect caused by superposition of two systems of waves.

INTERFERENCE ANGLE Difference between angles where valve seat and face.

INTERFLOW Lateral motion of water through upper layers until it enters stream channel.

INTERFLUVE Region of higher land between two connected river valleys.

INTERGLACIAL Occurring between glacial periods.

INTERIM STATUS Period during which treatment, storage and disposal facilities are temporarily permitted to operate while awaiting permanent permit.

INTERIOR PLAIN Vast region that spreads across stable core of North America.

INTERLINE Between two airlines or railroads.

INTERLOCK 1. Electrical switch, usually magnetically operated; 2. To fit or clasp together securely.

INTERLOCKING Binding particles together.

INTERMEDIATE BELT Residual part of zone of aeration occurring between belt of soil water and capillary fringe.

INTERMEDIATE COVER Fill placed on areas that have received waste and completed layer.

INTERMEDIATE GROUNDWATER Water in zone of aeration between periphery and groundwater.

INTERMEDIATE HANDLER Facility that either treats regulated medical waste to reduce or eliminate its infectious nature or destroys it by rendering it unrecognizable, but does not do both.

INTERMEDIATE SEED EXAMINATION Onsite examination performed approximately every 3 years or more frequently if conditions dictate, to evaluate operation, performance and existing condition of all features.

INTERMEDIATE SYNOPTIC TIMES Times of 0300, 0900, 1500 and 2100 UTC.

INTERMEDIATE TREATMENT Removal from wastewater of higher portion of suspended solids and colloidal matter, but comparatively little dissolved matter.

INTERMEDIATE VADOSE WATER Water held due to its intermolecular forces against pull of gravity.

INTERMEDIATE WATER Water between soil water belt and upper surface of capillary fringe.

INTERMEDIATE ZONE Subsurface water zone below root zone and above capillary fringe.

INTERMEDIATE-LEVEL WASTES Waste containing higher radioactivity levels than low level waste.

INTERMITTENT CAPACITY Load carrying capability of generator having less than full availability for meeting loads over specific periods of time.

INTERMITTENT CONTROL SYSTEM Dispersion technique which alters speed where pollutants are emitted to atmosphere in accordance with meteorological conditions or ambient concentrations of pollutant.

INTERMITTENT FILTER Filter for effluent that requires treatment prior to surface or subsurface discharge.

INTERMITTENT HORSEPOWER Maximum momentary output of full-load horsepower.

INTERMITTENT INTERRUPTED STREAM Stream with occasional stretches divided by ephemeral stretches.

INTERMITTENT PERIODIC SPRING Periodic spring that discharges intermittently.

INTERMITTENT SAMPLING Collects desired sample volume by cycling flow (or pump) on and off during the sampling period.

INTERMITTENT SPRING Springs that flow at intervals and are not dependent upon precipitation.

INTERMITTENT STREAM One which flows only at certain times of year when it receives water from springs or from some surface source (e.g., rainfall runoff, melting snow).

INTERMITTENT VAPOR PROCESSING SYSTEM Vapor processing system that employs intermediate vapor holder to accumulate total organic compound vapors and treats accumulated vapors only during automatically controlled cycles.

INTERMODAL 1. Relating to more than one mode of transport; 2. Transfer point to load or unload trailers or railcars.

INTERMODAL CAR Double-stack or stack rail vehicle.

INTERMODAL FACILITY Location that provides mode to change transportation.

INTERMONTANE Situated between or surrounded by mountains, mountain ranges or mountainous regions.

INTERNAL COMBUSTION PLANT Plant where prime mover is internal combustion engine.

INTERNAL COMPACTION TRANSFER SYSTEM Reciprocating activity of hydraulically powered bulkhead arrested within enclosed trailer that carries and compresses solid waste.

INTERNAL CONTROL Implementation of segregation of duties.

INTERNAL DIAMETER Diameter of inside of tube, pipe or other object.

INTERNAL DOSE Amount of substance penetrating absorption barriers (e.g., skin, lung tissue, gastrointestinal tract) of organism through either physical or biological processes.

INTERNAL DRAINAGE 1. Movement of water down through soil to porous aquifers or to surface outlets at lower elevations; 2. Drainage within basin that has no outlet (e.g., ocean).

INTERNAL EROSION Formation of voids within soil or soft rock caused by mechanical or chemical removal of material by seepage.

INTERNAL FLOATING ROOF Floating cover on waste management liquid surface that has fixed roof.

INTERNAL INSPECTION Inspection performed from inside of system.

INTERNAL LINK Roadway, at interchange ramp, between two signalized intersections.

INTERNAL PIPEWORK Piping for building services (e.g., hot water, cold water, gas, sanitary).

INTERNAL TENDON High-strength, steel wire used to prestress concrete.

INTERNAL THREAD Thread of screw on inner surface.

INTERNAL TREATMENT Conditioning (i.e., water treatment) of impurities within boiler system.

INTERNAL WATER Waters on landward side of baseline of nation's territorial waters, except in archipelagic states.

INTERNATIONAL BOUNDARY COMMISSION SURVEY DATUM Geodetic datum established at numerous monuments along United States-Canada boundary by International Boundary Commission.

INTERNATIONAL DATE LINE Line of longitude located at 180° East or West (with few local deviations) where date changes by day where west of line is one day later than east of line.

INTERNATIONAL ROUGHNESS INDEX Standard that quantifies road surface roughness where continuous profile of road is measured to determine pavement surface deviations impacting vehicle suspension movement.

INTERNATIONAL SYSTEM OF UNITS 1. System of measurement with several base units; 2. Modern metric system.

INTERPLANETARY MAGNETIC FIELD Magnetic field carried with solar wind.

INTERPOLATE To estimate value within interval between two known values.

INTERROGATOR Device that requests data from another device.

INTERRUPTED FLOW Traffic flow regulated by traffic signals.

INTERRUPTED STREAM One which contains alternating reaches, that are either perennial, intermittent or ephemeral.

INTERRUPTED WATERTABLE Pronounced drop of watertable near groundwater dam.

INTERRUPTIBLE DEMANDS Demands that, by contract, can be interrupted if supplying system lacks capacity.

INTERRUPTIBLE LOAD Program activities that can interrupt consumer load at times of peak load.

INTERSECTION 1. Junction of two (or more) paths, streets, highways or other thoroughfares; 2. Point or set of points common to two geometrical objects.

INTERSECTION ANGLE Angle between any two intersecting legs at point centerlines intersect.

INTERSECTION CONTROL BEACON One or more beacon signal faces directed toward each intersection approach.

INTERSECTION DELAY Additional travel time experienced by vehicle after it enters intersection and before it reaches free-flow speed.

INTERSECTION LEG Any roadway radiating from and forming part of intersection.

INTERSTATE CARRIER WATER SUPPLY Source of water for drinking and sanitary use on planes, buses, trains and ships operating in more than one state.

INTERSTATE HIGHWAY SYSTEM Network of controlled-access highways that forms part of National Highway System in United States.

INTERSTATE WATERS Waters that flow across or form part of state or international boundaries (e.g., Great Lakes, Mississippi River, coastal waters).

INTERSTICE Opening or space in rock or soil.

INTERSTITIAL FLOW Portion of surface water that infiltrates streambed and moves through pores in subsurface.

INTERSTITIAL ICE Ice formed in narrow spaces between small rocks and sediment in soil.

INTERSTITIAL MONITORING Continuous surveillance of space between walls of underground storage tank.

INTERSTITIAL WATER Subterranean water in pores of rocks, soils and bottom sediments.

INTERSTREAM GROUNDWATER RIDGE Residual ridge in water table that develops between two effluent streams.

INTERTIDAL Alternately flooded and exposed by tides.

INTERTROPICAL CONVERGENCE ZONE Region where northeasterly and southeasterly trade winds converge, forming often continuous band of clouds or thunderstorms near equator.

INTERVAL 1. Distance in space; 2. Period of time.

INTERVAL SEQUENCE Signal order during cycle.

INTERZONAL TRIP Trip between two zones.

INTOLERANT ORGANISMS Organisms that are not adaptable to human alterations to environment and thus decline in numbers where human alterations occur.

INTRADOS Curved downstream surface of horizontal arch elements of arch dam.

INTRAPERMAFROST WATER Water occurring in unfrozen zones within permafrost.

INTRASEASONAL OSCILLATION Oscillation with variability on timescale less than season.

INTRASTATE PRODUCT Pesticide products once registered by states for sale and use only in state.

INTRINSIC ENERGY VALUE Energy consumed in fabrication of product or to create feedstock.

INTRINSIC PERMEABILITY Relative ease with which porous medium can transmit liquid under hydraulic gradient or potential gradient.

INTRINSIC RATE OF INCREASE Maximum theoretical rate of increase of population per individual (i.e., maximum population growth rate).

INTRINSICALLY SAFE EQUIPMENT Equipment and wiring incapable of releasing sufficient energy to cause ignition.

INTRUSION Feature (e.g., landform, vegetation, structure) generally considered out of context because of excessive contrast and disharmony with characteristic landscape.

INUNDATE To cover with large amounts of water.

INUNDATION Process of covering normally dry areas with flood waters.

INUNDATION MAP Map delineating area that would be flooded by particular flood event (e.g., dam failure).

INVENTORY Detailed list of all of items.

INVENTORY RATING Load that can safely utilize bridge for unlimited time.

INVERSE CONDEMNATION Taking of private property by government without providing compensation, thus forcing previous owner to file lawsuit against government to recover value of property taken.

INVERSE METHOD Method of calibrating model using computer code to systematically vary inputs or input parameters to minimize residuals or residual statistics.

INVERSE MODEL Computer code for determination of selected unknown parameters and stresses in groundwater system, given that response of system to all stresses is known and that information is available regarding certain parameters and stresses.

INVERSE SLUDGE INDEX Reciprocal of sludge volume index multiplied by 100.

INVERSION 1. Increase of air temperature with increase in altitude; 2. Layer of warm air that prevents rise of cooling air and traps pollutants beneath it.

INVERT Lowest point of underground excavation or lowest point inside pipe at certain point.

INVERT LEVEL Base interior level of conduit, trench, tunnel, etc.

INVERTEBRATE Animal having no backbone or spinal column.

INVERTEBRATES Animals without vertebral column (e.g., spiders, crabs, worms).

INVERTED CAPACITY Maximum rate where recharge well can dispose of water admitted to it.

INVERTED DRAINAGE WELL Well that drains water at or near surface.

INVERTED SIPHON Conduit, "U" shaped, that allows stormwater or wastewater sewers to pass under obstruction in flow path.

INVERTED WELL Well that recharges water back into aquifer.

INVESTOR OWNED UTILITY Company owned by stockholders for profit that provides utility services.

INVITATION FOR BID Method to submit offer on specific project to be realized or product and/or service to be furnished.

INVITATION FOR BIDS All documents utilized for soliciting bids.

INVITED BIDDERS Bidders chosen for sole bids to be received.

INVOICE Document issued by seller to buyer indicating products, quantities and agreed prices for products or services that seller has already provided buyer with.

INVOLUNTARY LIEN Liens placed by government revenue authorities for unpaid duties or taxes.

INWARD-FLOW TURBINE Turbine that combines radial and axial flow concepts where water enters horizontally in spiral shaped pipe, wrapped around outside of turbine's rotating runner and exits down vertically through center of turbine.

IODINE-131 Radioactive isotope of iodine having half-life of 8 days used as medical tracer.

IODOMETRIC CHLORINE TEST Determination of residual chlorine in water or wastewater by addition of potassium iodide and titration of liberated iodine with standard solution of sodium thiosulfate, with starch solution as colorimetric indicator.

ION Electrically charged atom, group of atoms or molecule that has lost or gained one or more electrons.

ION EXCHANGE Reversible process whereby ions are exchanged between solution and solid and/or gel.

ION EXCHANGE RESINS Insoluble matrix of small beads having charged surface.

ION EXCHANGE TREATMENT Water-softening method where water passed through resin containing exchangeable ions, often found on large scale at water purification plants that remove some organics and radium by adding calcium oxide or calcium hydroxide to increase pH to level where metals will precipitate out.

ION-SELECTIVE ELECTRODE Electrode whose potential is dependent on concentration of specific ion in solution where it is immersed.

IONIC CONCENTRATION Concentration of any ion in solution, usually expressed in moles/liter.

IONIZATION Any process that leads to dissociation of neutral atom or molecule into charged particles ions.

IONIZATION CHAMBER Device that measures intensity of ionizing radiation.

IONIZING RADIATION Radiation that can strip electrons from atoms (e.g., alpha, beta, gamma radiation).

IONOSPHERE Complex atmospheric zone of ionized gases that extends between 50 and 400 miles (80 to 640 kilometers) above Earth's surface that's located between mesosphere and exosphere and is included as part of thermosphere.

IONOSPHERIC STORM Disturbance in F region of ionosphere, which occurs in connection with geomagnetic activity.

IOWA VANE Flow deflector that will divert water in way calculated to attract fish.

IRIDESCENCE Brilliant spots or borders of colors in clouds, usually red and green, caused by diffraction of light by small cloud particles.

IRIDESCENT CLOUDS Clouds that exhibit brilliant bright spots, bands or borders of colors, usually red and green, observed up to about 30° from Sun.

IRON Metallic chemical element having atomic number 26.

IRON BACTERIA Any of group of bacteria that oxidize ferrous iron to ferric iron.

IRON PIPE THREAD Regular system of threads for linking numerous kinds of rigid piping.

IRON VITRIOL Green, crystalline salt of divalent iron and sulfuric acid used as pigment and in water treatment.

IRRADIATED FOOD Food subject to brief radioactivity, usually gamma rays, to kill insects, bacteria and mold and to permit storage without refrigeration.

IRRADIATION Having been exposed to radiation of wavelengths shorter than those of visible light (gamma, x-ray or ultra- violet).

IRREGULAR WEIR Weir with non-standard cross-section.

IRRETRIEVABLE Commitments that are lost for period of time.

IRREVERSIBLE Commitments that cannot be reversed.

IRREVERSIBLE EFFECT Effect characterized by inability of body to partially or fully repair injury caused by toxic agent.

IRRIGABLE ACRE FOR SERVICE Acreage classified as irrigable for which project works have been constructed and project water is available.

IRRIGABLE AREA Maximum area which could be irrigated in reference year using equipment and quantity of water normally available on holding.

IRRIGABLE AREA NOT FOR SERVICE Part of total irrigable area for which service eventually will be available but for which project facilities have not yet been constructed.

IRRIGABLE LAND Arable land under specific plan for which water supply is or can be made available and that is provided with or planned to be provided with irrigation, drainage, flood protection and other facilities as necessary for sustained irrigation.

IRRIGATED ACREAGE Irrigable acreage actually irrigated in any one year (i.e., irrigated cropland harvested, irrigated pasture, cropland planted but not harvested and acreage in irrigation rotation used for soil-building crops).

IRRIGATED AREA Amount of outdoor vegetated area supplied water regularly.

IRRIGATING HEAD 1. Flow for irrigation of particular tract of land; 2. Flow or water distributed at single irrigation.

IRRIGATING STREAM 1. Flow for irrigation of particular tract of land; 2. Flow or water distributed at single irrigation.

IRRIGATION Supply of water (not via rainfall) to land in order to grow crops, grass, trees and other plants.

IRRIGATION CANAL Main waterways that bring irrigation water from water source to areas to be irrigated.

IRRIGATION CHECK Small dike or dam used in furrow alongside irrigation border to make water spread evenly across border.

IRRIGATION DISTRICT Cooperative, self-governing public corporation having taxing power to obtain and distribute water for irrigation of lands within district.

IRRIGATION DITCH Channels that deliver water to homes, farms, industries and other human uses.

IRRIGATION EFFICIENCY 1. Ratio (expressed as percent) of average depth of irrigation water beneficially used to average depth of irrigation water applied; 2. Amount of water stored in crop root zone compared to amount of irrigation water applied.

IRRIGATION REQUIREMENT Quantity of water (not via rainfall) required to grow crops, grass, trees and other plants.

IRRIGATION RETURN FLOW Surface and subsurface water which leaves field following application of irrigation water.

IRRIGATION ROTATION Rotation method among irrigation groups to counter water-shortage problems.

IRRIGATION STRUCTURE Structure used to conduct water for irrigation.

IRRIGATION WATER Water applied to lands to meet growing needs.

IRRIGATION WATER USE Water application on lands to assist in growing of crops and pastures or to maintain vegetative growth.

IRRIGATION WITHDRAWALS Withdrawals of water for application on land to assist in growing of crops and pastures or to maintain recreational lands.

IRRIGATION-WATER REQUIREMENTS Quantity of water necessary for crop growth.

IRRITANT 1. Substance that can cause irritation of skin, eyes or respiratory system.; 2. Chemicals which inflame living tissue by chemical action at site of contact, causing pain or swelling.

ISENTROPIC ANALYSIS Looking at atmosphere in 3-dimensions instead of looking at constant pressure surfaces in 2-dimensions.

ISENTROPIC LIFT Lifting of air traveling along upward-sloping isentropic surface.

ISENTROPIC SURFACE Two-dimensional surface containing points of equal potential temperature.

ISLAND 1. Mid-channel bar that has permanent woody vegetation, is flooded once year, on average and remains stable except during large flood events; 2. Contiguous area of land, smaller than continent, totally surrounded by water.

ISO-PERCENTILE Line, on chart or graph, of identical percentages.

ISOBAR Line connecting points of equal or constant pressure.

ISOBARIC CHART Weather map representing conditions on surface of equal atmospheric pressure.

ISOBARIC PROCESS Any thermodynamic change of state of system that takes place at constant pressure.

ISOBATH Imaginary line on Earth's surface or line on map connecting all points which are same vertical distance above upper or lower surface of water-bearing formation or aquifer.

ISOCHION Line of equal water equivalent of snow or equal snow depth.

ISOCHLOR Line joining all points of equivalent concentrations of chlorides.

ISOCHROME 1. Line on chart connecting equal times of occurrence of event; 2. Curve showing distribution of excess hydrostatic pressure at given time during process of consolidation.

ISOCHRONE Isoline on map or chart connecting points that have same value of quantity that has dimension time.

ISOCONCENTRATION More than one sample point exhibiting same isolate concentration.

ISODOP Contour of constant Doppler velocity values.

ISODROSOTHERM Line connecting points of equal dew point temperature.

ISOELECTRIC POINT pH of colloidal suspension or of ampholyte where solute does not move in electrophoretic field.

ISOHEIGHT Contour depicting vertical height of some surface above datum plane.

ISOHEL Line on weather map connecting points receiving equal or constant solar radiation.

ISOHYET Line that connects points of equal or constant precipitation or rainfall intensity.

ISOHYETAL LINE Line drawn connecting points that receive same amount of precipitation.

ISOHYETAL MAP Map depicting contours of equal precipitation amounts recorded during given time period.

ISOLATED National Weather Service convective precipitation descriptor for 10% chance of measurable precipitation.

ISOLATED INTERSECTION Intersection not influenced by upstream signalized intersections.

ISOLATED INTERSTICE Little opening in homogeneous stone.

ISOLATED SOLAR GAIN Utilizing solar energy to passively move heat from or to living space using fluid by convection.

ISOLATING SWITCH Switch that isolates electric circuit from source of power.

ISOLATING VALVE Valve that stops flow.

ISOLATION JOINT Separation between adjacent sections of concrete to allow relative movement.

ISOLATOR UNIT Horizontally flexible and vertically stiff bearing that permits large lateral deformation under seismic load.

ISOMERS 1. Any of two or more compounds with same molecular formula but with different structure; 2. Any of two or more atomic nuclei with same mass number and atomic number but with different radioactive properties.

ISONIVAL Line of equal water equivalent of snow or equal snow depth.

ISOPHYTOCHRONE Line, on map, that joins points where growing season for plants is same time.

ISOPLETH 1. Line connecting points with equal values of particular atmospheric variable (temperature, dew point); 2. Line or area represented by isoconcentration.

ISOSEISMIC LINE Line, on map, bounding points of equal intensity for particular earthquake.

ISOTACH Line connecting points of equal or constant wind speed.

ISOTHERMAL Having constant temperature.

ISOTHERMAL COMPRESSION Compression at constant temperature.

ISOTOPE Variation of element that has same atomic number of protons but different weight because of number of neutrons.

ISOTROPIC Having same characteristics or properties in all directions.

ISOTROPIC MASS Mass having same property (or properties) in all directions.

ISOTROPIC PLATE Plate with same structural properties in two directions.

ISOTROPIC SOIL Soil having over 90% similarity in all directions.

ISOTROPY Condition where hydraulic or other properties of aquifer are same in all directions.

ISSUANCE TIME Time product is transmitted.

ITEM Line of text having legal or other meaning.

J

JACKET Removable or replaceable protective or insulating cover for object.

JACKETED PUMP Pump with jackets that permits handling of materials exhibiting high viscosity.

JACKETING When section of existing structural member is restored to original dimensions or increased in size by encasement using suitable materials.

JACKING FORCE Vertical force exerted by tension device.

JACKING PIPE Technique for installing finished pipeline as tunnel is excavated.

JAM DENSITY Traffic density when flow stops completely.

JANUARY THAW Period of mild weather popularly supposed to recur each year in late January.

JAR TEST Laboratory procedure that simulates water treatment plant's coagulation/flocculation units with differing chemical doses, mix speeds and settling times to estimate minimum or ideal coagulant dose required to achieve certain water quality goals.

JAW CRUSHER Primary crushers in mine or ore processing plant.

JEOPARDY OPINION Action is likely to jeopardize continued existence of listed species or result in destruction or adverse modification of critical habitat.

JET 1. Fast-moving stream, spurt or flow of liquid or gas; 2. Engine that propels vehicle using stream of fluid as propulsion; 3. Part of carburetor that controls amount of fuel mixed with air.

JET ACTION Valve where flow is controlled by location of nozzle.

JET HEIGHT Vertical distance unconfined jet rises.

JET MAX Point or area of relative maximum wind speeds within jet stream.

JET PIERCING DRILL Device for directed destruction of solid mineral mediums through thermal and mechanical action.

JET PUMP Pump where small jet of steam, air, water or other fluid in rapid motion lifts or otherwise moves by its impulse large quantity of fluid with which it mingles.

JET ROUTES High altitude airways.

JET STREAK Point or area of relative maximum wind speeds within jet stream.

JET STREAM Relatively strong winds concentrated in narrow stream in atmosphere, normally referring to horizontal, high-altitude winds.

JET STREAM CIRRUS Filamentous cirrus that appears to radiate from point in sky and exhibits characteristics associated with strong vertical wind shear (e.g., twisted or curved filaments).

JET WIND SPEED PROFILE Vertical wind speed profile characterized by relatively narrow current of high winds with slower moving air above and below.

JET-FLOW GATE High pressure gate consisting of wheel-mounted leaf moved vertically by motor-driven screw hoist designed for regulating flow with minimal cavitation damage.

JETTING 1. Method of compacting soil using hose or other device, with high velocity stream of water, worked down through depth of soil placed; 2. Drilling with high pressure water or air jets.

JETTING PUMP Water pump that develops very high discharge pressure.

JETTY Structure extended into ocean to maintain coastal inlets position.

JITNEY Small vehicle that typically operates service on fixed route.

JOB ACCOUNT Account giving all specific details of work.

JOB HAZARD ANALYSIS Study of job or activity to identify hazards or potential accidents associated with each step or task and develop solutions that will eliminate, nullify or prevent such hazards or accidents.

JOB LAYOUT Scaled drawing of proposed construction site.

JOB SITE Location of job where work is performed.

JOB SPECIFICATION Statement of essential components of job class.

JOB SUPERINTENDENT Supervises every aspect of project.

JOB-MIX FORMULA Mixture of aggregate and asphalt binder.

JOBSITE INSTRUCTIONS Written terms, conditions or specifications within contract.

JOGGING Frequent starting and stopping of electric motor.

JOINT 1. Any break or rupture formed in ice cover or floe due to deformation; 2. Break in rock formation due to structural stresses (e.g., faults, shears, joints, planes of fracture cleavage); 3. Point where two things are joined, but are still able to rotate.

JOINT ACCEPTANCE TEST Procedure for testing joints of pipelines to demonstrate integrity of joint and construction procedures.

JOINT AND SEVERAL LIABILITY Liability for Superfund site cleanup and other costs on part of more than one potentially responsible party.

JOINT GAP Distance between ends of contiguous rails measured below top of rail.

JOINT INFORMATION CENTER Single location where public information officials gather to collaborate on and coordinate release of, emergency public information.

JOINT METERS Embedded instrument that uses electrical principles to measure movement across joint or crack.

JOINT RING Band used to create push-fit joint in drain pipe.

JOINT RUNNER Incombustible materials for packing to be caulked joint with lead.

JOINT SEAL Device designed to prevent infiltration and increase capacity in underground pipes.

JOINT SEALANT Substance that excludes water and materials from joints.

JOINT TAPE Tape used on threads of pipe fittings.

JOINT USE CAPACITY Reservoir capacity assigned to flood control purposes during certain periods of year and to conservation purposes during other periods of year.

JOINT-SEALING COMPOUND Paste that stays permanently flexible for filling joints.

JOINTED REINFORCED CONCRETE PAVEMENT Concrete pavement that uses contraction joints and reinforcing steel to control cracking.

JOINTING COMPOUND Substance applied to pipe threads to prevent leaks.

JOKULHLAUP Icelandic term meaning glacier dammed lake outburst flood.

JOULE Unit of energy or work equal to work done by force of 1 newton when its point of application moves 1 meter in direction of action of force equal to 1⁄3600 watt-hour.

JOULE PER KELVIN Unit of heat capacity required to raise unit mass of substance by unit temperature.

JOULE PER KILOGRAM Unit of specific latent heat required to raise unit mass of substance by unit temperature interval under specified conditions.

JOULE PER KILOGRAM KELVIN Unit of specific latent heat required to raise unit mass of substance by unit temperature interval under specified conditions.

JOULE PER MOLE Unit of molar internal energy per amount of material.

JUNCTION Where roads or pipes meet.

JUNCTION CHAMBER Accessibility manhole where sewers join into main sewer.

JUNCTION MANHOLE Accessibility manhole formed over junction of sewers.

JUNCTION POINT Location where branch line track connects with main line track.

JUNK FILL Refuse fill that has been dumped over long time.

JURISDICTION 1. Boundary of authorization; 2. Level of management responsibility entity has for specific area using its rules and regulations.

JUST COMPENSATION Compensation provided for property taken via eminent domain.

JUVENILE Young fish older than 1 year but not capable of reproduction.

JUVENILE GROUNDWATER Water brought to surface or added to underground supplies from magmas that's not potable.

JUVENILE WATER Water formed chemically within Earth and brought to surface in intrusive rock.

K

K CORONA White-light corona that portion that is caused by sunlight scattered by electrons in hot outer atmosphere of Sun.

K-FACTOR 1. Proportion of annual average daily traffic that occurs during peak hour; 2. Ratio between material thickness and neutral fiber axis.

K-INDEX Measure of thunderstorm potential based on vertical temperature lapse rate, moisture content of lower atmosphere and vertical extent of moist layer.

KAME Round hill or short ridge of sand or gravel deposited by melting glacier.

KANAT Underground conduit, between vertical shafts, that leads water from interior of hill to villages in valley.

KAOLIN Fine clay, rich in kaolinite, used in ceramics, papermaking, etc.

KAPLAN TURBINE Propeller-type water turbine that has adjustable blades.

KARST Geologic formation that results from dissolution and collapse of carbonate rocks (e.g., limestone, dolomite, gypsum) and characterized by closed depressions or sinkholes, caves and underground drainage.

KATABATIC WIND Wind created by high density air flowing from higher elevation downhill, due to force of gravity.

KATAFRONT Front where warm air descends frontal surface (except in low layers of atmosphere).

KEEL 1. Material similar to chalk or crayon used to mark item; 2. two lowest petals of corolla of papilionaceous flower, united and enclosing stamens and pistil.

KEETCH-BYRUM DROUGHT INDEX Index used to gauge severity of drought in deep duff and organic soils.

KELVIN Unit of thermodynamic temperature equal to fraction 1⁄273.16 of thermodynamic temperature of triple point of water.

KELVIN TEMPERATURE SCALE Absolute temperature scale where change of 1 Kelvin equals change of 1 °C.

KELVIN WAVES Fluctuations in wind speed at ocean surface at Equator result in eastward propagating waves.

KELVIN-HELMHOLTZ WAVES Vertical waves in air associated with wind shear across statically-stable regions.

KESSENER BRUSH Device used to provide both aeration and circulation in oxidation ditches in activated sludge process.

KETTLE Steep-sided hole or depression, commonly without surface drainage, formed by melting of large detached block of stagnant ice that had been buried in glacial drift.

KETTLE LAKE Body of water occupying kettle.

KEY EVENT Phase of work progress in planning and scheduling.

KEY INTERLOCK Ability to lock equipment in desired position allowing removal of key, that in turn, can be used to lock or unlock related equipment in sequentially predetermined manner.

KEY OPERATED SWITCH Switch that features keyed lock within its own mechanical operation.

KEY SWITCH Switch that can be activated only by use of key.

KEY VALVE Valve controlled with removable key.

KEYWAY Excavation in foundation of dam, usually located upstream of dam axis or centerline crest which extends to bedrock or to impervious stratum, later to be filled with impervious material to limit seepage beneath dam.

KILL Stream or creek.

KILO Prefix symbolizing 10^3.

KILOGRAM Unit of mass equal to 1,000 grams.

KILOGRAM METER PER SECOND Unit of momentum equal to 1 kg x m⁄s.

KILOGRAM METER SQUARED Unit of moment of inertia equal to kg x m^2.

KILOGRAM METER SQUARED PER SECOND Unit of angular momentum equal to kg x m^2⁄s.

KILOGRAM PER CUBIC METER Unit of density (mass density) equal to kg⁄m^3.

KILOGRAM PER LITRE Unit of density equal to kg⁄L.

KILOGRAM PER SECOND Unit of mass flow rate equal to kg⁄s.

KILOHERTZ Unit of frequency equal to 1,000 hertz.

KILOJOULE Unit of energy equal to 1,000 joules.

KILOKELVIN Unit of temperature equal to 1,000 °K.

KILOLITER Unit of volume equal to 1,000 liters.

KILOMETER Unit of length equal to 1,000 meters.

KILOMETER PER HOUR Unit of speed equal to 1,000 m⁄hr.

KILONEWTON Unit of force equal to 1,000 Newtons.

KILOPASCAL Unit of pressure equal to 1,000 Pascals.

KILOVOLT Unit of electric potential to 1,000 volts.

KILOVOLT AMPS REACTIVE 1,000 volt amps reactive.

KILOVOLT-AMPERE Unit of power equal to 1,000 volt-amperes and approximately 89/100 of kilowatt.

KILOWATT Electrical unit of work or power equal to 1,000 watts.

KILOWATT HOUR Unit of energy equal to 3.45 megajoules.

KILOWATT-HOUR Unit of energy equivalent to 1,000 watt-hours.

KINEMATIC SIMILARITY Fluid flow of both model and prototype undergo similar time rates of change motions.

KINEMATIC VELOCITY Ratio of dynamic viscosity by fluid density.

KINEMATIC VELOCITY COEFFICIENT Ratio of absolute viscosity coefficient of fluid to its unit weight.

KINEMATIC VISCOSITY Ratio of absolute viscosity to density.

KINETIC COMPRESSOR Compressor using rotating vane and/or screw to move and/or compress gas.

KINETIC ENERGY Energy of body with respect to motion of body.

KINETIC FLOW FACTOR Measure of turbulence or tranquility that prevails in flowing water.

KINETIC FRICTION COEFFICIENT Ratio of frictional force between object and surface to normal force action on object.

KINETIC HEAD Energy of flowing water that's function of its velocity.

KINETIC RATE COEFFICIENT Number that describes rate at which water constituent (e.g., biochemical oxygen demand, dissolved oxygen) rises or falls or at which air pollutant reacts.

KIP Unit of mass equal to 1,000 lb.

KISS AND RIDE Car park at transportation center (e.g., railway station, airport) for dropping off and picking up passengers.

KJELDAHL NITROGEN Method for quantitative determination of nitrogen contained in organic substances plus nitrogen contained in inorganic compounds ammonia and ammonium.

KLYSTRON Electron tube used as low-power oscillator or high-power amplifier at ultrahigh frequencies.

KNIFE SWITCH Switch used to control flow of electricity in circuit, consisting of hinge which allows metal lever or knife, to be lifted from or inserted into slot or jaw.

KNIFE-BLADE SWITCH Switch where moving contact is moved between fixed contact.

KNOT Unit of speed used in navigation equal to 1 nautical mile/hour.

KNUCKLE BEND Conduit bend that turns dramatically.

KNUCKLES Lumpy protrusions on edges and sometimes underside, of thunderstorm anvil.

KOLK Underwater vortex similar to whirlwind, capable of dislodging, picking up and moving boulders.

KRAUS PROCESS Return sludge bypasses reaeration unit and is delivered to mixed liquid aeration tank.

KRIGING Geostatistical interpolation procedure for estimating spatial distributions of model inputs from scattered observations.

L

LA NIÑA Periodic cooling of surface ocean waters in eastern tropical Pacific along with shift in convection in western Pacific further west than climatological average.

LAB PROCESS WATER Water used for laboratory experiments and procedures.

LABELLED Ticket or sign giving information about something.

LABOR Price of services of workers performing particular task.

LABOR CONSTANT Amount of labor necessary to perform unit of work.

LABORATORY ANIMAL STUDIES Investigations using animals as surrogates for humans.

LABORATORY REPORTING LEVEL Twice yearly determined long-term method detection level.

LABORATORY SAMPLE Sample of material to be tested or analyzed prepared from gross sample and retains latter's composition.

LACCOLITH Igneous intrusion that squeezes between sedimentary layers and domes overlying layers.

LACING Tied or threaded laces that form net like pattern.

LACUSTRINE Pertaining to, produced by or formed in lake.

LACUSTRINE HABITAT Lake or reservoir wetland habitat.

LACUSTRINE PLAIN Lakes that get filled by deposited sediments when water drains from lake.

LADDER 1. Frame, usually portable, of wood, metal or rope, used for ascent and descent, consisting of two side pieces to which are fastened rungs; 2. Digging-boom of hydraulic dredge or chain bucket-type excavator.

LADDER DITCHER Machine that digs trenches by means of bucket-ladder excavator.

LADDER DREDGE Dredger that uses chain with specialized cutting links to loosen material to be dredged.

LADING Shipment, cargo or freight.

LAG 1. Time from beginning to peak; 2. Center of mass of rainfall to hydrograph peak; 3. Time it takes flood wave to move downstream; 4. Wrap pipes with insulation.

LAG TIME Delay between dependent tasks.

LAGGING 1. In tunneling, planking placed against dirt or rock walls and ceiling, outside ribs; 2. Covering of something with strips of felt, wood etc, either as insulation or for protection.

LAGOMORPHS Order of mammals including rabbits, hares and pikas.

LAGOON 1. Shallow pond where sunlight, bacterial action and oxygen work to purify wastewater; 2. Storage of wastewater or spent nuclear fuel rods; 3. Shallow body of water, often separated from sea by coral reefs or sandbars.

LAGOONING Natural treatment technique that consists in accumulation of wastewater in ponds or basins, where series of biological, biochemical and physical processes take place.

LAHAR Volcanic mudflow.

LAID LENGTH Span of finished pipeline.

LAKE Large, landlocked stretch of water.

LAKE BREEZE Thermally produced wind blowing during day from surface of large lake to shore, caused by difference in rates of heating of surfaces of lake and of land.

LAKE EFFECT SNOW Snow showers that are created when cold, dry air passes over large warmer lake (e.g., one of Great Lakes) and picks up moisture and heat.

LAKE EFFECT SNOW ADVISORY Issued by National Weather Service when pure lake effect snow may pose hazard or it is life threatening.

LAKE EFFECT SNOW SQUALL Local, intense, narrow band of moderate to heavy snow squall that can extend long distances inland.

LAKE EFFECT SNOW WARNING Issued by National Weather Service when pure lake effect snow may pose hazard or it is life threatening.

LAKE EFFECT STORM Fall or winter storm that produces heavy but localized precipitation as result of temperature differences between air over snow-covered ground and air over open waters of lake.

LAKE SAND Substance having waterworn borders.

LAKESHORE FLOOD ADVISORY Minor flooding is possible (i.e., over and above normal high tide levels).

LAKESHORE FLOOD WARNING Flooding that will pose serious threat to life and property is occurring, imminent or highly likely.

LAKESHORE FLOOD WATCH Flooding with significant impacts is possible.

LAMBERT Unit of brightness equal to 1 $lumen/cm^2$.

LAMINAR 1. Smooth; 2. Non-turbulent.

LAMINAR FLOW Streamline flow where successive flow particles follow similar path lines and head loss is proportional to first power of velocity.

LAMINAR VELOCITY Speed in particular flat stream of water, where streamline or turbulent flow may happen.

LAMINATION Small scale sequence of fine layers that occurs in sedimentary rocks.

LAMP HOLE Small, vertical pipe or shaft extending from surface of ground to sewer.

LAND Part of Earth which is not covered by oceans or other bodies of water.

LAND APPLICATION Discharge of wastewater onto ground for treatment or reuse.

LAND ASSEMBLY Process of forming single site from number of lands, usually for eventual development or redevelopment.

LAND BAN Phasing out of land disposal of most untreated hazardous wastes.

LAND BANKING Practice of buying land as investment, holding it for future use and making no specific plans for its development.

LAND BATTURE Part of river bank that might be inundated at relatively infrequent periods.

LAND BOUNDARY Line that divides one territory or political unit from another.

LAND BREEZE Coastal breeze at night blowing from land to sea, caused by difference in rates of cooling of their respective surfaces.

LAND CLASSIFICATION Bureau of Reclamation systematic placing of lands into classes based on their suitability for sustained irrigated farming with Class 1 being most productive.

LAND DEVELOPMENT Altering landscape via work.

LAND DISPOSAL Placement of waste into or on land.

LAND DISPOSAL RESTRICTIONS Rules that require hazardous wastes to be treated before disposal on land to destroy or immobilize hazardous constituents that might migrate into soil and groundwater.

LAND DRAIN Drain that rapidly drains away water or relieves hydrostatic pressure.

LAND DRAINAGE Removes water from top or inside earth.

LAND FARMING OF WASTE Disposal process where hazardous waste deposited on or in soil is degraded naturally by microbes.

LAND FILTRATION Filtering wastewater where grass uses nutrients and bacteria in soil breaks down pollutants.

LAND IMPROVEMENT Enhancements to land to make it more usable.

LAND MANAGEMENT AGENCY Agencies that manage for purpose of preservation, recreation and development of natural resources.

LAND OWNERSHIP Land held in title.

LAND RECLAMATION Process of creating new land from submerged sediments.

LAND RETIREMENT Permanent removal of land from agricultural production.

LAND RIGHTS Rights of indigenous peoples to land, either individually or collectively.

LAND SUBSIDENCE General lowering or sudden sinking of Earth's surface owing to subsurface movement of earth materials.

LAND TREATMENT Disposal of something from sewage treatment plants by distribution over arable laud.

LAND TREATMENT FACILITY Unlined facility where hazardous waste is either applied on soil surface or incorporated into upper layers of soil.

LAND USE Management and modification of land to produce, modify or maintain land and its resources.

LAND USE ANALYSIS Study and categorization of different types of activities relating to how land is used.

LAND USE CONTROL Non-engineered instruments, such as administrative and legal controls or engineered and physical barriers, such as fences and security guards.

LAND USE DESIGNATION Planned established for area of land defined by authority.

LAND USE INTENSITY Extent to which land is developed and used in conformity with zoning ordinances.

LAND USE MAP Map that shows types and intensities of different land uses in particular area.

LAND USE PLAN Documents that provide officials with set of guidelines for development patterns and other land use issues that are important to community.

LAND USE PLANNING Process where lands are evaluated and assessed to become basis for decisions involving land disposition and utilization.

LAND USE REGULATION Regulates land use activities through permit process in accordance with local rules.

LAND USE SURVEY Record to identify boundaries and features of land to determine ownership at point in time.

LAND-CLEARING RAKE Equipment with long, curved teeth for gathering and piling brush, debris, etc.

LAND-SURFACE DATUM Datum plane approximately at land surface at each groundwater observation well.

LAND-USE STUDY Network of existing shallow wells in area having relatively uniform land use.

LANDBRIDGE Transport of material between either side of land mass.

LANDFALL Intersection of surface center of tropical cyclone with coastline.

LANDFILL 1. Open area where trash is buried; 2. Facility where solid waste from municipal and/or industrial sources is disposed.

LANDFILL CELL Waste placed, compacted and covered in landfill daily.

LANDFILL GAS Mix of different gases created by action of microorganisms within landfill.

LANDFILL MACHINE Machine (e.g., compactor, dozer, loaders, tractor) used at landfill.

LANDFORM Physical feature of Earth's surface from man-made or natural processes.

LANDHOLDER Individual or legal entity owning or leasing land subject to acreage limitation provisions.

LANDING 1. Coming to surface; 2. Place on shoreline where boat lands; 3. Level area at top of flight of stairs.

LANDSCAPE 1. Traits, patterns and structure of specific geographic area, including its biological composition, its physical environment and its anthropogenic or social patterns; 2. Area where interacting ecosystems are grouped and repeated in similar form.

LANDSCAPE CHARACTERIZATION Documentation of traits and patterns of essential elements of landscape.

LANDSCAPE ECOLOGY Study of distribution patterns of communities and ecosystems, ecological processes that affect those patterns and changes in pattern and process over time.

LANDSCAPE INDICATOR Measurement of landscape, calculated from mapped or remotely sensed data used to describe spatial patterns of land use and land cover across geographic area.

LANDSCAPING Act of improving landscape.

LANDSLIDE Failure of sloped bank of soil or rock where movement of mass takes place along surface of sliding.

LANDSLIDE SPRING Spring at foundation of landslide.

LANDSPOUT Tornado that does not arise from organized storm-scale rotation and therefore is not associated with wall cloud (visually) or mesocyclone (on radar).

LANE 1 Farthest left or fast lane, on roadway.

LANE 2 Highway lane left and adjacent to Lane 1.

LANE Road, street or similar thoroughfare intended for single line of vehicles.

LANE BALANCE When number of departing and approaching traffic lanes are equal.

LANE DISTRIBUTION Average annual daily traffic in lane along roadway.

LANE DROP Location on highway where number of traffic lanes decrease.

LANE GROUP Set of lanes established at intersection for separate analysis.

LANE LINE Traffic lines used to delineate separation of traffic lanes for travel in same direction.

LANE LINE MARKINGS Pavement marking lines that indicate if traffic is traveling in one or two directions.

LANE UTILIZATION Distribution of traffic volume per lane.

LANE WIDTH Width measured from center of edgeline to center of centerline or lane line.

LANE-USE CONTROL SIGNAL Overhead signals placed over certain lane or lanes.

LANGELIER INDEX Index reflecting equilibrium pH of water with respect to calcium and alkalinity that's used in stabilizing water to control both corrosion and scale deposition.

LANGELIER'S INDEX Calculated number used to predict whether or not water will precipitate, be in equilibrium with or dissolve calcium carbonate.

LAP WELD PIPE Pipe made by welding together overlapping edges or ends.

LAPSE RATE Rate of change of atmospheric variable (e.g., temperature) with height.

LARGE AIRCRAFT Any aircraft of more than 12,500 pounds maximum certificated takeoff weight.

LARGE CONSTRUCTION ACTIVITY Includes clearing, grading and excavating resulting in land disturbance that will disturb equal to or more than five acres of land or will disturb less than five acres of total land area but is part of larger common plan of development or sale that will ultimately disturb equal to or more than five acres.

LARGE QUANTITY GENERATOR Person or facility generating more than 2200 pounds of hazardous waste per month.

LARGE SCALE Size scale in meteorology referring generally to weather systems with horizontal dimensions of several hundred miles or more.

LARGE WATER SYSTEM Water system serving more than 50,000 people.

LARGE-CAPACITY SEPTIC SYSTEMS Class V wells that dispose of sanitary waste through septic tank and are used by multiple dwellings, business establishments, communities and regional business establishments for injection of wastes.

LARVAL FISH Immature stage that develops from fertilized egg before assuming characteristics of adult.

LASER INDUCED FLUORESCENCE Method for measuring relative amount of soil and/or groundwater with in-situ sensor.

LAST UPDATE Time and date where forecast was issued or updated.

LATE FINISH In Critical Path Method, latest date that task can finish without delaying finish of project.

LATE START DATE In Critical Path Method, last date upon which given activity can be started without delaying completion date of project.

LATENCY Time from first exposure of chemical until appearance of toxic effect.

LATENT ENERGY Thermal energy released or absorbed, by body or thermodynamic system, during constant-temperature process.

LATENT HEAT Heat absorbed or released during change of phase at constant temperature and pressure.

LATENT HEAT FLUX Amount of heat energy that converts water from liquid to vapor (evaporation) or from vapor to liquid (condensation) across specified cross-sectional area per unit time.

LATENT SENSIBLE HEAT-FLUX DENSITY 1. Amount of heat energy that moves by turbulent transport through air across specified cross-sectional area per unit time and goes to heating (cooling) air; 2. Flux of heat from Earth's surface to atmosphere not associated with phase changes of water that's component of surface energy budget.

LATERAL 1. Channel that conveys water from canal to farm, municipality, etc.; 2. Of or pertaining to side.

LATERAL BRACING COMPONENT Lateral bracing system component to prevent lateral buckling of components and/or to resist lateral loads.

LATERAL CANAL Canal running parallel to stream.

LATERAL CLEARANCE Smallest distance laterally between tire and nearest fixed point of vehicle.

LATERAL EARTH PRESSURE Pressure that soil exerts in horizontal direction.

LATERAL EROSION Widening of stream, channel or valley via erosion.

LATERAL MORAINE Ridge-like pile of sediment along side of glacier.

LATERAL MOTION Moving in sideways direction.

LATERAL PRESSURE Pressure that soil exerts in horizontal direction.

LATERAL SEWER Sewer pipeline connecting property to main sewer line.

LATERAL SEWERS Pipes that run under city streets and receive sewage from homes and businesses.

LATERAL-FLOW SPILLWAY Spillway that discharges overflow into channel perpendicular to inflow direction.

LATERAL-TORSIONAL BUCKLING Unrestrained beams that experience both lateral displacement and twisting.

LATEST EVENT OCCURRENCE TIME In Critical Path Method, final deadline where particular item of work must be completed in order to avoid delaying entire project.

LATEST FINISH DATE In Critical Path Method, absolute latest date task or activity can finish without delaying completion of project.

LATEST START DATE In Critical Path Method, latest possible point in time where activity must be started if project is not to be delayed.

LATITUDE Location north or south in reference to equator, that is designated at 0 °.

LATRINE Open trench or pit used for urination and defecation.

LAUNCHING Process over time (e.g., release material downstream).

LAUNCHING APRON Flexible apron that comes down or launches to cover face of scour.

LAUNDER 1. Gutter; 2. Trough used by miners to receive powdered ore from box where it is beaten or for carrying water to stamps or other apparatus for sorting ore.

LAUNDERING WEIR Sedimentation basin overflow weir.

LAVA Fluid, molten igneous rock erupted on Earth's surface.

LAVATORY Vessel or fixture for washing.

LAY BARGE Vessel that submerged pipeline is assembled and increasingly lowered.

LAYER Rock or soil with thickness of 6 inches or less.

LAYERED HAZE Haze produced when air pollution from multiple line, area or point sources is transported long distances to form distinguishable layers of discoloration in stable atmosphere.

LAYERED SYSTEM SOIL Two or more distinctly different soil or stone types ordered in layers.

LAYOUT OF CONSTRUCTION PLANT Site physical properties, materials and necessary for project.

LAYOVER TIME Time between one flight and another flight.

LC_{50} Amount of material, given all at once, which causes death of 50% (one half) of group of test animals.

LD_{50} Dose of toxicant that will kill 50% of test organisms within designated period of time.

LEACH 1. To remove components from soil by action of water trickling through; 2. Quantity of wood ashes, where water passes and thus imbibes alkali.

LEACH BED Piping from septic system that allow waters to flow out (disperse) of piping and into surrounding soils.

LEACHATE Water that collects contaminants as it trickles through wastes, pesticides or fertilizers.

LEACHATE COLLECTION SYSTEM System that gathers leachate and pumps it to surface for treatment.

LEACHATE RECIRCULATION Recycling tool in managing leachate on-site and minimizing long-term impact of disposed waste to human health and environment.

LEACHING 1. Process by which soluble constituents are dissolved and filtered through soil by percolating fluid; 2. Removal of materials in solution from soil or rock; 3. Movement of pesticides or nutrients from land surface to groundwater.

LEACHING BASIN Drainage pit with sand or gravel sides constructed to allow water to dissipate.

LEACHING CESSPOOL Lined or partially lined underground pit into which raw household water (sewage) is discharged and from which liquid seeps into surrounding soil.

LEACHING FIELD Lined or partially lined underground pit into which raw household water (sewage) is discharged and from which liquid seeps into surrounding soil.

LEACHING REQUIREMENT Quantity of irrigation water required for transporting salts through soil profile to maintain favorable salt balance in root zone for plant development.

LEAD Heavy, pliable, inelastic metal element with atomic number 82 that's hazardous to health if breathed or swallowed.

LEAD ARSENATE Inorganic insecticide (white, heavy powder) used primarily against potato beetle.

LEAD CAULKED JOINT Cast-iron pipework joint where molten lead is poured in to seal or join pipes.

LEAD LINE Instrument used in navigation to measure water depth.

LEAD POISONING Chronic intoxication produced by absorption of lead into body, characterized by severe colicky pains, dark line along gums and local muscular paralysis.

LEAD SERVICE LINE Service line made of lead which connects water to building inlet and any lead fitting connected to it.

LEAD TIME 1. Time available to determine if conditions at dam and reservoir or in basin, warrant declaration of specific emergency classification level; 2. Amount of time between initiation of some process and its completion.

LEAD WOOL Thin strands of lead metal that can be used to cold-caulk cast iron and steel pipes.

LEADER 1. Streamer that initiates first phase of each stroke of lightning discharge; 2. Pipe for conducting rain water from roof to cistern or to ground; 3. Fast-growing terminal shoot of woody plant.

LEADER SPOT In magnetically bipolar or multipolar sunspot group, western part precedes and main spot in that part is called leader.

LEAK DETECTOR Apparatus effective at discovering existence of leaks.

LEAK-DETECTION SYSTEM System effective at discovering failure of containment structure.

LEAKAGE 1. Flow of water, through hole or crack, from one hydrogeologic unit to another; 2. Natural loss of water from artificial structures as result of hydrostatic pressure; 3. Undesirable flow of electric current through insulation.

LEAKAGE CURRENT Current that flows through protective ground conductor to ground.

LEAKER Hole or crack that enables fluid to escape.

LEAKY AQUIFER Aquifers that lose or gain water through adjacent less permeable layers.

LEAN COMBUSTION Decrease in exhaust emissions from internal combustion engines.

LEAPING WEIR Weir before aqueduct or sewer intake that causes flood water to overshoot opening.

LEASE OF POWER PRIVILEGE Contractual right given to non-federal entity to utilize, consistent with project purposes, water power head and storage from Reclamation projects for electric power generation.

LEASEHOLD Tenure of property held by lessee under lease.

LEAT Artificial watercourse, canal or aqueduct, but especially millrace.

LEDGE DRAIN Drainage system installed in upstream face of dam.

LEE 1. Side or part sheltered or turned away from wind (e.g., with mountain); 2. Facing away from flow of fluid.

LEE WAVE Wavelike effect, characterized by severe updrafts and downdrafts, that occurs in lee of mountain range when rapidly flowing air is lifted up steep front of mountain range.

LEESIDE LOW Extratropical cyclones that form on downwind (lee) side of mountain chain.

LEEWARD Side away from wind.

LEFT ABUTMENT Abutment on respective side of observer looking downstream.

LEFT BANK Bank to left when facing downstream.

LEFT DESIGNATION Made with observer looking in downstream direction.

LEFT EXIT REGION Area downstream from and to left of upper-level jet max (as would be viewed looking along direction of flow).

LEFT FRONT QUADRANT Area downstream from and to left of upper-level jet max (as would be viewed looking along direction of flow).

LEFT OVERBANK Part of channel to left of overbank.

LEG Side post in tunnel timbering.

LEGAL WATER LEVEL Water stage where shoreline defines riparian boundaries.

LEGEND Key to symbols and color codes on map, chart, etc.

LEGIONELLA Genus of bacteria, some species of which have caused type of pneumonia called Legionnaires Disease.

LEGIONNAIRE'S DISEASE Type of pneumonia caused by Legionella bacteria.

LENGTH Distance in direction of flow between two specific points along river, stream or channel.

LENGTH OF DAM Measured length of dam along axis, centerline crest (e.g., top) or roadway surface on crest.

LENGTH OF NEED Total length of longitudinal barrier needed upstream to shield area of concern.

LENGTH OF RUN 1. Time that one user is permitted to work with irrigating head in rotational irrigation system; 2. Space that water must flow between ditches.

LENGTH ON WATERLINE Waterline where ship will float when fully loaded.

LENGTH OVERALL Length of ship from bow to stern.

LENGTH-OF NEED POINT Point on barrier where it will contain and redirect impacting vehicle along face of barrier.

LENS 1. Signal lens in front of lamp and colored glass; 2. Geologic deposit of variable thickness, that disappears laterally in all directions and cannot be correlated to adjacent borings.

LENTIC Standing waters (e.g., lakes, ponds, marshes).

LENTIC SYSTEM Non-flowing or standing body of fresh water (i.e., lake, pond).

LENTICULAR CLOUD Very smooth, round or oval, lens-shaped cloud often seen, singly or stacked in groups, near or in lee of mountain ridge.

LESS THAN CARLOAD Shipments less than container load size.

LESS THAN CONTAINER LOAD Ocean freight shipments not requiring ocean container full capacity.

LETHAL CONCENTRATION 50 Concentration of pollutant or effluent at which 50% of test organisms die that's common measure of acute toxicity.

LETHAL CONCENTRATION Point estimate of toxicant concentration that would be lethal to given percentage of test organisms during specified period.

LETHAL DOSE 50 Dose of toxicant that will kill 50% of test organisms within designated period of time.

LETHAL DOSE Dose of substance or of radiation likely to cause death.

LETHAL DOSE LOW Lowest dose in animal study at which lethality occurs.

LETTER OF AGREEMENT Formal document that states intentions and/or responsibilities of signatory parties.

LEVEE 1. Natural or manmade earthen barrier (i.e., embankment) along edge of stream, lake or river that helps prevent banks from overflowing and land from flooding; 2. Steep bank of river; 3. Border of irrigated field; 4. Pier or other landing place on river.

LEVEE GRADE Crown slope of levee centerline.

LEVEL To cause to conform to specified grade.

LEVEL CONTROL Apparatus that senses and keeps water level in storage container.

LEVEL I RESPONSE First and least serious, of three response levels that dam operating organization will declare after analyzing potentially threatening event that may be perceived as emergency or may be of general interest to public, but does NOT pose hazard, either at dam or to downstream populations at risk when observed.

LEVEL II RESPONSE Second of three response levels dam operating organization will declare after analyzing threatening event that means emergency conditions are such that populations at risk should prepare to leave predetermined inundation areas for higher ground and safe shelter.

LEVEL III RESPONSE Third and most serious, response level dam operating organization will declare after analyzing threatening events that indicates life-threatening flood waters, as result of high operational releases or dam failure, present imminent danger to public located downstream from dam.

LEVEL OF CONCERN Concentration in air of extremely hazardous substance above which there may be serious immediate health effects to anyone exposed to it for short periods.

LEVEL OF FREE CONVECTION Level at which parcel of saturated air becomes warmer than surrounding air and begins to rise freely.

LEVEL OF PROTECTION Level of loading selected that corrective actions will be designed to prevent dam failure.

LEVEL OF SERVICE Qualitative measure used to relate quality of motor vehicle traffic service.

LEVEL OF SERVICE SCORE Numerical output of support needs for various parts of their everyday routine and their level of independence.

LEVEL RECORDER Tool used to measure and record level or height of fluids.

LEVEL TERRAIN Condition where highway sight distances, governed by both horizontal and vertical restrictions, are generally long or could be made so without construction difficulty or major expense.

LEVELING COURSE Thin layer of asphalt mixture applied over existing pavement to correct surface irregularities.

LICHENS Any of many symbiotic organisms, being associations of fungi and algae.

LID 1. Layer of clay or other impermeable material, installed over top of closed landfill to prevent entry of rainwater and minimize leachate; 2. Layer of warm air several thousand feet above Earth's surface which suppresses or delays development of thunderstorms.

LIDAR Optical analogue of radar, using intense pulses of laser light to measure composition and structure of atmosphere.

LIEN Legal claim; charge upon real or personal property for satisfaction of some debt or duty.

LIEN RELEASE When holder of lien or security interest in piece of property, lifts or waives lien, rendering property free to purchase.

LIEN WAIVER Document commonly exchanged among parties in construction industry as part of payment process.

LIFE CYCLE Various stages animal passes through from egg fertilization to death.

LIFE CYCLE OF PRODUCT All stages of product's development, from extraction of fuel for power to production, marketing, use and disposal.

LIFE EXPECTANCY Amount of time one is expected to live.

LIFE TEST Lab process used to ascertain resistance of item into particular group of damaging forces or states.

LIFE ZONE 1. Major area of plant and animal life; 2. Region characterized by particular plants and animals and distinguished by temperature differences.

LIFE-CYCLE ANALYSIS Systematic analysis of environmental impact of products during their entire life cycle.

LIFE-CYCLE COST Cost to acquire and sustain product or system over its useful life.

LIFE-CYCLE COST ANALYSIS Method for assessing total cost of facility ownership.

LIFELINE 1. Line to which drowning or falling victim may cling to; 2. Means or route for transporting indispensable supplies; 3. Source of salvation in crisis; 4. System or structure of vital importance to community; 5. On deck of boat, line to which one can attach oneself to stay aboard on rough seas.

LIFETIME AVERAGE DAILY DOSE Figure for estimating excess lifetime cancer risk.

LIFT 1. Height of column or body of fluid below certain point; 2. Perpendicular space between static water source as well as suction chamber of pump; 3. Step or bench in multiple layer excavation or fill layer of solid wastes in sanitary landfill and upper layer of cover material; 4. Perpendicular space boat increases or drops when passing by means of lock.

LIFT BRIDGE Movable bridge where span rises vertically while remaining parallel with deck.

LIFT DEPTH Perpendicular depth of solid waste volume and cover material in sanitary landfill.

LIFT LINE Horizontal construction joint created when new concrete is placed on previously placed concrete.

LIFT LOCK canal lock that lifts boat by flotation from one level to higher level when water flows into lock.

LIFT PUMP Pump (i.e., suction) that discharges above or below suction inlet elevation.

LIFT STATION Wastewater facilities designed to move wastewater from lower to higher elevation.

LIFT-AND-CARRY WASTE CONTAINER Container lifted onto service vehicle and transported to disposal site.

LIFTED INDEX Measure of atmospheric instability obtained by computing temperature that air near ground would have if it were lifted to some higher level and comparing that temperature to actual temperature at that level.

LIFTING CONDENSATION LEVEL Level at which parcel of moist air becomes saturated when it is lifted dry adiabatically.

LIFTING STATION Mechanical device installed in sewer or water system or other liquid-carrying pipelines to move liquids to higher level.

LIGHT BRIDGE Observed in white light, bright tongue or streaks penetrating or crossing sunspot umbrae.

LIGHT DELIVERY TRUCK Motor vehicle designed to carry payload not to exceed 2,000 pounds.

LIGHT ENGINE Locomotive running on its own or with other powered locomotives, with no wagons or carriages attached.

LIGHT FIRE HAZARD Where quantity and/or combustibility of materials is small.

LIGHT GUN Green, red or white spotlight to signal airplane.

LIGHT MATERIAL BUCKET Bucket for moving, excavating and loading large amounts of snow and other light material.

LIGHT NON-AQUEOUS PHASE LIQUID Non-aqueous phase liquid with specific gravity less than 1.0.

LIGHT RAIL MODE Transit vehicles that run on tracks and are powered from overhead wires.

LIGHT RAIL TRANSIT Rail public transportation system with lower capacity and speed than heavy rail/metro system, but higher capacity speed than traditional tram system.

LIGHT RAIL VEHICLE Urban rail transit vehicle (e.g., tram, streetcar or trolley) that uses rolling stock.

LIGHT RAIN Light precipitation that falls from sky with intensity between trace and 0.10 in/hr.

LIGHT-ATTENUATION COEFFICIENT Measure of water clarity.

LIGHT-DUTY TRUCK US classification for vehicles with gross vehicle weight up to 8,500 pounds and payload capacity up to 4,000 pounds.

LIGHT-EMITTING DIODE Semiconductor light source that requires very little power and emits light when current flows through it.

LIGHT-GAUGE COPPER TUBE Copper conduit thin-walled and used for domestic plumbing.

LIGHT-WATER REACTOR Thermal- neutron reactor that utilizes normal water as opposed to heavy water, form of water that contains larger amount of hydrogen isotope deuterium.

LIGHTED BUOY Floating, navigation channel marker with light.

LIGHTER ABOARD SHIP Loading barges (i.e., lighters) aboard bigger vessel for transport.

LIGHTNING Visible electrical discharge produced by thunderstorm.

LIGHTNING ACTIVITY LEVELS 1. LAL 1 - No thunderstorms; 2. LAL 2 - Few building cumulus with isolated thunderstorms; 3. LAL 3 - Much building cumulus with scattered thunderstorms with light to moderate rain; 4. LAL 4 - Thunderstorms common with moderate to heavy rain reaching ground; 5. LAL 5 - Numerous thunderstorms with moderate to heavy rain reaching ground; 6. LAL 6 - Dry lightning (same as LAL 3 but without rain).

LIGHTNING CHANNEL Irregular path through air along which lightning discharge occurs.

LIGHTNING DISCHARGE Series of electrical processes by which charge is transferred along channel of high ion density between electrical charge centers of opposite sign.

LIGHTNING STROKE Any of series of repeated electrical discharges comprising single lightning discharge (strike).

LIGHTWEIGHT CONCRETE Concrete mixture made with lightweight coarse and fine aggregates.

LIGNIN Complex non-carbohydrate aromatic polymer present in all wood.

LIKELY In probability of precipitation statements, equivalent of 60 or 70% chance.

LIMB Edge of solar disk.

LIMB FLARE Solar flare seen at edge (Limb) of Sun.

LIME 1. Any inorganic material containing calcium, usually calcium oxide (e.g., quicklime) or calcium hydroxide (e.g., slaked lime); 2. Deciduous tree of genus Tilia.

LIME AND SODA-ASH PROCESS Process for softening water by addition of lime and soda ash to form insoluble compounds of calcium carbonate and magnesium hydroxide.

LIME BLOOM Occurs when free lime in highly alkaline cement based product is taken into solution by inherent moisture content of unit, resulting in efflorescence.

LIME POWDER Calcium carbonate and small amounts of other minerals such as magnesium.

LIME ROCK Calcium carbonate that contains varying levels of silica and hardens on exposure to atmosphere.

LIME STABILIZATION Addition of binder product to host material that reduces moisture dramatically turning unusable material into valid construction material.

LIME-SODA SOFTENING Process used in water treatment to remove hardness from water.

LIMESTONE Sedimentary rock composed mostly of mineral calcite and often containing marine fossils.

LIMESTONE SCRUBBING Use of limestone and water solution to remove gaseous stack-pipe sulfur before it reaches atmosphere.

LIMIT CONTROL Switch that opens when limits gets above setpoint.

LIMIT OF DETECTION Minimum concentration of substance being analyzed test that has 99% probability of being identified.

LIMIT OF WORKS Area or point where work starts or finishes.

LIMIT STATE Condition of structure beyond where it no longer fulfills design criteria.

LIMIT SWITCH Switch preventing travel of object in mechanism that's mechanically operated by motion of object itself.

LIMITED AUTOMATIC REMOTE COLLECTOR Electronic device that interfaces river or precipitation gauge with telephone line making it possible for remote computers to call gaging site and retrieve data.

LIMITED DEGRADATION Environmental policy permitting some degradation of natural systems but terminating at level well beneath established health standard.

LIMITED WATER-SOLUBLE SUBSTANCES Chemicals that are soluble in water at less than 1,000 mg/l.

LIMITING FACTOR 1. Condition whose absence or excessive concentration, is incompatible with needs or tolerance of species or population and which may have negative influence on their ability to thrive; 2. Constraint that limits business from achieving higher level of profitability.

LIMITS 1. Maximum allowed measurements of component; 2. Value to which sequence converges; 3. Border or edge.

LIMITS OF OSCILLATION Distance stream has ranged over time.

LIMNETIC Of or pertaining to living in deeper, open waters of lake or pond (greater than 2 meters deep).

LIMNIC Relating to freshwater.

LIMNOLOGICAL CONDITIONS Conditions on freshwater lakes.

LIMNOLOGY Study of physical, chemical, hydrological and biological aspects of fresh water bodies.

LIMOUSINE Automobile body with seats and permanent top like coupe and with top projecting over driver and projecting front.

LINDANE Pesticide that causes adverse health effects in domestic water supplies and is toxic to freshwater fish and aquatic life.

LINE 1. Hose or pipe for transferring fluid; 2. Path through two or more points.

LINE ECHO WAVE PATTERN Radar echo pattern formed when segment of line of thunderstorms surges forward at accelerated rate.

LINE HAUL 1. Movement of freight by road between distant cities or of passengers or cargo between terminals; 2. Designated truck route between distant cities.

LINE ITEM Specified amount in budget to spend on particular activity.

LINE LOSS Energy waste resulting from transmission of electrical energy across power lines.

LINE MANHOLE Sewer manhole for reasons other than reaching branch sewer.

LINE PIPE Steel pipe used for transporting petroleum, natural gas, oil and water.

LINE SOURCE Array of pollutant sources along defined path that can be treated in dispersion models as aggregate uniform release of pollutants along line.

LINE SPINNING System of rotating metal conduit.

LINE SQUALL Violent local storm occurring as one of number along cold front.

LINEAL In direction of line.

LINEAMENT Rectilinear topographic feature.

LINEAR ALKYLATE SULFONATE Biodegradable surfactant that's part of artificial, household detergents.

LINEAR DESIGN AND CONSTRUCTION System where each phase follows in logical sequence and doesn't commence until preceding phase is finished.

LINEAR ENERGY TRANSFER Amount of energy transferred to local environment in form of ionizations and excitations.

LINEAR FORCE Force applied in straight line.

LINEAR GROWTH Increase in value by same amount at each time step.

LINEAR MODEL Mathematical model based entirely on linear equations.

LINEAR SHRINKAGE Decrease in single dimension of specimen of soil mass when moisture content is reduced from liquid limit to shrinkage limit of soil.

LINEARITY Relationship between several quantities which can be considered as proportional.

LINED CANAL Impermeable layer provided at bed and sides of canal to improve life and discharge capacity.

LINER 1. Relatively impermeable barrier designed to keep leachate inside landfill; 2. Insert or sleeve for sewer pipes to prevent leakage or infiltration; 3. Cover on inside surface.

LINING Any protective material used to line interior surface of conduit, pipe or tunnel to provide water-tightness, prevent erosion, reduce friction or support periphery of structure.

LINK 1. Any intermediate rod or piece for transmitting force or motion; 2. Length of one joint of Gunter's chain; 3. Bond of affinity or unit of valence between atoms applied to unit of chemical force or attraction.

LINK LENGTH Urban street segment length minus width of upstream boundary intersection.

LINKED SWITCH Two or more electric switches which are mechanically connected by operating arms or levers.

LINKED TRIP Journey from origin to destination.

LIP UNION In plumbing, pipe union having lip to prevent gasket from being squeezed into pipe.

LIPID One of family of compounds that are insoluble in water and that make up one of principal components of living cells.

LIPID SOLUBILITY Maximum concentration of chemical that will dissolve in fatty substances.

LIPOIDAL MATTER Wastewater material consisting of fatty acids, salts, oils and animal fat.

LIPOPHILIC High affinity for lipids (fats).

LIQUEFACTION Changing solid into liquid.

LIQUEFIED PETROLEUM GAS Mixture of propane and butane with traces of some other gases, kept in liquid state at high pressure in metal bottles and used as source of gas for cooking, heating and as fuel in internal combustion engines in vehicles.

LIQUID Substance flowing and keeping no shape.

LIQUID CHLORINE Solution of sodium hypochlorite that produces hypochlorous acid instantly when it contacts water.

LIQUID COLLECTOR Solar collector where liquid is heat transfer medium.

LIQUID INJECTION INCINERATOR System that relies on high pressure to prepare liquid wastes for incineration by breaking them up into tiny droplets to allow easier combustion.

LIQUID LIMIT Moisture content corresponding to arbitrary limit between liquid and plastic states of consistency of soil.

LIQUID SLUDGE Watered sludge that flows by gravity or pumping.

LIQUID SPECIFIC GRAVITY Ratio of density of liquid compared to density of water.

LIQUID TANK Transport barge for liquids or liquefied gases.

LIQUID TRAP Trap that prevents liquids from entering pump.

LIQUID WASTE Wastewater, fats, oils or grease, used oil, liquids, solids, gases or sludges and hazardous household liquids.

LIQUID WASTE HAULER Individual or firm that collect, pump, transport or dispose of liquid wastes.

LIQUID WATER EQUIVALENT Liquid content of solid precipitation that has accumulated on ground (snow depth).

LIQUID WEIGHT Water content where soil passes from plastic to liquid state.

LIQUID-METAL FAST-BREEDER REACTOR Nuclear reactor that uses plutonium dioxide or uranium dioxide as fuel and molten sodium and potassium as coolant.

LIQUID-MOUNTED SEAL Foam or liquid-filled primary seal mounted in contact with liquid between tank wall and floating roof continuously around circumference of tank.

LIQUIDATED DAMAGES PROVISION Predetermined money amount that must be paid for failure to perform under contract.

LIQUOR 1. Liquid where desired reaction takes place; 2. Liquid obtained by cooking meat or vegetables (or both).

LISTED Protected from demolition or alteration.

LISTED SPECIES Any species of fish, wildlife or plant officially designated by agency as being endangered or threatened.

LISTED WASTE Wastes listed as hazardous under RCRA but which have not been subjected to Toxic Characteristics Listing Process because dangers they present are considered self-evident.

LITER Unit of volume equal to cubic decimeter.

LITHIFICATION Compaction and cementation of sediment into rock.

LITHIUM HYPOCHLORITE Colorless, crystalline lithium salt of hypochlorous acid that when dissolved in water, releases active chlorine.

LITHOLOGY Description of rocks, based on color, mineral composition and grain size.

LITHOMETEOR Dry particles that hang suspended in atmosphere (e.g., dust, smoke, sand, haze).

LITHOSPHERE Rocky portion of Earth's crust and upper mantle.

LITMUS Dyestuff extracted from certain lichens, that changes color when exposed to pH levels greater than or less than certain critical levels (red for pH values below 4.5 and blue for pH values above 8.3).

LITTER 1. Highly visible portion of solid waste carelessly discarded outside regular garbage and trash collection and disposal system; 2. Leaves and twigs fallen from forest trees; 3. Offspring of mammal born in one birth; 4. Covering of straw for plants.

LITTORAL Of or relating to shore or shallow-water zone, especially seashore (less than 2 meters deep).

LITTORAL CURRENT Current generated by waves breaking at angle to shoreline.

LITTORAL DEPOSIT Marine deposit laid down between low- and high-water marks.

LITTORAL DRIFT Transport of non-cohesive sediments along foreshore and shoreface due to action of breaking waves and longshore current.

LITTORAL ZONE 1. Area on or near shore of body water; 2. Zone or strip of land along shoreline between high and low water marks; 3. Portion of body of fresh water extending from shoreline lakeward to limit of occupancy of rooted plants.

LIVE BOTTOM BIN Bin that receives mass and provides metered feed.

LIVE BOTTOM PIT Endless belt-like structure receiving truck-unloaded material with controlled discharge or retrieval of material.

LIVE CAPACITY That part of total reservoir capacity which can be withdrawn by gravity that's equal to total capacity less dead capacity or sum of active-and inactive storage.

LIVE LOAD Temporary load (e.g., people, snow, ice) that roof structure must be designed to support.

LIVE MAIN Conduit (e.g., gas, water) under pressure.

LIVE STORAGE That part of total reservoir capacity which can be withdrawn by gravity that's equal to total capacity less dead capacity or sum of active-and inactive storage.

LIVE-BOTTOM TRAILER Trailer used for hauling loose material that uses hydraulic conveyor belt to exit material from rear of trailer.

LIVESTOCK WATER USE Water for livestock watering, feed lots, dairy operations, fish farming and other on-farm needs.

LIXIVIANT Solution of water or steam, possibly mixed with other chemicals, injected through injection well into formation to extract minerals in solution mining operations.

LOAD 1. Amount of electrical capacity or energy delivered or required at given point; 2. Power output of engine or power plant under given circumstances; 3. Material moved or carried by streams; 4. General term that refers to material or constituent in solution, in suspension or in transport.

LOAD ALLOCATION Maximum load each discharger of waste pollutants is permitted to release into waterway.

LOAD ALLOCATIONS Portion of receiving water total maximum daily load attributed to either one of its existing or future non-point sources of pollution or to natural background sources.

LOAD CURVE Graph of variation in load over specific time.

LOAD FACTOR 1. Average load carried by engine, machine or plant, expressed as percentage of its maximum capacity; 2. Ratio of average or actual amount of some quantity and maximum possible or permissible.

LOAD FOLLOWING Generating capacity having fast response time to varying loads over wide range of unit loading by automatic generation control from load frequency control equipment.

LOAD HEIGHT Height of load hauler to help avoid hitting obstructions.

LOAD LINE Line on outside of vessel indicating depth where it sinks in water when loaded.

LOAD PATH Direction where each consecutive load will pass through connected members.

LOAD-BEARING ATTACHMENT attachment to main member where both stress of main member and transverse stress are considered separately.

LOAD-CAPACITY RATING Load carrying capacity of bridge.

LOAD-DIVIDING PRESSURE VALVE Valve that proportions pressure between two pumps in series.

LOAD-TRANSFER DEVICE Mechanical means designed to transfer wheel loads across joint.

LOADED FILTER Graded filter located at foot of earth dam.

LOADED GUN SOUNDING Sounding characterized by extreme instability but containing cap.

LOADING 1.Quantity of waste material that can be transported or processed at one time; 2. Volume of work required to be performed; 3. Force exerted on structural component; 4. Electrical current or power delivered by device.

LOADING AREA Stop where carriers are loaded and unloaded.

LOADING CAPACITY Maximum concentration of solute that water can contain without violating water quality standards.

LOADING PIPE Filling conduit to prevent distortion or failure of conduit wall when bending.

LOAM 1. Soft, easily worked soil containing sand, silt and clay; 2. Top layer of soil.

LOCAL AUTHORITIES Personnel authorized by election or job title to carry out operation, planning, mitigation, preparedness, response and recovery functions of emergency management programs at local levels.

LOCAL AVERAGE VELOCITY Local discharge intensity divided by flow depth.

LOCAL BUS Bus that makes frequent stops that links neighborhoods with urban centers.

LOCAL CAPABILITIES ASSESSMENT Evaluation and report performed by to assess warning and evacuation capabilities of local jurisdictions located downstream from dams.

LOCAL CLIMATOLOGICAL DATA Publication that summarizes temperature, relative humidity, precipitation, cloudiness, wind speed and direction observation.

LOCAL CONVECTIVE WIND Local thermally driven winds arising over comparatively small area and influenced by local terrain.

LOCAL EMERGENCY OPERATIONS PLAN General planning document, required by law, that describes responsibilities and actions to be performed in event of emergency and/or disaster.

LOCAL EMERGENCY PLANNING COMMITTEE Committee made up of local officials, citizens and industry representatives charged with development and maintenance of emergency response plans for local emergency planning district.

LOCAL FLOOD WARNING SYSTEM General designator for network of stream and rain gauges implemented by community or local government to monitor hydrologic events as they occur.

LOCAL FLOODING Flooding conditions over relatively limited (localized) area.

LOCAL INFLOW/OUTFLOW POINT Points along any river segment at which water and sediment enter or exit that segment as local flow.

LOCAL LIMITS Conditional discharge limits imposed by municipalities upon industrial or commercial facilities that discharge to municipal sewage treatment system.

LOCAL MAGNITUDE Magnitude of earthquake measured as common logarithm of displacement amplitude, in microns, of standard Wood-Anderson seismograph located on firm ground 100 km from epicenter and having magnification of 2800, natural period 0.8 second and damping coefficient of 80 percent.

LOCAL OFFICIALS Personnel authorized by election or job title to carry out operation, planning, mitigation, preparedness, response and recovery functions of emergency management programs at local levels.

LOCAL ORGANIZATION Local government agency or office having principal or lead role in emergency planning and preparedness.

LOCAL ROAD OR STREET Road or street that provides immediate access.

LOCAL RUNOFF Water running off local area (e.g., rainfall draining into nearby creek).

LOCAL SCOUR Erosion caused by abrupt change in flow direction or velocity.

LOCAL STORM REPORT Product to inform users of reports of severe and/or significant weather-related events.

LOCAL STREET Road that provides local traffic access and circulation.

LOCAL TRAFFIC Traffic by motorists who reside or have business in particular area.

LOCAL TRANSIT SERVICE Transport system for local passengers that makes frequent stops.

LOCALIZER Component of instrument landing system that provides runway centerline guidance to aircraft, but not glideslope information.

LOCATOR TONE Repeating sound that informs approaching pedestrians to push button for crossing signal.

LOCK 1. Segment of canal or other waterway enclosed by gates, used for raising and lowering boats between levels; 2. Fastening together or interlacing.

LOCK CHAMBER Enclosure on waterway used to raise and lower vessels as they pass from one level to.

LOCK CUT Part of waterway that has been modified to provide locations for boats to moor while awaiting lock gates to open or allow people to board or disembark.

LOCK GATE One of pair of gates at each end of lock chamber on canal.

LOCK PADDLE Sluice that fills or empties lock chamber.

LOCK RAY Water space in lock chamber.

LOCK SEAM Joint between two pieces of sheet metal formed by folding overlapping edges against one other and then folding them over one side.

LOCK SILL Raised floor of lock chamber where lock gates sit when shut.

LOCKAGE Amount of elevation and descent made by locks of canal.

LOCKOUT 1. Placement of lockout device on energy isolating device, in accordance with established procedure, ensuring that energy isolating device and equipment being controlled cannot be operated until lockout device is removed; 2. Clearance procedure where physical locks replace safety tags and prevent operating switches, controls, etc.

LOCKOUT DEVICE Device that utilizes positive means (e.g., lock, either key or combination type) to hold energy isolating device in safe position and prevent energizing of machine or equipment.

LOCKOUT RELAY Electrical trip and hold out of service that requires inspection and/or repair before zone may be safely placed back in service.

LOCOMOTIVE Power unit of train that pulls rail cars.

LOCOMOTIVE HORN Air horn that serves as audible warning device.

LOESS Homogeneous, fine-grained sediment made up primarily of silt and clay and deposited over wide area.

LOESSIAL Medium-textured materials (usually silt or very fine sand) transported and deposited by wind action.

LOFTING Pattern of plume dispersion in stable boundary layer topped by neutral layer, where upper part of plume disperses upward while lower part of plume undergoes little dispersion.

LOG BOOM Barrier installed in reservoir, upstream of principal spillway, to prevent logs, debris and boaters from entering water discharge facility or spillway.

LOG CHUTE Channel in dam for passing logs.

LOGBOOK Dated, written record of performed operation and maintenance items or observations pertinent to structure.

LOGIC BOARD Primary circuit board of personal computer.

LOGIC DEVICE Chip that provides processing capability.

LOGIC PROBE Probe used for determining and troubleshooting logical states of digital circuit.

LOGIC STATE Two levels that binary digital signal can possess: logic 1 (i.e., high state) and logic 0 (i.e., low state).

LOGIC SYMBOL Symbol that indicates operation of device.

LOGICAL RELATIONSHIP Established and existing reliance or dependency between two particular elements of project.

LOGO Visual symbol or emblem that acts as trademark or means of identification of company or organization.

LONG NIPPLE Conduit nipple having unthreaded span at its center.

LONG PIPE Ratio of pipeline length to diameter greater than 500.

LONG RANGE NAVIGATION System of long range navigation whereby latitude and longitude are determined from time displacement of radio signals from two or more fixed transmitters.

LONG SCREW Threaded connector 6 inches or greater.

LONG TON Unit of mass equal to 2240 lb.

LONG TUBE Ratio of tube length to diameter greater than 3, that's inserted in orifice.

LONG-QUARTER BEND 90° pipe fitting where one end is longer than other.

LONG-RADIUS ELBOW Pipe fitting having bigger radius than standard for 90° bend.

LONG-SPAN CULVERT Culvert, composed of corrugated metal liner and compacted soil envelope, that exceeds standard sizes.

LONG-SWEEP FITTING Longer than standard pipe fitting radius curve.

LONG-TERM METHOD DETECTION LEVEL Detection level derived by determining standard deviation of minimum of 24 method detection limit spike sample measurements over extended period of time.

LONG-TERM MONITORING Data collection over period of years or decades to assess changes in selected hydrologic conditions.

LONG-TERM RETENTION Retention of data for 5 years to satisfy requirements for local studies and to support litigation.

LONG-TERM STORAGE DAMS Reservoirs used for recreational use or storage of irrigation, municipal or industrial water.

LONG-TUBE EVAPORATOR Evaporator tube heat exchanger.

LONGITUDE Location east or west in reference to Prime Meridian, that is designated as 0° longitude.

LONGITUDINAL 1. Pertaining to or extending along long axis or length, of structure; 2. Lengthwise.

LONGITUDINAL BARRIER Barrier with primary function to prevent penetration and safely redirect errant vehicle away from roadside or obstacle.

LONGITUDINAL CHANNELIZER Device that delineates edge of roadway.

LONGITUDINAL CRACKING Cracks that occur parallel to pavement centerline.

LONGITUDINAL DRAINAGE When conduits are placed parallel to land incline that's being drained.

LONGITUDINAL FORCE Force in direction of main plane.

LONGITUDINAL JOINT Joint between two slabs.

LONGITUDINAL MARKINGS Pavement markings, placed parallel and adjacent to traffic flow, that provide continuous guidance.

LONGITUDINAL PROFILE 1. Average stream slopes and depths of riffles, pools, runs, glides, rapids, etc.; 2. Perpendicular deviations of pavement surface from established reference parallel to lane direction.

LONGSHORE CURRENTS Current that moves parallel to shore.

LONGSHORE DRIFT Process whereby sediment moves along beach shoreline when waves approach shore at angle.

LONGWAVE RADIATION Term used to describe infrared energy emitted by Earth and atmosphere at wavelengths between about 5 and 25 micrometers.

LONGWAVE TROUGH Trough in prevailing westerly flow aloft that is characterized by large length and (usually) long duration.

LOOP 1. Path that starts and ends at same point; 2. Length of thread, line or rope doubled over to make opening; 3. Complete circuit for electric current.

LOOP APRON Heavy material (e.g., stones) placed to stop erosion on embankment berms.

LOOP DETECTOR Device to detect vehicular traffic traveling or stopping overhead in roadway.

LOOP PROMINENCE SYSTEM System of loop prominences associated with major flares.

LOOP RAMP Ramp that converts stop or signal controlled left-turn movement into merge or yield-controlled movement.

LOOP VENT Vent that loops, to allow gas to flow up pipe as water flows down pipe.

LOOPED WATER MAIN Water main that provides water supply from more than one direction.

LOOPING Chimney plume that loops when there has been superadiabatic lapse rate and solar heating.

LOOSE CUBIC YARD 1 yd^3 of soil after excavation.

LOOSE FIT Fit between parts with sufficient clearance.

LOOSE GROUND Material that tends to slough from sidewalls into borehole.

LOOSE MATERIAL BUCKET Bucket designed for free flowing materials.

LOOSE YARD 1. Unit of volume of excavation; 2. Cubic yard loose measurement.

LOOSE YARDS Measurement of soil or rock after it has been loosened by digging or blasting.

LOOSE-BOUNDARY HYDRAULICS Study of movement of material due to flow of fluid along unstable boundaries.

LOOSE-MEASURE VOLUME Measure of volume of material after it's been excavated.

LOOSE-ROCK DAM Dam built with loosely placed rock and no mortar.

LORAN Long range hyperbolic radio navigation system.

LOSING STREAM Stream or reach of stream that contributes water to zone of saturation.

LOSS Difference between volume of rainfall and volume of runoff.

LOSS OF EARNINGS EXPENDITURE Reduction from environmentally-caused harm or illness.

LOSS OF HEAD Dissipation of energy in any moving fluid because of friction.

LOSS-OF-HEAD GAUGE Gauge that measures head loss through filter.

LOST ENERGY Energy converted to heat by friction.

LOST RIVER Stream losses via aridity increase.

LOST TIME 1. To fall behind schedule; 2. Sum of clearance start-up time missed.

LOT 1. Distinct portion or plot of land; 2. Large quantity or number.

LOTIC Flowing body of fresh water (i.e., rivers, streams).

LOUDNESS Strength of sound pressure level.

LOUVER Series of sloping overlapping slats (i.e., slits) that admit air and light, but exclude rain, etc.

LOW Region of low pressure, marked as "L" on weather map.

LOW ALTITUDE Elevation less than or equal to 4,000 feet.

LOW ALTITUDE CONDITIONS Test altitude less than 1,800 feet.

LOW DAM Dam up to 100 feet high.

LOW DENSITY MATERIAL Material having low weight per unit volume either as it occurs in its natural state or after compacting it.

LOW DENSITY POLYETHYLENE Plastic material used for both rigid containers and plastic film applications.

LOW DOSE 1/10 of high dose.

LOW DUTY OF WATER Quantity of water required to irrigate land.

LOW EMISSIVITY WINDOWS Window technology that lowers amount of energy loss through windows by inhibiting transmission of radiant heat while still allowing sufficient light to pass through.

LOW FLOOR BUS Bus that has no steps between ground and floor of bus.

LOW FLOW CHANNEL Natural or artificial watercourse channel occupied during periods of low flow.

LOW FREQUENCY Portion of radio frequency spectrum from 30 to 300 kHz.

LOW GROUND PRESSURE Heavy equipment that has been specially designed to spread weight of equipment over larger area.

LOW HAZARD Downstream hazard classification for dams where no lives are in jeopardy and minimal economic loss (undeveloped agriculture, occasional uninhabited structures or minimal outstanding natural resources) would occur as result of failure of dam.

LOW IMPACT DEVELOPMENT Sustainable landscaping approach that can be used to replicate or restore natural watershed functions and/or address targeted watershed goals and objectives.

LOW LEVEL JET Region of relatively strong winds in lower part of atmosphere.

LOW LEVEL OUTLET Opening at low level from reservoir generally used for emptying or for scouring sediment and sometimes for irrigation releases.

LOW NOX BURNERS One of several combustion technologies used to reduce emissions of nitrogen oxides.

LOW PLATFORM Platform at top of rail elevation.

LOW PRESSURE SYSTEM Area of relative pressure minimum that has converging winds and rotates in same direction as Earth.

LOW RAIL Inner rail of curve that has at-grade alignment.

LOW TIDE Minimum height reached by each falling tide.

LOW VOLUME WASTE SOURCE Wastewater collected from multiple sources except any above restriction limitations.

LOW WATER 1. Location of low tide in coastal area; 2. Lowest stage of river.

LOW WATER ADVISORY Advisory to describe water levels which are significantly below average levels over Great Lakes, coastal marine zones and any tidal marine area, waterway or river inlet within or adjacent to marine zone that would potentially be impacted by low water conditions creating hazard to navigation.

LOW-FLOW FREQUENCY CURVE Graph showing magnitude and frequency of minimum flows for period of given length.

LOW-HEAD HYDROELECTRIC Hydroelectric power that operates with head of 66 feet or less.

LOW-LEVEL GROIN Slowly eroding natural outcroppings or accretion from stream bed.

LOW-LEVEL WASTE Nuclear waste (e.g., cloth, bottles, plastic, wipes, etc. that have contact radiation) that represents 90% of all radioactive waste.

LOW-MAINTENANCE/SELF-RESTORING CRASH CUSHIONS Traffic barriers that either suffer no damage or partial rebound after impact.

LOW-PRECIPITATION STORM Supercell thunderstorm characterized by relative lack of visible precipitation.

LOW-PRESSURE AIR TEST Test that detects damaged piping or improper jointing by measuring rate at which air under pressure escapes from isolated section of sewer.

LOW-PRESSURE STEAM HEATING SYSTEM Steam heat system with pressures between 0 and 15 psig.

LOW-PRESSURE STORAGE TANK Storage tank with pressures between 0.51 and 14.5 psig.

LOW-RATE TRICKLING FILTER Dripping filter that removes BOD to produce effluent.

LOW-RELAXATION STEEL Steel wire for use in prestressed concrete.

LOW-VELOCITY ZONE Zone within upper mantle beneath lithosphere and asthenosphere.

LOW-WATER CROSSING Bridge when water flow is low.

LOW-WATER LINE Line representing intersection of land and water at mean low water elevation.

LOW-WATER LUNITIDAL INTERVAL Interval between transit of moon and following low water.

LOW-WATER MARK Lowest elevation of water at low tide or during drought or dry season.

LOW-WATER REGULATION Flows adapted to some desired, essential or imposed state.

LOW-WATER SLACK Short period of time in tidal water when there is no current velocity in either direction of tidal stream.

LOWER ATMOSPHERE STABILITY INDEX Index composed of stability term and moisture term that indicates potential for large fire growth.

LOWER BANK Streambank area with elevation less than stream mean water level.

LOWER DETECTION LIMIT Smallest signal above background noise instrument can reliably detect.

LOWER EXPLOSIVE LIMIT Concentration of compound in air below which mixture will not catch on fire.

LOWERING VALVE Valve that enables supported load to go down.

LOWEST ACCEPTABLE DAILY DOSE Largest quantity of chemical that will not cause toxic effect, as determined by animal studies.

LOWEST ACHIEVABLE EMISSION RATE Rate of emissions that reflects most stringent emission limitation in implementation plan of any state for such source unless owner or operator demonstrates such limitations are not achievable or most stringent emissions limitation achieved in practice, whichever is more stringent.

LOWEST OBSERVED ADVERSE EFFECT LEVEL Lowest concentration of effluent or toxicant that results in statistically significant adverse health effects as observed in chronic or subchronic human epidemiology studies or animal exposure.

LOWEST-OBSERVED-ADVERSE-EFFECT LEVEL Lowest dose where there was observed toxic or adverse effect in animal species.

LOWEST-OBSERVED-EFFECT LEVEL Lowest dose of substance that has been reported to cause adverse health effects.

LOWLAND FLOODING Inundation of low areas near river, often rural, but may also occur in urban areas.

LUGGED PIPE BRACKET Wrap-around pipe bracket with lug to fix pipes to wall.

LUGGER BODY Container loaded by crane like boom and chains with 6 to 20 yd^3 capacity.

LUMEN Unit of luminous flux equal to amount of light emitted per second in unit solid angle of 1 steradian from uniform source of 1 candela.

LUMEN SECOND Unit of luminous energy.

LUMINAIRE Electrical device that contains electric lamp that provides illumination.

LUMP SUM BID Single bid that covers all costs.

LUMP SUM CONTRACT Contract that requires supplier to provide specified services for stipulated or fixed price.

LUNAR DAY Time interval of time between two successive crossings of Earth's meridian by moon.

LUX Unit of illumination of 1 $lumen/m^2$.

LUX-SECOND Unit of light exposure.

LYSIMETER Device to measure quantity or rate of downward water movement through block of soil usually undisturbed or to collect such percolated water for analysis as to quality.

M

M 3000 Optimum high frequency radio wave with 3,000 km range, which reflects only once from ionosphere (single hop transmission).

M1 AND M2 CURVES Curves representing mild sloping water surface profiles.

MACADAM Surface of road consisting of layers of crushed stone.

MACHINE Device that directs and controls energy, often in form of movement or electricity, to produce certain effect.

MACHINE AVAILABILITY TIME Percentage of operating time machine isn't under service or repair.

MACHINE BASE Mechanical equipment foundation.

MACHINE BEAM Flat steel beam that supports machines.

MACHINE DOWNTIME Time machine is unable to work.

MACHINE DRAWING Drawing that furnishes all information required for manufacture and assembly of components of machine.

MACHINE EFFICIENCY Ratio of horsepower output energy or work to horsepower input energy.

MACHINE LANGUAGE Set of instructions that particular computer is designed to execute.

MACHINE MECHANICAL DELAY TIME Scheduled operating time machine cannot function due to repair or service.

MACHINE OPERATING EFFICIENCY Ratio of real work time to accessible work time.

MACHINE OPERATING TIME Productive time machine works.

MACHINE OPERATIONAL LOST TIME Time machine production is stopped.

MACHINE OTHER PRODUCTIVE TIME Time machine performs tasks other than primary.

MACHINE OUT-OF-SHIFT REPAIR TIME Time during which no production is scheduled and machine is being repaired.

MACHINE OUT-OF-SHIFT SERVICE TIME Time during which no production is scheduled and machine is being serviced.

MACHINE PERSONNEL TIME Time that machine lacks operator or another machine team members.

MACHINE PRODUCTIVE MACHINE HOUR Time machine works during scheduled working hours.

MACHINE PRODUCTIVE TIME Scheduled operating time that machine is works.

MACHINE RATE Cost for possessing and running machine and labor cost for operator.

MACHINE RATING Mechanical output power machine can produce.

MACHINE REPAIR TIME Time spent servicing machine.

MACHINE RESISTING MOMENT Moment of dead weight of crane or derrick, less boom weight, about tipping fulcrum.

MACHINE ROOM Room in building dedicated to HVAC equipment and other machinery.

MACHINE SCHEDULED MACHINE HOUR Hour machine is scheduled to work that including down time.

MACHINE SCHEDULED NON-OPERATING TIME Time when machine is not scheduled to work.

MACHINE SCHEDULED OPERATING TIME Time machine is scheduled to work.

MACHINE SERVICE TIME Time machine is scheduled to be serviced.

MACHINE TIME Time spent mechanical equipment works.

MACHINE UTILIZATION TIME Ratio of machine working time to completion time.

MACHINE WAITING REPAIR TIME Time machine is expecting service.

MACHINED END-END Conduit milled on both ends and not in center.

MACHINED OVER ALL Conduit milled end to end.

MACKERAL SKY Cirrocumulus clouds with small vertical extent and composed of ice crystals.

MACROBURST Convective downdraft with affected outflow area of at least 2 miles wide and peak winds lasting between 5 and 20 minutes.

MACROHABITAT Extensive habitat presenting considerable variation of environment, containing variety of ecological niches and supporting large number and variety of complex flora and fauna.

MACROPHYTES Plant large enough to be seen by naked eye.

MACROPORES Secondary soil features (e.g., root holes, desiccation cracks) that can create significant conduits for movement of non-aqueous phase liquid and dissolved contaminants or vapor-phase contaminants.

MACROSCALE Large scale, characteristic of weather systems several hundred to several thousand kilometers in diameter.

MACROSCOPIC MODEL Mathematical traffic model that formulates relationships among traffic flow characteristics.

MADDEN-JULIAN OSCILLATION Tropical rainfall exhibits strong variability on time scales shorter than seasonal El Niño-Southern Oscillation (ENSO).

MADE GROUND Land that has been man-made.

MAGFLARE Geomagnetic and/or cosmic storm has been associated with this flare.

MAGMA Molten or fluid rock material from which igneous rock is derived.

MAGMATIC WATER Water that exists with magma or water-rich volatile fluids.

MAGNESIUM Light, flammable, silvery metal, with atomic number of 12.

MAGNETIC BAY Relatively smooth excursion of H (horizontal) component of geomagnetic field away from and returning to quiet levels.

MAGNETIC FRACTION Ferrous scrap remaining following removal of non-magnetic contaminants.

MAGNETIC LEVITATION Method of supporting and transporting objects and vehicles by using magnetic field to counterbalance gravitational pull.

MAGNETIC SEPARATION Use of magnets to separate ferrous materials from mixed municipal waste stream.

MAGNETIC SEPARATOR Apparatus that uses magnetic field to separate ferromagnetic particles.

MAGNETIC SWITCH Switch controlled by electomagnet.

MAGNETOGRAM Solar magnetograms are graphic representation of solar magnetic field strengths and polarity.

MAGNETOPAUSE Boundary layer between solar wind and magnetosphere.

MAGNETOSPHERE Magnetic cavity surrounding Earth, carved out of passing solar wind by virtue of geomagnetic field, which prevents or at least impedes, direct entry of solar wind plasma into cavity.

MAGNITUDE 1. Rating of given earthquake, independent of place of observation; 2. Measure of strength of earthquake or strain energy released by it, as determined by seismographic observations; 3. How much of pollutant (or pollutant parameter) expressed as concentration or toxic unit is allowable.

MAGNITUDE, LOCAL Magnitude of earthquake measured as common logarithm of displacement amplitude, in microns, of standard Wood-Anderson seismograph located on firm ground 100 km from epicenter and having magnification of 2800, natural period 0.8 second and damping coefficient of 80 percent.

MAGNITUDE, RICHTER Magnitude of earthquake measured as common logarithm of displacement amplitude, in microns, of standard Wood-Anderson seismograph located on firm ground 100 km from epicenter and having magnification of 2800, natural period 0.8 second and damping coefficient of 80 percent.

MAGNITUDE, SURFACE WAVE Magnitude of earthquake measured as common logarithm of resultant of maximum mutually perpendicular horizontal displacement amplitudes, in microns, of 20 second period surface waves.

MAIN Principal pipe artery that branches may be connected.

MAIN CANAL Main irrigation conduit that receives its water in supply and provides it to laterals.

MAIN CHANNEL Deepest or central part of bed of stream, containing main current.

MAIN CHANNEL POOL Reach of stream or river with low bed elevation, relative to rapids or riffles.

MAIN LINE Principal route or line of railway.

MAIN PHASE That period when horizontal magnetic field at middle latitudes is generally decreasing.

MAIN SEWER Sewer that carries sewage from branches to treatment plant.

MAIN STEM 1. Main course or channel of river or stream; 2. Main street of city or town.

MAIN SYNOPTIC TIMES 1. Times of 0000, 0600, 1200 and 1800 UTC; 2. Standard synoptic times.

MAIN VENT Stack in plumbing system that links system to atmosphere.

MAINLINE Principal route or line of railway.

MAINLINE METER Water meter of water distribution system installed on primary supply line.

MAINLINE OUTPUT Maximum number of vehicles that can depart freeway node.

MAINSTEM 1. Reach of river/stream formed by tributaries that flow into it; 2. Principal trunk of river or stream; 3. Main course of stream where current is strongest.

MAINSTREAM Main course of stream where current is strongest.

MAINTENANCE 1. Natural process which keeps organism alive; 2. All routine and extraordinary work necessary to keep facilities in good repair and reliable working order to fulfill intended designed project purposes.

MAINTENANCE CONTRACT Contract to maintain something.

MAINTENANCE DECISION SUPPORT SYSTEM Effort to mitigate challenges associated with highway winter maintenance decisions.

MAINTENANCE EXPENDITURE EXPENDITURE Price of keeping capital structures to fix or halt deterioration as result of unfavorable environmental impact.

MAINTENANCE MANAGEMENT SYSTEM Any organized system used to ensure that all operations and maintenance activities (e.g., maintenance, inspection, operational testing) at facility is accomplished and documented.

MAINTENANCE PERIOD Time following completion of construction project where contractor is responsible for certain maintenance issues.

MAINTENANCE SCHEDULE Comprehensive record of care processes needed to be finished according to established schedule.

MAINTENANCE STANDARD Officially recognized criteria for particular process.

MAJOR CONTRACT ITEMS Items listed in both bid schedule and special provisions.

MAJOR DISASTER Any natural catastrophe (e.g., hurricane, tornado, storm, tidal wave, tsunami, earthquake).

MAJOR DIVERGE AREA Where two primary roadways, each having multiple lanes, diverge from single freeway segment.

MAJOR FACILITY Larger publicly owned treatment works with flows equal to at least 1 million gal/day or servicing population equivalent to 10,000 persons.

MAJOR FLARE This flare is basis for forecast of geomagstorm, cosmic storm and/or protons in Earth's vicinity.

MAJOR FLOODING General term including extensive inundation and property damage.

MAJOR GEOMAGNETIC STORM Storm for which Ap index was greater than 49 and less than 100.

MAJOR HURRICANE Hurricane which reaches Category 3 (sustained winds greater than 110 mph) on Saffir/Simpson Hurricane Scale.

MAJOR INTERCHANGE 1. Interchanges with other expressways or freeways; 2. Interchanges with high-volume multi-lane highways, principal urban arterials or major rural routes.

MAJOR IONS Constituents commonly present in water in concentrations exceeding 1.0 mg/l.

MAJOR MERGE AREA Where two multilane freeway segments combine to form single freeway segment.

MAJOR MODIFICATION Term used to define modifications of major stationary sources of emissions.

MAJOR STREET Street that carries higher vehicular traffic than local service street.

MAJOR WEAVING SEGMENT Freeway segment with at least three entry and exit legs.

MAJORS Larger publicly owned treatment works with flows equal to at least 1 million gal/day or servicing population equivalent to 10,000 persons.

MAKEUP WATER Water supplied to compensate for losses (e.g., evaporation, leakage).

MALATHION Insecticide in chemical family known as organophosphates.

MALE COUPLING Eternally threaded hose nipple which fits in thread of female.

MALE THREAD External helical ridge on end of conduit.

MALFUNCTION Failure to function as designed.

MALFUNCTION MANAGEMENT UNIT Device used to detect improper and conflicting traffic signals.

MALLEABLE CAST IRON Cast as white iron with structure being metastable carbide in pearlitic matrix.

MAMMATUS CLOUDS Rounded, smooth, sack-like protrusions hanging from underside of cloud (usually thunderstorm anvil).

MAN HOUR Unit of work that can be done by one worker in one hour.

MAN LOCK Airlock that allows workers to pass spaces with differing air pressures.

MAN-MADE BETA PARTICLE AND PHOTON EMITTERS All radionuclides emitting beta particles and/or photons listed in Maximum Permissible Body Burdens and Maximum Permissible Concentrations of Radionuclides in Air and Water for Occupational Exposure.

MANAGED LANE Highway lane operated with management scheme.

MANAGEMENT INDICATOR SPECIES Any species, group of species or species habitat element selected to focus management attention for purpose of resource production, population recovery, maintenance of population viability or ecosystem diversity.

MANAGEMENT INDICATORS Plant and animal species, communities or special habitats that are selected for emphasis in planning and that are monitored during forest-plan implementation to assess effects of management activities on their populations and populations of other species with similar habitat needs that they may represent.

MANAGEMENT INFORMATION SYSTEM Computerized information-processing system designed to support activities of company or organizational management.

MANAGERIAL CONTROLS Methods of non-point source pollution control based on decisions about managing agricultural wastes or application times or rates for agrochemicals.

MANDATORY RECYCLING Programs which by law require consumers to separate trash so that some or all recyclable materials are recovered for recycling rather than going to landfills.

MANDATORY STANDARD Makes particular safety or information features on products compulsory and legally enforced.

MANEUVER TIME Time it takes to complete action.

MANGANESE Metallic chemical element with atomic number of 25.

MANGANESE BACTERIA Non-pathogenic bacteria that occur in soil, shallow aquifers and surface waters and feed on iron and manganese.

MANHOLE 1. Hole in ground used to access sewers or other underground vaults and installations; 2. Hole providing access to inside of boiler, tank, etc.

MANHOLE COVER Removable plate forming lid over opening of manhole.

MANHOLE HEAD Assembly for mounting cover onto manhole frame.

MANIFEST One-page form used by haulers transporting waste that lists EPA identification numbers, type and quantity of waste, generator it originated from, transporter that shipped it and storage or disposal facility that it is being shipped.

MANIFEST SYSTEM Procedure where hazardous materials are identified and followed as they are produced, treated, transported and disposed of by series of permanent, linkable, descriptive documents.

MANIFOLD Large pipe that series of smaller pipes are connected.

MANNING ROUGHNESS COEFFICIENT Roughness or friction applied to flow by channel.

MANNING'S EQUATION Empirical equation commonly applied in water surface profile calculations that defines relationship between surface roughness, discharge, flow geometry and rate of friction loss for given stream location.

MANNING'S N-VALUE Coefficient of boundary roughness that accounts for energy loss due to friction between bed and water.

MANNING'S ROUGHNESS COEFFICIENT Coefficient of roughness that accounts for energy loss due to friction between bed and water that's used to describe relative roughness of channel and overbank areas.

MANOMETER Instrument for measuring pressure.

MANTLE 1. Thick layer of rock deep within Earth that separates Earth's crust above from Earth's core below; 2. Zone of hot gases around flame.

MANUAL BLOCK SIGNAL Manually controlled signal.

MANUAL COUNT Counting by human observation.

MANUAL LANE Toll lane with attendant to accept payment.

MANUAL SEPARATION Hand sorting of recyclable or compostable materials in waste.

MANUAL TRAFFIC CONTROL Train movement is controlled by motorman.

MANUFACTURER'S FORMULATION List of substances or component parts as described by maker of coating, pesticide or other product containing chemicals or other substances.

MANUFACTURERS' IDENTIFICATION Code that indicates manufacturer.

MANUFACTURING USE PRODUCT Any product intended (labeled) for formulation or repackaging into other pesticide products.

MANUFACTURING WASTES Wastes (e.g., solid and liquid) generated during production.

MAP Two-dimensional representation of all or part of Earth's surface showing selected natural or manmade features or data.

MAP CRACKING Shallow or hairline cracks that extend only through upper surface of concrete.

MARBLE Metamorphic rock formed by "baking" and recrystallization of limestone.

MARC VELOCITY SIGNATURE Doppler radar-velocity based precursor towards forecasting initial onset of damaging straight-line winds in linear Quasi-Linear Convective System or bowing convective system.

MARE'S TRAIL Name given to thin, wispy cirrus clouds composed of ice crystals that appear as veil patches or strands, often resembling horse's tail.

MARGIN OF EXPOSURE Ratio of no-observed adverse-effect-level to estimated exposure dose.

MARGIN OF SAFETY Maximum amount of exposure producing no measurable effect in animals (or studied humans) divided by actual amount of human exposure in population.

MARGINAL LAND Land which, in its natural state, is not well suited for particular purpose (e.g., raising crops).

MARGINAL VISUAL FLIGHT RULES General weather conditions pilots can expect at surface with ceiling between 1,000 and 3,000 feet and/or 3 to 5 miles visibility.

MARIGRAPH Device used to measure tidal levels.

MARINA Harbour for small boats.

MARINE BAYS AND ESTUARIES Partially enclosed coastal waters.

MARINE CLIMATE Relatively mild climate moderated by presence of large body of water.

MARINE DEPOSIT Sedimentary deposit laid down in ocean water.

MARINE ENVIRONMENT Oceans, seas, bays, estuaries and other major water bodies.

MARINE INVERSION Temperature inversion produced when cold marine air underlies warmer air.

MARINE OPTICAL BUOY Measures solar radiation to calibrate satellite ocean color instruments.

MARINE PUSH Replacement of current air mass with air from off ocean.

MARINE REPORT Voluntary marine observation program of National Weather Service whose goal is to solicit meteorological and oceanographic observations in plain language from recreational and small commercial mariners who are not part of Voluntary Observing Ship program.

MARINE SANITATION DEVICE Any equipment or process installed on board vessel to receive, retain, treat or discharge sewage.

MARINE SOIL Soil formed from substances deposited by or in marine waters.

MARINE TERMINAL Port area where goods and cargo can be loaded and unloaded from ship.

MARINE WEATHER STATEMENT Product to provide mariners with details on significant or potentially hazardous conditions not otherwise covered in existing marine warnings and forecasts.

MARINE WETLAND Wetlands that are exposed to waves and currents of open ocean and to water having salinity greater than 30 parts/1,000.

MARINE ZONE Specific, defined over-water areas contained in various National Weather Service marine forecasts.

MARITIME AIR MASS Air mass influenced by sea.

MARITIME PLANTS Plants that live on saltwater tidal flats.

MARITIME POLAR AIR MASS Air mass characterized by cold, moist air.

MARITIME TROPICAL AIR MASS Air mass characterized by warm, moist air.

MARKET Geographical area where certain commercial demand exists.

MARKET VALUE Price seller or insurer might reasonably expect to fetch for goods, services or securities on open market.

MARKET-BASED RATES Rates for power or electric service that are established in unregulated, competitive market.

MARKETER Agent for generation projects who markets power on behalf of generator.

MAROB Voluntary marine observation program of National Weather Service in early stages of development whose goal is to solicit meteorological and oceanographic observations in coded format from recreational and small commercial mariners who are not part of more in-depth Voluntary Observing Ship program.

MARS Voluntary marine observation program of National Weather Service whereby U.S. Coast Guard Sector Stations report marine weather conditions from several shore locations within their operating area.

MARSH Type of wetland that does not accumulate appreciable peat deposits and is dominated by aquatic and grasslike vegetation.

MASKING AGENT Reagent that reacts with chemical species that may interfere in analysis.

MASONRY CHECK Obstacle constructed with cement mortar and non-disintegrating stones.

MASONRY DAM Dam constructed mainly of stone, brick or concrete blocks pointed with mortar.

MASS BALANCE Accounting for material entering and leaving system.

MASS CARE CENTER Facility for providing emergency lodging and care for people made temporarily homeless by emergency or disaster.

MASS CONCRETE Large volume of concrete cast-in-place, generally as monolithic structure.

MASS CURVE Graph of cumulative values of hydrologic quantity (i.e., precipitation, inflow, outflow, runoff), generally as ordinate, plotted against time or date as abscissa.

MASS DIAGRAM Plots of cumulative volumes of cut and fill along alignment.

MASS EXCAVATION BUCKET Pail created for volume truck loading.

MASS MOVEMENT Movement of land surface caused by gravity.

MASS PER SQUARE INCH Unit of force equal to 1 lb/in^2.

MASS RUNOFF Overall volume of overflow from certain place and time.

MASS SPECTROMETER Instrument that identifies molecular composition and concentrations of various chemicals in water and soil samples.

MASS SPECTROMETRY Analytical technique that measures mass or charge ratio of ions formed when molecule or atom is ionized, vaporized and introduced into vacuum.

MASS TRANSPORTATION Method of transport that carries people in greater quantities than typical private method.

MASS-BASED STANDARD Discharge limit measured in mass unit (e.g., lbs/day).

MASSIF Compact portion of mountain range, containing one or more summits.

MASSIVE Over 3 ft thick.

MASSIVE HEAD BUTTRESS DAM Buttress dam where buttress is greatly enlarged on upstream side to span gap between buttresses.

MAST ARM Horizontal post that attaches light fixture to streetlight.

MASTER SCHEDULE Summary-level project schedule used to identify all of major deliverables for project as well as all individual work breakdown structure components.

MASTER SPECIFICATIONS Collection of standardized, prewritten technical or guide specifications.

MASTER SWITCH Switch controlling supply of electricity or fuel to system.

MASTIC Soft sealing material.

MATCH Portion of state or local funds needed for federally funded project.

MATERIAL Matter which may be shaped or manipulated.

MATERIAL CATEGORY In asbestos program, broad classification of materials into thermal surfacing insulation, surfacing material and miscellaneous material.

MATERIAL DENSITY Amount of mass contained per unit volume.

MATERIAL HANDLER Perform physical and administrative tasks involved in receiving, order fulfillment, storing and distributing of materials, parts, etc.

MATERIAL SAFETY DATA SHEET Document that contains information on potential hazards and how to work safely with chemical product.

MATERIAL SPECIFICATION Material composition required to satisfy performance standards.

MATERIAL TYPE Classification of suspect material by its specific use or application (e.g., pipe insulation, fireproofing, floor tile).

MATERIALLY UNBALANCED BID Bid that generates reasonable doubt that award to bidder will result in lowest ultimate cost.

MATERIALS AND METHOD SPECIFICATIONS Direct contractor to use specified material, specific type of equipment and methods to place.

MATERIALS RECOVERY Specialized plant that receives, separates and prepares recyclable materials for marketing to end-user manufacturers.

MATERIALS RECOVERY FACILITY Facility that processes residentially collected mixed recyclables into new products available for market.

MATHEMATICAL MODEL Model that uses mathematical expressions (i.e., set of equations) to represent physical process.

MATHEMATICALLY UNBALANCED BID Bid with high price on some items and low price on other items in unit price contract.

MATTRESS Form of retaining wall used to support foundations or embankment.

MATURE RIVER River with gradient less steep and more discharge than youthful rivers fed by many tributaries.

MATURE SHORELINE Stable coastline not exposed to wave erosion.

MATURE VALLEY Valley that's wider, less steep sided and deeper than young valley.

MATURITY STREAM Stage in development of stream at which it has reached its maximum efficiency, when velocity is just sufficient to carry sediment delivered to it by tributaries, where flat uplands have been widely dissected by deep river valleys.

MAX OUT To reach maximum or point at which no more growth or improvement is possible.

MAX PARCEL LEVEL Highest attainable level that convective updraft (e.g., thunderstorm) can reach.

MAXIMALLY EXPOSED INDIVIDUAL Person with highest exposure in given population.

MAXIMUM 30-DAY AVERAGE Maximum average of daily values for 30 consecutive days.

MAXIMUM ACCEPTABLE CONCENTRATION Level that has been established for certain substances that are known or suspected to cause adverse health effects.

MAXIMUM ACCEPTABLE PRESSURE Maximum water pressure that will not cause premature or accelerated damage.

MAXIMUM ACCEPTABLE TOXIC CONCENTRATION Range (or geometric mean) between No Observable Adverse Effect Level and Lowest Observable Adverse Effects Level.

MAXIMUM ACCEPTABLE TOXICANT CONCENTRATION Value calculated through aquatic toxicity tests to help set water quality regulations for protection of aquatic life.

MAXIMUM ALLOWABLE CONCENTRATION Highest concentration of chemical or toxin allowed by law that worker may be exposed to during period of time.

MAXIMUM ALLOWABLE COST Payment model contractually agreed to in marketplace by all participants.

MAXIMUM AVAILABLE CONTROL TECHNOLOGY Emission standard for sources of air pollution requiring maximum reduction of hazardous emissions, taking cost and feasibility into account.

MAXIMUM AVAILABLE WATER 1. Amount of water which can be removed from aquifer without lowering groundwater table; 2. Field capacity minus wilting point.

MAXIMUM COMPUTED FLOOD Largest computed momentary flood discharge from watershed.

MAXIMUM CONTAMINANT LEVEL Maximum permissible level of contaminant in water delivered to any user of public water system.

MAXIMUM CONTAMINANT LEVEL GOAL Level of contaminant in drinking water below which there is no known or expected risk to health.

MAXIMUM CONTAMINATION LEVEL Maximum concentration of chemicals which are allowed in public drinking water systems.

MAXIMUM CONTROLLABLE WATER SURFACE Highest reservoir water surface elevation at which gravity flows from reservoir can be completely shut off.

MAXIMUM CREDIBLE EARTHQUAKE Earthquake(s) associated with specific seismotectonic structures, source areas or provinces that would cause most severe vibratory ground motion or foundation dislocation capable of being produced at site under currently known tectonic framework.

MAXIMUM DEMAND Greatest of all demands of load that has occurred within specified period of time.

MAXIMUM DENSITY Highest attainable density of substance under given conditions.

MAXIMUM DESIGN EARTHQUAKE Postulated seismic event, specified in terms of specific bedrock motion parameters at given site, that is used to evaluate seismic resistance of manmade structures or other features at site.

MAXIMUM DISCHARGE Maximum rate of flow conduit, stream, channel or structure is effective at passing.

MAXIMUM EXCURSION After unexpected disruption, maximum pressure deviation from operating pressure.

MAXIMUM EXPOSURE LIMIT Highest allowable chemical-concentration in period-limited worker exposure.

MAXIMUM EXPOSURE RANGE Estimate of exposure or dose level received by individual in defined population greater than 98th percentile dose for all individuals in that population, but less than exposure level received by person receiving highest exposure level.

MAXIMUM FLOOD CONTROL LEVEL Highest elevation of flood control storage.

MAXIMUM FLOOD FLOW Peak discharge of flood water.

MAXIMUM GREEN Maximum time green signal can be displayed in presence of conflicting demand.

MAXIMUM GROSS WEIGHT Weight of vehicle laden with heaviest load which it is constructed or adapted to carry.

MAXIMUM HIGHWAY TRAFFIC SIGNAL PREEMPTION TIME Maximum amount of time needed following initiation of preemption sequence for highway traffic signals to complete timing of right-of-way transfer time, queue clearance time and separation time.

MAXIMUM INDIVIDUAL RISK Maximum risk to exposed person.

MAXIMUM INLET PRESSURE Maximum rated gage pressure applied to pump inlet.

MAXIMUM LOAD POINT Point on route where total number of passengers carried is maximum.

MAXIMUM MINING YIELD Maximum volume of groundwater that can be extracted.

MAXIMUM POOL 1. Highest acceptable water surface elevation with all factors affecting safety of structure considered; 2. Highest water surface elevation resulting from computed routing of inflow design flood through reservoir under established operating criteria.

MAXIMUM PROBABLE FLOOD Flood that may be expected from most severe combination of critical meteorological and hydrologic conditions (i.e., largest flood for which there is any reasonable expectancy in this climatic era) that are reasonably possible in drainage basin under study.

MAXIMUM PROBABLE PRECIPITATION Greatest depth of precipitation for given duration that's physically possible over given size storm area at particular geographical location at certain time of year.

MAXIMUM PROBABLE RAINFALL Maximum precipitation likely to happen over confirmed span in drainage basin depending on historical data.

MAXIMUM RATED RPM Engine speed measured in revolutions per minute (rpm) where peak net brake power is developed.

MAXIMUM RECALL Phase when local detection is not present or inoperative.

MAXIMUM RESIDUE LEVEL Enforceable limit on food pesticide levels in some countries.

MAXIMUM SOUND LEVEL Highest time-weighted sound level measured.

MAXIMUM SPILLWAY DISCHARGE Spillway discharge when reservoir is at maximum designed water surface elevation.

MAXIMUM STREAM FLOW Maximum speed of discharge from stream at point over given interval.

MAXIMUM SUSTAINED SURFACE WIND Highest one-minute average wind (at elevation of 10 meters with unobstructed exposure) associated with that weather system at particular point in time.

MAXIMUM SUSTAINED YIELD Largest yield that can be taken from source over indefinite period.

MAXIMUM TEMPERATURE Highest temperature recorded during specified period of time.

MAXIMUM TOLERATED DOSE Maximum dose that animal species can tolerate for major portion of its lifetime without significant impairment or toxic effect other than carcinogenicity.

MAXIMUM UNAMBIGUOUS RANGE Range from radar at which echo can be known unquestionably as being at that range.

MAXIMUM UNAMBIGUOUS VELOCITY Highest radial velocity that can be measured unambiguously by pulsed Doppler radar.

MAXIMUM UNIT WEIGHT Dry unit weight defined by peak of compaction curve.

MAXIMUM WATER SURFACE 1. Highest acceptable water surface elevation with all factors affecting safety of structure considered; 2. Highest water surface elevation resulting from computed routing of inflow design flood through reservoir under established operating criteria.

MAXIMUM WATER-HOLDING CAPACITY Approximate volume of water which can be permanently kept in ground in opposition to gravity.

MAXIMUM WAVE Highest wave in wave group.

MAXIMUM WIND Most severe wind for generating waves reasonably possible at particular reservoir.

MAXIMUM WORKING PRESSURE Maximum pressure for which system is designed.

MEAN Arithmetic average of set of observations.

MEAN ANNUAL FLOOD Mean average of maximum flood discharges experienced in particular waterway.

MEAN ANNUAL FLOW Average flow for individual year or multi-year period of interest for waterway.

MEAN ANNUAL PRECIPITATION Average precipitation for year based on whole period of record or for selected period.

MEAN ANNUAL RUNOFF Average amount of water that flows per year, expressed either as depth across entire drainage basin or as volume of water flowing past given point.

MEAN ANNUAL TEMPERATURE Average temperature for entire year at any given location.

MEAN AREAL PRECIPITATION Average rainfall over given area expressed as average depth over area.

MEAN CELL RESIDENCE TIME Average time that microorganism will spend in activated sludge process or how long sludge is in system.

MEAN CONCENTRATION OF SUSPENDED SEDIMENT Time-weighted concentration of suspended sediment passing stream cross-section during given time period.

MEAN DAILY DISCHARGE Average discharge of any specified calendar day.

MEAN DAILY TEMPERATURE Average of highest and lowest temperatures during 24-hour period.

MEAN DEPTH Average depth of water in stream channel or conduit that's equal to cross-sectional area divided by surface width.

MEAN DISCHARGE Arithmetic mean of individual daily mean discharges of stream during specific period.

MEAN DOPPLER VELOCITY Reflectivity-weighted average velocity of targets in given pulse resolution volume.

MEAN FILTRATION RATING Measurement of average size of openings in filter element.

MEAN FLOW Typical discharge for certain interval.

MEAN GRADIENT Average gradient or inclination.

MEAN HIGH TIDE Average altitude of all high tides recorded at given place over certain period of time (e.g., 19-year period).

MEAN HIGH WATER Average of all high water heights observed over National Tidal Datum Epoch.

MEAN LAYER CAPE Calculated using parcel consisting of Mean Layer values of temperature and moisture from lowest 100 mb above ground level.

MEAN LAYER LIFTED INDEX Lifted Index calculated using parcel consisting of Mean Layer values of temperature and moisture from lowest 100 mb above ground level.

MEAN LOW TIDE Average altitude of all low tides recorded at given place over certain period of time (e.g., 19-year period).

MEAN LOW WATER Average of all low water heights observed over National Tidal Datum Epoch.

MEAN LOWER LOW WATER Average of lower low water height of each tidal day observed over National Tidal Datum Epoch.

MEAN NUMBER Average number.

MEAN PANEL RATING Ride quality, of individual highway pavement site, measured in panel rating study.

MEAN RANGE OF TIDE Average difference in height between mean high water and mean low water.

MEAN ROUGHNESS INDEX Roughness statistic calculated by averaging international roughness index (IRI) values computed for left and right wheel path profiles.

MEAN SEA LEVEL 1. Arithmetic mean of hourly heights observed over National Tidal Datum Epoch; 2. Arithmetic mean of hourly water elevations observed at given place over certain period of time (e.g., 19-year period).

MEAN SLOPE VARIANCE Average slope variance for inner and outer pavement wheel paths.

MEAN TIDE LEVEL Arithmetic mean of mean high water and mean low water.

MEAN VELOCITY 1. Mean velocity is discharge divided by area of water at cross-section; 2. Time average of velocity of flow at fixed point over interval of time.

MEAN VELOCITY POSITION Point on vertical section where real rate is equivalent to mean rate for flows using free surface.

MEAN WAVE DIRECTION Direction corresponding to energy of dominant wave period.

MEANDER 1. Big bend and loops in river channel as river snakes through flat land area; 2. Winding, crooked, complicated path or curve.

MEANDER AMPLITUDE Distance between points of maximum curvature of successive meanders of opposite phase in direction normal to general course of meander belt, measured between centerlines of channels.

MEANDER BELT 1. Part of valley bottom where stream sometimes shifts its channel; 2. Area between lines drawn tangential to extreme limits of fully developed meanders.

MEANDER BREADTH Distance between lines used to define meander belt.

MEANDER LENGTH Distance of meanders between corresponding points of successive meanders.

MEANDERING CHANNEL Channel that takes twists and turns over its length.

MEANDERING LINE Line of high water mark where constant vegetation ends.

MEANDERING STREAM Alluvial stream characterized in planform by series of pronounced alternating bends.

MEASURE OF EXPOSURE Measurable characteristic of stressor (e.g., specific amount of mercury in body of water) used to help quantify exposure of ecological entity or individual organism.

MEASURED CEILING Ceiling classification applied when ceiling value has been determined by instrument, (e.g., ceilometer, ceiling light) or by known heights of unobscured portions of objects, other than natural landmarks, near runway.

MEASURED VARIABLE Physical quantity, property or condition to be measured.

MEASUREMENT CONTRACT Where payment of contract is founded on finished work measurement.

MEASUREMENT ENDPOINT Measurable characteristic of ecological entity that can be related to assessment endpoint.

MEASUREMENT FOR PAYMENT Determination of work performed, materials or services for contract payment.

MEASUREMENT STANDARD Physical quantity that serves as basis for measurement of quantity.

MEASURING Ascertain size, amount or degree of something.

MEASURING POINT Arbitrary permanent reference point from which distance to water surface in well is measured to obtain water level.

MEASURING WEIR Device for measuring rate of flow of water where height of water above weir crest is used to determine rate of flow.

MECHANIC'S LIEN Security interest in title to property for benefit of those who have supplied labor or materials that improve property.

MECHANICAL AERATION Use of mechanical energy to inject air into water to cause water (e.g., waste stream) to absorb oxygen.

MECHANICAL AERATOR Machine that mixes air and water so that oxygen can be absorbed into water.

MECHANICAL AGITATION Mechanical device to add atmospheric oxygen into liquid.

MECHANICAL ANALYSIS Analysis by screening soil to calculate grain size percentage composition.

MECHANICAL AND THERMAL INTEGRITY Ability of converter to continue to operate at its previously determined efficiency and light-off time and be free from exhaust leaks when subject to thermal and mechanical stresses representative of intended application.

MECHANICAL COLLECTOR Apparatus that divides entrained debris from gas via of inertial and/or gravitational forces.

MECHANICAL CONTROL 1. Controlling pests and diseases with help of mechanical measures; 2. Using man-made constructions to reduce erosion.

MECHANICAL EFFICIENCY Ratio of input energy to output energy.

MECHANICAL ENGINEER Engineers that design power-producing and power-using machines.

MECHANICAL ENGINEERING Subfield of engineering concerned with designing and building machines and mechanical systems.

MECHANICAL FOAM Fire fighting foam produced by mixing foam concentrate with water at appropriate concentration and then aerating and agitating solution to form bubble structure.

MECHANICAL INTEGRITY Absence of significant leakage within injection tubing, casing or packer (known as internal mechanical integrity) or outside of casing (known as external mechanical integrity).

MECHANICAL OPERATOR Apparatus that causes some other mechanism to actuate.

MECHANICAL PIPE JOINT Joint that's flexible.

MECHANICAL RAKE Apparatus that clean debris from conduit intake racks.

MECHANICAL SEPARATION Separation of waste into various components using mechanical methods.

MECHANICAL SERVICE PIPE Welded steel pipe.

MECHANICAL SERVICES Building services (e.g., heating, ventilating, air conditioning, plumbing, electrical distribution, gas installations).

MECHANICAL SPECIFICATIONS Product description related to installation, operation and maintenance.

MECHANICAL TURBULENCE Random irregularities of fluid motion in air caused by buildings or other non-thermal processes.

MECHANICAL UNIT Mechanical systems such as air conditioning, fire prevention, air compressor, elevator, building automation system, sewage plant, water pump and other mechanical systems.

MECHANICALLY STABILIZED EARTH WALL Soil constructed with artificial wall reinforcing.

MECHANICALLY-CLEANED SCREEN Reduce operating and maintenance problems and to increase efficiency of wastewater screening.

MECHANISTIC MODEL Model that assumes complex system can be understood by examining workings of its individual parts and manner they are coupled.

MEDIA 1. Specific environments (e.g., air, water, soil) that are subject of regulatory concern and activities; 2. Any of range of substances chosen and set to reach measured result.

MEDIA MIGRATION Effluent stream material composed of material for filter medium.

MEDIAN 1. Middle value in distribution, above and below which lie equal number of values; 2. 50th percentile; 3. Line segment joining vertex of triangle to midpoint of opposing side.

MEDIAN BARRIER Barriers that physically separate opposing traffic streams and help stop vehicles travelling into opposing traffic lanes.

MEDIAN DIAMETER Midpoint of grain-size distribution (i.e., 50% (by weight) of sediment is coarser and 50% is finer than median grain size).

MEDIAN OPENING Opening in restrictive median for u-turn, crossing and turning traffic.

MEDIAN STREAM FLOW Middle rate of flow of water past given point for which there have been several greater and lesser rates of flow occurring during specified period.

MEDIAN TOLERANCE LIMIT Concentration of pollutants where 50% of animals in test are able to survive for specified period of exposure.

MEDICAL EMERGENCY Acute injury or illness that poses immediate risk life or long-term health.

MEDICAL SURVEILLANCE Periodic comprehensive review of worker's health status.

MEDICINAL SPRING Water perceived to have therapeutic properties.

MEDIUM Environmental vehicle by which pollutant is carried to receptor (e.g., air, surface water, soil, groundwater).

MEDIUM FREQUENCY Portion of radio frequency spectrum from 0.3 to 3 MHz.

MEDIUM GRAINED Grain diameters between 0.016 inch to 0.079 inch for sedimentary rocks or 0.039 inch to 0.197 inch for igneous or metamorphic rocks.

MEDIUM GRAINED SAND-SIZED Less than 10% of fine and coarse grained sand sizes are present.

MEDIUM RANGE In forecasting, three to seven days in advance.

MEDIUM SAND Sediment particles with diameters between 0.01 and 0.02 in.

MEDIUM-BEDDED 0.3 ft to 1 ft thick.

MEDIUM-HEIGHT DAM Dam between 100 and 300 feet high.

MEDIUM-SIZE PLANT Processes between 8,200 and 23,000 lbs/day of raw materials.

MEDIUM-SIZE WATER SYSTEM Water system that serves 3,300 to 50,000 customers.

MEDIUM-TEMPERATURE WATER HEATING SYSTEM Water heating system with boiler that heats water to between 250 °F and 350 °F before it is supplied to heating system.

MEDIUM-THICK ARCH DAM Arch dam with base thickness to structural height ratio between 0.2 and 0.3.

MEETING Place or instance of junction or intersection.

MEGA 1. Prefix symbolizing 10^6 (i.e., one million); 2. Very large.

MEGAELECTRON VOLT Unit of energy used to describe total energy carried by particle or photon.

MEGAGRAM Unit of mass equal to 1,000,000 grams.

MEGAHERTZ Unit of frequency equal to 1,000,000 hertz.

MEGAJOULE Unit of energy equal to 1,000,000 joules.

MEGALOPOLIS Large conurbation, where two or more large cities have sprawled outward to meet, forming something larger than metropolis; megacity.

MEGANEWTON Unit of force equal to 1,000,000 newtons.

MEGASCOPIC Visible to naked eye.

MEGAWATT Unit of power equal to 1,000,000 watts.

MEGAWATT-HOUR One million watt-hours of electrical energy.

MEGOHM Unit of resistance equal to 1,000,000 ohms.

MEGOHMMETER Instrument that measures extremely high resistances, in megaohms.

MELTDOWN Severe overheating of core of nuclear reactor resulting in core melting and radiation escaping.

MELTING LEVEL Altitude which ice crystals and snowflakes melt as they descend through atmosphere.

MELTING POINT 1. Temperature where solid material changes to liquid; 2. Temperature where solid and liquid phases of substance are in equilibrium.

MELTWATER Water from melting ice or snow.

MEMBRANE Sheet or thin zone or facing, made of flexible impervious material (e.g., asphaltic concrete, plastic concrete, steel, wood, copper, plastic).

MEMBRANE BARRIER Thin material that prevents flow of gas or water.

MEMBRANE FILTER Thin micro-porous material of specific pore size used to filter bacteria, algae and other very small particles from water.

MEMBRANE POISONING Low mobility ion absorption into membrane.

MEMBRANE PROCESSES Separation processes where characteristics of membrane separate components of solution or suspension.

MEMBRANE SELECTIVITY Ratio of permeability between two (2) species.

MEMBRANE WATERPROOFING Coating outside surfaces with waterproof material.

MEMORANDUM OF UNDERSTANDING Formal document that states intentions and/or responsibilities of signatory parties.

MENISCUS Curved top of column of liquid in small tube caused by surface tension.

MENSURATION Study of measurement (i.e., areas, volumes).

MERCAPTANS Any of class of organic compounds of sulphur.

MERCURY Silvery-colored, toxic, heavy metal, metallic chemical element, liquid at room temperature, with atomic number 80 that's highly toxic if breathed or swallowed.

MERCURY BAROMETER Instrument for measuring atmospheric pressure.

MERCURY CONTACT SWITCH Switch comprising two spaced apart electrical contacts and sealed mercury that opens or closes electrical circuit when making contact.

MERCURY GAUGE Column of mercury in glass tube that represents pressure of fluid.

MERGE To blend gradually into something else (e.g., traffic streams).

MERIDIAN Imaginary line on Earth's surface passing through both geographic poles and through any given point on planet that's line of longitude.

MERIDIONAL FLOW Large-scale atmospheric flow where north-south component (i.e., longitudinal, along meridian) is pronounced.

MEROMICTIC LAKE Lake where partial content remains unmixed with main body.

MEROMIXIS When water at specific depth does not circulate due to its high density.

MESA Flat area of land or plateau higher than other land, with one or more cliff like edges.

MESH 1. Network of computational points (nodes) linked together by finite element connection tables to form digital representation of modeled area's geometry; 2. Structure made of connected strands of material and evenly spaced openings between them.

MESH SCREEN Screen with small openings that can be used in windows or patios to keep bugs and outdoor elements out.

MESOCLIMATE Climate of small area of Earth's surface which may differ from general climate of district.

MESOCYCLONE Storm-scale region of rotation, typically around 2-6 miles in diameter and often found in right rear flank of supercell.

MESOHALINE Moderately salty or brackish.

MESOHIGH Relatively small area of high atmospheric pressure that forms beneath thunderstorm.

MESONET Regional network of observing stations (usually surface stations) designed to diagnose mesoscale weather features and their associated processes.

MESOPAUSE Top of mesosphere, corresponding to level of minimum temperature in atmosphere found at 70 to 80 km.

MESOPELAGIC Pelagic zone of oceans between photic epipelagic and aphotic bathypelagic zones.

MESOPHILIC BACTERIA Organism that thrives in reasonable temperature range between 68 °F and 113 °F.

MESOPHILIC DIGESTION Anaerobic digester that operates at mesophilic temperature range between 86 °F and 100 °F.

MESOPHILIC MICROORGANISMS Microorganisms with optimum development temperature between 68 °F and 113 °F.

MESOPHILIC RANGE Temperature range where mesophilic bacteria are most successful in digestion of sewage sludge (86 °F and 100 °F).

MESOPHILOUS Living and thriving at moderate temperatures.

MESOPHREATOPHYTE Phreatophyte (i.e., plants with long roots) not resistant to alkali or drought.

MESOPHYTE Any plant growing where moisture and aeration conditions have average supply of water.

MESOSAPROBIC Running water partially polluted.

MESOSCALE Size scale referring to weather systems smaller than synoptic-scale systems but larger than storm-scale systems.

MESOSCALE CONVECTIVE COMPLEX Large convective system generally round or oval-shaped, which normally reaches peak intensity at night.

MESOSCALE CONVECTIVE SYSTEM Complex of thunderstorms which becomes organized on scale larger than individual thunderstorms and normally persists for several hours or more.

MESOSCALE DISCUSSION When conditions actually begin to shape up for severe weather, Storm Prediction Center often issues statement anywhere from roughly half hour to several hours before issuing weather watch.

MESOSCALE HIGH WINDS High winds usually follow passage of organized convective systems and are associated with wake depressions or strong mesohighs.

MESOSCOPIC MODEL Model that includes combinations or approximations of elements from both macroscopic and microscopic models.

MESOSPHERE Atmospheric shell between about 20 km and about 70 to 80 km, extending from top of stratosphere (stratopause) to upper temperature minimum that defines mesopause (base of thermosphere).

MESOTROPHIC Area that provides (e.g., reservoir, lakes) moderate quantities of nutrients and minerals and are moderately productive in terms of aquatic animal and plant life.

METABOLISM Complete set of chemical reactions that occur in living cells.

METABOLITE Substance produced in metabolic reaction or by biological processes.

METACENTER Midway point between ship's center of buoyancy when upright and its center of buoyancy when tilted.

METACENTRIC HEIGHT Measurement of initial static stability of floating body.

METADATA Information that describes content, quality, condition and other characteristics of data.

METAL DETECTOR Apparatus that uses electromagnetic induction to detect metal.

METALIMNION Middle layer of thermally stratified lake or reservoir with rapid temperature decrease with depth.

METAMORPHIC 1. Rock compressed or changed by pressure, heat or water; 2. Rock formed from preexisting rock altered ("baked") by high temperatures and pressures, causing minerals to recrystallize but not melt.

METAMORPHIC ROCK Rock that has formed in solid state in response to pronounced changes of temperature, pressure and chemical environment.

METAMORPHIC STAGE Stage of development that organism exhibits during its transformation from immature form to adult form.

METAMORPHIC WATER Water driven out of rocks by metamorphism.

METAMORPHISM Process where rocks are changed into other forms by application of heat and/or pressure.

METAR International code used for reporting, recording and transmitting weather observations.

METASOMATISM Process where bulk chemical composition of rock is changed by introduction of components from external source, especially by hydrothermal fluid.

METEOGRAM Graphical depiction of trends in meteorological variables (e.g., temperature, dew point, wind speed and direction, pressure).

METEORIC WATER Water derived from precipitation.

METEOROGRAPH Device used to measure various meteorological conditions, such as temperature and humidity.

METEOROLOGIC ELEMENTS State of atmosphere, defined by measurement of six meteorologic elements (i.e., air temperature, barometric pressure, wind speed, humidity, cloudiness and precipitation).

METEOROLOGICAL HOMOGENEITY Climates and orographic influences that are alike or similar.

METEOROLOGICAL STORM Meteorological expression for precipitation, synoptic conditions, pressure condensation, pressure and specific humidity.

METEOROLOGICAL WATER Water from atmosphere.

METEOROLOGIST Person who studies meteorology.

METEOROLOGY Science dealing with atmosphere and its phenomena.

METER 1. Unit of length equal to 3.281 feet; 2. Device used for measuring amount of material consumed, passed or conveyed.

METER BOX Device used to house meter.

METER BYPASS Connection made prior to meter that redirects flow of water around meter and back into domestic plumbing.

METER RATE Cost based on quantity customer consumes as measured by service meter.

METER STOP Valve installed in water service pipe for control of flow of water.

METERED FLOW Controlled flow.

METERED SYSTEM Distribution system where meters can be found at all locations in system.

METERING PIN Needle-type, flow-control device in metering orifice used to vary fluid flow.

METHANE Colorless, non-poisonous, flammable, greenhouse gas emitted during production and transport of coal, natural gas and oil or from decomposition of organic wastes in municipal solid waste landfills and raising of livestock.

METHANE FERMENTATION Converts organic matter to methane and carbon dioxide under anaerobic conditions.

METHANOGENS Any of various archaebacteria capable of methanogenesis.

METHANOL Colorless, toxic, inflammable liquid, used as solvent, antifreeze, alternative fuel or as gasoline additive.

METHOD 18 EPA test method which uses gas chromatographic techniques to measure concentration of volatile organic compounds in gas stream.

METHOD 24 EPA reference method to determine density, water content and total volatile content (water and VOC) of coatings.

METHOD 25 EPA reference method to determine VOC concentration in gas stream.

METHOD CODE One-character code that identifies analytical or field method used to determine value stored in National Water Information System.

METHOD COMPARABILITY Characteristics that allow data produced by multiple methods to meet or exceed data-quality objectives of primary or secondary data users.

METHOD OF CUBATURES Method of computing discharge in tidal estuaries based on conservation of mass equation.

METHOD VALIDATION Process of substantiating method to meet certain performance criteria for sampling and/or analytical and/or data handling operations.

METHOXYCHLOR Pesticide that causes adverse health effects in domestic water supplies and is toxic to freshwater and marine aquatic life.

METHYL ORANGE ALKALINITY Measure of total alkalinity in water sample where color of methyl orange reflects change in level.

METHYL TERTIARY BUTYL ETHER Organic compound used to increase octane numbers of gasoline, through added oxygen, as anti-knock additive.

METHYL-ORANGE ALKALINITY Measure of total alkalinity in water sample.

METOLACHLOR Aniline derivative and chloroacetanilide herbicide, highly effective against grasses.

METRIBUZIN Herbicide found in drinking water supply with maximum acceptable concentration of 0.08 mg/l.

METRIC Biological attribute, some feature or characteristic of biotic assemblage, that reflects ambient conditions, especially influence of human actions on these conditions.

METRIC TON Unit of weight equivalent to 2205 lb.

METROPOLIS Large, busy city.

METROPOLITAN PLANNING ORGANIZATION Policy board of organization created and designated to carry out metropolitan transportation planning process.

METROPOLITAN STATISTICAL AREA Geographical region with relatively high population density at its core and economic ties throughout area.

MICA Group of minerals that form thin, platy flakes, typically with shiny surfaces, especially common in metamorphic rocks.

MICRO 1. Prefix symbolizing 10^{-6}; 2. Prefix symbolizing one millionth; 3. Small.

MICRO-FILTRATION Uses membrane that only allows particles smaller than 0.1 microns to pass through it.

MICROBAROGRAPH Instrument designed to continuously record barometer's reading of very small changes in atmospheric pressure.

MICROBE Any microorganism.

MICROBIAL ACTIVITY Microbiological processes of microorganisms is potential indicator of quality.

MICROBIAL FILM Film of growths covering filtration medium.

MICROBIAL GROWTH Amplification or multiplication of microorganisms (e.g., bacteria, algae, diatoms, plankton, fungi).

MICROBIAL METALLURGY Bacterium that pulls precious metals ions out of solution.

MICROBIAL PESTICIDE Microorganism used to kill pest, but is of minimum toxicity to humans.

MICROBIOLOGICAL GUIDELINES Criteria indicating microbiological condition when there are no established microbiological standards.

MICROBIOLOGY Branch of biology that deals with microorganisms, especially their effects on man and other living organisms.

MICROBIOTA Microbial flora harbored by normal, healthy individuals.

MICROBURST Convective downdraft with affected outflow area of less than 2.5 miles wide and peak winds lasting less than 5 minutes.

MICROCHEMICAL Of or pertaining to microchemistry.

MICROCLIMATE 1. Climate of small area, particularly that of living space of certain species, group or community; 2. Climate around tree or shrub or stand of trees.

MICROCLIMATOLOGY Branch of climatology that studies microclimates.

MICROENVIRONMENTAL METHOD Method for sequentially assessing exposure for series of microenvironments that can be approximated by constant concentrations of stressor.

MICROENVIRONMENTS Well-defined surroundings (e.g., home, office, kitchen) that can be treated as uniform in terms of stressor concentration.

MICROFARAD One millionth of farad.

MICROGRAMS PER GRAM Unit expressing concentration of chemical constituent as mass (micrograms) of element per unit mass (gram) of material analyzed.

MICROGRAMS PER KILOGRAM Unit expressing concentration of chemical constituent as mass (micrograms) of constituent per unit mass (kilogram) of material analyzed that's equivalent to 1 part per billion.

MICROGRAMS PER LITER Unit expressing concentration of constituents in solution as weight (micrograms) of solute per unit volume (liter) of water that's equivalent to 1 part per billion.

MICROHABITAT Small, specialized and effectively isolated location.

MICROHENRY One millionth of henry.

MICROMETEOROLOGY Portion of meteorology dealing with observation and explanation of small scale weather and weather forecasting for local area up to several kilometers in diameter.

MICROMETER Unit of length equal to one millionth of meter.

MICRON Unit of length equal to one thousandth of a millimeter.

MICRONUTRIENT Mineral, vitamin or other substance essential, even in very small quantities, for growth or metabolism.

MICROORGANISM Organism too small to be seen by unaided eye.

MICRORELIEF Very small relief (i.e., difference of elevations on surface).

MICROSAND Aggregate, basically free of clay and shale, that passes No. 100 (0.0059 in.) sieve.

MICROSCALE Pertaining to meteorological phenomena (e.g., wind circulations, cloud patterns) that are less than 2 km in horizontal extent.

MICROSCOPIC So small that it can only be seen using microscope.

MICROSCOPIC MODEL Models of vehicular traffic dynamics.

MICROSCREEN Screen for filtering water from wastewater treatment plants.

MICROSIEMENS PER CENTIMETER Unit expressing amount of electrical conductivity of solution as measured between opposite faces of centimeter cube of solution at specified temperature.

MICROSITE Small area with distinctive ecological characteristics.

MICROSTRAINER Wire cloth mounted on revolving drum partially submerged in water.

MICROSWITCH Electrical switch that operates with very little travel of (or pressure on) actuator.

MICROSYSTEM IRRIGATION Method of precisely applying irrigation water to immediate root zone of target plant at very low rates.

MICROTUNNEL Microscale tunnel.

MICROTUNNELLING Digging technique used to construct small tunnels.

MICROWAVE Type of electromagnetic radiation with wavelengths between those of infrared radiation and radio waves.

MICROWAVE BURST Radiowave signal associated with optical and/or X-ray flares.

MICROWAVE LANDING SYSTEM Radio guidance system to assist aircraft in landing at large airports.

MID-FLAME WIND Wind measured at midpoint of flames, considered to be most representative of speed of wind affecting fire behavior.

MID-LEVEL CLOUDS Term used to signify clouds with bases between 6,500 and 23,000 feet.

MID-LEVEL COOLING Local cooling of air in middle levels of atmosphere (roughly 8,000 to 25,000 feet), which can lead to destabilization of entire atmosphere if all other factors are equal.

MIDBLOCK CROSSING Locations between intersections with marked crosswalks.

MIDBLOCK STOP Transit stop located at location away from intersections.

MIDDLE CLOUDS Term used to signify clouds with bases between 6,500 and 23,000 feet.

MIDDLE LATITUDE AREAS Latitude belt roughly between 30° and 60° North and South of Equator.

MIDGE Small fly in family Chironomidae where larval (juvenile) life stages are aquatic.

MIE SCATTERING Any scattering produced by spherical particles whose diameters are greater than 1/10 wavelength of scattered radiation.

MIGMATITE Rock composed of complex mixture of metamorphic rock and igneous granitic rock.

MIGRATION Movement from one place to another.

MIGRATORY Moving from one area to another on seasonal basis.

MIL Unit of length equal to 0.001 in.

MILD SLOPE Conduit slope less than critical slope.

MILDEW Growth of minute powdery or webby fungi, whitish or of different colors, found on various diseased or decaying substances.

MILE Unit of length equal to 5280 feet or 1760 yards.

MILE PER HOUR Unit of velocity equal to 5280 ft/hr.

MILE YARD Excavation payment for 1 yd^3 over 1 mile horizontal distance.

MILE, NAUTICAL Unit of length equal to 6076 feet.

MILEAGE BLOCK Daily distance traveled in block.

MILESTONE Measurable event, action, state or goal that marks point of achievement in person's life or career, in history of nation, etc.

MILESTONE SCHEDULE Summary level schedule that allows project team leader to review and identify significant milestones.

MILITARY CREST Ridge that interrupts view between valley and hilltop.

MILITARY OPERATIONS AREA Airspace established outside Class A airspace to separate or segregate certain nonhazardous military activities from IFR Traffic and to identify for VFR traffic where these activities are conducted.

MILL Monetary cost and billing unit used by utilities that's equal to 1/1,000 of U.S. dollar (or 1/10 of one cent).

MILLED REFUSE Solid waste that has been grinded, pressed or milled.

MILLI Prefix symbolizing 10^{-3}.

MILLIBAR Unit of pressure equal to 1,000 $dynes/cm^2$.

MILLIFARAD Unit of electrical capacitance equal to 1/1,000 of farad.

MILLIGRAM Unit of mass equal to 1/1,000 of gram.

MILLIGRAM PER LITER Equivalent to 1 part per million.

MILLIGRAMS PER CUBIC METER Mass in micrograms of substance contained within m^3 of another substance or vacuum.

MILLIGRAMS PER LITER Unit expressing concentration of chemical constituents in solution as weight (milligrams) of solute per unit volume (liter) of water (e.g., 0.001 g/L) that's equivalent to one part per million (ppm).

MILLIHENRY Unit of inductance equal to 1/1,000 of henry.

MILLILITER Unit of volume equal to 1/1,000 of litre.

MILLIMETER Unit of length equal to 1/1,000 of meter.

MILLIMETER OF MERCURY Unit of pressure equal to 133.32 pascals.

MILLIMICRON 1. Unit of length equal to 1/1,000 of micron or 1/1,000,000 of millimeter; 2. Nanometer.

MILLION ACRE-FEET Volume of water that would cover 1 million acres to depth of 1 foot.

MILLION GALLONS PER DAY Rate of flow of water commonly used for wastewater discharges that's equivalent to 1,000,000 gal/day, 1.547 ft^3/s, 133,680.56 ft^3/day or 3.0689 acre-ft/day.

MILLIPASCAL Unit of pressure equal to 1/1,000 of pascal.

MILLISECOND DELAY Type of detonating delay cap, with definite but extremely short interval between initiation or passing of current and explosion hat fires from 0.02 to 0.5 s after firing current passes through it.

MIMIC DIAGRAM Line diagram of pipe system or items of equipment.

MINE Excavation from which ore or solid minerals are taken.

MINE BACKFILL WELLS Class V wells that inject water, sand, mill tailings or other mining byproducts in order to control subsidence caused by mining, to dispose of mining byproducts or to fill sections of mine.

MINE DEWATERING Action of removing groundwater from mine.

MINE DRAINAGE 1. Water removed from mine; 2. Formed when pyrite, iron sulfide, is exposed and reacts with air and water to form sulfuric acid and dissolved iron.

MINE TAILINGS DAM Industrial waste dam where waste materials come from mining operations or mineral processing.

MINE WATER Water that collects in mine and brought to surface.

MINER'S INCH Rate of discharge through orifice 1 in^2 under specific head.

MINERAL Any naturally occurring inorganic material that has definite chemical composition and characteristic physical properties.

MINERAL DEPOSIT Natural concentration of minerals in Earth's crust.

MINERAL FIBER INSULATION Insulating material composed of stone, slag or glass.

MINERAL SOIL Soil composed predominantly of mineral rather than organic materials that's less than 20% organic material.

MINERAL SPRING Naturally occurring springs that produce water containing minerals.

MINIBUS Small bus, especially one used as transport for small groups.

MINIMIZATION Comprehensive program to minimize or eliminate wastes, usually applied to wastes at their point of origin.

MINIMUM ACCEPTABLE PRESSURE Lowest water pressure permitting operation at most hydraulically remote point or component.

MINIMUM ANNUAL FLOOD Lowest elevation flood recorded.

MINIMUM BURST PRESSURE Lowest pressure conduit can handle before rupturing.

MINIMUM CLEARANCE WIDTH Smallest distance in floor space (e.g., wheelchair clearance).

MINIMUM DISCERNIBLE SIGNAL 1. Smallest input signal that will produce detectable signal at output; 2. Minimal amount of back scattered energy required to produce target on radar screen.

MINIMUM ENERGY LINE Energy line of critical flow.

MINIMUM FACTOR Limits tree distribution and growth.

MINIMUM FLOW Flow of water in stream, that will sustain aquatic population of agreed-upon levels, during prolonged dry weather.

MINIMUM GRADE Slope of gravity sewer to maintain minimum self-cleaning velocity.

MINIMUM GREEN First timed portion of green interval.

MINIMUM LEVEL 1. Level at which entire analytical system gives recognizable mass spectra and acceptable calibration points when analyzing for pollutants of concern; 2. Lowest point at which calibration curve is determined.

MINIMUM OPERATING LEVEL Lowest level that reservoir is drawn down under normal operating conditions.

MINIMUM RECALL Parameter that results in phase being called and timed for at least its minimum green time whether vehicle is present or not.

MINIMUM REPORTING LEVEL Smallest measured concentration of constituent that may be reliably reported using given analytical method.

MINIMUM TEMPERATURE Lowest temperature recorded during specified period of time.

MINIMUM TURNING RADIUS Circular arc formed by turning path of front outside vehicle tire.

MINIMUM WARNING TIME Minimum time required for warning system.

MINIMUM YIELD STRENGTH Tensile stress required to produce total elongation of 0.5%.

MINING Removal of soil or rock having value because of its chemical composition.

MINING OF AQUIFER Withdrawal over period of time of groundwater that exceeds rate of recharge of aquifer.

MINING WASTE Residues resulting from extraction, beneficiation and processing of raw materials from Earth.

MINING WATER USE Water use during quarrying rocks and extracting minerals from land.

MINOR CHANGE Change that does not require creation of new record.

MINOR FLOODING Indicating minimal or no property damage but possibly some public inconvenience.

MINOR GEOMAGNETIC STORM Storm for which Ap index was greater than 29 and less than 50.

MINOR INTERCHANGE Interchanges where traffic is local and very light.

MINOR MOVEMENT Vehicular movement performed at unsignalized intersections.

MINOR SOURCE New emissions sources or modifications to existing emissions sources that do not exceed NAAQS emission levels.

MINOR STREET Street that carries lower vehicular traffic volume.

MINOR TIDAL OVERFLOW Minor flooding caused by high tides that results in little if any damage.

MINORS Publicly owned treatment works with flows less than 1 million gal/day.

MINUTE Unit of time equal to 60 seconds.

MINUTE OF ARC Unit of angle equal to 60 seconds of arc.

MISCELLANEOUS ACM Interior asbestos-containing building material or structural components, members or fixtures (i.e., floor and ceiling tiles).

MISCELLANEOUS MATERIALS Interior building materials on structural components (i.e., floor and ceiling tiles).

MISCELLANEOUS SITE Site where streamflow, sediment, and/or water-quality data or water-quality or sediment samples are collected once or more often on random or discontinuous basis to provide better areal coverage for defining hydrologic and water-quality conditions over broad area in river basin.

MISCIBLE LIQUIDS Two or more liquids that can be mixed and will remain mixed under normal conditions.

MISOSCALE Scale of meteorological phenomena that ranges in size from 40 meters to about 4 kilometers.

MISSED APPROACH Failure of pilot to properly land aircraft due to various factors, requiring go-around for next attempt.

MISSED DETECTION Situation that occurs when test indicates that tank is "tight" when in fact it is leaking.

MISSION OF HYDROLOGIC SERVICES PROGRAM To provide river and flood forecasts and warnings for protection of life and property and to provide basic hydrologic forecast information for Nation's economic and environmental well-being.

MISSION OF NATIONAL WEATHER SERVICE To provide weather and flood warnings, public forecasts and advisories for all of United States, its territories, adjacent waters and ocean areas, primarily for protection of life and property.

MIST Water or other liquid, finely suspended in air, measuring 40 to 500 micrometers.

MISUSE OF LAND Land usage that might adversely affect community interests.

MITIGATION Action of reducing severity, seriousness or painfulness of something.

MITIGATION MEASURES 1. Methods or plans to reduce, offset or eliminate adverse project impacts; 2. Actions taken to avoid, reduce or compensate for effects of human-induced environmental damage.

MIX-IN-PLACE Stabilization technique where soil is mixed via mobile mixer.

MIXED BOUNDARY Linear combination of head and flux at boundary.

MIXED FACE Digging in dirt and rock in same heading at same time.

MIXED FUNDING Settlements where potentially responsible parties and EPA share cost of response action.

MIXED GLASS Recovered container glass not sorted into categories (e.g., color, grade).

MIXED LAYER Atmospheric layer, usually layer immediately above ground, where pollutants are well mixed by convective or shear-produced turbulence.

MIXED LIQUOR Mixture of activated sludge and water containing organic matter undergoing activated sludge treatment in aeration tank.

MIXED LIQUOR VOLATILE SUSPENDED SOLIDS Concentration of suspended solids, in aeration tank during activated sludge process, that occurs during treatment of wastewater.

MIXED METALS Recovered metals not sorted into categories.

MIXED MUNICIPAL WASTE Solid waste that has not been sorted into specific categories.

MIXED PAPER Recovered paper not sorted into categories.

MIXED PLASTIC Recovered plastic not sorted into categories.

MIXED PRECIPITATION Any of following combinations of freezing and frozen precipitation: snow and sleet, snow and freezing rain or sleet alone.

MIXED-FLOW PUMP Pump that mixes features of radial flow and axial flow pumps.

MIXER 1. Machine outfitted with blades with which it mixes or beats ingredients in bowl below; 2. Device for combining hot and cold water before it emerges from single spout or shower head.

MIXING BASIN Container where agitation is applied to liquid.

MIXING CHAMBER Chamber where several gases, often at different temperatures, converge before going through air pollution control system.

MIXING CHANNEL Channel where two or more flows or chemicals are mixed.

MIXING CYCLE 1. In mixing concrete in batch mixer, time taken for complete cycle; 2. Pattern of seasonal variations in temperature conditions and mixing status in large body of freshwater.

MIXING DEPTH Vertical distance between ground and altitude that pollutants are mixed by turbulence caused by convective currents or vertical shear in horizontal wind.

MIXING FAUCET VALVE Faucet for mixing hot and cold water.

MIXING HEIGHTS Height that parcel of air or column of smoke, will rise, mix or disperse.

MIXING LAYER Zone of flow where exist mixing of fluid by turbulence and strong inflection of mean velocity profile when two or many flows are different velocity.

MIXING RATIO 1. In specified volume, ratio of weight of water vapor to dry air; 2. Number of single material relative to another.

MIXING SPEED Turning speed of mixer drum or paddles.

MIXING TANK Machine container used to blend various components together.

MIXING VALVE Valve that combines flows from several inlets.

MIXING ZONE Area where effluent discharge undergoes initial dilution and is extended to cover secondary mixing in ambient water body.

MIXTURE Something that consists of diverse elements.

MIXTURE DESIGN Mixture where product under investigation is made up of several components or ingredients.

MOBILE BELT Linear or arcuate zone, on regional scale, which has undergone compressional tectonics.

MOBILE COMPACTOR Vehicle that carry and compress solid waste.

MOBILE CONVEYOR Mechanical handling equipment that moves materials.

MOBILE INCINERATOR SYSTEMS Hazardous waste incinerators that can be transported from one site to another.

MOBILE PACKER REFUSE TRUCK Vehicle for loading, compressing and spreading garbage.

MOBILE SOURCE Any non-stationary source of air pollution (i.e., cars, trucks, motorcycles, buses, airplanes, locomotives).

MOBILE WATER Water free to go in almost any direction.

MOBILITY 1. Ability of goods to travel from destination to destination; 2. Degree to which particles of liquid or gas are in movement.

MODAL SPLIT Percentage of travelers using particular type of transportation or number of trips using transportation system.

MODE 1. Most frequently occurring value in distribution; 2. Means of transportation.

MODE SHIFT Change of people or freight to improve overall efficiency of transportation system.

MODEL 1. Assembly of concepts in form of mathematical equations that portray understanding of natural phenomenon; 2. Physical or mathematical representation of process that can be used to predict some aspect of process.

MODEL APPLICATION Physical configuration and operational conditions where traffic analysis tool is used.

MODEL CONSTRUCTION Process of transforming conceptual model into parameterized mathematical form.

MODEL GRID System of connected nodal points superimposed over problem domain to spatially discretize problem domain into cells (finite difference method) or elements (finite element method) for purpose of numerical modeling.

MODEL INPUT Constitutive coefficients, system parameters, forcing terms, auxiliary conditions and program control parameters required to apply computer code to particular problem.

MODEL OUTPUT STATISTICS Short range (6 to 60 hours) guidance package generated from NGM, GFS and ETA models for over 300 individual stations in continental United States that use model output to forecast probability of precipitation, high and low temperature, cloud cover and precipitation amount for many cities across USA.

MODEL PLANT Hypothetical plant design used for developing economic, environmental and energy impact analyses as support for regulations or regulatory guidelines.

MODEL VERIFICATION Procedure of determining if (site-specific) model's accuracy and predictive capability lie within acceptable limits of error by tests independent of calibration data.

MODELING Process of formulating model of system of process by using mathematical equations to simulate and predict real events and processes.

MODELING OBJECTIVES Purpose(s) of model application.

MODERATE EXPOSURE CONDITION Environment where concrete will only occasionally be exposed to moisture and will not be saturated before freezing and where no deicing agents or other aggressive chemicals are used.

MODERATE FLOODING Inundation of secondary roads where transfer to higher elevation necessary to save property.

MODERATE FREQUENCY FLOOD Flood of lesser magnitude than inflow design flood, used for service spillway design when supplemented by separate auxiliary spillway.

MODERATE RAIN Rain with intensity between 0.11 and 0.30 in/hr.

MODERATE RISK OF SEVERE THUNDERSTORMS Severe thunderstorms are expected to affect between 5 and 10% of area.

MODERATELY FRACTURED Spacing 1 ft to 3 ft.

MODERATELY HARD Can be scratched easily with knife, but cannot be scratched by fingernail.

MODERATELY OPEN 0.020 inch to 0.098 inch.

MODERATELY WEATHERED Discoloration is evident and surface is pitted and altered with alterations penetrating well below rock surfaces with 10% to 50% of rock altered.

MODERATOR Substance used to decrease speed of fast neutrons in nuclear reactor and hence increase likelihood of fission.

MODIFICATION 1. Alteration or adjustment to something; 2. Change to organism as result of its environment that's not transmissible to offspring.

MODIFICATION DECISION ANALYSIS Process of determining with confidence whether dam safety deficiencies exist for maximum loading conditions.

MODIFIED AERATION Tank equipped with mechanical surface aerator where biological growth is settled from wastewater.

MODIFIED BIN METHOD Way of calculating required heating or cooling for building based on determining how much energy system would use if outdoor temperatures were within certain temperature interval and then multiplying energy use by time temperature interval typically occurs.

MODIFIED HOMOGENEOUS EARTHFILL DAM Homogeneous earthfill dam that uses pervious material specially placed in embankment to control seepage.

MODIFIED MERCALLI SCALE Earthquake intensity scale that has twelve divisions to assess effect of earthquake on Earth's surface.

MODIFIED SOURCE Enlargement of major stationary pollutant sources is often referred to as modification, implying that more emissions will occur.

MODIFIED VELOCITY Drift and angularity corrected flow velocity.

MODIFIER Additive that changes aspects of substance.

MODULATE To regulate, adjust or adapt.

MODULATED CONTROLS Control rules that provide for continuous degree of control applied to pump or flow regulator as determined by value of some controller variable, such as water depth at node or by time.

MODULUS OF RUPTURE Mechanical parameter indicating material's ability to resist deformation under load.

MODULUS OF SUBGRADE REACTION Reaction pressure sustained by soil sample under rigid plate of standard diameter per unit settlement measured at specified pressure or settlement.

MOHR CIRCLE Graphical representation of stresses acting on various planes at given point.

MOIST ADIABAT Line on Skew T-Log P chart that depicts change in temperature of saturated air as it rises and undergoes cooling due to adiabatic expansion.

MOIST ADIABATIC LAPSE RATE Rate at which temperature of parcel of saturated air decreases as parcel is lifted in atmosphere.

MOIST SOIL Soil perpetually damp.

MOIST-ADIABATIC PROCESS Adiabatic process for which air is saturated and may contain liquid water.

MOISTURE 1. Water diffused in atmosphere or ground; 2. Moderate degree of wetness.

MOISTURE ADVECTION Transport of moisture by horizontal winds.

MOISTURE CAPACITY 1. Amount of water held in soil against pull of gravity; 2. Depth of water retained in soil after ample irrigation or heavy rain when rate of downward movement has substantially decreased.

MOISTURE CONTENT 1. Ratio of weight of water in soil sample to weight of dry soil, expressed as percentage; 2. Quantity of water contained in material.

MOISTURE CONTENT OF SOIL Quantity of water soil contains.

MOISTURE CONVERGENCE Measure of degree that moist air is converging into given area, taking into account effect of converging winds and moisture advection.

MOISTURE DENSITY Mass of water per unit volume of space.

MOISTURE EJECTION VALVE Apparatus for draining-off liquids.

MOISTURE EQUIVALENT Percentage of water that soil can retain in opposition to centrifugal force 1,000 times that of gravity.

MOISTURE FILM COHESION Resistance of particles to being pulled apart, due to surface tension of moisture film surrounding each particle.

MOISTURE GRADIENT Difference in moisture content between surface and inner portion.

MOISTURE HOLDING CAPACITY Ability of soil within soil profile to retain water.

MOISTURE MOVEMENT Process where moisture moves through porous medium.

MOISTURE PENETRATION Depth to which moisture permeates soil.

MOISTURE PERCENTAGE Ratio of water content of semi-liquid substance to decrease in weight after drying to initial weight of sample.

MOISTURE REGAIN Percentage of water present in material to dry mass of material.

MOISTURE RIDGE Axis of relatively high dew point values.

MOISTURE TENSION Measure of suction (i.e., negative pressure).

MOISTURE-DENSITY CURVE Curve showing relationship between dry density (dry unit weight) and moisture content of soil for given compactive effort.

MOLD 1. Growth of fungi that occurs in moist, warm conditions; 2. Hollow shape/frame for shaping/forming.

MOLE 1. Self-propelled excavator used to form underground drains or to clear underground pipelines; 2. Type of underground drain used in farm fields, where mole plow creates unlined channel through clay subsoil.

MOLE DRAIN Subsurface method of draining clay soils that experience regular waterlogging from irrigation or high rainfall with drainage channel.

MOLE FRACTION Ratio of number of moles of one component of mixture to total number of moles.

MOLE PER CUBIC METER 1. Unit of concentration; 2. Number of atoms of substance per unit volume.

MOLE PER KILOGRAM Unit of molarity.

MOLE PLOUGH Subsoiler with torpedo-shaped tip.

MOLECULAR OXYGEN Diatomic molecule composed of two oxygen atoms held together by covalent bond.

MOLECULAR WEIGHT Sum of atomic weights of every atom in molecule.

MOLECULE Smallest particle of substance that retains properties of substance and is composed of one or more atoms.

MOLING Trenchless method used to lay pipes where pneumatically-driven machine forces its way through soil along desired path of pipe.

MOLTEN SALT REACTOR Thermal treatment unit that rapidly heats waste in heat-conducting fluid bath of carbonate salt.

MOMENTARY-CONTACT SWITCH Switch that returns to its resting state once actuating force is removed.

MONITOR Tool holder, as for lathe, shaped like low turret and capable of being revolved on vertical pivot so as to bring several tools successively into position.

MONITOR STAGE Stage which, when reached by rising stream, represents level where appropriate officials are notified of threat of possible flooding.

MONITORING 1. Periodic or continuous surveillance or testing to determine level of compliance with statutory requirements and/or pollutant levels in various media or in humans, plants and animals; 2. Repeated observation, measurement or sampling at site, on scheduled or event basis, for particular purpose.

MONITORING DEVICE Display produced by device that takes signals and displays them on monitor.

MONITORING SYSTEM System used to monitor resources and performance in computer system.

MONITORING WELL Well designed for measuring water levels and testing groundwater quality.

MONOCHLORAMINE Combination of chlorine and ammonia widely used as drinking water disinfectant.

MONOCHLOROBENZENE Colorless, flammable liquid, aromatic organic compound.

MONOCLINE Bend or steplike fold in rock layers where all strata are inclined in same direction.

MONOCLONAL ANTIBODIES 1. Man-made (anthropogenic) clones of molecule, produced in quantity for medical or research purposes; 2. Molecules of living organisms that selectively find and attach to other molecules that their structure conforms exactly.

MONOCOQUE GATE Thin-shell radial gate where usual skin plate and cross-beam framework are replaced by hollow shell having approximate elliptical cross-section.

MONOCYCLIC AROMATIC HYDROCARBONS Single-ring aromatic compounds.

MONOFILAMENT Single strand of man-made fiber.

MONOMER Relatively small molecule which can be covalently bonded to other monomers to form polymer.

MONOMICTIC Lakes and reservoirs which are relatively deep, do not freeze over and undergo single stratification and mixing cycle during year.

MONOMOLECULAR LAYER Layer that's one molecule thick.

MONORAIL Railroad system where trains run on one rail.

MONOSTATIC RADAR Radar that uses common antenna for both transmitting and receiving.

MONSOON Thermally driven wind arising from differential heating between land mass and adjacent ocean that reverses its direction seasonally.

MONTANE Of, pertaining to or inhabiting cool upland slopes below timber line.

MONTHLY AVERAGE Average parameter for given month of year.

MONTHLY REPORT OF RIVER AND FLOOD CONDITIONS (E-5)
Monthly narrative report covering flooding which occurred over past month.

MOOR Extensive waste covered with patches of heath and having poor, light soil, but sometimes marshy and abounding in peat.

MOORINGS Act of securing vessel with cables, anchors, etc.

MOPED Lightweight, two-wheeled vehicle equipped with small motor and pedals.

MORAINE Mound, ridge or other distinct accumulation of unsorted, unstratified glacial drift, predominantly till, deposited chiefly by direct action of glacier ice.

MORATORIUM During negotiation process, period during which agency and potentially responsible parties may reach settlement but no site response activities can be conducted.

MORBIDITY Rate of disease incidence.

MORNING GLORY Elongated cloud band, visually similar to roll cloud, usually appearing in morning hours, when atmosphere is relatively stable.

MORNING GLORY SPILLWAY Vertical or inclined shaft into which flood water spills and then is conducted through, under or around dam by means of conduit or tunnel.

MORPHOLOGY 1. Study of form and structure of animals and plants; 2. Study of structure of rocks and landforms.

MORTALITY Death rate.

MORTALITY RATE Number of deaths per given unit of population over given period of time.

MOSS 1. Small, leafy-stemmed, flowerless plant of class Musci; 2. Bog.

MOST PROBABLE NUMBER 1. Estimate of microbial density per unit volume of water sample; 2. Method of getting quantitative data on concentrations of discrete items from positive or negative data.

MOST UNSTABLE LIFTED INDEX Lifted Index calculated using parcel from pressure level that results in Most Unstable (lowest) value of Lifted Index possible.

MOSTLY CLEAR When $1/8^{th}$ to $2/8^{th}$ of sky is covered by with opaque (not transparent) clouds.

MOSTLY CLOUDY When $6/8^{th}$ to $7/8^{th}$ of sky is covered by with opaque (not transparent) clouds.

MOSTLY SUNNY When $1/8^{th}$ to $2/8^{th}$ of sky is covered by with opaque (not transparent) clouds.

MOTHER LIQUOR Solution remaining after product has been removed by crystallization, etc.

MOTILE Having power to move spontaneously.

MOTOR Machine or device that converts any form of energy into mechanical energy or imparts motion.

MOTOR CONTROL CENTER Assembly to control some or all electric motors in central location.

MOTOR EFFICIENCY Ratio of energy delivered by motor to energy supplied to it during fixed period or cycle.

MOTOR GENERATOR Device consisting of mechanically coupled electric motor and generator that may be used to control electrical supply.

MOTOR-CIRCUIT SWITCH Switch rated in horsepower, capable of interrupting maximum operating overload current of motor of same horsepower rating as switch at rated voltage.

MOTOR-GENERATOR SET Device for converting electrical power to another form.

MOTORCYCLE Open-seated motor vehicle handlebars instead of steering wheel and having two (three for sidecar) wheels.

MOTORIZED VALVE Valve opened by electric motor.

MOUND BREAKWATER Structure, built of rocks, to protect structure against wave attack.

MOUNT WILSON MAGNETIC CLASSIFICATIONS 1. Classification system for sunspots: Alpha - denotes unipolar sunspot group; 2. Beta - sunspot group having both positive and negative magnetic polarities, with simple and distinct division between polarities; 3. Beta-Gamma - sunspot group that's bipolar but where no continuous line can be drawn separating spots of opposite polarities; 4. Delta - complex magnetic configuration of solar sunspot group consisting of opposite polarity umbrae within same penumbra; 5. Gamma - complex active region where positive and negative polarities are so irregularly distributed as to prevent classification as bipolar group.

MOUNTAIN WAVE Wavelike effect, characterized by updrafts and downdrafts, that occurs above and behind mountain range when rapidly flowing air encounters mountain range's steep front.

MOUNTAIN WIND SYSTEM System of diurnal winds that forms in complex terrain area, consisting of mountain-plain, along-valley, cross-valley and slope wind systems.

MOUNTAIN-PLAIN WIND SYSTEM Closed, large-scale, thermally driven circulation between mountains and surrounding plain.

MOUNTAINADO Vertical-axis eddy produced in downslope windstorm by vertical stretching of horizontal roll vortices produced near ground by vertical wind shear.

MOUNTAINOUS TERRAIN 1. FAA obstacle clearance required in certain areas; 2. Containing many mountains.

MOUNTING HEIGHT Distance from center of signal lamp to surface where traffic vehicle stands.

MOUTH 1. Opening/entrance point; 2. Exit point of discharge.

MOVABLE BED Portion of waterway channel cross-section considered to be subject to erosion or deposition.

MOVABLE BED LIMITS Lateral limits of movable bed that define where scour or deposition occur.

MOVABLE BED MODEL Model where bed and/or side material is erodible and transported in manner similar to prototype.

MOVABLE BRIDGE Bridge that may be moved (e.g., raised) in order to open safe waterway for watercraft.

MOVABLE BRIDGE RESISTANCE GATE Provides physical deterrent, downstream of movable bridge, to road users when placed in appropriate position.

MOVABLE BRIDGE SIGNAL Signal installed to notify traffic to stop when bridge is opened for waterborne traffic.

MOVABLE BRIDGE WARNING GATE Traffic gate designed to warn, but not block, vehicle and/or pedestrian traffic.

MOVABLE BULKHEAD REFUSE TRUCK Side-loading, enclosed compactor truck.

MOVABLE DAM Obstacle which will be opened partially, creating adjustable weir.

MOVABLE DISTRIBUTOR Apparatus consisting of rotating or reciprocating conduits or troughs to release liquid.

MOVABLE GRATE Grate that feeds solid fuel or wastes into furnace.

MOVABLE WEIR Temporary weir that may be taken out of channel.

MOVE LATERALLY IN SOIL Transfer through soil in horizontal plane.

MOVE-UP TIME Time for vehicle to move from second to first position at stop or yield line.

MOVEABLE BED STREAMS Streams where steep slopes and lack of vegetation result in lot of erosion.

MOVEMENT GROUPS Traffic movement organization at signalized intersection.

MOVING VANE Vane pushed by flow from stationary nozzle.

MOVING WALKWAY Slow conveyor belt that transports people horizontally or on incline in similar manner to escalator.

MS CONNECTORS Surface delays for use when shooting with detonating cord.

MUCK 1. Mud rich in humus; 2. Stone, dirt, debris or useless material; 3. Dark, finely divided, well-decomposed, organic matter forming surface deposit in some poorly drained areas; 4. Finely blasted rock, particularly from underground.

MUCK SOILS Earth made from decaying plant materials.

MUD Any soil containing enough water to make it soft.

MUD BLANKET Layer of flocculent forming on sand filter surface.

MUD BOIL Upsurge of watery subsoil via opening in soil due to soil pressure change.

MUD CAKE Deposit of mud on walls of drill hole.

MUD FLAT Flat expanse of mud at edge of body of water.

MUD FLOW Flowing mass of soft, wet, unconsolidated earth and debris material.

MUD LINE Where water meets shoreline.

MUD PUMP Reciprocating piston or plunger pump designed to circulate drilling fluid under high pressure.

MUD SLIDE Fast moving soil, rocks and water that flow down mountain slopes and canyons during heavy downpour of rain.

MUD VALVE Valve in base of settling tank where collected sediment is emptied.

MUD WAVE Mud particles that move when subsoil of reclamation area is soft, with low shear strengths and can result in inadequate fill mass properties.

MUDBALLS Round material that forms in filters and gradually increases in size when not removed by backwashing.

MUDFLAT 1. Mud-covered, gently sloping tract of land alternately covered and left bare by water; 2. Muddy, nearly level bed of dry lake.

MUDSTONE Fine-grained sedimentary rock formed from hardened clay and silt that lacks thin layers typical of shale.

MUFFLER Part of exhaust pipe of car that dampens noise engine produces.

MUGGY Term for warm and excessively humid conditions.

MULCH Decaying material spread around to enrich soil, insulate plant roots, prevent moisture evaporation and control weeds.

MULTI-LANE Having two or more lanes for traffic in same direction.

MULTI-LEG INTERSECTION Intersection with five or more legs.

MULTI-MEDIA APPROACH Joint approach to several environmental media (i.e., air, water, land).

MULTI-SECTOR GENERAL PERMIT Authorizes discharge of stormwater from industrial facilities, consistent with terms of permit, in areas of United States where EPA manages NPDES permit program.

MULTI-STAGE PUMP Pump that has more than one impeller.

MULTICELL THUNDERSTORM Thunderstorms organized in clusters of at least 2-4 short-lived cells.

MULTIDIRECTIONAL Involving or moving in multiple directions.

MULTILANE HIGHWAY Highway, with more than one lane, for traffic flow in both directions.

MULTILANE ROUNDABOUT Roundabouts that have at least one approach leg with two or more entry lanes.

MULTILEVEL INTAKE Hydraulic structure that permits water to enter from various depths below water surface.

MULTIMEDIA FILTER Multibed filter used to reduce level of silt and suspended solids in incoming feed water.

MULTIMODAL Having or employing multiple modes of transportation.

MULTIPASS FILTER PERFORMANCE TEST Standard test created to give ratio between number of particles that filter allows to pass and total amount supplied to filter.

MULTIPHASE FLOW Simultaneous flow of materials with different states or phases or different chemical properties but in same state or phase.

MULTIPLE ARCH DAM Buttress dam comprised of series of arches for upstream face.

MULTIPLE CHEMICAL SENSITIVITY Diagnostic label for people who suffer multi-system illnesses as result of contact with or proximity to, variety of airborne agents and other substances.

MULTIPLE CORRELATION Relationship between random variable and its regression function.

MULTIPLE DOPPLER ANALYSIS Use of more than one radar to reconstruct spatial distributions of 2D or 3D wind field, which cannot be measured from single radar alone.

MULTIPLE DOSING TANKS More than one dosing tank where flow is passed.

MULTIPLE HEARTH INCINERATOR Incinerator with circular hearths vertically stacked.

MULTIPLE OUTLETS Branching conduits leading to outlets where wastewater or effluent is released.

MULTIPLE USE Use of water or land for more than one purpose.

MULTIPLE VORTEX TORNADO Tornado where two or more condensation funnels or debris clouds are present at same time, often rotating about common center or about each other.

MULTIPLE WEAVING SEGMENT Series of closely spaced merge and diverge areas that creates overlapping weaving movements and share same segment of roadway.

MULTIPLE-ARCH DAM Dam featuring repeated or contiguous arches that slope at about 45° to horizontal on parallel buttress wall.

MULTIPLE-CHAMBER INCINERATOR Any incinerator consisting of three (3) or more refractory lined combustion furnaces in series, physically separated by refractory walls, interconnected by gas passage ports or ducts and employing adequate design parameters necessary for maximum combustion.

MULTIPLE-DOME DAM Buttress dam where dome arched horizontally and longitudinally is supported by widely-spaced buttresses.

MULTIPLE-EFFECT EVAPORATOR Apparatus for efficiently using heat from steam to evaporate water.

MULTIPLE-LOAD-PATH STRUCTURE Structure that will support specific load after loss of main load-carrying component.

MULTIPLE-PLATE SAMPLERS Artificial substrates of known surface area used for obtaining benthic invertebrate samples.

MULTIPLE-STAGE PUMP Pump with two or more impellers.

MULTIPLE-STAGE SLUDGE DIGESTION Digestion of waste sludge in two or more tanks arranged in series.

MULTIPLE-TRAY CLARIFIER Clarifier consisting of vertical series of trays to collect suspended solids.

MULTIPLE-UNIT ELECTRIC LOCOMOTIVE Locomotive composed of two or more multiple-unit electric motive-power unit connected for simultaneous operation of all such units from single control station.

MULTIPLE-UNIT RAIL CAR Two or more connected rail cars equipped so that all cars in train may be operated from any car that has operator's station.

MULTIPLE-USE MANAGEMENT Management of public land resources for variety of uses (e.g., timber harvesting, energy development).

MULTIPLE-USE RESERVOIR Water storage reservoir with greater than one purpose.

MULTIPURPOSE BUCKET Allows machine operator to grab larger material and doze smaller material with single attachment.

MULTIPURPOSE DAM Dam constructed for two or more purposes (e.g., storage, flood control, navigation, power generation, recreation, fish and wildlife enhancement).

MULTIPURPOSE PROJECT Project designed for irrigation, power, flood control, municipal and industrial, recreation and fish and wildlife benefits, in any combinations of two or more (contrasted to single-purpose projects serving only one need).

MULTIPURPOSE RESERVOIR 1. Reservoir constructed and equipped to provide storage and release of water for two or more purposes (i.e., flood control, power development, navigation, irrigation, recreation, pollution abatement, domestic water supply); 2. Reservoir planned to operate for more than one purpose.

MULTISTAGE COMPRESSOR Compressor that compresses (i.e., pressurizes) gas at each stage more than previous stage.

MULTISTAGE REMOTE SENSING Strategy for landscape characterization that involves gathering and analyzing information at several geographic scales, ranging from generalized levels of detail at national level through high levels of detail at local scale.

MUNICIPAL DISCHARGE Discharge of effluent from waste water treatment plants which receive waste water from households, commercial establishments and industries in coastal drainage basin.

MUNICIPAL ENGINEERING Field of engineering concerned with municipal infrastructure (e.g., streets, water supply, sewers, electricity distribution, solid waste management and public parks).

MUNICIPAL SERVICES Services provided to residents of city in exchange for paying taxes.

MUNICIPAL SEWAGE Wastes (mostly liquid) originating from community that may be composed of domestic wastewaters and/or industrial discharges.

MUNICIPAL SLUDGE Semi-liquid residue remaining from treatment of municipal water and wastewater.

MUNICIPAL SOLID WASTE Common garbage or trash generated by industries, businesses, institutions and homes.

MUNICIPAL SOLID WASTE LANDFILL Discrete area of land or excavation that receives household waste.

MUNICIPAL SOLID WASTES Everyday items we use and then throw away.

MUNICIPAL STOP Stop valve on communication pipe nearest Municipal main, controlling supply of water.

MUNICIPAL USE OF WATER Various uses of water in developed urban areas (i.e., domestic use, industrial use, street sprinkling, fire protection).

MUNICIPAL UTILITY Utility owned and operated by city.

MUNICIPAL WASTE COMBUSTOR Apparatus that burns municipal (i.e., solid, liquid or gasified) waste.

MUNICIPAL WATER SYSTEM Water system that has at least five service connections or which regularly serves 25 individuals for 60 days.

MUNICIPAL-TYPE SOLID WASTE Household, commercial, retail, and/or institutional waste.

MUNICIPALIZATION Process by which municipal entity assumes responsibility for supplying utility service to its constituents.

MUNICIPALLY-OWNED ELECTRIC SYSTEM Electrical utility owned and operated by city.

MUSHROOM Thunderstorm with well-defined anvil rollover.

MUSHROOM VALVE Valve typically used to control timing and quantity of gas or vapor flow into engine.

MUSKEG 1. Large expanses of peatlands or bogs in subarctic zones; 2. Terrain composed of peat bog with tussocky meadow and woody vegetation including spruce.

MUTAGEN Chemical or physical agent that induces permanent change in genetic material.

MUTAGENIC Capable of causing mutation.

MUTAGENICITY Capacity of chemical or physical agent to cause such permanent changes.

MUTATION Any heritable change of base-pair sequence of genetic material.

N

NACREOUS CLOUDS Clouds of unknown composition that have soft, pearly luster and that form at altitudes about 25 to 30 km above Earth's surface.

NADIR Point on any given observer's celestial sphere diametrically opposite of one's zenith.

NAME PLATE CAPACITY Equipment rating as required by purchaser.

NAMEPLATE Power generation capacity of generator that can be guaranteed under continuous operation.

NAMEPLATE RATING Maximum power-generating capacity of generator or power-generating facility.

NANO Prefix symbolizing 10^{-9}.

NANOGRAMS PER LITER Unit expressing concentration of chemical constituents in solution as mass (nanograms) of solute per unit volume (liter) of water.

NANOMETER 1. One millionth of millimeter; 2. Millimicron.

NANOPLANKTON Plankton between 2 and 20 microns in size.

NANOTESLA Unit of magnetism equal to 10^{-9} tesla that's equivalent to gamma (10^{-5} gauss).

NAO INDEX Index measures anomalies in sea level pressure between Icelandic low pressure system and Azores high pressure system in North Atlantic Ocean.

NAPPE Profile of body of water flowing over obstruction in vertical drop.

NARROW Railway with track gauge narrower than standard 4 ft 8.5 in.

NARROW-BASE TERRACE Terrace with base width between 4 to 8 ft.

NARROWS Narrow part of navigable waterway.

NATIONAL AMBIENT AIR QUALITY STANDARDS National standards for ambient concentrations in air of different air pollutants designed to protect human health and welfare.

NATIONAL CENTERS FOR ENVIRONMENTAL PREDICTION Part of National Weather Service which provides nationwide computerized and manual guidance to Warning and Forecast Offices concerning forecast of basic weather elements.

NATIONAL CLIMATIC DATA CENTER Agency that archives climatic data from National Oceanic and Atmospheric Administration, as well as other climatological organizations.

NATIONAL DIGITAL FORECAST DATABASE National Weather Service's access to gridded forecasts of sensible weather elements (e.g., wind, wave height).

NATIONAL DISASTER MEDICAL SYSTEM System designed to deal with extensive medical care needs in very large disasters or emergencies.

NATIONAL EMISSIONS STANDARDS FOR HAZARDOUS AIR POLLUTANTS Emissions standards set by EPA for air pollutant not covered by NAAQS that may cause increase in fatalities or in serious, irreversible or incapacitating illness.

NATIONAL ENVIRONMENTAL PERFORMANCE PARTNERSHIP AGREEMENTS System that allows states to assume greater responsibility for environmental programs based on their relative ability to execute them.

NATIONAL ENVIRONMENTAL POLICY ACT Act requiring analysis, public comment and reporting for environmental impacts of federal actions.

NATIONAL ENVIRONMENTAL SATELLITE, DATA, AND INFORMATION SERVICE Collects, processes, stores, analyzes and disseminates various types of hydrologic, meteorological and oceanic data.

NATIONAL ESTUARY PROGRAM Program established to develop and implement conservation and management plans for protecting estuaries and restoring and maintaining their chemical, physical and biological integrity, as well as controlling point and non-point pollution sources.

NATIONAL FIRE DANGER RATING SYSTEM Uniform fire danger rating system used in United States that focuses on environmental factors that impact moisture content of fuels.

NATIONAL FLOOD SUMMARY National Weather Service daily product that contains nationwide information on current flood conditions.

NATIONAL GEODETIC VERTICAL DATUM As corrected in 1929, vertical control measure used as reference for establishing varying elevations.

NATIONAL HIGHWAY SYSTEM Network of strategic highways within United States serving major airports, ports, railway stations, terminals and transport facilities.

NATIONAL HURRICANE CENTER One of three branches of Tropical Prediction Center that maintains continuous watch on tropical cyclones over Atlantic, Caribbean, Gulf of Mexico and Eastern Pacific from 15 May through November 30.

NATIONAL HURRICANE OPERATIONS PLAN Issued annually by federal coordinator for Meteorological Services and Supporting Research that documents interdepartmental agreements relating to tropical cyclone observing, warning and forecasting services.

NATIONAL INSTITUTE FOR OCCUPATIONAL SAFETY AND HEALTH Federal agency responsible for conducting research and making recommendations for prevention of work-related disease and injury.

NATIONAL INTERAGENCY FIRE CENTER Nation's support center for wildland firefighting.

NATIONAL MUNICIPAL PLAN Policy created to bring all publicly owned treatment works into compliance with Clean Water Act requirements.

NATIONAL OCEANIC AND ATMOSPHERIC ADMINISTRATION Mission is to describe and predict changes in Earth's environment and conserve and wisely manage Nation's coastal and marine resources.

NATIONAL OIL AND HAZARDOUS SUBSTANCES CONTINGENCY PLAN Federal regulation that guides determination of sites to be corrected under both Superfund program and program to prevent or control spills into surface waters or elsewhere.

NATIONAL OPERATIONAL HYDROLOGIC REMOTE SENSING CENTER Organization under National Weather Service Office of Hydrology that mainly deals with snow mapping.

NATIONAL PARK SERVICE Agency within U.S. Department of Interior that preserves unimpaired natural and cultural resources and values of national park system for enjoyment, education and inspiration of this and future generations.

NATIONAL POLLUTANT DISCHARGE ELIMINATION SYSTEM Permitting program of Clean Water Act required for all point sources discharging pollutants into waters of United States to protect human health and environment.

NATIONAL PRETREATMENT STANDARD Any regulation promulgated that applies to specific category of industrial users and provides limitations on introduction of pollutants into publicly owned treatment works.

NATIONAL PRIORITIES LIST List of most serious uncontrolled or abandoned hazardous waste sites identified for possible long-term remedial action under Superfund.

NATIONAL REGISTER OF HISTORIC PLACES Federally maintained register of districts, sites, buildings, structures, architecture, archeology and culture.

NATIONAL RESPONSE CENTER Operated by U.S. Coast Guard, federal operations center that receives notifications of all releases of oil and hazardous substances into environment.

NATIONAL RESPONSE TEAM Consists of representatives from several federal agencies that provides program direction, planning and preparedness guidance and review of regional response activities to nationally significant incidents of pollution.

NATIONAL SEVERE STORMS LABORATORY Internationally known as Environmental Research Laboratories that lead way in investigations of all aspects of severe weather.

NATIONAL STANDARD HOSE COUPLING Fire hose coupling having number of threads per inch.

NATIONAL WARNING SYSTEM Dedicated, commercially leased, nationwide voice telephone warning system operated on 24-hour basis, with National Warning Center and Alternate National Warning Center staffed by attack warning officers.

NATIONAL WATER QUALITY ASSESSMENT PROGRAM Long term USGS program to assess occurrence and distribution of water-quality conditions Nationwide.

NATIONAL WEATHER AND CROP SUMMARY 1. Product of National Agricultural Statistics Service, Agricultural Statistics Board and U.S. Department of Agriculture that contains weekly national agricultural weather summaries, including weather's effect on crops and farm progress for 44 states and New England area.

NATIONAL WEATHER SERVICE Agency of Federal Government within Department of Commerce, National Oceanic and Atmospheric Administration, that provides weather, hydrologic and climate forecasts and warnings for United States, its territories, adjacent waters and ocean areas, for protection of life and property and enhancement of national economy.

NATIONAL WEATHER SERVICE RIVER FORECAST MODEL System of data entry, data preprocessing and forecast programs which are used by River Forecast Centers.

NATIONWIDE RIVERS INVENTORY Listing of more than 3,400 free-flowing river segments in United States that are believed to possess one or more "outstandingly remarkable" natural or cultural values judged to be of more than local or regional significance.

NATIVE ELEMENT Any chemical element which occurs in nature uncombined with other elements.

NATIVE SPECIES Any animal and plant species originally in United States.

NATRIC Soil containing sodium.

NATURAL ALLUVIAL CHANNEL Movable bed in loose sedimentary materials.

NATURAL CONDITIONS Phenomena that's naturally occurring and reduces visibility.

NATURAL CONTROL Stream gaging control that is natural to stream channel, in contrast to artificial control structure by man.

NATURAL FLOODWAY Channel of water course and those portions of adjoining flood plain which are reasonably required to carry selected probability flood.

NATURAL FLOW Unimpaired flow.

NATURAL FREQUENCY Frequencies of free vibration.

NATURAL GAS Mixture of gaseous hydrocarbons associated with petroleum deposits.

NATURAL GAS LIQUIDS Condensable hydrocarbons (e.g., natural gas, crude oil).

NATURAL GRADE Soil profile unaffected by construction techniques.

NATURAL GROUND Elevation of original ground level before excavation.

NATURAL HARBOR When part of body of water is protected by land.

NATURAL LEVEE Embankment built by stream on its flood plain along one or both banks of its channel.

NATURAL MOISTURE CONTENT Ratio of weight of water to weight of solids in given mass.

NATURAL OUTLET Outlet into pond, ditch, lake, groundwater, etc.

NATURAL PERIOD OF VIBRATION Time required for one cycle of simple harmonic motion in one of these characteristic patterns (shapes).

NATURAL PURIFICATION Natural processes occurring in body of water as water percolates through soil layers, removing large pieces of debris as bacteria or soil micro-organisms break down nutrients or contaminants.

NATURAL RESOURCES Naturally occurring water, minerals, fossil fuels, land, timber, fish, biota, animals, etc.

NATURAL SLOPE Maximum angle soil will hold without slipping.

NATURAL SUB-IRRIGATION Leakage of water from channels to irrigate crops via capillarity action.

NATURAL SUBSTRATE Any naturally occurring immersed or submersed solid surface (e.g., rock, tree) upon which organism lives.

NATURAL TURNOVER Yearly rate of transfer of material through active part of cycle.

NATURAL VENTILATION Supplying/removing air from indoor space without using mechanical devices.

NATURAL WATERCOURSE Watercourses, with headwaters, that have not been significantly altered from historical flow path or floodplain.

NATURALLY DEVELOPED WELL Well with screen in direct contact with formation materials and no filter pack.

NATURALNESS Area which generally appears to have been affected primarily by forces of nature, with imprint of man's work substantially unnoticeable.

NAUTICAL DAWN 1. Time at which Sun is 12° below horizon in morning; 2. Time at which there is just enough sunlight for objects to be distinguishable.

NAUTICAL DUSK 1. Time at which Sun is 12° below horizon in evening; 2. Time at which objects are no longer distinguishable.

NAUTICAL MILE Unit of length used in marine navigation and marine forecasts, equal to 1.15 statute miles, 1852 meters, 6076 feet or 1 minute of latitude.

NAUTICAL TWILIGHT Time after civil twilight, when brighter stars used for celestial navigation have appeared and horizon may still be seen that ends when center of Sun is 12° below horizon and it is too difficult to perceive horizon, preventing accurate sighting of stars.

NAVIGABLE AIRSPACE Airspace at or above minimum altitudes of flight that includes airspace needed to ensure safety in takeoff and landing of aircraft.

NAVIGABLE DEPTH Physical depth waterway where ships can still safely maneuver.

NAVIGABLE WATER Waters sufficiently deep and wide for navigation by all or specified vessels.

NAVIGABLE WATERS Waters that are subject to ebb and flow of tide and/or are presently used or have been used in past or may be susceptible for use to transport interstate or foreign commerce.

NAVIGABLE WATERWAY Body of water that's deep, wide and slow enough for vessel.

NAVIGATION CANAL Canal constructed for navigation or transportation by vessels.

NAVIGATION CHANNEL Deeper channel cut into sea or river bed, to enable larger ships to pass through to port.

NAVIGATION LIGHT One of two or more lights fitted to ship or aircraft that indicate position and direction taken.

NAVIGATION MODEL Model to study maneuverability of vessels under currents, waves, wind, etc. for design of navigable waterways.

NAVIGATIONAL AID Marker that aids traveler in navigation.

NAVTEX International automated medium frequency (518 kHz) direct-printing service for delivery of navigational and meteorological warnings and forecasts, as well as urgent marine safety information to ships.

NAVTEX FORECAST National Weather Service marine forecast combining various Coastal Waters and Offshore forecasts.

NAVY OPERATIONAL GLOBAL ATMOSPHERIC PREDICTION SYSTEM 144-hour numerical model of atmosphere run by U.S. Navy twice daily.

NEAP TIDE Minimum tide occurring at first and third quarters of Moon.

NEAP TIDES Tide which occurs just after first and third quarters of moon, when there is least difference between high tide and low tide.

NEAR FIRST SERVICE CONNECTION At one of 20% of all service connections in entire system that are nearest water supply treatment facility, as measured by water transport time within distribution system.

NEAR-CRITICAL ACTIVITY Schedule activity that has low total float.

NEAR-SIDE STOP Transit stop located on approach side of intersection.

NEARSHORE FORECAST National Weather Service seasonal marine forecasts for areas of Great Lakes extending from line approximating mean low water datum along coast or island, including bays, harbors and sounds, out to 5 nm.

NEATLINES OF STRUCTURE 1. Line which defines limits of work (e.g., excavation, cut stone); 2. True face line of building regardless of projections of stones; 3. Line back of or inside of incidental projections.

NECK 1. Long narrow tract of land projecting from main body; 2. Narrow channel connecting two bodies of water.

NECROSIS Death of plant or animal cells or tissues.

NEEDLE 1. Object to regulate flow of water through hydraulic structure; 2. Fine measurement indicator on dial or graph.

NEEDLE DAM Weir designed to maintain level or flow through use of thin "needles" of wood.

NEEDLE VALVE Any of family of valves which regulate flow through use of needle moving into and out of orifice.

NEEDLE WEIR Frame weir of hefty vertical timbers.

NEGATIVE AIR PRESSURE TEST Test to identify leaks using air at pressure less than atmospheric.

NEGATIVE CONFINING BED Confining bed that retards or prevents downward motion of groundwater when overlying water has adequate head to create downward pressure.

NEGATIVE HEAD Condition of negative pressure or vacuum that occurs when pump outlet is at or above level of storage tank.

NEGATIVE PRESSURE 1. Pressure within pipe that's less than atmospheric pressure; 2. Tension in liquid that's thermodynamically metastable with respect to vapor phase.

NEGATIVE PRESSURE VALVE Atmospheric relief valve that discharges before it collects amounts needed to actuate automatic air valve.

NEGATIVE VORTICITY ADVECTION Advection of lower values of vorticity into area.

NEGATIVE WELL Shaft or well driven through impermeable stratum to allow surface water to drain through to permeable stratum.

NEGATIVE-TILT TROUGH Upper level system that is tilted to west with increasing latitude (i.e., with axis from southeast to northwest).

NEGLIGENCE Duty of reasonable care was breached, causing damage.

NEGLIGIBLE RESIDUE Tolerance set on food or feed crop permitting minuscule amount of pesticide at harvest as result of indirect contact with chemical.

NEGOTIATED CONTRACT Contract awarded via direct agreement, without going through competitive bidding process.

NEKTON Consumers in aquatic environment and consist of large, free-swimming organisms that are capable of sustained, directed mobility.

NEMATICIDES Pesticides designed to kill nematodes (i.e., roundworms).

NEMATOCIDE Chemical agent that is destructive to nematodes.

NEOPRENE Synthetic rubber with superior resistance to oils.

NEOTROPICAL MIGRANT Bird that migrates to tropical regions during winter.

NEPHELOMETER Instrument for measuring various aspects of suspended particles in fluid.

NEPHELOMETRIC Method of measuring turbidity in water sample by passing light through sample and measuring amount of light deflected.

NEPHELOMETRIC TURBIDITY UNIT Unit of measure for turbidity of water.

NERITIC Describing marine environment of shallow waters.

NESSLER TUBES Glass tubes for filling with standard solution colors for visual color comparison with similar tubes filled with solution samples.

NESSLERIZATION Procedure for ascertaining existence of ammonia via solution of mercuric iodide in potassium iodide and potassium hydroxide.

NET ALL-WAVE RADIATION Net or resultant value of upward and downward longwave and shortwave radiative fluxes through plane at Earth-atmosphere interface that's component of surface energy budget.

NET AVAILABLE HEAD Difference in pressure elevation between water entering and exiting hydraulic structure, minus friction losses.

NET CAPABILITY Maximum load-carrying ability of equipment, exclusive of station use, under specified conditions for given time interval, independent of characteristics of load.

NET CONSUMPTIVE USE OF WATER Water applied in irrigation minus rainfall contribution.

NET CUT Cut required, less fill required, at particular station or part of road.

NET DEPTH Depth of concrete excluding that in form work.

NET DUTY OF WATER Water delivered to irrigate crops that's measured at point of delivery.

NET ECONOMIC BENEFITS Economic benefits less economic costs.

NET FILL Fill required, less cut required, at particular station or part of road.

NET HEAD Difference in pressure elevation between water entering and exiting hydraulic structure, minus friction losses.

NET METERING Method of crediting customers for electricity that they generate on site in excess of their own electricity consumption.

NET PEAK FLOW Maximum stream flow after subtracting base flow.

NET POSITIVE SUCTION HEAD Either absolute pressure at suction port of pump (i.e., available) or minimum pressure required at suction port of pump to prevent cavitation (i.e., required).

NET POWER Effective power output by engine.

NET RAINFALL Portion of rainfall which reaches stream channel or concentration point as direct surface flow.

NET REGISTERED TONNAGE Measured capacity of inside of ship for carrying goods not including fuel space, engines, etc.

NETTING Concept where all emissions sources in same area that owned or controlled by single company are treated as one large source, thereby allowing flexibility in controlling individual sources in order to meet single emissions standard.

NETWORK Any interconnected group or system.

NETWORK PATH Constant group of linked tasks in project network diagram.

NETWORK PLANNING Includes forecasting and capacity planning of activities to balance performance.

NETWORK PLANNING TECHNIQUE Method to analyze, describe, structure, plan, control and projects and processes, taking into consideration time, cost, resources, etc.

NEUTRAL AREA Paved triangular space between ramp and lane of highway.

NEUTRAL ATMOSPHERE Term used in ionospheric context to describe non-ionized component of atmosphere.

NEUTRAL DEPTH Depth of uniform flow or hypothetical depth of steady, non-uniform flow.

NEUTRAL LINE Line that separates longitudinal magnetic fields of opposite polarity.

NEUTRAL PRESSURE Hydrostatic pressure of water in soil pore space.

NEUTRAL SHORELINE Shoreline with features not dependent of either submergence or emergence of adjacent land.

NEUTRAL SOIL Soil ranging from slightly acid to slightly alkaline.

NEUTRAL STABILITY Atmospheric condition that exists in unsaturated air when environmental lapse rate equals dry adiabatic rate or in saturated air when environmental lapse rate equals moist adiabatic rate.

NEUTRAL ZONE Zone between underside of surface scum and top of deposited sediments in sedimentation or settling tank.

NEUTRALIZATION Decreasing acidity or alkalinity of substance by adding alkaline or acidic materials, respectively.

NEUTRON Subatomic particle forming part of nucleus of atom and having no charge that's combination of up quark and two down quarks.

NEUTRON PROBE Instrument used to estimate soil moisture by relating rate of attenuation in pulsed neutron emissions to soil water content.

NEVE 1. Term for firn; 2. Granular form of snow that's been partially melted, refrozen, compacted, but not yet ice.

NEW SOURCE 1. Any stationary source built or modified after publication of final or proposed regulations that prescribe given standard of performance; 2. Discharges of pollutants for specified categories of sources.

NEW SOURCE PERFORMANCE STANDARDS Uniform national EPA air emission and water effluent standards which limit amount of pollution allowed from new sources or from modified existing sources.

NEW SOURCE REVIEW Clean Air Act requirement that State Implementation Plans must include permit review that applies to construction and operation of new and modified stationary sources in non-attainment areas to ensure attainment of national ambient air quality standards.

NEW WATER Groundwater artificially brought to surface that would have otherwise flowed to underground reservoir or source.

NEW YORK ROD Leveling rod marked with narrow lines, ruler-fashion.

NEWTON 1. Unit of force equal to force that would give mass of 1 kg acceleration of 1 m/s^2; 2. 100,000 dynes.

NEWTON METER 1. Unit of moment of force; 2. Energy/work by force of 1 Newton acting through distance of 1 meter.

NEWTON PER METER 1. Unit of surface tension; 2. Energy to move object.

NEWTONIAN BEHAVIOR Fluid property where shear speed or flow rate is proportional to shear stress or pressure.

NEXT GENERATION RADAR National Weather Service network of Doppler radars operating nationwide.

NICKEL Silvery, white, elemental metal with atomic number of 28.

NIEVE PENITENTE Spike or pillar of compacted snow, firn or glacier ice, caused by differential melting and evaporation.

NIGHT Period of day between dusk and dawn.

NIGHT SOIL Human feces.

NIMBOSTRATUS 1. Cloud of class characterized by formless layer almost uniformly dark gray; 2. Rain cloud of layer type, of low altitude, usually below 8,000 feet.

NIPHER SHIELD Conically shaped, copper, rain-gauge shield.

NIPPLE 1. Short piece of pipe with male threads on each end; 2. Mechanical device where liquids or gases can be passed in regulated manner.

NITRATE Compound containing nitrogen that can exist in atmosphere or as dissolved gas in water that can have harmful effects on humans and animals but is plant nutrient.

NITRIC OXIDE Gas formed by combustion under high temperature and high pressure in internal combustion engine.

NITRIFICATION 1. To treat or react with nitrogen or nitrogen-containing compound; 2. Process whereby ammonia in wastewater is oxidized to nitrite and then to nitrate by bacterial or chemical reactions.

NITRILOTRIACETIC ACID Polyamino carboxylic acid used as chelating agent used to replace phosphates in detergents.

NITRITE 1. Intermediate in process of nitrification; 2. Nitrous oxide salts used in food preservation.

NITRITES Any salts or esters of nitrous acid.

NITROBACTERIA BACTERIA Nitrite to nitrate that's oxidized by bacteria.

NITROGEN Colorless, odorless, gaseous chemical element with atomic number of 7.

NITROGEN CYCLE Natural circulation of nitrogen where atmospheric nitrogen is converted to nitrogen oxides by lightning and deposited in soil by rain, where it's assimilated by plants and either eaten by animals or decomposed back to elemental nitrogen by bacteria.

NITROGEN DIOXIDE Result of nitric oxide combining with oxygen in atmosphere that's major component of photochemical smog.

NITROGEN FIXATION Conversion of atmospheric nitrogen into ammonia and organic derivatives by natural means.

NITROGEN OXIDE Result of photochemical reactions of nitric oxide in ambient air that's major component of photochemical smog.

NITROGEN OXIDES Any oxides of nitrogen.

NITROGEN-FIXING PLANTS Plants that take nitrogen from atmosphere and convert it into form useable by plant roots in soil.

NITROGENOUS Of, relating to or containing nitrogen.

NITROGENOUS WASTES Animal or vegetable residues that contain significant amounts of nitrogen.

NITROPHENOLS Synthetic organopesticides containing carbon, hydrogen, nitrogen and oxygen.

NITROSOMONAS Rod-shaped chemoautotrophic bacteria that oxidizes ammonia into nitrite as metabolic process.

NITROUS FUMES Toxic, reddish fumes generated by burning nitroglycerine explosives.

NITROUS OXIDE Clear, colorless gas, with slightly sweet odor.

NO ACTION ALTERNATIVE 1. Projected baseline condition or future without; 2. Expected future condition if no action is taken.

NO DETECTABLE EMISSIONS When detection device measures less than 500 ppm above background levels.

NO FURTHER REMEDIAL ACTION PLANNED Determination following preliminary assessment that site does not pose significant risk and so requires no further activity.

NO JEOPARDY OPINION Opinion that action is not likely to jeopardize continued existence of listed species or result in destruction or adverse modification of critical habitat.

NO OBSERVABLE ADVERSE EFFECT LEVEL Exposure level at which there are no statistically or biologically significant increases in frequency or severity of adverse effects between exposed population and its appropriate control.

NO-BUILD ALTERNATIVE Baseline for comparison of future conditions and impacts that assumes existing remains.

NO-FLOW BOUNDARY Model boundary where assigned flux is equal to zero.

NO-HUB PIPE Conduit with smooth ends.

NO-MIGRATION PETITIONS Demonstrations that fluids injected into Class I hazardous waste well will remain in injection zone for 10,000 years or as long as fluid remains hazardous or until waste decomposes or otherwise is attenuated to nonhazardous levels before migrating from injection zone.

NO-OBSERVED-ADVERSE-EFFECT LEVEL Denotes level of exposure of organism that there is no biologically or statistically significant increase in frequency or severity of any adverse effects in exposed population when compared to control.

NO-OBSERVED-ADVERSE-EFFECT-LEVEL Tested dose of effluent or toxicant below which no adverse biological effects are observed, as identified from chronic or subchronic human epidemiology studies or animal exposure studies.

NO-OBSERVED-EFFECT LEVEL Greatest concentration or amount of substance that causes no detectable adverse alteration of morphology, functional capacity, growth, development or lifespan of target organism under defined conditions of exposure.

NO-OBSERVED-EFFECT-CONCENTRATION Highest tested concentration of effluent or toxicant at which no adverse effects are observed on aquatic test organisms at specific time of observation.

NO-OBSERVED-EFFECT-LEVEL Exposure level at which there are no statistically or biological significant differences in frequency or severity of any effect in exposed or control populations.

NO-PASSING ZONE Segment where passing is prohibited on roadway.

NOAA WEATHER RADIO Broadcasts National Weather Service warnings, watches, forecasts and other hazard information 24 hours day.

NOAA WEATHER WIRE Mass dissemination via satellite of National Weather Service products to media and public.

NOBLE METAL Chemically inactive metal (e.g., gold) that does not corrode easily.

NOCTILUCENT CLOUDS Wavy, thin, bluish-white clouds that are best seen at twilight in polar latitudes.

NOCTURNAL 1. Related to nighttime; 2. Occurring at night.

NOCTURNAL INVERSION Increase in temperature with increasing height above earth's surface that develops during night as result of radiational cooling of surface.

NOCTURNAL JET Maximum wind speed that occurs just above nocturnal inversion at night.

NOCTURNAL THUNDERSTORMS Thunderstorms that develop after sunset.

NODAL POINT In numerical model, location in discretized model domain where dependent variable (i.e., hydraulic head) is computed.

NODE 1. In numerical model, location in discretized model domain where dependent variable (i.e., hydraulic head) is computed; 2. Location in numerical network where computations are performed and/or output is requested; 3. Point where orbit of planet, as viewed from Sun, intersects ecliptic; 4. Similar point on surface, where there is more than one tangent-plane; 5. Point along standing wave where wave has minimal amplitude.

NOISE 1. Sound or signal generated by random fluctuations; 2. Product-level or product-volume changes occurring during test that are not related.

NOISE ABATEMENT PROCEDURE Flight by pilot to limit noise around airport.

NOISE ABATEMENT ZONE Control of noise source in area.

NOISE AND NUMBER INDEX Subjective measure of aircraft noisiness.

NOISE CERTIFICATION Noise levels aircraft should never surpass on takeoff or landing.

NOISE CONTROL Strategies to reduce noise pollution or impact.

NOISE CONTROL BOUNDARY Perimeter around noise abatement zone.

NOISE FOOTPRINT Area below aircraft takeoff or landing experiencing maximum noise levels.

NOISE INDICES Measures of sound-loudness.

NOISE RATING Curve developed for rating acceptability of indoor environments for purposes of hearing preservation, speech communication and annoyance.

NOISE REDUCTION COEFFICIENT Coefficient of sound energy absorbed upon striking particular surface.

NOISE REDUCTION RATING Unit of measurement to determine effectiveness of hearing protection.

NOISE ZONING Regions classified by noise levels.

NOISE-INDUCED HEARING LOSS Hearing loss due to long or repeated exposure to sounds at or above 85 decibels.

NOMINAL Of or relating to presumed or approximate value, rather than actual value.

NOMINAL CONCENTRATION Number (percent by weight) of ingredient that's anticipated to show up in average sample of pesticide product when manufactured.

NOMINAL DIMENSION Dimension material is generally called or sold in trade.

NOMINAL DOLLARS Dollar value in indicated year, not adjusted for inflation.

NOMINAL FILTER RATING Filter rating indicating majority micrometer value that will not pass through filter.

NOMINAL FILTRATION RATING Arbitrary value determined by manufacturer and expressed in terms of percentage retention of given size.

NOMINAL LOAD Magnitude of load specified by applicable code.

NOMINAL MAXIMUM SIZE Smallest sieve opening that entire amount of aggregate is permitted to pass.

NOMINAL RESISTANCE Average of minimum and maximum allowed resistance values.

NOMINAL SIZE Approximate dimension(s) of standard materials.

NOMINAL STRENGTH Capacity of structure or component to resist effects of loads.

NOMINAL VALUE Accepted value as opposed to real value.

NOMINALLY HORIZONTAL At angle less than 45° with horizontal.

NOMINATED SUBCONTRACTOR Contractor who contracts with main contractor.

NOMINATED SUPPLIER Supplier that supplies goods or services selected by owner, not general contractor.

NOMOGRAM Diagram where relationship between three variables is represented by straight line or curve for each variable.

NOMOGRAPH Diagram where relationship between three variables is represented by straight line or curve for each variable.

NON-ABSTRACTIVE USE Use that does not consume resources.

NON-ALLUVIAL CHANNEL Channel of non-erodible material (e.g., bedrock, pipe).

NON-AQUEOUS PHASE LIQUID Contaminants that remain undiluted as original bulk liquid in subsurface(e.g., spilled oil).

NON-ARTESIAN WELL Non-flowing well where water does not rise to surface without aid of pump.

NON-ATTAINMENT AREA Area that does not meet ambient air quality standards.

NON-BINDING ALLOCATIONS OF RESPONSIBILITY Process to propose way for potentially responsible parties to allocate costs among themselves.

NON-BIODEGRADABLE Any material that's not susceptible to being decomposed by biological activity.

NON-CARBONATE HARDNESS Portion of total hardness in excess of alkalinity.

NON-CLOGGING IMPELLER Impeller with substantial passageways for passing solids.

NON-COMBUSTIBLE WASTE Inorganic solid waste.

NON-COMMUNITY WATER SYSTEM Public water system that's not community water system (e.g., water supply at national park).

NON-COMPLIANCE COAL Any coal that emits greater than 3.0 pounds of sulfur dioxide per million BTU when burned.

NON-CONCUSSIVE ACTION Slow-closing valve to prevent water hammer.

NON-CONDENSABLE Gas not easily condensed by cooling.

NON-CONFORMING USE Use of property that was allowed under zoning regulations at time use was established but that, because of subsequent changes in regulations, is no longer permitted.

NON-CONFORMING WORK Work that departures from policies, procedures or requirements.

NON-CONSERVATIVE POLLUTANT Pollutants that are mitigated by natural biodegradation or other environmental decay or removal processes in receiving stream after in-stream mixing and dilution have occurred.

NON-CONSUMPTIVE WATER USES Water uses that do not substantially deplete water supplies (i.e., swimming, boating, water skiing, fishing).

NON-CONTACT COOLING WATER Water used for cooling which does not come into direct contact with any raw material, product, byproduct or waste.

NON-CONTACT COOLING WATER WELLS Class V wells that are used to inject non-contact cooling water that contains no additives and has not been chemically altered.

NON-CONTACT WATER RECREATION Recreational activities (i.e., fishing, boating) that do not include direct contact with water.

NON-CONTRIBUTING AREA Natural formed area that does not contribute surface-water runoff to hydrologic unit outlet.

NON-CONVENTIONAL POLLUTANT Any pollutant not statutorily listed or that is poorly understood by scientific community (e.g., nitrogen, phosphorus).

NON-CORRODIBLE Resists corrosion.

NON-CRITICAL ACTIVITIES OR WORK ITEMS Activities with positive float.

NON-DEGRADATION Environmental policy which disallows any lowering of naturally occurring quality regardless of pre-established health standards.

NON-DEPLETABLE ENERGY SOURCE Sustainable energy source (e.g., solar or wind energy).

NON-DEPOSITING VELOCITY Water velocity that maintains suspended sediment during flow.

NON-DETECTABLE EMISSIONS Less than 500 parts per million by volume above background levels.

NON-DIRECTIONAL BEACON Low frequency, ground-based radio transmitter used as instrument approach for airplanes.

NON-DISCHARGING TREATMENT PLANT Treatment plant that does not discharge treated wastewater into any stream or river.

NON-ENCAPSULATING CHEMICAL PROTECTIVE SUIT Not gas or vapor tight.

NON-ERODING VELOCITY Velocity that will not cause scouring.

NON-FERROUS METALS Non-magnetic metals (e.g., aluminum, lead, copper).

NON-FILTERABLE Portion of total residue retained by filter.

NON-FIRM COMMERCIAL ENERGY Energy periodically available for sale at lower than firm power rates.

NON-FIRM POWER Power not available continuously and may be interruptible.

NON-FLOWING WELL Well that doesn't release water at surface without aid of pumping.

NON-FRIABLE ASBESTOS-CONTAINING MATERIALS Any material containing more than 1% asbestos that, when dry, cannot be crumbled, pulverized or reduced to powder by hand pressure.

NON-FRONTAL PRECIPITATION Precipitation from lifting of air that converges into low pressure region.

NON-GOVERNMENTAL ORGANIZATION Voluntary group of individuals or organizations that's formed to provide services or advocate.

NON-HOME-BASED TRIP Neither end of trip is origin or destination.

NON-IONIC SURFACTANT Surfactant where molecules stay connected when in solution.

NON-LOCAL TRANSIT SERVICE Transit service with stop spacing greater local service.

NON-MANIPULATIVE JOINT Conduit compression joint.

NON-METHANE HYDROCARBON Sum of all hydrocarbon air pollutants except methane.

NON-METHANE ORGANIC GASES Sum of all organic air pollutants except methane.

NON-OVERFLOW DAM Dam not designed to be overtopped.

NON-OVERFLOW SECTION Section of dam not designed to be overtopped.

NON-PARAMETRIC Free of assumptions about frequency distributions of variables being assessed.

NON-PERFORMANCE Failure to perform task.

NON-PERSISTENT EMERGENT PLANTS Emergent plants whose leaves and stems break down at end of growing season from decay or by physical forces of waves and ice.

NON-POINT SOURCE Pollution source (of any water-carried material) that cannot be defined as originating from discrete points (e.g., pipe discharge).

NON-POINT SOURCE CONTAMINANT Substance that pollutes or degrades water that comes from lawn or cropland runoff, atmosphere, roadways and other diffuse sources.

NON-POINT SOURCE POLLUTION Pollution discharged over wide land area, not from one specific location.

NON-POINT-SOURCE WATER POLLUTION Water contamination that originates from broad area (e.g., leaching of agricultural chemicals from crop land) and enters water resource diffusely over large area.

NON-POSITIVE DISPLACEMENT PUMP Pump that moves fluid where flow is not constant and varies in pressure.

NON-POTABLE Water unsafe or unpalatable to drink because it may contain objectionable pollution, contamination, minerals or infective agents.

NON-PREFERENCE CUSTOMER Customer excluded by law from preference in purchase of federally produced electrical energy.

NON-PRESSURE DRAINAGE Gravity flow drainage system.

NON-PRESSURE PIPE Conduit intended for gravity flow only.

NON-RECOVERABLE SLOPE Slope that vehicle may retain or regain control by slowing or stopping but unlikely to return to roadway.

NON-REGENERATIVE CARBON Chain of non-regenerative carbon beds.

NON-REIMBURSABLE Cost of constructing, operating or maintaining project that's borne by taxpayer and is not reimbursed by any other individual, entity or organization.

NON-RISING-STEM VALVE Valve stem that doesn't rise when opening.

NON-ROAD EMISSIONS Pollutants emitted by combustion engines on farm and construction equipment, gasoline-powered lawn and garden equipment and power boats and outboard motors.

NON-SELECTIVE HERBICIDE Kills or significantly retards growth of most higher plant species.

NON-SETTLEABLE MATTER Matter suspension for specific period of time.

NON-SETTLEABLE SOLIDS Suspended matter (e.g., biological organisms, bacteria, viruses, organic matter).

NON-SILTING VELOCITY Flow velocity keeping silt in suspension during flow.

NON-SIPHON TRAP Plumbing trap that prevents back siphonage where water is delivered under pressure directly from water supply.

NON-SPLITTABLE ACTIVITY Activity that needs to be finished once it has begun.

NON-TERRITORIAL COMMUNITIES Networks of associations around shared goals, values and norms (i.e., agricultural, environmental, recreational community).

NON-THRESHOLD EFFECTS Associated with exposure to chemicals that have no safe exposure levels (i.e., cancer).

NON-TIDAL CURRENT Current not induced by tidal action.

NON-TRANSIENT, NON-COMMUNITY WATER SYSTEM Public water system which supplies water to 25 or more of same people at least six months per year in places other than their residences (e.g., schools, office buildings).

NON-UNIFORM CHANNEL Channel with irregular streamlines and varying velocity.

NON-UNIFORM FLOW Flow where velocity varies with position for given time.

NON-UNIFORM SKY CONDITION Localized sky condition which varies from that reported in body of report.

NON-UNIFORM VISIBILITY Localized visibility which varies from that reported in body of report.

NON-UTILITY GENERATORS Facilities for generating electricity that are not owned exclusively by electric utility and which operate connected to electric utility system.

NON-UTILITY POWER PRODUCER Corporation, person, agency, authority or other legal entity or instrumentality that owns electric generating capacity and is not electric utility.

NON-VESSEL OPERATING COMMON CARRIER Ocean carrier that transports goods, licensed by Federal Maritime Commission, without operating ocean transportation vessels.

NON-WEAVING FLOW Traffic that does not weave in weaving roadway segment.

NONLINEAR MODEL Mathematical model using one or more nonlinear equations.

NONRENEWABLE RESOURCE Resource of economic value that cannot be readily replaced by natural means on level equal to its consumption.

NONUSE OF LAND Proclamation of failure to use land.

NOR'EASTER Strong low pressure system that affects Mid-Atlantic and New England States with continuously strong northeasterly winds blowing in from ocean ahead of storm and over coastal areas.

NORMAL 1. Long-term average value of meteorological parameter (i.e., temperature, humidity) for certain area; 2. According to norms or standards; 3. Perpendicular to tangent of curve or derivative of surface.

NORMAL AMBIENT VALUE Concentration of chemical species reasonably anticipated to be present in water column, sediments or biota in absence of disposal activities at disposal site in question.

NORMAL CONSOLIDATION Consolidation of soil in equilibrium with overburden pressure.

NORMAL DEPLETION CURVE Mean values plotted from procured of groundwater-depiction curves.

NORMAL DEPTH Depth of flow that would exist for steady-uniform flow condition.

NORMAL DISTRIBUTION Any of family of continuous probability distributions such that probability density function is Gaussian function.

NORMAL EROSION Natural land erosion caused by environment.

NORMAL FLOW Mean flow that predominates.

NORMAL HAUL Haul cost that's included in excavation cost.

NORMAL LOAD DURATION Load that fully stresses member to allowable design using full design load for ten year duration.

NORMAL LOADING CONDITIONS Loading conditions that occur or are anticipated to occur with some degree of regularity or frequency.

NORMAL MOISTURE CAPACITY Water in soil after all gravitational water is drained.

NORMAL POOL LEVEL 1. For reservoir with fixed overflow sill, lowest crest level of that sill; 2. For reservoir whose outflow is controlled wholly or partly by moveable gates, siphons or other means, maximum level that water may rise under normal operating conditions, exclusive of any provision for flood surcharge.

NORMAL POSITION Default location.

NORMAL PRESSURE Atmospheric pressure at sea level.

NORMAL RESERVOIR LEVEL 1. For reservoir with fixed overflow sill, lowest crest level of that sill; 2. For reservoir whose outflow is controlled wholly or partly by moveable gates, siphons or other means, maximum level that water may rise under normal operating conditions, exclusive of any provision for flood surcharge.

NORMAL SOLUTION Solution that contains as stated amount of reagent per liter.

NORMAL STAGE Water elevation during most of year.

NORMAL TEMPERATURE AND PRESSURE Gas industry reference base temperature of 21 °C and pressure of 1013 hPa.

NORMAL THREAD ENGAGEMENT Thread overlap required to produce tight joint.

NORMAL VELOCITY Normal depth flow velocity.

NORMAL WATER SURFACE 1. Highest elevation at which water is normally stored or that elevation which reservoir should be operated for conservation purposes (i.e., top of active conservation capacity); 2. Maximum elevation that reservoir may rise under normal operating conditions exclusive of flood control capacity.

NORMAL WATER SURFACE ELEVATION 1. For reservoir with fixed overflow sill, lowest crest level of that sill; 2. For reservoir whose outflow is controlled wholly or partly by moveable gates, siphons or other means, maximum level that water may rise under normal operating conditions, exclusive of any provision for flood surcharge.

NORMAL WEAR AND TEAR Physical depreciation that occurs naturally over time.

NORMAL YEAR Year when streamflow or precipitation approximates average for long period of record.

NORMALLY CONSOLIDATED SOIL Soil that's never experienced vertical effective stress that's greater than its present vertical effective stress.

NORMATIVE MODEL Model that asks what should be answer to problem.

NORTH AMERICAN DATUM OF 1927 Horizontal control datum that was defined by location and azimuth on Clarke spheroid of 1866.

NORTH AMERICAN ELECTRIC RELIABILITY COUNCIL Principal organization for coordinating and promoting reliability for North America's electric utilities.

NORTH AMERICAN VERTICAL DATUM OF 1988 Fixed reference adopted as official civilian vertical datum for elevations determined by federal surveying and mapping activities in United States established in 1991 by minimum-constraint adjustment of Canadian, Mexican and United States first-order terrestrial leveling networks.

NORTH ATLANTIC OSCILLATION Large-scale fluctuation in atmospheric pressure between subtropical high pressure system located near Azores in Atlantic Ocean and sub-polar low pressure system near Iceland and is quantified in NAO Index.

NORTH PACIFIC HIGH Semi-permanent, subtropical area of high pressure in North Pacific Ocean.

NORTH WALL North side boundary of Gulf Stream generally extending northeast from Cape Hatteras where Gulf Stream turns northeast.

NORTHERN LIGHTS Luminous, radiant emission from upper atmosphere over middle and high latitudes and centered around Earth's magnetic poles.

NOTCH 1. V-shaped cut in measuring weir; 2. Mountain pass.

NOTCH PLATE Thin plates with V- or U- shape opening that act as weirs to measure water flow.

NOTCHED WEIR Weir having substantial width and notch of known hydraulic characteristics.

NOTICE OF AWARD Legal document awarding contract to favorable bidder.

NOTICE OF DEFICIENCY Request to facility owner or operator requesting additional information before preliminary decision on permit application can be made.

NOTICE OF INTENT TO CANCEL Notification sent to registrants when EPA decides to cancel registration of product containing pesticide.

NOTICE OF INTENT TO DENY Notification by EPA of its preliminary intent to deny permit application.

NOTICE OF INTENT TO SUSPEND Notification sent to pesticide registrant when EPA decides to suspend product sale and distribution because of failure to submit requested data in timely and/or acceptable manner or because of imminent hazard.

NOTICE TO AIRMEN Notice filed with aviation authority to alert aircraft pilots of potential hazards along flight route.

NOTICE TO BIDDERS Legal notice given that bids will be accepted.

NOTICE TO PROCEED Notification letter to contractor stating date contractor can begin project work.

NOTIFICATION 1. To inform appropriate individuals about emergency condition so they can take appropriate action; 2. Third of five Early Warning System components consisting of communicating alerts and warnings about emergency condition at dam to appropriate local officials so they can take proper action(s).

NOWCAST Short-term weather forecast, generally out to six hours or less.

NOWCASTING Weather forecast predicting weather for very short upcoming period (i.e., few hours).

NOZZLE Short tube forming vent of hose or pipe.

NOZZLE AERATOR Short tube that sprays water into atmosphere.

NOZZLE JET Concentrated stream of water issued from nozzle.

NPDES PERMIT National Pollutant Discharge Elimination System (NPDES) permit to regulate point sources that discharge pollutants into waters of United State.

NUCLEAR ENERGY Energy released by nuclear reaction (i.e., fission or fusion).

NUCLEAR FISSION Nuclear reaction where large nucleus splits into smaller ones with simultaneous release of energy.

NUCLEAR FUSION Combining of nuclei of small atoms to form nuclei of larger ones, with resulting release of large quantities of energy.

NUCLEAR POWER Power obtained using nuclear fission or nuclear fusion.

NUCLEAR RADIATION Energy given off from radioactive element as it decays.

NUCLEAR REACTOR Any device where controlled chain reaction is maintained for purpose of creating heat or for creating neutrons and other fission products.

NUCLEAR REACTOR WASTES Waste products generated from nuclear reactions.

NUCLEAR REACTORS AND SUPPORT FACILITIES Uranium mills, commercial power reactors, fuel reprocessing plants and uranium enrichment facilities.

NUCLEAR WINTER Prediction that smoke and debris rising from massive fires of nuclear war could block sunlight for weeks or months, cooling Earth's surface and producing climate changes.

NUCLEIN Phosphorus-rich protein found in nucleus of cell.

NUCLEUS 1. Positively charged central part of atom made up of protons and neutrons; 2. Core, central part of something, around which other elements are assembled.

NUCLIDE Atom characterized by number of protons, neutrons and energy in nucleus.

NUISANCE Anything harmful or offensive to community or to member of it, that legal remedy exists.

NUISANCE SPECIES Undesirable plants and animals.

NUISANCE THRESHOLD Concentration where no more than small proportion of population experiences annoyances.

NULL Non-existent or empty value or set of values.

NUMBER Abstract entity used to describe quantity.

NUMERICAL EXPERIMENTS Varying input data or internal parameters of numerical model to ascertain impact on output of those variations.

NUMERICAL FORECASTING Computer forecast or prediction based on equations governing motions and forces affecting motion of fluids.

NUMERICAL METHODS Set of procedures used to solve equations where applicable partial differential equations are replaced by set of algebraic equations written in terms of discrete values of state variables at discrete points in space and time.

NUMERICAL MODEL Mathematical model that uses numerical methods to solve governing equations of applicable problem.

NUMERICAL SOLUTION Approximate solution of governing (partial) differential equation derived by replacing continuous governing equation with set of equations in discrete points of model's time and space domains.

NUMERICAL WEATHER PREDICTION Computer forecast or prediction based on equations governing motions and forces affecting motion of fluids.

NUMEROUS National Weather Service convective precipitation descriptor for 60 or 70% chance of measurable precipitation (0.01 inch).

NUTRIENT Any substance, element or compound essential for animal and plant growth.

NUTRIENT CYCLE Movement and exchange of organic or inorganic matter back into production of living matter.

NUTRIENT POLLUTION Contamination of water resources by excessive inputs of nutrients.

NUTRIENTS Sources of nourishment that can be metabolized by organism to give energy and build tissue.

NWR NOAA WEATHER RADIO NOAA Weather Radio broadcasts National Weather Service warnings, watches, forecasts and other hazard information 24 hours day.

NYMPH Larva of certain insects.

O

O-RING Rubber seal used around stems of some valves to prevent water from leaking past.

OAKUM Loosely woven hemp rope that has been treated with oil or other waterproofing agent used to caulk joints in bell and spigot pipe and fittings.

OBJECT MARKER Mark obstructions within or adjacent to roadway.

OBJECTIVE FUNCTION Mathematical expression that allows comparison between calculated result and specified goal.

OBLIGATE AEROBES Organism that requires oxygen to grow.

OBLIGATE RIPARIAN SPECIES Species that depends completely upon habitat along body of water.

OBLIGATION Legal agreement stipulating specified payment or action.

OBSCURATION Any atmospheric phenomenon, except clouds, that restricts vertical visibility (e.g., dust, rain, snow).

OBSCURING PHENOMENA Any atmospheric phenomenon, except clouds, that restricts vertical visibility (e.g., dust, rain, snow).

OBSERVATION WELL Non-pumping well (hole) used for observing water level, water temperature or water quality at atmospheric pressure within soil or rock.

OBSERVED EFFECT CONCENTRATION Lowest concentration of material used in aquatic toxicity test, with controls, that has statistically significant adverse effect on exposed population of test organisms compared.

OBSTRUCTION LIGHT Light that helps mark tall structures that occupy airspace.

OCCASIONAL STORM Precipitation intensity expected once every 10 to 25 years.

OCCLUDED FRONT Composite of two fronts, formed as cold front overtakes warm or quasi-stationary front.

OCCLUDED MESOCYCLONE Mesocyclone where air from rear-flank downdraft has completely enveloped circulation at low levels, cutting off inflow of warm unstable low-level air.

OCCUPANCY REQUIREMENT Any restriction that specifies minimum number of persons in vehicle.

OCCUPANT Person who occupies office, position, place or vehicle.

OCCUPANT IMPACT VELOCITY Theoretical impact velocity of occupant with vehicle interior.

OCCUPATIONAL EXPOSURE LIMIT Maximum concentration of toxic substance in air where worker can be exposed over period of time without suffering any harmful consequences.

OCCUPATIONAL EXPOSURE STANDARD Regulations mandated by government that limit airborne levels of substance in workplace that cause adverse effect to majority of workers.

OCCUPATIONAL SAFETY AND HEALTH ACT OF 1970 Law designed to protect health and safety of industrial workers.

OCCURRENCE AND DISTRIBUTION ASSESSMENT Characterization of broad-scale spatial and temporal distributions of water-quality conditions in relation to major contaminant sources and background conditions for surface water and groundwater.

OCEAN One of large bodies of water separating continents.

OCEAN CURRENT Any more or less permanent or continuous, directed movement of water that flows in one of Earth's oceans.

OCEAN DISCHARGE WAIVER Variance from Clean Water Act requirements for discharges into marine waters.

OCEANIC Of or relating to ocean.

OCEANIC CLIMATE Climate determined by ocean.

OCEANIC TRENCH Hemispheric-scale long but narrow topographic depression of seabed.

OCEANOGRAPHY Scientific study of oceans and ocean floor.

OCTANE NUMBER Measure of antiknock properties of automobile fuel.

OCTAVE 1. Pitch octave higher than given pitch; 2. Subjectional vibration of planet.

ODOR Any smell, whether fragrant or offensive.

ODOR PANEL Members that evaluate smell of items.

ODOR THRESHOLD 1. Lowest concentration of certain odor compound detected by smell; 2. Minimum odor of water sample that can just be detected after successive dilutions with odorless water.

ODORANT Any substance that has distinctive smell.

OFF-GAS Gas produced as side effect of chemical reaction.

OFF-HIGHWAY TRUCK Truck with size, weight or dimensions that it cannot be used on public highways.

OFF-PEAK DIRECTION During peak periods, off-peak direction traffic.

OFF-PEAK ENERGY Electric energy supplied during periods of relatively low system demand.

OFF-PEAK POWER Electrical energy generated during times when most are not using energy.

OFF-PEAK, ELECTRIC HEATING SYSTEM System where electric heat source is its main component with supplemental heating source for times of peak electrical demand.

OFF-PREMISE SIGN Sign displaying advertising that pertains to business, person, place, etc.

OFF-RAMP Segment of roadway that directs vehicular traffic from freeway onto ordinary roads.

OFF-SITE FACILITY Hazardous waste treatment, storage or disposal area located away from generating site.

OFFENSIVE INDUSTRY Industry that requires to be isolated from other buildings, when all measures to reduce or minimise impact have been employed.

OFFICE OF EQUAL OPPORTUNITY AND DIVERSITY MANAGEMENT Advises and assists responsibilities relative to Civil Rights laws, Executive Orders, regulatory guidelines and other nondiscrimination laws within Federal Government by fostering inclusive workforce, building environment that respects individual and offering opportunities for all employees to develop to their full potential.

OFFICE OF GLOBAL PROGRAMS Sponsors focused scientific research, within approximately eleven research elements, aimed at understanding climate variability and its predictability.

OFFICE PAPER High grade papers (e.g., copier paper, computer printout) and stationary almost entirely made of uncoated chemical pulp, although some ground wood is used.

OFFSET 1. Short distance measured at right angles from line actually run to some point in irregular boundary or to some object; 2. Abrupt bend in object, such as rod, where one part is turned aside out of line, but nearly parallel, with rest; 3. Short prostrate shoot that takes root and produces tuft of leaves, etc.; 4. Anything that acts as counterbalance.

OFFSET ELBOW S-shaped conduit fitting.

OFFSET FITTING Conduit fitting with two direction changes.

OFFSETS Concept whereby emissions from proposed new or modified stationary sources are balanced by reductions from existing sources to stabilize total emissions.

OFFSHORE BAR Mounds of sand deposited by currents and waves.

OFFSHORE BREEZE 1. Wind that blows from land towards body of water; 2. Land breeze.

OFFSHORE FLOW Occurs when air moves from land to sea and is usually associated with dry weather.

OFFSHORE TERRACE Sand mound in offshore zone.

OFFSHORE WATERS That portion of oceans, gulfs and seas beyond coastal waters extending to specified distance from coastline, to specified depth contour or covering area defined by specific latitude and longitude points.

OFFSHORE WATERS FORECAST National Weather Service marine forecast product for that portion of oceans, gulfs and seas beyond coastal waters extending to specified distance from coastline, to specified depth contour or covering area defined by specific latitude and longitude points.

OFFSTREAM USE Water withdrawn from surface or groundwater sources for use at another place.

OFFTRACKING Path of outside of outer tire on rear that deviates inward toward center of turn from circular path of outside of outer front tire while vehicle is making turn.

OGEE Reverse curve, shaped like elongated letter S.

OGEE CREST Shape of concrete spillway crest that represents lower profile of undernappe of jet of water flowing over sharp-crested weir at design depth.

OGEE SPILLWAY Spillway that has control weir ogee-shaped (S-shaped) in profile.

OHM Unit of electric resistance between two points of conductor when constant potential difference of 1 volt applied between these two points produces current of 1 ampere.

OHMMETER Portable device for measuring relatively small values of electrical resistance.

OHMS-PER-VOLT Evaluation revealing sensitivity of electric meter.

OIL AND GAS WASTE Gas and oil drilling muds, oil production brines and other waste associated with exploration for, development and production of crude oil or natural gas.

OIL DESULFURIZATION Pre-combustion method for reducing sulfur dioxide emissions from oil-burning power plants where oil is treated with hydrogen, which removes some of sulfur by forming hydrogen sulfide gas.

OIL FIELD Area under which are reserves of petroleum.

OIL FINGERPRINTING Method that identifies sources of oil and allows spills to be traced to their source.

OIL SHALES Sedimentary rock containing kerogen from which petroleum-like shale oil can be produced by pyrolysis, hydrogenation or thermal dissolution.

OIL SLICK Thin film of oil floating on surface of water.

OIL SPILL Accidental or intentional discharge of oil which reaches bodies of water.

OIL SPILL CONTINGENCY FUND Revolving fund for spill control efforts has been authorized in cases where federal government has taken over containment and cleanup operations.

OIL-WATER SEPARATOR Device that separates oil and suspended solids from wastewater effluents.

OILY WASTEWATER Wastewater generated during refinery process containing oil, emulsified oil or other hydrocarbons.

OKTA Used for measurement of total cloud cover where one okta of cloud cover is equivalent of $1/8$ of sky covered with cloud.

OLD RIVER Dry river bed due to change of course.

OLD VALLEY Very wide stream valley with flat bottom and low sloping sides.

OLIGOTROPHIC Reservoirs and lakes that are nutrient poor, little organic matter (aquatic plant or animal life) and high dissolved-oxygen level.

OLIGOTROPHIC LAKE Lake with low primary productivity due to low nutrient content.

OMBROMETER Instrument for measuring rainfall.

OMEGA Term used to describe vertical motion in atmosphere determined by amount of spin (i.e., large scale rotation) and warm (i.e., cold) advection present in atmosphere.

OMEGA HIGH Warm high aloft which has become displaced and is on polarward side of jet stream.

OMNIVORE Animal that eats both vegetable and animal substances.

ON-AND-OFF RIDE CHECK Passenger record (recorded by checker on bus) of number who board and alight at each stop on route.

ON-DISTRICT STORAGE Small water storage facilities located within boundaries of irrigation entity, including re-regulating reservoirs, holding ponds or other new storage methods that allow for efficient water use.

ON-FARM Activities (especially growing crops and applying irrigation water) that occur within legal boundaries of private property.

ON-FARM IRRIGATION EFFICIENCY Ratio of volume of water used for consumptive use and leaching requirements in cropped areas to volume of water delivered to farm (applied water).

ON-HIGHWAY TRUCK Paved surfaces designed vehicle.

ON-LINE REPLACEMENT Provides replacement pipe within deteriorated section.

ON-PEAK ENERGY Electric energy supplied during periods of relatively high system demand.

ON-PREMISE SIGN Sign that advertises products or services that are sold, produced, manufactured or furnished on property where sign is located.

ON-RAMP Segment of roadway that directs vehicular traffic from local roads onto freeway.

ON-SCENE COORDINATOR Federal official to coordinate and direct federal removal efforts at scene of oil or hazardous substance discharge.

ON-SITE FACILITY Hazardous waste treatment, storage or disposal area located on generating site.

ON-STREET TRANSITWAY Street right-of-way, segregated from other traffic, dedicated to transit.

ON/OFF-HIGHWAY TRUCK Vehicle made for paved and rough surfaces.

ONBOARD CONTROLS Devices placed on vehicles to capture gasoline vapor during refueling and route it to engines when vehicle is starting so that it can be efficiently burned.

ONCE-THROUGH OPERATION Treatment where flow is processed once.

ONCONOGENICITY Capacity to induce cancer.

ONE HUNDRED YEAR FLOOD Flood level with 1% chance of magnitude being equaled or exceeded in any given year.

ONE HUNDRED-YEAR FLOODPLAIN Land area susceptible to being inundated by water from any source during 100-year flood.

ONE PERCENT CHANCE FLOOD Statistic that indicates magnitude of flood that can be expected to be equaled or exceeded during any year.

ONE TWO THREE RULE Means of avoiding winds associated with tropical cyclone by taking into account forecast track error of National Weather Service over 10 year period that is approximately 100 nautical miles in 24 hours, 200 nautical miles for 48 hours and 300 nautical miles in 72 hours where forecast track error is added to 34 knot wind radii to compute danger area.

ONE-DIMENSIONAL MODEL Model defined with one space coordinate (i.e., variables are averaged over other two directions.)

ONE-HIT MODEL Mathematical model based on biological theory that single "hit" of some minimum critical amount of carcinogen at cellular target (e.g., DNA can start irreversible series events leading to tumor).

ONE-HUNDRED-YEAR FLOOD Flood event that has 1% probability of occurring in any given year.

ONE-PIPE SYSTEM Two perpendicular conduits are connected to same pipe, that acts as both flow pipe and return pipe.

ONE-SIDED WEAVING SEGMENT Weaving segment where no weaving maneuver requires more than two lane changes.

ONE-TO-MANY From single source to multiple recipients.

ONSHORE 1. Moving from sea towards land; 2. Positioned on or near shore.

ONSHORE BREEZE 1. Wind that blows from body of water towards land; 2. Seabreeze.

ONSHORE FLOW Occurs when air moves from sea to land and is usually associated with increased moisture.

ONSITE DISPOSAL Volume decrease of refuse onsite before removal for final disposal.

ONSITE INCINERATOR Incinerator that burns waste onsite.

ONSITE RENEWABLE ENERGY GENERATION Electricity generated by renewable resources using system or device located at site where power is used.

OPACITY 1. Amount of light obscured by particulate pollution in air; 2. Measure of relative impenetrability to electromagnetic radiation such as light; 3. Indicator of changes in performance of particulate control systems.

OPAQUE Condition where material (e.g., cloud) blocks passage of radiant energy, especially light.

OPEN Not drawn together, closed or contracted.

OPEN ACCESS SAME-TIME INFORMATION SYSTEM Electronic information system that allows users to instantly receive data on current operating status and transmission capacity of transmission provider.

OPEN BIDDING/TENDERING System of public promotion of building or material supply conditions.

OPEN BURNING 1. Burning of unwanted materials where smoke or emissions are released directly into air; 2. Uncontrolled fires in open dump.

OPEN CENTER VALVE Valve that connects all lines directly back to reservoir, bypassing pump.

OPEN CENTRIFUGAL PUMP Pump having impeller constructed with separate vanes.

OPEN CHANNEL Type of flow where one surface is not restricted.

OPEN CHANNEL FLOW Liquid flow within conduit with free surface.

OPEN CUT Surface mining technique of extracting material from earth from open pit or borrow.

OPEN DUMP Uncovered site used for disposal of waste without environmental controls.

OPEN GEAR Gear subjected to environmental surroundings.

OPEN GEARING Gears not sealed.

OPEN GRADED AGGREGATE 1. Aggregate containing little or no mineral filler; 2. When void spaces in compacted aggregate are relatively large.

OPEN GRID FLOOR Raised, perforated floor.

OPEN IMPELLER Impeller where fluid enters eye of impeller where turning vanes add energy to fluid.

OPEN INTERVAL Length of unscreened opening or of well screen where water enters well, in feet below land surface.

OPEN JOINT In pipe, flat ends that meet but do not overlap.

OPEN RIB Rib in orthotropic deck parallel to each other in longitudinal direction.

OPEN SPOOL VALVE Valve that allows flow of hydraulic oil when in neutral.

OPEN TOP TRUCK Vehicle with no top assembly.

OPEN TRANSMISSION ACCESS Enables all participants in wholesale market equal access to transmission service, as long as capacity is available, with objective of creating more competitive wholesale power market.

OPEN WELL Cylindrical shaft dug into ground with opening not covered by permanent cap.

OPEN-BOTTOMED WELL Well where bottom of casing is open to allow inflow of water.

OPEN-CHANNEL DRAINAGE Ditch drainage.

OPEN-CHANNEL FLOW Fluid flow with at least one free surface subjected to surroundings.

OPEN-CIRCUIT APPARATUS Breathing apparatus where wearer's exhalations are vented to atmosphere.

OPEN-CUT Method of excavation where working area is kept open to sky.

OPEN-CUT MINE Surface mining technique of extracting rock or minerals from earth by their removal from open cut mine.

OPEN-CUT TRENCHING Excavating down to and exposing existing pipe so that it can be repaired or replaced and backfilled.

OPEN-PIT INCINERATOR Device used to burn waste for primary purpose of reducing volume by removing combustible matter.

OPEN-ROAD ETC LANE Non-attended lane designed to allow electronic toll payments from vehicles traveling at normal speeds.

OPEN-ROAD TOLLING Collection of tolls without use of toll booths.

OPEN-ROAD TOLLING POINT Location along Open-Road ETC lane where detection and receiving equipment are placed to electronically assess toll.

OPEN-TOP CULVERT Culvert that channels water off road and can withstand normal vehicle traffic.

OPEN-WORK MATERIALS Poorly-graded (uniform or gap-graded gradation) gravels, cobbles and boulders with few fines in matrix, resulting in deposit containing large amount of interconnected void space where seepage water (and soil particles) can easily move.

OPERABLE TREATMENT WORKS Treatment works that will treat, transport or dispose of wastewater.

OPERABLE UNIT Term for each of number of separate activities undertaken as part of Superfund site cleanup.

OPERATING BAND Pressure ranges above and below operating pressure.

OPERATING BASIS EARTHQUAKE Earthquakes for which structure is designed to resist and remain operational.

OPERATING CHARACTERISTICS CURVE Curve that plots probabilities of accepting lot versus fraction defective.

OPERATING CONDITIONS Conditions specified in permit that dictate how incinerator must operate as it burns different waste types.

OPERATING COSTS Expenses related to operation of business, device, component, piece of equipment or facility.

OPERATING DEVICE Device used to actuate control.

OPERATING ENGINEER Heavy equipment operators, heavy equipment mechanics or service oilers.

OPERATING EXPENSE 1. Any expense incurred in running business, such as sales and administration, as opposed to production; 2. Money entity spends in order to turn inventory into throughput.

OPERATING FLOOR Floor in building where controls may be used and signaling devices are found.

OPERATING GATE Gate or valve that operates under full pressure flow conditions to regulate rate of discharge, flow through outlet works or spillway.

OPERATING INSTRUCTIONS Details how something should be operated.

OPERATING LOG Dated, written record of performed operation and maintenance items or observations pertinent to structure.

OPERATING POLICY Rule that specifies how water is managed throughout water resource system.

OPERATING PRESSURE Amount of internal force applied to walls of some type of pressure vessel during normal conditions.

OPERATING RATING Maximum permissible live load that can be placed on bridge.

OPERATING RATIO Percentage of operating expenses to revenue.

OPERATING RESERVE Generation capacity dedicated to maintaining adequate and reliable system operation during sudden reduction in system capacity.

OPERATING RULE Rule that specifies how water is managed throughout water resource system.

OPERATING SPEED Speed that motor vehicles generally operate on road.

OPERATION SELECTOR SWITCH Switch that can be set to chosen mode of functioning.

OPERATIONAL LOSSES Losses of water resulting from evaporation, seepage and spills.

OPERATIONAL PRODUCTS Product that has been fully tested and evaluated and is produced on regular and ongoing basis.

OPERATIONAL WASTE Water lost or otherwise discarded from irrigation system after having been diverted into it as part of normal operations.

OPERATIONS AND MAINTENANCE Activities related to performance of routine, preventive, predictive, scheduled and unscheduled actions aimed at preventing equipment failure or decline with goal of increasing efficiency, reliability and safety.

OPERATIONS MANUAL LETTER Serve as updates to policy and procedure for National Weather Service Operations Manual.

OPERATIONS-LEVEL ANALYSIS Application where user all required inputs to procedure.

OPERATOR 1. Mechanism that works another apparatus; 2. Function or other mapping that carries variables defined on domain into another variable or set of variables in defined range.

OPERATOR CERTIFICATION Certification of operators of community and non-transient non-community water systems, asbestos specialists, pesticide applicators, hazardous waste transporter and other such specialists.

OPERATOR PROTECTIVE STRUCTURE Device on articulating or telescopic boom lifts that provides secondary protection for operator.

OPERATOR'S MANUAL Instructional book t supplied with almost all technologically advanced products.

OPPOSING APPROACH Approach opposite of subject approach.

OPPOSING FLOW RATE Flow rate opposite of travel direction.

OPPOSING TRAFFIC Vehicles traveling in opposite direction.

OPTICAL DENSITY Degree to which refractive medium retards transmitted rays of light.

OPTIMAL CORROSION CONTROL TREATMENT Erosion control treatment that minimizes lead and copper concentrations at users' taps while also ensuring that treatment does not cause water system to violate any national primary drinking water regulations.

OPTIMUM AIR SUPPLY Quantity of air necessary to produce maximum efficiency.

OPTIMUM MOISTURE CONTENT 1. Moisture content to obtain maximum dry unit weight; 2. One water content (percent of dry weight of total material) of given soil and given compactive effort that will result in maximum dry density of soil.

OPTIMUM POINT OF COAGULATION Hydrogen-ion concentration in coagulation process where greatest floc appears in least time.

OPTIMUM POPULATION Population that Earth can support sustainably.

OPTIMUM WATER CONTENT One water content (percent of dry weight of total material) of given soil and given compactive effort that will result in maximum dry density of soil.

OPTION One of set of choices that can be made.

OPTION VALUE 1. Value associated with people who know they can visit area in future if they so desire; 2. Reversible decision or option to develop at some time in future have option value.

ORAL TOXICITY Ability of pesticide to cause injury when ingested.

ORDINARY FIRE HAZARD, GROUP 1 Quantity of combustibles is moderate, combustibility of contents is low, storage heights are 8 feet or less and moderate rates of heat release expected.

ORDINARY FIRE HAZARD, GROUP 2 Quantity of combustibles is moderate to high, combustibility of contents is moderate to high, storage heights do not exceed 12 feet and moderate to high rates of heat release expected.

ORDINARY FIRE HAZARD, GROUP 3 Combustibility of contents is high with fires of high rates of heat release are expected.

ORDINARY HIGH WATER Average high water mark that defines lateral limits of Federal jurisdiction.

ORDINARY STORM Rain intensity that could be expected to happen once every 5 years.

ORDINATE Number whose absolute value (modulus) is perpendicular distance of point from horizontal axis (i.e., X-axis).

ORE Naturally-occurring material from which metal or valuable mineral can be profitably extracted.

ORE BODY Large deposit of mineral or minerals, below earth's crust.

ORE DRESSING Any treatment of ore used to concentrate its valuable constituents (minerals) into products (concentrate) of smaller bulk and simultaneously to collect worthless material (gangue) into discardable waste (tailings).

ORGAN 1. Larger part of organism, composed of tissues that perform similar functions; 2. Species of cactus.

ORGANIC 1. Any chemical containing element carbon, referring to or derived from living organisms (e.g., from animal or plant sources).

ORGANIC CARBON Measure of organic matter present in aqueous solution, suspension or bottom sediment.

ORGANIC CHEMICALS Naturally occurring (animal or plant-produced or synthetic) substances containing mainly carbon, hydrogen, nitrogen and oxygen.

ORGANIC COMPOUNDS Chemicals that contain carbon.

ORGANIC CONTAMINANTS Carbon-based chemicals (e.g., solvents, pesticides) that can get into water through runoff from cropland or discharge from factories.

ORGANIC DEPOSIT Deposit formed by organisms or their remains.

ORGANIC DETRITUS Any loose organic material in streams (i.e., leaves, bark, twigs) removed and transported by mechanical means (e.g., disintegration, abrasion).

ORGANIC FARMING Farming based on biological methods that avoid use of synthetic crop or livestock production inputs.

ORGANIC MASS OF LIVING SUBSTANCE Difference between dry mass and ash mass and represents actual mass of living matter.

ORGANIC MATTER Plant and animal residues or substances made by living organisms that are based upon carbon compounds.

ORGANIC MATTER DEGRADATION Conversion of organic to inorganic matter.

ORGANIC NITROGEN Nitrogen combined in organic molecules.

ORGANIC SILT Particles varying in size from 0.002 to 0.003 inches containing organic substances.

ORGANIC SOIL Soil that contains more than 20% organic matter in upper 16 inches.

ORGANIC WASTE Biodegradable material that comes from plants or animals.

ORGANISM Any form of animal or plant life.

ORGANISM COUNT AREA Number of organisms collected and enumerated in sample and adjusted to number per area habitat.

ORGANISM COUNT VOLUME Number of organisms collected and enumerated in sample and adjusted to number per sample volume.

ORGANIZATION OF ECONOMIC AND COOPERATIVE DEVELOPMENT GUIDELINES Testing guidelines of United Nations that assist in preparation of protocols for studies of toxicology, environmental fate, etc.

ORGANOCHLORINE Any of very many chlorine substituted organic compounds (e.g., insecticides).

ORGANOCHLORINE COMPOUND Synthetic organic compounds containing carbon and chlorine.

ORGANOCHLORINE INSECTICIDE Class of organic insecticides containing high percentage of chlorine.

ORGANOCHLORINE PESTICIDE Class of organic pesticides containing high percentage of chlorine.

ORGANOMERCURY FUNGICIDES Toxic compounds that have compounds of mercury.

ORGANONITROGEN HERBICIDES Group of herbicides consisting of nitrogen ring with associated functional groups and including such classes as triazines and acetanilides (e.g., atrazine, cyanazine, alachlor, metolachlor).

ORGANOPHOSPHATE INSECTICIDES Class of insecticides derived from phosphoric acid that tend to have high acute toxicity to vertebrates.

ORGANOPHOSPHATES Pesticides that contain phosphorus that are short-lived, but some can be toxic when first applied.

ORGANOPHOSPHORUS COMPOUNDS Primarily used as pesticides (e.g., malathion, diazonon), but many are chemical warfare agents.

ORGANOPHOSPHORUS INSECTICIDES Insecticides derived from phosphoric acid and generally most toxic of all pesticides to vertebrate animals.

ORGANOPHYLLIC Substance that easily combines with organic compounds.

ORGANOTINS Chemical compounds used in anti-foulant paints to protect hulls of boats and ships, buoys and pilings from marine organisms (e.g., barnacles).

ORIENTATION EXERCISE Activity designed to introduce, discuss and update emergency planning documents, organization structure or early warning system component to familiarize key personnel with emergency procedures and their responsibilities.

ORIFICE 1. Opening with closed perimeter that water may flow, used for purpose of measurement or control of water; 2. End of small tube (i.e., pitot tube, piezometer).

ORIFICE FEED TANK Apparatus that receives chemical solution applied to flow solution.

ORIFICE FLOW Flow through orifice of conduit when inlet is completely submerged.

ORIFICE METER Flow meter used to measure rate of flow of liquid or gas using differential pressure measurement principle.

ORIFICE PLATE Device used for measuring flow rate, reducing pressure or restricting flow.

ORIFICE RESTRICTOR Device that controls flow rates.

ORIGIN 1. Source of river, information, goods, etc.; 2. Beginning of something (e.g., trip).

ORIGINAL AHERA INSPECTION Examination of school buildings arranged by Local Education Agencies to identify asbestos-containing-materials, evaluate their condition and take samples of materials suspected to contain asbestos that's typically performed by EPA-accredited inspectors.

ORIGINAL GENERATION POINT Where regulated medical or other material first becomes waste.

ORIGINAL GROUND SURFACE Surface of Earth as it exists in unaltered state (i.e., prior to any earthwork).

OROGENY Process of mountain building by upward folding of Earth's crust.

OROGRAPHIC Related to or caused by, physical geography (e.g., mountains, sloping terrain) and their general effect on storm path and generation of rainfall.

OROGRAPHIC LIFTING Occurs when air is forced to rise and cool due to terrain features (e.g., hills, mountains).

OROGRAPHIC PRECIPITATION Precipitation that is caused by hills or mountain ranges deflecting moisture-laden air masses upward, causing them to cool and precipitate their moisture.

OROGRAPHIC UPLIFT Occurs when air is forced to rise and cool due to terrain features (e.g., hills, mountains).

OROGRAPHIC WAVES Wavelike airflow produced over and in lee of mountain barrier.

ORPHAN ANVIL Anvil from dissipated thunderstorm, below which no other clouds remain.

ORTHOPHOSPHATE Any salt or ester of orthophosphoric acid; ordinary phosphate.

ORTHOTOLIDINE Colorimetric indicator of chlorine residual.

ORTHOTOLIDINE REAGENT Solution to ascertain colorimetrically available chlorine in solution.

ORTHOTOLIDINE-ARSENITE TEST Technique to determine amount of free chlorine residual.

ORTHOTROPIC 1. Growing vertically; 2. Having material properties that differ along three mutually orthogonal twofold axes of rotational symmetry.

ORTHOTROPIC DECK Bridge deck typically comprised of structural steel deck plate stiffened either longitudinally or transversely or in both directions.

OSCILLATING GRATE Oscillating surface grate that transfers fuel and deposits from feed to discharge.

OSCILLATION 1. Regular, periodic fluctuation in value about some mean; 2. Shift in position of various high and low pressure systems that in climate terms is usually defined as index.

OSMOSIS Passage of liquid from weak solution to more concentrated solution across semipermeable membrane that allows passage of solvent (e.g., water) but not dissolved solids.

OSMOTIC PRESSURE Hydrostatic pressure exerted by solution across semipermeable membrane from pure solvent; pressure needed to counteract osmosis.

OSMOTIC WATER TRANSPORT Movement of water between detached compartments by way of membrane.

OS&Y VALVE Valve that provides fire protection that has handle to raise and lower gate of valve by interacting directly with stem of valve.

OTHER FERROUS METALS Recyclable metals from strapping, furniture and metal found in tires and consumer electronics.

OTHER GLASS Recyclable glass from furniture, appliances and consumer electronics.

OTHER NON-FERROUS METALS Recyclable non-ferrous metals (e.g., lead, copper, zinc) from appliances, consumer electronics and non-packaging aluminum products.

OTHER PAPER Recyclable paper from books, third-class mail, commercial printing, paper towels, plates and cups, posters, photographic papers, cards and games, milk cartons, folding boxes, bags, wrapping paper and paperboard.

OTHER PLASTICS Recyclable plastic from appliances, eating utensils, plates, containers, toys and various kinds of equipment.

OTHER SOLID WASTE Recyclable non-hazardous solid wastes, other than municipal solid waste.

OTHER WOOD Recyclable wood from furniture, consumer electronics cabinets and other non-packaging wood products.

OTTAWA SAND Naturally rounded grains, obtained from deposits near Ottawa, IL, of nearly pure quartz, produced by processing silica sand.

OUNCE Unit of mass equal to 1/16 lb.

OUT-OF-POCKET Of or pertaining to spending of cash rather than using credit.

OUTBOUND TRIP Travelling to another business district or country.

OUTBURST BANK Central section of slope of sea embankment, between footing and swash bank.

OUTBURST FLOOD Icelandic term meaning glacier dammed lake outburst flood.

OUTCROP 1. Piece of land that stands out (usually into water) from land surrounding; 2. Coming out of bedrock or of unconsolidated deposit to surface of ground.

OUTDOOR AIR SUPPLY Air brought into building from outside.

OUTER CONVECTIVE BAND Bands in hurricane that occur in advance of main rain shield and up to 300 miles from eye of hurricane.

OUTER MARKER BEACON Beacon located 4 to 7 nautical miles from threshold to mark point where glide slope altitude is verified or where descent without glide slope is initiated.

OUTER SEPARATION Area between traveled ways of highway for through traffic and frontage road or service road.

OUTFALL 1. Outlet, structure, point or place where effluent (e.g., sewer, drain, stream, culvert) is discharged into receiving waters.

OUTFALL SEWER Sewer that receives wastewater from collection system or treatment plant and carries it to point of ultimate or final discharge in environment.

OUTFLOW 1. Amount of water passing given point downstream of structure; 2. Water flowing out of body of water; 3. Air that flows outward from thunderstorm.

OUTFLOW BOUNDARY Storm-scale or mesoscale boundary separating thunderstorm-cooled air (outflow) from surrounding air.

OUTFLOW CHANNEL Natural stream channel which transports reservoir releases.

OUTGO Conduit tail from sanitary fitting that joins to drain.

OUTGOING LONGWAVE RADIATION Polar satellite derived measurement of radiative character of energy radiated from warmer Earth surface to cooler space.

OUTHOUSE Outbuilding containing toilet or seat over cesspit.

OUTLET 1. Opening where water can be freely discharged from reservoir; 2. River that runs out of lake; 3. Vent or similar passage to allow escape of something; 4. Wall-mounted device such as socket or receptacle connected to electrical system.

OUTLET CAPACITY Amount of water that can be safely released through outlet works.

OUTLET CHANNEL 1. Channel downstream from terminal structure that conveys releases back to "natural" stream or river; 2. System that carries water far from storage facility.

OUTLET CONTROL When culvert barrel is not capable of conveying as much flow as inlet opening will accept.

OUTLET DISCHARGE STRUCTURE Structure that protects downstream end of outlet pipe from erosion or designed to slow down velocity of released water to prevent erosion of stream channel.

OUTLET GATE Gate controlling flow of water through reservoir outlet.

OUTLET HOSE NOZZLE Relatively little hose connection on hydrant (e.g., 2.5 inch in diameter).

OUTLET PIPE 1. Conduit that carries effluent from sewage treatment plant to its point of release; 2. Section of drain pipe between drain outlet of fixture and trap serving it.

OUTLET WORKS 1. Combination of structures and equipment required for safe operation and control of water released from reservoir to serve various purposes; 2. Series of components (e.g., pipe), mainly to supply downstream demands, located in dam where normal releases from reservoir are made.

OUTLET WORKS TOWER Tower within reservoir that contains mechanisms to open entrance to outlet works.

OUTLOOK Broad discussion of weather pattern expected across any given area, generally confined to forecast periods beyond 48 hours used to indicate that hazardous weather or hydrologic event may develop.

OUTPUT All information produced by computer code.

OUTPUTS Data sent out of model.

OUTSIDE AIR Atmospheric air not previously circulated through HVAC system.

OUTSIDE SCREW-AND-YOKE VALVE Control valve for sprinkler system where position of center screw indicates whether valve is open or closed.

OUTSIDE STANDPIPE Fire departments siamese connection on riser on exterior of building.

OUTWARD-FLOW TURBINE Turbine where discharge stage of flow is away from axis.

OUTWASH Soil material washed down hillside by rainwater and deposited upon more gently sloping land.

OUTWASH DEPOSIT Silt, sand and/or gravel carried by running water from melting ice of glacier and laid down in stratified deposits.

OUTWASH PLAIN Plain formed of glacial sediments deposited by meltwater outwash at end of glacier.

OVALITY 1. Measurement of deviation from circularity of oval or approximately elliptical shape; 2. Difference between maximum and minimum dimensions of outside conduit diameter.

OVER CALIBRATION Achieving artificially low residuals by inappropriately adjusting model input parameters without field data to support adjusted model parameter value.

OVER-BANK FLOW Flows greater than channel can handle.

OVER-CONSOLIDATED SOIL Stress acting on soil versus highest stress experienced by soil.

OVER-CONSOLIDATION RATIO Ratio of highest stress to current stress in soil.

OVER-SPEED SWITCH Switch that eliminates electricity from electric motor when preset speed is surpassed.

OVER-YEAR STORAGE Slow accumulation of water kept in reservoir during times of excessive supply.

OVERALL EFFICIENCY Efficiency of pump and motor together.

OVERALL REDUCTION Percent decrease in quantifiable component between start and finish.

OVERALL SAFETY OF DAMS CLASSIFICATION 1. Classifications assigned to dam following onsite examination and subsequent analyses using available data and state-of-the-art knowledge: Satisfactory - no existing or potential dam safety deficiencies are recognized; 2. Fair - no existing dam safety deficiencies are recognized for normal loading conditions; 3. Conditionally Poor - Potential dam safety deficiency is recognized for unusual loading conditions that may realistically occur during expected life of structure; 4. Poor - Potential dam safety deficiency is clearly recognized for normal loading conditions; 5. Unsatisfactory - Dam safety deficiency exists for normal loading conditions.

OVERAMPING Exceeding rated capacity of system.

OVERBAND MAGNET Electromagnet suspended over conveyor belts, transporting materials, to reclaim ferrous metals.

OVERBANK Surface area between bank on main channel and limits of floodplain.

OVERBANK FLOW natural diversion where stream surpasses bankfull stage and excess flows into nearby channel draining to different hydrologic unit.

OVERBREAK Moving or loosening of rock as result of blast, beyond intended line of cut.

OVERBURDEN Rock and soil cleared away before mining.

OVERBURDEN PRESSURE Stress imposed on layer of soil or rock by weight of overlying material.

OVERCAST When sky is completely covered by aloft obscuring phenomenon.

OVERCHUTE Overhead flume.

OVERCONSOLIDATION State that exists in case ground deposit continues to be subjected to powerful stress greater compared to present overburden pressure.

OVERDEVELOPMENT State where volume of water pumped from groundwater supply always surpasses discharge rate.

OVERDRAFT Pumping of water from groundwater basin or aquifer in excess of supply flowing into basin.

OVERDREDGING Additional depth dredged beyond minimum dredging depth used to provide sufficient navigational depth, to minimize re-dredging and to help compensate for sloughing off and resettling of sediment after dredging occurs.

OVEREXCAVATION 1. Excavation beyond specified or directed excavation; 2. Removing unsuitable foundation material.

OVERFALL Turbulent section of body of water caused by strong currents passing over submerged ridges.

OVERFALL DAM Dam that allows water to overflow dam.

OVERFALL WEIR Impounding structure over which water flows.

OVERFIRE AIR Air forced into top of incinerator or boiler to fan flames.

OVERFIRE AIR FAN Apparatus providing air, over fuel bed, to incinerator combustion chamber.

OVERFLOW 1. Spillage resultant from overflow; 2. Outlet for escape of excess material.

OVERFLOW CHANNEL Artificial hydraulic structure for conducting water away from overflowing structure.

OVERFLOW DAM Dam designed to be overtopped.

OVERFLOW DAM SECTION Section or portion of dam designed to be overtopped.

OVERFLOW QUEUE At signalized intersection, queued vehicles remaining after green phase.

OVERFLOW RATE Volume of water flow per unit of time divided by surface area of storage container.

OVERFLOW SIPHON Siphon that removes particles that are floating on water surface in tank.

OVERFLOW SPILLWAY Spillway on dam that functions like dam, but allows water to safely flow over it, typically using control weir ogee-shaped (S-shaped) in profile.

OVERFLOW STAND Standpipe where water rises to overflow at hydraulic gradient.

OVERFLOW STREAM Stream containing water that has overflowed banks of another water body.

OVERFLOW TOWER Exit from pressurized water conduit, placed at elevational peaks to restrict pressure on pipe.

OVERFLOW TUBE Vertical conduit in that prevents overflow.

OVERFLOWED LAND Land covered by unnavigable waters by tidal or seasonal flows.

OVERHANG 1. Projecting parts of face or bank; 2. Radar term indicating region of high reflectivity at middle and upper levels above area of weak reflectivity at low levels.

OVERHAUL 1. Movement of material far enough so that payment, in addition to excavation pay, is made for haulage; 2. Distance in excess of that given as stated haul distance to haul excavated material.

OVERHEAD Expense not directly assigned to goods or services provided.

OVERHEAD SIGN Structures built over high-traffic roads with several lanes.

OVERIRRIGATION Water to agricultural land over what is necessary.

OVERLAND FLOW 1. Flow of rainwater or snowmelt over land surface where there are no defined channels, flood water spreads out over large area at uniform depth, it becomes runoff; 2. Land application technique that cleanses waste water by allowing it to flow over sloped surface.

OVERLAND RUNOFF Precipitation runoff over landscape (i.e., surface runoff).

OVERPASS Section of road or path that crosses over obstacle.

OVERPOUR HEAD GATE Canal head gate where excess water discharges.

OVERPUMPING To pump groundwater excessively (i.e., extracting more water than it receives).

OVERRIDE PRESSURE Difference between cracking pressure of valve and pressure reached when valve is passing its rated flow.

OVERRIDE SWITCH Switch that operator can switch off automatic function.

OVERRUNNING Weather pattern where relatively warm air mass is in motion above another air mass of greater density at surface.

OVERSATURATED FLOW Loss in freedom of movement with respect to traffic condition.

OVERSHOOTING TOP Dome-like protrusion above thunderstorm anvil, representing very strong updraft and hence higher potential for severe weather with that storm.

OVERSTORY Upper portion of community of plants, above understory.

OVERTOPPING Flow of water over top of dam or embankment.

OVERTOPPING FLOOD Inundation of roadway where traffic services provided by facility are impacted.

OVERTURN One complete cycle of top to bottom mixing of previously thermally stratified water masses.

OVERWINDING Rope or cable wound and attached so that it stretches from top of drum to load.

OWNER One who owns something.

OWNER DEFAULT Owner failure to compensate contractor under contract conditions.

OWNER OF RECORD Entity named in public records as owner of particular property.

OWNER-CONTRACTOR AGREEMENT Agreement to perform detailed job between owner and contractor.

OWNING COSTS Machine ownership costs (e.g., equipment depreciation, interest, insurance, taxes and permits).

OXBOW 1. Bow-shaped lake formed in abandoned meander of river; 2. Land enclosed by meandering river.

OXBOW CHANNEL Natural U-shaped channel in river as viewed from above.

OXBOW LAKE Crescent-shaped lake formed near flat-valleyed river when one of its meanders becomes pinched off and isolated.

OXIDANT Collective term for some of primary constituents of photochemical smog.

OXIDATION Combination of substance with oxygen, used to break down pollutants or organic waste (i.e., destruction of chemicals in sewage by bacterial and chemical means).

OXIDATION DITCH Process that utilizes long solids retention times, in channels, to remove biodegradable organics.

OXIDATION LAGOON Pond with artificial aeration to promote biological oxidation of wastewater.

OXIDATION POND 1. Man-made (anthropogenic) body of water where waste is consumed by bacteria, used most frequently with other waste-treatment processes; 2. Sewage lagoon.

OXIDATION PROCESS Procedures designed to remove organic materials in water and wastewater by oxidation.

OXIDATION RATE Speed where organic fraction within wastewater is stabilized.

OXIDATION TREATMENT Procedures for removal and/or destruction of persistent organics in water and wastewater.

OXIDATION-REDUCTION POTENTIAL Electric potential required to transfer electrons from one compound or element (oxidant) to another compound (reductant) that's used as qualitative measure of state of oxidation in water treatment systems.

OXIDE FUEL Nuclear fuel made in oxides of fissile material.

OXIDIZABLE SALT Salt in groundwater that may be oxidized and deposited.

OXIDIZED ORGANICS Organic materials that have been broken down in biological process.

OXIDIZED SEWAGE Sewage where atmospheric or other oxygen has acted upon putrescible organic matter and rendered it stable.

OXIDIZED SLUDGE Liquid and solid sewage where organic matter has been combined with oxygen and stabilized.

OXIDIZED WASTEWATER Organics in wastewater stabilized by dissolved or atmospheric oxygen.

OXIDIZER Provider of oxygen.

OXIDIZING To combine with oxygen.

OXIDIZING AGENT Any substance that oxidizes or receives electrons from.

OXIDIZING SUBSTANCE Material that may cause or enhance combustion of other materials by yielding oxygen.

OXYGEN Colorless, odorless gas at room temperature with atomic number of 8.

OXYGEN BALANCE 1. Relationship involving biochemical oxygen demand of liquid and also dissolved oxygen accessible receiving stream; 2. Amount of dissolved oxygen at any given point in stream.

OXYGEN CONSUMED Oxygen uptake capability of organic and inorganic matter within water.

OXYGEN DEFICIENCY Insufficient oxygen to support life or fire.

OXYGEN DEFICIT Difference, in water, between saturation dissolved oxygen level and actual dissolved oxygen concentration.

OXYGEN DEMAND Need for molecular oxygen to meet needs of biological and chemical processes in water.

OXYGEN DEPLETION Reduction in concentration of dissolved oxygen in water, especially as result of pollution.

OXYGEN SAG Reduction in dissolved oxygen plotted over distance along water body from point where sewage or other pollutants have been discharged.

OXYGEN SATURATION Maximum quantity of dissolved oxygen that liquid can contain.

OXYGEN UTILIZATION Oxygen used in aerobic biological process or combustion.

OXYGENATED FUELS Gasoline which has been blended with alcohols or ethers that contain oxygen in order to reduce carbon monoxide and other emissions.

OXYGENATED SOLVENT Organic solvent containing oxygen as part of molecular structure (e.g., alcohols, ketones).

OXYGENATION CAPACITY Aerator capacity to provide oxygen to liquid.

OZONATION Application of ozone to water for disinfection or for taste and odor control.

OZONATOR Device used to purify air, water and other environmental factors with ozone (i.e., ozonation).

OZONE 1. Toxic blue gas, bubbled through water, generated from oxygen by electrical discharge, used to disinfect potable water; 2. Gas found in two layers of atmosphere, stratosphere and troposphere and major component of photochemical smog.

OZONE ACTION DAY Message when ozone levels may reach dangerous levels next day.

OZONE ADVISORY Issued when ozone levels reach 100 where levels above 100 are unhealthy for people with heat and/or respiratory ailments.

OZONE CRACKING Cracks formed by exposure to ozone.

OZONE DEPLETION Destruction of stratospheric ozone layer which shields Earth from ultraviolet radiation harmful to life caused by breakdown of certain chlorine and/or bromine containing compounds (chlorofluorocarbons or halons), which break down when they reach stratosphere and then catalytically destroy ozone molecules.

OZONE DEPLETION POTENTIAL Relative amount of degradation to ozone layer.

OZONE HOLE 1. Severe depletion of stratospheric ozone caused by chemical reaction involving ozone and chlorine, primarily from human produced sources, cloud particles and low temperatures; 2. Thinning break in stratospheric ozone layer; 3. Designation is made when detected amount of depletion exceeds fifty percent.

OZONE LAYER Protective layer in atmosphere that contains high proportion of oxygen that exists as ozone, about 15 miles above ground, that absorbs some of Sun's ultraviolet rays, thereby reducing amount of potentially harmful radiation that reaches Earth's surface.

OZONE RESISTANCE Ability of material to resist deteriorating effects of ozone exposure.

OZONOMETER Instrument for measuring how much ozone is present in mixture of gases.

P

P-TRAP Sanitary trap that captures water to prevent sewer gases from entering buildings, while allowing waste to pass through.

PACE Speed or velocity.

PACIFIC DECADAL OSCILLATION Pattern of climate variation on timescale of decades that primarily affects weather patterns and sea surface temperatures in Pacific Northwest, Alaska and northern Pacific Islands.

PACIFIC FLYWAY Established air route of migratory birds along west coast of United States.

PACIFIC TSUNAMI WARNING CENTER Center has international warning responsibility for entire Pacific and regional warning responsibility for State of Hawaii.

PACKAGE BUILDER Contractor who supplies wide selection of services toward conclusion of job.

PACKAGE PLANT Decentralized wastewater treatment system that uses biological extended aeration principle of operation to process given volume per hr/day.

PACKAGING Assembly of one or more containers and any other components necessary to ensure minimum compliance with program's storage and shipment packaging requirements.

PACKED BED SCRUBBER Air pollution control device where emissions pass through alkaline water to neutralize hydrogen chloride gas.

PACKED TOWER Pollution control device that forces dirty air through tower packed with crushed rock or wood chips while liquid is sprayed over packing material where pollutants in air stream either dissolve or chemically react with liquid.

PACKER Ring of packing or special device to render gastight and watertight space between tubing and bore of oil well.

PACKER TRUCK Refuse or recycling collection vehicle that compacts materials.

PACKING Material used to fill in space around something.

PACKING GLAND Stuffing used to seal reciprocating or rotating shaft against fluid.

PACKING NUT Nut on valve stem that compresses packaging.

PAD FOOT Tapered-shank pad using circular cross-section that projects from roller of earth compactor.

PADDLE In sluice, panel that controls flow of water.

PADDLE AERATOR Device, similar to paddle wheel shape, used in aeration of water.

PADDLE HOLE Hole that permits water to or from lock.

PADDLE WHEEL Large wheel, fitted with paddles, used to propel vessel.

PALATABLE WATER Water, at desirable temperature, free from objectionable tastes, odors, colors and turbidity.

PALEO Studies of geologic past.

PALEOFLOOD Peak discharges estimated using geology, fluvial, geomorphology and stratigraphic records where geologic information is used to determine flood depths, carbon 14 dating techniques are typically used to determine time frame when these depths were reached and hydraulic models (e.g., step backwater techniques) are used to determine associated flow given depth.

PALEOFLOOD DATA Distinguished from both historical and systematic (conventional) flood data by lack of human observation, regardless of time of occurrence and include two broad categories: fluvial geomorphic evidence and botanical evidence.

PALEOFLOOD HYDROLOGY 1. Study of past or ancient floods which occurred prior to time of human observation or direct measurement by modern hydrologic procedures; 2. Study of movements of water and sediment in channels before time of continuous hydrologic records or direct measurements.

PALEOHYDROLOGY Study of hydrologic processes and events, using geological, botanical and cultural evidence, that occurred before beginning of systematic collection of hydrologic data and observations.

PALLET Portable platform where goods can be stacked for transport or storage.

PALMER DROUGHT SEVERITY INDEX Index used to gauge severity of drought conditions by using water balance equation to track water supply and demand taking into account precipitation, potential and actual evapotranspiration, infiltration of water into soil and runoff.

PALUSTRINE HABITAT 1. Marsh habitat; 2. All non-tidal wetlands that are dominated by trees, shrubs, persistent emergents, emergent mosses or lichens and all such wetlands in tidal areas where salinity owing to ocean-derived salts is below 0.5 parts/1,000.

PALUSTRINE WETLANDS Freshwater wetlands including open water bodies of less than 20 acres where water is less than 2 meters deep that includes marshes, wet meadows, fens, playas, potholes, pocosins, bogs, swamps and shallow ponds.

PAN COEFFICIENT Ratio of amount of evaporation from large body of water to that measured in evaporation pan.

PAN EVAPORATION Evaporative water losses from standardized pan sometimes used to estimate crop evapotranspiration and assist in irrigation scheduling.

PAN PAN Headline within National Weather Service high seas forecasts transmitted via GMDSS to indicate that hurricane or hurricane force winds are forecast.

PANAMETRIC ESTIMATING Technique for estimating cost and duration that uses relationship between variables to calculate cost or duration.

PANCAKE ICE Circular flat pieces of ice with raised rim where shape and rim are due to repeated collisions.

PANDEMIC Widespread epidemic throughout area, nation or world.

PANEL 1. Rectangular section of surface; 2. Soft pad beneath saddletree to prevent chafing.

PANEL HEATING SYSTEM System consisting of wall or floor or baseboard or ceiling panels containing electric conductors or heating pipes.

PANEL MOUNTING Panel where numerous elements might be mounted.

PANEL POINT Point where members of truss intersect.

PANHANDLE HOOK Low pressure systems that originate in panhandle region of Texas and Oklahoma which initially move east and then "hook" or recurve more northeast toward upper Midwest or Great Lakes region.

PAPER In recycling business, refers to products and materials, including newspapers, magazines, office papers, corrugated containers, bags and some paperboard packaging that can be recycled into new paper products.

PARABOLIC WEIR Weir having notch formed as parabola having perpendicular axis.

PARADOX GATE Similar to ring follower gate except gate leaf is supported along either side by endless trains of rollers gate that seals by rollers disengaging from support of leaf when gate is completely closed, allowing hydrostatic forces to seal gate.

PARALIC Laid down on landward side of coast.

PARALLEL CURB RAMP Two ramps joined in middle by landing that's level with roadway.

PARALLEL OPERATION 1. Connection of two or more power sources of same size and output voltage to obtain higher output current; 2. Functioning of multistage pump when each of its own impellers receives water from common source and brings volume straight to discharge.

PARALLEL PATH FLOW Refers to flow of electric power on electric system's transmission facilities resulting from scheduled electric power transfers between two other electric systems.

PARALLEL THREAD Screw thread with uniform diameter.

PARAMETER Value kept constant during experiment, equation, calculation, etc.

PARAMETER ESTIMATION Selection of parameter value based on results of analysis and/or engineering judgment.

PARAMETER IDENTIFICATION MODEL Computer code for determination of selected unknown parameters and stresses in groundwater system, given that response of system to all stresses is known and that information is available regarding certain parameters and stresses.

PARAMETRIC DATA Data (e.g., rating curves, unit hydrographs, rainfall/runoff curves) which define hydrologic variables in models.

PARAPET WALL Solid wall built along top of dam (upstream and/or downstream edge) used for ornamentation, for safety of vehicles and pedestrians or to prevent overtopping caused by wave runup.

PARAQUAT Standard herbicide used to kill various types of crops (e.g., marijuana) that causes lung damage if smoke from crop is inhaled.

PARASITE Organism that lives on or in another organism, deriving benefit from living on or in that other organism.

PARASITIC BACTERIA Bacteria that normally live off another living organism.

PARATHION Powerful, dangerously toxic insecticide where maximum acceptable concentration in drinking water is 0.05 mg/l.

PARATRANSIT SYSTEM Special transportation services for people with disabilities unable to use fixed-route public transportation system.

PARCEL Volume of air small enough to contain uniform distribution of its meteorological properties and large enough to remain relatively self-contained and respond to all meteorological processes.

PARCEL PLATE Map of property needed for roadway purposes.

PARENT MATERIAL Underlying geological material where soil horizons form.

PARHELION 1. Either of two colored luminous spots that appear at roughly 22° on both sides of Sun at same elevation caused by refraction of sunlight passing through ice crystals; 2. Scientific name for Sun dogs.

PARK AND RIDE LOT Parking facility with public transport connections.

PARKING AREA Any area designated for parking of motor vehicles.

PARKING LANE Auxiliary lane reserved for parallel parking of vehicles.

PARKWAY 1. Scenic freeway; 2. Divided highway.

PARSHALL FLUME 1. Device used to measure flow of water in open channel; 2. Flume with specially shaped open-channel flow section that may be installed in drainage lateral or ditch to measure rate of flow of water.

PART PER MILLION Unit of concentration equal to 1 mg/kg or 1 mg/l.

PART TRIP Trip added to start or end of standard transit run.

PARTIAL ACCEPTANCE Agreement by someone to pay part of value of bill of exchange, with other acceptors responsible for paying rest.

PARTIAL BEAM FILLING Limitation of rainfall estimation techniques where reflectivity and associated rainfall rates are underestimated.

PARTIAL CLOVERLEAF INTERCHANGE Variation on cloverleaf interchange where there are fewer loop ramps.

PARTIAL CONTROL OF ACCESS When through traffic is given priority to access connections.

PARTIAL DIAMOND INTERCHANGE Diamond interchange with relatively low traffic but is too close to another interchange on one side.

PARTIAL DIVERSION Taking or removal of waters from any point in water body.

PARTIAL PRESSURE Pressure that one component of mixture of gases would contribute to total pressure.

PARTIAL STOP When vehicle slows in queue but doesn't completely stop.

PARTIAL TAKING Taking of part of property under power of eminent domain for public use.

PARTIAL VACUUM Region of low atmospheric pressure, especially one where most of air has been pumped out.

PARTIAL-DURATION FLOOD SERIES List of all flood peaks that exceed chosen base stage or discharge, regardless of number of peaks occurring in year.

PARTIAL-RECORD STATION Site where discrete measurements of one or more hydrologic parameters are obtained over period of time without continuous data being recorded or computed.

PARTIALLY PRESTRESSED CONCRETE Concrete with both prestressing and reinforcing.

PARTICIPATION RATE Portion of population participating in recycling program.

PARTICLE Minute portion of matter.

PARTICLE ACCELERATION Time rate of change of particle velocity.

PARTICLE COUNT Results of microscopic examination of treated water with special "particle counter" that classifies suspended particles by number and size.

PARTICLE COUNT BLANK Allowance for determinable background pollution.

PARTICLE DISPLACEMENT Difference between initial position of soil particle and any later temporary position during shaking.

PARTICLE SIZE DISTRIBUTION List of values or mathematical function that defines relative amount, typically by mass, of particles present according to size.

PARTICLE TRAJECTORY MODEL Computer sub-model that tracks trajectories of multiple particles that are released into atmospheric flow model.

PARTICLE VELOCITY Time rate of change of particle displacement.

PARTICLE-SIZE ANALYSIS Collective name that determines size range, and/or average or mean size of particles in powder or liquid sample.

PARTICULARLY DANGEROUS SITUATION WATCH Wording used in rare situations when long-lived, strong and violent tornadoes are possible.

PARTICULATE CONTAMINATION Unintended presence of extraneous, mobile and undissolved particles in parenteral solution.

PARTICULATE LOADING Mass of particulates per unit volume of air or water.

PARTICULATE MATTER Complex mixture of extremely small particles and liquid droplets made up of number of components, including acids (e.g., nitrates, sulfates), organic chemicals, metals and soil or dust particles.

PARTICULATE SAMPLING Procedures where to ascertain volume of particles present in gas flow.

PARTICULATES Fine liquid or solid particles (i.e., dust, smoke, mist, fumes, smog) found in air or emissions.

PARTITION COEFFICIENT Measure of sorption phenomenon, whereby pesticide is divided between soil and water phase.

PARTITIONING FUNCTION Mathematical relation describing distribution of reactive solute between solution and other phases.

PARTLY CLOUDY Between 3/8 and 5/8 of sky is covered by clouds.

PARTLY SUBMERGED ORIFICE Orifice with top above and bottom below water discharge surface.

PARTLY SUNNY Between 3/8 and 5/8 of sky is covered by clouds during daylight hours.

PARTS PER BILLION Parts of substance equal to 1/1,000,000,000.

PASCAL Unit of pressure (stress) resulting from force of 1 Newton acting uniformly over area of 1 m^2.

PASCAL SECOND Unit of dynamic viscosity equal to 1 kg/m x s.

PASCAL'S LAW Principle in fluid mechanics that states that pressure change occurring anywhere in confined incompressible fluid is transmitted throughout fluid such that same change occurs everywhere.

PASS 1. To move or be moved from one place to another; 2. To eliminate from body by natural processes; 3. Opening, road or track, available for passage of excavating, grading or compaction machine; 4. Channel connecting river or body of water to sea.

PASS THROUGH Wastewater treatment plant discharge of volume of effluent that violates quality levels.

PASSAGE TIME Maximum allowable duration of time between vehicle calls on phase before phase ends.

PASSENGER Person who rides or travels in vehicle, but does not operate it and is not member of crew.

PASSENGER CAR Railroad car or road vehicle that carries passengers.

PASSENGER CAR EQUIVALENT Unit to represent large vehicle by expressing it as number of equivalent passenger vehicles.

PASSENGER CONTROLS System for collecting fares and controlling passenger movements.

PASSENGER COUNT Number of passengers on vehicle or who use facility.

PASSENGER FARES FOR LINE SERVICE Payment for carrying passengers in transit.

PASSENGER FLOW Speed and dwell-time of passengers at location during specific period.

PASSENGER LOAD Measures number of passengers that utilize public transport.

PASSENGER LOCOMOTIVE High speed train that carries passengers.

PASSENGER MILES Basic unit used by transport companies for calculating profit levels determined by multiplying total distance travelled in given period by number of passengers.

PASSENGER RIDING COUNT Number of passengers on vehicle on particular run or who use particular facility.

PASSENGER STANDING COUNT Number of people estimated aboard vehicle when passing checkpoint.

PASSENGER TRIP LENGTH Average distance traveled by passenger from origin to destination.

PASSENGER-CAR EQUIVALENT Impact mode of transport has on traffic variables compared to single passenger car.

PASSING LANE Lane on multi-lane highway or motorway closest to median of road.

PASSING SIGHT DISTANCE Minimum sight distance required on highway that will allow driver to pass another vehicle without colliding with vehicle in opposing lane.

PASSIVE EARTH PRESSURE Maximum value of earth pressure that exists when soil mass is compressed sufficiently to cause its internal shearing resistance along potential failure surface to be completely mobilized.

PASSIVE PRESSURE Condition where retaining system exerts force on soil.

PASSIVE PRIORITY Traffic signal strategy that favors roads with significant transit use.

PASSIVE SMOKING/SECONDHAND SMOKE Inhalation of others' tobacco smoke.

PASSIVE SOLAR ENERGY Sun's energy used for power.

PASSIVE TREATMENT WALLS Technology where chemical reaction takes place when contaminated groundwater comes in contact with barrier (e.g., limestone, wall containing iron filings).

PATH CONVERGENCE Point in network diagram where one or more parallel paths come together.

PATHOGEN Any living organism (e.g., bacteria, viruses, bacteria, fungi, parasites) that causes disease.

PATHOGENIC BACTERIA Bacteria that can cause disease and/or infection.

PATHOGENS Any organism or substance, especially microorganism, capable of causing disease (e.g., bacteria, viruses, protozoa, fungi).

PATHWAY Physical course chemical or pollutant takes from its source to exposed organism.

PATTERN CRACKING Fine cracks in form of pattern on concrete surface.

PAVED Covered in bituminous surface treatment, mixed bituminous concrete or Portland cement concrete.

PAVEMENT CLASSIFICATION INDEX Number that indicates significant overloading of pavement by current aircraft.

PAVEMENT CONDITION General condition of pavement section.

PAVEMENT CONDITION INDICATOR Time measure of existing pavement sections condition.

PAVEMENT CONDITION RATING Quantifies road surface's overall performance regarding defects and ride quality.

PAVEMENT DISTRESS Indication on performance of unfavorable pavement and signals pending failure.

PAVEMENT MACROTEXTURE Deviation of pavement surface from true planar surface.

PAVEMENT MANAGEMENT Process of planning maintenance and repair of network of roadways or other paved facilities to optimize pavement conditions over entire network.

PAVEMENT MANAGEMENT SYSTEM Set of defined procedures for collecting, analyzing, maintaining and reporting pavement data, to assist decision makers in finding optimum strategies for maintaining pavements in serviceable condition over given period of time for least cost.

PAVEMENT MARKINGS Indicators (e.g., white or yellow lines) used to convey messages to roadway users.

PAVEMENT MICROTEXTURE Deviation of pavement surface from true planar surface.

PAVEMENT PERFORMANCE Ability of pavement to serve traffic over period of time.

PAVEMENT PERFORMANCE PREDICTION MODEL Predict future pavement conditions, trigger rehabilitation activities and their impact on pavement conditions.

PAVEMENT PRESERVATION Long-term strategy that enhances pavement performance by using cost effective practices that extends pavement life, improves safety and meets motorist expectations.

PAVEMENT STRUCTURE Combination of surface, base and sub-base layers or courses.

PAY ADJUSTMENT Change in employee's pay rate.

PAY ADJUSTMENT SCHEDULE FOR QUALITY Schedule for assigning pay factors associated with estimated quality levels.

PAY FACTOR Proportion of whole payment to be paid to Contractor.

PAY FORMATION Layer or deposit of soil or rock whose value is sufficient to justify excavation.

PAY-AS-YOU-THROW/UNIT-BASED PRICING Systems under which residents pay for municipal waste management and disposal services by weight or volume collected, not fixed fee.

PAYBACK Benefit, reward or form of recompense.

PAYLINE Lines of excavation, backfill, compacted backfill or embankment which are described in specifications or shown on drawings which describe or show limits that earthwork is paid for.

PAYLOAD 1. Part of cargo that produces revenue; 2. Part of rocket, missile, propelled stinger or torpedo not concerned with propulsion or guidance; 3. Excavated load that payment is created.

PAYLOAD AND BODY ALLOWANCE Payload capability of truck, allowing for truck body weight.

PAYMENT REQUEST Request created with contractor or provider for payment.

PAYMENT SCHEDULE Schedule defining dates and amounts of payments to be made for financial instrument.

PAYNE'S PROCESS Process for preserving timber and making it incombustible by impregnating it successively with solutions of iron sulfate and calcium chloride in vacuum.

PEA GRAVEL Uniformly graded gravel with particle size of approximately 3/16 of inch.

PEAK 1. Highest value reached by some quantity (e.g., flood wave) in time period; 2. Top or one of tops, of hill, mountain or range; 3. Sharp end or top of anything that terminates in point; 4. Maximum crest.

PEAK DEMAND 1. Period of simultaneous, strong demand for resources (i.e., energy, water); 2. Historically high point in sales record.

PEAK DIRECTION Principal direction of traffic during period of time.

PEAK DISCHARGE Highest rate of discharge of volume of water passing given location during given period of time.

PEAK ELECTRICITY DEMAND Maximum electricity used to meet cooling load of building or buildings in given area.

PEAK FLOOD LEVEL Design flood's water elevation.

PEAK FLOW 1. Maximum instantaneous flow in specified period of time at given location; 2. Point of hydrograph that has highest flow.

PEAK GUST Highest instantaneous wind speed observed or recorded.

PEAK HORSEPOWER Horsepower output of motor.

PEAK HOUR Part of day where traffic congestion is highest.

PEAK HOUR FACTOR Hourly volume during maximum-volume hour of day divided by peak 15-minute flow rate within peak hour.

PEAK LEVELS Levels of airborne pollutant contaminants much higher than average or occurring for short periods of time in response to sudden releases.

PEAK LOAD Maximum power load in stated period of time.

PEAK LOAD PLANT Power plant that normally is operated to provide power during maximum load periods.

PEAK PERIOD Time of day when traffic demand is highest.

PEAK PULSE Amount of power transmitted by radar during given pulse.

PEAK SERVICE Maximum number of vehicles operating during peak period.

PEAK SOUND PRESSURE LEVEL Measure of maximum instantaneous sound pressure at specified location.

PEAK STAGE Maximum height of water surface above established datum plane.

PEAK TORQUE Maximum torque that machine or motor produce to achieve revolutions per minute.

PEAK WIND SPEED Maximum instantaneous wind speed since last observation that exceeded 25 knots.

PEAK-LOAD PLANT Plant that meets demand or load required during peak hours.

PEAK/BASE RATIO Ratio of number of vehicles operating in passenger service during peak versus base time period.

PEAKEDNESS Describes rate of rise and fall of hydrograph.

PEAKING CAPACITY Capacity of generating equipment normally reserved for operation during hours of highest daily, weekly or seasonal loads.

PEAKING POWER Power plant capacity typically used to meet highest levels of demand in utility's load or demand profile.

PEAT Highly organic soil, composed of partially decomposed vegetable matter.

PEBBLE Sediment particles with diameter between 0.06 and 2.3 in.

PECELT NUMBER Relationship between advective and diffusive components of solute transport expressed as ratio of product of average interstitial velocity, times characteristic length, divided by coefficient of molecular diffusion where small values indicate diffusion is dominant transport process, large values indicate advection dominance.

PEDDLER 1. Itinerant seller of small goods; 2. local freight train.

PEDESTAL BASIN Floor mounted basin that hides pipework behind basin.

PEDESTAL POLE Traffic signal head pole that attaches directly to pole.

PEDESTAL URINAL Urinal that's not connected to wall but mounted on single pedestal.

PEDESTRIAN Of or intended for those who are walking or by wheelchair.

PEDESTRIAN ACCESS ROUTE Minimum accessible route of passage within walkway that may not comprise full width of pedestrian circulation route.

PEDESTRIAN CHANGE INTERVAL Time when flashing UPRAISED HAND signal is indicated.

PEDESTRIAN CLEAR INTERVAL Time needed for pedestrians to walk from curb to curb during signal WALK indication.

PEDESTRIAN CLEARANCE TIME Time required by pedestrians who enter crosswalks at end of green indication to complete crossing before conflicting vehicular traffic movements are released.

PEDESTRIAN CROSSWALK Facility for pedestrians to cross roadway.

PEDESTRIAN DENSITY Average number of pedestrians per unit of area within walkway or queuing area.

PEDESTRIAN FACILITIES Components to accommodate walking.

PEDESTRIAN FLOW RATE Pedestrian count passing point per time.

PEDESTRIAN HYBRID BEACON Traffic control device to increase motorists' awareness of pedestrian crossings at uncontrolled marked crosswalk locations.

PEDESTRIAN MODE Mode that optimizes route to your destination and estimated time of arrival depending on your walking speed.

PEDESTRIAN OVERPASS Bridge designed for pedestrians over roadway.

PEDESTRIAN PLAZA Open, urban public space.

PEDESTRIAN QUEUING AREA Location where pedestrians temporarily stand.

PEDESTRIAN RECALL Walk mode programmed into signal controllers.

PEDESTRIAN SERVICE TIME Time from when first pedestrian departures until last pedestrian arrives.

PEDESTRIAN SIGNAL HEAD Signal head installed to direct pedestrian traffic.

PEDESTRIAN SPACE Areas reserved for pedestrians only.

PEDESTRIAN START-UP TIME Difference between time pedestrian steps off curb until arrival at opposite curb.

PEDESTRIAN STREET Road for pedestrian use only.

PEDESTRIAN UNDERPASS Facility for pedestrians to pass under roadways.

PEDESTRIAN WALKWAY Portion of public right-of-way that provides separated area for people traveling on foot.

PEDIMENT Broad, gently sloping expanse of rock debris extending outward from foot of mountain slope.

PEDOGENESIS Process of formation of soil.

PEDON Three-dimensional body of soil with dimensions large enough to permit study of individual soil horizons.

PEER-REVIEWED LITERATURE Referable, obtainable, published document reviewed by minimum of two technical reviewers who are located external to author's organization.

PEGMATITE Very coarse-grained intrusive igneous rock of granitic composition that typically fills fractures to form veins.

PELAGIC Of or pertaining to oceans.

PELLETED FUEL Waste, combustible material, compressed into solid pellets for fuel.

PELLICULAR FRONT Even front, grown in pervious stuff, that pellicular water, depleted by evaporation, transpiration or chemical action, is regenerated by influent seepage.

PELLICULAR WATER Thin films of water that cling to soil and rock particles above water-bearing material.

PELLICULAR ZONE Maximum depth from surface up to which water will evaporate.

PENALTY FOR DELAY Financial penalty contractor must pay if contract isn't finished by date stipulated in contract files.

PENDANT ECHO Radar signature generally similar to hook echo, except that hook shape is not as well defined.

PENDANT SWITCH Electric wiring switch suspended overhead to control devices mounted overhead.

PENETRATING TOP Dome-like protrusion above thunderstorm anvil, representing very strong updraft and hence higher potential for severe weather with that storm.

PENSTOCK 1. Conduit used to convey water under pressure to turbines of hydroelectric plant; 2. Sluice or pipe that allows controlled flow of water from behind dam.

PENTACHLOROPHENOL Fully chlorinated derivative of phenol used as fungicide and wood preservative.

PENTOSANS Polysaccharides that produce only pentoses on hydrolysis and occur in plant tissue.

PENUMBRA Sunspot area that may surround darker umbra or umbrae that consists of linear bright and dark elements radial from sunspot umbra.

PEOPLE MOVER Fully automated, grade-separated mass transit system.

PER CAPITA USE Average amount of water used per person during standard time period (e.g., per day).

PER HECTARE FACTOR Conversion number from sample plot to per hectare information.

PERCEIVED NOISE DECIBELS Unit for measuring noise by comparison with reference sound.

PERCENT COMPLETE Approximation of services performed to date versus overall performance of activity.

PERCENT COMPOSITION Unit for expressing ratio of particular part of sample or population to total sample or population, in terms of types, numbers, weight, mass or volume.

PERCENT DEFECTIVE Percentage falling outside of specified tolerance limits.

PERCENT FINES 1. Quantity of substance in aggregate finer when compared to specified sieve, generally No. 200 (0.0029 in.) expressed as percent; 2. Amount of fine aggregate in concrete mixture expressed as percentage of total amount of aggregate.

PERCENT FREE-FLOW SPEED Ratio of average travel speed to free-flow speed.

PERCENT GRADE Slope (rise divided by run) written as percent.

PERCENT HUMIDITY Percentage of current absolute humidity to maximum possible humidity at given temperature.

PERCENT LOAD Fraction of maximum available torque at given engine speed.

PERCENT OF SLOPE Ratio between vertical and horizontal distances expressed in percent.

PERCENT OF TOTAL Unit for expressing ratio of particular part of sample or population to total sample or population, in terms of types, numbers, weight, mass or volume.

PERCENT REMOVAL Removal efficiency across treatment plant, expressed as percent.

PERCENT SATURATION Amount of substance dissolved in solution compared to amount that could be dissolved in it.

PERCENT SHADING Measure of amount of sunlight potentially reaching stream.

PERCENT TIME-SPENT-FOLLOWING Ratio of time spent in platoons versus total travel time.

PERCENT VOLATILE Percentage of chemical that will evaporate at ordinary temperatures.

PERCENT WITHIN LIMITS Percentage of variable between specified upper and lower limits.

PERCENTAGE REDUCTION Amount of decrease, expressed as percent.

PERCENTILE Any of ninety-nine points that divide ordered distribution into one hundred parts, each containing one per cent of population.

PERCHED Either losing stream or insulated stream separated from underlying groundwater by zone of aeration.

PERCHED GROUNDWATER Unconfined groundwater separated from underlying body of groundwater by unsaturated zone.

PERCHED SPRING Spring whose source of water is body of perched groundwater.

PERCHED WATER Zone of unpressurized water held above water table by impermeable rock or sediment.

PERCHED WATER TABLE Water table of relatively small groundwater body supported above general groundwater body.

PERCOLATING FILTER Treatment bed where sewage or waste is sprayed intermittently and trickles through coarse aggregate.

PERCOLATING WATER Water that passes through rocks or soil under force of gravity.

PERCOLATION Flow, seepage or filtration of liquid through porous substance.

PERCOLATION PATH Course followed by water moving or percolating through any other permeable material or under dam which rests upon permeable foundation.

PERCOLATION RATE Rate (velocity) at which water moves through porous media (e.g., soil).

PERCOLATION TEST Evaluation to quantify how quickly soil will absorb and drain.

PERCUSSION DRILL Drill with hammering action initiated by rotation of chuck.

PERENNIAL INTERRUPTED STREAM All-year-flowing stream having stretches where there is no surface flow.

PERENNIAL SPRING Spring that flows continuously all year round.

PERENNIAL STREAM Stream that flows continually throughout year.

PERENNIAL YIELD Maximum quantity of water that can be annually withdrawn from groundwater basin over long period of time (during which water supply conditions approximate average conditions) without developing overdraft condition.

PERENNIALLY-FROZEN SOIL Land constantly frozen for over 3 years.

PERFLUOROCARBONS Potent greenhouse gases that accumulate in atmosphere and remain there for thousands of years.

PERFORATED CASING Well casing with openings to permit inflow of water.

PERFORATED PIPE Pipe designed to discharge water through small, multiple, closely spaced orifices or nozzles, placed in segment of its circumference for irrigation purposes.

PERFORMANCE Act of doing what is required by contract.

PERFORMANCE BOND Cash or securities deposited before landfill operating permit is issued, which are held to ensure that all requirements for operating and subsequently closing landfill are faithful performed.

PERFORMANCE CURVE Graphical portrayal of operating feature of item.

PERFORMANCE DATA FOR INCINERATORS Information collected, during trial burn, on concentrations of designated organic compounds and pollutants found in incinerator emissions.

PERFORMANCE EVALUATION SAMPLE Benchmark sample of method where performance is documented and evaluated.

PERFORMANCE FACTOR Numerical ratio used to measure actual output capacity to input of system.

PERFORMANCE MEASURE Measurement of outcomes and results to determine effectiveness and efficiency.

PERFORMANCE SPECIFICATION Document that specifies operational requirements of component or installation.

PERFORMANCE SPECIFICATIONS Document that specifies operational requirements.

PERFORMANCE STANDARD Approved expression of performance threshold(s), requirement(s) or expectation(s) that must be met to be appraised at particular level of performance.

PERFORMANCE STANDARDS 1. Regulatory requirements limiting concentrations of designated organic compounds, particulate matter and hydrogen chloride in emissions from incinerators; 2. Operating standards established by EPA for various permitted pollution control systems, asbestos inspections and various program operations and maintenance requirements.

PERFORMANCE-BASE SPECIFICATIONS Desired performance level or performance target to be achieved.

PERFORMANCE-BASED METHODS SYSTEM System that permits use of any appropriate measurement methods that demonstrates ability to meet established performance criteria and that complies with specified data-quality needs.

PERFORMANCE-RELATED SPECIFICATION Quality assurance specifications that describe desired levels of key material and construction quality characteristic that have been found to correlate with fundamental engineering properties that predict performance.

PERICLINE Form of albite exhibiting elongated prismatic crystals.

PERIGEE Closest distance between Moon and Earth or Earth and Sun.

PERIGLACIAL Of or pertaining to area around edge of glacier.

PERIHELION Point on annual orbit of body (about Sun) closest to Sun.

PERIMETER HEATING SYSTEM Warm air heating system where ducts are located in concrete slab of building without basement.

PERIOD Length of time for completion of event.

PERIODIC Relative to period or periods.

PERIODIC APPLICATION OF COVER MATERIAL Application and compaction of soil or other suitable material over disposed solid waste at end of each operating day or at such frequencies and in such manner as to reduce risk of fire and to impede vectors access to waste.

PERIODIC FACILITY REVIEW Field review performed on high- or significant-hazard dam every 6 years that entails thorough examination from both operation and maintenance and dam safety perspectives.

PERIODIC SPRING Spring where flow of water varies over fairly regular time-scale.

PERIODIC TABLE Tabular chart of chemical elements according to their atomic numbers so that elements with similar properties are in same column.

PERIODIC-RECORD STATION Site where stage, discharge, sediment, chemical, physical or other hydrologic measurements are made one or more times during year but at frequency insufficient to develop daily record.

PERIPHERAL FLOW Flow parallel to circumference of tank.

PERIPHERAL PUMP Centrifugal pumps whose impeller rotates in largely concentric casing channel with inlet and outlet opening.

PERIPHERAL WEIR Weir around inside edge of circular tank.

PERIPHYTON Microorganisms (e.g., bacteria, fungi, protozoa, rotifers) attached to and living upon submerged solid surfaces.

PERISTALTIC PUMP Positive displacement pump for pumping fluids contained within flexible tube fitted inside circular pump casing.

PERM Water vapor transmission for differential in partial pressures on either side of material or membrane.

PERMAFROST Permanently frozen ground.

PERMANENT CONTROL Stream gaging control that is substantially unchanging and is not appreciably affected by scour, fill or backwater.

PERMANENT DEFLECTION Inelastic action where deflection remains after load removed.

PERMANENT HARDNESS Hardness in water due to presence of chlorides, nitrates and sulphates of calcium and magnesium.

PERMANENT LOAD Forces, deformations or accelerations applied to structure components.

PERMANENT MONUMENT Fixed monuments placed away from dam which allow movements in horizontal and vertical control points on dam to be monitored by using accurate survey procedures.

PERMANENT SNOW LINE 1. In summer, altitude above which or latitude beyond which snow does not melt; 2. Line to which snow extends at given point in time.

PERMANENT STREAM Streams that flow throughout year.

PERMANENT WILTING POINT Soil water content below which plants cannot readily obtain water and permanently wilt.

PERMANENTLY ABSORBED WATER Water that percolates past root depth to enter groundwater inside saturation zone.

PERMANGANATE VALUE Concentration of material in waste stream oxidizable by potassium permanganate under acidic conditions.

PERMEABILITY Rate that which gasses, liquids or plant roots penetrate or pass through layer of soil or porous media.

PERMEABILITY COEFFICIENT Rate of flow of fluid through cross-section of porous mass under unit hydraulic gradient.

PERMEABLE Having pores or openings that permit liquids or gasses to pass through.

PERMEAMETER 1. Laboratory instrument for determining permeability by measuring discharge through sample of material (e.g., soil) when known hydraulic head is applied; 2. Instrument used to measure electromagnetic permeability of material.

PERMEANCE 1. Measure of degree to which material allows fluid to permeate it; 2. Reciprocal of reluctance in magnetic circuit; 3. Analogue of conductance in electrical circuit.

PERMEATE To penetrate and pass through.

PERMEATING TUBE Sealed polymer tube comprising liquefied sample of gas useful for calibrating devices as well as in analytic means of quantifying concentrations of pollutants in atmosphere.

PERMEATION GROUTING Chemical grouting that permeates pore spaces in granular soils that hardens to create cemented mass.

PERMISSIBLE DOSE Dose of chemical that may be received by individual without expectation of significantly harmful result.

PERMISSIBLE EXPOSURE LIMIT Federal limits for workplace exposure to contaminants as established by OSHA.

PERMISSIBLE EXPOSURE LIMITS Guide to acceptable levels of chemical exposure.

PERMISSIBLE VELOCITY Highest velocity where water may be carried safely conduit before damage.

PERMIT Authorization, legal license or equivalent control document needed in order to operate.

PERMIT STATUS Period during which treatment, storage and disposal facilities are temporarily permitted to operate while awaiting permanent permit.

PERMIT TO WORK Authorization used to ensure that work is done safely and efficiently.

PERMIT VEHICLE Authorization granted for vehicle movement.

PERMITTED TURN Turning movement at signalized intersection where driver must yield to opposing traffic.

PEROXYACETYL NITRATES Thermally unstable, secondary pollutant present in photochemical smog.

PERPENDICULAR CURB RAMP Two ramps joined in middle by landing level with roadway.

PERSISTENCE 1. Length of time compound stays in environment, once introduced; 2. Continuation of existing conditions.

PERSISTENCE FORECAST Forecast that current weather condition will persist and that future weather will be same as present.

PERSISTENT PESTICIDES Pesticides that do not break down chemically or break down very slowly and remain in environment after growing season.

PERSISTENT POLLUTANT Not subject to decay, degradation, transformation, volatilization, hydrolysis or photolysis.

PERSON CAPACITY Maximum number of persons that pass given point during period of time.

PERSON TRIP Trip by person(s) in any mode of transportation.

PERSONAL AIR SAMPLES Air samples taken with pump directly attached to worker with collecting filter and cassette placed in worker's breathing zone.

PERSONAL MEASUREMENT Measurement collected from individual's immediate environment.

PERSONAL PROPERTY Any property that's movable (i.e., non-real estate).

PERSONAL PROTECTIVE EQUIPMENT Equipment worn to minimize exposure to hazards that cause serious workplace injuries and illnesses.

PERSONAL RAPID TRANSIT Public transport mode featuring small automated vehicles operating on network of specially built guideways.

PERSONAL SAMPLER Worn apparatus that samples atmosphere to record pollutant exposure.

PERT CHART Project management tool used to schedule, organize and coordinate tasks within project.

PERT SCHEDULE Project management tool used to chart of actions or events expected within project.

PERTURBATION MODEL Computer model used to calculate air pollution concentrations.

PERVIOUS Permeable, having openings that allow water to pass through.

PERVIOUS BED Stratum voids where water flows.

PERVIOUS SOIL Soil that offers minimum resistance to flow through it.

PERVIOUS ZONE Part of cross-section of embankment dam comprising material of high permeability.

PERVIOUSNESS Property of allowing passing of another body or substance.

PEST 1. Insect, rodent, nematode, fungus, weed or other form of terrestrial or aquatic plant or animal life that's injurious to health or environment; 2. Plague, pestilence, epidemic.

PEST CONTROL 1. Regulation or management of another species defined as pest; 2. Service that removes pests.

PEST CONTROL OPERATOR Person or company that applies pesticides as business (e.g., exterminator).

PEST PROBLEM Infestation of species that impacts adversely on human activities.

PESTICIDE Substance used to control or kill undesirable organisms (e.g., plant or animal pests).

PESTICIDE REGULATION NOTICE Formal notice to pesticide registrants about important changes in regulatory policy, procedures, regulations.

PESTICIDE TOLERANCE Amount of pesticide residue allowed by law to remain in or on harvested crop.

PESTILENCE Any epidemic disease highly contagious, infectious, virulent and devastating.

PETA Prefix symbolizing 10^{15}.

PETCOCK 1. Small drain valve; 2. Small ground key type valve used with soft copper tubing.

PETROCHEMICAL Any compound derived from petroleum or natural gas.

PETROLEUM Crude oil or any fraction thereof that's liquid under normal conditions of temperature and pressure, occurring naturally in deposits under Earth's surface.

PETROLEUM DERIVATIVES Chemicals formed when gasoline breaks down in contact with groundwater.

PETROLOGY Study of origin, composition and structure of rock.

pH 1. Measure of acidity or basicity of aqueous solution, equal to negative logarithm of concentration of hydronium ions in molars, from 0 to 14, of how acidic or basic (alkaline) material is, where pH of 7 is neutral, pH less than 7 is acidic and pH greater than 7 is basic; 2. Symbol for logarithm of reciprocal of hydrogen ion concentration in gram atoms per liter; 3. Symbol for picohenry, equal to 10^{-12} Henry's.

PHARMACOKINETICS Study of way that drugs move through body after they are swallowed or injected.

PHASE 1. Distinguishable part of sequence or cycle occurring over time; 2. Component in material system distinguished by chemical composition and/or physical state (solid, liquid or gas) and/or crystal structure; 3. Particular appearance or state in regularly recurring cycle of changes with respect to quantity of illumination or form or absence, of its enlightened disk.

PHASE RECALL Function that cancels conflicting green phases and starts green phase of public transport vehicle approach.

PHASE SEQUENCE Order where traffic phases occur during complete signal cycle.

PHASED DISPOSAL Method of uranium byproduct material or tailings management and disposal that uses lined impoundment that are filled and then immediately dried and covered.

PHASING Timing of flows from main stem of river and its tributaries.

PHENOL Caustic, poisonous, white crystalline compound derived from benzene and used in resins, plastics and pharmaceuticals and in dilute form as disinfectant and antiseptic.

PHENOL FORMALDEHYDE Synthetic polymer resin obtained by reaction of phenol or substituted phenol with formaldehyde.

PHENOLPHTHALEIN ALKALINITY Alkalinity in water sample measured by amount of standard acid needed to lower pH to level of 8.3 as indicated by change of color of phenolphthalein from pink to clear.

PHENOMENOLOGICAL MODEL Computer model used to calculate air pollution concentrations.

PHIEZOMETER 1. Instrument used to measure pressure head; 2. Used in dams to measure level of saturation.

PHILADELPHIA ROD Leveling rod where hundredths of feet or eights of inches, are marked by alternate bars of color width of measurement.

PHOSPHATE Any salt or ester of phosphoric acid.

PHOSPHATES Certain chemical compounds containing phosphorus.

PHOSPHOGYPSUM PILES Principal byproduct generated in production of phosphoric acid from phosphate rock.

PHOSPHOGYPSUM STACKS Principal byproduct generated in production of phosphoric acid from phosphate rock.

PHOSPHORUS Chemical element with atomic number of 15, that can play key role in stimulating aquatic growth in lakes and streams and contribute to eutrophication of lakes and other water bodies.

PHOSPHORUS PLANTS Facilities using electric furnaces to produce elemental phosphorous for commercial use, (e.g., high grade phosphoric acid, phosphate-based detergent, organic chemicals use).

PHOTIC ZONE Part of near-surface body of ocean or lake that receives enough sunlight for photosynthesis to be possible.

PHOTO INDEX Map showing layout of aerial photographs.

PHOTO MAP Detailed map that was copied from satellite images.

PHOTO-IDENTIFICATION Identity document that includes photograph of holder.

PHOTO-INTERPRETATION Determination of features in remote sensing image through visual interpretation.

PHOTOCHEMICAL Of, relating to or produced by photochemistry or by photochemical reactions.

PHOTOCHEMICAL OXIDANTS Air pollutants formed by action of sunlight on oxides of nitrogen and hydrocarbons.

PHOTOCHEMICAL SMOG Air pollution containing ozone and other reactive chemical compounds formed by reaction of nitrogen oxides and hydrocarbons in presence of sunlight.

PHOTOCONDUCTIVE CELL Light-sensitive resistors where resistance decreases with increase in light intensity when illuminated.

PHOTODEGRADABLE Capable of being chemically broken down as result of photochemical reaction.

PHOTODIODE Semiconductor two-terminal component whose electrical characteristics are light-sensitive.

PHOTOELASTICITY Technique for studying and measuring stresses and strains in object by means of mechanical birefringence.

PHOTOELECTRIC CELL Transducer that generates electric current proportional to light intensity.

PHOTOELECTRIC EFFECT Emission of electrons from surface of material following absorption of electromagnetic radiation.

PHOTOGRAMMETRY Techniques of obtaining precise measurements from images.

PHOTOMETER Any of several instruments used to measure various aspects of intensity of light.

PHOTOMETRY Measurement of intensity and spectrum of light (e.g., stars).

PHOTOMICROGRAPH Photograph taken using microscope.

PHOTON Quantum of light and other electromagnetic energy, regarded as discrete particle having zero rest mass, no electric charge and indefinitely long lifetime.

PHOTOPERIODISM Growth, development and other responses of plants and animals according to length of day and/or night.

PHOTOSPHERE Intensely bright portion of Sun visible to unaided eye considered "surface" of Sun.

PHOTOSYNTHESIS Process that plant chlorophyll-containing cells manufacture carbohydrates and oxygen (chemical energy) from carbon dioxide in presence of sunlight, forming organic compounds from inorganic compounds.

PHOTOSYNTHETIC BACTERIA Bacteria that convert light energy into chemical free energy.

PHOTOSYNTHETIC EFFICIENCY Fraction of light energy converted into chemical energy during photosynthesis in plants and algae.

PHOTOTHEODOLITE Arrangement of two cameras, whose plates may be brought into exactly same plane, used in surveying and map-making.

PHOTOTROPIC Having tendency to move in response to light.

PHOTOVOLTAIC System that converts sunlight directly into electricity using cells made of silicon or other conductive materials.

PHOTOVOLTAIC ARRAY Linked collection of solar modules to supply usable solar power by means of photovoltaics.

PHOTOVOLTAIC CELL Device that absorbs radiant energy and converts it into electrical energy.

PHOTOVOLTAIC EFFECT Conversion of electromagnetic radiation into electric power through absorption by semiconducting material.

PHREATIC 1. Of or pertaining to groundwater; 2. Involving explosively rapid heating of groundwater by magma.

PHREATIC CYCLE Period of time during which water table rises and then falls.

PHREATIC DECLINE Lowering upper level of watertable.

PHREATIC DIVIDE Underground watershed.

PHREATIC FLUCTUATION Upward and downward movement of water table.

PHREATIC LINE Top flow line of saturated soil mass below where seepage takes place.

PHREATIC SURFACE Free surface of groundwater at atmospheric pressure.

PHREATIC WATER Groundwater in zone of saturation where all openings in rocks and soil are filled, upper surface of which forms water table.

PHREATIC ZONE 1. Locus of points below water table where soil pores are filled with water; 2. Zone of saturation.

PHREATICOLOUS Organisms that live in subterranean fresh water.

PHREATOPHYTE Plant that habitually obtains its water supply from zone of saturation, either directly or through capillary fringe.

PHTHALATES Class of organic compounds containing phthalic acid esters and derivatives that's used as plasticizers in plastics.

PHYSICAL AND CHEMICAL TREATMENT Processes generally used in large-scale wastewater treatment facilities where physical processes may include air-stripping or filtration and chemical treatment includes coagulation, chlorination or ozonation.

PHYSICAL DETERIORATION Loss in physical efficiency of asset as it ages.

PHYSICAL GORE Unpaved area created between highway mainline and ramp that merges into or diverges from mainline.

PHYSICAL INVENTORY Weight, measurement, volume and count of items in company's inventory.

PHYSICAL LIFE Length of time that asset lasts.

PHYSICAL MAINTENANCE Preservation and upkeep to maintain or improve original condition.

PHYSICAL MODEL 1. Model using physical properties and behavior of modeling materials to represent prototype; 2. Scale model of prototype.

PHYSICAL PROPERTIES Properties that can be observed or measured without changing composition of matter.

PHYSICAL RESTRAINT Physical task that has to be finished before following tasks in sequence can start.

PHYSICAL STABILITY Ability to maintain physical dimensions and properties when exposed to conditions normally encountered in its service environment.

PHYSICAL SURVEY Legal document drawn to scale containing map of your property, showing existing condition and improvements, signed and sealed by professional land surveyor.

PHYSICAL TESTING Qualitative or quantitative procedure that consists of determination of one or more characteristics of given product, process or service according to specified procedure.

PHYSICAL WASTE TREATMENT PROCESS Wastewater treatment processes that includes use of racks, screens, clarifiers (i.e., sedimentation and flotation) and filtration.

PHYSICOCHEMICAL EFFLUENT TREATMENT Non-biological effluent treatment.

PHYSIOGRAPHIC EQUILIBRIUM Stability in hydrologic environment, regions and land forms, vegetative cover, precipitation and erosion.

PHYSIOGRAPHIC PROVINCE Region where landforms are distinctive and differ significantly from those of adjacent regions.

PHYSIOGRAPHY Description of nature or natural phenomenon in general (e.g., description of surface features of Earth, with emphasis on origin of landforms).

PHYTOMETER Any plant used as phytometric standard.

PHYTOPHAGOUS Plant eating.

PHYTOPLANKTON Portion of plankton community comprised of small, usually microscopic, plants (e.g., algae, diatoms) found in lakes, reservoirs and other bodies of water.

PHYTOREMEDIATION Low-cost remediation option for sites with widely dispersed contamination at low concentrations.

PHYTOTOXIC Harmful to plants.

PHYTOTREATMENT Cultivation of specialized plants that absorb specific contaminants from soil through their roots or foliage.

PICK Tool used for digging.

PICKING TABLE OR BELT Table or belt on which solid waste is manually sorted and certain items removed.

PICLORAM Herbicide 4-amino-3,5,6-trichloropicolinic acid, where maximum acceptable concentration in drinking water is 0.19 mg/l.

PICO Prefix symbolizing 10^{-12} (i.e., one trillionth).

PICOCURIE One-trillionth of amount of radioactive nuclide represented by curie.

PICOCURIES PER LITER Unit of measure for levels of radon gas that's equal to Becquerel's per m^3.

PICTOGRAPH Picture that represents word or idea.

PIEDMONT ALLUVIAL DEPOSIT Group of impinging alluvial cones built up by streams originating from mountain range.

PIEDMONT ALLUVIAL PLAIN Chain of alluvial fans merging one into other.

PIER 1. Structure that projects tangentially from shoreline to accommodate ships; 2. Structure supporting junction between two spans of bridge; 3. Rectangular pillar that supports arch, wall or roof or gate hinges.

PIER SHAFT High-capacity deep foundation or footing.

PIESTIC INTERVAL Difference in static levels between two lines of equal pressure.

PIEXOELECTRIC EFFECT Ability of certain materials to generate electric charge in response to applied mechanical stress.

PIEXOMETRIC HEAD Measurement of liquid pressure above geodetic datum.

PIEXOMETRIC LEVEL Level where water in confined aquifer would rise were it completely pierced with wells.

PIEZOMETER Instrument for measuring water levels or pore water pressures in embankments, foundations, abutments, soil, rock, concrete, tanks and soil.

PIEZOMETER NEST Group of two or more piezometers which are set very close to each other but are screened at different depths.

PIEZOMETER TUBE 1. Device used to measure liquid pressure in system by measuring height to which column of liquid rises; 2. Device that measures piezometric head of groundwater at specific point.

PIEZOMETRIC HEAD 1. Sum of pressure head and datum head at any point; 2. Liquid pressure above vertical datum.

PIEZOMETRIC MAP Map depicting isopiestic lines revealing piezometric surface of aquifer.

PIEZOMETRIC SURFACE Imaginary surface of piezometric pressure or hydraulic head throughout all or part of confined or semi-confined aquifer.

PIGGYBACK Transportation of goods where one transportation unit is carried on back of something else.

PILE Relatively slender structural element that is driven or otherwise introduced, into soil, usually for purpose of providing vertical or lateral support.

PILE BENT Driven or placed pile with cap to hold position.

PILE SHOE Pointed metal casting that's fitted on lower end of pile to prevent damage.

PILLOW Piece of metal or wood, forming support to equalize pressure.

PILOT 1. Person who steers object (e.g., vessel, aircraft); 2. V-shaped device on front of locomotive (i.e. cowcatcher); 3. Short plug at end of counterbore to guide tool.

PILOT BALLOON Small helium-filled meteorological balloon tracked as it rises through atmosphere to determine how wind speed and direction change with altitude.

PILOT CIRCUIT Controller for electrical power.

PILOT IN COMMAND Person responsible for aircraft operation and safety during flight.

PILOT LIGHT 1. Small, persistent flame used to ignite gas; 2. Small telltale lamp on dynamo or battery circuit whose brightness indicates voltage of current.

PILOT PLANT Small-scale chemical engineering facility designed to research full-scale production of product.

PILOT PRESSURE Pressure required to create force greater than opposing force in device, thus allowing control device to function.

PILOT REPORT Report of inflight weather by aircraft pilot or crew member.

PILOT TESTS Testing cleanup technology under actual site conditions to identify potential problems prior to full-scale implementation.

PILOT TUNNEL Small tunnel excavated over entire length or over part of larger tunnel, to explore ground conditions and/or to assist in final excavation.

PILOT VALVE Small hand-operated valve to let in liquid to operate valve difficult to turn by hand.

PILOT-OPERATED VALVE Valve actuated by pilot pressure.

PILOTAGE Use of landmarks to guide vessel or aircraft to its destination.

PIN CONNECTION Joint that transfers vertical and horizontal shear loads that cannot resist any bending or rotational forces.

PINCH-OUT Location where porous, permeable formation located between overlying and underlying confining formations thins to zero thickness and confining formations are in contact with each other.

PINGO Large frost mound of more than one-year duration.

PINION GEAR 1. Smaller gear in pair of meshing gears; 2. Cylindrical gear used in rack and pinion mechanism.

PINNED END Allows member to rotate around given point but not shift laterally or vertically.

PIONEER PLANT Herbaceous annual and perennial seedling plants that colonize bare areas as first stage in secondary succession.

PIONEER PLANTS Plants that grow anywhere.

PIONEERING First working over of rough or overgrown areas.

PIPE 1. Circular conduit constructed of any one of number of materials that conveys water by gravity or under pressure; 2. Cylindrical conduit or conductor with wall thickness sufficient to receive standard pipe thread.

PIPE AND FITTINGS CLASS Working pressure rating of specific pipe (i.e., cast iron, ductile iron, asbestos cement and some plastic pipe) for use in water distribution systems that includes allowances for surges.

PIPE BRACE Brace for securing pipe passing through frame of wall.

PIPE BRACKET Brace for securing pipe to wall or ceiling.

PIPE BUFFER Tradesperson who buffs ends of conduit to be joined just before welding.

PIPE BURSTING Replaces brittle pipes (e.g., clay, asbestos cement, concrete and cast iron) by pulling through host pipe static or dynamic bursting head to fragment existing pipe and expander cone to push fragments outwards into surrounding ground.

PIPE CLIP Conduit support having removable section.

PIPE CLOSER Non-combustible sleeve that closes gap between conduit and wall opening to permit passing.

PIPE COUPLING Short length of conduit that allows two conduits to be joined, welded or soldered.

PIPE COVERING Wrapping around conduit that serves as insulation or barrier.

PIPE CROSS Pipe fitting with four openings at right angles on same axis plane.

PIPE CUTTER Tool or machine for cutting pipe.

PIPE DIAMETER Nominal, bored or designated inside conduit diameter.

PIPE DRILL Hollow, piping used on drilling rigs to allow drilling fluid to be pumped down hole through drill bit and back up annulus.

PIPE DUCT Fittings that connect straight duct pieces together.

PIPE ENAMEL Asphaltic coating for buried conduits.

PIPE FINDER Device that locates metal such as pipes, cables or steel within wall.

PIPE FITTER Industrial tradesman who installs and maintains piping systems.

PIPE FITTING Occupation of installing or repairing conduit systems that convey liquid, gas and occasionally solid materials.

PIPE FLOW Type of liquid flow within closed conduit.

PIPE FREEZING Method of freezing liquid pipe contents so that it forms ice plug.

PIPE GALLERY Space allowed for installing conduit runs.

PIPE HANGER Apparatus designed to transfer load from pipe to supporting structures.

PIPE HOOK Hook used to support run of conduit.

PIPE INSPECTION CHAMBER Manhole used for inspection of pipe.

PIPE JACKING Tunnelling method for installing underground pipelines with minimum surface disruption.

PIPE KOOK Spiked fastener driven into stud or masonry joint with curved end.

PIPE LAYER Tradesman who lays pipe.

PIPE LINER Flexible tube used to rehabilitate deteriorating sanitary drain and storm sewer lines.

PIPE LINING Application to restore pipe from corrosion, leaks or cracks from roots.

PIPE PLUG Used to close ends of internally threaded pipes and fittings.

PIPE RAMMING Trenchless method for installation of steel pipes and casings through ground by percussive hammer.

PIPE RUN Path followed by piping system.

PIPE SCHEDULE Standard pipe sizes.

PIPE SECTION Single unit of pipe.

PIPE SLEEVE Insert placed in form for concrete wall at position where pipe is to penetrate in order to prevent flow of concrete into opening.

PIPE STOPPER Drain plug.

PIPE STRAP Attachment used to hold pipe directly to structure.

PIPE TAIL Short length of conduit for connection into branch.

PIPE THREAD Conduit screw threads.

PIPE TRENCH Temporary excavation where pipe is placed and eventually covered with soil material.

PIPE VISE Vise used to hold conduit.

PIPELINE Conduit made of pipes used to convey water, gas, petroleum, etc.

PIPELINE TRANSPORT Mode of transportation of goods or material through pipe.

PIPEWORK Any system of pipes or tubes.

PIPEWRAP Tape engineered to bond to virtually any pipe to shield against leaking, corroded and/or damaged pipework.

PIPING 1. System of pipes that compose structure; 2. Piece cut off to be set or planted; 3. Movement of soil particles as effect of unbalanced seepage forces; 4. Process of animal just beginning to break out of its egg.

PIPING SYSTEM System of pipes used to convey fluids from one location to another.

PISTON PUMP Positive displacement pump where high-pressure seal reciprocates with piston.

PIT Any mine, quarry or excavation area worked by open-cut method to obtain material of value.

PIT RUN GRAVEL 1. Natural gravelly material taken from excavation; 2. Gravelly material not processed.

PITCH 1. Elevation of surface or surface slope in reference to horizontal plane (i.e., steepness of slope); 2. Inclination (slope) of pipeline, conduit, stream channel or culvert; 3. Sound governed by producing vibration; 4. Dark viscous substance obtained as residue in distillation of organic (e.g., tar) materials.

PITCHBLENDE Naturally-occurring uranium oxide, variety of mineral uraninite.

PITCHER FITTING Pipe fitting using gradually curved turn as opposed to sharp angles on standard plumbing fittings.

PITCHING Squared masonry, precast blocks or embedded stones laid in regular fashion with dry or filled joints on upstream slope of embankment dam, reservoir shore or sides of channel as protection against wave and ice action.

PITLESS ADAPTER Two-piece pipe fitting that allows well pipe to exit well casing below frost line.

PITOMETER Apparatus used to gauge speed of moving fluids.

PITOT TUBE Device for measuring velocity of flowing water using velocity head of stream as index of velocity.

PITTED Voids 0.039 inch to 0.236 inch in diameter.

PLACED To put in particular place or in particular state.

PLACER Surficial mineral deposit formed by mechanical concentration of mineral particles from weathered debris.

PLACER MINING Mining of stream bed deposits for minerals.

PLAGE Extended emission feature of active region that exists from emergence of first magnetic flux until widely scattered remnant magnetic fields merge with background.

PLAGE CORRIDOR Space in chromospheric plage lacking plage intensity, coinciding with polarity inversion line.

PLAGENIL Spotless disc free of calcium plage.

PLAIN Expanse of land with relatively low relief.

PLAIN SEDIMENTATION Settling or storage of water without aid of chemicals.

PLAIN SETTLING TANK Tank with adequate volume that keeps flow for time required for settleable solids to precipitate without mechanical or chemical assistance.

PLAN POSITION INDICATES NO ECHOES Refers to radar detects no precipitation within its range.

PLAN POSITION INDICATOR Displays radar data horizontally using map projection.

PLANE OR SATURATION Earth inside watertable.

PLANETARY BOUNDARY LAYER Layer within atmosphere between 1 km and Earth's surface where friction affects wind speed and wind direction.

PLANETARY SET GEAR Gear set consisting of inner (sun) gear, outer ring with internal teeth and two or more small (planet) gears meshed with both sun gear and ring.

PLANFORM Shape and size of channel and overbank features as viewed from directly above.

PLANIMETER Device that measures area.

PLANKTON Tiny plants and animals that float and live in water.

PLANNING BASIS Guidance concerning size (distance) of planning area, time dependence of releases and flood characteristics of releases.

PLANNING MODELS Large scale models that describe how planning may proceed.

PLANNING STANDARDS Standards which emergency plans of dam operating organization and local authorities should conform to.

PLANNING-LEVEL ANALYSIS Measures adequacy of number of lanes based on travel demand rather than density of traffic volumes.

PLANNING-PROGRAMMING-BUDGETING SYSTEM Relates to developing detailed resource requirements and actions needed to implement plans.

PLANOSOL Soil with light-colored, coarse-textured surface horizon that shows signs of periodic water stagnation and abruptly overlies dense, slowly permeable subsoil with significantly more clay than surface horizon.

PLANS Drawing showing technical details of building, machine, etc.

PLANT 1. Organism that's not animal; 2. Station where mechanical, chemical, and/or nuclear energy is converted into electric energy.

PLANT AND EQUIPMENT Physical or tangible assets that are long-term assets that typically have life of more than one year.

PLANT FACTOR Ratio of average load on plant for period of time considered to be aggregate rating of all generating equipment installed in plant.

PLANT-USE ELECTRICITY Electric energy used in operation of plant.

PLANTING EASEMENT Easement for reshaping roadside areas and establishing, maintaining and controlling plant growth thereon.

PLAQUE 1. Piece of flat material with writing on it; 2. Clearing in bacterial lawn caused by virus.

PLASMA Any ionized gas (i.e., any gas containing ions and electrons).

PLASMA ARC REACTORS Devices that use electric arc to thermally decompose organic and inorganic materials at ultra-high temperatures into gases and vitrified slag residue.

PLASMID Circular piece of DNA that exists apart from chromosome and replicates independently of it.

PLASTIC 1. Soil property that enables it to be deformed or modeled without cracking or appreciable volume change; 2. Synthetic, solid, hydrocarbon-based polymer.

PLASTIC LIMIT 1. Water content corresponding to arbitrary limit between plastic and semisolid state of consistency of soil (e.g., will just begin to crumble when rolled into thread approximately 1/8 inches in diameter); 2. Minimum amount of water in terms of percent of oven-dry weight of soil that will make soil plastic.

PLASTIC SOIL Soil that exhibits plasticity.

PLASTICITY Property of soil or rock which allows it to be deformed beyond point of recovery without cracking or appreciable volume change.

PLASTICITY INDEX 1. Numerical difference between liquid limit and plastic limit; 2. Measure of plasticity of soil (i.e., size of range of water contents where soil exhibits plastic properties).

PLASTICIZER Any of various substances added to material in order to make it more pliable.

PLASTICS Non-metallic chemoreactive compounds molded into rigid or pliable construction materials, fabrics, etc.

PLASTICS PROCESSOR Intermediate facility where recovered plastic products and materials are sorted, decontaminated and prepared for final recycling.

PLATE 1. Flat material of uniform thickness; 2. Layer of material on surface of something.

PLATE COUNT Determination of degree of bacterial contamination of sample made by enumeration after period of incubation.

PLATE PRESS Filter press with flat carriage and roller, used for printing from engraved steel or copper plates.

PLATE SCREEN Display consisting of one or maybe more perforated plates where liquid flows.

PLATE TECTONICS Large-scale movement of tectonic plates that contributes to continental drift.

PLATE TOWER SCRUBBER Air pollution control device that neutralizes hydrogen chloride gas by bubbling alkaline water through holes in series of metal plates.

PLATFORM 1. Generic radar term often used to encompass pedestal and antenna assembly; 2. Flat expanse of rock, often result of wave erosion.

PLATFORM LABOR Cost for trainmen and bus operator services.

PLATFORM TIME Total time operator is behind wheel of vehicle in revenue and non-revenue service.

PLATFORM TRUCK Truck without sides for transporting heavy material.

PLATOON Group of self-driving vehicles travelling in close convoy and communicating electronically with each other.

PLATOON RATIO Measure of individual phase progression performance derived from percentage arrivals on green.

PLAYA Intermittent shallow lake formed at bottom of undrained desert basin after heavy rains.

PLAYA LAKE Shallow, temporary lake in arid or semiarid region, covering or occupying playa in wet season but drying up in summer.

PLOT Area of land studied or used for experimental purpose, where sample areas are often located.

PLOUGH Device pulled through ground in order to break it open into furrows for planting.

PLUG 1. Any substance used to stop or fill hole; 2. Pronged connecting device which fits into mating socket.

PLUG COCK VALVE Valve where fluid passes by way of hole in tapered stopper which is shut by turning plug.

PLUG FLOW 1. Type of flow occurs in tanks, basins or reactors when slug of water moves through without ever dispersing or mixing with rest of water flowing through; 2. Flow of body of ice or viscous fluid with no shearing between adjacent layers.

PLUG PRESSURE Fire hydrant water pressure.

PLUG VALVE Tapering valve which turns in case like plug of faucet.

PLUGGING Act or process of stopping flow of water, oil or gas into or out of formation through borehole or well penetrating that formation.

PLUMB 1. Exactly vertical; 2. At right angles to horizontal.

PLUMB BOB Pointed weight hung from string used for vertical alignment.

PLUMBER One who furnishes, fits and repairs pipes and other apparatus for conveyance of water, gas or drainage.

PLUMBER'S DOPE Pipe-joint compound.

PLUMBER'S FURNACE Portable heater for melting solder and lead.

PLUMBER'S RULE Measure using typical scale on single side as well as scale for measuring span of 45° counterbalance on another.

PLUMBER'S SOIL Mixture of adhesive and lampblack to prevent adhesion of solder.

PLUMBER'S SOLDER Soft solder, used in making wiped joints and seams, containing approximately two parts of lead to one part of tin.

PLUMBING Pipes, together with joints, tanks, stopcocks, taps and other fixtures of water, gas or sewage system in house or other building.

PLUMBING FIXTURE Device which can be connected to plumbing system to deliver and drain water.

PLUMBING INSPECTOR Individual authorized to inspect pipes works for compliance having regulatory code.

PLUMBING SYSTEM Any system that conveys fluids for wide range of applications.

PLUMBING TILE Tile for under-level drains.

PLUMBLINES Measures movement of concrete dam due to applied reservoir water pressures and temperature changes.

PLUME 1. Visible or measurable discharge of contaminant from given point of origin; 2. Area of radiation leaking from damaged reactor; 3. Area downwind within which release could be dangerous for those exposed to leaking fumes; 4. Upward spray of water or mist.

PLUME BLIGHT Visibility impairment caused by air pollution plumes aggregated from individual sources.

PLUME IMPINGEMENT Collision of plume with topography that rises above plume altitude.

PLUME RISE Flow of gases increases with higher buoyancy or momentum and decreases with increasing wind speed or vertical temperature.

PLUME-DOMINATED FIRE Fire whose behavior is governed primarily by local wind circulation produced in response to strong convection above fire rather than by general wind.

PLUNGE POINT Point where wave curls over and breaks.

PLUNGE POOL Natural or artificially created pool that dissipates energy of free falling water.

PLUNGER Device used to remove blockages from drain of basin or tub, by suction.

PLUNGER PUMP Positive displacement pump where high-pressure seal is stationary and smooth cylindrical plunger slides through seal.

PLUTONIUM Radioactive metallic element with atomic number of 94.

PLUVIAL Of, pertaining to or produced by precipitation.

PLUVIAL INDEX Amount of precipitation falling in specified period likely to be equaled or exceeded at given location.

PLUVIOGRAPH Instrument for measuring amount of water that has fallen with feature to register data in real time to demonstrate rainfall over short period, often automated graphing instrument.

PLUVIOMETER Scientific instrument for measuring amount of rainfall.

PLUVIOMETRIC COEFFICIENT For any month at given station, ratio of monthly normal precipitation to one-twelfth of annual normal precipitation.

PLY One of several layers of fabric or of other strength-contributing material.

PM_{10} Measure of particles in atmosphere with diameter of less than ten or equal to nominal 10 micrometers.

$PM_{2.5}$ Measure of particles smaller than PM_{10} in atmosphere.

PNEUMATIC Powered or inflated by compressed air.

PNEUMATIC EJECTOR Receiver that allows liquids or solids to enter without restriction and when it becomes filled, compressed air displaces contents up to higher discharge line.

PNEUMATIC PUMPING Airlift via pumping.

PNEUMATIC VALVE Valve where gas-flow direction is switched.

PNEUMATICS Branch of mechanics that deals with mechanical properties of gases.

POCOSIN Local term along Atlantic coastal plain, from Virginia south, for shrub-scrub wetland located on relatively flat terrain, commonly between streams.

PODZOL Typical soil of coniferous or boreal forests.

POINT Tapered, sharp end of tool, weapon or other object.

POINT BAR Deposits of sediment that occur on convex side or inside curve of meander (i.e., inside of channel bends).

POINT DEVIATION Connector between origin/destination and fixed-route service.

POINT DISCHARGE Instantaneous rate of discharge, in contrast to mean rate for interval of time.

POINT GAUGE Device used to measure water levels.

POINT OF APPLICATION Location where force is applied to body.

POINT OF CONTRAFLEXURE In flexing beam, location where bending moment is zero.

POINT OF DISINFECTANT APPLICATION Point where disinfectant is applied and water downstream of that point is not subject to recontamination by surface water runoff.

POINT OF INJECTION Last accessible sampling point prior to waste fluids being released into subsurface environment through Class V injection well.

POINT OF ORIGIN Location where something comes from.

POINT OF WASTE GENERATION Place where samples of waste are collected to calculate flow rate.

POINT PRECIPITATION Precipitation at particular site, in contrast to mean precipitation over area.

POINT RAINFALL Rainfall measured at particular site.

POINT SOURCE Any single identifiable source of pollution, radiation, waves, fluid or other substance that has one specific location.

POINT-BEARING PILE Pile that's driven down to hard ground to carry full load at its point.

POINT-OF-CONTACT MEASUREMENT OF EXPOSURE Estimating exposure by measuring concentrations over time (while exposure is taking place) at or near place where it is occurring.

POINT-OF-DISINFECTANT APPLICATION Point where disinfectant is applied and water downstream of that point is not subject to recontamination by surface water runoff.

POINT-OF-ENTRY TREATMENT DEVICE Device installed to treat water entering house or building for purpose of reducing contaminants in water distributed throughout house or building.

POINT-OF-USE TREATMENT DEVICE Treatment device applied to single tap to reduce contaminants in drinking water at one faucet.

POINT-SOURCE CONTAMINANT Any substance that degrades water quality and originates from discrete locations (e.g., discharge pipes, drainage ditches, wells, concentrated livestock operations, floating craft).

POINT-SOURCE POLLUTION Pollution coming from single point.

POISE 1. State of balance, equilibrium or stability; 2. Unit of dynamic viscosity.

POISON Chemical that, in relatively small amounts, is able to produce injury by chemical action when it comes in contact with susceptible tissue.

POLAR CAP ABSORPTION Anomalous condition of polar ionosphere whereby HF and VHF (3-300 MHz) radiowaves are absorbed and LF and VLF (3-300 kHz) radiowaves are reflected at lower altitudes than normal.

POLAR FRONT Semi-permanent, semi-continuous front that separates tropical air masses from polar air masses.

POLAR JET Boundary between polar air and subtropical air, marked by concentration of isotherms and strong vertical shear.

POLAR ORBITING SATELLITE Weather satellite which travels over both poles each time it orbits Earth.

POLARIZATION 1. Production of polarized light; 2. Direction where electric field of electromagnetic wave points; 3. Separation of positive and negative charges in nucleus, atom, molecule or system.

POLARIZATION RADAR Radar which takes advantage of ways where transmitted waves' polarization affect backscattering.

POLAROGRAPHY Electrochemical technique for analysis of redox reactions.

POLICY 1. Principle of behaviour, conduct, etc. thought to be desirable or necessary; 2. Document containing or certifying contract.

POLLEN 1. Fertilizing element of flowering plants; 2. Background air pollutant.

POLLUTANT Any substance that, when present in system at sufficient concentrations, degrades or contaminates air, water or soil or could become harmful to human and/or ecological health or that impair use of water for recreation, agriculture, industry, commerce or domestic purposes.

POLLUTANT PATHWAYS Avenues for distribution of pollutants.

POLLUTANT STANDARD INDEX Indicator of one or more pollutants that may be used to inform public about potential for adverse health effects from air pollution in major cities.

POLLUTANT, CONSERVATIVE Pollutants that do not readily degrade in environment and which are mitigated primarily by natural stream dilution after entering receiving bodies of waters.

POLLUTANT, NON-CONSERVATIVE Pollutants that are mitigated by natural biodegradation or other environmental decay or removal processes in receiving stream after in-stream mixing and dilution have occurred.

POLLUTION 1. Resource out of place; 2. Physical contamination of environment by harmful substances or by disruptive levels of noise, light, etc.

POLLUTION CONTROL Reduces amount of pollution generated by industries, agriculture or consumers.

POLLUTION CONVERSION Transformation of single pollutant into another or removal of pollutant.

POLLUTION EPISODE Air pollution incident in given area caused by concentration of atmospheric pollutants under meteorological conditions that may result in significant increase in illnesses or deaths that may also describe water pollution events or hazardous material spills.

POLLUTION HAZARD APPRAISAL Assessment of pollutant to discover its risk.

POLLUTION INDEX Formula to determine air quality scale that uses measured or predicted concentrations of several criteria pollutants.

POLLUTION INDICATORS Organism or organisms that react naturally to identified pollutants.

POLLUTION LOAD Tradable permit that essentially allows for rights to pollute to be sold in market.

POLLUTION PERMIT Practice that reduces, eliminates or prevents pollution at its source.

POLLUTION PREVENTION 1. Identifying areas, processes and activities which create excessive waste products or pollutants in order to reduce or prevent them through, alteration or eliminating process; 2. Number used to describe air quality and pollution measured.

POLLUTIONAL INDEX Quantity of pollutants present in environment.

POLYBUTYLENE PIPE Conduit formed of light colored polyolefin or saturated polymer.

POLYCHLORINATED BIPHENYLS 1. Any of class of aromatic organic compounds that are mixtures of chlorinated biphenyl compounds having various percentages of chlorine used as industrial chemicals; 2. Synthetic, toxic, persistent chemicals used in electrical transformers and capacitors for insulating purposes and in gas pipeline systems as lubricant.

POLYCHLORINATED NAPHTHALENES Industrial chemicals that are mixtures of chlorinated naphthalene compounds.

POLYCYCLIC AROMATIC HYDROCARBON Class of organic compounds with fused-ring aromatic structure.

POLYELECTROLYTE Electrolytic materials that are widely used in batteries, fuel cells and other applications.

POLYELECTROLYTES Synthetic chemicals that help solids to clump during sewage treatment.

POLYETHYLENE Polymer consisting of many ethylene monomers bonded together.

POLYETHYLENE PIPE Thermoplastic conduit made from high-molecular-weight organic compound.

POLYETHYLENE TEREPTHALATE Thermoplastic material used in plastic soft drink and rigid containers.

POLYMER Natural or synthetic chemical structure where two or more like molecules are joined to form more complex molecular structure.

POLYMERIZATION Chemical process, normally with aid of catalyst, to form polymer by bonding together multiple identical units.

POLYPROPYLENE Thermoplastic resin made by polymerization of propylene and used for films, fibres or moulding materials.

POLYPROPYLENE PIPE Conduit formed by lightweight, thermo-plastic.

POLYSACCHARIDE Polymer made of many saccharide units linked by glycosidic bonds.

POLYSAPROBIC Of, pertaining to or inhabiting body of water containing decomposing organic matter and thus little dissolved oxygen.

POLYSTYRENE 1. Alkane chain of benzene molecules; 2. Vinylic polymer of styrene.

POLYSYNTHESIS Act or process of combining many separate elements into whole.

POLYUNSATURATED Of or relating to long chain organic compounds that have multiple double bonds; polyunsaturated fatty acids are essential to human nutrition.

POLYURETHANE Any of various polymeric resins containing urethane links.

POLYVINYL Designating any polymer derived from vinyl compound.

POLYVINYL CHLORIDE Type of tough, environmentally indestructible plastic (i.e., thermoplastic resin) produced by polymerization of vinyl chloride used to make pipe and fittings for water distribution and irrigation that releases hydrochloric acid when burned.

POLYVINYL CHLORIDE PIPE Conduit of polymer of vinyl chloride.

POND 1. Inland body of standing water smaller than lake; 2. Water impounded by diversion dam; 3. Inland body of standing water of any size fed by springs rather than by river.

POND WATER SURFACE AREA Water surface area of pond created by impoundment for storage of processing waste water at average operating levels.

PONDAGE 1. Holding back of water for later release; 2. Water in reservoir.

PONDAGE FACTOR Index of amount pondage needed when stream flow is constant and plant is operational during specific periods.

PONDING 1. In flat areas, runoff collects in depression and cannot drain out; 2. Method for compacting soil using water added after soil is in place until free water stands on surface.

PONTOON 1. Flat-bottomed boat used as support for temporary bridge; 2. Floating structure supporting bridge or dock.

POOL 1. Deep reach of stream where current is slow with little velocity; 2. Reach of stream between two riffles; 3. Small and rather deep collection of water supplied by spring or occurring in course of stream; 4. Reservoir for water; 5. Elevation of surface of body of water (e.g., lake).

POOL HEIGHT Height of water behind dam.

POOL PRICE Cost of megawatt hour of electricity at end of given hour paid to electricity generators.

POP-OFF VALVE Secondary pressure regulating device used to back up primary pressure regulating device.

POP-UP WASTE Drain plug found at bottom of sink basin or tub.

POPCORN CONVECTION Slang for showers and thunderstorms that form on scattered basis with little or no apparent organization, usually during afternoon in response to diurnal heating.

POPPET VALVE Valve that consists of sliding shaft with disk on end and which is operated by springs and possibly cams interacting with shaft.

POPULATION 1. People living within political or geographical boundary; 2. Group of interbreeding organisms occupying particular space.

POPULATION AT RISK 1. Population subgroup more likely to be exposed to chemical or is more sensitive to chemical, than is general population; 2. Population potentially affected by flood waters as result of large operational releases or dam failure.

POPULATION DENSITY Number per unit area of individuals of any given species at given time.

POPULATION DYNAMICS Branch of life sciences that studies size and age composition of populations as dynamical systems and biological and environmental processes driving them.

POPULATION EQUIVALENT Means of expressing strength of organic material in wastewater.

PORCELAIN ENAMEL Thin layer of ceramic or glass to protect surfaces from chemical attack, physical damage, modify structural characteristics and improve appearance.

PORE Small to minute opening or interstice in rock or soil that allow passage of fluid.

PORE PRESSURE Interstitial pressure of fluid (e.g., air, water) within mass of soil, rock or concrete.

PORE SIZE DISTRIBUTION Relative abundance of each pore size in representative volume of soil.

PORE SPACE Open spaces in soil, sediments and rocks that are filled (i.e., volume of pores) by air or water.

PORE TREATMENT Chemicals applied to exterior of porous materials to reduce infiltration of moisture.

PORE WATER Water contained in pores in soil or rock.

PORE-WATER PRESSURE Internal hydrostatic pressure in embankment caused by level of water in reservoir acting through pressure-transmitting paths between soil particles in fill.

PORE-WATER-PRESSURE CELL Device for measuring pore water pressures on account of load changes.

POROSITY 1. Ratio of volume of void space to total volume of undisturbed sample; 2. Degree that soil, gravel, sediment or rock is permeated with pores or cavities where water (i.e., water-bearing capacity) or air can move; 3. Degree of perviousness.

POROUS 1. Full of tiny pores that allow fluids or gasses to pass through; 2. Voids less than 0.039 inch in diameter.

POROUS PIPE Conduit, used for drainage, that lets in water through holes or substance it's made from.

POROUS TUBE Tube whose permeability can be changed by adjusting porosity.

PORT 1. Opening where connection (i.e., pipe) is made; 2. Entryway or gate.

PORT AUTHORITY Public authority that operates port or other transportation infrastructure.

PORT CAPACITY Measure of how much port and its terminals can handle over given period.

PORTABLE CONTAINER Reusable container.

PORTABLE TANK Container that stores large amount of fluid.

PORTABLE TRAFFIC CONTROL SIGNAL Transportable and reusable temporary traffic control signal.

PORTAL Opening into tunnel.

PORTAL BRACING System of struts and ties that are in plane of portal inclined braces.

PORTAL FRAME Two-dimensional rigid frames that have rigid joint between column and beam.

PORTAL-OF-ENTRY EFFECT Local effect produced in tissue or organ of first contact between toxicant and biological system.

PORTLAND CEMENT CONCRETE PAVEMENT Pavement surface of aggregate mixed with portland cement.

POSITION HEAD Fluid elevation head.

POSITIVE AREA 1. Area on sounding representing layer where lifted parcel would be warmer than environment; 2. Measure of energy available for convection.

POSITIVE CLOUD TO GROUND LIGHTNING Lightning that delivers positive charge to ground, as opposed to more common negative charge.

POSITIVE CONFINING BED Confining bed of aquifer whose head is above upper surface of saturation zone.

POSITIVE CONTROL Separation of all air traffic within designated airspace.

POSITIVE DISPLACEMENT PUMP Pump which operates by periodically changing volume of cavity.

POSITIVE HEAD Energy per unit weight of fluid above datum.

POSITIVE PRESSURE Pressure greater than atmospheric pressure.

POSITIVE ROTARY PUMP Displacement pump designed for either clockwise or counterclockwise rotation.

POSITIVE VORTICITY ADVECTION Advection of higher values of vorticity into area.

POSITIVE-DISPLACEMENT PUMP Pump that makes fluid move by trapping fixed amount and forcing trapped volume into discharge pipe.

POSITIVE-PRESSURE FABRIC FILTER Fabric filter with fans on upstream side of filter bags.

POSITIVE-TILT TROUGH Upper level system that is tilted to east with increasing latitude (i.e., from southwest to northeast).

POST INDICATING VALVE Freestanding post that displays position of valve.

POST-CHLORINATION Addition of chlorine to plant effluent for disinfectant purposes after effluent has been treated.

POST-CLOSURE Time period following shutdown of waste management or manufacturing facility (for monitoring purposes, often considered to be 30 years).

POST-CONSUMER RECYCLING Use of materials generated from residential and consumer waste for new or similar purposes.

POST-CONSUMER WASTE Materials or finished products that have served their intended use and have been diverted or recovered from waste destined for disposal, having completed their lives as consumer items.

POST-FLARE LOOPS Loop prominence system often seen after major two-ribbon flare, which bridges ribbons.

POST-INJECTION SITE CARE Appropriate monitoring and other actions (including corrective action) needed following cessation of injection to assure that USDWs are not endangered.

POST-MOUNTED SIGN Roadway sign with no portion of it supported above.

POST-STORM REPORT Report summarizing impact of storm on it's forecast area.

POST-TENSIONING To apply tension after casting concrete.

POST-TENSIONING DUCT Form device to apply tension after casting concrete.

POSTCHLORINATION Treatment process that occurs before water leaves treatment plant with addition of chlorine to prevent microbial growth.

POSTCONSUMER RECOVERED PAPER Paper, paperboard and fibrous wastes from retail stores, office buildings, homes and so forth, after they have passed through their end-usage as consumer item.

POSTCONSUMER RECYCLING Process of collecting material used by consumers and turning waste into more products.

POSTCONSUMER WASTE Waste type produced by end (i.e., final) consumer.

POSTED OPERATING INSTRUCTIONS Operations and maintenance instructions taken from standing operating procedures that pertain to mechanical features in immediate area of dam and which are posted adjacent to those features.

POSTED SPEED LIMIT Maximum speed limit posted on section of highway using regulatory sign.

POSTEMERGENCE HERBICIDE Herbicide applied to foliage after crop has sprouted to kill or significantly retard growth of weeds.

POSTPROCESSING Using computer programs to assist in preparing data sets for use with generic simulation codes.

POTABLE WATER Water safe and satisfactory for drinking and cooking.

POTAMOLOGY Study of rivers.

POTASSIUM PERMANGANATE Readily available commercial permanganate used as oxidizing agent both in industry and in analytical laboratory.

POTENTIAL 1. Gravitational potential is radial component of gravitational field; 2. Work (energy) required to move reference particle from reference location to specified location in presence of force field.

POTENTIAL CAPACITY Maximum capacity STOP-controlled intersection could theoretically handle.

POTENTIAL DAM SAFETY DEFICIENCY Condition that currently does not significantly affect safety of dam, but is capable of becoming dam safety deficiency.

POTENTIAL DOSE Amount of compound contained in material swallowed, breathed or applied to skin.

POTENTIAL ENERGY Energy of body with respect to its position (e.g., gravitational or electric field) or its condition (e.g., stretched or compressed spring).

POTENTIAL EVAPOTRANSPIRATION Rate at which water, if available, would be removed from soil and plant surfaces by evapotranspiration.

POTENTIAL HABITAT Habitat suitable for, but currently unoccupied by, species or community in question.

POTENTIAL HEAD Measurement of liquid pressure above geodetic datum.

POTENTIAL NATURAL WATER LOSS Water loss during years when annual precipitation greatly exceeds average water loss.

POTENTIAL RATE OF EVAPORATION Amount of evaporation that would occur if sufficient water source were available.

POTENTIAL SPILL Accident or other circumstances which threatens to result in discharge of oil or other hazardous substances.

POTENTIAL SWITCH Contactor whose coil is in series using security circuit.

POTENTIAL TEMPERATURE Temperature parcel of dry air would have if brought adiabatically (i.e., without transfer of heat or mass) to standard pressure level of 1,000 mb.

POTENTIALLY RESPONSIBLE PARTY Any individual or company.

POTENTIATION Ability of one chemical to increase effect of another chemical.

POTENTIOMETRIC SURFACE Surface that water in aquifer can rise by hydrostatic pressure (i.e., imaginary surface that represents total head in aquifer).

POUND Unit of mass equal to 16 ounces.

POUND PER CUBIC FOOT Unit of density equal to 1 lb/ft^3.

POUND PER CUBIC INCH Unit of density equal to 1 lb/in^3.

POUND PER SQUARE INCH Unit of force/pressure equal to 1 lb/in^2.

POUNDS PER SQUARE INCH Pressure designation for pounds per in^2 that may be pounds per in^2 gauge (psig) or absolute (psia).

POUNDS PER SQUARE INCH, ABSOLUTE Relative to vacuum, atmospheric pressure plus gauge pressure.

POUNDS PER SQUARE INCH, GAUGE Pressure relative to atmospheric pressure 0 at sea level.

POUR POINT Temperature below which liquid loses its flow characteristics.

POWDER FACTOR Term expressing amount of explosives used to break quantity of rock.

POWDER SNOW Dry, loose, unconsolidated snow.

POWER 1. Physical force or strength; 2. Electrical energy generated, transferred or used; 3. Rate to magnify optical image by lens or mirror; 4. Measure of rate of doing work or transferring energy.

POWER CAPACITY OF STREAM Stream hydroelectric creating potential.

POWER DEMAND Rate at which electric energy is required and delivered to or by system over any designated period of time.

POWER FACTOR Ratio of real to total power.

POWER HEAD Deep-well pump actuating mechanism.

POWER POOL Two or more interconnected electric systems which operate as single system to supply power to meet combined load requirements.

POWER UNIT Engine for use in portable operation of mechanical equipment.

POWERPLANT Structure that houses turbines, generators and associated control equipment.

PRACTICAL QUANTIFICATION LIMIT Lowest level that can be reliably achieved within specified limits of precision and accuracy during routine laboratory operating conditions.

PRACTICAL QUANTITATION LIMIT Minimum concentration of analyte that can be measured to determine if analyte is present at or above concentration.

PRAIRIE Extensive area of relatively flat grassland with few, if any, trees.

PRAIRIE POTHOLE Shallow depression, generally containing wetlands, occurring in outwash plain, recessional moraine or till plain.

PRE-AERATION Addition of air at initial stages of treatment to freshen wastewater, remove gases, add oxygen, promote flotation of grease and aid coagulation.

PRE-ALARMS Warning prior to actual signal.

PRE-CHARGE PRESSURE Accumulator compressed gas pressure before entry of liquid.

PRE-CHLORINATION Adding chlorine to water supply collection system for potable water treatment, corrosion and odor control.

PRE-CONSUMER WASTE Materials generated in manufacturing and converting processes (e.g., manufacturing scrap, trimmings, cuttings).

PRE-DAMPENING Adding water to reduce dust, rebound and static electricity while improving compaction of material on construction surface.

PRE-DRAINAGE Dewatering of site by lowering groundwater levels in advance of excavation.

PRE-EXCAVATION 1. Removal of soil by driving or cleaning out open ended pipe; 2. Advance excavation of site.

PRE-FILTER Device that removes unwanted matter from air or water.

PRE-FRONTAL SQUALL LINE Line of thunderstorms that precedes advancing cold front.

PRE-FRONTAL TROUGH Elongated area of relatively low pressure preceding cold front usually associated with shift in wind direction.

PRE-HARVEST INTERVAL Time between last pesticide application and harvest of treated crops.

PRE-HURRICANE SQUALL LINE First serious indication that hurricane is approaching generally straight line and resembles squall-line that occurs with mid-latitude cold front.

PRE-SETTLING Sedimentation period prior to additional wastewater treatment.

PRE-SIGNAL Traffic signals in advance of intersection to improve efficiency.

PRE-TENSIONING Stressing wires or strands (i.e., tendons) to predetermined amount by stretching them between two anchorages prior to placing concrete.

PRECAST CONCRETE/PRECAST MEMBER Concrete structural member cast in different location than its final position.

PRECAST DAM Dam constructed mainly of large precast concrete blocks or sections.

PRECAUTIONARY PRINCIPLE When information about potential risks is incomplete, basing decisions about best ways to manage or reduce risks on preference for avoiding unnecessary health risks instead of on unnecessary economic expenditures.

PRECEDENCE DIAGRAMMING METHOD Graphical representation technique that demonstrates inter-dependencies among various project activities.

PRECEDENCE OF DOCUMENTS Order of contract file precedence if there is conflict between files.

PRECEDENCE PROGRAMMING Project network diagram with nodes representing activities and arrows representing dependencies.

PRECEDENCE RELATIONSHIP Diagramming method where activities are placed in boxes and connected with arrows.

PRECHLORINATION Addition of chlorine at headwork of treatment plant prior to other treatment processes.

PRECIPITABLE WATER Measure of depth of liquid water at surface that would result after precipitating all of water vapor in vertical column over given location.

PRECIPITATE 1. Water in air falling to ground (e.g., rain, snow, sleet or hail); 2. Substance separated from solution or suspension by chemical or physical change.

PRECIPITATION 1. Discharge of water, in liquid or solid state, out of atmosphere, generally onto land or water surface; 2. Total measurable amount of water received in form of snow, rain, drizzle, hail and sleet; 3. Removal of hazardous solids from liquid waste to permit safe disposal; 4. Removal of particles from airborne emissions as in rain (e.g., acid precipitation); 5. Act or process of producing solid phase within liquid medium.

PRECIPITATION ATTENUATION Loss of energy that radar beam experiences as it passes through area of precipitation.

PRECIPITATION GAUGE Instrument to measure amount of liquid precipitation over set period of time.

PRECIPITATION MODE Standard or default, operational mode of WSR-88D.

PRECIPITATION PROCESSING SYSTEM WSR-88D system that generates 1-hour running, 3-hourly and running storm total precipitation accumulations.

PRECIPITATOR Pollution control device that collects particles from air stream.

PRECISE SETTLEMENT GAUGE Device that tracks land settlement.

PRECISION 1. Accuracy with which number can be represented (i.e., number of digits used to represent number); 2. Ability of measurement to be reproduced consistently.

PRECISION APPROACH RADAR Radar guidance system designed to provide lateral and vertical guidance to aircraft pilot for landing.

PRECISION WORK State/quality of work required to be exact.

PRECOAT To coat in advance of another process.

PRECURSOR Compound antecedent to pollutant.

PREDECESSOR ACTIVITY Activity that logically occurs before dependent activity in schedule.

PREDECESSOR WORK ITEM Work thing that immediately precedes one or even more work things in sense sequence of precedence network.

PREDEVELOPMENT HYDROLOGY Combination of runoff, infiltration and evapotranspiration rates and volumes that typically existed on site before human-induced land disturbance occurred (e.g., construction of infrastructure on undeveloped land).

PREDOMINANT PERIOD Period(s) at which maximum spectral energy is concentrated.

PREDOMINANT WIND 1. Wind that prevails and generates local component of significant sea conditions across forecast area where in marine forecast products this is 10-meter wind and in nearshore marine zones this is 3-meter wind.

PREEMERGENCE HERBICIDE Herbicide applied to bare ground after planting crop but prior to crop sprouting above ground to kill or significantly retard growth of weed seedlings.

PREEMPTION Purchase of something before it's offered for sale to others.

PREFERENCE Priority access to federal power by public bodies and cooperatives.

PREFERENCE CUSTOMERS Publicly owned systems and non-profit cooperatives which, by law, have preference over investor-owned systems for purchase of power from federal projects.

PREFERENTIAL LANE Lane separated from general purpose lanes for exclusive use .

PREFERENTIAL PARKING Spaces to assist in finding parking near desired locations.

PREFERENTIAL TREATMENT Special privileges given to specific transportation mode.

PRELABORATORY Methods that include all activities involved in collecting, preparing and delivering sample to place of analysis.

PRELIMINARY ASSESSMENT Process of collecting and reviewing available information about known or suspected waste site or release.

PRELIMINARY FILTER Filter that protects pump against fine to larger debris to permit uninterrupted pumping.

PRELIMINARY REPORT Report summarizing life history and effects of storm.

PRELIMINARY TREATMENT Removal of coarse solids and large materials in raw wastewater.

PRELOAD To load in advance.

PREPARATION TIME Time between reaching target and movement initiation.

PREPOSITIONING To place in location before some other event occurs.

PREQUALIFICATION Evaluation of expected bidder prior to bid invitation.

PREQUALIFICATION QUESTIONNAIRE Forms bidder is required to submit to help assess ability to execute and finance work.

PRESCRIBED ELEVATION Elevation as called out or dictated in notes, drawings or specifications paragraphs.

PRESCRIBED FIRE Management ignited or natural wildland fire that burns under specified conditions where fire is confined to predetermined area and produces fire behavior and fire characteristics required to attain planned fire treatment and resource management objectives.

PRESCRIBED PROCESS Regulation that defines how waste is to be treated.

PRESCRIBED SUBSTANCE Regulation that control release of air, land or water.

PRESCRIPTIVE Water rights which are acquired by diverting water and putting it to use in accordance with specified procedures.

PRESENT MOVEMENT Best estimate of movement of center of tropical cyclone at given time and given position.

PRESENT SERVICEABILITY Ability of specific section of pavement to serve traveling public.

PRESENT SERVICEABILITY INDEX Measure of current overall rating of section of highway based upon visual observation.

PRESENT SERVICEABILITY RATING Rating on 0 to 5 scale that represents qualitative ride quality of pavement from user perspective.

PRESENT VALUE METHOD Calculates present value of costs by converting present and future expenses to present.

PRESENT WEATHER Type of weather observed at reporting time.

PRESERVATIVE Any natural or artificial agent that acts to preserve.

PRESERVATIVE TREATMENT Wood surface treatment for resistance to decay damage and insect attack.

PRESSED WOOD PRODUCTS Materials used in building and furniture construction that are made from wood veneers, particles or fibers bonded together with adhesive under heat and pressure.

PRESSURE 1. Force per unit area; 2. Exertion of force upon surface by fluid (e.g., atmosphere) in contact with it.

PRESSURE ALTIMETER Aneroid barometer calibrated to indicate altitude in feet instead of units of pressure.

PRESSURE ALTITUDE Altitude in standard atmosphere at which given pressure will be observed.

PRESSURE BALANCE System or container pressure equivalent to outside pressure.

PRESSURE CELL Instrument for measuring pressure within mass of soil, rock or concrete or at interface between one and other.

PRESSURE CHANGE Net difference between barometric pressure at beginning and ending of specified interval of time.

PRESSURE CHARACTERISTIC Pattern of pressure change during specified period of time using three categories: falling, rising or steady.

PRESSURE CONTROL VALVE Valves that safeguard against excessive system pressure or limit operation pressure.

PRESSURE COUPLET Area where you have high pressure area located adjacent to low pressure area.

PRESSURE CUTOUT Detector that functions at set pressure.

PRESSURE DROP Difference in total pressure between two points of fluid flow.

PRESSURE EQUALIZING Balancing high and low pressure together.

PRESSURE FALLING RAPIDLY Decrease in station pressure at rate of 0.06 inch of mercury or more per hour which totals 0.02 inch or more.

PRESSURE FILTER Semi-continuous machines that enter wash and cake discharge mode at end of filtration cycle.

PRESSURE FLOW Experienced when flow moves from high to low pressure in system.

PRESSURE GAUGE Device for registering pressure of solids, liquids or gases.

PRESSURE GRADIENT Amount of pressure change occurring over given distance.

PRESSURE GRADIENT FORCE Three-dimensional force vector operating in atmosphere that accelerates air parcels away from regions of high pressure and toward regions of low pressure in response to air pressure gradient.

PRESSURE GROUTING Injecting grout into inaccessible but interconnected space.

PRESSURE HEAD Internal energy of fluid created by depth of one unit of fluid that's exerted on its container.

PRESSURE ICE Floating sea, river or lake ice that has been deformed, altered or forced upward in pressure ridges by lateral stresses of any combination of wind, water currents, tides, waves and surf.

PRESSURE INTENSITY Pressure applied at point.

PRESSURE JUMP Sudden, sharp increase in atmospheric pressure, typically occurring along active front and preceding storm.

PRESSURE LINE Pressurized fluid line to apparatus.

PRESSURE LOSS Decrease in pressure as measured between two points in flowing fluid system.

PRESSURE METER Device used to determine absolute and relative pressures for gases and/or liquids.

PRESSURE POTENTIAL Component of water potential which represents hydrostatic pressure exerted on water in cell.

PRESSURE RANGE Force magnitude of wind load effect in limit-state.

PRESSURE REGULATING VALVE Control valve that reduces input pressure to desired output pressure.

PRESSURE REGULATOR Control valve that maintains constant pressure.

PRESSURE REGULATOR VALVE Control valve that reduces input pressure to desired output pressure.

PRESSURE RELEASE Propagation of fractures near surface of rock due to release and expansion of confining pressure.

PRESSURE RELIEF PIPES Pipes used to relieve uplift or pore water pressure in dam foundation or in dam structure.

PRESSURE RIDGE Ridge resulting from interaction sea ice expanses.

PRESSURE RISING RAPIDLY Increase in station pressure at rate of 0.06 inch of mercury or more per hour which totals 0.02 inch or more.

PRESSURE SENSING VALVE Device used to prevent explosion or implosions of pressurized tanks.

PRESSURE SEWERS System of pipes where water, wastewater or other liquid is pumped to higher elevation.

PRESSURE STRAINER Apparatus designed to protect pipeline and valves from large foreign objects.

PRESSURE SWITCH Electric switch controlled by pressure.

PRESSURE TANK Tank where liquid or gas is stored at pressure greater than atmospheric.

PRESSURE TENDENCY 1. Character and amount of atmospheric pressure change during specified period of time; 2. Sign (plus or minus) and amount of pressure change (hPa) for three hour period ending at time of observation on buoy report.

PRESSURE TRANSDUCER Transducer that converts pressure into analog electrical signal.

PRESSURE UNLOADING VALVE Control valve used to dump excess fluid to tank at little or no pressure.

PRESSURE UNSTEADY Pressure that fluctuates by 0.03 inch of mercury or more from mean pressure during period of measurement.

PRESSURE VESSEL Container designed to maintain volume at significantly different pressure to that of ambient.

PRESSURE ZONE Area of water distribution system that provides constant hydraulic gradient.

PRESSURE, STATIC 1. Pressure of fluid on body when body is at rest relative to fluid; 2. In flowing air, total pressure minus velocity pressure, pushing equally in all directions.

PRESSURE, TOTAL 1. Sum of static pressure, dynamic pressure and gravitational head; 2. Force per unit area of soil; 3. Sum of each gas' partial pressure; 4. In flowing air, sum of static and velocity pressures.

PRESSURE, VELOCITY 1. Due to motion, increase in moving fluid's pressure compare to its static pressure; 2. In flowing air, pressure due to velocity and density of air.

PRESSURE-ACTUATED SEAL Seal that activates by fluid pressure.

PRESSURE-DIFFERENTIAL SWITCH Electric switch controlled by means of difference in pressure.

PRESSURE-DRIVEN CHANNELING Channeling of wind in valley by synoptic-scale pressure gradients superimposed along valley's axis.

PRESSURE-LIMITING DEVICE mechanism built that prevents output pressure from exceeding pressure capacity.

PRESSURE-PRESERVATIVE TREATED Impregnating wood with preservatives by pressure process.

PRESSURE-REDUCING VALVE Safety valve used to control or limit pressure in system.

PRESSURE-RELIEF CONE Depression in piezometric surface of confined artesian aquifer created when well is employed to take out water at fixed speed.

PRESSURE-RELIEF DAMPER Damper, in system, that relieves excess pressure.

PRESSURE-RELIEF VALVE Safety control valve used to limit pressure in system.

PRESTRESSED CONCRETE Concrete with stresses induced in it before use so as to counteract stresses that will be produced by load.

PRESTRESSED DAM Dam, stability of which depends in part on tension in steel wires, cables or rods that pass through dam and are anchored into foundation rock.

PRETIMED CONTROL Preset signal timings non-responsive to real-time fluctuations in traffic demand.

PRETIMED OPERATION Traffic control signal will alternate green, yellow and all-red time without regard to traffic flow.

PRETREATMENT 1. Any treatment received before some other process; 2. Processes used to reduce, eliminate or alter nature of wastewater pollutants from non-domestic sources before they are discharged into publicly owned treatment works.

PREVAILING CONDITION Conditions during analysis.

PREVAILING VISIBILITY Visibility considered representative of conditions at station that's greatest distance that can be seen throughout at least half horizon circle, not necessarily continuous.

PREVAILING WESTERLIES Westerly winds that dominant in middle latitudes.

PREVAILING WIND Wind from direction that's predominant at particular place or season.

PREVAILING WINDS Wind that consistently blows from one direction more than from any other.

PREVALENT LEVEL SAMPLES Air samples taken under normal conditions (i.e., ambient background samples).

PREVALENT LEVELS Levels of airborne contaminant occurring under normal conditions.

PREVENTION OF SIGNIFICANT DETERIORATION Program, specified in Clean Air Act, whose goal is to prevent air quality from deteriorating significantly in areas of country that are presently meeting ambient air quality standards.

PREVENTIVE EXPENDITURE Sum paid to prevent or reduce unwanted effects.

PREVENTIVE MAINTENANCE Maintenance performed regularly on equipment to reduce likelihood of breakdown.

PRICE CURRENT METER Current meter with series of conical cups from which velocity of water may be computed.

PRIMACY Having primary responsibility for administering and enforcing regulations.

PRIMARY AIR Amount of air required for complete combustion of fuel.

PRIMARY AMBIENT AIR QUALITY STANDARDS Air quality standards designed to protect human health.

PRIMARY BURNER Burner installed in primary combustion chamber to dry out and ignite fuel.

PRIMARY CONTROL SYSTEM Air pollution control system designed to remove gaseous and particulate matter.

PRIMARY CONTROL TIDE STATION Tide station where continuous observations have been made for minimum of 19 years.

PRIMARY CRUSHER Machine for breaking large stones into pieces.

PRIMARY DRAIN Main, sloping drain connection, located on system riser, to main drain.

PRIMARY DRINKING WATER REGULATION Applies to public water systems and specifies contaminant level, which, in judgment of EPA Administrator, will not adversely affect human health.

PRIMARY EFFECT Effect where stressor acts directly on ecological component of interest, not on other parts of ecosystem.

PRIMARY ELEMENT Device that measures physical condition or variable of interest.

PRIMARY EMISSION CONTROL SYSTEM Combination of equipment used for capture and collection of primary emissions.

PRIMARY EXCAVATION Digging in undisturbed soil.

PRIMARY FILTER Filter used for coarse pre-filtration in multistage filter systems.

PRIMARY INFLOW POINTS Always at upstream most end of tributary or main stem segment.

PRIMARY MANUFACTURING RESIDUE Waste byproducts from manufacturing of raw materials.

PRIMARY MINERALS Minerals that form as magma solidifies.

PRIMARY POLLUTANT Substances that are pollutants immediately on entering atmosphere.

PRIMARY PRODUCTION Creation of organic matter by photosynthesis.

PRIMARY PRODUCTIVITY 1. Measure of rate at which new organic matter is formed and accumulated through photosynthetic and chemosynthetic activity of producer organisms (chiefly, green plants); 2. Rate of primary production is estimated by measuring amount of oxygen released (oxygen method) or amount of carbon assimilated (carbon method) by plants.

PRIMARY RECYCLING When recyclable materials are recovered or reused without being altered and used for same purpose.

PRIMARY SETTLING TANK Sewage treatment tank designed to reduce velocity of wastewater flow for organic solids to settle.

PRIMARY SEWAGE TREATMENT Physical process (filtering and settling) that mainly removes solids from wastewater.

PRIMARY SIGNAL FACE Signal face mounted on post at or near left of stop line of approach.

PRIMARY SLUDGE Sludge procured from capture of suspended solids and organics via settling in primary wastewater treatment process.

PRIMARY SPILLWAY Structure located on or adjacent to storage or detention dam over or where surplus or floodwaters which cannot be contained in allotted storage space are discharged and at diversion dams to bypass flows exceeding those which are turned into diversion system.

PRIMARY STANDARDS National ambient air quality standards designed to protect human health with adequate margin for safety.

PRIMARY SWELL DIRECTION Prevailing direction of swell propagation.

PRIMARY TREATMENT First stage of wastewater treatment practice of removing some portion of suspended solids and organic matter are removed by screening and settling.

PRIMARY TRIBUTARY Tributary directly connected to or that joins with main river segment.

PRIME 1. To prepare mechanism (e.g., action of filling pump casing with water to remove air); 2. First in time, order or sequence.

PRIME COAT Coating applied directly to prepared base before additional layers.

PRIME CONTRACT Contract entered into directly between client and contractor.

PRIME CONTRACTOR Contractor hired to manage project and to hire subcontractors.

PRIME MOVER Machine that receives and modifies energy as supplied by some natural source or fuel and transforms it into mechanical work.

PRIMING Carrying over of matter to initiate reaction.

PRIMING PUMP Pump used to prime main pump.

PRIMITIVE WATER Water inside inner part of Earth's lithosphere.

PRINCIPAL LEGEND Information (e.g., street names, routes) placed on guide signs.

PRINCIPAL ORGANIC HAZARDOUS CONSTITUENTS Hazardous compounds monitored during incinerator's trial burn, selected for high concentration in waste feed and difficulty of combustion.

PRINCIPAL ORGANIZATIONS Federal, state and local agencies or departments or executive offices having major or lead roles in emergency planning and preparedness for emergency incidents at dams.

PRINCIPAL SPILLWAY Spillway which would be used first during normal inflow and flood flows.

PRIONS Microscopic particles made of protein that can cause disease.

PRIOR APPROPRIATION DOCTRINE Water law that allocates right to use water on first-come first-serve basis.

PRIORITY CIRCUIT Electrical circuit that delivers power.

PRIORITY CONTROL Process of communicating start and completion dates in order to execute plan.

PRIORITY LANE Lane that provides preferential treatment to traveler.

PRIORITY WATER QUALITY AREAS Water body areas near potential pollutant sources and areas.

PRISTINE Earliest condition of quality of water body that's unaffected by human activities.

PRIVATE CAR 1. Vehicle that's for personal use and owned privately; 2. Railroad passenger car that's either privately owned or used privately.

PRIVATE ROAD OPEN TO PUBLIC TRAVEL Private toll roads and roads (including any adjacent sidewalks that generally run parallel to road) within shopping centers, airports, sports arenas and other similar business and/or recreation facilities that are privately owned, but where public is allowed to travel without access restrictions. Roads within private gated properties (except for gated toll roads) where access is restricted at all times, parking areas, driving aisles within parking areas and private grade crossings shall not be included in this definition.

PRIVATE SEWAGE DISPOSAL SYSTEM Sewage treatment and disposal system independently owned.

PRIVATE SEWER Sewers owned and cared for by private property owners.

PRIVATE SIDING Railway siding owned by company, also possibly served by their own locomotive, which is connected to and served by railway, but is not part of railway's infrastructure.

PRIVATE UTILITY Private business organization that provides utility services to public.

PRIVATE WATER Systems composed of private groundwater residential wells, cisterns and larger private water systems that serve more than one residence.

PRIVATE WATER SUPPLY Water supply not provided by public company or government.

PRIVY Outdoor facility for urination and defecation.

PROBABILISTIC MODEL Mathematical model where behavior of one or more of variables is either completely or partially subject to probability laws.

PROBABILISTIC QUANTITATIVE PRECIPITATION FORECAST Form of quantitative precipitation forecast that includes assigned probability of occurrence for each numerical value in forecast product.

PROBABILITY Chance or likelihood, that certain event might happen.

PROBABILITY CURVE Curve showing probability distribution of continuous random variable.

PROBABILITY FORECAST Forecast of probability that one or more of mutually exclusive set of weather conditions will occur.

PROBABILITY OF DETECTION Likelihood, expressed as percentage, that test method will correctly identify leaking tank.

PROBABILITY OF FAILURE Likelihood that, given loading event and failure mode, structure responds with necessary adverse occurrences to ultimately result in uncontrolled reservoir release.

PROBABILITY OF HAIL Product from NEXRAD hail detection algorithm that estimates likelihood that hail is present in storm.

PROBABILITY OF LOAD Likelihood that certain loading event (e.g., static, seismic, hydrologic) occurs at structure site.

PROBABILITY OF PRECIPITATION Likelihood that precipitation will be reported at certain location during specified period of time.

PROBABILITY OF THUNDERSTORMS Likelihood, based on climatology, that thunderstorm will be reported at that location during specified period of time.

PROBABILITY OF TROPICAL CYCLONE CONDITION Probability, in percent, that cyclone center will pass within 50 miles to right or 75 miles to left of listed location within indicated time period when looking at coast in direction of cyclone's movement.

PROBABILITY SAMPLE Sample selected by method involving knowledge of likelihood of any unit being selected.

PROBABLE 1. Likely to occur; 2. Reasonably expected.

PROBABLE MAXIMUM FLOOD Largest flood that may be expected from most severe combination of critical meteorological and hydrologic conditions that are reasonably possible in particular drainage area in this climatic era.

PROBABLE MAXIMUM PRECIPITATION Largest precipitation for which there is any reasonable expectancy in this climatic era.

PROBABLE RESERVES Unproved reserves that analysis of data suggests are likely to be recoverable.

PROBE Small device used to explore, investigate or measure something by penetrating or being placed in it.

PROBING Action of investigating/exploring.

PROCESS 1. Set of procedures used to produce product; 2. Structure that arises above surface.

PROCESS VARIABLE Physical or chemical quantity that is usually measured and controlled in operation of water treatment plant or industrial plant.

PROCESS VERIFICATION Verifying that process raw materials, water usage, waste treatment processes, production rate and other facts relative to quantity and quality of pollutants contained in discharges are substantially described in permit application and issued permit.

PROCESS WASTE Waste pollutants due to manufacturing process.

PROCESS WASTEWATER Any water that comes into contact with any raw material, product, byproduct or waste.

PROCESS WATER Water that cannot be classified as drinking water and used in connection with industrial processes.

PROCESS WEIGHT Total weight of all materials, including fuel, used in manufacturing process.

PROCTOR CURVE Curve showing relationship between dry density (dry unit weight) and moisture content of soil for given compactive effort.

PRODUCERS Plants that perform photosynthesis and provide food to consumers.

PRODUCT LEVEL Level of product in storage tank.

PRODUCT PIPE Pipe used to transport product from one location to another.

PRODUCT RESOLUTION Smallest spatial increment or data element distinguishable in given Doppler radar product.

PRODUCT STANDARD Standard that specifies product requirements to be fulfilled.

PRODUCT WATER Water that has passed through water treatment plant and is ready to be delivered to consumers.

PRODUCTION Act or process of producing electrical energy from other forms of energy, expressed in kilowatt-hours (kWh).

PRODUCTION WELL well used to retrieve liquid or gas from underground.

PRODUCTION-BASED STANDARD Discharge standard expressed in terms of pollutant mass allowed in discharge per unit of product manufactured.

PRODUCTS OF COMBUSTION Result of combustion of material (e.g., carbon dioxide, particulate matter).

PRODUCTS OF INCOMPLETE COMBUSTION Organic compounds formed by combustion.

PROFILE 1. Vertical or longitudinal section through object; 2. Plot of elevation versus distance.

PROFILE DEPTH Difference between largest profile peak height and largest profile valley depth.

PROFILE GRADE Elevation change along centerline of roadway.

PROFILE RECORD Record of surface elevation changes over length.

PROFILER Instrument designed to measure horizontal winds directly above its location and thus measure vertical wind profile.

PROFILOMETER Device designed to measure profile (i.e., roughness) of surface.

PROFUNDAL Deepest part of ocean or lake where light does not penetrate.

PROGRAM 1. Set of structured activities; 2. Software application or collection of software applications, designed to perform specific task.

PROGRAM EVALUATION AND REVIEW TECHNIQUE Method for diagramming and analyzing flow of dependent tasks and other events in project.

PROGRAMMABLE LOGIC CONTROLLER Programmable electronic device used in industrial automation to provide logic and sequencing controls for machinery.

PROGRAMMATIC ENVIRONMENTAL IMPACT STATEMENT Statement that addresses proposal to implement specific policy, to adopt plan for group of related actions or to implement specific statutory program.

PROGRESS CHART Graph or chart showing amount of work done on major project.

PROGRESS PAYMENT Partial payment to cover completed work, up to point of invoicing.

PROGRESS REPORT Report to inform about progress made on project over certain period of time.

PROGRESS SCHEDULE Tool utilized to track progress and performance of projects through all phases until completion.

PROGRESSION 1. Act of moving forward or proceeding in course; 2. Sequence obtained by adding or multiplying each term by constant.

PROGRESSIVE DERECHO Derecho characterized by short curved squall line oriented nearly perpendicular to mean wind direction with bulge in general direction of mean flow.

PROGRESSIVE SIGNAL SYSTEM Traffic control signals coordinated to promote optimum traffic movement.

PROGRESSIVE SYSTEM System that dewaters ahead of excavation.

PROGRESSIVE WAVE Water wave where particles move in circles or ellipses.

PROHIBITED AREA Area of airspace within which flight of aircraft is not permitted.

PROJECT Single financial entity which can be composed of several units or divisions, integrated projects or participating projects.

PROJECT ACCOUNTANT Accountant responsible for managing all financial information related to project.

PROJECT AGREEMENT Contract setting forth mutual obligations with regard to portion or all of specific project.

PROJECT ANALYSIS Assessment of all costs pr problems of project before work commences.

PROJECT BRIEF Capsulated description that defines scope and requirements for project.

PROJECT BUDGET Financial and personnel disposal limits that are needed for project.

PROJECT CHARTER Statement of scope, objectives and project members for project.

PROJECT CLOSEOUT Assesses project, ensures completion and derives any lessons learned and best practices to be applied to future projects.

PROJECT COMPLETION Period where job is finished according to contract.

PROJECT CONTROL Data gathering, management and analytics used to predict and influence time pr cost of project.

PROJECT COST Cost to procure all products, services and resources to complete job.

PROJECT COST MANAGEMENT Management method that uses technology to measure cost and productivity through life cycle of project.

PROJECT DEFINITION Statement of scope, objectives and participants in project.

PROJECT DEPTH Depth, authorized by Congress, of federal navigation channel.

PROJECT DRAWING Drawing that details project and contract standards.

PROJECT ENGINEER Engineer who designs and coordinates engineering aspects of construction project.

PROJECT EVALUATION AND REVIEW TECHNIQUE Method for diagramming and analyzing flow of dependent tasks and other events in project.

PROJECT HANDOVER At end of project, where client formally accepts: asset, work area, and/or relevant information from contractor.

PROJECT HAZARDOUS WASTE COORDINATOR Individual designated by Area Manager to coordinate records and maintain proper management of hazardous wastes on each project that acts as primary contact with regional hazardous waste management coordinator.

PROJECT HUMAN RESOURCES MANAGEMENT Identifies, documents and assigns project roles, responsibilities and reporting relationships for project.

PROJECT INSPECTOR Inspector focused on monitoring work of contractors to ensure compliance with contract specifications and plans.

PROJECT INTEGRATION MANAGEMENT Takes numerous processes that are being used in project and makes ensures coordination.

PROJECT LIFE CYCLE Step process (i.e., initiation, planning, execution and closure) followed when moving through phases of project completion.

PROJECT MANAGEMENT Discipline of organizing and managing people in such way that project is completed within defined scope, quality, time and cost constraints.

PROJECT MANAGEMENT BODY OF KNOWLEDGE Terminology and guidelines for project management.

PROJECT MANAGEMENT PROFESSIONAL Professional designation offered by Project Management Institute.

PROJECT MANAGEMENT SOFTWARE Type of computer applications designed to facilitate project delivery and efficiency.

PROJECT MANAGEMENT TEAM Group of individuals teamed together to achieve specific business task or goal.

PROJECT MANAGER One given special responsibilities to manage projects.

PROJECT MANUAL Defines and recommends both procedures and protocols for project.

PROJECT NETWORK DIAGRAM Graphic or tabular schedule depicting sequence that project's elements are to be completed.

PROJECT PHASE Set of project activities to prepare product.

PROJECT PLAN Document that contains intended scope of project, facilitate communication and tracks planning assumptions or decisions.

PROJECT PLAN DEVELOPMENT Takes outputs of other planning processes to create document that can be used to guide project execution and control.

PROJECT PLAN EXECUTION Primary process for carrying out project plan by performing tasks contained therein.

PROJECT PLANNING Use of schedules to develop, plan and report progress of project.

PROJECT PROCUREMENT MANAGEMENT Creation of relationships with outside vendors and suppliers for goods and services needed to complete project.

PROJECT QUALITY MANAGEMENT Processes and activities needed to determine and achieve project quality.

PROJECT RISK MANAGEMENT Process of identifying, analyzing and responding to risks that arises over life cycle of project.

PROJECT SCHEDULE Document specifying work needed to deliver project on time.

PROJECT SCOPE MANAGEMENT Act of defining and controlling all processes involved in project.

PROJECT SPECIFICATIONS Comprehensive description of objectives for development project containing all goals, functionality and details established by client.

PROJECT TEAM MEMBERS Individuals who report to project manager.

PROJECT TIME MANAGEMENT Component of project management where timeline is analyzed and developed for completion of project or deliverables.

PROJECT XL EPA initiative to give states and regulated community flexibility to develop comprehensive strategies as alternatives to multiple current regulatory requirements in order to exceed compliance and increase overall environmental benefits.

PROJECT-USE ENERGY Primarily power and energy used for project operations (e.g., pumping, miscellaneous uses).

PROJECT-USE POWER Power for main conveyance pumping, designated drainage pumping and other designated miscellaneous electric loads directly associated with operation of project.

PROJECTION RATIO Vertical distance pipe projects above ground divided by outside vertical pipe height.

PROLIFERATION 1. Process where organism produces others of its kind; 2. Spread of biochemical, nuclear and other weapons of mass destruction to countries not originally involved in developing them.

PROMINENCE Term identifying cloud-like features in solar atmosphere.

PROOF PRESSURE Amount of over-pressure sensor can be subjected to without damage or malfunction.

PROPAGATE 1. To cause to continue or multiply by generation; 2. To spread.

PROPAGATION 1. Movement; 2. Transmission of energy as waves through or along medium.

PROPELLANT Liquid product that expels active ingredient from its container.

PROPELLER FAN Propeller mounted on axle or directly-mounted to shaft of motor.

PROPELLER PUMP Centrifugal pump that develops head by propelling or lifting action of vanes upon liquid.

PROPERLY SHREDDED GARBAGE wastes from preparation, cooking and dispensing of food that have been shredded to such degree that all particles will be carried freely under flow conditions normally prevailing in public sewers, with no particle greater than ½in (1.27 cm) in any dimension.

PROPERTY PROTECTION Measures that are undertaken usually by property owners in order to prevent or reduce flood damage.

PROPORTIONAL CONTROL Control system technology based on response in proportion to difference between desired process variable and current value.

PROPORTIONAL SAMPLING Sampling method where investigator divides finite population into sub-populations and applies random sampling techniques to each sub-population.

PROPORTIONAL WEIR Weir where discharge (i.e., flow) is directly proportional to head.

PROPORTIONATE MORTALITY RATIO Number of deaths from specific cause in specific period of time per 100 deaths from all causes in same time period.

PROPORTIONER Basic pumping unit consisting of two positive displacement pumps designed to dispense two components at precisely controlled ratio.

PROPOSAL Offer by party of what he has in view as to intended business transaction.

PROPOSAL GUARANTY Security furnished with proposal to guarantee that bidder will enter into contract if department awards contract to bidder.

PROPOSED PERMIT Permit prepared after close of public comment period (and when applicable, any public hearing and administrative appeals) that is sent for review before final issuance.

PROPOSED PLAN Plan for site cleanup available to public for comment.

PRORATABLE WATER That portion of total water supply available subject to proration in times of water shortage.

PRORATE 1. To divide proportionately; 2. Organophosphorus insecticide and acaricide.

PROSPECTING Search, by drilling and excavation, for deposits within Earth's surface and crust.

PROTECTED MODE Traffic control signal operation where left or right turns are permitted when left or right GREEN ARROW signal is indicated.

PROTECTED PLUS PERMITTED Left turn at signalized intersection that displays protected phase before permitted phase.

PROTECTED TURN When opposing traffic faces red light and that pedestrians crossing lane used by protected left-turn movement will face DON'T WALK signal.

PROTECTED WASTE PIPE Waste pipe not directly linked to soil, drain, vent or waste pipe.

PROTECTION Means of keeping or remaining safe.

PROTECTIVE ACTION Action or measure taken to avoid or reduce exposure to hazard.

PROTECTIVE ACTION DECISION MAKING Process whereby local authorities select one of more actions to protect threatened populations.

PROTEIN 1. Any of numerous large, complex naturally-produced molecules composed of one or more long chains of amino acids held together by peptide bonds; 2. One of three major classes of food or source of food energy abundant in animal-derived foods (i.e., meat) and vegetables (i.e., legumes).

PROTEINS Complex nitrogenous organic compounds of high molecular weight made of amino acids that are essential for growth and repair of animal tissue.

PROTEOLYSIS Hydrolysis of proteins into peptides and amino acids.

PROTEOLYTIC ENZYMES Enzymes that break long chain-like molecules of proteins into shorter fragments (i.e., peptides) and eventually into their components (i.e., amino acids).

PROTOCOL Series of formal steps for conducting test.

PROTON 1. Positively charged subatomic particle forming part of nucleus of atom and determining atomic number of element; 2. Solar activity levels with at least one high energy event (i.e., Class X Flares).

PROTON EVENT Any flare producing significant fluxes of greater than 10 million electron volts protons in vicinity of Earth.

PROTOPLASM Entire contents of cell comprising nucleus and cytoplasm.

PROTOPLAST Membrane-bound cell from which outer wall has been partially or completely removed.

PROTOTYPE Full-sized structure, system, process or phenomenon being modeled.

PROTOZOA One-celled animals that are larger and more complex than bacteria.

PROTOZOAN Any of diverse group of eukaryotes that are primarily unicellular.

PROVEN RESERVES Quantity of natural resources that company reasonably expects to extract from given formation.

PROXIMATE ANALYSIS 1. Process of quantifying portion of free moisture, volatile matter, ash and given carbon in solid fuel; 2. Set of methods to get information about nutritional value of feed or food.

PROXIMITY SWITCH Sensor able to detect presence of nearby objects without any physical contact.

PSEUDO-COLD FRONT Boundary between supercell's inflow region and rear-flank downdraft.

PSEUDO-WARM FRONT Boundary between supercell's inflow region and forward-flank downdraft.

PSI Measurement of pressure in system, device or appliance.

PSIA Atmospheric pressure plus gauge pressure.

PSIG Pressure above atmospheric pressure 0 at sea level.

PSYCHODA Genus of moth flies in family Psychodidae.

PSYCHOLOGICAL VALUE Concept that existence of wilderness provides condition that could allow individual to achieve control over stressful conditions, thus contributing to psychological health of many off-site users.

PSYCHROMETER Any of several instruments used to measure relative humidity of atmosphere (e.g., wet-and-dry-bulb hygrometer).

PSYCHROMETRIC CHART Graph or chart showing relationships of thermodynamic properties of moist air.

PSYCHROMETRY Measurement or study of thermodynamic properties of moist air.

PSYCHROPHILIC BACTERIA Group of bacteria that have preferential temperature for growth at less than 59 °F.

PUBLIC AFFAIRS OFFICER Organization's designated person responsible for public affairs.

PUBLIC ALERT AND NOTIFICATION SYSTEM System for obtaining attention of public and providing appropriate emergency information.

PUBLIC AUTHORITY Any authority which has legal mandate to govern, administrate part or aspect of public life.

PUBLIC COMMENT PERIOD Time allowed for public to express its views and concerns regarding action.

PUBLIC CONTRACT Legally enforceable commitment of party to undertake work or improvement desired by public authority.

PUBLIC HEALTH Science and practice of community hygiene.

PUBLIC HEALTH APPROACH Regulatory and voluntary focus on effective and feasible risk management actions at national and community level to reduce human exposures and risks, with priority given to reducing exposures with biggest impacts in terms of number affected and severity of effect.

PUBLIC HEALTH CONTEXT Incidence, prevalence and severity of diseases in communities or populations and factors that account for them, including infections, exposure to pollutants and other exposures or activities.

PUBLIC HEARING Formal meeting wherein officials hear public's views and concerns about action or proposal.

PUBLIC INFORMATION OFFICER Person on team who is in charge of public information affairs.

PUBLIC INVOLVEMENT PLAN Document that presents procedural plans to inform and gather information from project beneficiaries and general public.

PUBLIC LAND ORDER Action on federal lands to withdraw it from public use for specified purpose.

PUBLIC LANDS HIGHWAY Forest road under jurisdiction of and maintained by public authority and open to public.

PUBLIC NOTICE Notification informing public of actions.

PUBLIC NOTIFICATION Advisory that requires water system to distribute to affected consumers when system has violated MCLs or other regulations that advises consumers what precautions, if any, they should take to protect their health.

PUBLIC RELATIONS OFFICER Person on team who is in charge of public information affairs.

PUBLIC ROAD Any road or street under jurisdiction of and maintained by public authority and open to public travel.

PUBLIC SERVICE COMMISSION Regulatory agency that governs retail utility rates and practices and, in many cases, issues approvals for construction of new generation and transmission facilities.

PUBLIC SERVICE VEHICLE Motor vehicle used for carrying passengers for hire or reward.

PUBLIC SEWER Sewer systems that are maintained by local public works department.

PUBLIC SUPPLY Water withdrawn by public and private water suppliers and delivered to users.

PUBLIC TRANSPORTATION Transportation service that can be used by member of public.

PUBLIC USE OF WATER Acquisition, treatment and distribution of raw or potable water for public use via agency, corporation or alternative organization.

PUBLIC UTILITIES COMMISSION Regulatory agency that governs retail utility rates and practices and, in many cases, issues approvals for construction of new generation and transmission facilities.

PUBLIC UTILITY Organization supplying community with electricity, gas, water or sewerage.

PUBLIC UTILITY HOLDING COMPANY ACT Federal legislation, enacted in 1935, which regulates corporate structure and financial operations of certain utility holding companies.

PUBLIC UTILITY REGULATORY POLICIES ACT Bill aimed at expanding use of cogeneration and renewable energy resources.

PUBLIC WATER Water accessible to public.

PUBLIC WATER MAIN Water supply pipe for public utilization controlled by public authority.

PUBLIC WATER SUPPLY Provides water to public, subject to regulation.

PUBLIC WATER SYSTEM System that provides piped water for human consumption to at least 15 service connections or regularly serves 25 individuals for at least 60 days annually.

PUBLIC WATER USE Water supplied from public-water supply and used for such purposes as firefighting, street washing and municipal parks and swimming pools.

PUBLIC WORKS Technical projects carried out by government on behalf of community.

PUBLIC-SUPPLY DELIVERIES Water provided to users through public-supply distribution system.

PUBLIC-SUPPLY WITHDRAWALS Water withdrawn by public and private water suppliers for use within general community.

PUBLICLY OWNED TREATMENT WORKS Waste-treatment works owned by state, unit of local government or Indian tribe, usually designed to treat domestic wastewaters.

PUBLICLY-OWNED TREATMENT WORKS Sewage treatment plant owned and operated by government agency.

PUBLICS Any interested group or individual, including federal, state and local agencies, interest groups, ad hoc groups and general public.

PUDDLE 1. Small pool of water; 2. To compact loose soil by soaking it and allowing it to dry; 3. Act of compacting earth, soil clay, etc., by mixing them with water and rolling or tamping mixture.

PUDDLE CLAY Clay used to make waterproof layer for dam or canal on permeable ground.

PUDDLED CORE Earth dam core constructed of puddle clay.

PUDDLING Method of compacting soil with soil deposited into pool of water and then stirred or rodded.

PUFF PIPE Antisiphon conduit.

PUG MILL Mill for grinding and mixing clay for art or brick making.

PULL IN To approach or drive up to place and come to stop.

PULL OUT To leave place or begin trip.

PULL SHOVEL Shovel with hinge-and-stick mounted bucket that digs while being pulled inward.

PULL SWITCH Switch actuated by chain or string.

PULL-ON CONTAINER System where detachable container is pulled on vehicle and taken to disposal site.

PULMONARY IRRITANTS Airborne pollutants that irritate mucous lining of respiratory tract.

PULSE Short burst of electromagnetic energy that radar sends out in straight line to detect precipitation target.

PULSE DURATION Time over which radar pulse lasts.

PULSE LENGTH Linear distance in range occupied by individual pulse from radar.

PULSE RADAR Radar, designed to facilitate range (i.e., distance) measurements, where energy is transmitted in periodic, brief transmission.

PULSE REPETITION FREQUENCY Amount of time between successive pulses or bursts, of electromagnetic energy transmitted by radar.

PULSE REPETITION TIME Time elapsed between pulses by radar.

PULSE RESOLUTION VOLUME Discrete radar sampling volume.

PULSE SEVERE THUNDERSTORMS Single cell thunderstorms which produce brief periods of severe weather (3/4 inch hail, wind gusts in excess of 58 miles hour or tornado).

PULSE STORM Thunderstorm within which brief period (i.e., pulse) of strong updraft occurs, during and immediately after which storm produces short episode of severe weather.

PULSE WIDTH Linear distance in range occupied by individual pulse from radar.

PUMP Machine that uses suction/pressure to forcefully move liquid or gas.

PUMP CAN Hand-operated, five-gallon water can used to supply drinking water or extinguish small fires.

PUMP CHARACTERISTIC CURVE Graphic representation showing total dynamic head, brake horsepower and net positive suction head and efficiency plotted over capacity range of pump.

PUMP DISCONNECT Mechanism that disconnects pump service.

PUMP EFFICIENCY Ratio of water horsepower output from pump to shaft horsepower input for pump.

PUMP HOUSE Building containing pumping equipment.

PUMP LIFT Vertical distance that pump will raise waters (i.e., distance water must be lifted in well from pumping level to ground surface).

PUMP PIT Sump or subterranean hole used to remove water via pump.

PUMP PRIMER Vacuum pump that primes suction pump.

PUMP SETTING Distance well pump is located below pumping head.

PUMP SLIPPAGE Difference between suction water and discharged water from pump.

PUMP STAGES Centrifugal pump with series of impellers, where stage is determined by number of impellers.

PUMP STRAINER Apparatus that filters out matter before entering pump.

PUMP STROKE Distance piston, of reciprocating pump, travels.

PUMP SUBMERGENCE Distance between pump suction bottom and water surface elevation.

PUMP VALVES Devices that control flow of liquid through pump.

PUMP WELL Pump used in extracting fluid from well.

PUMP-BACK SYSTEM Return-flow system where tailwater is pumped back to head of irrigation ditch for reuse.

PUMP-VALVE CAGE Water valves mounted on end of pump suction.

PUMPABILITY Quality or degree of being pumpable.

PUMPAGE 1. Volume raised by pumps; 2. Amount of work done by pumps.

PUMPED CONCRETE Concrete transported to heights by means of pumping via concrete pumps.

PUMPED STORAGE RESERVOIR Reservoir filled completely with pumped water.

PUMPED-STORAGE PLANT Power plant designed to generate electric energy for peak load use by releasing water previously pumped into elevated storage reservoir, usually during off-peak periods.

PUMPER OUTLET NOZZLE Fire hydrant outlet 4 inches in diameter or larger.

PUMPING Mechanical transfer of liquids or gases.

PUMPING ENERGY Energy required to pump water from lower reservoir to upper reservoir of pumped-storage plant.

PUMPING HEAD 1. Energy given to fluid by pump; 2. Pressure pump must generate for water to flow from outlet at desired rate of flow.

PUMPING LEVEL Level of water in well when pumping.

PUMPING LINE Pump discharge line.

PUMPING MAIN Conduit on pump's outlet side.

PUMPING PLANT Facility that lifts water up and over hills.

PUMPING STATION 1. Mechanical device installed in sewer or water system or other liquid-carrying pipelines to move liquids to higher level; 2. Building that houses one or more pumps.

PUMPING TEST Test conducted to determine aquifer or well characteristics.

PUNCHOUT Distress that typically occurs between closely spaced transverse cracks in concrete.

PURCHASED POWER Represents purchase cost of energy for firming up power supply.

PURE OXYGEN ACTIVATED SLUDGE PROCESS Sludge treatment procedure that uses pure oxygen as alternative to surrounding atmosphere.

PURGE 1. To remove by cleansing or washing away; 2. Cleansing of pipes; 3. To remove air.

PURGING Removing stagnant air or water from sampling zone or equipment prior to sample collection.

PURIFICATION Removal of impurities.

PURVEYOR Agency or person that supplies water.

PUSHBUTTON Can be initiated or activated at push of button.

PUSHBUTTON INFORMATION MESSAGE Recorded message that provides street name and intersection associated with pushbutton.

PUSHBUTTON LOCATOR TONE Repeating locator tone that provides information to blind pedestrians about presence and location of pedestrian pushbutton.

PUSHER ENGINE Railway locomotive temporarily assisting train that requires additional power or traction.

PUTREFACTION Biological decomposition of organic matter associated with anaerobic conditions.

PUTRESCIBILITY Likely to become putrid.

PYCNOMETER Device used to determine specific gravity of liquid.

PYRETHROIDS Any of several synthetic insecticides having structure based on pyrethrin.

PYRHELIOGRAPH Device for measuring and recording Sun's radiation strength.

PYROCLASTIC Ash mixed with larger igneous rock fragments produced by explosive volcanic eruptions.

PYROLYSIS Decomposition of material, chemical or compound due to extreme heat, in absence of oxygen or other reagents.

PYROMETER 1. Thermometer designed to measure high temperatures; 2. Instrument for measuring thermal expansion of solids.

PYRRHOPHYTA Free-swimming unicells characterized by red pigment spot.

Q

Q_{10} Amount growth or activity of organism or enzymic reaction increases for every 10 °C rise in temperature.

QUAD Prefix symbolizing 10^{15} (i.e., one quadrillion).

QUADRAT Square or rectangular plot of land marked off for study of plants, animals soil, natural processes, etc.

QUADRATURE Component of complex signal 90° out of phase with in-phase component.

QUAGMIRE Swampy, soggy area of ground.

QUALIFICATION OF BIDDER Provisions to determine ability of bidder to perform work of contract.

QUALIFIED BIDDER Bidder that meets requirements that enables them to bid on contract.

QUALIFIED GROUNDWATER SCIENTIST Scientist or licensed professional engineer who has received baccalaureate or post-graduate degree in groundwater hydrology or related fields.

QUALIFIED PERSON One who, by possession of recognized degree, certificate or professional standing or who by extensive knowledge, training and experience, has successfully demonstrated his ability to solve or resolve problems relating to subject matter, work or project.

QUALITATIVE 1. Having to do with quality or qualities; 2. Descriptive of kind, type or direction, as opposed to size, magnitude or degree.

QUALITATIVE USE ASSESSMENT Report summarizing major uses of pesticide including percentage of crop treated and amount of pesticide used on site.

QUALITY 1. Level of excellence; 2. Property or attribute that differentiates thing or person; 3. Being of good worth, well made, fit for purpose; 4. In two-phase liquid-vapor mixture, ratio of mass of vapor present to total mass of mixture.

QUALITY ASSURANCE Systematic process of checking to see whether product or service being developed is meeting specified requirements.

QUALITY ASSURANCE SPECIFICATIONS Statistically based specifications that let contractor know if operations are producing acceptable product.

QUALITY CHARACTERISTIC Inherent characteristic of product that says something about aspect of product quality.

QUALITY CHARACTERISTICS Inherent characteristic of product that says something about as aspect of quality of product.

QUALITY CONFORMANCE INSPECTION OR TEST Examination of samples from production run of hose to determine adherence to given specifications, for acceptance of that production.

QUALITY CONTROL System of procedures, checks, audits and corrective actions to ensure that all research design and performance, environmental monitoring and sampling and other technical and reporting activities are of highest achievable quality.

QUALITY FACTOR Measure of performance of coil, capacitor inductor in terms of its losses and resonator bandwidth.

QUALITY INDEX Statistical measure that assesses key dimensions of operational and service quality.

QUALITY LEVEL ANALYSIS Statistical procedure for estimating percent compliance to specification.

QUALITY MEASURE Numeric quantification of quality for designated accountable entity.

QUALITY OF SERVICE Overall performance of telephony or computer network, based on error rates, throughput, transmission delay, etc.

QUALITY OF SNOW Amount of ice in snow sample expressed as percent of weight of sample.

QUALITY STANDARD Details of requirements, specifications, various guidelines and characteristics to be able to meet its quality by product in order to meet product purpose, process or service.

QUALITY STANDARDS Details of requirements, specifications, guidelines and characteristics that products, services and processes should consistently meet.

QUANTITATIVE 1. Having to do with quantity, capable of being measured; 2. Descriptive of size, magnitude or degree as opposed to kind, type or direction.

QUANTITATIVE ANALYSIS 1. Technique that seeks to understand behavior by using mathematical and statistical modeling, measurement and research; 2. Determination of amounts of element or compound in sample.

QUANTITATIVE PRECIPITATION FORECAST Spatial and temporal precipitation forecast that will predict potential amount of future precipitation for specified region or area.

QUANTITY OF SERVICE Measure of number of units of service during particular period.

QUANTITY STANDARD Established criterion that should be used in production of single unit.

QUANTITY TAKEOFF Detailed measurement of materials and tasks needed to complete project.

QUANTUM Amount or quantity observably present.

QUARANTINE Sanitary measure to prevent spread of contagious plague by isolating those believed to be infected.

QUARRY Open cut mine in rock (i.e., rock pit) chosen for physical rather than chemical characteristics.

QUARRY LANDFILLING Landfilling where cover material is obtained from inside pit depression.

QUARRY RUN STONE Stone supplied directly from source with little or no processing, without regard to gradation requirements.

QUARTER BEND Conduit bend changing direction 90°.

QUARTERING 1. Division into four parts; 2. At right angles that are in planes forming right angle with each other.

QUARTZ Very hard, clear or translucent mineral with chemical composition of silicon dioxide.

QUARTZITE 1. Metamorphic rock composed of sand-sized quartz grains that have fused together by heat and pressure; 2. Hard, well-cemented quartz-rich sandstone.

QUASI-STATIONARY Describes low or high pressure area or front nearly stationary.

QUASI-STATIONARY FRONT Front that is nearly stationary or moves very little since last synoptic position.

QUASI-STEADY-STATE MODEL Model where time-dependent variables are simulated by sequence of steady-states.

QUASI-THREE-DIMENSIONAL MODEL Combination of two-dimensional models used to simulate variations in three dimensions.

QUATERNARY Geologic period of last 2 million years and includes Pleistocene and Holocene (last 10,000 years) epochs.

QUATERNARY SEWAGE TREATMENT Chemical treatment to promote precipitation, oxidation or conversion of various undesirable chemical fractions.

QUENCH TANK Water-filled tank used to cool incinerator residues or hot materials during industrial processes.

QUEUE Line of people, vehicles or other objects organized into first-come first-served order.

QUEUE CLEARANCE TIME Time it takes for queue to clear after beginning of green signal.

QUEUE DELAY Time vehicle spends in queued state.

QUEUE DISCHARGE FLOW Traffic flow that passes congestion.

QUEUE JUMP Roadway geometry that provides preference to buses at intersections.

QUEUE JUMPER Roadway geometry that provides preference to buses at intersections.

QUEUE LENGTH Distance between queue ends.

QUEUE SPILLBACK When queue from downstream intersection occupies all storage length that prevents vehicles from entering upstream traffic.

QUEUE STORAGE RATIO Ratio of predicted queue length to available storage length.

QUEUED STATE When vehicle is stopped and within one car length of another stopped vehicle.

QUICK-OPERATING VALVE Valve with lever to quickly open or close.

QUICKLIME Calcium oxide, which is produced by heating limestone and gives slaked lime on treatment with water.

QUICKSAND Any wet inorganic soil, where friction between particles is reduced, that will not support any load.

QUIESCENT PROMINENCE Long, sheet-like prominences nearly vertical to solar surface.

QUIET Solar activity levels with less than one chromospheric event per day.

QUIET ZONE Section of rail line where locomotive horns are not routinely sounded.

QUOTATION Act of naming price.

R

R VALUE Measure of insulation where higher R-value indicates resistance to transfer of heat.

RABBLE ARM Rod where stirring or agitating action is produced in ore under treatment in furnace.

RABBLING Action of raking, stirring or skimming molten material in furnace using rabble.

RACE 1. Fast-moving current of water; 2. Bushings of rolling element bearing which contacts rolling elements.

RACHION Line of lake where maximum wave action shows influence.

RACK 1. Series of one or more shelves, stacked one above other; 2. Bar with teeth on its face or edge; 3. 2. Waterway waste screen.

RACK GEAR Bar with teeth.

RACK RAKE Apparatus to clear collected debris on face of trash screen.

RAD Measure of absorbed dose of radiation equivalent to 0.01 joules/kg.

RADAR Radio device or system for locating object by means of ultrahigh-frequency radio waves reflected from object and received, observed and analyzed by receiving part of device in such way that characteristics (as distance and direction) of object may be determined.

RADAR ADVISORY SERVICE Radar used to assist in providing services to ensure separation between aircraft.

RADAR BEACON Transmitter-receiver that automatically returns distinctive signal that appears on display of triggering radar.

RADAR BEAM Straight line that radar pulse travels along that as radar beam gets further away from radar, it gets wider and wider.

RADAR CODED MESSAGE Alphanumeric coded message which will be used in preparation of national radar summary chart.

RADAR CROSS-SECTION Area of fictitious, perfect reflector of electromagnetic waves (e.g., metal sphere) that would reflect same amount of energy back to radar as actual target (e.g., lumpy snowflake).

RADAR DATA ACQUISITION Hardware component of NEXRAD system that consists of radar antenna, transmitter, receiver, tower and controlling computer that collects unprocessed, analog voltages from radar antenna and converts signal to base reflectivity, base velocity and spectrum width (in polar coordinate form).

RADAR METEOROLOGY Branch of meteorology that uses radars for weather observations and forecasts.

RADAR MONITORING Studying current weather via radar.

RADAR MOSAIC Radar product that combines information from multiple radars to give regional or national view of reflectivity or precipitation.

RADAR PRODUCT GENERATOR Computer in NEXRAD system that receives polar-coordinate base radar data from RDA and processes these data into end-user products.

RADAR RANGE Distance from radar antenna.

RADAR REFLECTIVITY Sum of all backscattering cross-sections (e.g., precipitation particles) in pulse resolution volume divided by that volume.

RADAR SEPARATION Radar spacing to ensure two aircraft are minimum horizontal distance from each other.

RADAR VECTOR When plane is visible to air traffic controller on radar screen and controller tells pilot to fly to specific heading.

RADIAL Lines converging at single center.

RADIAL DRAINAGE Drainage pattern where streams diverge outward from high central area.

RADIAL EXPANSION Communities experiencing rapid growth radially away from center.

RADIAL FLOW Constrained fluid that moves at right angles to axis of rotation.

RADIAL GATE Dam gate with curved upstream plate, flat pivot axis and radial arms hinged to piers or other supporting structure.

RADIAL HIGHWAY Arterial highway leading to or from urban center.

RADIAL IMPELLER Impeller where flow leaves in radial direction that's perpendicular to pump shaft.

RADIAL VELOCITY Component of motion toward or away from given location.

RADIAL WELL System of well strainers, driven horizontally into water-bearing stratum, to extract water.

RADIAL-FLOW TANK Circular tank where flow is to or from center.

RADIAN Unit of plane angle equal to measure of plane angle with its vertex at center of circle and subtended by arc equal in length to radius.

RADIAN PER SECOND Unit of angular velocity equal to 1 rad/s.

RADIAN PER SECOND SQUARED Unit of angular acceleration equal to 1 rad/s^2.

RADIANCE Measure of intensity of radiant energy flux emitted by body in given direction.

RADIANT HEATING SYSTEM Systems that supplies heat directly to floor or to panels in wall or ceiling of house.

RADIATION Transmission of energy though space or any medium through radiating waves or particles (i.e., alpha, beta, gamma or X-rays).

RADIATION DOSE EQUIVALENT Measure of result of ionizing radiation on material or tissue that consumes it.

RADIATION FOG Fog that forms when outgoing longwave radiation cools near-surface air below its dew point temperature.

RADIATION LAWS Four physical laws which fundamentally describe behavior of blackbody radiation: Kirchhoff's law, Planck's law, Stefan-Boltzmann law and Wien's displacement law.

RADIATION SICKNESS Any illness produced by ionizing radiation with symptoms ranging from nausea through to death.

RADIATION STANDARDS Regulations that set maximum exposure limits for protection of public from radioactive materials.

RADIATION WINDOW Radiation spectrum band where little absorption by water vapor happens.

RADIATIONAL COOLING Cooling of Earth's surface (i.e., at night, Earth suffers net heat loss to space due to terrestrial cooling).

RADIATIONAL INVERSION Temperature inversion that develops during night as result of radiational cooling of surface.

RADICAL 1. Group of atoms, joined by covalent bonds, that take part in reactions as single unit; 2. Pertaining to root.

RADIDE Plant embryo part that grows into primary root.

RADIO ALTIMETER Altimeter that uses narrow radio or radar beam directed at ground.

RADIO CONTROL Use of radio signals to remotely control device.

RADIO EMISSION Emissions of Sun in radio wavelengths from centimeters to dekameters, under both quiet and disturbed conditions.

RADIO EVENT Flares with centimetric bursts and/or definite ionospheric event.

RADIO FREQUENCY RADIATION 1. Radiation that does not change structure of atoms but does heat tissue and may cause harmful biological effects; 2. Microwaves, radio waves and low-frequency electromagnetic fields from high-voltage transmission lines.

RADIO-CONTROLLED ENGINE Unmanned engine situated within train, separated by cars from lead unit that's controlled by radio signals.

RADIOACTIVE BURIAL GROUND Disposal site for radioactive waste materials that uses earth or water as shield.

RADIOACTIVE DECAY Spontaneous change in atom by emission of charged particles and/or gamma rays.

RADIOACTIVE GRAVEYARD Disposal site for radioactive waste materials that uses earth or water as shield.

RADIOACTIVE MATERIAL Material with atoms that will spontaneously decay to form other types of atoms by emitting radiation.

RADIOACTIVE SPRING Spring where water has high radioactivity.

RADIOACTIVE WASTE Any waste that emits energy as rays, waves, streams or energetic particles.

RADIOACTIVE-SOLUTION GAUGING Radioactive measurement technique where solution is added to stream and dilution measured downstream.

RADIOACTIVITY Spontaneous emission of ionizing radiation as consequence of nuclear reaction or directly from breakdown of unstable nucleus.

RADIOACTIVITY HALF-LIFE Period of time where half nuclei of species of radioactive substance would decay.

RADIOACTIVITY SHIELDING Material or obstruction that absorbs radiation and provides protection.

RADIOBIOLOGY 1. Study of effects of ionizing radiation on living organisms; 2. Use of radioactive labels to study biological processes.

RADIOCHEMISTRY 1. Chemistry of radioactive substances; 2. Use of radioisotopes to study kinetics of chemical reactions.

RADIOFACSIMILE Broadcasting graphic weather maps and other graphic images via HF radio.

RADIOISOTOPE SNOW GAUGE Snow water equivalent gauge based on absorption of gamma radiation by snow that can measure up to 55 inches water equivalent with 2 to 5% error.

RADIOISOTOPES 1. Chemical variants of radioactive elements with potentially oncogenic, teratogenic and mutagenic effects on human body; 2. Isotopic forms of elements that exhibit radioactivity.

RADIOMETRIC AGE Approximate age determined by tendency that radioactive atoms in geologic materials decay at known rates.

RADIONUCLIDE Radioactive particle, man-made (anthropogenic) or natural, with distinct atomic weight number.

RADIOSONDE Instrument carried aloft by balloon to send back information on atmospheric temperature, pressure and humidity by means of small, expendable radio transmitter.

RADIUS Horizontal distance from center of rotation.

RADIUS OF INFLUENCE 1. Radial distance from center of wellbore to point where there is no lowering of water table or potentiometric surface (edge of cone of depression); 2. Radial distance from extraction well that has adequate air flow for effective removal of contaminants when vacuum is applied to extraction well.

RADIUS OF MAXIMUM WINDS Distance from center of tropical cyclone to location of cyclone's maximum winds.

RADIUS OF VULNERABILITY ZONE Maximum distance from point of release of hazardous substance where airborne concentration could reach level of concern under specified weather conditions.

RADON Naturally occurring, colorless, odorless, radioactive chemical element (i.e., gas) with atomic number 86, formed by radioactive decay of radium atoms in soil or rocks that's damaging to human lungs when inhaled.

RADON DAUGHTERS Short-lived radioactive decay products of radon that decay into longer-lived lead isotopes that can attach themselves to airborne dust and other particles and, if inhaled, damage linings of lungs.

RADON DECAY PRODUCTS Short-lived radioactive decay elements of radon that decay into longer-lived lead isotopes that can attach themselves to airborne dust and other particles and, if inhaled, damage linings of lungs.

RADON PROGENY Short-lived radioactive decay products of radon that decay into longer-lived lead isotopes that can attach themselves to airborne dust and other particles and, if inhaled, damage linings of lungs.

RAFFINATE PONDS Liquid resulting from extractions with solvent.

RAIL 1. Metal bar that makes track for railroad; 2. Horizontal bar extending between supports and used for support or barrier.

RAIL BANKING Method where rail lines proposed for abandonment can be preserved for future rail use through interim conversion to trail use.

RAIL JOINT Fastener connecting adjacent rail ends.

RAIL MOTOR Railway car with small steam, diesel or petrol engine.

RAIL RAPID TRANSIT CAR Bi-directional rail car for transit systems.

RAIL TRAFFIC Vehicles, persons or property that's moved by railway.

RAILBANK PROTECTION Rock-filled, wire baskets supported by rails or posts in streambed.

RAILROAD Permanent road consisting of two fixed metal steel rails to drive trains.

RAILROAD GRADE CROSSING Intersection where roadway crosses railroad tracks at same level.

RAIN Liquid precipitation that falls to Earth in drops more than 0.5 mm in diameter.

RAIN CHAINS Water feature used as alternative to downspout.

RAIN FOOT Horizontal bulging near surface in precipitation shaft, forming foot-shaped prominence.

RAIN FOREST Forest which grows in region of heavy annual precipitation with two major types: tropical and temperate.

RAIN GARDEN Depressed area of ground planted with vegetation, allowing runoff from impervious surfaces (e.g., parking lots, roofs) opportunity to be collected and infiltrated into groundwater supply or returned to atmosphere through evaporation and evapotranspiration.

RAIN GAUGE Instrument for measuring quantity of rainfall.

RAIN INDUCED FOG When warm rain falls through cooler air, water evaporates from warm rain that subsequently condenses in cool air forming fog.

RAIN SHADOW Fry region on lee side of topographic obstacle, usually mountain range, where rainfall is noticeably less than on windward side.

RAIN SHIELD Solid or nearly solid area of rain that typically becomes heavier as one approaches eye of hurricane.

RAIN-FREE BASE Dark, horizontal cloud base with no visible precipitation beneath it.

RAINBOW Luminous arc featuring all colors of visible light spectrum (i.e., red, orange, yellow, green, blue, indigo, violet) created by refraction, total reflection and dispersion of light.

RAINBOW REPORT Comprehensive document giving status of all pesticides now or ever in registration or special reviews.

RAINFALL 1. Quantity of water that falls as rain only; 2. Weighted average of current and antecedent precipitation "effective" in correlating with runoff; 3. Part of precipitation falling on irrigated area effective in meeting consumptive use requirements.

RAINFALL AREA Surface rain has fallen around.

RAINFALL DISTRIBUTION COEFFICIENT Ratio of maximum rainfall at any point to mean rainfall for drainage area or basin.

RAINFALL ESTIMATES Series of NEXRAD products that employ Z-R relationship to produce accumulations of surface rainfall from observed reflectivity.

RAINFALL EXCESS Volume of rainfall available for direct runoff that's equal to total rainfall minus interception, depression storage and absorption.

RAINFALL INDEX Reflects how much precipitation is received relative to long-term average for specified area and timeframe.

RAINFALL INTENSITY Ratio of total amount of rain (i.e., depth) falling during given period to duration of period in in/hr.

RAINFALL INTENSITY CURVE Illustrates rates of rainfall to duration for confirmed interval.

RAINFALL SIMULATION Method to assess overland flow, infiltration, erosion, etc. from rainfall.

RAINFALL, EXCESSIVE Rainfall where rate of fall is greater than certain adopted limits, chosen with regard to normal precipitation (excluding snow) of given place or area.

RAINFED STREAM Stream fed mostly by rain runoff.

RAINOUT Radioactive fallout deposited by means of rain.

RAINWASH 1. Washing action of rain, capable of erosion and transporting soil; 2. Deposit formed by rain.

RAINWATER Water sourced from precipitation which has not joined spring, stream, river, pond, lake or sea.

RAISE Pile of stones.

RAISED BEACH Relatively flat beach lying above water level due to geological changes since formation.

RAISED PAVEMENT MARKER Safety device markers, slightly above road surface, that enhance delineation and driver awareness.

RAKE 1. Toothed machine for collecting; 2. Set of coupled rail vehicles; 3. Fissure or mineral vein traversing strata vertically; 4. Slant to incline from vertical.

RAKINGS Matter in wastewater removed with bar screens.

RAM PUMP Cyclic water pump powered by hydropower via gravity.

RAMP 1. Inclined surface that connects two levels; 2. Road that connects freeway to surface street or another freeway; 3. Length of steeply laid drainpipe.

RAMP CONTROL SIGNAL Highway traffic control signal that controls traffic flow onto freeway ramp.

RAMP FLUME Flume calibrated to measure flow of liquid in open channel.

RAMP LANDFILLING Landfill waste is disposed in slightly inclined surfaces, compacted, forming ramps.

RAMP METER Device that regulates traffic flow entering freeways according to current conditions.

RAMP METERING System to regulate vehicular volume entering freeway to optimize operations.

RAMP RATE Rate of change in instantaneous output from power plant.

RAMP-FREEWAY JUNCTION Area between ramp and freeway that connects traffic.

RANDOM CRACKING Irregular cracking.

RANDOM EARTHQUAKE Earthquakes not associated with known geological structures.

RANDOM ERROR Error in measurement caused by factors that vary from one measurement to another.

RANDOM FILL Fill of various substances.

RANDOM INCIDENT FIELD Sound field where angle of arrival of sound at given point in space is random in time.

RANDOM NOISE Random signal having equal intensity at different frequencies.

RANDOM SAMPLE Sample randomly taken from investigated population.

RANGE 1. Geographic region where given plant or animal normally lives or grows; 2. Distance from radar antenna; 3. Line or series of mountains, buildings, etc.; 4. Length of smallest interval which contains all data in sample.

RANGE FOLDING Occurs when radar receives signal return from pulse other than most recent pulse.

RANGE GATE Discrete point in range along single radial of radar data at which received signal is sampled.

RANGE HEIGHT INDICATOR Radar display where radar scans vertically, with antenna pointing at specific azimuth or radial.

RANGE MANAGEMENT Professional natural science that centers around study of rangelands and conservation.

RANGE NORMALIZATION Receiver gain function in radar which compensates for effect of range (distance) on received power for equivalent reflectivity.

RANGE OF TIDE Vertical difference between high tide and succeeding low tide.

RANGE RESOLUTION Ability of radar to distinguish two targets along same radial but at different ranges.

RANGE UNFOLDING Process of removing range ambiguity in apparent range of multi-trip target on radar.

RANK 1. One's position in list sorted by shared property such as physical location, population or quality; 2. Size of any basis of given matroid.

RANKINE Unit of temperature where freezing point is 491.6°R, boiling point is 671.6°R and absolute zero is 0°R.

RANKINE VORTEX Velocity profile for symmetric circulation where inner core is in solid rotation and tangential winds outside core vary inversely with radial distance from center.

RAPID 1. Section of river where current is very fast moving, caused by steep descent in riverbed through constriction of main channel; 2. Very swift or quick.

RAPID DEEPENING Decrease in minimum sea-level pressure of tropical cyclone of 1.75 mb/hr or 42 mb for 24 hours.

RAPID DRAWDOWN When submerged slopes experience rapid reduction of external water level.

RAPID FILTER Sand or pressure filter used in water purification.

RAPID FLOW 1. State of flow where water depth is below critical depth, velocities are greater than critical, Froude number is greater than one and where inertial forces dominate gravitational forces; 2. Flow is described as rapid or shooting.

RAPID SAND FILTER Filter, consisting of layers of increasingly smaller media, used in water purification.

RAPID UPDATE CYCLE MODEL Numerical model run by NCEP that focuses on short-term forecasts out to 12 hours.

RAPIDLY INTENSIFYING Any maritime cyclone whose central pressure is dropping or is expected to drop, at rate of 1 MB/hr for 24 hours.

RAPTORS Birds of prey.

RASP Machine that grinds waste into manageable material and helps prevent odor.

RASPING SYSTEM Provides initial sorting, grinding, roller crushing and ballistic and/or gravity separation.

RATE 1. Proportional relationship between one amount, value, etc. and another; 2. Set price or charge; 3. Relative speed of change or progress; 4. Wage calculated in relation to unit of time.

RATE BASE Value of property on which public utility is permitted to earn specified rate of return.

RATE MAKING Determination of rates charged by insurance companies.

RATE OF ANNUAL DEPLETION Typical yearly rate where groundwater is removed from reservoir.

RATE OF FLOW Rate where liquid flows pass given location.

RATE OF RISE Temperature increase per time.

RATE OF ROUGHNESS Total roughness per distance.

RATE-OF-FLOW CONTROLLER Apparatus that controls fluid rate of flow.

RATE-OF-FLOW INDICATOR Apparatus that signals or records fluid rate of flow.

RATE-OF-FLOW RECORDER Apparatus that recorders fluid rate of flow.

RATED CAPACITY 1. Capacity which hydrogenerator can deliver without exceeding mechanical safety factors or nominal temperature rise; 2. Nameplate rating except where turbine power under maximum head is insufficient to deliver nameplate rating of generator.

RATED FATIGUE PRESSURE Pressure component can endure 10 million times without failure.

RATED HEAD Water depth for which hydroelectric generator and turbines were designed.

RATED HORSEPOWER Normal maximum, allowable, continuous power output of engine, turbine motor or other prime mover.

RATED POWER Amount of power that something was designed for.

RATED PRESSURE Maximum pressure that manufacturer assigns to specific component as maximum desirable pressure where device will function properly.

RATED SPEED Speed of motor (that produces its rated (i.e., maximum) power at specified (i.e., rated) voltage and rated load.

RATED STATIC PRESSURE Pressure, prior to failure, component can withstand.

RATING 1. Position on scale; 2. Relationship between water level and volume of water that can be removed from stream, well or aquifer.

RATING CURVE Graph showing relation between water height (ordinate) and volume of water flowing in channel (abscissa) per unit of time.

RATING FLUME Flume used for purpose of calibrating flow measuring devices.

RATING TABLE 1. Table showing relation between gauge height and discharge of stream or conduit at given gaging station; 2. Table showing relation between two mutually dependent quantities or variable over given range of magnitude.

RATING VEHICLE Vehicular loads applied to structure to determine operating ratings.

RATIO Number representing comparison between two named things.

RATIO OF REDUCTION Relationship between maximum size of stone which will enter crusher and size of its product.

RATIO REGULATOR Apparatus used to maintain precise pressure ratio in pipeline.

RATIONAL METHOD Technique for estimating design discharge from small watershed.

RATIONAL RUNOFF FORMULA Formula used to predict stream discharges.

RAVELLING Decomposition of road surface.

RAVINE Deep narrow valley or gorge in earth's surface worn by running water.

RAW AGRICULTURAL COMMODITY Unprocessed human food or animal feed crop (e.g., raw carrots, apples, corn, eggs).

RAW DATA Data that have not been analyzed or processed.

RAW LAND Land in its natural state with no man-made improvements.

RAW MATERIAL Material in its unprocessed, natural state considered usable for manufacture.

RAW SEWAGE Untreated wastewater and its contents.

RAW SEWAGE SLUDGE Suspended or dissolved organic or inorganic materials that result from wastewater treatment.

RAW WASTEWATER Plant influent or wastewater before any treatment.

RAW WATER Water in its natural state, prior to any treatment for drinking or use.

RAWINSONDE Radiosonde tracked to measure winds.

RAWINSONDE OBSERVATION Radiosonde observation which includes wind data.

RAYLEIGH SCATTERING Changes in directions of electromagnetic energy by particles whose diameters are 1/16 wavelength or less responsible for sky being blue.

RE-ENTRY Air exhausted from building immediately brought back into system through air intake and other openings.

RE-ENTRY DELAY Time vehicle is delayed as it enters into traffic stream after departing previous stop.

RE-ENTRY INTERVAL Period of time immediately following application of pesticide during which unprotected workers should not enter field.

RE-REGISTRATION Re-evaluation and re-licensing of existing pesticides originally registered prior to current scientific and regulatory standards.

RE-REGULATING RESERVOIRS Reservoir for reducing diurnal fluctuations resulting from operation of upstream reservoir for power production.

REACH Length of channel,between two specified gaging stations, control points or computational points, uniform with respect to discharge, depth, area and slope.

REACTION 1. Transformation where one or more substances is converted into another by combination or decomposition; 2. Action or statement in response to stimulus or other event.

REACTION PATH MODELING Simulation approach to studying chemical evolution of (natural) system.

REACTION TANK Tanks used to mix water, corrosive chemicals, solvents, etc. apart from chemicals.

REACTION VELOCITY Speed of chemical reaction.

REACTIVE 1. Material that readily takes part in reactions; 2. Characterized by induction or capacitance rather than resistance.

REACTIVE POWER 1. Portion of power produced by load inductances or capacitances; 2. Time average of instantaneous product of voltage and current, with current phase shifted 90 °.

REACTIVITY Those hazardous wastes that are normally unstable and readily undergo violent chemical change but do not explode.

READILY ACCESSIBLE Capable of being reached quickly for operation, renewal or inspections without requiring those to whom ready access is requisite to actions such as to use tools, to climb over or remove obstacles or to resort to portable ladders and so forth.

READILY WATER-SOLUBLE SUBSTANCES Chemicals that are soluble in water at concentration equal to or greater than 1,000 mg/l.

READILY-AVAILABLE WATER Water that plant can easily extract from soil.

READOUT Display that presents numerical data.

READY BIODEGRADABILITY Tests that detect growth-linked biodegradation.

READY CONDITION In suitable condition that can be used without further preparation or adjustment.

READY RESERVE Generation capacity that can be synchronized and ready to serve load in short time period, usually 10 minutes or less.

REAERATION 1. Replenishment of oxygen in water from which oxygen has been removed; 2. Exchange of gases between atmosphere and water.

REAERATION CONSTANT Expression of speed where oxygen is dissolved into oxygen-deficient water.

REAGENT Compound or mixture of compounds used to treat materials, samples, other compounds or reactants in laboratory or sometimes industrial setting.

REAL DENSITY Mass divided by volume occupied by material excluding pores and voids.

REAL PROPERTY Real estate.

REAL-TIME Refers to rapid retrieval, processing and transmission of data.

REAL-TIME DATA Data collected by automated instrumentation and telemetered and analyzed quickly enough to influence decision that affects monitored system.

REAL-TIME MEASUREMENT Measurement of reactants and end-products during chemical, physical or dynamic process.

REAL-TIME MONITORING Monitoring and measuring environmental developments with technology and communications systems that provide time-relevant information to public in easily understood format people can use in day-to-day decision-making about their health and environment.

REAR FLANK DOWNDRAFT Region of dry air subsiding on back side of and wrapping around, mesocyclone.

REAR LOADER, DETACHABLE CONTAINER Detachable container system with roll-out containers.

REAR-DUMP SCRAPER Scraper that discharges rear load.

REAR-DUMP TRUCK Vehicle body that's lifted where load slides out back.

REASONABLE FURTHER PROGRESS Annual incremental reductions in air pollutant emissions as reflected in State Implementation Plan that EPA deems sufficient to provide for attainment of applicable national ambient air quality standards by statutory deadline.

REASONABLE MAXIMUM EXPOSURE Maximum exposure reasonably expected to occur in population.

REASONABLE WORST CASE Estimate of individual dose, exposure or risk level received by individual in defined population greater than 90th percentile but less than that received by anyone in 98th percentile in same population.

REASONABLY AVAILABLE CONTROL MEASURES Technological and other measures for pollution control.

REASONABLY AVAILABLE CONTROL TECHNOLOGY Control technology reasonably available and both technologically and economically feasible.

REATTACHMENT DEPOSIT Sand deposit located where downstream flow meets channel bank at downstream end of recirculating zone.

REAUMUR Temperature scale where boiling point of water is 80° and its freezing point is 0°.

REBAR Reinforcing steel bar.

RECALCINATION Act or process of calcining again.

RECARBONATION Process where carbon dioxide is bubbled into water being treated to lower pH.

RECARBONIZATION Process where carbon dioxide is bubbled into water being treated to lower pH.

RECEIVER Electronic device which detects backscattered radiation, amplifies it and converts it to low-frequency signal that is related to properties of target.

RECEIVING BASES Container where effluent is released.

RECEIVING LANES Intersection departing lanes.

RECEIVING WATER Body of water that wastewater or treated effluent is discharged into.

RECEPTION PIT Pit used for collection of discharged.

RECEPTION SHAFT Opposite end of launch shaft.

RECEPTOR Ecological entity exposed to stressor.

RECESSION CONSTANT Constant used to reduce Antecedent Precipitation Index (API) value daily in API method of estimating runoff.

RECESSION CURVE Portion of hydrograph showing decreasing rate of runoff, where runoff is predominantly produced from basin storage (subsurface and small land depressions), that's separated from falling limb of hydrograph by inflection point.

RECESSION VELOCITY Mean rate of water flowing in channel or conduit immediately downstream.

RECESSIONAL MORAINE End moraine built during temporary but significant pause in final retreat of glacier.

RECHARGE 1. Amount of water added to aquifer; 2. Charge electric battery after its power has been consumed.

RECHARGE AREA Land area where water infiltrates ground and reaches zone of saturation.

RECHARGE BASIN Device used to manage stormwater runoff, prevent flooding and downstream erosion and improve water quality in river, stream, lake, etc.

RECHARGE RATE Quantity of water per unit time that replenishes or refills aquifer.

RECHARGE WELL Well that returns surface water into groundwater system.

RECIPROCATING COMPRESSOR Positive-displacement compressor that uses pistons to compress air.

RECIPROCATING DRILL Drill with spiral ratchet that rotates its bit.

RECIPROCATING GRATE Ash discharge and over feed type stoker.

RECIRCULATED COOLING WATER Systems that remove heat picked up in plant by evaporative cooling by passing water through condensers.

RECIRCULATION Act or process to circulate again.

RECIRCULATION ZONE Area of flow composed of one or more eddies immediately downstream from constriction in channel (e.g., debris fan, rock outcrop).

RECLAIMED Processed material retrieved from its original application for purposes of subsequent use.

RECLAIMED WASTEWATER Treated wastewater that can be used for beneficial purposes.

RECLAMATION 1. Act of reclaiming or state of being reclaimed; 2. Recovery of wasteland or of flooded land so it can be cultivated.

RECLAMATION AREA Process of creating new land or returning it to its original state.

RECLAMATION REFORM ACT OF 1982 Limits amount of owned land eligible to receive project (reclamation) irrigation water and addresses rate paid for such water delivered to owned and leased land.

RECOMBINANT BACTERIA Microorganism whose genetic makeup has been altered by deliberate introduction of new genetic elements.

RECOMBINANT DNA New DNA formed by combining pieces of DNA from different organisms or cells.

RECOMMENDED MAXIMUM CONTAMINANT LEVEL Maximum level of contaminant in drinking water at which no known or anticipated adverse effect on human health would occur and that includes adequate margin of safety.

RECONNAISSANCE CODE Aircraft weather reconnaissance code that has come to refer primarily to in-flight tropical weather observations, but actually signifies any detailed weather observation or investigation from aircraft in flight.

RECONSTRUCTED SOURCE Facility where components are replaced to such extent that fixed capital cost of new components exceeds 50% of capital cost of constructing comparable brand-new facility.

RECONSTRUCTION OF DOSE Estimating exposure after it has occurred by using evidence within organism (e.g., chemical levels in tissue or fluids).

RECORD DRAWING Set of drawings that records any changes made since first delivery or completion.

RECORD EVENT REPORT Issued by National Weather Service to report meteorological and hydrological events that equal or exceed existing records.

RECORD KEEPING Maintenance log of activities.

RECORD OF DECISION Public document that explains which alternative(s) will be used.

RECORDER Apparatus for recording operation or procedure.

RECORDING GAUGE Apparatus that continuously plots value it's monitoring.

RECORDING RAIN GAUGE Apparatus that gathers and measures amount of liquid precipitation over set period of time.

RECOVERABLE Amount of given constituent in solution after representative water sample has been extracted or digested.

RECOVERABLE FILTERED Amount of given constituent in solution after part of representative water suspended sediment sample that has passed through filter has been extracted.

RECOVERABLE RESOURCE Any resource that can be recovered and reused.

RECOVERABLE SLOPE Road slope where motorist may slow or stop to regain control of vehicle.

RECOVERABLE SUSPENDED SOLIDS Amount of given constituent in solution after part of representative water-suspended sediment sample retained on 0.45-micrometer filter has been extracted or digested.

RECOVERABLE UNFILTERED Amount of given constituent in representative water-suspended sediment sample that has been extracted or digested.

RECOVERED MATERIALS Waste materials and byproducts that have been recovered or diverted from solid waste, but does not include materials and byproducts generated from and commonly reused within, original manufacturing process.

RECOVERED SOLVENT Solvent backed out of process waste stream.

RECOVERY Extraction of substances from source or waste.

RECOVERY AREA Clear zone, adjacent to highway, available for errant vehicle usage.

RECOVERY DEVICE Equipment to recover for use, reuse or sale.

RECOVERY PHASE That period when depressed northward field component returns to normal levels.

RECOVERY RATE Percentage of usable recycled materials that have been removed from total amount of municipal solid waste generated in specific area or by specific business.

RECREATION REPORT National Weather Service product that reports on conditions for resorts and recreational areas and/or events.

RECREATIONAL BENEFIT Value of recreational activity to recreationist, usually measured in dollars above cost of participating in recreational activity (e.g., travel, entrance fees).

RECREATIONAL VEHICLE Self-propelled vehicle or trailer designed for use as living quarters on overnight recreational activities and travel.

RECRUITMENT Survival of young plants and animals from birth to life stage less vulnerable to environmental change.

RECTANGULAR WEIR Contracted or suppressed weir with horizontal crest, rectangular in shape, having vertical sides.

RECTIFIER ELECTRIC MOTOR Rail car that collects propulsion power from alternating-current distribution system and converts it to direct current for direct-current motors.

RECTILINEAR TIDAL CURRENT Where direction of flow is inverted between tides in tidal estuaries.

RECURRENCE INTERVAL Average period in years between storm events equal to or larger than given amount that's reciprocal of probability of that storm event being equaled or exceeded in any year.

RECURRENCE PERIOD Time between incidents of repeating events.

RECYCLABLE PAPER Waste paper turned into new paper products.

RECYCLE Minimizing waste generation by recovering and reprocessing into reusable material (i.e., recycling of aluminum cans, paper, bottles).

RECYCLED MATERIAL Materials (e.g., glass, paper, cardboard, metal, plastic, tires, textiles, electronics) that can be properly used in place of raw materials.

RECYCLED WATER Water used more than one time before it passes back into natural hydrologic system.

RECYCLING Practice of sorting and collecting waste materials for reuse.

RECYCLING AGENT Added to recycled mixtures to restore to desired specifications.

RECYCLING BODY Multi-compartmented truck used for recyclable garbage collection.

RECYCLING ECONOMIC DEVELOPMENT ADVOCATES Individuals hired by state or tribal economic development offices to focus financial, marketing and permitting resources on creating recycling businesses.

RECYCLING MILL Facility where recovered materials are remanufactured into new products.

RECYCLING TECHNICAL ASSISTANCE PARTNERSHIP NATIONAL NETWORK National information-sharing resource designed to help businesses and manufacturers increase their use of recovered materials.

RECYCLING TRAIN Series of mobile processing units that recycle paving.

RED BAG WASTE Hazardous waste capable of causing infections in humans (e.g., contaminated animal waste, human blood and blood products, isolation waste, pathological waste, needles, scalpels, broken medical instruments).

RED BOX Slang for Tornado Watch.

RED CLEARANCE INTERVAL Period of time following yellow signal indication where red signal indication is displayed to all conflicting movements at intersection.

RED FLAG Fire weather program which highlights onset of critical weather conditions conducive to extensive wildfire occurrences.

RED FLAG WARNING Term used by fire-weather forecasters to call attention to limited weather conditions of particular importance that may result in extreme burning conditions.

RED TIDE Proliferation of marine plankton toxic and often fatal to fish, perhaps stimulated by addition of nutrients where tide can be red, green or brown, depending on coloration of plankton.

RED TIME Period during signal cycle that signal is red.

RED WATCH Slang for Tornado Watch.

REDD Depression (i.e., nest) in riverbed or lakebed dug by spawning female fish to deposit eggs.

REDEMPTION PROGRAM Program where consumers are monetarily compensated for collection of recyclable materials, generally through prepaid deposits or taxes on beverage containers.

REDOX POTENTIAL Numerical measure of intensity of oxidation or reducing conditions where positive potential indicates oxidizing conditions and negative potential indicates reducing conditions.

REDUCE 1. To bring down size, quantity, quality, value or intensity of something; 2. To add electrons or hydrogen or to remove oxygen; 3. To produce metal from ore by removing non-metallic elements in smelter.

REDUCER 1. Hydraulic device for reducing pressure and hence increasing movement; 2. Component used to connect two pipes of different bores; 3. Breaks down dead organic matter into simpler compounds.

REDUCING AGENT Any substance that reduces or donates electrons to, another and becomes oxidized.

REDUCING TEE T-shaped pipe fitting with different sized openings.

REDUCING VALVE Valve that reduce higher, inlet pressure to lower, downstream pressure.

REDUCTION Addition of hydrogen, removal of oxygen or addition of electrons to element or compound.

REDUCTION CONTROL SYSTEM System that reduces sulfur recovery plant emissions to hydrogen sulfide.

REDUNDANT MEMBER Framed structure member not necessary for support.

REED BEDS Aquatic plant based systems that allow bacteria, fungi and algae to digest sewage and clean water.

REEF Chain or range of rocks, sand or coral lying at or near surface of water.

REEF ECOSYSTEM Diverse collection of species that interact with each other and coral red.

REEL CARRIER Self-loading reel to support and discharge conduit or cable.

REFERENCE Measurement one can compare to.

REFERENCE AMBIENT CONCENTRATION Concentration of chemical in water which will not cause adverse impacts to human health expressed in units of mg/l.

REFERENCE CONDITIONS Characteristics of water body segments least impaired by human activities.

REFERENCE DOSE Estimate of daily exposure to human population not likely to cause harmful effects during lifetime.

REFERENCE MARK Relatively permanent point of known elevation tied to benchmark.

REFERENCE PHASE Phases coordinated through signal system.

REFERENCE SITE NAWQA sampling site selected for its relatively undisturbed conditions.

REFERENCE STANDARDS Standardized and validated reference material used to calibrate working standards.

REFERENCE TISSUE CONCENTRATION Concentration of chemical in edible fish or shellfish tissue which will not cause adverse impacts to human health when ingested expressed in units of mg/kg.

REFERENCE VALUE Chemical, physical or biological quality or condition exhibited at either single site or aggregation of sites that represent least impacted or reasonably attainable condition at least impacted reference sites.

REFILL Material which had previously been excavated as result of construction activities and again placed to lines as shown on drawings.

REFILL TUBE Tube goes into overflow tube to fill toilet bowl with water.

REFLECTED WAVE Wave reflected from surface, discontinuity or junction.

REFLECTION Process whereby radiation (or other waves) incident upon surface is directed back into medium where it traveled.

REFLECTION CRACKING Breaks in underlying layers due to movement at original crack.

REFLECTIVITY Measure of efficiency of radar target in intercepting and returning electro magnetic energy that's sum of all backscattering cross-sections (e.g., precipitation particles) in pulse resolution volume divided by that volume.

REFLECTIVITY CROSS-SECTION WSR-88D radar product displays vertical cross-section of reflectivity on grid with heights up to 70,000 feet on vertical axis and distance up to 124 nautical miles on horizontal axis.

REFLECTIVITY FACTOR Result of mathematical equation that converts analog power received by radar antenna into more usable quantity.

REFLUX Backwards flow of any fluid.

REFLUX VALVE Valve that allows fluid to flow in only one direction.

REFORMULATED GASOLINE Gasoline with different composition from conventional gasoline (e.g., lower aromatics content) that cuts air pollutants.

REFRACTION 1. Changes in direction of energy propagation as result of density changes within propagating medium; 2. Part of wave advancing in shallower water moves more slowly than that part still advancing in deeper water, causing wave crest to bend toward alignment with underwater contours.

REFRACTIVE INDEX Measure of amount of refraction equal to ratio of wave velocity in vacuum to wave speed in medium.

REFRACTORY MATERIALS Materials difficult to remove entirely from wastewater.

REFRIGERANT Substance used in heat cycle that undergoes phase change between gas and liquid to allow cooling.

REFRIGERATOR CAR Rail car designed to carry perishable freight at specific temperatures.

REFUELING EMISSIONS Emissions released during vehicle re-fueling.

REFUGE COMPATIBILITY REQUIREMENTS Requirements under Refuge Administration Act that all uses of national wildlife refuge must be compatible with purpose for which refuge was established.

REFUGE ISLAND Raised platform or other protective barrier in middle of street crossing that permits pedestrians to pause safely in middle of road.

REFUSE Non-liquid, non-soluble materials ranging from municipal garbage to industrial wastes (e.g., sewage sludge, agricultural refuse, demolition wastes, mining residues) that contain complex and sometimes hazardous substances.

REFUSE BURNER Apparatus that reduces solid waste via burning.

REFUSE CHUTE Transports waste materials by chute.

REFUSE COMPACTOR Apparatus that reduces size of garbage through compaction.

REFUSE HANDLING Turning wastes into product.

REFUSE RECLAMATION Conversion of solid waste into useful products.

REFUSE TRUCK Truck designed to collect and transport waste.

REFUSE-DERIVED FUEL Fuel produced from municipal, industrial or commercial wastes.

REGAIN OF MOISTURE Moisture absorbed by means of substance as percentage of its own initial weight.

REGELATION Fusion of two pieces of ice by pressure (i.e., lowers freezing point).

REGENERATE To construct or create anew.

REGENERATED WATER Water which comes from secondary treatment outflow in wastewater treatment stations.

REGENERATION 1. Manipulation of cells to cause them to develop into whole plants; 2. Rebuilding or restructuring.

REGENERATION LEVEL Amount of regenerant used in ion exchange procedure.

REGIME THEORY Theory of forming of channels in material carried by streams.

REGIMEN State in stream where erosion and deposition are in equilibrium.

REGIMEN OF STREAM System or order characteristic of stream.

REGIONAL FLOOD OF RECORD Maximum flood known or recorded in drainage area.

REGIONAL HAZARDOUS WASTE MANAGEMENT COORDINATOR Individual designated by Regional Director to assess and coordinate records and maintain proper management of hazardous waste activities in each region.

REGIONAL HAZE Haze mixed uniformly between surface and top of convective boundary layer.

REGIONAL RESPONSE TEAM Representatives of federal, local and state agencies who may assist in coordination of activities at request of On-Scene Coordinator before and during significant pollution incident (e.g., oil spill, major chemical release, Superfund response).

REGIONAL TRANSMISSION GROUP Voluntary organization of transmission owners, transmission users and other entities interested in coordinating transmission planning, expansion, operation and use within region.

REGISTER 1. Device that automatically records quantity; 2. Grille at outflow of ventilation duct.

REGISTERED LAND Land governed by system for registering title to land.

REGISTERING FAREBOX Device that compares amount inserted by passenger versus fare.

REGISTRANT Any manufacturer or formulator who obtains registration for pesticide active ingredient or product.

REGISTRATION Formal listing with EPA of new pesticide before it can be sold or distributed.

REGISTRATION STANDARDS Published documents which include summary reviews of data available on pesticide's active ingredient, data gaps and Agency's existing regulatory position on pesticide.

REGOLITH Layer or mantle of fragmented and unconsolidated rock material, residual or transported, that nearly everywhere forms surface of land and overlies or covers bedrock.

REGOSOL Weakly developed mineral soil in unconsolidated materials.

REGRESSION Analytic method to measure association of one or more independent variables with dependent variable.

REGRESSION MODEL Model for estimating relationships between dependent variable and one or more independent variables.

REGULAR RUN Trip or group of trips designated by number on timetable, arranged for selection by or assignment to, crew as day's work.

REGULAR SERVICE Auto maintenance required for vehicle on scheduled basis.

REGULAR TRAIN Train that runs using timetable schedule.

REGULATED ASBESTOS-CONTAINING MATERIAL Friable asbestos material or non-friable ACM that will be or has been subjected to sanding, grinding, cutting or abrading or has crumbled or been pulverized or reduced to powder in course of demolition or renovation operations.

REGULATED FLOW Streamflow that's subject to control through hydraulic works.

REGULATED MEDICAL WASTE Any solid waste generated in diagnosis, treatment or immunization of human beings or animals, in research pertaining thereto or in production or testing of biologicals.

REGULATING DAM Dam impounding reservoir from which water is released to regulate flow downstream.

REGULATING GATE Gate used to regulate rate of discharge through outlet works or spillway that operates under full pressure flow conditions.

REGULATING SLEEVE VALVE Valve for regulating high pressure outlets and ensuring energy dissipation.

REGULATING VALVE Valve used to regulate rate of discharge through outlet works or spillway that operates under full pressure flow conditions.

REGULATION Law or administrative rule, issued by organization, used to guide or prescribe conduct of members of that organization.

REGULATION OF STREAM Artificial manipulation of flow of stream.

REGULATOR Device that controls or limits something.

REGULATORY FLOOD Flood having one percent chance (i.e., 100-year flood) of being equaled or exceeded in any given year.

REGULATORY FLOODWAY Area where construction regulations require special provisions to account for this extra hazard.

REGULATORY SIGN Signs used for traffic laws, regulations or requirements.

REHABILITATION Process of renovating facility or system whose performance is failing to meet original criteria and needs of project.

REIMBURSABLE Costs of constructing, operating or maintaining project that are repaid by some other individual, entity or organization.

REINFORCED CONCRETE Portland cement concrete with matrix of steel bars or wires (i.e., rebar) to increase its tensile strength.

REINFORCED EARTH RETAINING WALL Structure that's combination of earth and linear reinforcing strips to bear large tensile stresses.

REINFORCED REVETMENT Stream bank protection using stone toe-fill.

REINFORCEMENT Steel embedded is concrete to resist forces.

REINFORCING STEEL Steel bar or mesh used as tension device in reinforced concrete to strengthen structure.

REJECTABLE QUALITY LEVEL Largest defective rate or defect rate tolerated.

REJUVENATED STREAM Stream that, due to uplift, has reverted back to erosion cycle.

REJUVENATED WATER Water that has been returned to Earth's water supply as result of compaction and metamorphism.

RELATIVE ABUNDANCE Number of organisms of particular kind present in sample relative to total number of organisms in sample.

RELATIVE COMPACTION Percentage ratio of field density to maximum density as determined by standard compaction.

RELATIVE DENSITY 1. Ratio of density of substance to density of water at same temperature; 2. Ratio of difference between void ratio of cohesionless soil in loosest state and any given void ratio to difference between its void ratios in loosest and in densest states.

RELATIVE ECOLOGICAL SUSTAINABILITY Ability of ecosystem to maintain relative ecological integrity indefinitely.

RELATIVE ERROR Errors due to precision and accuracy.

RELATIVE EVAPORATION Ratio of rate of evaporation from land or water surface in contact with atmosphere, to evaporability under existing (i.e., potential) atmospheric conditions.

RELATIVE HUMIDITY Dimensionless ratio, expressed in percent, of amount of moisture in air to maximum amount of moisture air could hold under same conditions.

RELATIVE IONOSPHERIC OPACITY METER Radio receiver specially designed for continuous monitoring of cosmic noise.

RELATIVE PERMEABILITY Ratio of effective permeability of particular fluid at particular saturation to absolute permeability of that fluid at total saturation.

RELATIVE RISK ASSESSMENT Estimating risks associated with different stressors or management actions.

RELATIVE SETTLEMENT Settlement in proportion to distance.

RELATIVE STABILITY Stability of system with respect to certain parameter or conditions.

RELATIVE TRANSPIRATION Transpiring power of plant that yields amount of water lost per time per area.

RELATIVE UTILIZATION FACTOR Ratio of use efficiency of fuel to usage efficiency of base fuel.

RELATIVE VELOCITY Velocity of object in rest frame of another object.

RELATIVE VORTICITY Sum of rotation of air parcel about axis of pressure system and rotation of parcel about its own axis.

RELATIVE WIND Wind with reference to moving point.

RELAXATION Reduction of stress in prestressing steel tendons.

RELAY 1. Series of vehicles travelling in sequence; 2. Electrical actuator that allows relatively small electrical voltage or current to control larger voltage or current.

RELAY EMERGENCY VALVE Valve that acts as relay station to speed up application and release of trailer brakes.

RELAY VALVE Valve to remotely control brakes at rear of truck.

RELEASE 1. Amount of water released after use; 2. Process where chemical substance is set free; 3. Any spilling, leaking, pumping, pouring, emitting, emptying, discharging, injecting, escaping, leaching, dumping or disposing into environment of hazardous or toxic chemical or extremely hazardous substance; 4. Device adapted to hold or release device or mechanism as required.

RELIABILITY Probability that device will function without failure over specified time period or amount of usage.

RELICT Any species, population, etc., surviving in small local area and widely separated from closely related species, that's survivor of nearly extinct group.

RELICTION Gradual loss of water from river, lake, etc. to leave permanent dry land.

RELIEF 1. Path of least resistance where energy from explosions can be released; 2. Difference of elevations on surface.

RELIEF BRIDGE Embankment opening on floodplain for overbank flow passage.

RELIEF RUN Combination of other operators' schedule days off from regular runs to comprise week of work.

RELIEF SEWER Sewer designed to divert all flows in excess of certain quantity.

RELIEF VALVE Valve which will allow air or fluid to escape if its pressure becomes higher than valve setting (i.e., safety device that automatically provides protection against excessive temperatures, excessive pressures or both).

RELIEF VENT Vent pipe used in conjunction with circuit vent to provide additional air circulation between drainage system and venting system.

RELIEF WELL Well drilled to intersect natural gas and petroleum well that's experienced blowout.

RELIEF WELLS 1. Vertical wells downstream of or in downstream shell of embankment dam to collect and control seepage through and under dam where line of such wells forms "drainage curtain"; 2. Well drilled to carry excess water off agricultural fields.

RELIQUEFACTION Liquefaction again or anew.

REMAINING LIFE Number of years that pavement will be functionally and structurally acceptable.

REMEDIAL ACTION Actual construction or implementation phase of Superfund site cleanup that follows remedial design.

REMEDIAL DESIGN Phase of remedial action that follows remedial investigation/feasibility study and includes development of engineering drawings and specifications for site cleanup.

REMEDIAL INVESTIGATION In-depth study designed to: gather data needed to determine nature and extent of contamination, establish site cleanup criteria, identify preliminary alternatives for remedial action and support technical and cost analyses of alternatives.

REMEDIAL PROJECT MANAGER Official responsible for overseeing on-site remedial action.

REMEDIAL RESPONSE Long-term action that stops or substantially reduces release or threat of release of hazardous substances that's serious but not immediate threat to public health.

REMEDIATION Cleanup or other abatement methods used to remove or contain toxic spill or hazardous materials including evaluation, repair, enclosure, encapsulation or removal of greater than 3 linear feet or ft^2 of asbestos-containing materials from building.

REMOTE OBSERVATION SYSTEM AUTOMATION Type of automated data transmitter used by National Weather Service Cooperative Program observers.

REMOTE OPERATION Operation of mechanical features from on-site location other than at feature.

REMOTE SENSING Collection and interpretation of information about object without physical contact with object (e.g., satellite imaging, aerial photography, open path measurements).

REMOVAL ACTION Short-term immediate actions taken to address releases of hazardous substances that require expedited response.

RENEWABLE ENERGY CERTIFICATES Represent technology and environmental attributes of electricity generated from renewable sources that are referred to as Green Tags.

RENEWABLE ENERGY CREDITS Represent technology and environmental attributes of electricity generated from renewable sources that are referred to as Green Tags.

RENEWABLE ENERGY PRODUCTION INCENTIVE Incentive established for renewable energy power projects owned by state or local government or non-profit electric cooperative.

RENEWABLE ENERGY SOURCE Inexhaustible source of energy (i.e, sunlight, wind, rain, tides, geothermal heat).

RENEWABLE RESOURCES Naturally replenishable, but flow limited that are virtually inexhaustible in duration but limited in amount of energy available per unit of time (e.g., biomass, hydro, geothermal, solar, wind).

RENOVATION To revamp something to make it look new again.

REOXYGENATION 1. Stream oxygen replenishment by water entering; 2. Oxygenation subsequent to deoxygenation.

REPAIR To restore to good working order, fix or improve damaged condition.

REPEAT COMPLIANCE PERIOD Any subsequent compliance period after initial one.

REPEATABILITY Lack of variation amongst multiple measurements taken under same conditions.

REPLACEMENT 1. Person or thing that takes place of another; 2. Removal of edge of crystal, by one plane or more.

REPLACEMENT COST Amount to replace asset of company at same or equal value.

REPLACEMENT EXPENDITURE Asset volume multiplied by unit replacement cost.

REPLICATE 1. To make copy of; 2. To repeat (i.e., experiment or trial) with consistent result.

REPLICATE SAMPLES Group of samples collected in manner such that samples are thought to be essentially identical in composition.

REPORT Statement of actual conditions observed at specific time at specific site.

REPORT FOR FIRST FORMAL EXAMINATION Report of initial SEED onsite examination.

REPORT ON RIVER GAUGE STATION (E-19) Report to be completed every 5-10 years providing complete history of river station and all gauges that have been used for public forecasts since establishment of station that should be updated anytime significant change occurs at forecast point.

REPORT TIME Time operator must show up for work assignment.

REPORTABLE DISCHARGE OF OIL Harmful quantities of oil discharged or released to navigable waters of United States are any amount that violates state water quality standards and/or causes sheen or film (as little as 10 parts per million of oil can cause sheen), upon or discoloration of surface of water or adjoining shorelines or cause sludge or emulsion to be deposited beneath surface of water or upon adjoining shorelines.

REPORTABLE QUANTITIES Quantities of hazardous substances which require notification.

REPOWERING Rebuilding and replacing major components of power plant instead of building new one.

REPRESENTATIVE IMPORTANT SPECIES Species that are representative in terms of their biological needs, of balanced, indigenous community.

REPRESENTATIVE SAMPLE Portion of material or water that's as nearly identical in content and consistency as possible to that in larger body of material or water being sampled.

REPROCESSING Second or subsequent processing.

REPRODUCIBILITY Closeness of agreement among repeated measurements of variable made under same operating conditions over period of time or by different people.

REQUEST FOR PAYMENT Sending invoices and receiving corresponding payment.

REQUEST FOR PROPOSAL Formal publicly released document outlining need and inviting businesses to submit plans to fulfill that need.

REQUEST FOR PROPOSALS Document that announces and provides details about project and solicits bids from contractors.

REQUEST FOR QUALIFICATIONS Step in formal qualifications-based selection process.

REQUIRED FIRE FLOW Rate of flow at given residual pressure for firefighting.

REQUIRED STRENGTH Design consideration that must be attained and verified by properly evaluated test results.

REREGULATION DAM Dam located downstream from large hydroelectric power plant used to regulate discharges downstream.

RESEALING TRAP Plumbing trap designed so that discharge rate of flow seals trap but doesn't induce self-siphonage.

RESEARCH Diligent inquiry or examination to seek or revise facts, principles, theories, applications, etc.

RESERVATION All land within limits of any Indian reservation under jurisdiction of United States Government, notwithstanding issuance of any patent and including rights-of-way running through reservation.

RESERVE CAPACITY Extra treatment capacity built into solid waste and wastewater treatment plants and interceptor sewers to accommodate flow increases due to future population growth.

RESERVE GENERATING CAPACITY Extra generating capacity available to meet unanticipated capacity demand for power in event of generation loss due to scheduled or unscheduled outages of regularly used generating capacity.

RESERVES Kept back for future use.

RESERVOIR Pond, lake or basin, either natural or artificial, for storage, regulation and control of water.

RESERVOIR CAPACITY Sum of dead and live storage (i.e., complete volume) of reservoir.

RESERVOIR CLEARING Removal of vegetation from within proposed reservoir.

RESERVOIR DEPOSITION Collected sediment deposited in reservoir.

RESERVOIR INFLOW Amount of water entering reservoir.

RESERVOIR LINING Lining to prevent water loss from reservoir.

RESERVOIR OPERATING PROCEDURE Compilation of criteria, guidelines and specifications that govern storage and release function of reservoir.

RESERVOIR REGULATION PROCEDURE Operating procedures that govern reservoir storage and releases.

RESERVOIR RIM Boundary of reservoir including all areas along valley sides above and below water surface elevation associated with routing of IDF.

RESERVOIR RULE CURVE Compilation of criteria, guidelines and specifications that govern storage and release function of reservoir.

RESERVOIR SURFACE AREA Area covered by reservoir when filled to specified level.

RESERVOIR VOLUME Volume of reservoir when filled to specified level.

RESET To set back to initial state.

RESIDENCE 1. Place where one lives, corporation is established, etc.; 2. State of living in particular place or environment.

RESIDENCE TIME Average time for which particular molecule of water will remain in body of water.

RESIDENTIAL INCINERATOR Incinerator to be used in individual homes.

RESIDENTIAL SILENCER Exhaust muffler to quiet noise in residential areas.

RESIDENTIAL SOLID WASTE Any garbage, rubbish, yard trash or bulk item usual to housekeeping.

RESIDENTIAL USE Pesticide application in and around houses, office buildings, apartment buildings, motels and other living or working areas.

RESIDENTIAL WASTE Waste generated in single and multi-family homes (i.e., newspapers, clothing, disposable tableware, food packaging, cans, bottles, food scraps).

RESIDENTIAL WATER USE Water for household purposes (e.g., watering of livestock, drinking, food preparation, bathing, washing clothes and dishes, flushing toilets, watering lawns and gardens).

RESIDENTIAL/COMMERCIAL AREAS Places where individuals dwell or work.

RESIDUAL Amount of pollutant remaining in environment after natural or technological process has taken place that's difference between model-computed and field-measured values of variable (e.g., hydraulic head, groundwater flow rate) at specific time and location.

RESIDUAL CHLORINE Concentration of chlorine present in water after chlorine demand has been satisfied.

RESIDUAL CLAY Material formed in place by weathering of rock.

RESIDUAL DISINFECTANT CONCENTRATION Concentration (i.e., mg/l) of disinfectant in water sample.

RESIDUAL DRAWDOWN Water-level response measured, in one or more surrounding observation wells and control well, after pumping has stopped.

RESIDUAL FUEL OIL Remainder of crude oil after gasoline and distillate fuel oils have been extracted through distillation.

RESIDUAL LAYER Elevated portion of convective boundary layer that remains after stable boundary layer develops at ground (usually in late afternoon or early evening) and cuts off convection.

RESIDUAL MOISTURE Atmospheric moisture which lingers over area after main weather system has departed.

RESIDUAL OXYGEN Direct measurement of how much air is required for stoichiometric combustion of fuel.

RESIDUAL PRESSURE Force exerted on inside walls of pipe when any fluid is flowing through it.

RESIDUAL QUEUE Demand unmet at end of analysis period when demand exceeded capacity.

RESIDUAL RISK Extent of health risk from air pollutants remaining after application of Maximum Achievable Control Technology.

RESIDUAL SATURATION Saturation level below which fluid drainage will not occur.

RESIDUAL WASTE Material that remains after process of waste treatment has taken place.

RESIDUAL-MASS CURVE Graph of cumulative departures from given reference (e.g., arithmetic average plotted against time or date).

RESIDUE 1. Dry solids that remains after evaporation, distillation, filtration or any similar process; 2. Molecule released from polymer after bonds between neighbouring monomers are broken; 3. Whatever property or effects are left in estate after payment of all debts, other charges and deduction of what is specifically bequeathed by testator.

RESILIENCE Ability of any system to resist or to recover from stress or hardship.

RESILIENT CONNECTOR Flexible connector that joins pipe to another pipe or pump.

RESILIENT HANGER Anti-vibration hanger.

RESISTANCE 1. Act of resisting or capacity to resist; 2. Force that tends to oppose motion; 3. Shortened form of electrical resistance.

RESISTANCE FACTOR Factor that accounts for bending, tension and compression in design.

RESISTANCE LOSSES Head loss of flow due to conduit or channel resistance.

RESOLUTION Separation of constituent parts.

RESONANCE 1. Increase in amplitude of oscillation of system under influence of periodic force whose frequency is close to that of system's natural frequency; 2. Property of compound that can be visualized as having two structures differing only in distribution of electrons; 3. Influence of gravitational forces of one orbiting object on orbit of another, causing periodic perturbations; 4. Condition where inductive and capacitive reactances have equal magnitude.

RESONANT 1. Adjusted as to dimensions so that currents or electric surgings are produced by passage of electric waves of given frequency; 2. Resounding, echoing.

RESONANT FREQUENCY Natural frequency of vibration determined by physical parameters of vibrating object.

RESOURCE Something that one uses to achieve objective (e.g., raw material).

RESOURCE CONSERVATION AND RECOVERY ACT Federal statute which establishes framework for proper management and disposal of all hazardous wastes (i.e., generation, transportation, storage, treatment and disposal).

RESOURCE LEVELLING Process performed when availability of resources are less than required.

RESOURCE MANAGEMENT PLAN Written plan that addresses existing resources of area and provides future objectives, goals and management direction.

RESOURCE PLANNING Discovering what resources are essential to do job tasks.

RESOURCE RECOVERY Process of obtaining matter or energy from materials formerly discarded.

RESOURCE RECOVERY FACILITY Plant that processes residential, commercial or institutional solid wastes to regain useful products.

RESOURCE-LIMITED SCHEDULE project schedule whose schedule activities start dates and finish dates directly reflect availability of all resources in question.

RESOURCES Something that one uses to achieve objective (e.g., raw material).

RESPIRABLE DUST Particles hazardous when deposited in lower gas-exchange region of lung.

RESPIRATION Process of inhaling and exhaling that exchanges gases with its environment.

RESPIRATOR Device designed to allow breathing when it would otherwise be hindered (e.g., poisonous vapors).

RESPIRATORY QUOTIENT Ratio of volume of carbon dioxide evolved to that of oxygen consumed by organism, tissue or cell in given time.

RESPONSE Reaction to stimulus or provocation.

RESPONSE ACTION Actions taken (either short-term removal action or long-term removal response) in response to actual or potential health-threatening environmental events (e.g., spills, sudden releases, asbestos abatement/management problems).

RESPONSE INDICATOR Environmental indicator measured to provide evidence of biological condition of resource at organism, population, community or ecosystem level of organization.

RESPONSE SPECTRUM Plot of maximum values of acceleration, velocity, and/or displacement response of infinite series of single degree of freedom systems subjected to time history of earthquake ground motion.

RESPONSE TIME 1. Period of time between application of stimulus and associated response; 2. Time taken for display of instrument to move to new value following step change; 3. Amount of time where it will take watershed to react to given rainfall event.

RESPONSE, LEVEL I First and least serious, of three response levels that dam operating organization will declare after analyzing potentially threatening event that may be perceived as emergency or may be of general interest to public, but does NOT pose hazard, either at dam or to downstream populations at risk when observed.

RESPONSE, LEVEL II Second of three response levels dam operating organization will declare after analyzing threatening event that means emergency conditions are such that populations at risk should prepare to leave predetermined inundation areas for higher ground and safe shelter.

RESPONSE, LEVEL III Third and most serious, response level dam operating organization will declare after analyzing threatening events that indicates life-threatening flood waters, as result of high operational releases or dam failure, present imminent danger to public located downstream from dam.

RESPONSIBLE BIDDER Bidder that's able to perform work to satisfaction.

RESPONSIVE BID/PROPOSAL Offer meeting all contract requirements.

RESPONSIVENESS SUMMARY Summary of oral and/or written public comments received during comment period on documents and response to those comments.

REST AREA Place where one can stop to use restroom or to take break from travelling.

REST MASS Mass of body when it is not moving relative to observer.

REST-IN-WALK MODE Controller that displays WALK indication for minor street crossings from as long as there are no conflicting calls.

RESTORABLE SURFACE WATERS Surface waters and wetlands having function and values diminished by human impact that can be restored through various management techniques.

RESTORATION Measures taken to return site to pre-violation conditions.

RESTORATION CLAIM Claim for restoring, rehabilitating, replacing due to discharge of dangerous material.

RESTORATION EXPENDITURE Sum spent to restore.

RESTORE To reestablish or bring back into existence.

RESTRAINING RAIL Rail used in transit systems to reduce rail wear in sharp curves and increase track's resistance to flange climb derailment.

RESTRICTED AREA Airspace identified by area on surface of earth within which flight of aircraft, while not wholly prohibited, is subject to restrictions.

RESTRICTED ENTRY INTERVAL Time after pesticide application during which entry into treated area is restricted.

RESTRICTED SPEED Speed that will permit stopping within one-half range of vision.

RESTRICTED USE Pesticide may be classified (under FIFRA regulations) for restricted use if it requires special handling because of its toxicity, and, if so, it may be applied only by trained, certified applicators or those under their direct supervision.

RESTRICTED WASTES Solid waste with properties that make it dangerous to human health and/or environment that's prohibited from disposal in landfill.

RESTRICTED-ORIFICE SURGE TANK Storage container with restricted entry that increases damping action and reduces initial surge.

RESTRICTION Reduced cross-sectional area in conduit that reduces pressure.

RESTRICTION ENZYMES Enzymes that recognize specific regions of long DNA molecule and cut it at those points.

RESTRICTIVE MEDIAN Median of divided highway or driveway.

RESTRICTOR Apparatus that reduces cross-sectional area.

RESURFACING Application of new surface.

RESURGENT WATER Magmatic water.

RESUSCITATOR Device used to restore normal breathing to person by forcing oxygen or mixture of gases into lungs.

RETAINAGE Amount of money held back from payment as insurance against work not being completed.

RETAINED WATER Water in aeration zone.

RETAINING WALL Wall separating two levels.

RETARDATION FACTOR Used to simulate resistance of contamination to move through groundwater aquifer.

RETARDER Device for slowing down large trucks, buses and shunted railway wagons.

RETARDER YARD Switching yard for cars after they are released from locomotive.

RETARDING BASIN Excavated area to protect against flooding and downstream erosion by storing water for limited time.

RETARDING RESERVOIR Ungated reservoir for temporary storage of flood water.

RETENTION Total precipitation and total runoff difference for given area over specified interval.

RETENTION BASIN Water in storage is permanently obstructed from flowing downstream.

RETENTION CAPACITY Maximum capacity for retention for given conditions.

RETENTION CHAMBER Structure built to accumulate stormwater underground.

RETENTION PERIOD Time to displace volume of storage container for certain discharge rate.

RETENTION POND Stormwater control structure that holds stormwater runoff before slowly releasing.

RETENTION TIME 1. Theoretical calculated time required for small amount of water to pass through tank at given rate of flow; 2. Actual time that small amount of water is in storage reservoir (e.g., settling basin, flocculating basin, rapid-mix chamber).

RETREAT VELOCITY Average flow velocity downstream of hydraulic structure.

RETROFIT Addition of device on existing facility without making major changes to facility.

RETROGRADE MOTION Movement of weather system in direction opposite to that of basic flow where it is embedded.

RETROGRESSION 1. Movement of weather system in direction opposite to that of basic flow where it is embedded; 2. Lowering of river bed.

RETROREFLECTIVITY When reflected rays of light are preferentially returned in directions close to opposite of origin direction.

RETROSPECTIVE ANALYSIS Review and analysis of existing data in order to address objectives and to aid in design of studies.

RETURN BEND Pipe fitting for connecting contiguous ends of two nearly parallel pipes lying alongside or one above another.

RETURN FLOW 1. Drainage water from irrigated farmlands that reaches ground- or surface-water source after release from point of use and thus becomes available for further use that is not consumed in evaporation or transpiration and returns to surface stream or aquifer; 2. South winds on back (west) side of eastward-moving surface high pressure system.

RETURN OFFSET Double offset that permits pipeline to pass around obstruction.

RETURN PERIOD Average period in years between storm events equal to or larger than given amount that's reciprocal of probability of that storm event being equaled or exceeded in any year.

RETURN SEEPAGE Water that percolates to underlying strata that raises groundwater levels.

RETURN STROKE Electrical discharge that propagates upward along lightning channel from ground to cloud.

RETURN SYSTEM Conduits that return air or water from heated or cooled region to processing gear.

RETURN TIME Time it takes to travel from dump to load area or vice versa.

RETURN-FLOW SYSTEM System of pipelines or ditches to collect and convey surface or subsurface runoff from irrigated field for reuse.

RETURNED SLUDGE Solids returned to wastewater tank.

REUSABLE CRASH CUSHIONS Crash cushions for roadway hazards (e.g., ends of rigid concrete barriers, steel barrier, bridge piers, signs).

REUSABLE HOSE FITTING Hose fitting that can be removed and reapplied.

REUSE Minimizing waste generation by recovering and reprocessing usable products that might otherwise become waste (i.e., recycling of aluminum cans, paper, bottles).

REUSE SYSTEM System of pipelines or ditches to collect and convey surface or subsurface runoff from irrigated field for reuse.

REVENUE PASSENGER Passenger who has paid for trip.

REVENUE SERVICE Entity responsible for intake of revenue.

REVENUE TRACK MILES Total mileage of track in passenger-carrying service.

REVENUE VEHICLE Vehicles used by Contractor to perform customer transportation services under Contract.

REVERBERATION Violent oscillation or vibration.

REVERBERATION TIME Time required for sound to decay in closed space.

REVERBERATORY FURNACE Forge, used in smelting and refining, where material is melted by radiant heat and convection of hot gases.

REVERSE COMMUTING Round trip from urban area to suburban area and then returning.

REVERSE FLOW Flow in opposite direction to norm.

REVERSE OSMOSIS Treatment process whereby solution is forced through semipermeable membrane under pressure to generate drinkable water from seawater or to separate chemical compounds.

REVERSE OSMOSIS REJECT WATER Waste water released from reverse osmosis process.

REVERSE ROTARY DRILLING Drilling where waste is carried away by liquid forced up drill pipe interior.

REVERSE-ACTING DIAPHRAGM VALVE Valve that opens or closes when pressure is applied or released.

REVERSIBLE EFFECT Effect that is not permanent.

REVERSIBLE LANE Lane where traffic may travel in either direction.

REVERSIBLE MOTOR Motor where direction of rotation can be reversed.

REVERSING CURRENT Tidal current that flows alternately in approximately opposite directions with slack water at each reversal of direction.

REVETMENT Facing of stone, cement, sandbags, etc., to protect wall or embankment.

REVETMENT WALL Permanent slope protection structure designed to prevent erosion.

REVIEW OF OPERATION AND MAINTENANCE Periodic evaluation of operation and maintenance activities at particular facility.

REVOLUTIONS PER HOUR Unit of rotational speed equal to cycles/hr.

REVOLUTIONS PER MINUTE Unit of rotational speed equal to cycles/min.

REVOLUTIONS PER SECOND Unit of rotational speed equal to cycles/sec.

REVOLVING FUND Fund that's continually replenished after withdrawals.

REVOLVING SCREEN Circular trash rack.

REWIND Act of putting new copper insulated wire in armature windings of synchronous generator.

REX BLOCK Blocking pattern where there is upper level high located directly north of closed low.

REYNOLD'S CRITICAL VELOCITY Velocity where fluid flow changes from laminar to turbulent and friction ceases to be proportional to first power of velocity.

REYNOLD'S NUMBER Dimensionless number that yields measure of inertial forces to viscous forces ratio.

REYNOLDS NUMBER Dimensionless ratio of inertial force to viscous force that's defined as velocity multiplied by length divided by kinematic viscosity.

RHEOLOGY Branch of physics that studies deformation and flow of matter.

RHYOLITE Light-colored volcanic igneous rock that's extrusive equivalent of granite.

RIBBON LIGHTNING Succession of strokes, each blown bit to side of previous strokes by wind, but striking so fast that all strokes are seen at once as ribbon-like flash.

RIBONUCLEIC ACID Molecule that carries genetic message from DNA to cellular protein-producing mechanism.

RICHTER MAGNITUDE Magnitude of earthquake measured as common logarithm of displacement amplitude, in microns, of standard Wood-Anderson seismograph located on firm ground 100 km from epicenter and having magnification of 2800, natural period 0.8 second and damping coefficient of 80 percent.

RICHTER SCALE Scale of numerical values to describe magnitude of earthquake, ranging from 1 to 9, by measurements made in well-defined conditions and with given type of seismograph.

RIDE NUMBER Rideability of pavement section using scale of 0 to 5.

RIDE QUALITY RATING Quality of road's ride perceived by traveling public.

RIDEABILITY Condition of being rideable.

RIDEABILITY INDEX Estimate of current level of ride satisfaction for roadway lengths.

RIDEDOWN ACCELERATION Deceleration unrestrained occupant would experience when impacting vehicle interior.

RIDER One who rides in vehicle.

RIDER SEWER Shallow sewer receiving house connections.

RIDERSHIP People who ride form of transportation.

RIDESHARING Form of transportation that's not public transport where more than one person shares use of vehicle (e.g.,van) to make trip.

RIDGE 1. Line along which two sloping surfaces meet which diverge towards ground; 2. Chain of mountains or hills; 3. Long narrow elevation on ocean bottom; 4. Type of warm air that comes down on to land from mountains; 5. Highest point on roof, represented by horizontal line where two roof areas intersect, running length of area; 6. Elongated area of relatively high atmospheric pressure.

RIDGE ICE Ice piled haphazardly one piece over another in form of ridges or walls.

RIFFLE 1. Fast-flowing, shallow part of stream causing broken water; 2. Trough or sluice having cleats, grooves or steps across bottom for holding quicksilver and catching particles of gold when auriferous earth is washed; 3. Succession of small waves.

RIFFLE AND POOL COMPLEX Water habitat composed of riffles (water flowing rapidly over coarse substrate) and pools (deeper areas of water associated with riffles).

RIFT VALLEY Mid-oceanic ridge system with central valleys.

RIGGING 1. Supporting material used for construction work; 2. Tackle that support masts and spars of sailing vessel.

RIGHT ABUTMENT Abutment on respective side of observer looking downstream.

RIGHT ASCENSION Celestial longitude of Sun where value is 0° at vernal equinox, 90° at summer solstice, 180° at autumnal equinox and 270° at winter solstice.

RIGHT BANK Bank, of river, to right when facing downstream.

RIGHT DESIGNATION Made with observer looking in downstream direction.

RIGHT ENTRANCE REGION Area upstream from and to right of upper-level jet max.

RIGHT MOVER Thunderstorm that moves appreciably to right relative to main steering winds and to other nearby thunderstorms.

RIGHT OF ACCESS Legal right to pass through grounds or property belonging to another.

RIGHT OF IMMEDIATE POSSESSION Right to immediately enter upon and use property.

RIGHT OF WAY Right to proceed first in traffic, on land, on water or in air.

RIGHT OF WAY APPRAISAL Appraiser determines value of property before taking and then determines value of property after taking.

RIGHT OF WAY ESTIMATE Approximation of market value of property including damages in advance of appraisal.

RIGHT OF WAY LINE Line that bounds right of way at its outermost edge.

RIGHT OF WAY MAP Provide general alignments and standards for all transportation corridors and support development patterns.

RIGHT OVERBANK Part of channel to right of overbank.

RIGHT REAR QUADRANT Area upstream from and to right of upper-level jet max.

RIGHT TURN ON RED Permits vehicles at traffic light, showing red signal, to turn right.

RIGHT-OF-WAY Land on and immediately surrounding structure owned by entity holding title to structure.

RIGHT-OF-WAY APPRAISAL When appraiser determines value of property before taking and determines value after taking.

RIGHT-OF-WAY ASSIGNMENT Transfer of any right or interest in right-of-way grant or lease from assignor to assignee.

RIGHT-OF-WAY ESTIMATE Total estimated damages to each property impacted by project.

RIGHT-OF-WAY MAP Plot representing actual location and dimensions of all property, rights or franchises.

RIGHT-OF-WAY PHASE In design phase, revision or preparation of plans for new right-of-way.

RIGHT-OF-WAY TRANSFER TIME Maximum amount of time needed for worst case condition, prior to display of track clearance green interval.

RIGHT-TURN-ON-RED Permitting vehicles at traffic light, showing red signal, to turn right.

RIGID FRAME Load-resisting frame constructed with rigid connections that resist movements induced at joints of members.

RIGID GRAVITY AND SEMI-GRAVITY RETAINING WALL Structure that provides lateral support for mass of soil and that owes it stability primarily to its own weight and to weight of any soil located directly above its base.

RIGID PAVEMENT Pavement slabs placed on earth that takes load by itself though its strength and rigidity.

RIGID PIPE Pipe designed to transmit backfill load to foundation beneath pipe.

RIGID SLAB Portland cement concrete pavement bounded by joints and edges.

RILL Small grooves, furrows or channels in soil made by water flowing down over its surface.

RIM Edge around something.

RIME Ice formed by rapid freezing of cold water droplets of fog onto cold surface.

RIME ICE Opaque coating of tiny, white, granular ice particles caused by rapid freezing of supercooled water droplets on impact with object.

RING 1. Solid object in shape of circle; 2. Group of atoms linked by bonds to form closed chain in molecule; 3. Planar geometrical figure included between two concentric circles.

RING FOLLOWER GATE Gate consisting of rectangular leaf and opening equal in diameter to that of conduit that forms unobstructed passageway when leaf is in raised or open position.

RING GATE Ring- or annular-shaped steel drum operating in recess or gate chamber in spillway crest and controlled in manner similar to drum gate.

RING GEAR Ring, with teeth, fitted on periphery of flexplate or flywheel of engine.

RING NOZZLE Nozzle that reduces its diameter.

RING SEAL GATE Sealing by moveable seal extended by water pressure when gate is closed.

RING SYSTEM Well pointing system around construction area.

RINGELMANN CHART Scale for measuring apparent density or opacity of smoke.

RINGLEMANN CHART Series of shaded illustrations used to measure opacity of air pollution emissions, ranging from light grey through black.

RIP Strong outflow of surface water.

RIP CURRENT Relatively small-scale surf-zone current moving away from beach.

RIP TIDE Relatively small-scale surf-zone current moving away from beach.

RIPARIAN Of, living on, pertaining to or adjacent to water (e.g., river, lake, pond, ocean) with high density, diversity and productivity of plant and animal species relative to nearby uplands.

RIPARIAN AREAS Geographically delineable areas with distinctive resource values and characteristics that compose aquatic and riparian ecosystems.

RIPARIAN DEPENDENT RESOURCES Resources that owe their existence to riparian area.

RIPARIAN DOCTRINE OR RULE Precept that owner of adjacent property to surface water has first withdraw right.

RIPARIAN ECOSYSTEMS Transition between aquatic ecosystem and adjacent terrestrial ecosystem.

RIPARIAN HABITAT Transition zone between aquatic and upland habitat with differing density, diversity and productivity of plant and animal species relative to nearby uplands.

RIPARIAN LAND Terrain adjacent to rivers and streams.

RIPARIAN OWNER Person who owns land adjacent to water body.

RIPARIAN RIGHTS Entitlement of land owner to certain uses of water on or bordering property.

RIPARIAN WATER Water allocated to those who possess land along its path.

RIPARIAN ZONE Interface between land and water body.

RIPPLE 1. Small triangular-shaped bed forms; 2. Moving disturbance or undulation, in surface of fluid (i.e., pebble thrown into pond).

RIPRAP Layer of large uncoursed stone, precast blocks, bags of cement or other suitable material, generally placed on slope of embankment or along watercourse as protection against wave action, erosion or scour.

RIPTIDE Particularly strong tidal current that moves offshore.

RISE RATE Pressure rise to time ratio.

RISER Vertical utility conduit, pipe or path between floors of building for placement of cables or to convey fluids.

RISER PIPE Vertical pipe in building.

RISING LIMB Portion of hydrograph where runoff is increasing.

RISING MAIN Pipe where water from engine is delivered to elevated reservoir.

RISING SLUDGE Occurs when sludge settles to bottom of clarifier, compacted and rises to surface due to denitrification or anaerobic biological activity.

RISING STEM VALVE Valve stem that rises when valve opened.

RISK Relationship between consequences resulting from adverse event and its probability of occurrence.

RISK ANALYSIS Procedure to identify and quantify risks by establishing potential failure modes, providing numerical estimates of likelihood of event in specified time period and estimating magnitude of consequences.

RISK ASSESSMENT Process of identifying likelihood and consequences of failure (qualitative and quantitative evaluation) to provide basis for informed decisions that incorporates both risk analysis and risk evaluation phases.

RISK CHARACTERIZATION Last phase of risk assessment process that estimates potential for adverse effects to occur and evaluates uncertainty involved.

RISK COMMUNICATION Exchange of information about risks among risk assessors and managers, general public, news media, interest groups, etc.

RISK ESTIMATE Description of probability to develop adverse response.

RISK ESTIMATION Expressing duration, intensity, magnitude and reach of potential consequences of risk in quantifiable or monetary terms.

RISK FACTOR Characteristics (e.g., race, sex, age, obesity) or variables (e.g., smoking, occupational exposure level) associated with increased probability of adverse effect.

RISK FOR NON-ENDANGERED SPECIES Risk to species if anticipated pesticide residue levels are equal to or greater than LC_{50}.

RISK MANAGEMENT Process of evaluating and selecting alternative regulatory and non-regulatory responses to risk.

RISK MONITORING Regular updates and reporting of risk information to monitor progress during risk management process.

RISK-BASED TARGETING Direction of resources to those areas that have been identified as having highest potential or actual adverse effect on human health and/or environment.

RISK-SPECIFIC DOSE Dose associated with specified risk level.

RIVER Large and often winding stream which drains land mass, carrying water down from higher areas to lower point, ending at ocean or in inland sea.

RIVER BASIN Land area drained by river and its tributaries.

RIVER CAPTURE Geomorphological phenomenon when stream is diverted and flows down bed of neighbouring stream.

RIVER CORRIDOR River and strips of land adjacent to it, including talus slopes at bases of cliffs, but not cliffs themselves.

RIVER FLOOD STATEMENT Product is used by local National Weather Service Forecast Office to update and expand information in River Flood Warning.

RIVER FLOOD WARNING Product is issued by local National Weather Service Forecast Office when forecast points (those that have formal gaging sites and established flood stages) at specific communities or areas along rivers where flooding has been forecasted, is imminent or is in progress.

RIVER FLOODING Rise of river to elevation such that river overflows its natural banks causing or threatening damage.

RIVER FORECAST Contains stage and/or flow forecasts for specific locations based on existing and forecasted hydrometeorologic conditions.

RIVER FORECAST CENTER Centers that serve groups of Weather Service Forecast offices and Weather Forecast offices, in providing hydrologic guidance and is first echelon office for preparation of river and flood forecasts and warnings.

RIVER FORECASTING Forecasting estimate of river stage.

RIVER GAUGE Device for measuring river water surface above datum.

RIVER GAUGE DATUM Arbitrary zero datum elevation which all stage measurements are made from.

RIVER GRAVEL Gravel composed of small pieces of rounded stone of various colors.

RIVER HEAD River source.

RIVER ICE STATEMENT Public product containing narrative and numeric information on river ice conditions.

RIVER MILE Unit of measurement (in miles) used on rivers.

RIVER OBSERVING STATION Established location along river designated for observing and measuring properties of river.

RIVER OUTLET WORKS Combination of structures and equipment, located in dam, required for safe operation and control of water released from reservoir to serve various purposes.

RIVER REACH River or stream segment of specific length.

RIVER RECREATION STATEMENT Statement released to inform river users of current and forecast river and lake conditions.

RIVER SEGMENT 1. Specified portion of river with upstream inflow point and with downstream termination at control point; 2. Surface waters of approved planning area exhibiting common hydrological, natural, physical, biological or chemical processes.

RIVER STAGE Water level above datum.

RIVER STATEMENT Product issued to communicate notable hydrologic conditions which do not involve flooding.

RIVER SUMMARY Summary of river and/or crest stages for selected forecast points along river.

RIVER SYSTEM All of streams and channels draining river basin.

RIVER TERRACE Remnants of former floodplain of stream or river.

RIVER TRAINING Structural measures taken to improve river and its banks.

RIVER TRAINING STRUCTURE Structures that use river's energy to move sediment out of navigation channel and reduce dredging need.

RIVER TRASH WALL Walls constructed to deflect heavy floating debris away from upper ends of fishway.

RIVER VALLEY Valley eroded or followed by river.

RIVER WASH Matter deposited on land by flooding river.

RIVERBED 1. Path river runs or once ran; 2. Bottom earthen part of river, not including riverbanks.

RIVERINE 1. Riparian; 2. Of or pertaining to rivers.

RIVERINE HABITAT All wetlands and deep-water habitats within channel, with two exceptions - wetlands dominated by trees, shrubs, persistent emergents, emergent mosses or lichens and habitats with water that contains ocean-derived salt in excess of 0.5 parts/1,000.

RIVERINE WETLANDS Wetlands within river and stream channels or where ocean-derived salinity is less than 0.5 parts/1,000.

RIVERSIDE Bank or side of river.

RIVULET Small brook or stream.

ROAD Public way designed to accommodate many vehicles travelling in both directions.

ROAD LOCOMOTIVE Locomotive with broad wheels for use on roads.

ROAD METER Wheel that measures for professional surveying, highway and site work.

ROAD PRICING Charges levied for road use (e.g., tolls, distance fee, time fee).

ROAD USER Any person that makes use of road, whether they are driver, cyclist, motorcyclist, moped rider, pedestrian or horserider.

ROAD WEATHER INFORMATION SYSTEM System that provides surface weather data that improves maintenance efficiency and increases road safety.

ROADBED Prepared location for road, including its foundation.

ROADSIDE Area on either side of road.

ROADSIDE BARRIER Feature that protects traffic from roadside obstacles or hazards.

ROADSIDE OBSTRUCTION Object along road or in median that impacts traffic.

ROADSIDE SIGNS Displays that help passers-by, provide directions, announce upcoming events or display information.

ROADWAY Portion of bridge, road or railway used by traffic.

ROADWAY NETWORK Road segments that interconnect.

ROADWAY WIDTH Distance between barriers or curbs in road.

ROCK 1. Any naturally formed, consolidated or unconsolidated material (but not soil) consisting of two or more minerals; 2. Hard, firm and stable parts of Earth's crust.

ROCK ANCHOR Steel rod or cable placed in hole drilled in rock, held in position by grout, mechanical means or both.

ROCK BOLT Steel rod or cable placed in hole drilled in rock, held in position by grout, mechanical means or both.

ROCK CRUSHING AND GRAVEL WASHING Facilities that process crushed and broken stone, gravel and riprap.

ROCK DRILL Powerful drill that can cut through stone.

ROCK EXCAVATION 1. Hard and firm parts of Earth's crust that is dug out and removed from particular site or area; 2. Boulders or detached pieces of solid rock more than 1 yd^3 in volume.

ROCK FLOUR Fine-grained, silt-sized particles of rock generated by mechanical grinding of bedrock by glacial erosion or artificially.

ROCK FLOWAGE AND FRACTURE ZONE Part of lithosphere lying between zone of rock fracture and zone of stone flowage.

ROCK FLOWAGE ZONE Portion of lithosphere where all stone are under stresses surpassing their elastic limits and thus subject to long-term deformation.

ROCK FRAGMENT Detached pieces of rock which generally are not rounded.

ROCK REINFORCEMENT Placement of rock bolts, untensioned rock dowels, prestressed rock anchors or wire tendons in rock mass to reinforce and mobilize rock's natural competency to support itself.

ROCK WOOL INSULATION Building insulation made of melted rock fibers.

ROCK-RIPPING BUCKET Bucket designed for extreme rock excavation.

ROCKER BEARING Bearing that can accommodate rotations and movements widely used for railroad structures.

ROCKETSONDE Type of radiosonde shot into atmosphere by rocket, allowing it to collect data during its parachute descent from higher position in atmosphere than balloon could reach.

ROCKFILL DAM Embankment dam where more than 50% of total volume is comprised of compacted or dumped cobbles, boulders, rock fragments or quarried rock generally larger than 3-inch size.

ROCKING GRATE Form of firebars, where sections of grate may be rocked or tipped to break up clinker within fire or to drop fire after day's work.

ROD Graduated staff used in determining difference in elevation between two points.

ROD FLOAT Method used to determine velocities in open channels using rod or tube floats.

ROD-FLOAT Float that measures estimate of surface velocity.

RODDING Process of reinforcing concrete with metal rods.

RODDING EYE Conduit fitting with removable plug to provide inspection or cleaning access.

RODENTICIDE Chemical or agent used to destroy rats or other rodent pests or to prevent them from damaging food, crops, etc.

RODENTICIDES Toxic substance used to kill rodent pests.

ROENTGEN Unit of ionizing radiation that's amount producing one electrostatic unit of positive or negative ionic charge in one cubic centimeter of air under standard conditions.

ROGUE WAVE Unexpected wave of much greater height or steepness than other waves in prevailing sea or swell system.

ROLL CLOUD Low, horizontal tube-shaped arcus cloud associated with thunderstorm gust front (or sometimes with cold front).

ROLL CRUSHER Crusher with two rolls driven so that they rotate toward one another.

ROLL-OFF CONTAINER Open top dumpster utilizing wheels to facilitate rolling dumpster in place.

ROLL-ON/ROLL-OFF Vessel designed to carry wheeled cargo that's driven on and off using their own wheels or platform vehicle.

ROLL-OVER PROTECTION STRUCTURE Structure to protect equipment operators and motorists from injuries caused by vehicle roll overs.

ROLLED FILLED DAM Embankment dam of earth or rock where material is placed in layers and compacted by use of rollers or rolling equipment.

ROLLED-EARTH DAM Dam entirely constructed of one type of material.

ROLLER Any rotating cylindrical device that's part of machine.

ROLLER BEARING Bearing where load is carried by round (e.g., cylindrical or spherical) elements.

ROLLER COMPACTED CONCRETE Mixture of cement, water and aggregate compacted by rolling.

ROLLER COMPACTED CONCRETE DAM Concrete gravity dam constructed by use of dry mix concrete transported by conventional construction equipment and compacted by rolling, usually with vibratory rollers.

ROLLER DRUM GATE Movable crest or spillway gate, consisting of long hollow drum that may be held in its raised position by water pressure in flotation chamber beneath drum.

ROLLER GATE Gate rolled up or down inclined supporting rails by hoist through sprocket chains around ends of cylinder.

ROLLING Method of compacting soil using sheepsfoot roller or smooth drum roller.

ROLLING STOCK Vehicles that move on railway.

ROLLING TERRAIN Hilly slopes that consistently rise above and fall below road grade.

ROLLWAY Inclined slope or chute prepared for rolling logs into stream or into mill.

ROOF DRAIN Pipe designed to carry water away from roof to protect it.

ROOT 1. Part of plant that anchors and supports plant body, absorbs and stores water and nutrients; 2. Dam section that ties into ground.

ROOT CROPS Underground plant parts eaten as food.

ROOT ZONE 1. That depth of soil which plant roots readily penetrate and where predominant root activity occurs; 2. Area where low-angle thrust fault steepens and descends into crust.

ROPE CLOUD Narrow, rope-like band of clouds sometimes seen on satellite images along front or other boundary.

ROPE FUNNEL Narrow, often contorted condensation funnel usually associated with decaying stage of tornado.

ROPE STAGE Dissipating stage of tornado, characterized by thinning and shrinking of condensation funnel into rope (or rope funnel).

ROSSBY WAVES Series of troughs and ridges on quasi-horizontal surfaces in major belt of upper tropospheric westerlies where waves are thousands of kilometers long and have significant latitudinal amplitude.

ROTAMETER Device used to measure fluid flow.

ROTARY COMPRESSOR Screw compressors that uses two asymmetrical rotors to compress air.

ROTARY DRILL Drilling machine frequently used for oil, gas and geothermal exploration/production where drill bit is turned by rotary table and jetting liquid is used to cool bit and transport material to surface.

ROTARY DRYER Dryer to reduce or minimize liquid moisture content of material.

ROTARY GEAR PUMP Positive displacement pump that uses rotor and gear assembly to generate flow.

ROTARY INTERSECTION Enlarged intersection where moving vehicles cross roads or change direction.

ROTARY KILN INCINERATOR Incinerator with rotating combustion chamber that keeps waste moving, thereby allowing it to vaporize for easier burning.

ROTARY PLUG VALVE Valve where rotation of passage in transverse plug regulates flow of liquid or gas through pipes.

ROTARY PUMP Positive displacement pump to move fluid through pump housing.

ROTARY SWITCH Switch operated by rotation.

ROTARY TIDAL CURRENT Offshore ocean tidal current where direction of flow is not restricted by any barriers.

ROTARY VALVE Valve where rotation regulates flow of liquid or gas.

ROTATING BIOLOGICAL CONTACTOR Fixed-bed reactors, partially submerged, consisting of stacks of disks that rotate as wastewater flows through.

ROTATING-BALL FAUCET Faucet with single-handle that swings up and down and side to side.

ROTATION Spinning of body (e.g., Earth) about its axis.

ROTOR 1. Rotating part of mechanical device; 2. Quantity having magnitude, direction and position.

ROTOR CLOUD Turbulent altocumulus cloud formation found in lee of some mountain barriers when winds cross barrier at high speed.

ROTTEN ICE Ice in advanced stage of disintegration.

ROUGH FISH Fish not prized for sport or eating.

ROUGH SEAS Sea conditions associated with regionally defined wind thresholds over bays, inlets, harbors, inland waters and estuaries where larger waves are forming with whitecaps and spray everywhere.

ROUGHING FILTER Filters that provide pretreatment for turbid water.

ROUGHING TANK Settling tank to eliminate suspended matter and grease from wastewater.

ROUGHNESS Property of having uneven or irregular surface.

ROUGHNESS COEFFICIENT Represents roughness or friction applied to flow by channel or conduit.

ROUGHOMETER Device that measures roughness for roads.

ROUND TRIP Trip from one destination to another and then returning to starting location.

ROUND-CRESTED WEIR Weir having crest upwardly convex in direction of normal flow.

ROUND-HEADED BUTTRESS DAM Concrete dam constructed of thick parallel buttresses at upstream end until they touch.

ROUNDABOUT Road junction where traffic streams circularly around central island.

ROUNDHOUSE Circular building where locomotives are housed.

ROUNDING 1. Rounded surface; 2. Curve.

ROUNDOFF ERROR Cumulative error introduced by rounding of results from individual arithmetic operations because only finite number of digits can be retained after each operation of digital computer.

ROUTE 1. Course or way traveled or passed; 2. To direct or divert along particular course.

ROUTE ALERTING Method for alerting people in areas not covered by primary system or if primary system fails.

ROUTE DEVIATION When public transportation service area vehicle makes pick-up or drop-off around area.

ROUTE OF EXPOSURE Avenue by which chemical comes into contact with organism (e.g., inhalation, ingestion, dermal contact, injection).

ROUTING Technique used to compute effect of channel storage and conveyance on shape and movement of flood wave through river reach.

ROUTING MODEL Model for performing flood routing.

RUBBERNECKING Slowing down vehicle to view traffic accident.

RUBBISH Solid waste, excluding food waste and ashes, from homes, institutions and workplaces.

RUBBISH CHUTE Conduit where rubbish may be dropped for disposal.

RUBBLE 1. Broken remains of rock or masonry; 2. Mass or stratum of fragments of rock lying under alluvium and derived from neighbouring rock.

RUBBLE DAM Stone masonry dam where stones are unshaped or uncoursed.

RUBBLE DRAIN Trench lined with tile or perforated pipe backfilled with gravel and then covered to collect and channel water.

RUBBLE-MOUND BREAKWATER Breakwater built of heavy stones.

RULE AUTHORIZATION Allowing owner or operator to construct or operate as long as requirements are followed.

RUMBLE STRIP Pattern of variation in road's surface designed to alert inattentive drivers to potential danger by causing vibration and rumbling sound.

RUN 1. Relatively shallow part of stream with moderate velocity and little or no surface turbulence; 2. Seasonal upstream migration of anadromous fish; 3. One or more lengths of pipe that continue in straight line.

RUN CUTTING Scheduling work shifts for operators to minimize total cost of labor that meets service requirements.

RUN-OF-RIVER PLANTS Regulated inflow of one power plant is equal to outflow from power plant upstream.

RUN-THROUGH TRAIN Train not scheduled to add (i.e., pick up) or reduce (i.e., set out) rail cars enroute.

RUNNING AND QUICK-START CAPABILITY Net capability of generating units that carry load or have quick-start capability.

RUNNING GROUND Sand that flows downhill on incredibly small incline.

RUNNING LOSSES Evaporation of motor vehicle fuel from fuel tank while vehicle is in use.

RUNNING RAIL One of two rails of railway track where wheels of train make contact.

RUNNING SAND Flow of sand into excavation or void due to water pressure.

RUNNING SPEED Distance vehicle travels per time.

RUNNING TIME Time for vehicle to traverse roadway.

RUNNING TRAP Sanitary trap fitting located in drainage piping not directly connected to any fixture.

RUNOFF 1. That part of water yield (e.g., precipitation, snow melt, irrigation) that flows by overland flow, tile drains or groundwater toward streams; 2. Composed of baseflow and surface runoff; 3. Quantity of water discharged from drainage basin during given time period; 4. Dissolved chemicals.

RUNOFF COEFFICIENT Ratio of runoff to precipitation.

RUNOFF CYCLE Part of hydrologic cycle undergone by water between time it reaches land as precipitation and its subsequent evapotranspiration or discharge through stream channels.

RUNOUT DISTANCE GRADING Area where vehicle may travel after impacting terminal ahead of its length-of-need point.

RUNUP Vertical distance above setup that rush of water reaches when wave breaks.

RUNWAY Section of land, usually paved, for airplanes to land on or take off from.

RUNWAY THRESHOLD Beginning and end of designated space for landing and takeoff.

RUNWAY VISUAL RANGE System that measures runway visibility, luminance and light intensity to determine distance pilot should be able to see down runway.

RURAL Relating to countryside or agriculture.

RURAL AREA Predominantly agricultural, prairie, forest, range or undeveloped land where population is small.

RURAL ELECTRIC COOPERATIVE Customer-owned electric utility that was created to transmit and distribute power in rural areas.

RURAL HIGHWAY Low traffic volume highway that serve transportation purposes.

RURAL WATER USE Water used in suburban or farm areas for domestic and livestock needs.

RURAL WITHDRAWAL Water used in suburban or farm areas for domestic and livestock needs.

RUT 1. Furrow, groove or track worn in ground; 2. Roaring sound.

RUT DEPTH Depth of depression worn into road, measured at highway speeds with profilograph.

RUTTING Forming depressions or grooves in road by wheels or skis.

RYEGRASS GRASS Genus of tufted grasses in bluegrass subfamily of grass family.

S

S-TRAP Trap with 'S-bend' where water creates seal and prevents sewer gas from passing.

S1 CURVE Water surface profile that represents rise curve at upstream end of channel.

S2 CURVE Water surface profile that represents drawdown curve at downstream end of channel.

SACRED SITE Any specific, discrete, narrowly delineated location on federal land identified by Indian Tribe or Indian individual determined to be appropriately authoritative representative of Indian religion, as sacred.

SACRIFICAL ANODE Easily corroded material deliberately installed in pipe or intake to give it up (sacrifice it) to corrosion while rest of water supply facility remains relatively corrosion-free.

SACRIFICIAL ANODE Piece of corrodible metal, attached to metallic surface to be protected, preferentially consumed by electrolytic action.

SACRIFICIAL CRASH CUSHIONS Roadside safety barrier for vehicular impact.

SACRIFICIAL PROTECTION Protection of metal against corrosion by using more reactive metal.

SADDLE 1. Small tapered or sloped area structure that helps channel surface water to drains; 2. Low point, in shape of saddle, between two hills; 3. Conduit fitting.

SADDLE DAM Subsidiary dam of any type constructed across saddle or low point on perimeter of reservoir.

SADDLE FITTING Fitting to install branch from conduit run.

SADDLE REEF Opening at crest of sharp fold in sedimentary rocks, occupied by ore.

SAE NET HORSEPOWER Brake horsepower of engine measured at flywheel, on engine dyno, with all accessories installed.

SAFE Condition under which there is practical certainty that no harm will result.

SAFE BEARING CAPACITY Bearing capacity of soil with factor of safety.

SAFE DRINKING WATER ACT Main federal law that ensures quality of Americans' drinking water.

SAFE VELOCITY Flow speed that will keep solids in suspension without scouring.

SAFE WATER Water that does not contain harmful bacteria, toxic materials or chemicals and is considered safe for drinking even if it may have taste, odor, color and certain mineral problems.

SAFE WORKING PRESSURE Permitted maximum working pressure of vessel.

SAFE YIELD Annual amount of water that can be taken from source of supply over period of years without depleting that source beyond its ability to be replenished naturally in "wet years."

SAFE-POSITIONED Emergency vehicles located such that traffic flow through incident area is optimized.

SAFENER Chemical added to pesticide to keep it from injuring plants.

SAFETY BELT Belt or strap that attaches person to immovable object for safety.

SAFETY BOOM Floating structure (e.g., net, chain of logs, drums or pontoons) used to protect face of dam or reservoir that prevents or deflects floating debris and boaters from entering water discharge facility or obstructing spillways and intakes.

SAFETY CAN Container for safe storage of flammable liquids.

SAFETY DEVICE Any apparatus that reduces loss or damage.

SAFETY EVALUATION EARTHQUAKE Earthquake, expressed in terms of magnitude and closest distance from dam site or in terms of characteristics of time history of free-field ground motions, for which safety of dam and critical structures associated with dam are to be evaluated.

SAFETY EVALUATION FLOOD Largest flood for which safety of dam and appurtenant structure(s) are to be evaluated.

SAFETY EVALUATION OF EXISTING DAMS EXAMINATION On-site examination performed initially and at predetermined intervals where design, construction, operation, performance and existing condition of all features are evaluated in accordance with state-of-the-art criteria.

SAFETY EVALUATION OF EXISTING DAMS REPORT Compilation of independent technical reports that evaluate design, construction and performance of dam for its structural and hydraulic integrity using available data that identifies existing or potential dam safety deficiencies and recommends future actions appropriate for safety of dam.

SAFETY FENCING Highly visible barrier for construction sites, crowd control, road work, etc.

SAFETY FLOORING Floor materials that include skid-resistance and sterilization characteristics.

SAFETY GATE Protective barrier designed to prevent accessing areas.

SAFETY GEAR Helmet, clothing, goggles, etc. designed to protect body from injury or infection.

SAFETY GLASSES Spectacles designed to protect eyes against flying objects (e.g., dust).

SAFETY GOGGLES Protective eyewear that enclose or protect area surrounding eye against flying objects (e.g., dust).

SAFETY HARNESS Form of protective equipment designed to protect against falling.

SAFETY SHUTOFF VALVE Valve, that when tripped, stops flow quickly and indicates system interruption.

SAFETY SWITCH Switch primarily used to disconnect equipment and machinery from power source.

SAFETY VALVE Relief valve set to open at pressure below that where container (e.g., boiler) would burst.

SAFETY WORK SURFACE Surface finished to reduce foot slippage.

SAFETY-OPERATED SWITCH Switch to remove power from machine and brake on application for safety.

SAFFIR-SIMPSON HURRICANE WIND SCALE 1 to 5 categorization based on hurricane's intensity at indicated time.

SAG Difference in elevation of road, wire, cable, chain or rope suspended between two consecutive points.

SAG CULVERT Culvert entirely below hydraulic grade line, designed to run full and under pressure.

SALINA 1. Area where deposits of crystalline salt are formed (e.g., salt flat); 2. Body of saline water (e.g., saline playa, salt marsh).

SALINE Condition of containing dissolved or soluble salts.

SALINE CONTAMINATION Depreciation of useable water due to salt water infiltration.

SALINE ESTUARINE WATER Semi-enclosed coastal water that's seaward exchange with ocean water and similar salinity.

SALINE FORMATIONS Deep and geographically extensive sedimentary rock layers saturated with waters or brines that have high total dissolved solids content (i.e., over 10,000 mg/l).

SALINE INTRUSION BARRIER WELLS Class V wells that are used to inject fluids to prevent intrusion of salt water into aquifer.

SALINE SODIC LAND Soil that contains soluble salts in amounts that impair plant growth but not excess of exchangeable sodium.

SALINE SPRING Small pool of mineralized water.

SALINE WATER Water considered unsuitable for human consumption or for irrigation because of its contains significant amounts (i.e., more than 1,000 mg/l) of dissolved solids.

SALINITY Measure of concentration of salt in solution.

SALINOMETER Hydrometer used to measure concentration of salt in solution.

SALMONIDS Family of fish that includes salmon and steelhead.

SALT 1. Chemically consisting mainly of sodium chloride used extensively as condiment and preservative; 2. One of compounds formed from reaction of acid with base, where positive ion replaces hydrogen of acid.

SALT MARSH Marsh of saline water, found in intertidal zone between land and sea, subjected to periodic flooding with salt water.

SALT WATER INTRUSION Invasion of fresh surface or groundwater by salt water.

SALTATION Movement of sand or fine sediment by short jumps above streambed under influence of water current too weak to keep it permanently suspended in moving water.

SALTATION LOAD Portion of bed load that's moving along stream bed by flow or impact of bouncing particles.

SALTS Minerals that water picks up as it passes through air, over and under ground or from households and industry.

SALVAGE 1. Utilization of waste materials; 2. Similar rescue of property liable to loss.

SALVAGE AND RECLAMATION Refuse disposal procedure where cast-off stuff is divided into groups.

SALVAGE SCRAP Substances, goods or equipment beyond repair that must be sold or disposed of as scrap.

SALVAGE VALUE 1. Estimated book value of asset after depreciation; 2. Value of existing pavement as support layer for overlay after analysis period.

SALVAGING Act where discarded goods are put to use.

SALVINIA MOLESTA Aquatic fern prohibited in United States by federal law that's invasive, rapidly growing plant that covers surfaces of lakes and streams forming floating mats that shade and crowd out important native plants.

SAME LOCATION Property under common ownership or control.

SAMPLE 1. Subset of population selected for measurement, observation or questioning; 2. Part or snippet of something taken or presented for inspection or shown as evidence of quality of whole.

SAMPLE ERROR Random variation reflecting inherent variability within population being counted.

SAMPLER Device that takes samples.

SAMPLING FREQUENCY Interval between collection of successive samples (i.e., rate at which sensor data is read).

SAMPLING SPOON Cylinder with spoon-like cutting edge for taking earth samples.

SAMPLING TRAIN Order of sequence where personal health sampling equipment parts are assembled together to complete cycle of assimilated breathed atmosphere that person is exposed to during working time period.

SANCTIONS Actions taken for failure to provide or implement plan.

SAND Mineral grains that's ground more finely than gravel, but is not as fine as silt where size vary from No. 4 sieve to No. 200 sieve (i.e, 1/16 mm and 2 mm).

SAND ASPHALT Mixture of ungraded sand and liquid asphalt for roads.

SAND BACKFILL Material which has particle size which varies from No. 4 sieve to No. 200 sieve and is used for refilling excavation.

SAND BOIL Eruption of water through bed of sand.

SAND CATCHER Instrument used to remove sand and solid from high pressure gas well effluent to prevent erosion of downstream equipment.

SAND DRAIN Process of radial consolidation which increases rate of drainage in embankment by driving casing into embankment and making vertical boreholes.

SAND DUNE Large, semi-permanent mound of windblown sand.

SAND EQUIVALENT Measure of relative percentages of hazardous fine dust or clay in soils or fine aggregate.

SAND FILTER Filters, comprising layers from rough rock to fine quartz grains, used as step in water treatment process of water purification.

SAND FILTERS Devices that remove some suspended solids from sewage where air and bacteria decompose additional wastes filtering through sand so that cleaner water drains from bed.

SAND GATE Sluice gate where sand and sediment accumulated in sand trap may be ejected.

SAND ISLAND Island largely made of pumped sand.

SAND LINE Rope attached to well drilling bailer.

SAND POCKET Porous zone in concrete or mortar containing fine aggregate with deficient cement.

SAND PUMP Pump to remove sand, mud or silt.

SAND PUMP DREDGE Suction dredge that pumps sediment.

SAND TRAP Basin that prevents passage of sand into drainage system.

SAND WASHER Machine used to separate sand from water and to wash grit at same time in order to remove organic particles.

SAND WICK Strip added into small-diameter holes drilled vertically into water-bearing surface strata.

SAND-GRAIN METER Hydrographic device used to assess amount of sand in running water.

SAND:COARSE AGGREGATE RATIO Ratio of fine to coarse aggregate in batch of concrete by mass or volume.

SANDBAG Bag, filled with sand, that can be used to weigh something down or to make defensive wall against flooding.

SANDBAR Ridge of sand caused by action of waves along shore.

SANDSTONE Sedimentary rock composed of sand-sized grains (usually quartz) cemented together.

SANDSTORM Particles of sand carried aloft by strong wind.

SANDY CLAY Material with predominance of clay with sandy being modifier.

SANDY LOAM Soil made up of sand along with varying amounts of silt and clay.

SANITARY DRINKING FOUNTAIN Drinking fountain that ensures safe supply of clean water.

SANITARY ENGINEERING Application of engineering methods to improve sanitation of human communities.

SANITARY LANDFILL Sites where waste is isolated from environment until it is safe.

SANITARY LANDFILL COMPACTOR Vehicle designed with load concentrators to offer compaction as well as devastating effect.

SANITARY LANDFILLING Disposing of solid waste by spreading, compacting and applying cover material on land by end of every working day.

SANITARY PLUMBING Pipework for taking away releases from sanitary fittings by connecting to your drainage system.

SANITARY SEWAGE Liquid or waterborne waste including human body waste.

SANITARY SEWER Pipe or conduit (sewer) intended to carry wastewater to wastewater treatment plant.

SANITARY SEWER OVERFLOWS Untreated or partially treated sewage overflows from sanitary sewer collection system.

SANITARY SEWERS Underground pipes that carry off only domestic or industrial waste, not storm water.

SANITARY SURVEY On-site review of water sources, facilities, equipment, operation and maintenance of public water systems for purpose of evaluating adequacy of facilities for producing and distributing safe drinking water.

SANITARY WASTE Liquid or solid wastes originating solely from humans and human activities, (e.g., wastes collected from toilets, showers, wash basins, sinks used for cleaning domestic areas, food preparation, clothes washing operations, sinks or washing machines where food and beverage serving dishes, glasses and utensils are cleaned).

SANITARY WASTEWATER Wastewater generated by normal housekeeping activities.

SANITARY WATER Domestic wastewater composed of wash water from kitchen, bathroom and laundry sinks, tubs and washing machines.

SANITARY WATER USE Domestic water uses (e.g., faucets, toilets, showers).

SANITATION Control of physical factors in human environment that could harm development, health or survival.

SANTA ANA WIND Weather condition in southern California, where strong, hot, dust-bearing winds descend to Pacific Coast around Los Angeles from inland desert regions.

SAPROBE Organism that lives off of dead or decaying organic material.

SAPROBIC CLASSIFICATION Biological water quality classification.

SAPROGENIC Causing or resulting from putrefaction.

SAPROLITE Soft, clay-rich, thoroughly decomposed rock formed in place by chemical weathering of igneous or metamorphic rock.

SAPROPEL Aquatic sludge-like sediment rich in organic matter formed in oxygen-free areas at bottom of lakes, seas, etc.

SAPROPHAGOUS Feeding on dead or decaying organic matter.

SAPROPHYTES Organisms living on dead or decaying organic matter that help natural decomposition of organic matter in water.

SAPROPHYTIC BACTERIA Bacteria that do not develop in living organism and feed on waste generated within it.

SAPROZOIC Feeding on dead or decaying animal matter.

SAPWOOD Wood just under bark of stem or branch, different in color from heartwood.

SASTRUGI Ridges of snow formed on snow field by action of wind.

SATELLITE COMMUNITY Comparatively self contained town established in vicinity of city.

SATELLITE VEHICLE Solid waste collection vehicle that transfers its load to more substantial vehicle.

SATURATE Cause to become completely penetrated, impregnated or soaked.

SATURATED 1. Unable to hold or contain any more; 2. Having all available valence bonds filled; especially of any organic compound containing only single bonds between carbon atoms.

SATURATED AIR Air that holds water vapor at its highest level.

SATURATED FLOW Movement of water through soil that's temporarily saturated.

SATURATED LIQUID Liquid that contains so much of solute at that temperature that it can dissolve no more.

SATURATED ROCK Rock that contains only saturated minerals.

SATURATED SOIL Soil with micropore and macropore spaces filled with water.

SATURATED SOLUTION Solution where solvent can dissolve no more of specific solute at particular temperature.

SATURATED UNIT WEIGHT Wet unit weight of soil when saturated.

SATURATED ZONE Area below water table where all open spaces (i.e., voids) are filled with water under pressure equal to or greater than that of atmosphere.

SATURATION 1. Condition of liquid when it has taken into solution maximum possible quantity of given substance at given temperature and pressure; 2. Degree that voids in soil are filled with water; 3. Condition of being filled to capacity.

SATURATION CAPACITY Maximum water holding capacity of soil.

SATURATION DEFICIT 1. Difference involving total amount of dissolved oxygen within stream or body of water and amount of oxygen needed to generate saturation; 2. Amount where water vapor in air must be increased to achieve saturation without changing environmental temperature and pressure.

SATURATION FLOW RATE Number of passenger car units in dense traffic flow for specific intersection.

SATURATION HEADWAY Constant headway achieved once stable moving traffic queue is established.

SATURATION LINE Line by way of cross-section of earth construction used to hold or retain water that indicates uppermost limit of flow-through.

SATURATION PRESSURE Pressure of vapor in equilibrium with its liquid.

SATURATION TEMPERATURE 1. Atmosphere temperature where atmosphere is saturated; 2. Boiling point of liquid.

SATURATION VAPOR PRESSURE Vapor pressure of system, at given temperature, wherein vapor of substance is in equilibrium with plane surface of that substance's pure liquid or solid phase.

SATURATION ZONE Portion of soil profile where available water storage is completely filled.

SATURATION-ADIABATIC PROCESS Adiabatic process for which air is saturated and may contain liquid water.

SBCAPE CAPE calculated using "Surface Based" parcel.

SCABLAND High, flat land of igneous rock, with thin soil and deep channels formed by glaciers or glacial floods.

SCALE Ratio of parameter in model to corresponding parameter in prototype.

SCALE EFFECT Consequence of non-similarity between model and prototype resulting from fact that not all pertinent dimensionless numbers are same in model and prototype.

SCALE HOUSE Structure for loaded freight cars to be weighed on their way out.

SCALING 1. Prying loose pieces of rock off face or roof to avoid danger of their falling unexpectedly; 2. Adjustment to earthquake time history or response spectrum where amplitude of acceleration, velocity, and/or displacement is increased or decreased, usually without change to frequency content of ground motion.

SCALING LAWS Conditions that must be satisfied to achieve desired similarity between model and prototype.

SCALPING SCREEN Vibration screen.

SCARIFICATION Raking ground harshly to remove weeds, etc.

SCARIFIER Implement for stripping and loosening soil, without bringing up fresh surface.

SCARIFY To break up, loosen or roughen surface of field or road or hard surface.

SCARP Cliff at edge of plateau or ridge caused by erosion.

SCATTER Concentration of artifacts.

SCATTERED Area coverage of convective weather affecting 30% to 50% of forecast zone.

SCATTERED DEVELOPMENT Where commercial, residential and retail developments are not close together.

SCATTERING Process where beam of light is diffused or deflected by collisions with particles suspended in atmosphere.

SCAVENGING Act of searching through refuse for useful material.

SCENIC EASEMENT Easement for conservation and development of roadside views and natural features.

SCENIC OVERLOOK Elevated location to view scenery.

SCHEDULE 1. Written or printed table of information; 2. To plan activity at specific date or time in future.

SCHEDULE OF DEFECTS List of things remaining to be finished by contract at end of defects liability period.

SCHEDULE VARIANCE Indicator of whether project schedule is ahead or behind.

SCHEDULED FINISH DATE Date and time work was scheduled to be completed on specific activity or task.

SCHEDULED MAINTENANCE Predetermined maintenance actions carried out at regular time intervals that are aimed at prevention of breakdowns.

SCHEDULED OUTAGE Shutdown of generating unit or other facility, for inspection or maintenance, in accordance with advance schedule.

SCHEDULED START DATE Activity's planned start time.

SCHEDULER Person or device that determines schedule, that determines order that tasks are to be done.

SCHEDULING Function where events are timed to take place at most opportune time.

SCHEMATIC Simplified line-drawing generally used by engineers and technicians to describe and understand how system works at abstract level.

SCHEMATIZATION Representation of continuum by discrete elements (e.g., dividing real river into reaches with constant parameters).

SCHEME 1. Systematic program of action for solving governing equations of mathematical model; 2. Chart or diagram of system or object; 3. Representation of aspects of celestial bodies for any moment or at given event.

SCHIST Metamorphic rock composed of platy mica minerals aligned in same direction.

SCHOKLITSCH FORMULA Bed load equation that relates sediment flux with excess flow discharge rate.

SCHOOL 1. Group of fish or marine mammals; 2. Institution dedicated to teaching and learning.

SCHOOL ZONE Roadway segment or crosswalk near school that has likely presence of pedestrians.

SCIENCE ADVISORY BOARD Group of external scientists who advise on science and policy.

SCINTILLATION COUNTER Detector of particles and ionizing radiation which uses photomultiplier to generate flashes of light.

SCOPE 1 GHG EMISSIONS Direct GHG emissions from sources that are owned or controlled by reporting entity that include emissions from fossil fuels burned on site, emissions from agency-owned or agency-leased vehicles and other direct sources.

SCOPE 2 GHG EMISSIONS Indirect GHG emissions resulting from generation of electricity, heat or steam generated off site but purchased by reporting agency.

SCOPE 3 GHG EMISSIONS Indirect GHG emissions from sources not owned or directly controlled by reporting agency but related to agencies activities (e.g., vendor supply chains, delivery services, outsourced activities, employee travel and commuting).

SCOPE Determining and documenting specific goals, deliverables, tasks, costs and deadlines for project.

SCOPE CHANGE Official decision made by project manager and client to change project's cost or schedule.

SCOPE CHANGE CONTROL Defines procedures that project and product scope can be changed.

SCOPE OF WORK Area in agreement that describes work to be performed.

SCOPE PLANNING Project management process that includes project justification, deliverables and goals.

SCOPE VERIFICATION Process of formalizing acceptance of project scope by stakeholders.

SCORIA Rough masses of rock formed by solidified lava; Slag or dross that remains after smelting of metal from ore.

SCOUR Enlargement of flow section by removal of bed material through action of moving water.

SCOUR PROTECTION Mechanical protection of soil and submerged sediment against scour.

SCOURED DEPTH Elevation difference between water surface and lowest scoured point in bottom.

SCOURING SLUICE Opening in dam, controlled by gate, where accumulated silt, sand and gravel may be ejected.

SCOURING VELOCITY Water velocity necessary to dislodge solids from stream bed.

SCOW END Raised part of flooring in back of dump body.

SCRAP Fragment of material discarded.

SCRAP LOADER Machine designed having magnet or grapple accessory for managing metal in scrap yards.

SCRAP METAL PROCESSOR Intermediate operating facility where recovered metal is sorted, cleaned of contaminants and prepared for recycling.

SCRAPE Nest made from scratching in ground.

SCRAPER 1. Instrument used for scraping up earth; 2. Apparatus pushed or pulled to dislodge pipeline deposits; 3. Apparatus installed in base of sedimentation tank to move settled sludge; 4. Blade to remove collected matter from filter.

SCRATCH COAT First layer used to provide bonding surface for second layer.

SCREED PUMP Machine for blending and pumping screed.

SCREEFING Removing vegetation at ground level before digging hole.

SCREEN Mesh (e.g., filter) or bar surface used for separating pieces or particles of different sizes.

SCREEN CHAMBER Area where wastewater screen is installed.

SCREEN LINE Imaginary line on map composed of one or multiple straight segments.

SCREENED INTERVAL Length of unscreened opening or of well screen where water enters well, in feet below land surface.

SCREENED MATERIAL Recovered materials left over from screening at processing facility.

SCREENED WELL Water well with intake filtering screen.

SCREENING Process of using screens to remove coarse floating and suspended solids from water or wastewater.

SCREENING RISK ASSESSMENT Risk assessment performed with few data and many assumptions to identify exposures that should be evaluated more carefully for potential risk.

SCREENINGS Refuse left after screening sand, coal, ashes, etc.

SCREENINGS GRINDER Mechanism for reducing solids to smaller size.

SCREW AUGER Rotating helical blade, within tube, to move materials.

SCREW COMPRESSOR Gas compressor that uses rotary-type positive-displacement mechanism.

SCREW IMPELLER Helical impeller of pump.

SCREW-DOWN VALVE Valve with lid attached to end of screw that when rotated, lid moves towards valve seat to close and stop flow.

SCREW-FEED PUMP Pump with helix blade to transport or compress material.

SCREWED PIPE Conduit with male screw thread on both ends.

SCRUB Vegetation of inferior quality.

SCRUBBER Equipment for removing particulates from products of combustion from industrial gas output by sprayers and wet baffles/packing.

SCRUBBING Washing of particulates from products of combustion from industrial gas output.

SCS CURVE NUMBER Empirically derived relationship between location, soil-type, land use, antecedent moisture conditions and runoff.

SCUD Small, ragged, low cloud fragments that are unattached to larger cloud base and often seen with and behind cold fronts and thunderstorm gust fronts.

SCUM 1. Layer of impurities that accumulates at surface of liquid; 2. Topmost liquid layer of cesspool or septic tank.

SCUM BAFFLE Plank that scrapes surface scum and solids into outlet opening.

SCUM BOARD Strip of material partly immersed to hold back scum.

SCUM BREAKER Device to break up scum in sludge digestion tank.

SCUM CHAMBER Place where rising scum is gathered in sludge-digestion tank.

SCUM COLLECTOR Device that skims scum from wastewater treatment tank surface.

SCUM TROUGH Channel where scum is deposited for later processing.

SCUPPER Opening in wall or parapet that allows water to drain.

SEA Large body of salt water.

SEA BREEZE Thermally produced wind blowing during day from cool ocean surface onto adjoining warm land, caused by difference in rates of heating of surfaces of ocean and of land.

SEA BREEZE CONVERGENCE ZONE Zone at leading edge of sea breeze where winds converge.

SEA BREEZE FRONT Leading edge of sea breeze, whose passage is often accompanied by showers, wind shift or sudden drop in temperature.

SEA FOG Advection fog caused by transport of moist air over cold body of water.

SEA ICE Any form of ice found at sea which has originated from freezing of sea water.

SEA LEVEL Long-term average position of sea surface commonly referenced to national vertical datum (e.g., NAVD 1988).

SEA LEVEL PRESSURE Atmospheric pressure at sea level at given location.

SEA MOUNT Underwater mountain formed by volcanic activity.

SEA SURFACE TEMPERATURES Mean temperature of ocean in upper few meters.

SEAFLOOR SPREADING Outward movement of seafloor from central underwater ridges as result of plate tectonics.

SEAL 1. Depth of water held in trap to stop passing of air or gas; 2. Something designed to prevent liquids or gases from leaking through joint; 3. Tight closure, secure against leakage.

SEAL COAT Final coat applied for sealing surface.

SEALABLE EQUIPMENT Equipment enclosed in locking or sealing case.

SEALANT Material used to seal surface so as to prevent passage of fluid.

SEALANT RESERVOIR Channel or joint where sealant is set.

SEALED BEARING Bearing with seals on both sides.

SEALED BID Contract cost estimate where value is kept secret from competitors until all such bids have been made.

SEALED HYDRAULIC SYSTEM Permanently sealed hydraulic system designed to prevent leaks.

SEALED SYSTEM Any device with inner workings obscured.

SEALING WATER Water used to prevent wastewater or dirt from reaching moving parts.

SEAM Layer of rock, coal or ore.

SEAMLESS CARBON-STEEL PIPE Conduit with large carbon and manganese content that produces greater tensile strength but less ductability.

SEAMLESS PIPE Pipe without seam or weld-joint.

SEAPLANE Aircraft capable of taking off from and alighting on surface of water.

SEAS Combination of both wind waves and swell.

SEASON Each of four divisions of year: spring, summer, autumn (fall) and winter.

SEASONAL EFFICIENCY Measures true energy efficiency of heating and cooling devices during period of one year.

SEASONAL FACTOR Measures percentage that's amount above or below normal time period on average.

SEASONAL RECOVERY Cyclical recharge of groundwater after wet season and subsequent to seasonal depletion.

SEASONAL STORAGE Storage of water or energy in medium for later use.

SEASONAL STREAM Stream that flows only periodically or when it receives water from source (e.g., rainfall runoff, springs, melting snow).

SEAWALL Coastal structure to prevent erosion, flooding or direct wave damage.

SEAWATER Saltwater of sea or ocean.

SECCHI DEPTH Measure of water clarity.

SECCHI DISC Secchi disk used to measure water transparency or turbidity by lowering until no longer visible.

SECCHI DISK Opaque disk lowered into water to measure transparency by noting depth where disk ceases to be visible from surface.

SECOND 1. One-sixtieth of minute; 2. Unit of angle equal to one-sixtieth of minute of arc or one part in 3600 of degree.

SECOND OF ARC Unit of angle equal to 1/60 of arc minute.

SECOND STAGE BIOCHEMICAL OXYGEN DEMAND Oxygen demand related to biochemical oxidation of nitrogenous substances.

SECOND-DAY FEET Volume of water represented by flow of 1 ft^3/s for 24 hours that's equal to 84,000 ft^3.

SECOND-FOOT 1. Unit of flow equal to 1 ft^3/s; 2. One cusec.

SECOND-STAGE BIOCHEMICAL OXYGEN DEMAND Portion of oxygen demand associated with oxidation of nitrogenous substance.

SECONDARY AIR Air injected into exhaust stream for fuller combustion of exhaust gases.

SECONDARY AMBIENT AIR QUALITY STANDARDS Air quality standards designed to protect human welfare, including effects on vegetation and fauna, visibility and structures.

SECONDARY BRANCH Branch off main drain branch.

SECONDARY BURNER Burner installed in secondary combustion chamber to hold temperature to finish combustion of incompletely burned gases.

SECONDARY CIRCULATION Hot water pumped around system to ensure instantaneous hot water.

SECONDARY CONSOLIDATION Consolidation of sedimentary material, at essentially constant pressure, resulting from internal processes such as recrystallization.

SECONDARY CRUSHER Pulverizing machine after primary crushing.

SECONDARY CURRENTS Movement of water particles on cross-section normal to longitudinal direction of channel.

SECONDARY DRINKING WATER REGULATIONS Non-enforceable federal regulations applying to public water systems and specifying maximum contamination levels that are required to protect public welfare.

SECONDARY DRINKING WATER STANDARDS Non-enforceable federal guidelines regarding cosmetic effects (i.e., tooth, skin discoloration) or aesthetic effects (i.e., taste, odor, color) of drinking water.

SECONDARY EFFECT Action of stressor on supporting components of ecosystem, which in turn impact ecological component of concern.

SECONDARY EMISSION CONTROL SYSTEM System to control emission of electrons from surface of solid when bombarded with electrons or other charged particles.

SECONDARY EMISSIONS Emission of electrons from surface of solid when bombarded with electrons or other charged particles.

SECONDARY ENRICHMENT Enrichment of vein or mineral deposit by minerals that have been taken into solution from one part of vein or adjacent rocks and redeposited in another.

SECONDARY FILTER Second phase of multistage filter system.

SECONDARY FLOW Movement of water particles on cross-section normal to longitudinal direction of channel.

SECONDARY INDUSTRY CATEGORY Any industry category that is not primary industry category.

SECONDARY INTERSTICE Opening in rock developed by process that affected rock.

SECONDARY LIQUID FUEL Non-recyclable liquid wastes converted into alternative to fossil fuels used in industrial processes.

SECONDARY MATERIALS Materials that have been manufactured and used at least once and are to be used again.

SECONDARY MAXIMUM CONTAMINANT LEVEL Maximum level of contaminant or undesirable constituent in public water systems required to protect public welfare.

SECONDARY MEMBER Structural members not designed to carry primary loads.

SECONDARY POLLUTANT Pollutants generated by chemical reactions occurring within atmosphere.

SECONDARY PROCESS Parts separated from solid waste that be further processed to permit reuse.

SECONDARY RECYCLING Material reused in some other way without reprocessing.

SECONDARY SEDIMENTATION Sedimentation after secondary or biological treatment of sewage.

SECONDARY SETTLING TANK Tanks that allow microorganisms and other solids to settle after secondary or biological treatment.

SECONDARY SEWAGE TREATMENT Biological degradation of organic material by micro-organisms under controlled conditions.

SECONDARY SOILS Formerly existing organic soils and stone debris that have been transported through geologic and hydrologic processes.

SECONDARY STANDARDS National ambient air quality standards designed to protect welfare, damage to property, transportation hazards, economic values and personal comfort/well-being.

SECONDARY USE Use in human biological materials originally collected for purpose other than current research purpose.

SECONDARY WASTEWATER TREATMENT Second step in most publicly owned waste treatment systems involving biological process of reducing suspended, colloidal and dissolved organic matter in effluent from primary treatment systems and which generally removes 80 to 95% of biochemical oxygen demand and suspended matter where bacteria consume organic parts of waste.

SECRETION Any substance secreted by organism.

SECTION 304 CRITERIA Developed by EPA under authority of section 304(a) of Act based on latest scientific information on relationship that effect of constituent concentration has on particular aquatic species and/or human health.

SECTION 1. Part, piece, cross-section, subdivision of anything (e.g., area equal to 640 acres or 1 mi^2); 2. Sequence of rock layers; 3. Informal taxonomic rank below order ranks and above family ranks; 4. Thin slice of material prepared as specimen for research.

SECTIONAL TANK Storage container composed of molded panels.

SECTIONALIZING VALVE Valve installed at regular intervals to isolate pipeline sections during emergencies (e.g., leakage).

SECTOR BOUNDARY In solar wind, area of demarcation between sectors, which are large-scale features distinguished by predominant direction of interplanetary magnetic field, toward or away from Sun.

SECTOR GATE Roller gate in part-circle or sector form for dam crest.

SECTOR VISIBILITY Visibility in specific direction that represents at least 45° arc of horizontal circle.

SECTORIZED HYBRID SCAN Single reflectivity scan composed of data from lowest four elevation scans.

SECULAR Continuing over long period of time.

SECULAR CYCLE Long periods of time that define era.

SECURE MAXIMUM CONTAMINANT LEVEL Maximum permissible level of contaminant in water delivered to free flowing outlet of ultimate user or of contamination resulting from corrosion of piping and plumbing caused by water quality.

SECURITE Headline within National Weather Service high seas forecasts transmitted to indicate that no hurricane or hurricane force winds are forecast.

SECURITY FENCE Fence built to stop unauthorized entrance.

SEDENTARY SOIL Soils that hold its position of their formation, without transporting, just above their parent rock.

SEDIMENT 1. Solid material suspended in or settled from water; 2. Particles derived from biological materials that have been transported by fluid.

SEDIMENT CONCENTRATION Quantity of sediment relative to quantity of transporting fluid or fluid-sediment mixture.

SEDIMENT DISCHARGE Rate at which sediment passes stream cross-section in given period of time.

SEDIMENT GUIDELINE Threshold concentration above which there is high probability of adverse effects on aquatic life from sediment contamination, determined using modified U.S. Environmental Protection Agency procedures.

SEDIMENT LOAD 1. Material in suspension and/or in transport; 2. Amount of sediment passing through stream cross-section in specified period of time.

SEDIMENT PARTICLE Fragments of mineral or organic material in either singular or aggregate state.

SEDIMENT POOL Portion of storage pool intended to fill with sediment over time.

SEDIMENT QUALITY GUIDELINE Threshold concentration above which there is high probability of adverse effects on aquatic life from sediment contamination.

SEDIMENT SAMPLE Quantity of water-sediment mixture or deposited sediment collected to characterize some property or properties of sampled medium.

SEDIMENT STORAGE CAPACITY Volume of reservoir planned for deposition of sediment.

SEDIMENT TRANSPORT FUNCTION Formula/algorithm for calculating sediment transport rate given hydraulics and bed material at cross-section.

SEDIMENT TRANSPORT RATE Rate at which sediment passes stream cross-section in given period of time.

SEDIMENT TRANSPORT ROUTING Computation of sediment movement for selected length of stream (reach) for period of time with varying flows.

SEDIMENT TRANSPORTATION Complex processes of moving sediment particles from place to place where principal transporting agents are flowing water and wind.

SEDIMENT TRAP Any depression where sediments substantially accumulate over time.

SEDIMENT TRAP EFFICIENCY Proportion of sediment inflow to stream reach retained within that reach.

SEDIMENT YIELD 1. Amount of mineral or organic soil material in suspension, is being transported or has been moved from its site of origin and arriving at specific location; 2. Portion of eroded material that does travel through drainage network to downstream measuring or control point; 3. Ratio of dry weight of sediment to total weight of water-sediment mixture in sample or unit volume of mixture.

SEDIMENT YIELD RATE Sediment yield per unit of drainage area.

SEDIMENT-DISCHARGE RELATIONSHIP Tables that relate inflowing sediment loads to water discharge for upstream ends of main stem, tributaries and local inflows.

SEDIMENTARY Rock resulting from consolidation of loose eroded sediment, remains of organisms or crystals forming directly from water.

SEDIMENTARY GROIN Groin developed slowly from river or sea bed.

SEDIMENTARY ROCK One of major groups of rock that makes up crust of Earth, formed by deposition of either weathered remains of other rocks, results of biological activity or precipitation from solution.

SEDIMENTARY ROCKS 1. Rocks formed by consolidation of loose sediment that has accumulated in layers (e.g., sandstone, shale formed by fragments of other rock transported from their sources and deposited in water); 2. Rocks formed by or from secretions of organisms (e.g., most limestone).

SEDIMENTATION 1. Act or process of forming or accumulating sediment in layers due to decrease in velocity and corresponding reduction in size and amount of sediment that can be carried; 2. Letting solids settle out of wastewater by gravity during treatment.

SEDIMENTATION BASIN Temporary storage built to capture eroded or disturbed soil.

SEDIMENTATION DIAMETER Diameter of sphere of same specific weight and same terminal settling velocity as given particle in same fluid.

SEDIMENTATION TANK Structure where wastewater is filled and stored for some time to remove suspended particles present in water.

SEDIMENTATION TANKS Wastewater tanks where floating wastes are skimmed off and settled solids are removed for disposal.

SEDIMENTS Soil, sand and minerals washed from land into water.

SEED PROTECTANT Chemical applied before planting to protect seeds and seedlings from disease or insects.

SEEDING MATERIAL Digested sewage sludge used to 'prime' or 'seed' sludge digestion tanks.

SEEP Small area where groundwater percolates slowly to land surface, usually forming pool.

SEEPAGE 1. Process where liquid leaks through porous substance (e.g., soil, rock, structure); 2. Water that has seeped or oozed through porous soil.

SEEPAGE BOIL Uprising of water on landward side of water obstruction (e.g., levee) due to infiltration.

SEEPAGE COLLAR Projecting collar of concrete built around outside of tunnel or conduit, within embankment dam, to reduce seepage along outer surface of conduit.

SEEPAGE FACE Physical boundary segment of groundwater system along which groundwater discharges and that is present when water table surface ends at downstream external boundary of flow domain.

SEEPAGE FORCE Force transmitted to soil or rock grains by seepage.

SEEPAGE LOSS Water loss by capillary action and slow percolation.

SEEPAGE PIT Covered pit with open-jointed or perforated lining where septic tank effluent seeps into surrounding soil.

SEEPAGE SPRING Water flow that occurs when shallow groundwater oozes from ground over large area with no defined discharge point.

SEEPAGE TRENCH Small trenches excavated at slight gradient and filled with stone to distribute effluent.

SEEPAGE VELOCITY Rate of discharge of seepage water through porous medium per unit area of void space perpendicular to direction of flow.

SEGMENT 1. Straight path between two points that's shortest distance between them; 2. To divide; 3. Part of circle between its circumference and chord.

SEGMENT DELAY Difference between ideal and observed travel time.

SEGMENTAL CONCRETE BRIDGE Joining numerous cast-in-place or precast bridge elements to form continuous bridge span.

SEGMENTAL SLUICE GATE Radial hydraulic gate.

SEGREGATED STORMWATER SEWER SYSTEM Drainage and collection system designed and operated to collect rainfall runoff at facility segregated from all other individual drain systems.

SEGREGATION Separation of things.

SEICH Standing wave oscillating in body of water.

SEICHE Oscillating wave of water caused by landslide, earthquake or meteorological event (e.g., wind, large barometric pressure gradient).

SEISMIC Of or related to movement in Earth's crust caused by natural relief of rock stresses.

SEISMIC EVALUATION CRITERIA Guideline for determining which faults or seismic sources need to be assigned maximum credible earthquake.

SEISMIC INTENSITY Numerical index describing effects (i.e., shaking) of earthquake on man, manmade structures or other features of Earth's surface using descriptive scale.

SEISMIC MOMENT Measure of earthquake size containing information on rigidity of elastic medium in source region, average dislocation and area of faulting.

SEISMIC PARAMETERS Descriptors of earthquake loading or earthquake size (e.g., magnitude, peak acceleration, location, distance, focal depth, spectrum intensity).

SEISMOTECTONIC Of, relating to or designating structural features of Earth which are associated with or revealed by earthquakes.

SEISMOTECTONIC PROVINCE Geographic area characterized by combination of geology and seismic history.

SEISMOTECTONIC SOURCE AREA Area or areas of known or potential seismic activity that may lack specific identifiable seismotectonic structure.

SEISMOTECTONIC STRUCTURE Identifiable dislocation or distortion within Earth's crust resulting from recent tectonic activity or revealed by seismologic or geologic evidence.

SEIZE 1. To bind wire rope with soft wire; 2. To prevent rope from unraveling when cut.

SEIZURE LAWS Law that allows agencies to seize property for public interest.

SEL Watch cancellation statement issued to terminate watch before its original expiration time.

SELECT MATERIAL Backfill materials specially selected and segregated from excavated materials.

SELECTION CRITERIA Set of statements that describe suitable indicators or rationale for selecting indicators.

SELECTIVE BIDDING Process of soliciting competitive bids for award of contract for construction.

SELECTIVE DIGGING Distinguishing several varieties of soil while excavating them.

SELECTIVE HERBICIDE Compound that kills or significantly retards growth of unwanted plant species without significantly damaging desired plant species.

SELECTIVE MEMBRANE Sheet material that's preferentially selective to passage of either cations or anions in solution.

SELECTIVE PESTICIDE Chemical designed to affect only certain types of pests, leaving other plants and animals unharmed.

SELECTIVITY 1. Ability of device to respond to particular frequency without interference from others; 2. Characteristic of plants and insects to adapt to their environment by genetic selection; 3. Characteristic of radiation, toxins and heavy metals to affect some things and not others.

SELECTOR Environmental conditions that develop are intended to favor growth of certain organisms over others.

SELECTOR RECYCLE Recycling of return sludge or oxidized nitrogen to provide desired environmental conditions for microorganisms to perform desired function.

SELENIUM Non-metallic chemical element with atomic number of 34.

SELF-CLEANSING GRADIENT Slope where flow in pipe will carry away deposited solids.

SELF-CLEANSING VELOCITY Minimum flow in pipe that will carry away deposited solids.

SELF-CONTAINED BREATHING APPARATUS Apparatus worn to provide breathable air in immediately dangerous atmosphere.

SELF-LOADING Capacity of powered industrial truck to load automatically.

SELF-MONITORING Sampling and analyses performed by facility to determine compliance with permit or other regulatory requirements.

SELF-MULCHING SOIL Soil that mixes itself through shrinking and swelling.

SELF-POWERED SCRAPER Scraper built into single unit with tractor.

SELF-PRIMING CENTRIFUGAL PUMP Centrifugal pump that can use air-water mixture to reach fully-primed pumping condition.

SELF-PURIFICATION Purification by natural process.

SELF-SIPHONING State where partial vacuum in poorly ventilated plumbing system causes seal to be drawn from trap permitting passing of air or gases throughout fitting.

SELF-SUPPLIED WATER Water withdrawn from surface water or groundwater source by user rather than being obtained from public supply.

SELLER'S RISK Probability that acceptance plan will erroneously reject acceptable quality level material or construction.

SEMI-ACTUATED CONTROL Signal control where detection is provided for minor movements only.

SEMI-ACTUATED OPERATION Traffic control signal with capability for at least one signal phases to function in respond to presence of vehicles or pedestrians at intersection.

SEMI-ANALYTICAL MODEL Mathematical model where complex analytical solutions are evaluated using approximate techniques, resulting in solution discrete in either space or time domain.

SEMI-AQUATIC 1. Spending part of life cycle underwater; 2. Having roots partially or entirely underwater.

SEMI-CONFINED AQUIFER Aquifer partially confined by soil layers of low permeability where recharge and discharge can still occur.

SEMI-EMPIRICAL MODEL Representation of real system by mathematical description based on general physical laws but containing coefficients determined from experimental data.

SEMI-HYDRAULIC-FILL DAM Material for dam transported by hauling and dumped at edge of embankment.

SEMI-PERCHED Watertable with pressure head greater than underlying body of water not separated by unsaturated stone.

SEMIANNUAL Occurring twice year.

SEMIARID Receiving little rainfall but more than arid area would.

SEMIBOLSON Topographic basin with inward-flowing drainage basin.

SEMIDIURNAL Every 12 hours.

SEMINAR Activity designed to introduce, discuss and update emergency planning documents, organization structure or early warning system component to familiarize key personnel with emergency procedures and their responsibilities.

SEMIPERMANENT SNOW LINE Lower elevation limit of where snow accumulates and remains.

SEMIPERMEABLE MEMBRANE Membrane that will allow certain molecules or ions to pass through it in one direction by diffusion.

SEMIPERMEABLE MEMBRANE DEVICE Long strip of low-density, polyethylene tubing filled with thin film of purified lipid (e.g., triolein) that simulates exposure to and passive uptake of highly lipid-soluble organic compounds by biological membranes.

SEMIPERVIOUS ZONE Substantial part of cross-section of embankment dam comprising material whose grading is of intermediate size between that of impervious zone and that of permeable zone.

SEMITRAILER trailer without front axle.

SEMIVOLATILE ORGANIC COMPOUND Organic compounds that volatilize slowly at standard temperature (20 °C and 1 ATM pressure).

SENESCENCE Aging process that describes lakes or other bodies of water in advanced stages of eutrophication.

SENSIBLE HEAT Heat exchanged by thermodynamic system that changes temperature of system without changing some variables.

SENSIBLE HEAT FLUX 1. Amount of heat energy that moves by turbulent transport through air across specified cross-sectional area per unit time and goes to heating (cooling) air from Earth's surface to atmosphere not associated with phase changes of water (i.e., component of surface energy budget).

SENSING UNIT Apparatus that senses condition, state or location of something.

SENSITIVE SPECIES Plant and animal species for which population viability is concern, populations are small and widely dispersed or restricted to few localities or species whose numbers are declining so rapidly that official listing may be necessary on threatened and endangered list.

SENSITIVENESS Responsiveness of device to signal or condition.

SENSITIVITY 1. Variation in value of one or more output variables or quantities calculated from output variables due to changes in value of one or more inputs to model; 2. Ability to respond to external stimuli; 3. Proportion of individuals in population that will be correctly identified in binary classification test.

SENSITIVITY ANALYSIS Analysis where relative importance of one or more of variables thought to have influence on phenomenon under consideration is determined.

SENSITIVITY LEVEL Smallest concentration that can be quantified by device or method.

SENSOR Device or organ that detects certain external stimuli and responds in distinctive manner.

SEPARATE COLLECTION Depositing goods separately according to type of material they are made of.

SEPARATE SEWER Wastewater only sewer.

SEPARATE SLUDGE DIGESTION Digesting sludge process in tanks different from those where it was permitted to gather and settle.

SEPARATE SYSTEM Rainwater and sewage drainage system transported by individual conduits.

SEPARATE TURN SIGNAL FACE Signal face that exclusively controls turn movement and that displays signal indications that are applicable only to turn movement.

SEPARATION 1. Propensity for solids to separate from water by gravitational settlement; 2. Act of disuniting two or more things.

SEPARATION DEPOSIT Sand deposit located at upstream end of recirculation zone, where downstream flow becomes separated from channel bank.

SEPARATION EDDY Eddy that forms near ground on windward or leeward side of bluff object or steeply rising hillside.

SEPARATION LINE Clarified liquid and inherent sludge blanket interface.

SEPARATION TIME Component of maximum preemption time where minimum track clearance distance is clear of vehicular traffic prior to arrival of train.

SEPARATOR Object located between two or more things and hence separating them.

SEPARATOR ENTRAINMENT In desalinization apparatus, removal of vapor entrapped droplets from flow.

SEPARATOR TANK Apparatus to separate two immiscible liquids.

SEPTAGE Partially treated waste stored in septic tank.

SEPTIC BED Network of perforated pipes buried in gravel-filled trenches to spread effluent.

SEPTIC SEWAGE Sewage from dissolved oxygen consumed by organic matter.

SEPTIC SLUDGE Undigested sludge.

SEPTIC SYSTEM On-site system designed to treat and dispose of domestic sewage.

SEPTIC TANK Tank used to detain domestic wastes to allow settling of solids prior to distribution to leach field for soil absorption.

SEPTIC TREATMENT SYSTEM Set of components to receive, treat and dispose of sewage at residential or other property.

SEPTIC WASTEWATER Anaerobic wastewater experiencing putrefaction.

SEPTICITY Tendency to putrefaction.

SEQUENCE Ordered list of objects.

SEQUENCE FLASHING LIGHTS Lights that flash in sequence starting with farthest from runway and ending with closest.

SEQUENCE VALVE Pressure-operated valve that opens at adjustable set pressure.

SEQUENT DEPTHS Paired depths upstream and downstream of hydraulic jump.

SEQUESTERING AGENT Substance that removes ions from solution system by forming ring that does not have chemical reactions with ion removed.

SEQUESTRATION To prevent ion in solution from behaving normally by forming coordination compound.

SERIAL BONDS Bonds issued in series by public entity that are payable at different times.

SERIAL DERECHO Derecho that consists of extensive squall line that is oriented such that angle between mean wind direction and squall line axis is small.

SERIAL DISTRIBUTION System where all effluent is discharged to one lateral trench.

SERIES Number of things that follow on one after other or are connected one after other.

SERIES OPERATION Centrifugal pumps connected one after another (i.e., discharge of one pump is connected to suction side of second pump).

SERIOUS ACUTE EFFECTS Human injury or human disease processes that have short latency period for development and likely result in death or severe health issues.

SERIOUS CHRONIC EFFECTS Human injury or human disease processes that have long latency period for development and likely result in death or severe health issues.

SERULE Tiny sere of microhabitat.

SERVICE Utility (e.g., water, electric, sewer, gas) lines that link public services to building.

SERVICE AGE Time since physical unit began working.

SERVICE AREA Area built alongside limited-access road where one can stop to buy fuel, refreshments, etc.

SERVICE ATTRIBUTES placeholders for transportation service that applies to specific service only.

SERVICE CHARGE Extra charge assessed for service.

SERVICE CLAMP Conduit saddle.

SERVICE CONNECTION Water pipe connecting distribution water main to residential, commercial or industrial buildings.

SERVICE CONNECTOR Pipe that carries tap water from public water main to building.

SERVICE ENTRANCE SWITCH Panel where service conductors come into building and fed to fused distribution box.

SERVICE FLOW RATE Rate media can filter at maximum filtration efficiency.

SERVICE HYDROLOGIST Designated expert of hydrology program at WFO.

SERVICE LIFE Forecast life expectancy of products based on real world results.

SERVICE LIMIT STATE Boundary limit between acceptable and unacceptable performance of structure or its component.

SERVICE LINE SAMPLE One-liter sample of water that has been standing for at least 6 hours in service pipeline and is collected according to federal regulations.

SERVICE MEASURE Performance measure regarding aspect of transit service representing passenger's point of view.

SERVICE METER Recording apparatus installed to show quantity used.

SERVICE PIPE Pipeline extending from water main to building served or to consumer's system.

SERVICE RESERVOIR Water storage container that holds water after it has been treated, before it's piped to end users.

SERVICE ROOM Space supplied in building to adapt building service equipment.

SERVICE SPACE Space supplied in building to hide building service facilities.

SERVICE SPILLWAY Spillway designed to provide continuous or frequent regulated or unregulated releases from reservoir, without significant damage to either dam or its appurtenant structures.

SERVICE SWITCH Main electric switch located in service box.

SERVICE SYSTEMS Supply (e.g., ventilating, AC, heating, water, gas and electric) to building.

SERVICE TEST Evaluation of item under actual conditions.

SERVICE TIME Number of years something has been working.

SERVICE TRACK Railroad side tracks that are distant from platforms.

SERVICE VOLUME Maximum number of vehicles or passengers that can be accommodated under given conditions.

SERVICED LOT Surveyed lots with utilities (e.g., water, sanitary, sewer, gas).

SERVICES Systems that provide water, gas, drainage, electricity, etc. to structure.

SERVICING VALVE Valve for shutting off flow of water in pipe connected to fitting.

SERVO CIRCUIT Closed-loop circuit that's attached to position feedback device.

SERVO LOOP Hardware needed to remotely control motion of antenna dish.

SERVO PISTON Piston that transfers another part on command from servovalve.

SERVO VALVE Valve that controls hydraulic fluid sent to actuator.

SERVOCONTROL Control that uses servomechanism.

SET 1. Direction where flow is headed; 2. Punch for setting nails in wood; 3. Small tuber or bulb used instead of seed; 4. Piece placed temporarily upon head of pile when latter cannot otherwise be reached by weight or hammer.

SET POINT Setting or location fixed.

SET-BACK Setting thermometer to lower temperature when building is unoccupied to reduce consumption of heating energy or setting thermometer to higher temperature during unoccupied periods in cooling season.

SET-UP Vertical rise in stillwater level at upstream face of dam caused by wind stresses on water surface.

SETBACK LINE Minimum distance between edge of you property and where structure will be built.

SETTING TEMPERATURE Average temperature where material will set in place.

SETTLEABILTTY TEST Test to find amount of solids in aeration units.

SETTLEABLE SOLIDS Material heavy enough to sink to bottom of wastewater treatment tank.

SETTLED WASTEWATER Wastewater where nearly all settleable solids are removed via sedimentation.

SETTLEMENT Vertical downward movement of structure or its foundation because of subsurface compaction, usually occurring when moisture, added deliberately or by nature, causes reduction in void volumes.

SETTLEMENT SENSORS Monitors (e.g., pneumatic, vibrating-wire) difference in elevation between sensor unit and its reservoir.

SETTLING Gravitation procedure that encourages subsidence and deposition of suspended matter carried by water.

SETTLING CHAMBER 1. Mechanical chamber designed to remove solids via settlement before water is sent to biological filter; 2. Series of screens placed in way of flue gases to slow stream of air, thus helping gravity to pull particles into collection device.

SETTLING POND Open lagoon into which wastewater contaminated with solid pollutants is placed and allowed to stand where solid pollutants suspended in water sink to bottom of lagoon and liquid is allowed to overflow out of enclosure.

SETTLING TANK Holding area for wastewater, where heavier particles sink to bottom for removal and disposal.

SETTLING VELOCITY Falling or settling rate of particle in given medium.

SEVEN-DAY AVERAGE Arithmetic mean of values for samples collected seven successive days.

SEVEN-DAY, 10-YEAR LOW FLOW Lowest stream flow for seven consecutive days that would be expected to occur once in ten years.

SEVERANCE DAMAGES Compensation given to property owner for loss or decrease in property value after government takes via eminent domain.

SEVERE EXPOSURE States of high wind or heavy rain, frost or snowfall.

SEVERE EXPOSURE CONDITION Concrete exposure to freezing and thawing cycles.

SEVERE GEOMAGNETIC STORM Storm for which Ap index was 100 or more.

SEVERE ICING Rate of ice accumulation on aircraft is such that de-icing/anti-icing equipment fails to reduce or control hazard.

SEVERE LOCAL STORM Convective storm that usually covers relatively small geographic area or moves in narrow path and is sufficiently intense to threaten life and/or property.

SEVERE LOCAL STORM WATCH Alert issued by National Weather Service for contiguous U.S. and its adjacent waters of potential for severe thunderstorms or tornadoes.

SEVERE THUNDERSTORM Thunderstorm that produces tornado, winds of at least 58 mph, and/or hail at least 3/4 inches in diameter.

SEVERE THUNDERSTORM WARNING Warning issued by National Weather Service when either severe thunderstorm is indicated by WSR-88D radar or spotter reports thunderstorm producing hail 3/4 inches or larger in diameter and/or winds equal or exceed 58 miles hour.

SEVERE THUNDERSTORM WATCH Issued by National Weather Service when conditions are favorable for development of severe thunderstorms in and close to watch area.

SEVERE WEATHER POTENTIAL STATEMENT Statement issued by National Weather Service designed to alert public and state/local agencies to potential for severe weather up to 24 hours in advance.

SEVERE WEATHER PROBABILITY WSR-88D radar product algorithm displays numerical values proportional to probability that storm will produce severe weather within 30 minutes.

SEVERE WEATHER STATEMENT Statement issued by National Weather Service that provides follow up information on severe weather conditions which have occurred or are currently occurring.

SEVERE WEATHER THREAT INDEX Stability index developed by Air Force which incorporates instability, wind shear and wind speeds.

SEVERE-DUTY BUCKET Skid steer bucket for heavy construction.

SEVERITY INDEX Number system to measure probability of damage.

SEWAGE Waste and wastewater produced by residential and commercial sources and discharged into sewers.

SEWAGE CHARGE Wastewater treatment services charge for collection and/or treatment.

SEWAGE COLLECTION SYSTEM System of lateral sewers primarily installed to receive wastewaters directly.

SEWAGE DISPOSAL Process where sewage is transported through cities and inhabited areas to sewage treatment plants, where it's treated to remove contaminants.

SEWAGE DISTRIBUTOR Mechanism that applies sewage or sewage effluent to filter top.

SEWAGE EJECTOR Apparatus for lifting sewage to higher elevation.

SEWAGE GAS Gas by-product of sewage digestion.

SEWAGE LAGOON 1. Shallow pond where sunlight, bacterial action and oxygen work to purify wastewater; 2. Storage of wastewater or spent nuclear fuel rods; 3. Shallow body of water, often separated from sea by coral reefs or sandbars.

SEWAGE PUMP Pump used to transfer sewage liquids and solids from one place to another.

SEWAGE RATE Charge to users of wastewater collection and treatment system.

SEWAGE SLUDGE Sludge produced at publicly owned sewage treatment works.

SEWAGE TREATMENT Process that removes majority of contaminants from wastewater or sewage.

SEWAGE TREATMENT EFFLUENT WELLS Class V wells that are used by privately or publicly owned treatment works to inject treated or untreated domestic sewage through vertical well or leach field.

SEWAGE TREATMENT PLANT Facility designed to receive wastewater from domestic sources and to remove materials that damage water quality and threaten public health and safety when discharged into receiving streams or bodies of water.

SEWAGE TREATMENT RESIDUE Sludge and solids from coarse screening of domestic sewage.

SEWAGE TREATMENT WORKS Wastewater treatment facilities for collection, treatment and disposal of sewage.

SEWER System of underground pipes, channel or conduit that collect and deliver wastewater and storm-water runoff from source to treatment plant or receiving stream.

SEWER APPURTENANCES Accessories on sewerage system that are necessary for efficient operation.

SEWER CONNECTION Connection of conduits into sewer.

SEWER DISTRICT District created for intent of gathering, treating and disposing of wastewater.

SEWER GAS Mixture of toxic and non-toxic gases produced and collected in sewage systems by decomposition of organic household or industrial wastes.

SEWER LINE Conduit that carries wastewater to downstream elements of treatment system.

SEWER MANHOLE Top opening giving accessibility for inspection of sewer.

SEWER OUTFALL Discharge point of effluent from sewer.

SEWER OUTLET Hydraulic structure where effluent discharges.

SEWER PIPE Conduit used to transport sewage.

SEWER ROD Rod used in sewer to clean out or dislodge any blockage.

SEWER SYSTEM System for transporting sewage from residences and/or buildings to treatment facilities or disposal point.

SEWER SYSTEM EVALUATION SURVEY Systematic evaluation used to investigate reports of excessive inflow or infiltration sources in sewer system.

SEWER TAP Connection point between residence's main sewer line and municipal water system.

SEWER TILE Glazed waterproof clay pipe with bell joints.

SEWER UTILITY Business responsible for collecting, treating and disposing of wastewater.

SEWERAGE Entire system of sewage collection, treatment and disposal.

SFERIC Transient electric or magnetic field generated by any feature of lightning discharge (entire flash).

SHADING COEFFICIENT Amount of Sun's heat transmitted through given window compared with that of standard 1/8 inch thick single pane of glass under same conditions.

SHADOW Area protected by obstacle.

SHAFT 1. Round bar that rotates or provides axis of revolution; 2. Vertical or steeply inclined tunnel.

SHAFT HORSEPOWER Power delivered to propeller or turbine shaft.

SHAFT HOUSE House at top of access shaft to gate chamber that secures access to gates.

SHAFT SPILLWAY Vertical or inclined shaft into which flood water spills and then is conducted through, under or around dam by means of conduit or tunnel.

SHAKING SCREEN Material screen that moves back-and-forth.

SHALE Fine-grained sedimentary rock formed by consolidation of clay, silt or mud.

SHALLOW DRAFT WATERWAY All waterway segments with navigable depths of 12 feet or less.

SHALLOW FOG Fog where visibility at 6 feet above ground level is $5/8^{th}$ statute mile or more and apparent visibility in fog layer is less than $5/8^{th}$ statute mile.

SHALLOW FOUNDATION Building foundation where depth from ground surface to underside of foundation is less than five times width of foundation.

SHALLOW MANHOLE Manhole depth where access shaft to chamber is not needed.

SHALLOW WELL Well less than 10-50 feet deep.

SHALLOW-WATER DEPOSIT Ocean sediments up to 600-ft depths.

SHALLOW-WATER WAVE Water gravity wave with depth of up to one half of wavelength.

SHALLOWS Shallow place or area in body of water.

SHANNON INDEX Numerical expression of evenness of distribution of aquatic organisms.

SHAPE FACTOR Particle shape factor of perfect sphere is 1.0 and can be as low as 0.1 for very irregular shapes.

SHARED LANE Travel lane for both bicycle and motor vehicles.

SHARED ROADWAY Lane for both motor vehicle and bicycle travel.

SHARED TURN SIGNAL FACE Signal face, for controlling both turn movement and adjacent through movement, that always displays same color of circular signal indication that adjacent through signal face or faces display.

SHARED USE PATH Pathway that supports both recreation and transportation (e.g., walking, bicycling, handicapped, skating).

SHARED-LANE CAPACITY Capacity of lane for both motor vehicle and bicycle travel.

SHARED-USE PATH Lane for both motor vehicle and bicycle travel.

SHARP SAND Sand with angular shape and gritty texture.

SHARP-CRESTED WEIR Weir used to obtain open channel discharge by measuring upstream water head.

SHARPS Hypodermic needles, syringes (with or without attached needle), Pasteur pipettes, scalpel blades, blood vials, needles with attached tubing and culture dishes used in animal or human patient care or treatment or in medical, research or industrial laboratories.

SHAW-EDGED ORIFICE Orifice where water contacts just line of edge.

SHEAR 1. Variation in wind speed (speed shear) and/or direction (directional shear) over short distance within atmosphere; 2. Structural break where differential movement has occurred along surface or zone of failure.

SHEAR CONNECTOR Steel connection on top flange of steel composite bridge girders to provide shear transfer.

SHEAR FORCE 1. Unaligned forces pushing one part of body and another part in opposite directions; 2. Shear developed on wetted area of channel and it acts in direction of flow.

SHEAR GATE Slide used in water and sewage plants for filling or emptying tanks and low pressure sludge discharge lines.

SHEAR INTENSITY Dimensionless number taken from Einstein's bed load function that's inverse of Shield's parameter.

SHEAR KEY Structural element designed to resist lateral force created by lateral load from soil and seismic force due to earthquake.

SHEAR SLIDE Landslide produced by shear failure usually along plane of weakness.

SHEAR STRENGTH Maximum resistance of soil or rock to shearing stresses.

SHEAR STRESS 1. Component of stress that causes parallel layers of material to move relative to each other in their own planes; 2. Frictional force per unit of bed area exerted on bed by flowing water.

SHEAR VELOCITY Defined as square root of quantity shear stress divided by fluid density.

SHEAR WALL Vertical lateral-force-resisting element in structure assigned to resist wind or earthquake generated lateral forces.

SHEAR ZONE Area where rock mass has moved along plane of contact which often becomes channel for groundwater.

SHEAVE Wheel having groove in rim for rope to work pulley.

SHEEP'S FOOT Projection from compactor drum that leaves imprint.

SHEEP'S FOOT ROLLER Compacting soil roller with projected feet from its outer face.

SHEEPSFOOT ROLLER Tamping roller having lugs with feet extending at their outer tips.

SHEET Thin, flat layer of solid material.

SHEET ASPHALT Hot mixture of asphalt binder with graded sand and mineral filler.

SHEET DRAINAGE Water flowing in thin layer over surface of area.

SHEET EROSION Relatively even erosion of layer of soil without channel formation.

SHEET FLOW Flow that occurs overland in thin layer.

SHEET ICE Ice formed by freezing of liquid precipitation or freezing of melted solid precipitation.

SHEET PILING Steel strips shaped to interlock with each other when driven into ground.

SHEET WASH Flow of rainwater that covers entire ground surface with thin film and is not concentrated into streams.

SHEETED PIT Excavation with sheeting along sides.

SHEETERS Perpendicular steel planks driven before excavation to hold excavation sides.

SHEETING 1. Lining of planks or boards for protecting embankment; 2. Weathering process where rock breaks into sheets after compressional stress on it is removed.

SHEETING DRIVER Pile driver that fits on plank ends to be driven without splintering.

SHEETING JACK Turnbuckle that places trench bracing.

SHELF CLOUD Low, horizontal wedge-shaped arcus cloud, associated with thunderstorm gust front.

SHELL 1. Upstream and downstream parts of cross-section of embankment dam on each side of core or core wall; 2. Material composed of predominantly (¿75%) coarse-grained sand to gravel-sized whole or broken shell.

SHELL ICE Ice formed on water but water has receded.

SHELL PUMP Pump, used with water, to remove sand.

SHELTERBELT Natural or planned barrier of trees or shrubs to reduce erosion and provide shelter from winds or storms.

SHELTERING, IN-PLACE Protective action that involves taking cover in upper levels of building able to withstand high flood levels.

SHELVES Stream bank features extending nearly horizontally from flood plain to lower limit of persistent woody vegetation.

SHIELD 1. Wide and relatively low-profiled volcano; 2. To protect.

SHIELD'S DETERMINISTIC CURVE Curve of dimensionless tractive force plotted against grain Reynolds number used to help determine critical tractive force.

SHIELD'S PARAMETER Dimensionless number referred to as dimensionless shear stress where beginning of motion of bed material is function of this dimensionless number.

SHIELDING Crash cushion between vehicle and obstacle to reduce impact severity.

SHIFT DIFFERENTIAL Extra compensation for employees who work less desirable shift.

SHIFT PREMIUM Pay that compensates workers who are scheduled for second or third shift.

SHIFTING CONTROL Natural constriction of channel, long reach of channel, stretch of rapids or artificial structure downstream from gaging station that physically affects water-surface elevation and thereby determines stage-discharge relation at gauge (e.g., complete or partial constriction of channel, bedrock outcrop, gravel bar, artificial structure).

SHINGLE BEACH Beach armoured with pebbles or small cobble.

SHOAL Relatively shallow place in stream, lake or sea.

SHOCK LOAD 1. Arrival at water treatment plant of raw water containing unusual amounts of algae, colloidal matter, color, suspended solids, turbidity or other pollutants; 2. Sudden force exerted when object suddenly accelerates or decelerates.

SHOCK WAVE Powerful compression wave produced by movement of body through fluid or gas at velocity greater than local speed of sound.

SHORE Land adjoining non-flowing body of water (i.e., ocean, lake, pond).

SHORE CURRENT Ocean current due to oblique angled waves.

SHORE DRIFT Transportation of sediments (i.e, silt, sand) along shoreline due to currents.

SHORE EROSION Removal of sediment by action of water, wind or ice of land bordering water.

SHORE HEAD Brace for vertical shoring members.

SHORE ICE 1. Ice sheet in form of long border attached to bank or shore; 2. Border ice.

SHORE PROTECTION Protecting and preserving shore at waterline.

SHORELINE Divide between land and body of water.

SHORING Temporary bracing to hold sides of excavation from caving.

SHORING LAYOUT Drawing showing layout of shoring equipment.

SHORT CIRCUIT Unintentional connection of low resistance or impedance in circuit.

SHORT NIPPLE Conduit nipple with small unthreaded portion between threads.

SHORT PERIOD DELAY Type of delay cap with definite but extremely short interval between initiation or passing of current and explosion.

SHORT PIPE Conduit 500 times shorter than its diameter.

SHORT TERM FORECAST Product used to convey information regarding weather or hydrologic events in next few hours.

SHORT TON Unit of mass equal to 2,000 lb.

SHORT WAVE FADE Particular ionospheric solar flare effect under broad category of sudden ionospheric disturbances whereby short-wavelength radio transmissions, VLF, through HF, are absorbed for period of minutes to hours.

SHORT-CIRCUITING When some of water in tanks or basins flows faster than rest that may result in shorter contact, reaction or settling times than calculated or presumed.

SHORT-FUSE WARNING Warning issued by National Weather Service for local weather hazard of relatively short duration.

SHORT-LINE RAILROAD Railroad company that operates over relatively short distance.

SHORT-THROATED FLUMES Flumes that control and produce curvilinear flow (e.g., Parshall flume).

SHORTWAVE Disturbance in mid or upper part of atmosphere which induces upward motion ahead of it.

SHORTWAVE RADIATION Radiant energy emitted by Sun in visible and near-ultraviolet wavelengths (between about 0.1 and 2 micrometers).

SHORTWAVE TROUGH Meteorological wave of low atmospheric pressure in form of trough that has wave length of 600 to 1,500 miles and moves progressively through lower troposphere in same direction as that of prevailing current of air motion.

SHOTCRETE Concrete conveyed through hose and pneumatically projected at high velocity onto surface.

SHOULDER Upstream and downstream parts of cross-section of embankment dam on each side of core or core wall.

SHOULDER BIKEWAY Paved shoulder of road for bicyclists.

SHOULDER BYPASS LANE Paved shoulder for traffic to bypass slowing or stopped vehicles.

SHOULDER DITCH Ditch on slope to reduce erosion or transport runoff.

SHOULDER NIPPLE Conduit nipple with small unthreaded space between threads.

SHOVING Traffic load that causes deflection and bulging of road surface.

SHOWALTER INDEX Stability index used to determine thunderstorm potential.

SHOWER 1. Precipitation characterized by suddenness with which they start and stop, by rapid changes of intensity and usually by rapid changes in appearance of sky; 2. Device for bathing where water falls from tank or action of pump.

SHOWER HEAD Part of shower where water comes out.

SHOWER RECEPTOR Watertight floor unit for shower.

SHOWER ROOM Room with multiple shower heads.

SHOWER TRAY Watertight floor unit for shower.

SHREDDER Machine that tears up objects into smaller pieces.

SHREDDING Act of cutting or tearing into shreds.

SHRINKAGE Loss of bulk of soil when compacted in fill.

SHRINKAGE CRACKING Cracks caused by rapid loss of water from concrete surface before it sets.

SHRINKAGE INDEX Numerical difference between plastic and shrinkage limits.

SHRINKAGE LIMIT Maximum water content at which reduction in water content will not cause decrease in volume of soil mass.

SHRINKAGE RATIO Ratio of given volume change, expressed as percentage of dry volume, to corresponding change in moisture content above shrinkage limit, expressed as percentage of weight of oven-dried soil.

SHRUB Woody plant smaller than tree.

SHRUBLAND Land covered predominantly with shrubs.

SHUT-OFF DEVICE Valve coupled with control device which closes valve when set pressure or flow value is exceeded.

SHUTDOWN 1. Action of stopping operations; 2. Closing of computer, business, event, etc.

SHUTOFF VALVE Valves that turns off water to single fixture.

SHUTTER Swinging flashboard that opens due to water pressure.

SHUTTER WEIR Movable weir consisting of row of large panels hinged at bottom and inclined slightly downstream toward top when weir is closed.

SHUTTLE Back and forth motion.

SHUTTLE LOOP TRANSIT Vehicles that follow fixed paths with few or no switches.

SHUTTLE VALVE Valve that allows fluid to flow through it from one of two sources.

SHY DISTANCE Space between vehicles or pedestrians when they pass.

SI SYSTEM System of measurement with several base units that's modern metric system.

SIAL Rocks rich in silicon and aluminum that form upper layer of Earth's crust.

SIAMESE CONNECTION Pipe fitting that allows two or more hoses to be connected to single riser.

SICK BUILDING SYNDROME Building whose occupants experience acute health and/or comfort effects that appear to be linked to time spent therein, but where no specific illness or cause can be identified.

SIDE CHANNEL SPILLWAY Spillway whose crest is roughly parallel to channel immediately downstream of spillway.

SIDE DITCH Open drain to intercept and collect surface runoff or groundwater.

SIDE LOADING REFUSE TRUCK Truck for residential and commercial garbage collection using one person.

SIDE OUTLET Conduit fitting with opening in side.

SIDE VENT Drain vent with angle of 45° or less.

SIDE WATER DEPTH Depth of water at wall of basin.

SIDE-CHANNEL SPILLWAY Spillway where control weir is placed alongside and approximately parallel to upper discharge channel.

SIDE-DUMP BUCKET Bucket that can dump forward or side.

SIDE-DUMP TRUCK Vehicle with hydraulic rams that tilt dump body onto its side to spill contents.

SIDE-ENTRANCE MANHOLE Manhole with flat entrance shaft to one side of inspection chamber.

SIDECASTING Practice of dumping excavated material alongside line being excavated.

SIDEFILL Aggregate placed on sides of buried conduit.

SIDEHILL CUT Excavation in slope with bank on one side.

SIDELOBE Secondary energy maximum located outside main radar beam.

SIDEPATH Path for bicyclists located alongside roadway.

SIDESLOPE GRADIENT 1. Representative change in elevation in given horizontal distance perpendicular to stream (e.g., valley slope along line perpendicular to stream); 2. Layers of rock.

SIDESTREAM Stream of fluid taken from intermediate point in process (i.e., distillation).

SIDETRACKING Smaller tunnel or well drilled as auxiliary off main.

SIDEWALK Paved footpath.

SIDING Building material that covers and protects sides of house or other building.

SIEMEN Unit of electrical conductance that's equal to reciprocal of ohm.

SIEVE Device with mesh bottom to separate, in granular material, larger particles from smaller ones or to separate solid objects from liquid.

SIEVE ANALYSIS Test procedure where particles move vertically or horizontally through sieve mesh.

SIEVE CORRECTION Correction for deviation of sieve analysis operation.

SIEVE DIAMETER Smallest standard sieve opening size where given particle of sediment will pass.

SIEVE FRACTION Part of sample that passes by way of regular sieve of given size and is retained by some finer sieve of stated size.

SIEVE NUMBER Number of holes present in sieve within one inch length of sieve mesh.

SIEVE SHAKER Vibrated table where sieves are clamped.

SIEVE SIZE Number of openings in one square inch of screen.

SIEVE TEST Analyze used to determine quantity of sand, silt and clay.

SIEVERT Unit of radiation dose equal to 1 joule/kg of absorbing tissue.

SIGHT DISTANCE Length of roadway visible to driver.

SIGHT GLASS Transparent tube/window that permits level of liquid to be visually checked.

SIGHT LINE EASEMENT Easement for maintaining or improving sight distance.

SIGHT-LINE EASEMENT Easement for maintaining or improving sight distance.

SIGMA BANK System of collecting, storing and retrieving information pertaining to variance of measurement on samples of materials or items of construction.

SIGN Visible object bearing short message.

SIGN ASSEMBLY Signs that convey information to road users.

SIGN ILLUMINATION Sign lighted or exposed to artificial lighting.

SIGN LEGEND All words, numbers and symbols intended to convey specific meanings.

SIGNAL Volume or product-level change produced by leak in tank.

SIGNAL BACKPLATE Frame of signal head to highlight visibility.

SIGNAL COORDINATION Vehicles traveling along main road tend to progress or move through downstream signalized intersections without having to stop.

SIGNAL FACE Cluster of signal sections (e.g., red, yellow and green lamps) facing one traffic movement.

SIGNAL HEAD Signal face mounted to control vehicular or pedestrian movement.

SIGNAL HOUSING Structure that protects signal lights and components.

SIGNAL INDICATION Illumination of signal for direction.

SIGNAL LENS Part of signal that directs indication light.

SIGNAL LOUVER Device mounted inside signal visor to restrict visibility of signal indication.

SIGNAL PHASE Combination of vehicle movements and/or pedestrian movements that occur during same period of time.

SIGNAL PREEMPTION System that allows normal operation of traffic lights to be preempted.

SIGNAL SYSTEM Traffic control signals operating in coordination.

SIGNAL TIMING Amount of signal time allocated at signalized intersection.

SIGNAL VISOR Section of signal housing that protects signal viewer from obstruction by external light.

SIGNAL WORDS Words used on pesticide label (e.g., Danger, Warning, Caution) to indicate level of toxicity.

SIGNAL-TO-NOISE RATIO Ratio that measures comprehensibility of data, usually expressed as signal power divided by noise power, usually expressed in decibels.

SIGNIFICANT ADVERSE ENVIRONMENTAL EFFECTS Injury to environment by chemical substance.

SIGNIFICANT ADVERSE REACTIONS Reactions that may indicate substantial impairment to health or environment.

SIGNIFICANT BIOLOGICAL TREATMENT Use of aerobic or anaerobic biological treatment process in treatment works to consistently achieve 30-day average of at least 65% removal of BOD.

SIGNIFICANT CHANGE IN WORK Modification of major work item defined by contract.

SIGNIFICANT DETERIORATION Pollution resulting from new source in previously "clean" areas.

SIGNIFICANT ENCROACHMENT Action within base flood plain that involves significant potential for interruption or termination of transportation facility which is needed for emergency, significant risk or significant adverse impact on natural and beneficial flood plain values.

SIGNIFICANT ENVIRONMENTAL EFFECTS Irreversible damage to biological, commercial or agricultural resources of importance to society.

SIGNIFICANT HAZARD Hazard classification where lives are in jeopardy and appreciable economic loss (rural area with notable agriculture, industry or worksites or outstanding natural resources) would occur as result of failure.

SIGNIFICANT HAZARD TO PUBLIC HEALTH Any level of contaminant that causes or may cause aquifer to exceed any maximum contaminant level established in Primary Drinking Water Standard.

SIGNIFICANT INDUSTRIAL USER Indirect discharger that's focus of control efforts under national pretreatment program.

SIGNIFICANT MUNICIPAL FACILITIES Those publicly owned sewage treatment plants that discharge 1 million gal/day or more and are therefore considered by states to have potential to substantially affect quality of receiving waters.

SIGNIFICANT NON-COMPLIANCE Violations by point source dischargers of sufficient magnitude or duration to be regulatory priority.

SIGNIFICANT POTENTIAL SOURCE OF CONTAMINATION Facility or activity that stores, uses or produces compounds with potential for significant contaminating impact if released into source water of public water supply.

SIGNIFICANT SOURCE OF GROUND-WATER Aquifer saturated with water having less than 10,000 mgl of total dissolved solids; aquifer within 2500 feet of land surface; aquifer that has transmissibility greater than 200 gpd/ft; aquifer capable of continuously yielding at least 10,000 gpd to pumped or flowing well for period of at least year.

SIGNIFICANT VIOLATIONS Violations by point source dischargers of sufficient magnitude or duration to be regulatory priority.

SIGNIFICANT WAVE HEIGHT Average height of highest one-third of waves during 20 minute sampling period.

SIGNIFICANT WEATHER OUTLOOK Narrative statement issued by National Weather Service to provide information regarding potential of significant weather expected during next 1 to 5 days.

SIGNIFICANTLY MORE STRINGENT LIMITATION BOD_5 and suspended solids limitations necessary to meet percent removal requirements of at least 5 mg/l more stringent than otherwise applicable concentration-based limitations.

SILAGE Fermented green forage fodder stored in silo.

SILENCER Attachment to gas pipe that reduces sound it emits.

SILENCING PIPE Perpendicular tube in cistern from ball valve to underside where outlet is below water level.

SILICICLASTIC ROCKS Rocks (e.g., shale, sandstone) that are formed by compaction and cementation of quartz-rich mineral grains.

SILICOSIS Disease of lungs caused by inhalation of crystalline silica dust.

SILL 1. Horizontal member at base of any structure (e.g., line of dam, spillway, weir); 2. Tabular sheet of igneous rock intruded between and parallel with existing strata; 3. Submerged structure across river to control upstream water levels.

SILT 1. Sedimentary materials, intermediate between clay and sand that's non-plastic or very slightly plastic and that exhibits little or no strength when air dry, passes No. 200 United States Standard sieve (i.e., between 1/256 mm and 1/16 mm in diameter).

SILT BASIN Structure where solid particles floating in water can settle and be evacuated.

SILT LOAM Sandy substance having reasonable quantity of clay and sand.

SILT TRAP Settling hole that prevents waterborne soil from entering system.

SILTATION Deposition or accumulation of silt (i.e., small-grained material) in body of water.

SILTING Filling with soil or mud deposited by water.

SILTSTONE Fine-grained sedimentary rock composed mainly of silt-sized particles, having texture and composition of shale but lacking its fine lamination.

SILVA Forest trees of particular area.

SILVICULTURE Management of forest land for timber (i.e., cultivation of forest trees).

SIMA Lower layer of Earth's outer crust that underlies sial and is rich in silica, iron and magnesium.

SIMAZINE Herbicide and algicide.

SIMILARITY Correspondence between behavior of model and its prototype.

SIMILITUDE Correspondence between behavior of model and its prototype.

SIMPLE SURGE TANK Vertical standpipe connected to pipeline.

SIMPLE WEAVING SEGMENT Segment formed by single merge point followed by single diverge point.

SIMPLEX TELETYPE OVER RADIO Means of transmitting text broadcasts over radio.

SIMULATE To express physical system in mathematical terms.

SIMULATION One complete execution of modeling computer program, including input and output.

SIMULATION MODEL Mathematical model where detailed values of various parameters are computed, usually with respect to both space and time.

SIMULID Group of two-winged flying insects who live their larval stage underwater and emerge to fly about as adults.

SIMULTANEOUS PREEMPTION Notification of approaching train is forwarded to highway traffic control signal controller unit and railroad active warning devices at same time.

SINGLE CELL THUNDERSTORM Thunderstorm that develops in weak vertical wind shear environments.

SINGLE CENTRIFUGAL PUMP Pump where suction inlet accepts water at one side of impeller.

SINGLE ENTRY System of bookkeeping where each transaction is entered in one account.

SINGLE SUCTION PUMP Pump where water enters from one side.

SINGLE TRACK Having but one set of tracks for travel in one direction at time.

SINGLE-ACTING PUMP Pump where one side of piston functions.

SINGLE-ARCH DAM Dam that crosses between sidewalls within arc.

SINGLE-AXLE LOAD Equivalent load (e.g., 18,000 lb) used in pavement design.

SINGLE-BAG COMPACTOR Semiautomatic garbage compactor where garbage is smashed against front-opening door right into predetermined volume.

SINGLE-BREATH CANISTER Small one-liter canister designed to capture single breath.

SINGLE-CELL PROTEIN Edible unicellular microorganisms.

SINGLE-CHAMBER INCINERATOR Furnace charged via door in top area of chamber.

SINGLE-EFFECT EVAPORATOR Apparatus where steam condenses on tube surface and heat is transferred to solution.

SINGLE-HUB PIPE Conduit with bell at one end and spigot at other.

SINGLE-OCCUPANT VEHICLE Vehicle designed to accommodate one person (i.e., driver).

SINGLE-PASS COOLING To remove heat load, water is circulated once through piece of equipment and then disposed down drain.

SINGLE-PASS SOIL STABILIZER Machine with rapidly rotating toothed milling wheels in contact with soil.

SINGLE-PIECE CONVEYOR Single conveyor that carries and deposits material.

SINGLE-POINT URBAN INTERCHANGE Interchange between highway and street that allows traffic to proceed from either direction to other, governed by single set of traffic signals.

SINGLE-POLE SWITCH Switch to control electric circuit from only one location.

SINGLE-STACK DRAINAGE SYSTEM Sanitary conduit stack that conducts wastewater and serves as vent.

SINGLE-STAGE DIGESTION Sludge digestion reactions occur within single tank.

SINGLE-STAGE PUMP Pump with one dual suction impeller located on both sides.

SINGLE-UNIT RAIL CAR Rail car that operates alone with control cab at one or both ends.

SINK 1. Depression in land surface; 2. Place in environment where compound or material collects; 3. Basin used for holding water for washing; 4. Drain for carrying off wastewater; 5. Sinkhole; 6. Absorbs resources or energy.

SINK BIB Tap for kitchen sink.

SINKHOLE 1. Depression in Earth's surface caused by dissolving of underlying limestone, salt or gypsum; 2. Steep-sided depression formed when removal of subsurface embankment or foundation material causes overlying material to collapse into resulting void.

SINKING Controlling oil spills by using agent to trap oil and sink it to bottom of body of water where agent and oil are biodegraded.

SINKING PUMP Pump designed for use in well or to keep shaft dry.

SINUOSITY Ratio of channel length between two points on channel to straight-line distance between same two points (i.e., measure of river's meandering).

SIP CALL EPA action requiring state to resubmit all or part of its State Implementation Plan to demonstrate attainment of require national ambient air quality standards within statutory deadline.

SIPHON 1. Tube or pipe where water flows over high point by hydrostatic pressure exerted due to force of gravity moves liquid from one reservoir to another; 2. Tubelike organ found in animals or elongated cell found in plants.

SIPHON BREAK Little groove that prevents capillary action between adjoining surfaces.

SIPHON SPILLWAY Spillway with one or more siphons built at crest level.

SIPHON TUBE Relatively short, light-weight, curved tube used to convey water over ditch banks to irrigate furrows or borders.

SIPHONAGE Partial vacuum created by flow of liquids in pipes.

SIREN 1. Device that makes piercingly loud sound as alarm or signal; 2. Member of genus of aquatic salamanders of family Sirenidae.

SITE 1. Any location of past human activity; 2. Area or place within jurisdiction; 3. To situate or where building takes place.

SITE ACCOMMODATION Space provided on site during construction.

SITE ANALYSIS Design processes dedicated to study of geographical, historical, legal, climatic and infrastructural context of specific site.

SITE ASSEMBLY Put together on-site with parts created in another location.

SITE ASSESSMENT PROGRAM Means of evaluating hazardous waste sites through preliminary assessments and site inspections to develop Hazard Ranking System score.

SITE BOUNDARY Border of building site.

SITE CHARACTERIZATION 1. General term applied to investigation activities at specific location that examines natural phenomena and human-induced conditions, both in laboratory and in field, important to resolution of environmental, safety and water resources issues to establish conditions and ranges of those parameters of particular site relevant to program.

SITE CLEANUP Removal of unwanted building materials from site.

SITE CLEARING Process of clearing away vegetation and surface soil of construction site.

SITE CLOSURE Point/time, as determined by Director, at which owner or operator of site has completed their post-injection site care responsibilities.

SITE CONDITIONS Physical conditions on, under, over, at or in vicinity of site.

SITE CONSTRAINTS Restrictions, part of contract files, throughout building period.

SITE DESIGN Preparation for land development.

SITE DEVELOPMENT Improvements needed for site before construction can begin.

SITE DEVELOPMENT PLAN Comprehensive plan of proposed improvements to given lot.

SITE DIARY Record kept of what occurs on site everyday.

SITE DRAINAGE Movement of water to another area away from site.

SITE FENCE Temporary fence surrounding building site.

SITE INSPECTION Collection of information from site to determine extent and severity of hazards posed by site.

SITE INVESTIGATION Process of collecting information, assessment of data and reporting potential unknown hazards beneath site.

SITE MANAGER Representative that oversees daily operations, ensures work is done safely, on time and within budget to quality standards.

SITE MEETING Meetings between stakeholders on project to facilitate communication.

SITE OFFICE Temporary office of individual accountable for site.

SITE PREPARATION Demolition, clearing, blasting, test drilling, landfill, levelling, earth-moving, excavating, land drainage and other land preparation.

SITE REHABILITATION Restoring site to stable condition that will no longer require human intervention.

SITE SAFETY PLAN Crucial element in all removal actions, it includes information on equipment being used, precautions to be taken and steps to take in event of on-site emergency.

SITE SECURITY Measures and staff required to secure site.

SITE SERVICES Temporary public service connections of works proposed for site.

SITE SIGN Sign observable from outside construction site fence company information in charge of job.

SITE SPECIFIC Data taken in certain place.

SITE SPECIFIC HYDROLOGIC PREDICTION SYSTEM WFO hydrologic forecast model for small rivers and streams that uses RFC soil moisture state variables, stage and precipitation data.

SITE SUPERVISION Person responsible for construction management and oversight.

SITE-SPECIFIC Term used to convey fact that report is for individual location as opposed to general area.

SITE-SPECIFIC AQUATIC LIFE CRITERION Water quality criterion for aquatic life that has been derived to be specifically appropriate to water quality characteristics and/or species composition at particular location.

SITEWORK Process of getting construction site ready for actual construction.

SITING Process of choosing location for facility.

SIZE 1. Dimensions or magnitude of thing; 2. Rated capacity in tons per hour of storage bin.

SKATOLE Mildly toxic white crystalline organic compound of indole family occurring naturally in feces and coal tar.

SKETCH-PLANNING TOOLS Tools that produce general order-of-magnitude estimates of transportation and land use demand and impacts.

SKEW ANGLE Angle between normal/perpendicular to alignment/centerline of bridge and centerline of pier.

SKEWED On horizontal angle or in oblique course or direction.

SKEWNESS Numerical measure of lack of symmetry of asymmetrical frequency distribution.

SKID NUMBER Number used to report skid resistance (i.e., friction measurement) of roadway.

SKID RESISTANCE Force developed when tire, that cannot rotate, slides along surface.

SKIM 1. To clear liquid from scum or substance floating or lying on it; 2. To scrape off.

SKIMMED TREE Minimum travel time, cost or distance between each pair of zones obtained from tree records.

SKIMMER GATE Gate at spillway crest whose prime purpose is to control release of debris and logs with limited amount of water.

SKIMMING 1. Using machine to remove oil or scum from surface of water; 2. Diversion of water from stream or conduit by shallow overflow used to avoid diversion of sand, silt or other debris carried as bottom load.

SKIMMING DETRITUS TANK Tank made for settling of solids and skimming floating materials.

SKIMMING TANK Tank where floating matter rise and remain on surface of wastewater until removed.

SKIMMING WEIR Weir with adjustable crest depth that limits water overflow depth.

SKIMMINGS Removing floating materials from surface.

SKIN OF WATER Thin layer of water due to cohesion.

SKIP Non-digging bucket or tray that hoists material.

SKIP-GRADED AGGREGATE All particles are same size.

SKIP-STOP Train, bus or elevator that does not make all designated stops along its route.

SKIVING To dig in thin layers.

SKY CONDITION Used in forecast to describes predominant/average sky condition based upon octants (eighths) of sky covered by opaque (not transparent) clouds defined as Sky Condition - Cloud Coverage: Clear / Sunny - 0/8, Mostly Clear / Mostly Sunny - 1/8 to 2/8, Partly Cloudy / Partly Sunny - 3/8 to 4/8, Mostly Cloudy / Considerable Cloudiness - 5/8 to 7/8, Cloudy - 8/8, Fair (mainly for night) - Less than 4/10 opaque clouds (no precipitation, no extremes of visibility/temperature/wind).

SKYWARN Nationwide network of volunteer weather spotters who report to and are trained by National Weather Service.

SLAB Deck or floor of concrete bridge.

SLAB AND BUTTRESS DAM Buttress dam with buttresses which support flat slab of reinforced concrete which forms upstream face.

SLAB BRIDGE Short-span bridge with reinforced-concrete slab resting on abutments.

SLACK WATER 1. State of tide when turning; 2. Area of water with no current.

SLAG 1. Waste material from coal mine; 2. Scum that forms on surface of molten metal; 3. Scoria associated with volcano.

SLAKE Combine quicklime with water to produce calcium hydroxide.

SLAKED LIME Calcium hydroxide, soft white powder, produced by adding water to quicklime.

SLAKING Process of breaking up or sloughing when indurated soil is immersed in water.

SLANT 1. Sloped surface or line; 2. Sewer conduit joining residence to drain; 3. Heading driven diagonally between dip and strike of coal seam; 4. Sloping surface in culture medium.

SLAT BUCKET Primitive bucket made with vertical wood slats, metal strapping and side handles.

SLATE Fine-grained metamorphic rock formed by "baking" and recrystallizing shale or mudstone and which splits easily along flat, parallel planes.

SLAVE UNIT Machine controlled by another machine.

SLAVE VALVE Valve that uses pressure from pilot valve to move spool position.

SLEET Pellets of ice composed of frozen or mostly frozen raindrops or refrozen partially melted snowflakes.

SLEET WARNING Issued when accumulation of sleet in excess of 1/2 inch is expected.

SLEEVE 1. Covering or lining to protect piece of machinery; 2. Narrow channel of water.

SLICK Covering of liquid (e.g., oil); 2. Tool used to make something smooth or even.

SLICKENS Pulverised matter from quartz mill.

SLICKENSIDE Smooth, striated rock surface caused by friction of one mass sliding over another.

SLIDE 1. Falling of large amounts of rubble, earth and stones down slope of hill or mountain; 2. Small dislocation in beds of rock along line of fissure.

SLIDE COUPLING Slip joint.

SLIDE GATE Gate that can be opened or closed that slides on its bearings in edge guide slots.

SLIDING BEARING Bearing with just bearing surface and no rolling elements.

SLIDING GATE Valve that slides right on its bearings.

SLIDING-PANEL WEIR Frame weir with wooden panels that slide between grooved uprights.

SLIGHT CHANCE In probability of precipitation statements, usually equivalent to 20% chance.

SLIGHT RISK OF SEVERE THUNDERSTORMS Severe thunderstorms are expected to affect between 2 and 5% of area.

SLIGHTLY FRACTURED Spacing 3 ft to 6 ft.

SLIGHTLY WEATHERED Superficial discoloration, alteration and/or discoloration along discontinuities with less than 10% of rock volume is altered.

SLIME 1. Soft, moist earth or clay, having adhesive quality; 2. Any mucilaginous substance.

SLIMICIDES Any pesticide designed to kill organisms that produce slime.

SLING PSYCHROMETER Instrument used to measure water vapor content of atmosphere where wet and dry bulb thermometers are mounted on frame connected to handle at one end by means of bearing or length of chain.

SLIP 1. Thin, slippery mix of clay and water; 2. Difference between ideal and actual signal; 3. Component movements; 4. Smooth fracture.

SLIP BASE Structural element enables element to slip at base and fall should collision occur.

SLIP COUPLING Coupling that slips on to complete conduit connection.

SLIP JOINT Joint that allows extension and compression in linear structure (e.g., conduit).

SLIP NUT Threaded ring that can be tightened and loosened by hand.

SLIP PLANE Plane that offers minimum distortion or maximum atomic density.

SLIP RAMP Local road running parallel to higher-speed, limited-access road.

SLIP-ON FLANGE Flange that slips onto conduit and is soldered into place.

SLIPLINING Installing smaller pipe into larger, faulty pipe.

SLIPPAGE CRACKING Crescent shaped cracks having two ends pointed away from direction of traffic.

SLOP OIL Reclaimed petroleum waste mixture of oil, chemicals and water.

SLOP SINK Sink designed for dirty tasks.

SLOPE Ratio of vertical and horizontal distances between two points on line.

SLOPE ANGLE Inverse-tangent of rise divided by run.

SLOPE AREA DISCHARGE MEASUREMENT Uses channel surveys, estimates of water-surface slopes and estimates of channel roughness to calculate peak discharge.

SLOPE DISCHARGE CURVE Discharge graph in stream as function of slope of water surface.

SLOPE DISTANCE Length between two points of elevation.

SLOPE EASEMENT Permanent right to construct and maintain slope on land.

SLOPE GAUGE Tool for showing grade of terrain or water surface.

SLOPE OF FRONT Weather front gradient.

SLOPE PROTECTION Protection of slope against wave action or erosion.

SLOPE RATIO Comparison of rise of line to run of line.

SLOPE VARIANCE Variation of slope of pavement from mean slope value.

SLOPE-AREA METHOD Method of using slope of water surface in uniform reach of channel and average cross-sectional area of that reach to give rate of discharge.

SLOSH MODEL Computer model run by National Hurricane Center to estimate storm surge heights resulting from historical, hypothetical or predicted hurricanes by taking into account pressure, size, forward speed, track and winds.

SLOT DRAIN Hidden linear drain used to collect groundwater.

SLOUGH 1. Movement of soil mass downward along slope because of slope angle too great to support soil, wetness reducing internal friction among particles or seismic activity; 2. Small marshy tract lying in swale or other local shallow, undrained depression; 3. Sluggish creek or channel in wetland; 4. Release of contaminant from upstream side of filter element to downstream side of filter enclosure or into effluent stream; 5. Skin shed by snake or other reptile.

SLOUGHING Washing off media of trickling filters.

SLOW BEND Curve with large radius.

SLOW ORDER Rail line local speed restriction that's set below normal track speed limit.

SLOW SAND FILTER Filters used in water purification for treating raw water (i.e., removing turbidity and pathogenic organisms through various biological, physical and chemical processes) to produce potable water.

SLOW SAND FILTRATION Passage of raw water through bed of sand at low velocity, resulting in substantial removal of chemical and biological contaminants.

SLUDGE 1. Solids separated from suspension in liquid; 2. Residual semi-solid material left from industrial, water treatment or wastewater treatment processes; 3. Mass of small pieces of ice on surface of water; 4. Sediment of accumulated minerals in steam boiler; 5. Slop or drip slowly.

SLUDGE AGE Amount of time (i.e., days) that solids or bacteria are under aeration.

SLUDGE BLANKET Collected mat of sludge hydrodynamically suspended in wastewater.

SLUDGE BOIL Upwelling of sludge and water due to gas release in decomposing sludge.

SLUDGE BULKING Expanding sludge due to failure to separate out in sedimentation tanks.

SLUDGE CAKE Accumulation of sludge hardened.

SLUDGE CIRCULATION Overturning sludge digestion tank contents.

SLUDGE COLLECTOR Apparatus that transfers settling tank sludge to drawoff sump.

SLUDGE COMPARTMENT Segment of settling tank where collected sludge is digested.

SLUDGE CONCENTRATION Sequence to dewater and thicken sludge.

SLUDGE CONDITIONING Process where sludge solids are treated in preparation for dewatering processes.

SLUDGE DENSITY INDEX Describes settling characteristics of sludge in aeration tank, reciprocal of sludge volume index multiplied by 100.

SLUDGE DEWATERING Process to separate sludge into liquid and solid components.

SLUDGE DIGESTER Tank where complex organic substances like sewage sludges are biologically dredged.

SLUDGE DIGESTION Processes where microorganisms, in closed tanks, break down biodegradable material in absence of oxygen.

SLUDGE DIGESTION TANK Tank where active sludge digestion in wastewater is encouraged.

SLUDGE DRYER Apparatus to dehydrate sludge.

SLUDGE DRYING BED Shallow bed where sludge drys by drainage and evaporation.

SLUDGE EXCESS Excess sludge generated in activated sludge treatment process.

SLUDGE FILTER Apparatus where water is removed from wet sludge by pressure.

SLUDGE FOAMING Foam due to surplus amounts of gas in sludge blanket.

SLUDGE GAS Gas, due to decomposition of organics, collected from sewage sludge digesting tanks.

SLUDGE GAS HOLDER Tank used to store sludge digestion tank gas.

SLUDGE LAGOON Basin where sludge from sedimentation basins is pumped for drying.

SLUDGE MOISTURE CONTENT Net weight of water within sludge.

SLUDGE PRESS Apparatus to dewater sludge using pressure.

SLUDGE PUMP Centrifugal pump that increases pressure to move viscous substances.

SLUDGE REAERATION Continuous aeration of sludge after initial aeration in activated sludge process.

SLUDGE SEEDING Mixing microorganisms with sludge to rapidly begin biological decomposition.

SLUDGE SHREDDER Apparatus to reduce air dried sludge.

SLUDGE STRIPPER Apparatus that removes sludge from drying beds.

SLUDGE THICKENER Tank designed to continuously produce thick sludge.

SLUDGE THICKENING Raising solids content of sludge.

SLUDGE TREATMENT Processes to manage and dispose of sewage sludge.

SLUDGE UTILIZATION Alternative use of treated wastewater sludge.

SLUDGE VOLUME INDEX Computation used to describe settling characteristics of sludge in aeration tank.

SLUDGE VOLUME RATIO Volume of sludge blanket divided by daily volume of sludge pumped from thickener.

SLUFF 1. Sliding of overlying material; 2. Release of contaminant from upstream side of filter element to downstream side of filter enclosure or into effluent stream.

SLUFFING Washing off media of trickling filters.

SLUG 1. Imperial unit of mass; 2. Unit of mass which will undergo acceleration of 1 ft/s^2 when force of 1 pound is applied to it; 3. Terrestrial pulmonate gastropod mollusks; 4. Discrete mass of material that moves as unit through another material.

SLUG DISCHARGE Any discharge of non-routine, episodic nature, including but not limited to accidental spill or non-customary batch discharge, which has reasonable potential to cause interference or pass through or in any other way violate ordinances, local limits or permit conditions.

SLUGGING Pulsating and intermittent flow of shotcrete material due to improper use of delivery equipment and materials.

SLUGGISH STREAM Stream that gains deposition of silt and sediment due to low flow velocity.

SLUGS Occasional volumes discharged into stream.

SLUICE Opening for releasing water from below static head elevation.

SLUICE FLOW Flow in water channel controlled at its head by sluice gate.

SLUICE GATE Gate that can be opened or closed, to allow or prevent passage of water, by sliding in supporting guides.

SLUICEWAY Opening in diversion dam used to discharge heavy floating debris safely past dam.

SLUICING 1. Method of compacting soil where soil is washed into placed with high velocity stream of water; 2. Transferring something with running water.

SLUMP 1. Measure of fluidity of freshly mixed concrete; 2. Minor landslide.

SLURRY 1. Watery mixture of insoluble matter that is pumped beneath dam to form impervious barrier (i.e., grout) or pollution control techniques.

SLURRY SEAL Mixture of water, emulsified asphalt, well-graded fine aggregate and mineral filler.

SLURRY TRENCH Narrow excavation whose sides are supported by mud slurry filling excavation.

SLURRY TUNNEL-BORING MACHINE Underground excavator that removes excavated material hydraulically.

SLURRY WALL Deep, water-blocking wall structures surrounding tunnels, open cuts and used to lay foundations.

SLUSH 1. Half-melted snow or ice; 2. Mixture of white lead and lime, used as paint to prevent oxidation.

SMALL AIRCRAFT Aircraft with maximum gross takeoff weight of 12,500 lb or less.

SMALL CONCENTRATED ANIMAL FEEDING OPERATION Animal feeding operation designated as concentrated animal feeding operation and is not medium concentrated animal feeding operation.

SMALL CONSTRUCTION ACTIVITY Clearing, grading and excavating resulting in land disturbance that will disturb equal to or more than one acre and fewer than five acres of total land area but is part of larger common plan of development or sale that will ultimately disturb equal to or fewer than five acres.

SMALL CRAFT ADVISORY Advisory issued by coastal and Great Lakes Weather Forecast Offices for areas included in

Coastal Waters Forecast or Nearshore Marine Forecast products where any vessel that may be adversely affected.

SMALL CRAFT ADVISORY FOR HAZARDOUS SEAS Advisory for wind speeds lower than small craft advisory criteria, yet waves or seas are potentially hazardous due to wave height, wave period, steepness or swell direction.

SMALL CRAFT ADVISORY FOR ROUGH BAR Advisory for specialized areas near harbor or river entrances known as bars.

SMALL CRAFT ADVISORY FOR WINDS Advisory for wave heights lower than small craft advisory criteria, yet wind speeds are potentially hazardous.

SMALL CRAFT SHOULD EXERCISE CAUTION Precautionary statement issued to alert mariners with small, weather sensitive boats.

SMALL HAIL Snow pellets or graupel.

SMALL POWER PRODUCER Facility or small power producer that generates electricity using waste, renewable (e.g., water, wind, solar) or geothermal energy as primary energy source.

SMALL QUANTITY GENERATOR Persons or enterprises that produce 220-2200 lbs/month of hazardous waste.

SMALL STREAM FLOODING Flooding of small creeks, streams or runs.

SMELTER Facility that melts or fuses ore, often with accompanying chemical change, to separate its metal content.

SMOG Combination of smoke and fog.

SMOKE Particles suspended in air after incomplete combustion.

SMOKE ALARM Device that sounds alarm when it detects smoke.

SMOKE CONTROL AREA Place where smoke is regulated.

SMOKE DETECTOR Device that sounds alarm when it detects smoke.

SMOKE DISPERSAL Ability of atmosphere to ventilate smoke.

SMOKE MANAGEMENT Use of meteorology, fuel moisture, fuel loading, fire suppression and burn techniques to keep smoke impacts from prescribed fires within acceptable limits.

SMOKE ROCKET Device for testing tightness of drains by generating smoke within them.

SMOKE SHADE Color scale using amount of unburnt carbon in gases.

SMOKE STACK Chimney for discharging smoke.

SMOKE TEST Test for leaks involving blowing smoke into conduit.

SMOLTS Adolescent salmon 3 to 7 inches long.

SMOOTH RESPONSE SPECTRUM Response spectrum devoid of sharp peaks and valleys that specifies amplitude of spectral acceleration, velocity, and/or displacement to be used in analyses of structure.

SMOOTHED SUNSPOT NUMBER Average of 13 monthly RI numbers, centered on month of concern.

SMOTHER To extinguish by excluding oxygen.

SNAKE 1. Spring steel wire tool for unclogging plumbing or cable pulling; 2. Legless reptile of sub-order Serpentes with long, thin body and fork-shaped tongue.

SNOW Precipitation that consists of frozen flakes formed when water vapor accumulates on ice crystals, going directly to ice phase.

SNOW ACCUMULATION AND ABLATION MODEL Model which simulates snow pack accumulation, heat exchange at air-snow interface, areal extent of snow cover, heat storage within snow pack, liquid water retention and transmission and heat exchange at ground-snow interface.

SNOW ADVISORY Issued by National Weather Service when low pressure system produces snow that may cause significant inconveniences, but do not meet warning criteria and if caution is not exercised could lead to life threatening situations.

SNOW CORE Sample of either freshly fallen snow or combined old and new snow on ground obtained by pushing cylinder down through snow layer and extracting it.

SNOW CORNICE Mass of snow or ice projecting over mountain ridge.

SNOW COURSE Line or series of connecting lines along which snow samples are taken at regularly spaced points.

SNOW COVER Depth of snow that has fallen.

SNOW DENSITY Mass of snow per unit volume that is equal to water content of snow divided by its depth.

SNOW DEPTH Combined total depth of both old and new snow on ground.

SNOW FENCE Temporary barrier that forces windblown snow to accumulate in desired location.

SNOW FLAKE Single crystal of snow.

SNOW FLURRIES Intermittent light snowfall of short duration with no measurable accumulation.

SNOW GAUGE Apparatus used to gather and measure amount of solid precipitation over period of time.

SNOW GRAINS Precipitation consisting of white, opaque ice particles usually less than 1 mm in diameter.

SNOW LINE Altitude above where ground is covered with snow.

SNOW PACK Combined layers of snow and ice on ground at any one time.

SNOW PELLETS Precipitation, usually of brief duration, consisting of crisp, white, opaque ice particles, round or conical in shape and about 2 to 5 mm in diameter.

SNOW PILLOW 1. Instrument used to measure snow water equivalents; 2. Window of snow deposited in immediate lee of snow fence or ridge.

SNOW PLOW Apparatus mounted on vehicle used to transfer or remove snow.

SNOW RIPENING Period when melting snow spaces fill with water.

SNOW SAMPLE Core removed from snow accretion.

SNOW SAMPLER Apparatus that measures snow water equivalent.

SNOW SCALE Accumulated snow measuring rod.

SNOW SHED Structure that provides avalanche protection for roads and railroad tracks.

SNOW SHOE Footwear that distributes weight over large area to walk over snow.

SNOW SHOWER Short duration of moderate snowfall where some accumulation is possible.

SNOW SQUALL Intense, but limited duration, period of moderate to heavy snowfall, accompanied by strong, gusty surface winds and possibly lightning (generally moderate to heavy snow showers) where snow accumulation may be significant.

SNOW STAKE $1 - 3/4\ in^2$, semi-permanent stake, marked in inch increments to measure snow depth.

SNOW STICK Portable rod used to measure snow depth.

SNOW STORAGE Snow removed and stored for later availability.

SNOW STORM Weather involving blowing winds and heavy snowfall amount.

SNOW SURVEY Process of determining depth, water content and density of fallen snow.

SNOW TELEMETRY Automated network of snowpack data collection sites.

SNOW TRAP Clearing to gather snow.

SNOW WATER EQUIVALENT Amount of water content in snowpack or snowfall (i.e., obtained from melting accumulated snow).

SNOWBANK Heap of snow.

SNOWBOARD Flat, solid, white material, approximately 2 ft^2, that is laid on ground or snow surface by weather observers to obtain more accurate measurements of snowfall and water content.

SNOWCOVER Combined layers of snow and ice on ground at any one time.

SNOWFALL Amount of snow that falls on one occasion.

SNOWFIELD Large permanent expanse of snow on mountain or at head of glacier.

SNOWFLAKE Agglomeration of snow crystals falling as unit.

SNOWLINE General altitude that continuous snow cover of high mountains retreats in summer, chiefly controlled by depth of winter snowfall and by temperature of summer.

SNOWLINE, TEMPORARY Line sometimes drawn on weather map during winter showing southern limit of snow cover.

SNOWMELT FLOODING Flooding caused primarily by melting of snow.

SNOWPACK Total snow and ice on ground, including new snow, previous snow and ice which has not melted.

SNOWY Covered with snow.

SOAP 1. Substance able to mix with both oil and water, used for cleaning; 2. Metallic salt derived from fatty acid.

SOCIAL VALUE Concept that existence of wilderness provides condition that could allow individual to achieve control over stressful conditions, thus contributing to psychological health of many off-site users.

SOCKET 1. Hollow tool for grasping and lifting tools dropped in well-boring; 2. Opening where plug or other connecting part is designed to fit.

SOD Turf grown and cut specifically for establishment of lawns.

SODA ASH Industrial-grade sodium carbonate.

SODIC SOIL Contains sufficient exchangeable sodium to interfere with growth of most crop plants.

SODICITY 1. Exchangeable-sodium content of soil; 2. High sodium content.

SODIUM Soft, waxy, silvery reactive metal never found unbound in nature with atomic number of 11.

SODIUM ADSORPTION RATIO Expression of relative activity of sodium ions in exchange reactions within soil and is index of sodium or alkali hazard to soil.

SODIUM ALUMINATE Coagulating compound and softening agent used in water and wastewater treatment.

SODIUM ARSENITE Sodium salt of arsenous acid used to determine residual chlorine.

SODIUM CARBONATE 1. Salt formally derived from sodium hydroxide and carbonic acid; 2. Soda ash.

SODIUM HAZARD IN WATER Index that can be used to evaluate suitability of water for irrigating crops.

SOFFIT 1. Top point of inside open section of pipe or box conduit; 2. Visible underside of arch, balcony, beam, cornice, staircase, vault or any other architectural element.

SOFT Can be scratched with fingernail.

SOFT CONVERSION Changing measurement from U.S. customary units to equivalent metric units.

SOFT DETERGENT Biodegradable agent used for cleaning.

SOFT DETERGENTS Cleaning agents that break down in nature.

SOFT GROUND Earth unable to carry imposed load.

SOFT WATER Any water that does not contain significant amount of dissolved minerals (e.g., salts of calcium or magnesium) making it easier to lather with soap.

SOFTWARE Encoded computer instructions.

SOFTWOOD Wood from any conifer.

SOIL Mixture of sand and organic material used to support plant growth.

SOIL ABSORPTION CAPACITY Soil ability to absorb effluent flow.

SOIL ADHESION Sticking of soil to surfaces of other materials.

SOIL ADSORPTION FIELD Sub-surface area containing trench or bed with clean stones and system of piping where treated sewage may seep into surrounding soil for further treatment and disposal.

SOIL ANALYSIS Analysis of soil sample to determine nutrient content, composition, acidity and/or pH level.

SOIL AND WATER CONSERVATION PRACTICES Control measures consisting of managerial, vegetative and structural practices to reduce loss of soil and water.

SOIL BORING Shallow cores taken from sediment used to survey soil.

SOIL BRANCH Sewer conduit that receives toilet discharge.

SOIL BULK DENSITY Mass of dry soil per unit bulk soil.

SOIL CEMENT Tightly compacted mixture of water, Portland cement and pulverized soil, usually processed in tumble and mixed to specific consistency that, as cement hydrates, forms hard, durable, low-cost paving material usually placed in lifts and rolled to compact on upstream face of dam.

SOIL CLASSIFICATION Systematic arrangement of soils into classes of one or more categories or levels of classification for specific objective where broad groupings are made on basis of general characteristics and subdivisions are made on basis of more detailed differences in specific properties.

SOIL CLASSIFICATION SYSTEMS Systematic categorization of soils and rock.

SOIL COHESION Soil particles attraction via molecular forces and moisture.

SOIL COMPACTION Process where stress causes soil densification as air is displaced from voids.

SOIL COMPACTION TEST Evaluation to understand compaction characteristics of different soils with change in moisture content.

SOIL CONDITIONER Organic material like humus or compost that helps soil absorb water, build bacterial community and take up mineral nutrients.

SOIL CONSERVATION Protection of soil against physical loss by erosion and chemical deterioration by application of management and land-use methods that safeguard soil against all natural and human-induced factors.

SOIL CONSERVATION SERVICE Former name of branch of United States Department of Agriculture, renamed Natural Resources Conservation Service that has responsibilities in soil and water conservation and flood prevention.

SOIL CONSOLIDATION Technique to removal air and moisture from soil voids.

SOIL COVER Vegetation on surface of soil.

SOIL CREEP Slow ,downward progression of soil down low-grade slope.

SOIL DISCHARGE Groundwater evaporation at surface by capillarity action from saturation zone.

SOIL DRAINAGE Process where water moves across, through and out of soil due to gravity.

SOIL ENVELOPE Zone of controlled soil backfill around culvert.

SOIL ERODIBILITY Indicator of soil's susceptibility to raindrop impact, runoff and other erosive processes.

SOIL EROSION Displacement of soil by precipitation, wind, plant, animal and human activity.

SOIL FAILURE Permanent deformation of soil by shear and stress forces.

SOIL GAS Gaseous elements and compounds in small spaces between particles of earth and soil.

SOIL GROUPS Fundamental groups of soils (sand, silt, clay, loam).

SOIL HEAT FLUX DENSITY Amount of heat energy that moves by conduction across specified cross-sectional area of soil per unit time and goes to heating (or cooling) soil.

SOIL HORIZON Layer of soil distinguishable from adjacent layers by characteristic physical and chemical properties.

SOIL INFILTRATION RATE Velocity where water enters soil.

SOIL INVESTIGATION Investigation and report preparation of soil sample to determine nutrient content, composition, acidity and/or pH level.

SOIL LIQUID LIMIT Moisture content where soil changes from plastic to liquid state.

SOIL MAP Geographical representation of soil types and/or properties.

SOIL MECHANICS Branch of science concerned with properties and behavior of soil.

SOIL MOISTURE Water stored in soil.

SOIL MOISTURE ACCOUNTING Modeling process that accounts for continuous fluxes to and from soil profile where models can be event-based or continuous.

SOIL MOISTURE DEFICIT Difference between amount of water actually in soil and amount of water soil can hold.

SOIL ORGANIC MATTER Fraction of soil that consists of plant or animal tissue in various stages of decomposition.

SOIL PHASE Subdivision of soil type of other unit of soil type classification.

SOIL PIPE Conduit that carries off liquid wastes from toilet.

SOIL PLASTIC LIMIT Moisture content where soil begins to behave as plastic material.

SOIL PLASTICITY Property of soil to deform without rupture or breaking under external force.

SOIL PLASTICITY INDEX Difference between soil plastic and liquid limit.

SOIL POROSITY Amount of voids between soil particles.

SOIL PROFILE Uppermost layers of ground down to bedrock, subject to infiltration, evaporation and percolation fluxes that depicts all soil horizons.

SOIL REPORT Report prepared by professional that gives understanding of earth conditions affecting building.

SOIL SAMPLE Specimen of earth.

SOIL SAMPLER Tool used to pull subsurface soil samples.

SOIL SATURATION Soil voids full of water.

SOIL SHRINKAGE LIMIT Soil where all particles are in contact and dried below plastic limit.

SOIL STABILIZATION Soil treatment to enhance physical properties.

SOIL STACK Sanitation pipe that penetrates roof used to vent toilet fixtures.

SOIL STERILANT Chemical that temporarily or permanently prevents growth of all plants and animals.

SOIL SUBSIDENCE Lowering of normal level of ground, usually due to over pumping of water from wells.

SOIL SURFACE Surface of upper part of earth mass carrying load of structure.

SOIL TEST Analysis of soil sample to determine nutrient content, composition, etc.

SOIL TEST BORING Soil samples at depths for tests.

SOIL TEXTURE Classification to determine soil classes based on physical texture.

SOIL TYPES Soil classification into groups.

SOIL WATER 1. Water contained in upper regions near Earth's surface; 2. Water contained in pore space of unsaturated zone of soils; 3. Upper part of zone of aeration from which water is discharged by transpiration of plants or by soil evaporation.

SOIL pH Measure of acidity and alkalinity in soils.

SOIL, TYPE A Cohesive soils having unconfined compressive strength of 1.5 $tons/ft^2$ or greater.

SOIL, TYPE B Cohesive soil with unconfined compressive strength greater than 0.5 $tons/ft^2$ but less than 1.5 $tons/ft^2$.

SOIL, TYPE C Cohesive soil with unconfined compressive strength of 0.5 $tons/ft^2$ or less.

SOIL-CEMENT Highly compacted mixture of cement, soil/aggregate and water.

SOIL-CEMENT BEDDING Mixture of soil, portland cement and water placed for pipe bedding.

SOIL-OR-WASTE PIPE Conduit in sanitary drainage system.

SOIL-OR-WASTE STACK Vertical soil-or-waste conduit.

SOIL-WATER CONTENT Water lost from soil upon drying to constant mass at 105 °C expressed either as mass of water per unit mass of dry soil or as volume of water per unit bulk volume of soil.

SOLAR Of or pertaining to sun.

SOLAR ACCESS Exposure to sunlight across property lines without obstruction.

SOLAR BATTERY Rechargeable battery that integrates solar cell with battery power storage.

SOLAR CELL Semiconductor device that converts radiant energy of sunlight into electrical energy.

SOLAR COLLECTOR Device that collects and/or concentrates solar radiation.

SOLAR CONSTANT Amount of incoming solar radiation per unit area.

SOLAR COOLING SYSTEM Reduction in temperature by using solar energy to remove heat from it.

SOLAR COORDINATES Angular distance in solar longitude measured from central meridian.

SOLAR CYCLE Approximately 11-year quasi-periodic variation in frequency or number of solar active events.

SOLAR DAY Time between two Sun meridian passages.

SOLAR DEGRADATION Reduction in substance's properties due solar power exposure.

SOLAR ENERGY Energy in form of electromagnetic radiation emitted from Sun.

SOLAR ENERGY SYSTEM System to convert solar energy into power.

SOLAR FLARE Violent explosion in Sun's atmosphere.

SOLAR FLUX Measure of how much light energy is being radiated in given area by sun.

SOLAR FRACTION Solar energy provided divided by total energy required.

SOLAR FURNACE Structure that uses solar energy as heat source.

SOLAR GAIN Increase in temperature in space, object or structure due to solar radiation.

SOLAR HEATING Collect thermal energy from sun to use as heat.

SOLAR HOUSE House using heat-absorbing material to capture solar radiation for heating.

SOLAR INSOLATION Total solar radiation on surface of Earth.

SOLAR LIGHT Lighting system (i.e., lamp, solar panels, battery).

SOLAR LOAD Cooling load from Sun heat.

SOLAR MAXIMUM Month(s) during solar cycle when 12-month mean of monthly average sunspot numbers reaches maximum.

SOLAR MINIMUM Month(s) during solar cycle when 12-month mean of monthly average sunspot numbers reaches minimum.

SOLAR MONTH One-twelfth of solar year.

SOLAR NOON Time of day at which Sun is closest to its zenith (i.e., highest in sky).

SOLAR ORIENTATION Positioning of building in relation to seasonal variations of Sun's alignment.

SOLAR POWER Conversion of energy from solar radiation into electricity.

SOLAR PROTECTION Materials to completely or partly block ultraviolet radiation.

SOLAR PUMP Pump that operates due to solar radiation.

SOLAR RADIATION Electromagnetic radiation and subatomic particles radiated from Sun.

SOLAR REFLECTING SURFACE Exterior surface intended to reduce effects of solar heating.

SOLAR RIGHTS Right that prevents restriction on receiving direct sunlight.

SOLAR SECTOR BOUNDARY Apparent solar origin or base, of interplanetary sector boundary marked by larger-scale polarity inversion lines.

SOLAR STILL Distillation device powered by solar radiation.

SOLAR STORAGE Medium containing solar collector heat.

SOLAR SYSTEM Gear to collect solar radiation into energy.

SOLAR WIND Outward flux of solar particles and magnetic fields from Sun where velocities are typically near 350 km/s.

SOLAR YEAR Time it takes Earth to pass once around Sun.

SOLAR-CONTROL GLAZING Glass that reduces solar radiation through tinting and/or reflection.

SOLARIMETER Device used to measure solar radiation on surface.

SOLARIUM Room, with many windows, exposed to Sun.

SOLARIZE Exposing to sunlight.

SOLDER Metallic compound used to seal joints between pipes.

SOLE SOURCE AQUIFER Aquifer that supplies 50% or more of drinking water to particular human population (e.g., area).

SOLENOID Coil of wire that acts as magnet when electric current flows through it.

SOLID Absence of voids.

SOLID DRAWN PIPE Seamless conduit produced by extrusion.

SOLID HEAD BUTTRESS DAM Buttress dam where upstream end of each buttress is enlarged to span gap between buttresses.

SOLID SLEEVE Conduit fitting used for uniting two, straight line conduits of same nominal diameter.

SOLID WASTE Non-liquid, non-soluble materials ranging from municipal garbage to industrial wastes (e.g., sewage sludge, agricultural refuse, demolition wastes, mining residues) that contain complex and sometimes hazardous substances.

SOLID WASTE BOUNDARY Perimeter of solid waste disposal facility.

SOLID WASTE DISPOSAL Final placement of refuse not salvaged or recycled.

SOLID WASTE MANAGEMENT Supervised handling of waste materials from their source through recovery processes to disposal.

SOLID WASTE STORAGE CONTAINER Receptacle for solid waste temporary storage.

SOLID WASTE-DERIVED FUEL Fuel produced from solid waste, used with or in place of fossil fuel.

SOLID-PHASE EXTRACTION Procedure to isolate specific organic compounds onto bonded silica extraction column.

SOLIDIFICATION AND STABILIZATION Removal of wastewater from waste or changing it chemically to make it less permeable and susceptible to transport by water.

SOLIDITY 1. Solid contents of body; 2. Solid elevation area divide by total enclosed elevation area for truss.

SOLIDS CONCENTRATION Amount of solid material in wastewater sample.

SOLIDS CONTENT Proportion of non-volatile material contained in solution.

SOLIDS RETENTION TIME Average time activated-sludge solids are in system.

SOLIDS-CONTACT CLARIFIER Clarifier where liquid passes up by solids blanket and discharges near surface.

SOLSTICE Either of two times per year when Sun is at its greatest angular distance from celestial equator: about June 21 (Northern Hemisphere summer solstice), when Sun reaches its northernmost point on celestial sphere or about December 22 (Northern Hemisphere winter solstice), when it reaches its southernmost point.

SOLUBILITY Amount of mass of compound that will dissolve in unit volume of solution.

SOLUBILITY IN WATER Indicator of amount of chemical that can be dissolved in water, shown as percentage or as description.

SOLUBLE BOD BOD of water that has been filtered in standard suspended solids test.

SOLUM Upper layers of soil profile that are affected by climate.

SOLUTE Substance dissolved in another substance, thus forming solution.

SOLUTE CONCENTRATION Concentration of chemical species dissolved in groundwater.

SOLUTE TRANSPORT MODEL Application of model to represent movement of chemical species dissolved in groundwater.

SOLUTION Mixture of solvent and solute formed when solid, gas or another liquid in contact with liquid becomes dispersed homogeneously throughout liquid.

SOLUTION CHANNEL Void within rock enlarged by groundwater.

SOLUTION FEEDER Apparatus designed to feed water treatment chemicals into water system.

SOLUTION MINING Mining process used to recover minerals through well boreholes.

SOLUTION MINING WELLS Class V wells that inject leaching solutions (lixiviants) to remove mineral ores from their original geological settings.

SOLVENT Substance that dissolves other substances, thus forming solution (e.g., water is known as "universal solvent.")

SOLVENT EXTRACTION Separation of components of solution by partitioning between immiscible liquids.

SOMATIC MUTATION Cell genetic alteration that can be passed during cell division.

SONDE Device for testing physical conditions.

SONIC Of or relating to sound.

SONIC BOOM Audible effect of shock wave in air.

SOOT Fine particles of amorphous carbon and tar, produced by incomplete combustion.

SORB To take up and hold either by absorption or adsorption.

SORPTION 1. Process of absorption and adsorption; 2. Interaction (binding or association) of solute ion or molecule with solid; 3. All processes which remove solutes from fluid phase and concentrate them on solid phase of medium.

SORTED 1. Particles uniform in size; 2. Arranged into similar kinds, groups, classes or classifications.

SORTING Arranging into similar kinds, groups, classes or classifications.

SOUND 1. Determine water depth; 2. Body of water connecting larger water bodies; 3. Sensation perceived by ear caused by vibration of air or some other medium.

SOUND ABSORPTION COEFFICIENT Ratio of absorbed sound energy to incident energy.

SOUND ATTENUATING INSULATION Insulation designed to impede noise transmission.

SOUND ATTENUATION Reduction of energy loss in media.

SOUND BARRIER Structure designed to reduce noise.

SOUND BOARD Surface that gives distinctness and sonority to sound.

SOUND FOCUS Place where sound level is significantly higher than elsewhere in room.

SOUND INSULATION Material to reduce sound transmission through it.

SOUND ISOLATION Property of reducing sound through use of materials that reduce transmission.

SOUND LEVEL Logarithmic measurements of audible vibrations.

SOUND LOCK Barrier between room and outside.

SOUND POWER LEVEL Total sound power emitted by source in all directions.

SOUND PRESSURE Root mean square of instantaneous sound pressures over given time interval.

SOUND PRESSURE LEVEL Logarithmic measure of sound relative to reference value.

SOUND REDUCTION FACTOR Measure of sound transmission reduction in substance.

SOUND TRANSMISSION CLASS Rating of ability for building materials to attenuate airborne sound.

SOUND WAVE Longitudinal wave of pressure transmitted through any material at about 1120 ft/s.

SOUND-LEVEL METER Device built to record sound amplitude.

SOUNDING Set of data measuring vertical structure of parameter (e.g., depth, temperature, humidity, pressure, winds) at given time.

SOUNDING LINE Weighted line with distances marked off at regular intervals.

SOUNDING WELL Vertical hole in mass of coarse aggregate for obtaining characteristic data.

SOUNDNESS 1. Property of not only being valid, but also of having true premises; 2. Free of defects, cracks or differences.

SOUNDPROOFING Material to dampen sound from traveling through it.

SOUR GAS Any gas (e.g., natural) containing significant amounts of hydrogen sulfide.

SOURCE Person, place or thing where something comes or is acquired.

SOURCE AREA Location of liquid hydrocarbons or zone of highest soil or groundwater concentrations or both, of chemical of concern.

SOURCE CHARACTERIZATION MEASUREMENTS Measurements made to estimate rate of release of pollutants into environment from source (e.g., incinerator, landfill).

SOURCE LOADING Rate at which contaminant is entering groundwater system at specific source.

SOURCE OF CONTAMINANTS Physical location and spatial extent of source contaminating aquifer.

SOURCE REDUCTION Reducing amount of materials entering waste stream from specific source by redesigning products or patterns of production or consumption.

SOURCE ROCKS Rocks from which fragments and other detached pieces have been derived to form different rock.

SOURCE SEPARATION 1. Sorting of different waste materials at point of generation; 2. Separation of source signals from mixed signals.

SOURCE WATER Water in its natural state, prior to any treatment for drinking.

SOURCE-WATER PROTECTION AREA Area delineated by state for Public Water Supply or including numerous such suppliers, whether source is groundwater or surface water or both.

SOUTHERN OSCILLATION "See-saw" in surface pressure in tropical Pacific characterized by simultaneously opposite sea level pressure anomalies at Tahiti, in eastern tropical Pacific and Darwin, on northwest coast of Australia.

SPACE ENVIRONMENT CENTER Provides real-time monitoring and forecasting of solar and geophysical events, conducts research in solar-terrestrial physics and develops techniques for forecasting solar and geophysical disturbances.

SPACE MEAN SPEED Distance traveled divided by average travel time.

SPACING In blasting, distance between holes in row.

SPALL 1. Loss of surface concrete usually caused by impact, abrasion or compression; 2. To break off from surface in sheets or pieces (i.e., small fragment or chip).

SPALLING Process of reducing (e.g., stone blocks) to approximately level surface by hammering; 2. Fragment or chip of stone.

SPAN Spread of arch or between abutments or beam, girder, truss, bridge, etc. between supports.

SPAN WIRE 1. Wire attached between two poles; 2. Trolley-wire of electric road.

SPANDREL Space between two contiguous arches and straight feature above them.

SPARGE Injection of air below water table to strip dissolved volatile organic compounds and/or oxygenate groundwater to facilitate aerobic biodegradation of organic compounds.

SPARGER Vessel with perforated cover for sprinkling liquid.

SPARSELY FOSSILIFEROUS Less than 40% fossils.

SPATIAL CONCENTRATION Ratio of dry weight of sediment to total weight of water-sediment mixture in sample (i.e., unit volume of mixture).

SPATIAL STOP RATE Ratio of stop rate to facility length.

SPATIAL VARIABILITY Quantity measured at different spatial locations that exhibits different values across locations.

SPAWN To lay eggs (i.e., aquatic animals).

SPAWNING BEDS Places where eggs of aquatic animals lodge or are placed during or after fertilization.

SPEARHEAD ECHO Radar echo associated with downburst with pointed appendage extending toward direction of echo motion.

SPECIAL AQUATIC SITES Wetlands, sanctuaries and refuges, mud flats, vegetated shallows, coral reefs and riffle and pool complexes.

SPECIAL ASSESSMENT Tax levy related for purposes of paying for project or replenishing underfunded reserve.

SPECIAL AVALANCHE WARNING Issued by National Weather Service when avalanches are imminent or occurring in mountains.

SPECIAL DRAINAGE WELLS Class V wells that include potable water tank overflow, construction dewatering, swimming pool drainage and mine dewatering wells that receive fluids that cannot be classified as agricultural, industrial or storm water.

SPECIAL EVENTS Public event (e.g., ballgame, parade, festival) that requires transportation planning (i.e., roadway closure, permits).

SPECIAL EXAMINATION Field review performed on high- or significant-hazard dam to address identified visible dam safety deficiency or to investigate significant changes in operating or loading conditions.

SPECIAL FEATURES Area containing ecological, geological or other features of scientific, educational, scenic or historical value.

SPECIAL LOCAL-NEEDS REGISTRATION Registration of pesticide product by state agency for specific use that's not federally registered.

SPECIAL MARINE WARNING Issued for potentially hazardous weather conditions usually of short duration (up to 2 hours) producing sustained marine thunderstorm winds or associated gusts of 34 knots or greater; and/or hail 3/4 inches or more in diameter; and/or waterspouts affecting areas included in Coastal Waters Forecast, Nearshore Marine Forecast or Great Lakes Open Lakes Forecast not adequately covered by existing marine warnings that's used for short duration mesoscale events (e.g., strong cold front, gravity wave, squall line) lasting less than 2 hours and producing winds or gusts of 34 knots or greater.

SPECIAL POPULATIONS Those individuals or groups that may be institutionalized and have needs that require special consideration in emergencies.

SPECIAL PROVISIONS Additional standard requirements, prohibitions, exceptions, etc.

SPECIAL PURPOSE ROAD Road designed for purposes other than carriage of passengers or goods.

SPECIAL REVIEW Regulatory process where existing pesticides suspected of posing unreasonable risks to human health, non-target organisms or environment are referred for review by EPA.

SPECIAL TROPICAL DISTURBANCE STATEMENT Issued by National Hurricane Center furnishes information on strong and formative non-depression systems.

SPECIAL VFR CONDITIONS Weather conditions that require aircraft to maintain altitude of at least 500 feet above ground and 1,000 feet above nearest obstacle.

SPECIAL WASTE Items including household hazardous waste, bulky wastes (e.g., refrigerators, pieces of furniture) tires and used oil.

SPECIAL WASTES Waste that requires special handling, trained people, and/or special disposal methods.

SPECIAL WIND REGION Regions where wind-speed anomalies exist.

SPECIALLY-DESIGNATED LANDFILL Landfill where complete long term protection is provided for quality of surface and subsurface waters from pesticides, pesticide containers and pesticide-related wastes deposited therein and against hazard to public health and environment.

SPECIES 1. Basic category of biological classification intended to designate single kind of animal or plant; 2. Mineral with unique chemical formula whose crystals belong to unique crystallographic system; 3. Official mixture or compound powder of any kind.

SPECIES DIVERSITY Ecological concept that incorporates both number of species in particular sampling area and evenness with which individuals are distributed among various species.

SPECIES EMPHASIS Species of high public interest and demand.

SPECIES RICHNESS Number of species (taxa) present in defined area or sampling unit.

SPECIFIC ABSORPTION Material capacity to consume water after removing gravity water.

SPECIFIC AREA MESSAGE ENCODING Tone alert system which allows NOAA Weather Radio receivers equipped with SAME feature to sound alert for only certain weather conditions or within limited geographic area (e.g., county).

SPECIFIC CAPACITY Rate of discharge from well divided by drawdown of water level within well at specific time since pumping started (i.e., 1yield of well per unit of drawdown).

SPECIFIC CAPACITY OF WELL Discharge per unit drawdown in well.

SPECIFIC CONDUCTANCE Rapid method of estimating dissolved solid content of water supply by testing its ability to conduct electrical current.

SPECIFIC DISCHARGE 1. In surface water hydrology, discharge per unit area of upstream watershed; 2. In groundwater, volumetric flow rate per unit area of porous media.

SPECIFIC ENERGY Energy of substance measured per unit mass.

SPECIFIC GRADE Roadway grade expressed as percentage.

SPECIFIC GRAVITY 1. Ratio of mass of body to equal volume of water; 2. Comparison of weight of chemical to weight of equal volume of water; 3. Ratio of density of any substance to density of water.

SPECIFIC HEAT Ratio of amount of heat needed to raise temperature of unit mass of substance by unit degree to amount of heat needed to raise that of same mass of water by same amount.

SPECIFIC HUMIDITY Ratio of mass of water vapor to mass of air.

SPECIFIC IDENTIFICATION COST FLOW Identifies inventory units sold or remaining in ending inventory.

SPECIFIC LEVEL Given discharge water surface level.

SPECIFIC RETENTION Ratio of volume of water that material will hold vs. pull of gravity to volume of body.

SPECIFIC STORAGE Volume of water released from or taken into storage per unit volume of porous medium per unit change in head.

SPECIFIC VOLUME Volume of substance unit mass (reciprocal of its density).

SPECIFIC WEIGHT Weight per unit volume.

SPECIFIC YEAR FLOOD Stream flow (i.e., flood) is equalled or exceeded on average once during year.

SPECIFIC YIELD 1. Ratio of volume of water that will drain freely from material to total volume of that material (e.g., saturated rock, aquifer formation); 2. Volume of water released from storage by unconfined aquifer per unit surface area of aquifer per unit decline of water table.

SPECIFICATION explicit set of requirements to be satisfied by material, product or service.

SPECIFICATION LIMIT Values between where products or services should operate.

SPECIFICATION OF WORKS document describing materials and workmanship required.

SPECIFICATIONS Set of requirements defining exact description of object or process.

SPECIFIED CONCENTRATION BOUNDARY Boundary at which solute concentration is specified.

SPECIFIED FLUX BOUNDARY Model boundary condition where groundwater flux or mass flux is specified.

SPECIFIED STRENGTH OF CONCRETE Compressive strength of concrete based on 28-day test results.

SPECIMEN 1. Individual instance that represents class; 2. Sample used for diagnostic analysis.

SPECTRAL DENSITY Radar term for distribution of power by frequency.

SPECTRAL UNCERTAINTY Potential variation in exposure to sound spectra in office.

SPECTRAL WAVE DENSITY Energy for each frequency bin.

SPECTRAL WAVE DIRECTION Mean wave direction, in degrees from true North, for each frequency bin.

SPECTROGRAPH Machine for recording spectra, producing spectrograms.

SPECTROHELIOGRAM Photograph of Sun taken using spectroheliograph.

SPECTROHELIOGRAPH Device that produced monochromatic photographs of Sun using light at selected wavelength.

SPECTROHELIOSCOPE Instrument used for observing solar radiation.

SPECTROMETER Optical instrument for measuring absorption of light by chemical substances.

SPECTROPHOTOMETER Instrument used to measure intensity of electromagnetic radiation at different wavelengths.

SPECTROSCOPY Scientific study of spectra.

SPECTRUM 1. Pattern of absorption or emission of radiation produced by substance when subjected to energy (i.e., radiation, heat, electricity); 2. Range.

SPECTRUM INTENSITY Integral of pseudo-velocity response spectrum taken over range of significant structural vibration periods of structure being analyzed.

SPECTRUM OF TURBULENCE Turbulence at different frequencies.

SPECTRUM WIDTH WSR-88D radar product depicts full 360° sweep of spectrum width data indicating measure of velocity dispersion within radar sample volume.

SPECTRUM WIDTH CROSS-SECTION WSR-88D radar product displays vertical cross-section of spectrum width on grid with heights up to 70,000 feet on vertical axis and distance up to 124 nautical miles on horizontal axis.

SPEED 1. Rate of motion or magnitude of velocity; 2. Rate distance is traversed in given time.

SPEED HARMONIZATION Active traffic management strategy to delay onset of traffic flow breakdown and mitigate congestion.

SPEED LIMIT Maximum speed permitted by law for vehicles on road.

SPEED LIMIT SIGN BEACON Flashing signal beacon to indicate speed limit.

SPEED MEASUREMENT MARKINGS Speed measurement transverse marking placed on roadway to assist enforcement.

SPEED OF SOUND Rate sound is propagated through some medium under specified conditions.

SPEED SHEAR Component of wind shear that is due to change in wind speed with height.

SPEED ZONE Area where speed limit is in effect.

SPEED-CHANGE LANE Deceleration or acceleration lane for vehicles entering traveled roadway.

SPENT BRINE RETURN FLOW WELLS Class V wells that dispose of spent brine that results from extraction of minerals, halogens and other compounds from fluids.

SPENT NUCLEAR FUEL Nuclear fuel irradiated in nuclear reactor.

SPHAEROTILUS BULKING Sludge bulking that happens due to large amounts of Sphaerotilus.

SPHERE CALIBRATION Reflectivity calibration of radar by pointing dish at metal sphere of theoretically known reflectivity.

SPIGOT 1. Plain end of cast-iron pipe; 2. Plug of faucet or bell and spigot pipe joint; 3. Pin or peg used to stop vent in cask.

SPIKE 1. Large nail with offset head used to secure rails and base plates to railroad ties; 2. Surge in power.

SPILE Plank driven ahead of tunnel face for roof support.

SPILL To spread out or fall out.

SPILL AREA Place where observable hints of spill is found plus buffer zone.

SPILL BOUNDARIES Actual area of contamination as determined by post-cleanup verification sampling or by pre-cleanup sampling.

SPILL EVENT Discharge of contaminant in soil or water.

SPILL PREVENTION CONTROL AND COUNTERMEASURE PLAN Plan prepared by facility to minimize likelihood of spill and to expedite control and cleanup activities should spill occur.

SPILL PREVENTION, CONTAINMENT, AND COUNTERMEASURES PLAN Plan covering release of hazardous substances as defined in Clean Water Act.

SPILL THROUGH ABUTMENT Abutments that support bridge at intervals where water or roadways can move through abutments.

SPILLOVER 1. Something that overflows (e.g., pedestrian traffic); 2. Excess or side effect.

SPILLS Water releases that cannot be put to use for project purposes.

SPILLWAY 1. Structure that passes normal and/or flood flows in manner that protects structural integrity of dam; 2. Path designed to take away overflow safely.

SPILLWAY CAPACITY Maximum spillway outflow that dam can safely pass with reservoir at its maximum level.

SPILLWAY CHANNEL Open channel or closed conduit conveying water from spillway inlet downstream.

SPILLWAY CHUTE Steeply sloping spillway channel that conveys discharges at super-critical velocities.

SPILLWAY CREST Lowest level at which water can flow over or through spillway (i.e., elevation of highest point of spillway).

SPILLWAY GATE Gate on crest of spillway to control discharge or reservoir water level.

SPILLWAY, AUXILIARY Any secondary spillway designed to be operated infrequently, possibly in anticipation of some degree of structural damage or erosion to spillway that would occur during operation.

SPILLWAY, EMERGENCY Any secondary spillway designed to be operated infrequently, possibly in anticipation of some degree of structural damage or erosion to spillway that would occur during operation.

SPILLWAY, FUSE PLUG Form of auxiliary spillway consisting of low embankment designed to be overtopped and washed away during exceptionally large flood.

SPILLWAY, SERVICE Spillway designed to provide continuous or frequent regulated or unregulated releases from reservoir, without significant damage to either dam or its appurtenant structures.

SPILLWAY, SHAFT Vertical or inclined shaft into which water spills and then is conveyed through, under or around dam by means of conduit or tunnel.

SPIN-UP Small-scale vortex initiation (i.e., when gustnado, landspout or suction vortex forms).

SPINNING RESERVES Available capacity of generating facilities synchronized to interconnected electric system where it can be called upon for immediate use in response to system problems or sudden load changes.

SPIRAL BEVEL GEAR Round gear with helical teeth.

SPIRAL PIPE Round conduit formed from coil of metal into rigid tube with 4-ply lockseam.

SPIRAL REINFORCEMENT Reinforcement wrapped around main or longitudinal reinforcement.

SPIRAL WELD PIPE Conduit with seam running its entire length in spiral form.

SPIRAL-RIVETED PIPE Conduit formed sheets using overlapped and riveted edges.

SPIT 1. Small point or low tongue or narrow embankment of land having one end attached to mainland and other terminating in open water; 2. Depth spade goes in digging.

SPLASH PAD Structure that shields land from erosion by splashing water.

SPLIT 1. Crack or longitudinal fissure; 2. Segment of traffic signal cycle.

SPLIT FLOW 1. Flow that leaves main river flow and takes completely different path from main river; 2. Flow pattern high in atmosphere characterized by diverging winds.

SPLIT HEATING SYSTEM Heating system of radiators supplemented with mechanical circulation.

SPLIT PIPE Conduit cut along its length.

SPLIT RUN Combining two shifts for transportation driver.

SPLIT SAMPLE Sample prepared by dividing it into two or more equal volumes, where each volume is considered separate sample but representative of entire sample.

SPLIT SPOON SAMPLER Drill core for land investigation where tube splits into two equal halves lengthwise.

SPLIT WEB Cracks may run lengthwise along side, into or through web.

SPLIT-DIAMOND INTERCHANGE Interchange with ramps (i.e., entry and exit) separated between two crossroads.

SPLITTER BOX Structure to divide specific amount of flow.

SPLITTER ISLAND Raised or painted traffic island that separates opposing direction traffic.

SPLITTER WALL Wall or pier parallel to direction of flow in channel that separates flows released from different sources as means of energy dissipation.

SPLITTING STORM Thunderstorm which splits into two storms which follow diverging paths.

SPLITTING TENSILE STRENGTH Method to evaluate shear resistance and determine length of reinforcement for concrete.

SPOIL Overburden, dirt, rock or other waste material removed in mining, quarrying, dredging or excavating.

SPOON Tool, that splits, to get soil samples.

SPOON BLOW Blow of 140-lb hammer falling 30 inches onto 2 inch diameter O.D. by 1.375 inch I.D. split spoon sampler.

SPORADIC E Phenomenon occurring in E region of ionosphere, which significantly affects HF radiowave propagation.

SPORE Reproductive particle released by fungus, alga or plant that may germinate into another.

SPOTTING Outbreak of secondary fires as firebrands or other burning materials are carried ahead of main fire line by winds.

SPOUT Tube or lip where liquid or steam is poured or discharged.

SPOUT-DELIVERY PUMP Pump not capable of delivering water above its own elevation.

SPOUTING VELOCITY Velocity obtained during isentropic expansion of gas between entry and exit pressures of stage.

SPRAWL Unplanned development of open land.

SPRAY Fine, gentle, dispersed mist of liquid.

SPRAY AERATOR Apparatus with pressure nozzle that propels water into fine spray.

SPRAY DRYER Produces dry powder from liquid or slurry by hot gas rapid drying.

SPRAY IRRIGATION Irrigation method where water is shot from high-pressure sprayers onto crops and some water is lost to evaporation.

SPRAY NOZZLE Device that facilitates dispersion of liquid into spray.

SPRAY POND Reservoir where warmed water from cooling plant is sprayed to cool before reuse.

SPRAY TOWER Construction for purifying gases that pass up through tower where suitable liquid is sprayed from top.

SPRAY TOWER SCRUBBER Device that sprays alkaline water into chamber where acid gases are present to aid in neutralizing gas.

SPRAY-IN-PLACE INSULATION Sprayed on insulation.

SPREAD Depth of water in curb and gutter section.

SPREAD FOOTING Foundation with wider bottom section than load-bearing foundation walls it supports.

SPREAD TIME Total elapsed time between first report time and final turn-in time for transit run.

SPRING 1. Groundwater or oil seeping or flowing out of Earth where water table intersects ground surface; 2. Season of year comprising transition period from winter to summer occurring when Sun is approaching summer solstice.

SPRING MELT/THAW Process whereby warm temperatures melt winter snow and ice.

SPRING TIDE Tide higher than normal that occurs when moon is new or full.

SPRING WATER Water originating from spring.

SPRINGLINE Imaginary horizontal reference line located at mid-height or halfway point, of circular conduit, pipe, tunnel or point at which side walls are vertical on horseshoe-shaped conduit.

SPRINKLE Light covering with sprinkled substance.

SPRINKLER Device that sprays water into air.

SPRINKLER CONNECTION Siamese connection in sprinkler system to boost water pressure and supply.

SPRINKLER IRRIGATION Method of irrigation where water is sprayed or sprinkled, through air to ground surface.

SPRINKLER NOZZLE Irrigation spray nozzle actuated by water pressure.

SPRINKLER SYSTEM System comprising pipes and spray nozzles.

SPRINKLER TONGS Tool that stops flow of water from sprinkler head.

SPRINKLER WEDGE Wedge that stops water flow from sprinkler heads.

SPRINKLERED Having sprinkler system.

SPROCKET GEAR Profiled wheel with teeth that mesh with track, chain or other perforated material.

SPUD 1. To drill by moving drill bit and shaft up and down or by raising and dropping bit; 2. Vertical leg of jack-up barge.

SPUDDING Initial drilling of well.

SPUDDING DRILL Drill that lifts and drops bit to make hole.

SPUN PIPE Concrete pipe manufactured by spinning cage when pouring concrete.

SPUR DIKE Elongated structure having one end on bank of stream and other end projecting into current.

SPUR GEAR Gear with teeth projecting parallel to axis shaft.

SPUR VALLEY Little branch valley.

SQUALL 1. Strong wind characterized by sudden onset where wind speed increases at least 16 knots and is sustained at 22 knots or more for at least one minute; 2. Severe local storm considered as whole, that is, winds and cloud mass and (if any) precipitation, thunder and lightning.

SQUALL LINE Line of active thunderstorms, either continuous or with breaks, including contiguous precipitation areas resulting from existence of thunderstorms.

SQUARE Unit of measurement commonly applied to building (e.g., roofing) materials (e.g., 100 ft^2 = 10 ft. x 10 ft).

SQUARE CENTIMETER Unit of area equal to 1 cm^2.

SQUARE FOOT Unit of area equal to 1 ft^2.

SQUARE INCH Unit of area equal to 1 in^2.

SQUARE KILOMETER Unit of area equal to 1 km^2.

SQUARE METER Unit of area equal to 1 m^2.

SQUARE METER PER SECOND Unit of acceleration equal to 1 m^2/s.

SQUARE MILE Unit of area equal to 1 mi^2.

SQUARE YARD Unit of area equal to 1 yd^2.

ST. ELMO'S FIRE Glow on masthead produced by extreme buildup of electrical charge where lightning may strike mast within five minutes after it begins to glow.

STABILITY 1. Persistence of component of system; 2. Degree of resistance of layer of air to vertical motion; 3. Condition of structure or mass of material when it is able to support applied stress (e.g., soil cut or fill) for long time without suffering any significant deformation or movement not reversed by release of stress; 4. Resistant to change.

STABILITY INDEX Overall stability or instability of sounding is sometimes conveniently expressed in form of single numerical value.

STABILIZATION Conversion where wastes are converted to inert material.

STABILIZATION LAGOON Shallow ponds to reduce wastewater organic content and pathogen removal.

STABILIZATION POND Shallow Lagoon to reduce wastewater organic content and pathogen removal.

STABILIZATION PONDS 1. Shallow pond where sunlight, bacterial action and oxygen work to purify wastewater; 2. Storage of wastewater or spent nuclear fuel rods; 3. Shallow body of water, often separated from sea by coral reefs or sandbars.

STABILIZE To firm and limit soil movement.

STABILIZED Process of making something secure or in equilibrium.

STABILIZED CHANNEL Earth channel not eroding silt.

STABILIZED WASTE Waste that has been treated or decomposed to not cause impacts in receiving stream.

STABLE Atmospheric state with warm air above cold air which inhibits vertical movement of air.

STABLE AIR Motionless mass of air that holds, instead of dispersing, pollutants.

STABLE BOUNDARY LAYER 1. Stably-stratified layer that forms at surface and grows upward, usually at night or in winter, as heat is extracted from atmosphere's base in response to longwave radiative heat loss from ground; 2. Stably-stratified layer that forms when warm air is advected over cold surface or over melting ice.

STABLE CHANNEL Stream channel that does not change in planform or bed profile during particular period of time (i.e., years to tens of years).

STABLE CORE Post-sunrise, elevated remnant of temperature inversion that has built up overnight within valley.

STABLE EFFLUENT Effluent for treating wastewater that has no adverse impact on accepting water.

STABLE ISOTOPE RATIO PER MIL Unit expressing ratio of abundance of two radioactive isotopes.

STABLE ROCK Natural solid mineral matter that can be excavated with vertical sides and remain intact while exposed.

STACCATO LIGHTNING Cloud-to-ground lightning discharge which appears as single very bright, short-duration stroke, often with considerable branching.

STACK 1. Chimney, smokestack or vertical pipe that discharges gas; 2. Vertical drainpipe; 3. Coastal landform, consisting of large vertical column of rock in sea.

STACK EFFECT Air that moves upward because it is warmer than ambient atmosphere (e.g., chimney) creating positive pressure area at top and negative pressure area at bottom.

STACK EMISSIONS Solids and gases that come out of flue after incineration.

STACK GAS Combustion gas that passes through flue.

STACK PIPE Vertical conduit that removes sewage and waste from building.

STACK SAMPLING Method of collecting air or gas pollution samples at place of origin.

STACK VENT Continuation of soil or waste stack after final fixture has been tied into.

STAFF GAUGE Vertical staff graduated to indicate water depth.

STAGE 1. Height of water surface above established datum at given location; 2. Elevation; 2. Succession of rock strata laid down in single age on geologic time scale; 3. Floor of house.

STAGE AERATION Activated sludge process broken into phases allowing for intermediate settling of solids and sludge return.

STAGE DIGESTION Processing of waste sludge in two or more stages (i.e., primary and secondary).

STAGE II CONTROLS Systems placed on service station gasoline pumps to control and capture gasoline vapors during refueling.

STAGE TREATMENT Procedure where wastewater treatment stages are in series.

STAGE-CAPACITY CURVE Graph showing relation between surface elevation of water in reservoir (ordinate) plotted against volume below that elevation (abscissa).

STAGE-DISCHARGE CURVE Graph showing relation between water height (ordinate) and volume of water flowing (e.g., discharge) in channel (abscissa) per unit of time at given location.

STAGE-DISCHARGE RELATION Relationship between water level and discharge at certain location in stream.

STAGED TRICKLING FILTRATION System where effluent of dripping filter becomes influent.

STAGNATION Lack of motion in mass of air or water that holds pollutants in place.

STAGNATION POINT Point in flow field where local velocity of fluid is zero.

STAINLESS STEEL PIPE Versatile corrosion-resisting material comprised of steel alloy and small percentage of chromium.

STAIR STEPPING Process of continually updating river forecasts for purpose of incorporating effects rain that has fallen since previous forecast was prepared.

STAIRWAY Set of steps allowing one to walk up or down.

STAKE BODY TRUCK Vehicle with sockets along edge that hold upright stakes to form fence.

STAKEHOLDER Any organization, governmental entity or individual that has stake in or may be impacted by given approach to environmental regulation, pollution prevention, energy conservation, etc.

STALE WASTEWATER Dark colored wastewater with oxygen deficit and near putrefaction.

STALL TORQUE Maximum torque that can be applied to shaft and cause motor to stop rotating.

STAND 1. Particular grove or other group of trees or shrubs; 2. Contiguous group of trees sufficiently uniform in age-class distribution, composition and structure and growing on site of sufficiently uniform quality, to be distinguishable unit.

STAND-UP TIME Time unsupported tunnels and underground excavations can be held.

STANDARD Document that has been developed and established within consensus principles of American Society for Testing and Materials (ASTM) and that meets approval requirements of ASTM procedures and regulations.

STANDARD ABSORPTION TRENCH Trench 12 to 36 inches wide, containing 12 inches of clean, coarse aggregate and distribution pipe covered with minimum 12 inches of soil.

STANDARD AIR Air at standard temperature (68 °F) and pressure (29.92 inches of mercury).

STANDARD ATMOSPHERE Unit of pressure equal to 101.325 kPa.

STANDARD BIOCHEMICAL OXYGEN DEMAND Biochemical oxygen demand determined under normal laboratory processes.

STANDARD CONDITION Standard temperature and pressure conditions for experimental measurements.

STANDARD CUBIC FEET OF AIR PER MINUTE Cubic feet of air per minute at 70 °F (21 °C) and 1 atmospheric pressure.

STANDARD DEVIATION 1. Statistical measure of dispersion or scatter of series of values; 2. Square root of variance.

STANDARD DIMENSION Company's set dimensions.

STANDARD DRYING DAY Day that produces same net drying as experienced during 24-hour period under laboratory conditions where dry-bulb temperature is maintained at 90 °F and relative humidity at 20%.

STANDARD ERROR 1. Standard deviation of sample sizes to sample means.

STANDARD FALL DIAMETER Diameter of sphere having specific gravity of 2.65 and same standard fall velocity as particle in question.

STANDARD FALL VELOCITY Average rate of fall that particle would attain if falling alone in quiescent distilled water of infinite extend and at temperature of 75.2 °F.

STANDARD FORM OF CONTRACT Contract where terms and conditions are set by only one party.

STANDARD HYDROLOGIC EXCHANGE FORMAT Documented set of rules for coding data for operational day-to-day use in form for both visual and computer recognition.

STANDARD INDUSTRIAL CLASSIFICATION CODE Four digit code established by Office of Management and Budget to identify various types of industries.

STANDARD INSIDE DIAMETER DIMENSION RATIO Ratio of typical inside diameter to minimum specified wall thickness of conduit.

STANDARD INSTRUMENT DEPARTURE Published flight procedures followed by aircraft on IFR flight plan immediately after takeoff from airport.

STANDARD METHODS Best current practice by professionals.

STANDARD ORIFICE Orifice inserted into piping with bore hole diverging in flow direction to measure flow rate.

STANDARD OUTSIDE DIAMETER DIMENSION RATIO Ratio of typical outside diameter to minimum specified wall thickness of conduit.

STANDARD PENETRATION RESISTANCE Amount of strikes of 140-lb hammer, dropping 30 in. to move split-barrel sampler 12-in into soil.

STANDARD PENETRATION TEST In-situ testing method used to determine geotechnical engineering properties of subsurface soils.

STANDARD PLAN Repetitively used drawing.

STANDARD PLANS Drawings that provide consistent application of standard designs and details in preparation of construction contract documents.

STANDARD PRESSURE 1. One atmosphere; 2. Working pressure of 125 psi.

STANDARD PROJECT FLOOD Discharge expected to result from most severe combination of meteorological and hydrological conditions that are reasonably characteristic of geographic region involved.

STANDARD PROJECT STORM Most severe rainstorm that has actually occurred over region.

STANDARD PROVISION Common contract clause.

STANDARD RAIL Rail track gauge 4 ft 8½″.

STANDARD RATING Test ratings performed under standard conditions.

STANDARD SAMPLE 1. Part of finished drinking water examined for presence of coliform bacteria; 2. Small portion that conforms to specifications.

STANDARD SAND Silica sand, free of organic matter, graded to pass No. 20 (0.0331-in.) sieve and be retained on No. 30 (0.0232-in.) sieve.

STANDARD SHORT TUBE Conduit with length approximately three times its diameter.

STANDARD SIEVE Mesh screen for separating and characterizing particle size distribution of sample.

STANDARD SOLUTION Solution with precise concentration of substance or element.

STANDARD SPECIFICATIONS Requirements to be satisfied by material, design, product or service.

STANDARD STAGE BIOCHEMICAL OXYGEN DEMAND Biochemical oxygen demand as determined under normal laboratory processes for 5 days at 20 °C, expressed in mg/l.

STANDARD STEP METHOD Method where total distance is divided into reaches by cross-sections at fixed locations along channel and, starting from one control, profile calculations proceed in steps from cross-section to cross-section to next control.

STANDARD SYNOPTIC TIMES Times of 0000, 0600, 1200 and 1800 UTC.

STANDARD TEMPERATURE AND PRESSURE Temperature of 32 °F and pressure of 1 atmosphere.

STANDARD TERMINAL ARRIVAL ROUTE Flight procedure followed by aircraft on instrument flight rules (IFR) flight plan just before reaching destination airport.

STANDARD THREAD Allowance and tolerance for screw threads.

STANDARD URBAN BUS Motor vehicle to transport passengers over distance.

STANDARD WIPE TEST 1. For spills of PCBs, cleanup to numerical surface standards; 2. For sampling, to verify that numerical standards have been met.

STANDARD-RATE FILTER Trickling filter designed to receive small load of BOD per unit volume and have low dosage rate per unit area.

STANDARDISE To make to conform to standard.

STANDARDS Norms that impose limits on amount of pollutants or emissions produced.

STANDARDS OF PROFESSIONAL PRACTICE Ethical principles of professional organizations.

STANDBY 1. Sleep mode; 2. Remaining in preparation for; 3. Wait briefly.

STANDBY EQUIPMENT Gear that's on job and available for work, but not put into operation until necessary.

STANDBY POWER SUPPLY Power supply with circuit that's capable of determining disturbances in electricity.

STANDBY RESERVES Unused capacity in electric system in machines that are not in operation but that are available for immediate use if required.

STANDEE Passenger forced to stand up on crowded bus.

STANDING CROP Number of living matter present.

STANDING OPERATING PROCEDURES Comprehensive single-source document covering all aspects of dam and reservoir operation and maintenance and emergency procedures.

STANDING WATER LEVEL Height water rises up hole when reached by drill rig.

STANDING WAVE Wave form that occurs in limited, fixed medium where reflected wave coincides with produced wave.

STANDPIPE 1. Pipe or tank connected to closed conduit and extending to or above hydraulic grade line of conduit to afford relief from surges of pressure in pipelines; 2. Tank used for storage of water in distribution systems; 3. Water supply of building for firefighting use.

STAPHYLOCOCCUS Spherical gram-positive parasitic bacterium.

STARBOARD Right hand side of ship, boat or aircraft when facing front, fore or bow, respectively.

START DATE Actual start time of activity.

START OF RESPONSE ACTION Point in time when there is guarantee or set-aside of funding by federal agencies, states or Principal Responsible Parties in order to begin response actions.

START TIME Actual start time of activity.

START-TO-FINISH RELATIONSHIP Predecessor activity must finish before successor activity can start.

START-TO-START RELATIONSHIP Predecessor activity must start before successor activity can start.

START-UP Act or process of starting process or machine.

START-UP COST Non-recurring costs associated with setting up business, facility, etc.

START-UP LOST TIME Time lost when traffic signal changes from red to green.

START-UP TIME Time to achieve steady state condition.

STARTING MATERIAL Substance used in production of substance and incorporated into substance structure.

STARTING TORQUE Minimum torque required to get mechanical system moving from standstill.

STASIS Inactivity.

STATE 1. Set of circumstances applying at any given time; 2. Element of range of random variables that define random process.

STATE CLIMATE DIVISION Geographic area in state based primarily on crop-reporting districts where states can have 2 to 10 climate divisions.

STATE EMERGENCY RESPONSE COMMISSION Appointed by governor of each state for designation of emergency planning districts, appoint local emergency planning committees, supervise and coordinate their activities and review local emergency response plans as provided by SARA Title III.

STATE ENVIRONMENTAL GOALS AND INDICATION PROJECT Program to assist state environmental agencies by providing technical and financial assistance in development of environmental goals and indicators.

STATE FORECAST PRODUCT National Weather Service product intended to give good general picture of what weather may be expected in state during next 5 days.

STATE IMPLEMENTATION PLANS EPA approved state plans for establishment, regulation and enforcement of air pollution standards.

STATE MANAGEMENT PLAN State management plan required by EPA to allow states, tribes and U.S. territories flexibility to design and implement ways to protect groundwater from use of certain pesticides.

STATE ORGANIZATION State government agency or office having principal or lead role in emergency planning and preparedness.

STATE WEATHER ROUNDUP National Weather Service tabular product which provides routine hourly observations within state through National Weather Wire Service that gives current weather condition in one word (e.g., cloudy, rain, snow, fog), temperature and dew point in Fahrenheit, relative humidity, wind speed and direction and finally additional information (e.g., wind chill, heat index, secondary weather condition).

STATEMENT OF BASIS Document prepared for every draft NPDES permit for which fact sheet is not required that briefly describes how permit conditions were derived and reasons conditions are necessary for permit.

STATEMENT OF QUALIFICATIONS Written information discussing knowledge, skills and abilities applicant possesses in relation to desirable qualifications of advertisement.

STATEMENT OF WORK Document that lays foundation for planning and successful execution of project.

STATES Include: 50 States, District of Columbia, Guam, Commonwealth of Puerto Rico, Virgin Islands, American Samoa, Trust Territory of Pacific Islands and Commonwealth of Northern Mariana Islands.

STATIC DISCHARGE HEAD Vertical distance between pump centerline and point of free discharge or surface of liquid in discharge tank.

STATIC ELECTRICITY Electric charge that has built up on insulated body.

STATIC FLOW MODEL Model where inflow and outflow rates are same.

STATIC FORCE Constant force applied to stationary object.

STATIC FRICTION COEFFICIENT Index that shows relationship between friction force between two objects and normal reaction objects.

STATIC HEAD 1. Difference in elevation between pumping source and point of delivery; 2. Vertical distance between two points in fluid; 3. Static flow pressure in pipe, channel or duct.

STATIC PENETRATION TEST Test where testing device is pushed into soil with measurable force.

STATIC PRESSURE 1. Pressure of fluid on body when body is at rest relative to fluid; 2. In flowing air, total pressure minus velocity pressure, pushing equally in all directions.

STATIC RESERVE INDEX Ratio of current reserves to current consumption.

STATIC SUCTION HEAD Vertical distance from water line to centerline of impeller.

STATIC SUCTION LIFT Vertical distance from water line to centerline of impeller.

STATIC SYSTEM System where output depends only on present value of input.

STATIC WATER DEPTH Vertical distance from centerline of pump discharge down to surface level of free pool while no water is being drawn from pool or water table.

STATIC WATER LEVEL Elevation or level of water table in well when pump is not operating.

STATIC WATER SUPPLY Supply of water at rest that does not supply pressure head for fire fighting but may be used for fire pumps.

STATIC-REPLACEMENT TEST Test method where test solution is periodically replaced at specific intervals during test.

STATION Any one of series of stakes or points indicating distance from point of beginning or reference.

STATION CROSSING Crossing at railroad station platform.

STATION MODEL Specified pattern for plotting, on weather map, meteorological symbols that represent state of weather at particular observing station.

STATION PLATFORM Area adjacent to railway track providing train access.

STATION PRESSURE Absolute air pressure at given reporting station.

STATION RATING CURVE Plotted stage-discharge relationship of stream at gauging station.

STATION TIME Time where crews, passengers and cargo are on board for flight.

STATION USE Energy used in generating plant as necessary in production of electricity.

STATIONARY COMPACTOR Machine that compacts refuse into detachable container at site of generation.

STATIONARY DREDGER Dredge vessel that remains on same location and swings cutter head sideways to remove material.

STATIONARY EMISSION SOURCE Emission source that's not self-propelled.

STATIONARY FRONT Front between warm and cold air masses moving very slowly or not at all.

STATIONARY GAS TURBINE Turbine that recovers heat from gas turbine exhaust gases to generate steam, heat water or preheat combustion air inlet.

STATIONARY SOURCE Fixed-site producer of pollution (e.g., power plants, petroleum refineries, heavy industrial) that primarily use industrial combustion processes.

STATIONARY WHITE NOISE Random energy with statistical characteristics that do not vary with time.

STATISTICAL PROCESS CONTROL CHART Daily performance plot of operating plant.

STATISTICAL SIGNIFICANCE Measure of how unlikely it is that result has occurred by chance.

STATISTICAL SOUND LEVEL Sound level exceeded percent of time during given sample period.

STATISTICS Branch of mathematics dealing with collection, analysis, interpretation and presentation of masses of numerical data.

STATOR That portion of machine which contains stationary (i.e., non-moving) parts that surround moving parts (e.g., rotor).

STATOR WINDINGS Armature or stationary winding of synchronous generator.

STATUTORY SPEED LIMIT Speed limit specified by State's traffic code.

STEADY Continuous display of traffic signal.

STEADY FLOW 1. Open channel condition where flow velocity profile, pressure, cross-section do not vary with time; 2. No change occurs with respect to time.

STEADY NON-UNIFORM FLOW Flow conditions change from point to point but do not change with time.

STEADY STATE Any state that does not change over time (i.e., dynamic equilibrium).

STEADY STATE CONDITION 1. When model input values are nearly constant for defined period of time; 2. Condition where system inputs and outputs are in equilibrium so that there is no net change in system with time.

STEADY STATE FLOW Characteristic of groundwater or vadose zone flow system where magnitude and direction of specific discharge at any point in space are constant in time.

STEADY STATE MODEL Model where variables being investigated do not change with time.

STEADY UNIFORM FLOW Flow conditions do not change from point to point or with time.

STEADY-STATE BIOCONCENTRATION FACTOR Ratio of concentration of substance in organism to concentration in water.

STEADY-STATE OR APPARENT PLATEAU Situation where amount of chemical substance taken into test subject from water is equal to amount being eliminated from test subject.

STEAM Water vapor that rises from boiling water.

STEAM ENGINE 1. Engine driven by steam; 2. Heat engine that uses steam as its primary working fluid to do mechanical work.

STEAM FOG Fog formed when water vapor is added to air that is much colder than source of vapor.

STEAM GENERATING UNIT Apparatus used for combusting fuel for purpose of producing steam.

STEAM HEATING SYSTEM System where steam transports heat.

STEAM STRIPPING Steam distillation process where volatile organic matter is removed from water.

STEAM TRAP Valve that filters out condensate and noncondensable gases without letting steam escape.

STEAMER CONNECTION Pumper outlet nozzle for fire hose attachment.

STEEL PIPE Long, hollow conduits, either welded or seamless, produced from steel alloys.

STEENING Lining made of brick, stone or other hard material for well or soakpit.

STEEP SLOPE 1. Average incline of 20% or more in case of grassland or 15% or more in case of other land; 2. Slope greater than critical slope for particular value of discharge.

STEEPLY DIPPING 20° to 45°.

STEERING ANGLE Angle between front of vehicle and steered wheel direction.

STEERING CURRENTS Prevailing synoptic scale flow which governs movement of smaller features embedded within it.

STEERING WINDS Prevailing synoptic scale flow which governs movement of smaller features embedded within it.

STEM 1. Faucet part where handle is attached; 2. Slender supporting member of individual part of plant (e.g., flower, leaf).

STEMFLOW Precipitation that makes it to ground , sliding down trunk or stem of plants.

STEMMING Crushed stone, soil, sand or drill cuttings used to plug unloaded portion of drill hole.

STEP AERATION Activated sludge process where settled wastewater is added into aeration tank at multiple locations.

STEP IRON Metal rung built into walls to provide steps.

STEPPED LEADER Faint, negatively charged channel that emerges from base of thunderstorm and propagates toward ground in series of steps of about 1 microsecond duration and 50-100 meters in length, initiating lightning stroke.

STERADIAN Solid angle subtended at center of sphere radius by portion of surface of sphere area.

STERILIZATION Removal or destruction of all microorganisms (e.g., pathogenic and other bacteria, vegetative forms and spores) or prevent organism from reproducing.

STERILIZATION SEWAGE TREATMENT Sterilant injection to remove remaining pathogens.

STERILIZED WASTEWATER Wastewater effluent where all microorganisms have been eliminated.

STERILIZER One of three groups of anti-microbials registered by EPA for public health uses.

STERN Rear part or after end of ship or vessel.

STEVEDORE Person involved in supervising, loading and unloading cargo from vessel.

STEVENSON SCREEN Standard shelter to shield meteorological instruments against radiation and precipitation.

STICKY LIMIT Lowest moisture content at which soil will stick to metal blade drawn across surface of soil mass.

STIFF MIX Mix that can be adapted to wide variety of looks from solid to terrazzo look.

STIFFENER Anything added to something in order to stiffen it (e.g., support beam).

STIFFNESS Rigidity or inflexibility.

STILL WATER Water with no flow.

STILL-WATER LEVEL Level that sea surface would assume in absence of wind waves.

STILLING BASIN Basin constructed to dissipate energy of fast-flowing water (e.g., from spillway, bottom outlet) and to protect streambed from erosion.

STILLING POOL Pool, usually lined with reinforced concrete, located below spillway, gate or valve into which discharge dissipates energy to avoid downstream channel degradation.

STILLWATER LEVEL Elevation that water surface would assume if all wave actions were absent.

STIMULATION 1. Artificially boost well output; 2. Any action or condition that creates response.

STIPULATED PRICE CONTRACT Contract between owner and contractor that establishes single, pre-determined fixed price regardless of contractor costs.

STIPULATION Something that's stated as condition of agreement.

STOCHASTIC EFFECTS State that occurs by chance and can be compared to deterministic effects that result in direct effect.

STOCHASTIC MODEL Mathematical model for estimating probability distributions of potential outcomes by allowing for random variation in one or more inputs over time.

STOCK POND Spring excavation for impounding livestock drinking water.

STOCK RAIL Fixed rail in railroad track against that switch rail operates.

STOCK SOLUTION Concentrated solution diluted to lower concentration.

STOCKPILE Storage pile of materials.

STOICHIOMETRIC Quantitative relationship between reactants and products of specific reaction or equation.

STOKE 1. Measurement of resistance of liquid to flow (kinematic viscosity); 2. Stir up and add fuel to furnace or fire.

STOKER Device that feeds fuel into furnace.

STOKES' LAW Mathematical expression that expresses frictional force (i.e., drag) on spherical particles falling in fluid.

STOMA One of tiny pores in epidermis of leaf or stem where gases and water vapor pass.

STOMATAL TRANSPIRATION Loss of water through specialized pores in leaves.

STONE 1. Concretion of earthy or mineral matter; 2. Rock.

STONE DRAIN Trench filled with loose stones to drain water.

STONE MATRIX ASPHALT Coarse graded, rut resistant engineered hot mix asphalt surface layer.

STONE RIPRAP Rock material used to armor shorelines, streambeds, bridge abutments, pilings and shoreline structures against scour, water or ice erosion.

STONEY GATE Gate, for large openings, rolled up or down inclined supporting rails by hoist through sprocket chains around ends of cylinder.

STOP BEACON Signal in standard traffic signal face with flashing circular red signal light.

STOP LINE Solid white pavement surface markings that inform drivers where they should stop when approaching intersection.

STOP LOG Removable hydraulic structure elements to adjust water level or discharge.

STOP RATE Number of full stops versus number of vehicles served.

STOP SWITCH Safety switch used to shut off machinery in emergency.

STOP VALVE Valve that may be turned to regulate flow of fluid through pipe.

STOP WORK ORDER Orders issued to stop work when hazardous substances or conditions exist.

STOP-AND-WASTE COCK VALVE Prevent flow of water in conduit, permitting downstream water to be emptied by letting air into line.

STOPCOCK VALVE Valve for turning on/off water or gas supply.

STOPING Upward erosion/piping action into embankment or foundation (possibly leading to breach).

STOPLOGS Large logs, timbers or steel beams placed on top of each other with their ends held in guides on each side of channel or conduit so as to provide cheaper or more easily handled means of temporary closure than bulkhead gate.

STOPPED DELAY Time vehicle is stopped in queue while waiting to pass through intersection.

STOPPED STATE Vehicle traveling less than 5 mi/hr.

STOPPED TIME Time spent on trip stopped due to other traffic being stopped.

STOPWAY Area beyond runway that can be used for deceleration during aborted takeoff.

STORAGE 1. Water artificially impounded in surface or underground reservoirs for future use; 2. Water naturally detained in drainage basin (e.g., groundwater, channel storage, depression storage); 3. Temporary holding of waste pending treatment or disposal.

STORAGE CAPACITY Refers to how much volume reservoir can hold.

STORAGE COEFFICIENT Volume of water aquifer releases from or takes into storage per unit surface area of aquifer per unit change in head.

STORAGE EQUATION Equation for conservation of mass when calculating storage.

STORAGE LENGTH Length of turn lane designed to accommodate 1.5 to 2 times average number of vehicle queued per cycle.

STORAGE PIT Underground storage facility of waste materials.

STORAGE RATIO Ratio of net available storage of reservoir to annual mean inflow.

STORAGE RESERVOIR Reservoir for storing raw water for significant time interval.

STORAGE TANK Containers that hold liquids or compressed gases.

STORAGE, BANK Water that has infiltrated from reservoir into surrounding land where it remains in storage until water level in reservoir is lowered.

STORAGE, CONSERVATION Storage of water for later release for usual purposes (e.g., municipal water supply, power, irrigation) in contrast with storage capacity used for flood control.

STORAGE, DEAD Volume (e.g., storage) that lies below lowest outlet invert and cannot readily be withdrawn from reservoir by gravity.

STORAGE, DEPRESSION 1. Volume of water contained (stored) in elevation concavity; 2. Small low points in undulating terrain that can store precipitation that otherwise would become runoff.

STORAGE, TOTAL Volume (including dead storage) of storage below maximum designed water surface level.

STORAGE, USABLE Quantity of groundwater of acceptable quality that can be economically withdrawn from storage.

STORAGE-REQUIRED FREQUENCY CURVE Graph showing frequency with which storage equal to or greater than selected amounts will be required to maintain selected rates of regulated flow.

STORATIVITY Volume of water aquifer releases from or takes into storage per unit surface are of aquifer per unit change in head where for confined aquifer, storage coefficient is equal to product of specific storage and aquifer thickness and for unconfined aquifer, storage coefficient is approximately equal to specific yield.

STORET EPA's computerized STOrage and RETrieval water quality data base that includes physical, chemical and biological data measured in water bodies throughout United States.

STORM Disturbance of ordinary average conditions of atmosphere which, unless specifically qualified, may include any or all meteorological disturbances (e.g., wind, rain, snow, hail, thunder).

STORM CENTER Point of lowest barometric pressure in thunderstorm.

STORM DATA Monthly publication documents chronological listing, by states, of occurrences of storms and unusual weather phenomena.

STORM DRAIN Access point or conduit for rainwater that drains directly downstream in drainage basin without going through sewers or water treatment to prevent floods.

STORM FLOW Storm runoff resulting from storm precipitation.

STORM HYDROGRAPH Hydrograph representing total flow or discharge past point.

STORM MOTION Speed and direction at which thunderstorm travels.

STORM OVERFLOW Structures, in combined sewers, used to separate sewage from storm water.

STORM OVERFLOW SEWER Sewer that transports excess flows from combined sewer system.

STORM RELATIVE Measured relative to moving thunderstorm, usually referring to winds, wind shear or helicity.

STORM RELATIVE MEAN RADIAL VELOCITY MAP WSR-88D radar product depicts full 360° sweep of radial velocity data with average motion of all identified storms subtracted out.

STORM RELATIVE MEAN RADIAL VELOCITY REGI WSR-88D radar product depicts 27 nautical miles by 27 nautical miles region of storm relative mean radial velocity centered on point which operator can specify anywhere within 124 nautical miles radius of radar.

STORM SCALE Referring to weather systems with sizes on order of individual thunderstorms.

STORM SEEPAGE That part of precipitation which infiltrates surface soil and moves toward streams as ephemeral, shallow, perched groundwater above main groundwater level.

STORM SEWER System of pipes (separate from sanitary sewers) that carries surface runoff from buildings and land surfaces, street wash and snow melt from land.

STORM SURGE Abnormal and sudden rise of sea along shore as result of winds of storm.

STORM TIDE Actual level of sea water resulting from astronomic tide combined with storm surge.

STORM TOTAL PRECIPITATION Estimate of accumulated rainfall from radar since last time there was one-hour or more, break in precipitation.

STORM TRACKING INFORMATION WSR-88D radar product displays previous, current and projected locations of storm centroids (forecast and past positions are limited to one hour or less).

STORM WARNING Warning of sustained surface winds or frequent gusts, in range of 48 knots (55 mph) to 63 knots (73 mph) inclusive, either predicted or occurring and not directly associated with tropical cyclone.

STORM WATCH Watch for increased risk of storm force wind event for sustained surface winds or frequent gusts, of 48 knots

(55 mph) to 63 knots (73 mph), but its occurrence, location, and/or timing is still uncertain.

STORM WATER Water that originates from precipitation (i.e., rain, snow and ice melt).

STORM WATER DRAINAGE WELLS Shallow Class V wells that are designed for disposal of rain water and melted snow.

STORMFLOW Water that flows over ground surface (i.e., runoff) or through ground directly into streams, rivers or lakes after rainfall or snow melt.

STORMWATER Water not absorbed into soil and rapidly flows downstream from stormwater runoff, snow melt runoff, surface runoff and drainage.

STORMWATER CHANNEL Channel used to convey stormwater or floodwater.

STORMWATER DISCHARGE Precipitation that does not infiltrate into ground or evaporate due to impervious land surfaces but instead flows onto adjacent land or water areas and is routed into drain/sewer systems.

STORMWATER DISCHARGE-RELATED ACTIVITIES Activities that cause, contribute to or result in stormwater point source pollutant discharges, including excavation, site development, grading and other surface disturbance activities.

STORMWATER INFILTRATION Process where stormwater runoff penetrates into soil from ground surface.

STORMWATER OR WASTEWATER COLLECTION SYSTEM Equipment necessary to collect and transport flow of surface water runoff resulting from precipitation or domestic, commercial or industrial wastewater to and from retention areas or treatment areas.

STORMWATER OVERFLOW WEIR Weir structure that enables flows to be redirected.

STORMWATER POINT SOURCE System for collecting and conveying stormwater runoff.

STORMWATER SEWER SYSTEM Pipe network that collect and transport waste- and stormwater to treatment plant or receiving waters.

STORMWATER TANK Sedimentation tank for storm water inflow, prior to discharge.

STOSS Facing towards flow of ice or fluid.

STOSS-AND-LEE TOPOGRAPHY Glacier slopes where stoss slope is smoothly abraded and lee slope roughly plucked.

STRADDLE CARRIER Vehicle used in ports and yards for stacking and moving freight containers.

STRAIGHT TEE Pipe fitting to make 90 ° branch from main run of pipe.

STRAIGHT TIME Normal working hours that are paid at base rate.

STRAIGHT-IN APPROACH VFR Entry into traffic pattern by interception of extended runway centerline without executing any other portion of traffic pattern.

STRAIGHT-LINE HODOGRAPH Hodograph with straight line shape caused by steady increase of winds with height (i.e., vertical wind shear).

STRAIGHT-LINE WINDS Any wind not associated with rotation used to differentiate from tornadic winds.

STRAIN Change in length per unit of length in given direction.

STRAIN METERS Instrument (e.g., Carlson type, vibrating-wire) that uses electrical principles to measure strain at location of strain meter.

STRAINER 1. Device to where liquid is passed for purification, filtering or separation from solid matter; 2. Perforated screen used to prevent solid bodies from mixing in liquid stream or flow line.

STRAPPED ELBOW Plumbing fitting for drop elbow.

STRATA 1. Distinct layers of stratified rock; 2. Layer of sedimentary rock, visually separable from other layers above and below.

STRATEGIC PLAN Written plan outlining government agency's framework for management.

STRATIFICATION 1. Layering of water in lakes and streams; 2. Formation of separate layers (e.g., temperature, plant, sedimentary rocks, animal life); 3. Subareas that exhibit reasonably homogeneous environmental conditions; 4. Vertical layering of vegetation in forest.

STRATIFIED RESERVOIR Reservoir with several thermal layers of water.

STRATIFIED SOILS Soils characterized by abrupt porosity changes at various depths.

STRATIFORM Having extensive horizontal development, as opposed to more vertical development characteristic of convection.

STRATIFORM RINGS AND BANDS Occur between active convective bands of hurricane outside of eye wall.

STRATIGRAPHY 1. Geology that deals with origin, composition, distribution and succession of strata; 2. Study of formation, composition, layering and sequence of sediments, whether consolidated or not; 3. Description of layered or stratified rocks.

STRATOCUMULUS Low-level clouds, existing in relatively flat layer but having individual elements.

STRATOPAUSE Boundary between stratosphere and mesosphere in atmosphere.

STRATOSPHERE 1. Region of atmosphere extending from top of troposphere to base of mesosphere (i.e., 0 to 25 miles above Earth's surface); 2. Region of uppermost atmosphere where temperature increases along with altitude due to absorption of solar ultraviolet radiation by ozone; 3. Layers of Earth's crust primarily comprised of stratified deposits.

STRATOSPHERIC OZONE In stratosphere, ozone forms ozone shield that prevents dangerous radiation from reaching Earth's surface.

STRATUM 1. Layer of sedimentary rock, visually separable from other layers above and below; 2. Layer of tissue, vegetation or atmosphere.

STRATUS Low, generally gray cloud layer with fairly uniform base.

STRAW Dried stalk of grain.

STREAK CLOUDS Large gap of cloud drawn out in wind shear direction.

STREAM Body of flowing water in natural channel.

STREAM BANK EROSION Removal of bank material by force of flowing water and caving of stream banks.

STREAM CAPACITY Total volume of water that stream can carry within normal high water channel.

STREAM DEGRADATION Lowering of fluvial surface (i.e., stream bed, floodplain) through erosional processes.

STREAM DISCHARGE Volume of flow passing stream cross-section in unit of time.

STREAM FORCE Horizontal load that must be resisted caused by water flowing around bridge pier.

STREAM GAGING Process of measuring flow characteristics (e.g., depths, areas, velocities, rates).

STREAM GAUGE Device that measures and records flow characteristics (e.g., water discharge, water surface elevation) at specific location on stream.

STREAM GAUGING Process of measuring water discharge or flow at particular point on stream or river.

STREAM GRADIENT Slope of stream surface elevation.

STREAM LINE 1. Imaginary line within flow that is everywhere tangent to velocity vector; 2. Arrows on weather chart showing wind speed and direction where head of arrow points toward where wind is blowing and length of arrow is proportional to wind speed. Sometimes shows wind direction and trajectory only.

STREAM MILE Distance of 1 mile along line connecting midpoints of channel of stream.

STREAM ORDER Method of numbering streams as part of drainage basin network by ranking of relative sizes of streams within watershed based on nature of their tributaries where smallest unbranched tributary is called first order, etc.

STREAM POWER Product of bed shear stress and mean cross-sectional velocity at cross-section for given flow.

STREAM PROFILE Plot of elevation of stream bed versus distance along stream.

STREAM REACH Continuous part of stream between two specified points.

STREAM SEGMENT 1. Specified portion of river with upstream inflow point and with downstream termination at control point; 2. Surface waters of approved planning area exhibiting common hydrological, natural, physical, biological or chemical processes.

STREAM SIZE 1. Channel depth or width; 2. Volume of water flowing.

STREAM, CONTINUOUS Stream that does not have interruptions in space (i.e., its course).

STREAM, EPHEMERAL One that flows only in direct response to precipitation and whose channel is at all times above water table.

STREAM, GAINING Stream or reach of stream that receives water from zone of saturation.

STREAM, INTERMITTENT One which flows only at certain times of year when it receives water from springs or from some surface source (e.g., rainfall runoff, melting snow).

STREAM, INTERRUPTED One which contains alternating reaches, that are either perennial, intermittent or ephemeral.

STREAM, LOSING Stream or reach of stream that contributes water to zone of saturation.

STREAM, MATURITY Stage in development of stream at which it has reached its maximum efficiency, when velocity is just sufficient to carry sediment delivered to it by tributaries, where flat uplands have been widely dissected by deep river valleys.

STREAM, PERENNIAL Stream that flows continually throughout year.

STREAM, SEASONAL Stream that flows only periodically or when it receives water from source (e.g., rainfall runoff, springs, melting snow).

STREAM-AQUIFER INTERACTIONS Relations of water flow and chemistry between streams and aquifers that are hydraulically connected.

STREAM-GAGING STATION Gaging station where record of discharge of stream is obtained.

STREAMBANK EROSION Natural process that occurs when forces exerted by flowing water exceed resisting forces of bank materials and vegetation.

STREAMBANK FAILURE Sudden collapse of streambank due to unstable condition.

STREAMBANK PROTECTION Protection of streams by reducing bank erosion or failure.

STREAMBED AT DAM AXIS Lowest-point elevation in streambed at axis or centerline crest of dam prior to construction.

STREAMFLOW Water flowing (i.e., discharge) in stream channel, whether or not it is affected by diversion or regulation.

STREAMFLOW DEPLETION Sum of water that flows into stream channel minus water that flows out.

STREAMFLOW RECORD Stream flow recorded over time.

STREAMFLOW REGULATION Measures established to control water in stream.

STREAMFLOW SOURCE ZONE Upstream headwaters area that drains into recharge zone.

STREAMFLOW WAVE Travelling wave caused by sudden increase of fluid flow.

STREAMLINE Line on map parallel to direction of fluid flow and shows flow patterns.

STREAMLINE FLOW Parallel layer fluid flow with no turbulent velocity fluctuations.

STREET Paved part of public way (i.e., road).

STREET CORNER Where two roads meet at intersection.

STREET ELBOW Conduit fitting that turns corner and has male and female end connections.

STREET REFUSE Waste collected by street and sidewalk sweeping.

STREET SWEEPER Machine used to sweep streets.

STREET TEE Conduit tee with external and internal threads on run connections and side outlet.

STREET WASH Paved area surface runoff that's directed into storm or sewer drains.

STREET WASTES Materials picked up by manual or mechanical sweeping; wastes from public waste receptacles; and material removed from catch basins.

STREETCAR Tram or light rail vehicle operating on city streets.

STRENGTH LIMIT STATE Maximum load carrying capacity of structure.

STRESS Force per unit area.

STRESS RANGE Difference between maximum and minimum values of stress in member that result from different loading conditions.

STRESS TRANSFER Imparting force in pre-tensioning anchoring device to concrete.

STRESSOR INDICATOR Characteristic measured to quantify natural process, environmental hazard or management action that results in changes in exposure and habitat.

STRESSORS Physical, chemical or biological entities that can induce adverse effects on ecosystems or human health.

STRIATIONS 1. Scratches or grooves in bedrock caused by rocks within glacier grinding Earth's surface as glacier moves; 2. Grooves or channels in cloud formations, arranged parallel to flow of air and therefore depicting airflow relative to parent cloud.

STRIKE 1. Direction taken by bedding or fault plane as it intersects horizontal; 2. To be aligned or to trend in direction at right angles to direction of dip.

STRINGER 1. Horizontal timber that supports upright posts or hull of vessel; 2. Side rail supporting rungs of ladder or steps of flight of stairs.

STRIP CROPPING Growing crops in systematic arrangement of strips or bands that serve as barriers to wind and water erosion.

STRIP-MINING Process that uses machines to scrape soil or rock away from mineral deposits just under Earth's surface.

STRIPPED GASES Gases that are released from liquid by bubbling air through liquid or by allowing liquid to be sprayed or tumbled over media.

STRIPPED ODORS Odors that are released from liquid by bubbling air through liquid or by allowing liquid to be sprayed or tumbled over media.

STRIPPING 1. Removal of surface layer or deposit for purpose of excavating other material beneath it; 2. Physical separation process where one or more components are removed from liquid stream by vapor stream.

STRONTIUM-90 Radioactive isotope of strontium present in fallout from nuclear reactions.

STRUCK BUCKET Volume of material in bucket after being leveled with edge.

STRUCTURAL CAPACITY maximum load and/or number of repetitions structure can support.

STRUCTURAL DEFORMATION Distortion in walls of tank after liquid has been added or removed.

STRUCTURAL FILL Material placed and compacted in accordance with approved design criteria in order to improve land for intended use.

STRUCTURAL HEIGHT Vertical distance between lowest point in excavated foundation to top of dam.

STRUCTURAL LUMBER Lumber larger that 2X4 inches used in framing houses or buildings.

STRUCTURAL MASS CONCRETE Concrete footing at least 5 feet or other concrete placements at least 4 feet.

STRUCTURAL NUMBER Number expressing structural strength of pavement.

STRUCTURAL OVERLAY Overlay used to increase structural capacity.

STRUCTURAL SUPPORT Support providing stiffness and strength to resist forces.

STRUCTURALLY CONTINUOUS BARRIER Barrier only interrupted at deck joints.

STRUCTURE CONTOUR Imaginary line connecting points of equal elevation (i.e., contour) on single surface.

STRUT Inside brace.

STRUVITE Orthorhombic ammonium magnesium phosphate mineral.

STUD Bolt having one end firmly anchored.

STUDY To review materials already examined.

STUDY UNIT Major hydrologic system of United States where NAWQA studies are focused that's geographically defined by combination of ground- and surface-water features and generally encompass more than 4,000 mi^2 of land area.

STUDY-UNIT SURVEY Broad assessment of water-quality conditions of major aquifer systems of each NAWQA Study Unit.

STUFFING BOX Assembly containing seal or sealing material, around shaft.

SUB-ORGANIZATION Any organization (e.g., agencies, departments, offices, local jurisdictions) having supportive role in emergency planning and preparedness.

SUB-SURFACE LAYER Composed of well mixed sediments brought up from inactive layer plus sediment which has deposited from water column.

SUB-SYNOPTIC LOW Mesoscale low-pressure center.

SUBACUTE TOXICITY Adverse effects occurring after multiple or continuous exposure between 24 hours and 28 days.

SUBAQUEOUS Existing below water surface.

SUBAQUEOUS PIPE Submerged conduit.

SUBARTESIAN WELL Artesian well with insufficient head to raise water above ground level.

SUBBASE Layer used in pavement system between subgrade and base course or between subgrade and portland cement concrete pavement.

SUBCHRONIC 1. Of intermediate duration; 2. Describes studies or periods of exposure lasting between 5 and 90 days.

SUBCHRONIC DERMAL TOXICITY Adverse effects occurring as result of repeated daily exposure of experimental animals to chemical by dermal application for part of life span.

SUBCHRONIC EXPOSURE Multiple or continuous exposures lasting for approximately 10% of experimental species lifetime.

SUBCHRONIC INHALATION TOXICITY Adverse effects occurring as result of repeated daily exposure of experimental animals to chemical by inhalation for part of life span.

SUBCRITICAL FLOW State of flow where water depth is above critical depth and velocities are less than critical where gravity forces dominate inertial forces and flow is described as slow or tranquil.

SUBDRILLING 1. Overdrilling; 2. Drilling below final grade.

SUBDUCTION ZONE Region of Earth where one tectonic plate dives beneath another into interior of Earth.

SUBGRADE Soil prepared and compacted to support structure or pavement system.

SUBGRADE SURFACE Surface of earth or rock prepared to support structure or pavement system.

SUBHARMONIC Having frequency that's fraction of fundamental frequency.

SUBIRRIGATION Applying irrigation water below ground surface either by raising water table within or near root zone or by use of buried perforated or porous pipe system which discharge directly into root zone.

SUBJECT APPROACH Two-way and all-way STOP controlled intersection approach.

SUBLETHAL Stimulus below level that causes death.

SUBLIMATION Process of transformation directly between solid and gas, without passing through intermediate liquid phase.

SUBLITTORAL ZONE 1. Depth from below low tide to edge of continental shelf; 2. Portion of lake too deep for plant roots to grow.

SUBMEANDER Small meander contained with banks of main channel, associated with relatively low discharges.

SUBMERGED AQUATIC VEGETATION Vegetation (i.e., sea grasses) that cannot withstand excessive drying and therefore live with their leaves at or below water surface that provides important habitat for young fish and other aquatic organisms.

SUBMERGED ORIFICE Orifice that discharges completely under water.

SUBMERGED OUTLET Outlet completely under water.

SUBMERGED PIPE Conduit placed on bed in water body.

SUBMERGED SOIL Soils saturated with water for sufficiently long time.

SUBMERSED PLANT Plant that lies entirely beneath water surface, except for flowering parts in some species.

SUBMERSIBLE PUMP Enclosed pump and motor unit suitable for submerged installation.

SUBREFRACTION Bending of radar beam in vertical that is less than under standard refractive conditions.

SUBSEQUENT River, follow channels, cut naturally into readily erodible bed.

SUBSERE Secondary series of ecological communities beginning after succession has been interrupted.

SUBSIDENCE 1. Gradual downward settling or sinking of Earth's surface due to underground excavation (e.g., removal of groundwater, gas, oil) with little or no horizontal motion; 2. Compression of soft aquifer materials in confined aquifer due to pumping of water from aquifer; 3. Descending motion of air in atmosphere occurring over rather broad area.

SUBSIDENCE CONTROL WELLS Class V wells that are used to control land subsidence caused by groundwater withdrawal or over pumping of oil and gas.

SUBSIDENCE INVERSION Temperature inversion that develops aloft as result of air gradually sinking over wide area and being warmed by adiabatic compression, usually associated with subtropical high pressure areas.

SUBSOIL Layer of earth below topsoil.

SUBSOIL DRAIN Perforated conduit laid below ground, designed to remove excess water from surrounding soil.

SUBSOIL DRAINAGE PIPE Underground, perforated conduit installed to conduct and intercept groundwater.

SUBSOIL EXPLORATION Process of identifying layers of deposits that underlie proposed area and their physical characteristics.

SUBSTATION 1. Facility equipment that switches, changes or regulates electric voltage; 2. Location where observations are taken or other services are furnished by people not located at National Weather Service offices who do not need to be certified to take observations.

SUBSTRATE 1. Physical surface upon which organism grows, lives or is attached; 2. Underlying layer; 3. Surface to which substance adheres; 4. Metal, plated with another metal, that has different physical properties.

SUBSTRATE EMBEDDEDNESS CLASS Visual estimate of riffle streambed substrate larger than gravel surrounded or covered by fine sediment ($\leq$ 2 mm, sand or finer).

SUBSTRATE SIZE Diameter of streambed particles (e.g., clay, silt, sand, gravel, cobble, boulders).

SUBSTRATUM Layer that lies underneath another.

SUBSURFACE Something below layer on surface.

SUBSURFACE AIR Gases within pore space interstices in soil.

SUBSURFACE DRAIN Shallow drain installed in irrigated field to intercept rising groundwater level and maintain water table at acceptable depth below land surface.

SUBSURFACE FLUID DISTRIBUTION SYSTEM Assemblage of perforated pipes, drain tiles or other similar mechanisms intended to distribute fluids below surface of ground.

SUBSURFACE INTERFLOW Lateral motion of water through upper layers until it enters stream channel that usually takes longer to reach stream channels than runoff.

SUBSURFACE INVESTIGATION Appraisal of subsurface conditions at building site by analytic techniques of boring data.

SUBSURFACE IRRIGATION SYSTEM Irrigation by means of underground porous tile or its equivalent.

SUBSURFACE RUNOFF That part of precipitation which infiltrates surface soil and moves toward streams as ephemeral, shallow, perched groundwater above main groundwater level.

SUBSURFACE STORM FLOW Lateral motion of water through upper layers until it enters stream channel that usually takes longer to reach stream channels than runoff.

SUBSURFACE WATER Water, from precipitation, beneath Earth's surface.

SUBTERRANEAN STREAM Body of subsurface water flowing through interconnected caves or caverns or through large interconnected interstices.

SUBTIDAL 1. Continuously submerged; 2. Area affected by ocean tides.

SUBTROPICAL ANTICYCLONE Semi-permanent anticyclone located, on average, over oceans near 30° North and 30° South latitude.

SUBTROPICAL CYCLONE Non-frontal low pressure system that has characteristics of both tropical and extratropical cyclones.

SUBTROPICAL DEPRESSION Subtropical cyclone where maximum 1-minute sustained surface wind is 38 mph or less.

SUBTROPICAL JET Jet stream is usually found between 20° and 30° latitude at altitudes between 12 and 14 km.

SUBTROPICAL STORM Subtropical cyclone where maximum 1-minute sustained surface wind is 39 mph or more.

SUBURBAN COACH Bus line that generally connects suburban to downtown area.

SUBURBAN STREET Street adjacent to or surrounding city.

SUBURBAN TRANSIT BUS Bus line that generally connects suburban to downtown area.

SUBWATERSHED Topographic perimeter of catchment area of stream tributary.

SUCCESSION Directional, orderly process of community change where community modifies physical environment to eventually establish ecosystem that is as stable as possible at site in question.

SUCTION 1. Inlet (i.e., pull side) of pump; 2. Process of creating imbalance in pressure to draw matter from one place to another.

SUCTION BOOSTER Elevated pressure contributed to pumping system by booster pump.

SUCTION DREDGE Dredge vessel that sucks bed material (e.g., rocks, sand, gravel, silt) through suction hose and transports it discharge point.

SUCTION HOSE Hose located at inlet connection where fluid enters pump.

SUCTION LIFT 1. Pressure (i.e., negative) on suction side of pump; 2. Elevation from water surface to pump impeller center.

SUCTION LINE 1. Pump intake line; 2. Conduit between tank and pump inlet.

SUCTION PIPE Conduit between supply source and pump suction inlet.

SUCTION PIT Excavation dug where material is picked up by pump suction line.

SUCTION PRESSURE 1. Working pressure measured in compressor inlet suction line; 2. Pressure where cell absorbs water.

SUCTION PUMP Pump for drawing liquid through pipe.

SUCTION VALVE Suction conduit check valve.

SUCTION VORTEX Small but very intense vortex within tornado circulation.

SUCTION-CUTTER DREDGE Dredge with rotating screw cutter head that fragments material in suction line.

SUDDEN ACCIDENTAL OCCURRENCE Occurrence not continuous or repeated in nature.

SUDDEN COMMENCEMENT Abrupt increase or decrease in northward component of geomagnetic field, which marks beginning of geomagnetic storm.

SUDDEN DRAWDOWN Rapid drawdown in water level behind quarry, dam or embankment.

SUDDEN IMPULSE Sudden perturbation of several gammas in northward component of low-latitude geomagnetic field, not associated with following geomagnetic storm.

SUDDEN IONOSPHERIC DISTURBANCE HF propagation anomalies due to ionospheric changes resulting from solar flares, proton events and geomagnetic storms.

SUGAR Any of various small carbohydrates.

SULFATE Any salt or ester of sulfuric acid.

SULFATE ATTACK Damage to concrete caused by effects of chemical reaction between sulfates in soils or groundwater and hydrated lime and hydrated calcium aluminate in cement paste that results in considerable expansion and disruption of paste.

SULFATE-REDUCING BACTERIA Bacteria that perform anaerobic respiration utilizing sulfate and reducing it to hydrogen sulfide.

SULFITE Any salt or ester of sulfurous acid.

SULFONAMIDE Any amide of sulfonic acid.

SULFUR Chemical element with yellowish green color and atomic number of 16.

SULFUR BACTERIA Any of several bacteria which oxidize inorganic sulfur compounds.

SULFUR DIOXIDE Pungent, colorless, gas formed primarily by combustion of fossil fuels that's considered pollutant when present in large amounts and causes acid rain.

SULFUR HEXAFLUORIDE Potent greenhouse gas primarily used as electrical insulator in high voltage equipment.

SULFUR RECOVERY UNIT Apparatus that uses combustion and catalytic process to recover pure form of sulfur from acid gas streams containing hydrogen sulfide.

SULLAGE 1. Silt or sediment deposited from flowing water; 2. Liquid discharges from kitchens, washbasins, toilets, etc.

SUMMATION PRINCIPLE States that sky cover at any level is equal to summation of sky cover of lowest layer plus additional sky cover provided at all successively higher layers up to and including layer in question.

SUMMER Typically warmest season of year during which Sun is most nearly overhead.

SUMMER SOLSTICE Time at which Sun is farthest north in Northern Hemisphere, on or around June 21.

SUMMER SWITCH Electrical fan, manual switch of forced-air furnace.

SUMP 1. Pit, pool or tank for draining, collecting or storing liquid for drainage or disposal; 2. Chamber located at entrance to pump which provides water to pump; 3. Lowest part of mineshaft where water drains.

SUMP PUMP Pump used for removing collected water from sump.

SUN DOG Either of two colored luminous spots that appear at roughly 22° on both sides of Sun at same elevation caused by refraction of sunlight passing through ice crystals.

SUN GEAR Central gear in planetary gear box that spins planet gears.

SUN PILLAR Bright column above or below Sun produced by reflection of sunlight from ice crystals.

SUN POINTING Alignment of radar antenna by locating position of Sun in sky, which has exactly known position given radar's location and present time.

SUNK DRAIN Sink conduit for liquid wastes.

SUNLIGHT DIRECT AQUEOUS PHOTOLYSIS RATE CONSTANT First-order rate constant in units of day^{-1} and is measure of rate of disappearance of chemical dissolved in water body in sunlight.

SUNNY 1. When there are no opaque (not transparent) clouds; 2. Same as Clear.

SUNRISE Phenomenon of Sun's daily appearance on eastern horizon as result of Earth's rotation.

SUNSET Phenomenon of Sun's daily disappearance below western horizon as result of Earth's rotation.

SUNSPOT 1. Area seen as dark spot on photosphere of Sun; 2. Concentrations of magnetic flux, typically occurring in bipolar clusters or groups.

SUNSPOT GROUP CLASSIFICATION 1. A: small single unipolar sunspot or very small group of spots without penumbra; 2. B: bipolar sunspot group with no penumbra; 3. C: elongated bipolar sunspot group where one sunspot must have penumbra; 4. D: elongated bipolar sunspot group with penumbra on both ends of group; 5. E: elongated bipolar sunspot group with penumbra on both ends where longitudinal extent of penumbra exceeds 10° but not 15 °; 6. F: elongated bipolar sunspot group with penumbra on both ends where longitudinal extent of penumbra exceeds 15 °; 7. H: unipolar sunspot group with penumbra.

SUNSPOT NUMBER Daily index of sunspot activity.

SUPER TYPHOON Typhoon having maximum sustained winds of 130 knots (150 mph) or greater.

SUPERCELL Thunderstorm consisting of one quasi-steady to rotating updraft which may exist for several hours (potentially most dangerous of convective storm types) that possess structure to generate vast majority of long-lived strong and violent (F2-F5) tornadoes, as well as downburst damage and large hail.

SUPERCHLORINATION Chlorination with doses that are deliberately selected to produce water free of combined residuals so large as to require dechlorination.

SUPERCOOL To cool liquid below its freezing point without solidification or crystallization.

SUPERCOOLED LIQUID WATER Liquid water can survive at temperatures colder than 0 °C.

SUPERCOOLING Metastable state of liquid staying in liquid phase.

SUPERCRITICAL FLOW State of flow where water depth is below critical depth and velocities are greater than critical where inertial forces dominate gravitational forces that's described as rapid or shooting.

SUPERCRITICAL FLUID Fluid above its critical temperature and critical pressure.

SUPERCRITICAL OXIDATION Water warmed and pressurized, allowing it to break down complex organic compounds.

SUPERCRITICAL WATER Type of thermal treatment using moderate temperatures and high pressures to enhance ability of water to break down large organic molecules into smaller, less toxic ones.

SUPERELEVATION 1. Cant of railway track; 2. Difference in elevation (height) between its two edges (i.e., curve).

SUPERELEVATION RATE Rate of rise of pavement at full superelevation.

SUPERFLOOD Flood greater than 100-year flood.

SUPERFUND Program operated that funds and carries out EPA solid waste emergency and long-term removal and remedial activities.

SUPERFUND INNOVATIVE TECHNOLOGY EVALUATION PROGRAM EPA program to promote development and use of innovative treatment and site characterization technologies in Superfund site cleanups.

SUPERINTENDENT Person authorized to supervise, direct or administer.

SUPERNATANT 1. Lying above sediment or precipitate; 2. Floating on surface of liquid.

SUPERNATANT LIQUOR Liquid above solids from settling, precipitation or centrifugation.

SUPERPOSITION PRINCIPLE Addition or subtraction of two or more different solutions of governing linear partial differential equation to obtain composite solution of partial differential equation.

SUPERREFRACTION Bending of radar beam in vertical that is greater than sub-standard refractive conditions.

SUPERSATURATION 1. Condition of solution being more highly concentrated than is normally possible; 2. Condition of vapor having higher saturation vapor pressure than is normally possible.

SUPERSTRUCTURE 1. Any structure or edifice built on something else (i.e., raised on foundation or basis); 2. Structural parts of bridge that comprise horizontal span.

SUPERVISORY CONTROL System used to monitor conditions and operate mechanical features associated with facility from location other than at site.

SUPERVISORY INDICATOR Gauge (e.g., visual or audible) indicating state.

SUPERVISORY SWITCH Monitoring switch used to signal status of important settings.

SUPPLEMENTAL AGREEMENT Contract modification that's accomplished by mutual action of both parties.

SUPPLEMENTAL REGISTRATION Arrangement whereby registrant licenses another company to market its product under second company's registration.

SUPPLEMENTAL SIGNAL FACE Non-primary signal face used to provide enhanced visibility.

SUPPLEMENTAL SPECIFICATIONS Specifications prepared during final design that are based on standard specifications.

SUPPLEMENTARY CONTROL SYSTEM Any technique for inviting concentration of pollutant in ambient air by varying its emission.

SUPPLIER OF WATER Any person who owns or operates public water supply.

SUPPLY LINES Conduits joining source (e.g., water) to distribution system.

SUPPLY MAIN Primary piping or electrical system that brings water, gas or electricity into structure.

SUPPLY PIPE Service pipe connecting mains.

SUPPLY PUMP Pumps that transport liquid either directly to supply network or area.

SUPPRESSED WEIR Measuring weir with sides flush with channel sides (i.e., no end contractions).

SUPRAPERMAFROST WATER Free water in ground above permafrost.

SURCHARGE 1. To fill or load to excess; 2. Volume or space in reservoir between controlled retention water level and maximum water level; 3. Any load or vertical pressure applied to ground surface.

SURCHARGE CAPACITY Reservoir capacity between maximum water surface elevation and highest of following elevations: top of exclusive flood control capacity, top of joint use capacity or top of active conservation capacity (i.e., temporary storage) for use in passing inflow design flood through reservoir.

SURCHARGE STORAGE Reservoir capacity between maximum water surface elevation and highest of following elevations: top of exclusive flood control capacity, top of joint use capacity or top of active conservation capacity (i.e., temporary storage) for use in passing inflow design flood through reservoir.

SURETY Promise to pay sum of money in event that another person fails to fulfill obligation.

SURF ZONE Area of water between high tide level on beach and seaward side of breaking waves.

SURF ZONE FORECAST National Weather Service routine or event driven forecast product geared toward non-boating marine users issued for area extending from area of water between high tide level on beach and seaward side of breaking waves.

SURFACE AERATION Mix, circulate or dissolve air into liquid or another substance.

SURFACE AREA OF LAKE Area encompassed by boundary of lake as shown on maps or photographs.

SURFACE CASING Well casing to protect shallow zones and weak formations.

SURFACE COURSE Pavement that's flexible, provides friction for tires and protects underlying base from traffic and water.

SURFACE CURVE Longitudinal profile of channel water surface flow.

SURFACE DISTRESS Any indication or sign of pending pavement failure.

SURFACE DRAIN Drain to remove surface water around building foundations.

SURFACE ENERGY BUDGET Energy or heat budget at Earth's surface, considered in terms of fluxes through plane at Earth-atmosphere interface including radiative, sensible, latent and ground heat fluxes.

SURFACE EVAPORATION Vaporization that occurs on surface as it changes into gas phase.

SURFACE FILTRATION Solids retained on surface of filter.

SURFACE IMPOUNDMENT 1. Indented area in land's surface (e.g., pit, pond, lagoon); 2. Treatment, storage or disposal of liquid hazardous wastes in ponds.

SURFACE INVERSION Temperature inversion based at Earth's surface (i.e., temperature increase with height beginning at ground).

SURFACE LOADING Settling tank and clarifier parameter that determines how much water can be processed each day per area of tank.

SURFACE MOISTURE Absorbed water and moisture coating surface that's not chemically bound.

SURFACE PRESSURE CHART Charts that show surface pressure pattern using isobars (lines of equal pressure) and indicate areas of high and low pressure.

SURFACE PUMP Mechanism for removing water or wastewater from sump or wet well.

SURFACE RECYCLING Rehabilitation process that restores pavement in preparation for final thin wearing course.

SURFACE RUNOFF Precipitation, snow melt or irrigation that travels overland and is in excess of what can infiltrate soil surface and be stored.

SURFACE SHADOWS Dark areas on surface of hot mix asphalt.

SURFACE SLOPE Change of elevation per unit of surface slope length.

SURFACE SOIL Upper part of soil ordinarily moved in tillage or its equivalent in uncultivated soils, about 10 to 20 cm in thickness.

SURFACE TENSION Attraction of molecules to each other on liquid's surface.

SURFACE TREATMENT Processes that alters surface of manufactured item to achieve final property.

SURFACE URANIUM MINES Strip mining operations for removal of uranium-bearing ore.

SURFACE VEIL Surface that provides barrier between outermost layer of glass and composite surface.

SURFACE VIBRATION Method of compacting soil using vibrating plate or vibrating smooth drum roller used on surface of soil placed.

SURFACE WATER Water on surface of Earth naturally open to atmosphere (e.g., in stream, river, lake, reservoir, impoundment, sea, estuary).

SURFACE WATER DRAIN Conduit designed to drain excess rain and groundwater from impervious surfaces.

SURFACE WAVE MAGNITUDE Magnitude of earthquake measured as common logarithm of resultant of maximum mutually perpendicular horizontal displacement amplitudes, in microns, of 20 second period surface waves.

SURFACE WAVES Waves that travel along or near surface that also include Rayleigh and Love Waves of earthquake.

SURFACE WEATHER CHART Analyzed synoptic chart of surface weather observations that shows distribution of sea-level pressure (i.e., position of highs, lows, ridges, troughs) and location and nature of fronts and air masses.

SURFACE WIND Wind blowing 33 feet above Earth's surface.

SURFACE-ACTIVE Having ability to modify surface energy by altering properties and lowering surface tension between phases.

SURFACE-ACTIVE AGENT Substance that lowers surface tension of medium where it's dissolved, interfacial tension with other phases and is positively adsorbed at liquid-vapor interface and other interfaces.

SURFACE-BASED CONVECTION Convection occurring within surface-based layer (i.e., layer where lowest portion is based at or very near Earth's surface.).

SURFACE-BASED TEMPERATURE INVERSION Layer of atmosphere where air temperature increases with height.

SURFACE-WATER TREATMENT RULE 1. Rule that specifies maximum contaminant level goals for Giardia lamblia, viruses and Legionella and promulgates filtration and disinfection requirements for public water systems using surface-water or groundwater sources under direct influence of surface water; 2. Specify water quality, treatment and watershed protection criteria under which filtration may be avoided.

SURFACING ASBESTOS-CONTAINING MATERIAL Asbestos-containing material sprayed or troweled on or otherwise applied to surfaces.

SURFACING MATERIAL Material sprayed or troweled onto structural members for fireproofing, acoustical or decorative purposes.

SURFACTANT Detergent compound that promotes lathering and reduces surface tension of liquid.

SURFICIAL BED MATERIAL Upper surface (0.1 to 0.2 foot) of bed material sampled using U.S. Series Bed-Material Samplers.

SURGE 1. Sudden transient rush, flood or increase; 2. Jet of material from active regions that reaches coronal heights and then either fades or returns into chromosphere along trajectory of ascent; 3. Sudden electrical spike or increase of voltage and current; 4. Tapered part of windlass barrel or capstan.

SURGE CHAMBER Chamber or tank connected to pipe and located at or near valve that may quickly open or close or pump that may suddenly start or stop to allow water to flow into or out of pipe and minimize any sudden positive or negative pressure waves or surges in pipe.

SURGE CONTROL TANK Tanks or vessels used only for controlling pressure in pipeline system during surges or other variations from normal operations.

SURGE IRRIGATION Irrigation technique wherein flow is applied to furrows (or less commonly, borders) intermittently during single irrigation set.

SURGE PIPE Open-topped standpipe to discharge pressure from surge.

SURGE SUPPRESSOR Apparatus that minimizes surges in pipeline.

SURGE TANK Tanks installed in pipeline to relieve excess pressure caused by water hammer and to provide water supply to reduce negative pressure if valve suddenly opened.

SURROGATE Analyte that behaves similarly to target analyte, but highly unlikely to occur in sample.

SURROGATE DATA Data from studies of test organisms or test substance that are used to estimate characteristics or effects on another organism or substance.

SURVEILLANCE SYSTEM Series of monitoring devices designed to check on environmental conditions.

SURVEY 1. Measure external vertical and horizontal movement on surface; 2. Sampling of representative number of sites during given hydrologic condition; 3. Record of area and features examined to construct map, plan or description.

SUSCEPTIBILITY 1. Level organism is influenced by pesticide in specific degree of vulnerability; 2. Measure of how easily dielectric polarizes in response to external electric field.

SUSCEPTIBILITY ANALYSIS Analysis to determine whether Public Water Supply is subject to significant pollution from known potential sources.

SUSPECT MATERIAL Building material suspected of containing asbestos (e.g., surfacing material, floor tile, ceiling tile, thermal system insulation).

SUSPENDED 1. State of floating in water rather than being dissolved in it; 2. Amount (concentration) of undissolved material in water-sediment mixture.

SUSPENDED BED MATERIAL LOAD Portion of suspended load composed of particle sizes found in bed material.

SUSPENDED GROWTH PROCESS Wastewater treatment procedure where microorganisms are responsible for conversion of organic matter or other constituents in wastewater to gases and cell tissue.

SUSPENDED LOAD Sediment supported by upward components of turbulence, carried with flow of water and that stays in suspension for appreciable length of time.

SUSPENDED SEDIMENT Sediment (e.g., particles of rock, sand, soil, organic detritus) transported in suspension in water and remains in suspension for considerable period of time without contact with bottom due to upward components of turbulence and currents.

SUSPENDED SEDIMENT DISCHARGE Amount of sediment moved by stream, above bed layer, in given time.

SUSPENDED SEDIMENT LOAD Stream's total sediment load, transported by flow, that rarely contacts streambed.

SUSPENDED SOLIDS Small particles or solid pollutants that float on surface of or are suspended in, sewage or other liquids.

SUSPENDED SOLIDS, RECOVERABLE Amount of given constituent in solution after part of representative water-suspended sediment sample retained on 0.45-micrometer filter has been extracted or digested.

SUSPENDED SOLIDS, TOTAL Total amount of given constituent in part of water-sediment sample retained on 0.45-micrometer membrane filter.

SUSPENDED-SEDIMENT CONCENTRATION 1. Ratio of mass of dry sediment in water-sediment mixture to mass of water-sediment mixture; 2. Velocity-weighted concentration of suspended sediment in sampled zone (from water surface to point approximately 0.3 foot above bed).

SUSPENDED-SEDIMENT DISCHARGE Rate of sediment transport, as measured by dry mass or volume, that passing cross-section in unit of time.

SUSPENDED-SEDIMENT LOAD Given characteristic of material in suspension that passes point during specified period of time.

SUSPENSION 1. Method of sediment transport where air or water turbulence supports weight of sediment particles preventing them from settling out or being deposited; 2. Suspending use of pesticide when EPA deems it necessary to prevent imminent hazard resulting from its continued use.

SUSPENSION CULTURE Cells growing in liquid nutrient medium.

SUSPENSION OF WORKS Instructing contractor to suspend progress or any part of works.

SUSTAINABLE DEVELOPMENT Development that seeks to produce sustainable economic growth while ensuring future generation growth.

SUSTAINABLE YIELD Yield that can be extracted without reducing resource.

SUSTAINED OVERDRAFT Long-term withdrawal from aquifer of more water than is being recharged.

SUSTAINED WIND Wind speed determined by averaging observed values over 2-minute period.

SWABBING Removal of deposits from conduits using swab.

SWALE Slight depression that carries water.

SWAMP Area intermittently or permanently covered with water and having trees and shrubs.

SWAY BRACING Horizontal bridge bracing that prevents swaying.

SWEATING Water droplet formation on pipe or tank exterior due to condensation.

SWEEP T Pipe fitting where T branch is curved.

SWEET GAS Natural gas with small amounts of hydrogen sulfide.

SWEET WATER Solution of diluted glycerol.

SWELL 1. Increase of bulk in soil or rock when excavated; 2. Wind-generated waves that have travelled out of their generating area.

SWELL DIRECTION Compass direction from which swell wave are coming from.

SWELL HEIGHT Vertical distance between any swell crest and succeeding swell wave trough.

SWELL PERIOD Time that it takes successive swell wave crests or troughs pass fixed point.

SWELL PRESSURE Pressure required to maintain zero expansion.

SWEPT PATH WIDTH Difference in paths of front-most outside wheel and rear-most inside wheel of turning vehicle.

SWILL Mixture of solid and liquid food scraps.

SWING BRIDGE Movable bridge that has vertical locating pin and support ring as its primary structural support.

SWING CHECK VALVE Check valve to block reverse flow or allow forward flow.

SWING, CHECK VALVE Check valve to block reverse flow or allow forward flow.

SWITCH Apparatus to turn electric current on and off or direct its flow.

SWITCH STAND Device that manually operates switches.

SWITCHBOARD Device that directs electricity from one source to another.

SWITCHBOX Box with electrical switches mounted on it.

SWITCHGEAR Interrupting device used in connection with generation, transmission, distribution and conversion of electric power for controlling, metering protecting and regulating devices.

SWITCHING DEVICE Apparatus that opens and closes electrical circuit.

SWITCHING SPEED Time needed to change device power state.

SWITCHING STATION Facility equipment used to tie together two or more electric circuits through switches.

SYMBIOSIS Relationship of mutual benefit, especially among different species.

SYMBOL Any object meant to represent another.

SYMMETRIC DOUBLE EYE Concentrated ring of convection that develops outside eye wall in symmetric, mature hurricanes that propagates inward and leads to double-eye.

SYMMETRICAL DISTRIBUTION Equal distribution above and below mean.

SYNCHRONOUS CONDENSERS Synchronous machine running without mechanical load and supplying or absorbing reactive power.

SYNCHRONOUS DETECTION Radar processing that retains received signal amplitude and phase but that removes intermediate frequency carrier.

SYNCHRONOUS SATELLITE Satellite that maintains orbit location above point on Earth.

SYNCLINE Troughlike downward sag or fold in rock layers where strata dip inward from both sides toward axis that's opposite of anticline.

SYNERGISM 1. Interaction of two or more chemicals that results in effect greater than sum of their separate effects; 2. Benefits resulting from combining two different: groups, people, objects, substances, organs, organisms or processes.

SYNERGISTIC EFFECT Biological response to exposure to multiple chemicals that is greater than sum of effects of individual agents.

SYNGAS Gas obtained by heating coal and steam.

SYNOPSIS Broad discussion of weather pattern expected across any given area, generally confined to 0-48 hour time frame.

SYNOPTIC Relating to general weather pattern over wide region, (e.g., areas of high and low pressure, frontal boundaries) as opposed to mesoscale or smaller features (e.g., thunderstorm).

SYNOPTIC CODE Rules and procedures established by World Meteorological Organization for encoding weather observations.

SYNOPTIC SCALE Size scale in meteorology referring generally to weather systems (i.e., migratory high and low pressure systems of lower troposphere) with horizontal dimensions of several hundred miles or more.

SYNOPTIC SITES Sites sampled during short-term investigation of specific water-quality conditions during selected seasonal or hydrologic conditions to provide improved spatial resolution for critical water-quality conditions.

SYNOPTIC STUDIES Short-term investigations of specific water-quality conditions during selected seasonal or hydrologic periods to provide improved spatial resolution for critical water-quality conditions.

SYNOPTIC TRACK Weather reconnaissance mission flown to provide vital meteorological information in data sparse ocean areas as supplement to existing surface, radar and satellite data.

SYNTHETIC Produced by synthesis instead of being isolated from natural source.

SYNTHETIC DETERGENT Synthetic substance, other than soap, effective cleanser and functions equally well as surface-active agent in hard or soft water.

SYNTHETIC EARTHQUAKE Earthquake time history records developed from mathematical models that use white noise, filtered white noise and stationary and non-stationary filtered white noise or theoretical seismic source models of failure in fault zone.

SYNTHETIC FIBER Any fiber made by extruding synthetic substance through spinnerets to make thread.

SYNTHETIC HYDROGRAPH Empirical unit hydrograph that's based upon measurable physical basin characteristics.

SYNTHETIC ORGANIC CHEMICALS Man-made (i.e., anthropogenic) organic chemicals.

SYNTHETIC SLUDGE Sludge produced by wastewater organism seeding and aerating organic substances.

SYSTEM Physically connected generation, transmission and distribution facilities operated as integrated unit under one central management or operating supervision.

SYSTEM CAR Railroad car owned by railroad.

SYSTEM DISPLAY MONITOR Monitor that displays information relating to system status.

SYSTEM ELEMENTS Parts of transportation system.

SYSTEM PLANNING Method of analyzing, defining and designing information architecture of organizations.

SYSTEM PRESSURE Pressure in well pump system that must be kept to work.

SYSTEM WITH SINGLE SERVICE CONNECTION System that supplies drinking water to consumers via single service line.

SYSTEMATIC ERROR Experimental error that deviates by fixed amount from true value of measurement.

SYSTEMIC AGENT Agent that effects body in general but not specific organ.

SYSTEMIC PESTICIDE Chemical absorbed by organism that interacts with organism and makes organism toxic to pests.

SYSTEMIC TOXICANTS Chemical compounds that affect entire organ systems, often operating far from original site of entry.

SYSTEMS ENGINEERING Field of engineering focusing on development, analysis and organization of complex systems.

SYZYGY Instance when Earth, Sun and Moon (i.e., new moon, full moon) are all in straight line.

T

T INTERSECTION Three-way junction where three roads come together.

T-HEAD Braced, top horizontal member, forming T-shape.

T-NUMBER System used to subjectively estimate tropical cyclone intensity based solely on visible and infrared satellite images.

TABLETOP EXERCISE Informal activity involving discussions of actions to be taken on described emergency situations.

TABULATED DATA Processing information by putting it in table.

TACK COAT Thin layer applied between hot mix asphalt pavement lifts to promote bonding.

TACKLE Device for grasping object and attached means of moving it (e.g., rope and hook).

TACTICAL AIR NAVIGATION System used by aircraft that provides user with bearing and distance to ground or ship-borne station.

TACTILE WARNING Distinctive surface pattern of truncated domes, cones or bars detectable by long cane or underfoot used to alert visually impaired.

TAG 1. Small label; 2. Metallic binding, tube or point, at end of string or lace to stiffen it; 3. Any short peptide sequence artificially attached to proteins mostly in order to help purify, solubilize or visualize these proteins.

TAGOUT Placement of tagout device on energy isolating device, in accordance with established procedure, to indicate that energy isolating device and equipment being controlled may not be operated until tagout device is removed.

TAGOUT DEVICE Prominent warning device (e.g., tag and means of attachment) which can be securely fastened to energy isolating device in accordance with established procedure, to indicate that energy isolating device and equipment being controlled may not be operated until tagout device is removed.

TAIL 1. Short length of conduit that makes connection; 2. Part of distribution most distant from mode; 3. Visible stream of dust and gases blown from comet by solar wind; 4. Caudal appendage of animal.

TAIL BAY 1. Canal lock section located below lower gates; 2. Joist that rests on wall and girder.

TAIL CLOUD Horizontal, tail-shaped cloud (not funnel cloud) at low levels extending from precipitation cascade region of supercell toward wall cloud.

TAIL GATE Low-level gate at end of lock.

TAIL PIPE Pump's suction line.

TAIL WATER Water immediately downstream relative to structure.

TAIL-END CHARLIE Slang for thunderstorm at southernmost end of squall line or other line or band of thunderstorms.

TAILINGS Second grade or waste material separated from pay material during screening or processing (e.g., Rock that remains after processing ore).

TAILINGS DAM Industrial waste dam where waste materials come from mining operations or mineral processing.

TAILPIPE STANDARDS Emissions limitations applicable to mobile source engine exhausts.

TAILRACE Body of water immediately downstream from power plant, turbine or pumping plant.

TAILWATER 1. Water surface elevation immediately downstream from structure (i.e., dam, weir, drop structure) where elevation of water varies with discharge from reservoir; 2. Runoff of irrigation water from lower end of irrigated field.

TAILWATER HEIGHT Height of water immediately downstream of dam.

TAINTER GATE Pivoted crest gate, face of that is usually circular arc, with center of curvature at pivot about which gate swings and radial arms are hinged to piers or other supporting structure.

TAKEOFF Quantification of materials.

TAKEOFF DISTANCE Distance of ground run plus distance from where aircraft leaves ground to until it reaches 50 ft.

TAKEOFF RUN Amount of runway airplane uses to leave runway.

TALUS 1. Sloping accumulation of rock debris; 2. Rock fragments at base of cliff as result of slides or falls.

TALUS SPRING Spring occurring at base of talus slope.

TAMP To pound or press soil to compact.

TAMPER Tool for compacting soil in spots not accessible to rollers.

TAMPERING Adjusting, negating or removing pollution control equipment on motor vehicle.

TAMPING Method of compacting soil using impact of power or hand tamper on surface of soil placed.

TAMPING FOOT Pads attached to soil compactor drum.

TAMPING ROLLER Drum roller fitted with projection feet used to densify soil (e.g., sheepsfoot roller).

TANDEM Pair where one part follows other.

TANDEM AXLE LOAD Load imposed on highway by two or more consecutive axles.

TANDEM-CENTER VALVE Valve that lets pump unload while blocking ports.

TANGENT Line that touches circle and is perpendicular to its radius at point of contact.

TANGENTAL VELOCITY Velocity measured at any point tangent to circle.

TANK Storage chamber for liquids or gases.

TANK BODY Completely enclosed vehicle body to transport fluid.

TANK CAR Railroad car designed to transport liquids or gases.

TANK VEHICLE Any commercial motor vehicle designed to transport any liquid or gaseous materials within tank or tanks having individual rated capacity of more than 119 gallons and aggregate rated capacity of 1,000 gallons or more either permanently or temporarily attached to vehicle or chassis.

TANKER Vessel used to transport large quantities of liquid.

TAP 1. Connection made to electrical or fluid conductor without breaking it; 2. Device used to cut internal screw thread; 3. Device used to dispense liquids.

TAP BOLT Bolt with head on one end and thread on other end.

TAPER AREA Reduction in pavement width to traffic.

TAPER BEND Curved taper conduit.

TAPERED AERATION Air introduced at higher rate at head of tank than in subsequent sections during activated sludge process.

TAPERED INLET Inlet edge that decreases head loss as flow enters culvert.

TAPERED PIPE THREAD Pipe outside diameter thread cut on cone.

TAPERED REAMER Tool that tapers to point for cleaning burrs from drilled hole.

TAPERED THREAD Conical threaded pipe connection that's gas or liquid tight.

TAPPED T Cast iron fitting with branch tapped for screw pipe.

TAPPING MACHINE Machine that cuts hole and taps internal screw thread in metal.

TAPPING SLEEVE Split sleeve used to make branch outlet from main supply line.

TAPROOT Big root that grows downward from base of tree.

TAR SANDS Sand with oil deposits.

TARE WEIGHT Weight of empty container or vehicle.

TARGET Precipitation or other phenomena which produces echoes on radar display.

TARGET COMPLETION DATE Deadline date when all phases of specific project are to be completed.

TARGET FINISH DATE Date project is to be completed.

TARGET SPEED Speed intend for drivers to travel.

TARGET START DATE Date project is to be initiated.

TARIFF System of government-imposed duties levied on imported or exported goods.

TARN Relatively small and deep, steep-sided lake or pool occupying ice-gouged basin amid glaciated mountains.

TASK Work done as part of one's duties.

TASTE Sensation produced by tongue in response to certain chemicals.

TAUT Under tension (e.g., stretched bowstring, rope, sail).

TAXA RICHNESS Number of species (taxa) present in defined area or sampling unit.

TAXABLE HORSEPOWER Horsepower based on engine's bore size, number of cylinders and presumption of engine efficiency.

TAXI Vehicle hired for single journey by members of public.

TAXICAB Vehicle that passengers hire to take them between locations.

TAXON 1. Any identifiable group of taxonomically related organisms; 2. Plural taxa.

TAXONOMY Division of biology concerned with classification and naming of organisms.

TEAM TRACK Railroad siding where customers can load and unload freight cars.

TEARING Pulling of asphalt mix under paver.

TECHNICAL LIFE LENGTH Time from when machine goes into operation until it is no longer used in any operation.

TECHNICAL SUPPORT DOCUMENT Contains procedures for water quality-based limitation development.

TECHNICAL-GRADE ACTIVE INGREDIENT Pesticide chemical in pure form as it is manufactured prior to being formulated into end-use product (e.g., wettable powders, granules, emulsifiable concentrates).

TECHNOLOGY-BASED EFFLUENT LIMIT Permit limit for pollutant based on capability of treatment method to reduce pollutant to certain concentration.

TECHNOLOGY-BASED LIMITATIONS Industry-specific effluent limitations based on best available preventive technology applied to discharge when it will not cause violation of water quality standards at low stream flows.

TECHNOLOGY-BASED STANDARDS Industry-specific effluent limitations applicable to direct and indirect sources which are developed on category-by-category basis using statutory factors, not including water-quality effects.

TECHNOSPHERE Part of environment where techno-diversity extends its influence into biosphere.

TECTONIC Of, relating to or caused by large-scale movements of Earth's lithosphere (i.e., crust).

TECTONIC ACTIVITY Movement of Earth's crust resulting in formation of ocean basins, continents, plateaus and mountain ranges.

TECTONIC CREEP Slight, apparently continuous, movement along fault.

TECTONIC STRESS Stress due to deformation of Earth's crust.

TECTONICS Science that deals with structure of Earth's crust.

TEE Pipe fitting that has two threaded openings in line and third at right angles to them.

TELECONNECTION Linkage between changes in atmospheric circulation occurring in widely separated parts of globe.

TELEMETRY Science and associated technology, of automatic recording and transmission of data from remote source to receiving station for analysis.

TELESCOPE To slide one piece inside another.

TELLTALE Machine or contrivance for indicating or recording something.

TEMPERATURE Measure of internal energy (i.e., cold or heat) that substance contains.

TEMPERATURE APPROACH Difference between heat exchanger outlet and inlet temperature.

TEMPERATURE CHANGE 1. Mean lapse rate within layer of atmosphere, obtained by calculating difference between observed temperatures at bottom and top of layer; 2. Difference in temperature between surface of lake and 850mb, typically used to determine lake effect snow potential.

TEMPERATURE GRADIENT Variation of temperature change at particular location.

TEMPERATURE INVERSION Atmospheric condition where cooler layer of air lies below warmer layer of air that functions as "ceiling" for cooler layer.

TEMPERATURE RECOVERY Change in temperature over given period of time.

TEMPERATURE SCALES Calibrations (i.e., Fahrenheit, Celsius and Kelvin) to measure temperature quantitatively.

TEMPERATURE SENSOR Apparatus effective at measuring temperature.

TEMPORAL VARIABILITY Variant that's function of time.

TEMPORARILY ABSORBED WATER Water that only reaches belt of soil water or intermediate saturation zone, from which it's removed by evaporation, root absorption and/or transpiration.

TEMPORARY BARRIER Short-term road safety device installed with ongoing construction, short term emergencies, etc.

TEMPORARY IRRIGATION SERVICE LAND Irrigable land for which water supply is delivered under temporary arrangements.

TEMPORARY SNOWLINE Line sometimes drawn on weather map during winter showing southern limit of snow cover.

TEMPORARY SPRING Irregular spring.

TEMPORARY STRUCTURE Any structure that can be readily and completely dismantled and removed from site between periods of actual use.

TEMPORARY TRAFFIC CONTROL SIGNAL Limited-time traffic control signal.

TEMPORARY TRAFFIC CONTROL ZONE Section of roadway from first advance warning sign to last traffic control device.

TEN-YEAR LOW FLOW (SEVEN-DAY) Lowest stream flow for seven consecutive days that would be expected to occur once in ten years.

TEN-YEAR PRECIPITATION EVENT Maximum 24-hour precipitation event that can occur at least once in 10 years.

TENDER MIX Internally unstable mix that tends to displace laterally and shove rather than compact under roller loads.

TENDON Wire or bar used to strengthen prestressed concrete.

TENSILE FORCE Lateral force applied to solid body.

TENSIOMETER 1. Instrument used to measure moisture content of soil; 2. Instrument used to measure tension in wire; 3. Instrument used to measure surface tension of liquid.

TEPHIGRAM Thermodynamic diagram commonly used in weather analysis and forecasting.

TEPHRA Solid material thrown into air by volcanic eruption that settles on surrounding areas.

TERA Prefix symbolizing 10^{12}.

TERATOGEN Material that produces physical defect in developing embryo causing birth defects.

TERATOGENESIS Introduction of non-hereditary birth defects in developing fetus by exogenous factors (e.g., physical or chemical agents) acting in womb to interfere with normal embryonic development.

TERATOGENIC Causing malformations or defects to embryo or fetus.

TERBUFOS Organic thiophosphate, organothiophosphate insecticide and organosulfur compound with maximum acceptable concentration of 0.001 mg/l in drinking water.

TERMINAL 1. Building in airport where passengers transfer from ground transportation to facilities that allow them to board airplanes; 2. Harbour facility where ferries load and unload passengers and vehicles; 3. Rail station or town where service begins and ends; 4. Rate charged on all freight; 5. Storage tank for bulk liquids; 6. End of line where signals are either transmitted or received; 7. Electric contact on battery.

TERMINAL AERODROME FORECAST National Weather Service aviation product is concise statement of expected meteorological conditions at airport during specified period (usually 24 hours).

TERMINAL AREA Airspace surrounding major airport where there is high volume of traffic.

TERMINAL MORAINE Ridge of sediment piled at front edge of glacier or ice sheet.

TERMINAL STRUCTURE Portion of spillway downstream from conduit, chute, tunnel or conduit, which generally dissipates or stills releases.

TERMINAL TIME 1. Time allowed at terminal between arrival and departure vehicles; 2. Time required for passenger to pass through terminal and change mode.

TERMINAL VELOCITY Speed where object in free-fall and not in vacuum ceases to accelerate downwards because force of gravity is equal and opposite to drag force acting against it.

TERMINUS 1. End point of transportation system or town or city where it is located; 2. Border, post or stone marking boundary.

TEROTECHNOLOGY Engineering and technology related to design and maintenance of industrial plant machinery.

TERRACE Ridge, ridge and hollow or flat earth bench built along ground contour.

TERRACING Dikes built along contour of sloping farm land that hold runoff and sediment to reduce erosion.

TERRAIN 1. Ground surface; 2. Area having preponderance of particular rock or group of rocks.

TERRAIN FORCED FLOW Airflow modified or channeled as it passes over or around mountains or through gaps in mountain barrier.

TERRANE Area or surface over which particular rock type or group of rock types is prevalent.

TERRESTRIAL 1. Pertaining to, consisting of or representing Earth; 2. Living or growing on land; 3. Land-based.

TERRIGENOUS Derived from erosion of land-based rocks.

TERRITORIAL COMMUNITIES Populations that function within particular geographic area.

TERRITORIAL SEAS Sea and ocean areas belonging to territorial waters of country.

TERTIARY SEWAGE TREATMENT Dosing to remove suspended solids from secondary effluent, further reducing biochemical oxygen demand.

TERTIARY TREATMENT Advanced cleaning of wastewater that goes beyond secondary or biological stage, removing nutrients (e.g., phosphorus, nitrogen, most BOD and suspended solids).

TERTIARY WASTEWATER TREATMENT Selected biological, physical and chemical separation processes to remove organic and inorganic substances that resist conventional treatment practices that's additional treatment of effluent beyond that of primary and secondary treatment methods to obtain very high quality of effluent.

TERTIARY-TREATED SEWAGE Third phase of treating sewage that removes nitrogen and phosphorus before it is discharged.

TESLA 1. Unit of magnetic flux density; 2. Measurement to determine strength of magnetic field.

TEST Challenge or trial.

TEST CHAMBER Container where organism is preserved during test solution exposure.

TEST DATA Data that has been specifically identified for use in tests.

TEST PIT Pit dug for geologic investigation or inspection and testing.

TEST SOLUTION Solution of reagent in specified strength.

TEST SUBSTANCE Mixture added to test system in study.

TEST VEHICLE Vehicle used to test crash performance, comfort, reliability and safety.

TEST WELL Well to evaluate aquifer.

TESTED WELL CAPACITY Speed that well can produce water without substantial drawdown.

TETRAETHYL LEAD Toxic organometallic compound added to petroleum or gasoline as antiknock agent.

TETRAHEDRON Four triangular sided wind direction indicator that points direction wind is blowing.

TETRAPOD 1. Concrete structure to arrest wave energy; 2. Vertebrate with four limbs.

TEXAS HOOKER Low pressure systems that originate in panhandle region of Texas and Oklahoma which initially move east and then "hook" or recurve more northeast toward upper Midwest or Great Lakes region.

TEXTURE Feel or shape of surface or substance.

THALWEG Line that connects lowest points in valley or river channel.

THEODOLITE Instrument used in surveying to measure horizontal and vertical angles with small telescope that can move in horizontal and vertical planes.

THEORETICAL DISCHARGE Computed discharge from weir, conduit or pipeline compared to specific discharge.

THEORETICAL GORE Where broken white line becomes wide solid white "channelizing lines" that then becomes white triangle in front of gore.

THEORETICAL MAXIMUM RESIDUE CONTRIBUTION Theoretical maximum amount of pesticide in daily diet of average person.

THEORETICAL MODEL Representation of real system by mathematical description.

THEORY Possible or plausible explanation of phenomena based on available evidence.

THERAPEUTIC INDEX Ratio of dose required to produce toxic or lethal effects to dose required to produce non-adverse or therapeutic response.

THERM Unit of heat equal to 100,000 British thermal units (BTU).

THERMAL Relatively small-scale, rising air current produced when Earth's surface is heated.

THERMAL BELT Zone of high nighttime temperatures (and relatively low humidities) often experienced within narrow altitude range on valley sidewalls, especially evident during clear weather with light winds.

THERMAL COLUMN Column of rising air in lower altitudes of Earth's atmosphere.

THERMAL CONDUCTIVITY Measure of ability of material to conduct heat.

THERMAL CONVECTION STORM Storm due to local temperature inequalities with short, intense rainfall.

THERMAL CONVERSION Use of heat to convert biomass materials or feedstocks into other forms of energy.

THERMAL EFFICIENCY Efficiency of heat engine measured by ratio of work done by it to heat supplied to it.

THERMAL FORCE Change in temperature, due to thermal expansion and contraction, that creates stresses and strains.

THERMAL HIGH Area of high pressure that's shallow in vertical extent and produced primarily by cold surface temperatures.

THERMAL LOADING Amount of waste heat discharged to water body.

THERMAL LOW Area of low pressure that's shallow in vertical extent and produced primarily by warm surface temperatures.

THERMAL MAPPING Process where spatial variation of minimum night-time road surface temperature is measured.

THERMAL POLLUTION Reduction in water quality caused by increasing its temperature, often due to disposal of waste heat from industrial or power generation processes that can kill or injure aquatic organisms.

THERMAL SAND Sand used to dissipate heat away from buried electrical cables.

THERMAL SPRING Natural spring producing water above mean annual temperature.

THERMAL STRATIFICATION Formation of layers of different temperatures in lake or reservoir.

THERMAL SYSTEM INSULATION Asbestos-containing material applied to pipes, fittings, boilers, breeching, tanks, ducts or other interior structural components to prevent heat loss or gain or water condensation.

THERMAL TREATMENT Use of elevated temperatures to treat hazardous wastes.

THERMAL WIND Theoretical wind that blows parallel to thickness lines, for layer considered, analogous to how geostrophic wind blows parallel to height contours where closer thickness isopleths, stronger thermal wind.

THERMAL-ELECTRIC POWERPLANT Generating plant which uses heat to create steam driven electricity.

THERMALLY DRIVEN CIRCULATION Diurnally reversing closed cellular wind current resulting from horizontal temperature contrasts caused by different rates of heating or cooling over adjacent surfaces.

THERMIONIC CONVERTER Static device that converts heat into electricity by boiling electrons.

THERMISTOR Resistor whose resistance changes with temperature.

THERMOCLINE 1. Middle layer of lake, separating upper, warmer portion (epilimnion) from lower, colder portion (hypolimnion); 2. Layer within body of water or air where temperature changes rapidly with depth.

THERMOCOUPLE Heat-sensing device made of two conductors of different metals joined at their ends.

THERMODYNAMIC CHART Chart containing contours of pressure, temperature, moisture and potential temperature, all drawn relative to each other such that basic thermodynamic laws are satisfied.

THERMODYNAMIC DIAGRAM Diagram containing contours of pressure, temperature, moisture and potential temperature, all drawn relative to each other such that basic thermodynamic laws are satisfied.

THERMODYNAMICS Relationships between heat and other properties (e.g., temperature, pressure, density).

THERMOELECTRIC POWER Electrical power generated by use of fossil-fuel (e.g., coal, oil, natural gas), geothermal or nuclear energy.

THERMOELECTRIC POWER WATER USE Water used in process of generation of thermoelectric power.

THERMOGRAPH Instrument that measures and continuously records variations of temperature.

THERMOKARST Irregular land surface formed in permafrost region by melting ground ice and subsequent settling of ground.

THERMOMETER Instrument for measuring temperature.

THERMOPHILIC Living and thriving at relatively high temperatures.

THERMOPHILIC DIGESTION Sludge digestion between 113 °F and 145 °F.

THERMOPHILIC RANGE Favorable temperature (between 120 °F and 135 °F) for thermophilic bacteria.

THERMOPHONE Transducer that generates sound waves of precise power.

THERMOPLASTIC Plastic that softens when heated and hardens when cooled.

THERMOSIPHON Arrangement of siphon tubes for assisting circulation in liquid.

THERMOSPHERE Atmospheric shell extending from top of mesosphere to directly below exosphere (i.e., outer space).

THICK ARCH DAM Arch dam with base thickness to structural height ratio of 0.3 or greater.

THICK-BEDDED 1 ft to 3 ft thick.

THICKNESS OF DAM Thickness of dam as measured horizontally between upstream and downstream faces and normal to axis or centerline crest of dam.

THIEF HOLE Digester sampling well that allows sampling of digester contents without venting digester gas.

THIN ARCH DAM Arch dam with base thickness to structural height ratio of 0.2 or less.

THIN LINE ECHO Narrow, elongated, non-precipitating echo.

THIN PARTING Paper thin to 0.002 ft.

THIN-BEDDED 0.1 ft to 0.3 ft thick.

THIRD RAIL Electrified rail that runs beside or between train tracks to power electric trains.

THIRTY-DAY LIMITATION Limitation value that average of daily values for thirty consecutive days shall not exceed.

THIXOTROPHY Property of material that enables it to stiffen in relatively short time on standing but, upon agitation or manipulation, to change to very soft consistency or to fluid of high viscosity, process being completely reversible.

THIXOTROPY Reduction of viscosity with increased shear.

THREADED JOINT Parts joined using threads.

THREADER Device used to machine screw thread for conduits, etc.

THREAT ASSESSMENT Practice of determining credibility, seriousness and probability of potential threat.

THREATENED Legal classification for species that is likely to become endangered within foreseeable future.

THREATENED SPECIES Any species which has potential of becoming endangered in near future.

THREATENED WATERS Waters that fully support their designated uses, but may not support uses in future unless pollution-control action is taken because of anticipated sources or adverse pollution trends.

THREE C (3C) PROCESS Law that calls upon local officials to cooperate with states and public transportation providers in undertaking continuing, comprehensive and cooperative (i.e., 3C) multimodal transportation planning process.

THREE-DIMENSIONAL MODEL Model defined with three space coordinates.

THREE-HOUR RAINFALL RATE WSR-88D Radar product displays precipitation total (in inches) of current and past two clock hours as graphical image.

THREE-LEG INTERSECTION Where two roads join at intersection.

THREE-LEVEL DIAMOND INTERCHANGE Highway interchange where through traffic on both main roads is grade-separated from intersections that handle transferring traffic.

THREE-MINUTE MEAN CONCENTRATION Maximum permissible concentration of pollutant or mixture of pollutant, averaged over three minutes at ground level.

THREE-QUARTER S-TRAP Sanitary trap used primarily to join fixtures to Y-branch or into hub of pipe inclined at 45° in soil pipe where distance between trap and branch is short.

THREE-SECOND GUST WIND SPEED Highest average wind speed measured over three second time period to define wind loads.

THREE-WAY SWITCH Switch that can be arranged in pairs to control from two locations.

THREE-WAY VALVE Valve that can control fluid exchange between three separate transfer lines.

THRESHOLD 1. Lowest dose of chemical at which specified measurable effect is observed and below which it is not observed; 2. Quantitative point where action is triggered.

THRESHOLD EFFECTS Result from chemicals that have safe level (i.e., acute, subacute, chronic human health effects).

THRESHOLD LEVEL Time-weighted average pollutant concentration values, exposure beyond that is likely to adversely affect human health.

THRESHOLD LIMIT VALUE Concentration of airborne substance that average person can be repeatedly exposed without adverse effects.

THRESHOLD LIMITING VALUE Maximum concentration of airborne hazardous substances that healthy workers may be exposed to without ill effect over eight hour working day.

THRESHOLD ODOR Minimum odor of gas or water sample that can just be detected after successive dilutions with odorless gas or water.

THRESHOLD OF AUDIBILITY Level above where sound intensity, at any specified frequency, must rise in order to be detected by average human ear.

THRESHOLD PLANNING QUANTITY Quantity designated for each chemical on list of extremely hazardous substances that triggers notification by facilities that such facilities are subject to emergency planning requirements.

THRESHOLD RUNOFF Runoff in inches from rain of specified duration that causes small stream to slightly exceed bankfull (flood stage).

THRESHOLD TREATMENT Water treatment process of stopping precipitation-type reaction at threshold of precipitate formation.

THROPIC LEVELS Functional classification of species based on feeding relationships (i.e., generally aquatic and terrestrial green plants comprise first thropic level and herbivores comprise second.)

THROUGH CUT Excavation between parallel banks that begins and ends at original grade.

THROUGH RATE Single transportation rate for haul from origin to destination.

THROUGH ROUTING Route designed to transport people across downtown and out other side as different route.

THROUGH STREET Street designed to give preference to "pass through" movement of traffic.

THROUGH TRAIN Train that travels long distance without any stops.

THROUGH TRAVELED WAY Roadway portion set aside for movement of vehicles.

THROUGH VEHICLES Vehicles that do not turn and pass directly through street.

THROUGH-GIRDER SPAN Bridge roadway where girder is below top flange.

THROUGH-TRUSS SPAN Bridge roadway where floor elevation is nearly at bottom and traffic travels between supporting sections.

THROUGHFALL Excess precipitation that falls from foliage.

THROUGHPUT Rate vehicles and/or passengers pass point over time.

THRUST BLOCK 1. Massive block of concrete built to withstand thrust or pull; 2. Mass of concrete or similar material appropriately placed around pipe to absorb any axial thrust of pipeline.

THRUST BORER Gear for forcing hole through embankment for conduit or cable insertion.

THUNDER Sound caused by rapidly expanding gases in lightning discharge.

THUNDERSTORM Local storm produced by cumulonimbus cloud and accompanied by lightning and thunder.

TIDAL BASIN Enclosed body of water where tide changes water level.

TIDAL BORE Wave, in form of wall of water, formed by incoming tide funnelling into estuary.

TIDAL CURRENT Flow of water caused by rise and fall of tide.

TIDAL CYCLE Periodic changes in intensity of tides caused primarily by varying relations between Earth, Moon and Sun.

TIDAL FLAT Extensive, nearly horizontal, tract of land alternately covered and uncovered by tide and consists of unconsolidated sediment.

TIDAL LAG Delay between Moon's passage and tidal response.

TIDAL MARSH Low, flat marshlands traversed by channels and tidal hollows, subject to tidal inundation.

TIDAL MODEL Froude number model where tidal water levels and current fluctuations are reproduced.

TIDAL OUTLET Inlet where tide ebbs and flows.

TIDAL PILING Occurs when unusually high water levels occur as result of accumulation of successive incoming tides that do not completely drain due to opposing strong winds and/or waves.

TIDAL POWER Electric creation contained in tide flowing in or out of water body.

TIDAL RANGE Maximum difference in height between consecutive high and low waters over tide cycle.

TIDAL RIVER River that enters sea or estuary and is affected by tides.

TIDAL VALVE Check valve that drains at high tide.

TIDAL WAVES Series of long-period waves (on order of tens of minutes) that are usually generated by impulsive disturbance (e.g., earthquake, underwater volcanic activity, landslide) that displaces massive amounts of water.

TIDE 1. Periodic (occurring at regular intervals) variations in surface water level of Great Lakes, oceans, bays, gulfs and inlets from gravitational attraction between Sun and Moon on Earth; 2. Water level in feet above or below mean lower low water.

TIDE ANOMALY Actual water level minus prediction.

TIDE GATE Opening through where water may flow freely when tide moves in one direction, but closes automatically and prevents water from flowing in other direction.

TIDE GAUGE Device for showing height of tide.

TIDE PREDICTION Computation of tidal highs and lows at given location resulting from gravitational interactions between Sun and Moon on Earth.

TIDE TABLE Table giving time when tide will be high or low at particular place.

TIDEWATER PORT Port located in tidal water body.

TIE 1. Structural member firmly holding two pieces together; 2.Horizontal wooden or concrete structural member that supports and ties together rails.

TIE BAR Deformed steel bars or connectors that hold faces of abutting concrete slabs.

TIE LINES Transmission line connecting two or more power systems.

TIE PLATE Steel plate used on rail tracks between flanged T rail and crossties.

TIE-DOWN Rope, strap, chain or line used to secure load or item.

TIED ARCH Arch bridge where outward-directed horizontal forces of arch are borne as tension by chord tying arch ends.

TIER 1 Bureau of Reclamation public protection guideline dealing with loss of life considerations.

TIER 1 SEDIMENT GUIDELINE Threshold concentration above which there is high probability of adverse effects on aquatic life from sediment contamination, determined using modified USEPA procedures.

TIER 2 Bureau of Reclamation public protection guideline dealing with public trust responsibilities based on annual failure probability of structure.

TIERED PERMIT LIMITS Permit limits that only apply to discharge when certain threshold (e.g., production level), specific circumstance (e.g., batch discharge) or timeframe (e.g., after 6 months) triggers their use.

TIERED TESTING Any of series of tests that are conducted as result of previous test's findings.

TIERING Coverage of general matters in broad document with subsequent narrowly focused documents.

TIGHT Soil or rock formations lacking veins of weakness that's between 0.004 inch to 0.020 inch.

TIGHT COVER Cover for manhole lacking openings.

TIGHT URBAN DIAMOND INTERCHANGE Compressed diamond interchange, with signal-controlled intersection separation of less than 400 ft, used in urban and suburban areas where right-of-way is constrained.

TILE Pipe made of baked clay.

TILE DRAIN Buried perforated pipe designed to remove excess water from soils.

TILE SHOE Box that lays tile drains directly behind ditcher.

TILL 1. Deposit of sediment formed under glacier, consisting of unlayered mixture of clay, silt, sand and gravel ranging widely in size and shape; 2. To work, cultivate or plough soil.

TILLAGE Plowing, seedbed preparation and cultivation practices.

TILT Storm where line connecting centroid of mid-level storm component to centroid of lowest storm component is to right or rear of direction of motion.

TILT SEQUENCE Radar term indicating that radar antenna is scanning through series of antenna elevations in order to obtain volume scan.

TILTED STORM Thunderstorm or cloud tower that is not purely vertical but instead exhibits slanted or tilted character that's sign of vertical wind shear.

TILTED UPDRAFT Thunderstorm or cloud tower that is not purely vertical but instead exhibits slanted or tilted character that's sign of vertical wind shear.

TILTH 1. Soil condition in relation to lump or particle size; 2. Rich cultivated soil.

TILTING GATE Crest gates with flaps held by hinges that permit water to flow over top when lowered.

TILTMETER Instrument designed to measure very small changes from horizontal level.

TILTMETERS Instrument that monitors horizontal or vertical tilt of structures and rock masses.

TIMBER Wood beams larger than 4 inches by 6 inches.

TIMBERING Wood bracing in tunnel or excavation.

TIMBERING SET Tunnel support consisting of roof beam or arch and two posts.

TIME CYCLE Time needed for one sequence.

TIME FRAME Period of time where something is expected to occur.

TIME INTERVAL SCALE FACTOR Ratio of total freeway entrance demands to freeway exit counts in each time interval.

TIME LAG Period of time between two events.

TIME MEAN SPEED Average of vehicles passing point over time duration.

TIME OF COMPLETION Date stipulated in contract for end of work.

TIME OF CONCENTRATION 1. Time required for storm runoff to flow from hydraulically most remote point (i.e., furthest location) of catchment, drainage area or watershed to outlet or point under consideration; 2. Time from end of rainfall excess to inflection point on recession curve.

TIME OF FLOW Time for water to move from sewer and another location.

TIME OF OPENING Time between signal that stops and then releases traffic.

TIME OF OPERATION Time required to release span lock, wedges, etc. to open bridge.

TIME OF RISE Time from start of rainfall excess to peak of hydrograph.

TIME OF VESSEL PASSAGE Time required for bridge to remain open to permit vessels to pass through.

TIME OFF Period of time one is not required to work.

TIME ON Period of time where one is required to work.

TIME POINT Time or point where vehicles are scheduled to pass.

TIME TO PEAK Time from center of mass of rainfall excess to peak of hydrograph.

TIME-RESPONSE CURVE Two graphs comparing responses with different potency and efficacy that shows onset, peak and duration.

TIME-VARYING FLOW MODEL Simulation where flow changes with time.

TIME-WEIGHTED AVERAGE 1. Computed by multiplying number of days in sampling period by concentrations of individual constituents for corresponding period and dividing sum of products by total number of days; 2. Calculate workers daily exposure to hazardous substance, averaged over eight hour workday.

TIMED EXIT GATE OPERATING MODE Exit with predetermined gate interval time.

TIMER Device used to measure amounts of time.

TIMETABLE 1. Tabular schedule of events with times when they occur(e.g., arrivals and departures); 2. To arrange specific time for (event, class).

TINAJA 1. Pocket of water developed below waterfall; 2. Temporary pool.

TINKER'S DAM Tiny dam, built in pipe, to isolate work space flooded with solder.

TIPPING BAY Opening that enables waste vehicles to discharge loads directly into storage pits or hopper.

TIPPING FEE Charge for given quantity of waste received at waste processing facility.

TIPPING FLOOR Unloading area between garbage source and landfill.

TIPPING-BUCKET RAIN GAUGE Precipitation gauge where collected water is funneled into two compartment bucket where rain fills one compartment, tips bucket actuating electric circuit (emptying into reservoir) and moves second compartment beneath funnel.

TIRE BRAKING FORCE Longitudinal force from vehicle braking.

TIRE BRAKING FORCE COEFFICIENT Maximum value of tire braking force coefficient that occurs prior to wheel lockup as braking torque is progressively increased.

TIRE BRAKING FORCE COEFFICIENT SLIDE Braking force coefficient from wheel locked.

TIRE FORCES Forces generated between tire and ground (i.e., contact patch).

TIRE LOAD Maximum weight that tire can support when properly inflated.

TIRE PROCESSOR Intermediate operating facility where recovered tires are processed in preparation for recycling.

TISSUE STUDY Assessment of concentrations and distributions of trace elements and certain organic contaminants in tissues of aquatic organisms.

TITANIUM Chemical element that's strong, corrosion-resistant metal, with atomic number of 22.

TITER Concentration of substance as determined by titration.

TITLE Legal right to ownership of property (e.g., deed, certificate).

TITRATION Determination of concentration of some substance in solution by slowly adding measured amounts of some other substance until reaction is shown to be complete.

TOE 1. Projection from periphery of revolving piece; 2. Free side of dam foundation; 3. Underside of sanitary landfill working face.

TOE DRAIN System of pipe and/or pervious material along downstream toe of dam used to collect seepage from foundation and embankment and convey it to free outlet.

TOE FILTER Granular filter at low end of levee or earth dam.

TOE OF BANK Line formed by intersection of sloping side plane of watercourse with plane of bed of watercourse.

TOE OF DAM Point of intersection between bottom of slope or upstream or downstream face of dam and natural ground.

TOE WEIGHT Additional material placed at toe of embankment dam to increase its stability.

TOEBOARD Low board around edge of raised platform to prevent workers from accidentally falling.

TOGGLE SWITCH Switch actuated by moving small lever.

TOILET SEAL Wax ring used to seal connection between toilet and waste pipe.

TOKEN Voucher that can be exchanged for goods or services.

TOLERABLE DAILY INTAKE Threshold value of amount of substance in air, food or drinking water that can be taken in daily over lifetime without appreciable health risk.

TOLERANCE Variation or deviation from standard (i.e., maximum permitted variation in engineering measurement).

TOLERANCE PETITION Formal request to establish new tolerance or modify existing one.

TOLERANCES Permissible residue levels for pesticides in raw agricultural produce and processed foods.

TOLERANT SPECIES Those species that are adaptable to (tolerant of) human alterations to environment and often increase in number when human alterations occur.

TOLL BOOTH Booth on toll road or toll bridge where toll is collected.

TOLL FACILITY Toll structure to collect toll fee.

TOLL ISLAND Raised island where fee is collected as vehicles pass.

TOLL LANE Individual lane that provides drivers to pay toll.

TOLL PLAZA Row of toll booths on toll road or turnpike.

TOLL-TICKET SYSTEM When toll road user receives ticket from toll booth attendant or machine before entering toll system.

TON 1. Unit of weight (mass) equal to 2240 pounds (long ton) or 2,000 pounds (short ton) or 1,000 kilograms (metric ton); 2. Unit of thermal power equal to 12,000 BTU/h.

TON MILE Unit of measure expressed in number of short tons moved over specific distance of single mile.

TON, LONG Unit of mass equal to 2,240 lb.

TON, SHORT Unit of mass equal to 2,000 lb.

TONNAGE 1. Amount of waste that landfill accepts, usually expressed in tons/month; 2. Charge made on each ton of cargo when landed; 3. Number of tons of water that floating ship displaces.

TONNE Unit of mass equal to 1,000 kg.

TONNE-PER-DAY Mass flow of metric tons across threshold per unit time of day.

TONS PER ACRE-FOOT Dry mass (tons) of constituent per unit volume (acre-foot) of water that's equivalent to 2,000 lbs/day or 0.9072 metric ton/day.

TOO NUMEROUS TO COUNT Abbreviation for 300 or more bacterial colonies on culture plate (i.e., too many to view as individual colonies).

TOP OF ACTIVE CONSERVATION CAPACITY Reservoir water surface elevation at top of capacity allocated to storage of water for conservation purposes only.

TOP OF DAM Top surface of dam that water must rise before passing over structure excluding any parapet wall, railing, curb, etc.

TOP OF DEAD CAPACITY Lowest elevation in reservoir from which water can be drawn by gravity.

TOP OF EXCLUSIVE FLOOD CONTROL CAPACITY Reservoir water surface elevation at top of reservoir capacity allocated to exclusive use for regulation of flood inflows to reduce damage downstream.

TOP OF INACTIVE CAPACITY Reservoir water surface elevation below which reservoir will not be evacuated under normal conditions.

TOP OF JOINT USE CAPACITY Reservoir water surface elevation at top of reservoir capacity allocated to joint use.

TOP OF SURCHARGE CAPACITY Maximum water surface of reservoir.

TOP THICKNESS Thickness of dam at level of top of dam (excluding corbels or parapets).

TOP THICKNESS, DAM Thickness or width of dam at level of crest (top) of dam (excluding corbels or parapets) where term thickness is used for gravity and arch dams and width is used for other dams.

TOP WIDTH 1. Width of dam at level of top of dam (excluding corbels or parapets); 2. Width of stream section at water surface that varies with stage in most natural channels.

TOPOGRAPHIC DIVIDE Crest line that divides drainage basins.

TOPOGRAPHIC MAP Map showing natural and/or physical features of Earth's surface, including altitude contours.

TOPOGRAPHY Physical features of ground surface area (i.e., Shape of land) including relative elevations and position of natural and man-made (anthropogenic) features.

TOPOTYPE Specimen from type locality.

TOPSOIL Topmost layer of soil, usually containing organic matter, capable of supporting plant growth.

TORNADO Violently rotating column of air, usually pendant to cumulonimbus, with circulation reaching ground.

TORNADO EMERGENCY Issued when there is severe threat to human life and catastrophic damage from imminent or ongoing tornado.

TORNADO FAMILY Series of tornadoes produced by single supercell, resulting in damage path segments along same general line.

TORNADO VORTEX SIGNATURE Image of tornado on Doppler radar screen that shows up as small region of rapidly changing wind speeds inside mesocyclone.

TORNADO WARNING Issued by National Weather Service when tornado is indicated by WSR-88D radar or sighted by spotters.

TORNADO WATCH Issued by National Weather Service when conditions are favorable for development of tornadoes in and close to watch area.

TORQUE Moment of force defined as equivalent straight line force multiplied by distance from axis of rotation.

TORRICELLI'S THEOREM Statement in fluid dynamics relating speed of fluid flowing out of orifice to height of fluid above orifice.

TORSION 1. Twisting or wrenching of body by exertion of lateral force tending to turn one end or part of it about longitudinal axis; 2. Force where thread, wire or rod of any material, returns or tends to return, to state of rest after being twisted.

TORSION FORCE Load applied to material through torque (i.e., twisting).

TORT Wrongful act that causes injury and can be remedied in civil court usually through awarding of damages.

TOTAL Amount of given constituent in representative whole-water (unfiltered) sample, regardless of constituents physical or chemical form.

TOTAL APPLIED FORCE Total of all acting forces on area, body or material.

TOTAL CAPACITY 1. Maximum hydraulic flow that can be passed; 2. Maximum rate at which well or spring will yield water; 3. Largest amount that can be received or contained.

TOTAL COLIFORM BACTERIA Particular group of bacteria that are used as indicators of possible sewage pollution.

TOTAL CONCENTRATION Concentration of constituent regardless of its form (dissolved or bound) in sample.

TOTAL DDT Sum of DDT and its metabolites (breakdown products), including DDD and DDE.

TOTAL DISCHARGE Quantity of given constituent, measured as dry mass or volume, that passes stream cross-section per unit of time.

TOTAL DISSOLVED PHOSPHOROUS Total phosphorous content of all material that will pass through filter, that is determined as orthophosphate without prior digestion or hydrolysis.

TOTAL DISSOLVED SOLIDS 1. Quantitative measure of residual mineral dissolved in water that remains after evaporation of solution; 2. Measure of combined total of organic and inorganic substances contained in liquid.

TOTAL DYNAMIC HEAD 1. When pump is lifting or pumping water, vertical distance from elevation of energy grade line on suction side of pump to elevation of energy grade line on discharge side of pump; 2. Total amount of pressure when water is flowing in system.

TOTAL ENERGY 1. Sum of all different types of energies body can have (i.e., potential and kinetic energy).

TOTAL EVAPORATION Sum of water lost from given land area during any specific time by transpiration from vegetation, building of plant tissue, by evaporation from water surfaces, moist soil, snow and interception.

TOTAL GROSS RESERVOIR CAPACITY Total amount of storage capacity available in reservoir for all purposes from streambed to normal water or normal water or normal pool surface level.

TOTAL HEAD 1. Height above datum plane of column of water; 2. In groundwater system, it is composed of elevation head and pressure head; 3. In pumps, measure of pump's ability to push fluids through system (i.e., difference between discharge and suction heads plus friction head).

TOTAL LATERAL CLEARANCE Sum of highway lateral clearances of median and shoulder.

TOTAL LOAD 1. All of constituent in transport; 2. Sediment in transport including bed load, suspended bed material load and wash load.

TOTAL MAXIMUM DAILY LOAD Estimates of amount of specific pollutants that body of water (i.e., receiving water) can safely take without threatening beneficial uses and safely meet water quality standards.

TOTAL OPERATING REVENUE Total dollar amount received by company for selling goods and services.

TOTAL ORGANIC CARBON Measure of total amount of carbon in organic compounds.

TOTAL ORGANISM COUNT Number of organisms collected and enumerated in any particular sample.

TOTAL PETROLEUM HYDROCARBONS Measure of concentration or mass of petroleum hydrocarbon constituents present in given amount of soil or water.

TOTAL POPULATION Total number of humans currently living (i.e., population density).

TOTAL PRESSURE 1. Sum of static pressure, dynamic pressure and gravitational head; 2. Force per unit area of soil; 3. Sum of each gas' partial pressure; 4. In flowing air, sum of static and velocity pressures.

TOTAL PUMPING HEAD 1. Difference between total discharge and total suction heads; 2. Sum of total discharge head and total suction lift.

TOTAL RAMP DENSITY Total number of ramps within +/- 3 miles of midpoint of study segment divided by 6 miles.

TOTAL RECOVERABLE Amount of given constituent in whole-water sample after sample has been digested by method (usually using dilute acid solution) that results in dissolution of only readily soluble substances.

TOTAL RECOVERED PETROLEUM HYDROCARBON Method for measuring petroleum hydrocarbons in samples of soil or water.

TOTAL RESIDUAL CHLORINE Chlorine amount remaining after defined contact period.

TOTAL SEDIMENT DISCHARGE Total rate (e.g., tons/day) mass of suspended sediment plus bed-load transport (measured as dry weight) passes cross-section in given time.

TOTAL SEDIMENT LOAD Sediment in transport including bed load, suspended bed material load and wash load (i.e., all of constituent in transport).

TOTAL SOLIDS Measure of sum of suspended and dissolved solids in water.

TOTAL STORAGE Volume (including dead storage) of storage below maximum designed water surface level.

TOTAL SUSPENDED PARTICLES Method of monitoring airborne particulate matter by total weight.

TOTAL SUSPENDED SOLIDS Total amount of given constituent in part of water-sediment sample retained on 0.45-micrometer membrane filter.

TOTAL TOXIC ORGANICS Mass total of hazardous organic compounds in concentration greater than 0.010 mg/l.

TOTAL WATER DEMAND Water volume required.

TOTAL WELL CAPACITY Rate well water can be pumped after removing stored water.

TOTAL-TOTALS INDEX Stability index and severe weather forecast tool, equal to temperature at 850 mb plus dew point at 850 mb, minus twice temperature at 500 mb.

TOTALIZER Device or meter that continuously measures and calculates total flow.

TOTALLY ENCAPSULATING CHEMICAL PROTECTIVE SUIT Special protective suits made of material that prevents toxic or corrosive substances or vapors from coming in contact with body (i.e., gas and vapor tight suit).

TOTALLY ENCLOSED TREATMENT FACILITY Facility directly connected to industrial production process and is constructed and operated in manner that prevents release of any hazardous waste or any constituent thereof into environment.

TOTE BOX Container for storing and transporting industrial materials.

TOUCHLESS CONTROLS Electronic controls that operate by stimulus.

TOWAGE SERVICE When tug supplies power to draw another vessel.

TOWBOAT Powerful boat designed to push barges.

TOWERING CUMULUS Large cumulus cloud with great vertical development, usually with cauliflower-like appearance, but lacking characteristic anvil of towering cumulus.

TOXAPHENE Chemical that causes adverse health effects in domestic water supplies and is toxic to fresh water and marine aquatic life.

TOXIC Relating to harmful effects to biota caused by substance or contaminant.

TOXIC CHEMICAL RELEASE FORM Information form required of facilities that manufacture, process or use (in quantities above specific amount) chemicals listed.

TOXIC CHEMICAL USE SUBSTITUTION Replacing toxic chemicals with less harmful chemicals in industrial processes.

TOXIC CLOUD Airborne plume of gases, vapors, fumes or aerosols containing toxic materials.

TOXIC CONCENTRATION Concentration at which substance produces toxic effect.

TOXIC DOSE Dose level at which substance produces toxic effect.

TOXIC POLLUTANT Material that causes death, disease, behavioral abnormalities, cancer, genetic mutations, physiological malfunctions, (including malfunctions in reproduction) or physical deformations in organisms that ingest or absorb them.

TOXIC RELEASE INVENTORY Database of toxic releases in United States.

TOXIC SUBSTANCE Chemical or mixture that may present unreasonable risk of injury to health or environment.

TOXIC UNIT ACUTE Reciprocal of effluent concentration that causes 50% of organisms to die by end of acute exposure period.

TOXIC UNIT CHRONIC Reciprocal of effluent concentration that causes no observable effect on test organisms by end of chronic exposure period.

TOXIC UNITS Measure of toxicity in effluent as determined by acute toxicity units or chronic toxicity units measured.

TOXIC WASTE Waste that can produce injury if inhaled, swallowed or absorbed through skin.

TOXICANT Harmful substance or agent that may injure exposed organism.

TOXICITY Degree of danger posed by substance to cell, organism, animal or plant life.

TOXICITY ASSESSMENT Characterization of toxicological properties and effects of chemical, with special emphasis on establishment of dose-response characteristics.

TOXICITY REDUCTION EVALUATION Site-specific study conducted in stepwise process designed to identify causative agent(s) of effluent toxicity, isolate sources of toxicity, evaluate effectiveness of toxicity control options and then confirm reduction in effluent toxicity.

TOXICITY TEST Procedure to determine toxicity of chemical or effluent using living organisms.

TOXICOLOGICAL PROFILE Examination, summary and interpretation of hazardous substance to determine levels of exposure and associated health effects.

TOXICOLOGY Study of adverse effects of chemicals on biological systems and assessment of probability of their occurrence.

TOXIN Poisonous substance, generally from plant or animal.

TRACE Hydrograph or similar plot for extended-range time horizon showing one of many scenarios generated through ensemble forecast process.

TRACE ELEMENT Chemical element present in minute quantities in substance.

TRACE ELEMENTS Chemical elements present in sample in very small quantities.

TRACE OF PRECIPITATION Rainfall amount less than 0.01 of inch.

TRACER Stable, easily detected substance or radioisotope added to material to follow location of substance in environment or to detect any physical or chemical changes that it undergoes.

TRACER MODEL Fixed-bed model where patterns of erosion and deposition are estimated by using tracers of bed material or lightweight sediment proxies.

TRACK Path that storm or weather system follows.

TRACK APRON Railroad tracks along waterfront to transfer of cargo between vessel and vehicle.

TRACK CAR Any rail car that can be moved manually on or off track.

TRACK CROSSING Structure where one track crosses another at grade, consisting of four connected frogs.

TRACK DRILL Heavy drilling machine for quarrying, heavy construction and mining.

TRACK SCHEDULE Document describing track usage.

TRACK SIGNAL Device with display for control and management infrastructure of train movement.

TRACKING SOLAR COLLECTOR System that rotates and orientates equipment for solar energy collection.

TRACKWORK Tracks and associated equipment of railway line.

TRACTIVE FORCE Force developed in direction of water flow in channel.

TRACTOR 1. Vehicle used in farms for pulling equipment and preparing fields; 2. Truck for pulling trailer; 3. Airplane where propeller is located in front of fuselage.

TRACTOR GATE Rectangular gate supported and sealed similar to paradox gate.

TRACTOR TRAILER ANGLE Angle between adjoining units of tractor or semitrailer when turned.

TRADABLE RENEWABLE CREDITS Represent technology and environmental attributes of electricity generated from renewable sources that are referred to as Green Tags.

TRADE WINDS Persistent tropical, easterly winds that blow from subtropical high pressure centers towards equatorial low.

TRADITIONAL CULTURAL PROPERTY Site or resource eligible for inclusion in National Register of Historic Places because of its association with cultural practices or beliefs of living community.

TRAFFIC 1. Pedestrians or vehicles on roads; 2. Commercial transportation or exchange of goods.

TRAFFIC ASSIGNMENT Model of routes between origins and destinations in transportation networks.

TRAFFIC ASSIGNMENT ZONE Subdivision of traffic study area represented by centroid.

TRAFFIC BARRIER Barrier that keeps vehicles within road and prevent collision with dangerous obstacles.

TRAFFIC CONDITION Classification of transport by speeds, directions, trip times and vehicular queueing.

TRAFFIC CONTROL DEVICE Markers, signs and signals used to inform, control and guide traffic.

TRAFFIC CONTROL POINT Location staffed to ensure continued movement of traffic inside or outside area of risk during emergency or disaster.

TRAFFIC CONTROL SIGNAL Signalling devices positioned at road intersections, pedestrian crossings and other locations to control traffic flow.

TRAFFIC DELAY When transportation vehicles interact and reduce speed less than free-flow speed.

TRAFFIC INCIDENTS Non-recurring occurrences on roadway (i.e, crashes, debris).

TRAFFIC ISLAND Raised area in roadway or traffic circle, bordered by curbs, used to control traffic.

TRAFFIC OPERATION PLAN Developed study for evaluating and routing traffic and transportation vehicles.

TRAFFIC PATTERN Standard path followed by aircraft when taking off or landing while maintaining visual contact with airfield.

TRAFFIC PRESSURE Driving behavior of many drivers considered aggressive during traffic.

TRAFFIC SIGN Sign for control of traffic or information of drivers.

TRAFFIC SIGNAL Signalling device positioned at road intersection or pedestrian crossing to indicate when it may be safe to drive or walk.

TRAFFIC SIGNAL OPTIMIZATION TOOL Model applied to develop baseline signal phasing and timing plan.

TRAFFIC SIGNAL PREEMPTION System that permits normal operation of traffic lights to be preempted.

TRAFFIC SIGNAL PRIORITY Giving special treatment to transit vehicles at signalized intersections.

TRAFFIC VOLUME Number of pedestrian or vehicles passing that's conducted along particular road, path or intersection.

TRAFFIC-ACTUATED SIGNAL Traffic control signal that monitors intersection traffic conditions and dynamically adjusts maximum green times.

TRAILER 1. Unpowered wheeled vehicle towed behind another and used to carry equipment that cannot be carried in leading vehicle; 2. Furnished vehicle towed behind another that's used as dwelling when stationary.

TRAILER PUMP Mobile pumping system.

TRAIN 1. Line of connected railway cars or carriages considered overall as mode of transport; 2. Transient trail of glowing ions behind large meteor as it falls through atmosphere.

TRAINING Repeated areas of rain, typically associated with thunderstorms, that move over same region in relatively short period of time and are capable of producing excessive rainfall totals.

TRAINING WALL Wall built to confine or guide flow of water.

TRAMP Any ship that does not have fixed schedule or published ports of call.

TRAMP OIL Unwanted (i.e., contaminant) oil mixed with fluid.

TRANQUIL FLOW Fluid flow that exists when Froude number is less than one, control of flow depth is always downstream and surface waves can propagate upstream as well as downstream.

TRANSBOUNDARY POLLUTANTS Air pollution that travels from one jurisdiction to another, often crossing state or international boundaries.

TRANSCRIBED WEATHER BROADCAST Broadcasts of weather information providing continuous aeronautical and meteorological information.

TRANSCRIBED WEATHER BROADCASTS National Weather Service aviation product that forecast sky cover (height and amount of cloud bases), cloud tops, visibility (including vertical visibility), weather and obstructions to vision are described for corridor 25 miles either side of route.

TRANSDUCER Device that senses some varying condition and converts it to electrical signal for transmission to some other device (i.e., receiver) for processing or decision making.

TRANSECT 1. Line across stream perpendicular to flow and along which measurements are taken, so that morphological and flow characteristics along line are described from bank to bank; 2. Path researcher moves along, counts and records observations.

TRANSFER 1. To move or pass from one place, person or thing to another; 2. Design conveyed by contact from one surface to another (e.g., heat transfer); 3. To arrange for something to belong to or be officially controlled by somebody else.

TRANSFER CENTER Location for loading and unloading transportation passengers.

TRANSFER FEE Cost to change from one transportation mode to another.

TRANSFER LENGTH Distance from end of strand to point where effective stress is developed.

TRANSFER POINT Temporary point whose elevation is determined by additions and subtractions of backsights and foresights respectively.

TRANSFER STATION Facility where temporary deposition of waste is transferred to another transport mechanism.

TRANSFER SWITCH Switch that changes load between two power sources.

TRANSFER VALVE Valve that sets operation for multistage centrifugal pumps.

TRANSFORMATION Chemical alteration of compound by processes (e.g., reaction with other compounds, breakdown into component elements).

TRANSFORMED FLOW NET Flow net whose boundaries have been properly modified (transformed) so that net consisting of curvilinear squares can be constructed to represent flow conditions in anisotropic porous medium.

TRANSFORMER Device which through electromagnetic induction transforms alternating electric energy in one circuit into energy of similar type on another circuit, commonly with altered values of voltage and current.

TRANSIENT CONDITIONS Condition where system inputs and outputs are not in equilibrium so that there is net change in system with time.

TRANSIENT WATER SYSTEM Non-community water system that does not serve 25 of same non-residents per day for more than six months per year.

TRANSIENT, NON-COMMUNITY WATER SYSTEM Water system which provides water in place (e.g., gas station, campground) where people do not remain for long periods of time.

TRANSIT Surveying instrument that can measure both vertical and horizontal angles.

TRANSIT CAPACITY Number of passengers transportation mode can handle over time period.

TRANSIT CENTER Fixed facility where transit passengers transfer goods or passengers from one route to another.

TRANSIT DEPENDENT People unable to drive that reply on public transportation.

TRANSIT FREQUENCY 1. Time between consecutive transportation modes (e.g., buses, trains or ferries) that determines maximum wait time; 2. Number of transit vehicles and passengers stopping on street segment.

TRANSIT MODE Mode of travel where vehicles stop for passenger pick up and drop off.

TRANSIT RELIABILITY Measure of performance and regularity of transit vehicles used by customers in making transit decisions.

TRANSIT SHED Structure designed for storing goods before or after transportation.

TRANSIT SIGNAL PRIORITY Adjustments to signal timing that uses technology to reduce time at traffic signals.

TRANSIT SYSTEM Public transportation facility usually comprised of buses, subways and elevated trains.

TRANSITION Fitting that changes shape or size.

TRANSITION BELT Belt that carries material between loading point and main conveyor.

TRANSITION ZONE 1. Substantial part of cross-section of embankment dam comprising material whose grading is of intermediate size between that of impervious zone and that of permeable zone; 2. Stage in flow or fluid where it changes from laminar to turbulent; 3. Region between upper and lower mantle characterized by rapid increase in density.

TRANSITWAY Route designed for use by public transport.

TRANSLATION 1. Motion or compulsion to motion in straight line without rotation or other deformation; 2. Transfer of motion occurring within gearbox.

TRANSMISSIBILITY Capacity of rock to transmit water under pressure.

TRANSMISSION Act or process of transporting electric energy in bulk.

TRANSMISSION CONSTANT Capacity of material to transmit water.

TRANSMISSION LINE 1. Facility for transmitting electrical energy at high voltage from one point to another point; 2. Pipeline that transports raw water from its source to water treatment plant, then to distribution grid system.

TRANSMISSION LOSS Decrease in waveform energy along transmission path or through circuit.

TRANSMISSIVE BOUNDARY Boundary (cross-section) that will allow sediment that reaches it to pass without changing that cross-section.

TRANSMISSIVITY Volume of water at existing kinetic viscosity that will move (i.e., ability of aquifer to transmit water) in unit time under unit hydraulic gradient through unit width of aquifer that equals hydraulic conductivity multiplied by aquifer thickness.

TRANSMITTER Radar equipment used for generating and amplifying radio frequency (RF) carrier signal, modulating carrier signal with intelligence and feeding modulated carrier to antenna for radiation into space as electromagnetic waves.

TRANSPIRATION 1. Process by which water vapor escapes (i.e., discharges) from living plant surfaces, principally leaves and enters atmosphere; 2. Process of giving off water vapor through skin or mucous membranes.

TRANSPIRATION RATIO Ratio of water mass transpired to dry matter mass produced.

TRANSPONDER Radio or radar transceiver that transmits some signal in response to receiving predetermined signal.

TRANSPORT Hydrological, atmospheric or other physical processes that convey pollutants through and across media from source to receptor.

TRANSPORT CAPACITY Ability of river to transport given volume or weight of sediment material of specific size per time for given flow condition.

TRANSPORT POTENTIAL Rate at which river could transport sediment of given grain size for given hydraulic conditions if bed and banks were composed entirely of material of that size.

TRANSPORT WIND Average wind over specified period of time within mixed layer near surface of Earth.

TRANSPORTATION Act of transporting people and goods.

TRANSPORTATION BROKERAGE Transportation services of middle man between transporter and client seeking shipping services.

TRANSPORTATION CONTROL MEASURES Steps taken by locality to reduce vehicular emission and improve air quality by reducing or changing flow of traffic.

TRANSPORTATION DEMAND MANAGEMENT Strategies to reduce or redistribute travel demand in space or time.

TRANSPORTATION IMPROVEMENT PROGRAM Lists all transportation projects in metropolitan planning area that seek federal transportation funding within at least four-year horizon.

TRANSPORTATION PLAN Planning product that details long-term transportation needs and future demands.

TRANSPORTATION SYSTEM Multimodal transportation service that permits either vehicular movement or commodity flow.

TRANSPORTATION SYSTEM MANAGEMENT Transportation operations (e.g., tracking vehicles, maintenance, routing, fuel costs).

TRANSPORTER Hauling firm that picks up properly packaged and labeled hazardous waste from generators and transports it to designated facilities for treatment, storage or disposal.

TRANSVERSE Pertaining to or extending along short axis or width (i.e., perpendicular to or across long axis or length).

TRANSVERSE BANDS Bands of clouds oriented perpendicular to flow where they are embedded.

TRANSVERSE CRACKING Cracks perpendicular to pavement's centerline or laydown direction.

TRANSVERSE FISSURE Progressive crosswise fracture starting from inside head from where it spreads outward at right angle to length of rail.

TRANSVERSE LOAD Forces applied perpendicularly to longitudinal axis of member.

TRANSVERSE MARKINGS White transverse lines, marked across direction of traffic, spaced progressively closer together to create illusion of increasing speed.

TRANSVERSE PROFILE Pavement surface vertical deviation from horizontal reference perpendicular to travel direction.

TRANSVERSE REINFORCEMENT Reinforcement spanning along long direction.

TRANSVERSE ROLLS Elongated low-level clouds, arranged in parallel bands and aligned parallel to low-level winds but perpendicular to mid-level flow.

TRAP 1. Place in water pipe, pump, etc., where air accumulates for lack of outlet; 2. Bend in waste-pipe arranged so that liquid contents form seal that prevents escape of noxious gases, but permits flow of liquids; 3. Machine, device or depression designed to catch.

TRAP EFFICIENCY Proportion of sediment inflow to stream reach retained within that reach.

TRAP SEAL Retains small amount of water to create seal that prevents sewer gas from passing into occupied space.

TRAPEZOIDAL WEIR Contracted weir of trapezoidal shape.

TRAPPER Valley or basin where cold air becomes trapped or pooled.

TRASH 1. Material considered worthless or offensive that's thrown away; 2. Container where things are discarded.

TRASH COMPACTOR 1. Machine that reduces volume of trash by compression. 2. Nathaniel's favorite part of Star Wars - A New Hope movie where machine nearly crushes Rebels (i.e., Luke Skywalker, Han Solo and Chewbacca) during their rescue of Princess Leia Organa.

TRASH GATE Opening where floating waste is released.

TRASH RACK Grid structure designed to prevent floating and submerged debris from entering stormwater and wastewater systems while maintaining water flow.

TRASH SCREEN Screen structure designed to prevent floating and submerged debris from entering stormwater and wastewater systems while maintaining water flow.

TRASH-TO-ENERGY PLAN Burning trash to produce energy.

TRASHRACK Screen located at intake to prevent floating or submerged debris from entering intake.

TRASHRAKE Device used to remove debris which has collected on trash rack to prevent blocking associated intake.

TRAVEL DEMAND MODELS Models that forecast traffic flows on transportation system.

TRAVEL SPEED Distance divided by travel time.

TRAVEL TIME Time required for flood wave to travel from one location to subsequent location downstream.

TRAVEL TIME RATE 1. Time per distance traveled; 2. Rate of pay for travel.

TRAVEL TIME RELIABILITY Measure of extent of travel time dependability.

TRAVELED WAY Portion of road designated for movement of vehicles, excluding shoulders, parking areas, turn lanes, etc.

TRAVELER INFORMATION SYSTEMS Technology that updates drivers on current roadway conditions.

TRAVELER PERCEPTION MODEL Model indicating average rating travelers would give transportation facility.

TRAVELLING GRATE Moving grate in furnace covered with fuel.

TRAVELLING SCREEN Rotating device with screen to catch and remove debris.

TRAVERSABLE SLOPE Smooth terrain, sloped at 1:3 or flatter, where driver would unlikely be able to steer back into roadway but safely slow and stop.

TRAY AERATOR Aerator where water is released by slats or perforated trays.

TREATABILITY MANUAL Five-set library of EPA guidance manuals that contain information related to treatability of many pollutants.

TREATABILITY STUDIES Tests of potential cleanup technologies conducted in laboratory.

TREATABILITY STUDY Study or test that tells how wastewater might be treated.

TREATED REGULATED MEDICAL WASTE Medical waste treated to substantially reduce or eliminate its pathogenicity, but that has not yet been destroyed.

TREATED SEWAGE Sewage treated partially or completely.

TREATED WASTEWATER Wastewater that has been subjected to one or more physical, chemical and biological processes to reduce its potential of being health hazard.

TREATED WATER Water treated partially or completely.

TREATMENT 1. Any method, technique or process designed to remove solids and/or pollutants from solid waste, wastestreams, effluents and air emissions; 2. Use of substance or process to preserve or give particular properties to something.

TREATMENT PARAMETER Sewage characteristic that treatment is intended to address.

TREATMENT PLANT Structure built to treat wastewater before discharging it into environment.

TREATMENT TECHNIQUE Required process intended to reduce level of contaminant in drinking water.

TREATMENT WORKS Plant for sewage treatment.

TREATMENT WORKS TREATING DOMESTIC SEWAGE Includes all POTWs and other facilities that treat domestic wastewater and facilities that do not treat domestic wastewater, but that treat or dispose of sewage sludge.

TREATMENT, STORAGE, AND DISPOSAL FACILITY Site where hazardous substance is treated, stored or disposed of.

TREE Large plant with single trunk that grows in girth with age and branches.

TREMIE Device used to place concrete or grout under water.

TREMIE CONCRETE Concrete placement underwater by gravity through vertical pipe.

TREMIE SEAL Depth where discharge end of tremie pipe is kept embedded in concrete being placed.

TRENCH 1. Long, narrow excavation; 2. Constructed open channel for conducting water.

TRENCH BRACE Jack positioned between trench walls to support shoring.

TRENCH FORM Vertical sides and semicircular bottom of trench excavated through compacted soil to provide exterior form and base for concrete pipe that's cast-in-place.

TRENCH JACK Screw device used to hold shoring in place.

TRENCH LANDFILLING Trench excavated for wastes where excavated soil become cover material.

TRENCH SHIELD Trench shoring structure to protect workers within trench.

TRENCH SHORING Bracing trench walls to prevent collapse and cave-ins.

TRENCHER Machine for digging trenches.

TRENCHING BUCKET Narrow bucket used to dig trenches for cable or conduit.

TRENCHING OR BURIAL Placement of sewage sludge or septic tank pumpings in natural or man-made depression and subsequent covering with soil or suitable material at end of each operating day to prevent wastes from migrating to surface.

TRENCHLESS TECHNOLOGY Subsurface construction that requires little or no trenches.

TREND Statistical term referring to direction or rate of increase or decrease in magnitude of individual members of time series of data when random fluctuations of individual members are disregarded.

TRESTLE Bridge, usually of timber or steel, that has number of closely spaced supports between abutments.

TRIAL BURN Incinerator test where emissions are monitored for presence of specific organic compounds, particulates and hydrogen chloride.

TRIANGULAR NOTCH WEIR Weir with V-shaped opening to measure discharge.

TRIAXIAL SHEAR TEST Test where specimen of soil or rock encased by impervious membrane is subjected to confining pressure and then loaded axially to failure.

TRIAZINE HERBICIDE Class of herbicides containing symmetrical triazine ring (nitrogen-heterocyclic ring composed of three nitrogen's and three carbons in alternating sequence).

TRIBE Describes any Indian band, group or community recognized by Secretary of Interior and exercising governmental authority over Federal Indian reservation.

TRIBUTARY River or stream flowing into larger river, stream or lake.

TRICHADERMA VIRIDE Fungus and biofungicide that's used for soil and/or seed treatment for suppression of disease.

TRICHLOROETHYLENE Stable, low boiling-point colorless liquid, toxic if inhaled, used as solvent or metal degreasing agent and in other industrial applications.

TRICKLE Irrigation system where water is applied directly to root zone of plants by means of applicators (orifices, emitters, porous tubing, perforated pipe and so forth) operated under low pressure.

TRICKLE DRAIN Drain set vertically in water to transport excess water.

TRICKLE IRRIGATION Method where water drips to soil from perforated tubes or emitters.

TRICKLING FILTER Coarse treatment system where wastewater is trickled over bed of stones or other material covered with bacteria that break down organic waste and produce clean water.

TRIHALOMETHANE By-products of chlorination of drinking water that contains organic material.

TRIP 1. Journey; 2. Mechanical cutout device.

TRIP MATRIX Array of numbers that relates number of trips from originating zone to destination zone.

TRIP PURPOSE Main reason for taking trip.

TRIP SHEET Record of trip dates, driver name, beginning odometer reading, ending odometer reading, destination, etc.

TRIP TABLE Matrix that displays number of trips going from origin to destination.

TRIPLE HYDRANT Fire hydrant with three outlets (e.g., two 2.5 inch and one 4.5 inch).

TRIPLE POINT Intersection point between two boundaries (e.g., dry line, outflow boundary, cold front), often focus for thunderstorm development.

TRIPLE-TIPPER BODY Dump truck that can be raised by hydraulics to either side or rear.

TRIPOD Three-legged support for surveying instrument.

TRIPPER Short transportation run that's not part of regular scheduled run.

TRIPTON Inorganic particulate matter suspended in bodies of water.

TRITIUM Radioactive form of hydrogen, having one proton and two neutrons, with atoms of three times mass of ordinary hydrogen that can be used to determine age of water.

TROLLEY BUS Trolley vehicle, powered via overhead electric cables, that does not run on tracks.

TROLLEY POLE Pole of wood or metal, used to transfer electricity from overhead wire to power trolley vehicle.

TROPHIC LEVEL Position of organism in food chain.

TROPHOGENIC REGION Superficial layer of lake where organic production from mineral substances takes place on basis photosynthetic activity and light energy.

TROPICAL ADVISORY Official information issued by tropical cyclone warning centers describing all tropical cyclone watches and warnings in effect along with details concerning tropical cyclone locations, intensity and movement and precautions that should be taken.

TROPICAL ANALYSIS AND FORECAST BRANCH One of three branches of Tropical Prediction Center that provides year-round products involving marine forecasting, aviation forecasts and warnings and surface analyses.

TROPICAL CYCLONE Warm-core, non-frontal synoptic-scale cyclone, originating over tropical or subtropical waters with organized deep convection and closed surface wind circulation about well-defined center.

TROPICAL CYCLONE PLAN OF DAY Coordinated mission plan that tasks operational weather reconnaissance requirements during next 1100 to 1100 UTC day or as required, describes reconnaissance flights committed to satisfy both operational and research requirements and identifies possible reconnaissance requirements for succeeding 24-hour period.

TROPICAL CYCLONE POSITION ESTIMATE National Hurricane Center issues position estimate between scheduled advisories whenever storm center is within 200 nautical miles of U.S. land-based weather radar and if sufficient and regular radar reports are available to hurricane center.

TROPICAL CYCLONE UPDATE Brief statement issued by National Hurricane Center in lieu of or preceding special advisories to inform of significant changes in tropical cyclone or posting or cancellation of watches and warnings.

TROPICAL DEPRESSION Tropical cyclone where maximum 1-minute sustained surface wind is 33 knots (38 mph) or less.

TROPICAL DISTURBANCE Discrete tropical weather system of apparently organized convection originating in tropics or subtropics, having non-frontal migratory character and maintaining its identity for 24 hours or more.

TROPICAL PREDICTION CENTER Center which produces marine offshore and high seas forecasts south of 30° N in Eastern Pacific, Gulf of Mexico and Caribbean.

TROPICAL STORM Tropical cyclone where maximum 1-minute sustained surface wind ranges from 34 to 63 knots (39 to 73 mph) inclusive.

TROPICAL STORM WARNING Announcement that tropical storm conditions (sustained winds of 39 to 73 mph) are expected somewhere within specified coastal area within 36 hours.

TROPICAL STORM WATCH Announcement that tropical storm conditions (sustained winds of 39 to 73 mph) are possible within specified coastal area within 48 hours.

TROPICAL UPPER-TROPOSPHERIC TROUGH Semi-permanent trough extending east-northeast to west-southwest from about 35° N in eastern Pacific to about 15° to 20° N in central west Pacific.

TROPICAL WAVE Trough or cyclonic curvature maximum in trade wind easterlies.

TROPICAL WEATHER DISCUSSION Messages issued 4 times daily by Tropical Analysis and Forecast Branch to describe significant synoptic weather features in tropics.

TROPICAL WEATHER OUTLOOK Covers tropical and subtropical waters, discussing weather conditions, emphasizing any disturbed and suspicious areas which may become favorable for tropical cyclone development within next day to two.

TROPICAL WEATHER SUMMARY National Hurricane Center issues monthly summary of tropical weather is included at end of month or as soon as feasible thereafter, to describe briefly past activity or lack thereof and reasons why.

TROPICS Areas of Earth within 20° North and South of equator.

TROPOPAUSE Zone of transition between troposphere and stratosphere.

TROPOPAUSE JET Type of jet stream found near tropopause (e.g., subtropical, polar fronts).

TROPOPHYTE Any plant that lives in environment where heavy rainfall alternates with periods of drought.

TROPOSPHERE Lower levels of atmosphere extending from surface up to tropopause.

TROUGH 1. Elongated area of relatively low atmospheric pressure; 2. Opposite of ridge; 3. Elongated depression in potentiometric surface.

TROUGH OF WARM AIR ALOFT "Tongue" of relatively warm/moist air aloft that wraps around to north and west of mature cyclone.

TRUCK Vehicle designed for carrying cargo.

TRUCK APRON Apron between road portion of roundabout and inner circle of roundabout to allow large vehicles to navigate without striking other road or fixed objects.

TRUCK TRACTOR Motor vehicle designed to draw or pull semi-trailer.

TRUCK-MOUNTED ATTENUATOR Device attached to truck to absorb impact of crash, decrease damage made to vehicle and save lives.

TRUCKING PLATFORM Landing area where freight and baggage are moved to other vehicles.

TRUE WIND Wind relative to fixed point on Earth.

TRUMPET INTERCHANGE Three-way interchange where one loop ramp connects traffic entering or leaving roadway with far lanes of roadway.

TRUNCATED DOMES Ground surface indicators (e.g., yellow raised bumps) that assist and warn blind or visually impaired pedestrians.

TRUNCATION ERROR Error introduced by replacing derivative terms of differential equation by finite differences using Taylor series and then truncating after certain number of terms.

TRUNK 1. Upright part of tree between roots and branches; 2. Channel for flow of some kind.

TRUNK SEWER Sewer line that receives wastewater from tributary branches and sewer lines.

TRUNKLINE Main pipe that drains or supplies water to building.

TRUNNION Heavy horizontal hinge.

TRUSS Structure made up of one or more triangular units made from straight beams of wood or metal that are used to support structure (e.g., roof or bridge).

TRUSS BRIDGE Bridge with load-bearing superstructure is composed of truss.

TRUST FUND Fund set up under Comprehensive Environmental Response, Compensation and Liability Act to help pay for cleanup of hazardous waste sites and for legal action to force those responsible for sites to clean them up.

TSUNAMI Series of long-period waves (on order of tens of minutes) that are usually generated by impulsive disturbance (e.g., earthquake, underwater volcanic activity, landslide) that displaces massive amounts of water.

TSUNAMI ADVISORY Third highest level of tsunami alert issued to coastal populations within areas not currently in either warning or watch status when tsunami warning has been issued for another region of same ocean.

TSUNAMI INFORMATION STATEMENT Issued to inform emergency management officials and public that earthquake has occurred or that tsunami warning, watch or advisory has been issued for another section of ocean.

TSUNAMI WARNING Highest level of tsunami alert issued due to imminent threat of tsunami from large undersea earthquake or following confirmation that potentially destructive tsunami is underway.

TSUNAMI WATCH Second highest level of tsunami alert issued based on seismic information without confirmation that destructive tsunami is underway.

TUBE Anything that's hollow and cylindrical in shape.

TUBE SETTLER Device using bundles of tubes to let solids in water settle to bottom for removal by conventional sludge collection means that's sometimes used in sedimentation basins and clarifiers to improve particle removal.

TUBE VALVE Valve that is opened or closed by mechanically moving tube upstream or downstream by actuating screw.

TUBEAXIAL FAN Fan consisting of propeller or disk-type wheel within cylinder.

TUBERCLE 1. Round nodule, small eminence or warty outgrowth; 2. Small rounded wart-like protuberance of roots of some leguminous plants.

TUBERCULATION Development or formation of small mounds of corrosion products (i.e., rust blisters) on inside of iron pipe.

TUBING 1. Lightweight conduit; 2. Small-diameter pipe installed inside casing of well.

TUBULAR WELL Well where screen or casing is in direct contact with water bearing formation.

TUFF Igneous rock formed from hardened volcanic ash.

TUGBOAT Small, powerful boat used to push, pull or help maneuver barges or larger vessels.

TULE FOG Radiation fog that forms during night and morning hours in late fall and winter months following first significant rainfall.

TUMBLER SWITCH Lever-actuated switch where blades are actuated by lever being pushed up or pulled down.

TUNDRA Type of treeless ecosystem dominated by lichens, mosses, grasses and woody plants found at high latitudes (arctic tundra) and high altitudes (alpine tundra).

TUNDRA SOIL Arctic soil above permafrost.

TUNNEL 1. Covered portion of spillway between gate or crest structure and terminal structure, where open channel flow and/or pressure flow conditions may exist; 2. Portion of outlet works between upstream and downstream portals, excluding gate chamber that's generally located in dam abutments and are concrete lined or concrete/steel lined; 3. Enclosed channel, with two or more openings to surface constructed by excavating through natural ground that can convey water or house conduits or pipes.

TURBID Having cloudy or muddy appearance.

TURBIDIMETER Device that measures amount of suspended solids in liquid.

TURBIDITE Any sedimentary deposit formed by turbidity current.

TURBIDITY Measure of extent that light passing through water (i.e., transparency) is reduced due to suspended materials (e.g., suspended silt, organic matter).

TURBINE Machine for generating rotary mechanical power from energy of stream of fluid (e.g., water, steam, hot gas).

TURBOJET Jet engine that develops thrust solely from high-speed exhaust gases expelled from turbine that drives compressor.

TURBOPROP Gas turbine aircraft engine that drives and obtains essentially all thrust from external propeller.

TURBULENCE Irregular motion of medium (e.g., flowing fluid, atmosphere).

TURBULENT Violently disturbed or agitated.

TURBULENT FLOW Open channel flow characterized by random fluid motion where inertial forces are very much greater than viscous forces and Reynolds number is large.

TURBULENT MIXERS Apparatus that mix air bubbles and water and cause turbulence to dissolve oxygen in water.

TURBULENT VELOCITY Fluid flow where fluid undergoes irregular fluctuations (i.e., mixing).

TURKEY TOWER Narrow, individual cloud tower that develops and falls apart rapidly.

TURN ANGLES To measure angle between directions with surveying instrument.

TURN BAY Area where vehicles can pause to turn without risk of holding up traffic flow or being hit from behind another vehicle.

TURNDOWN RATIO Ratio of maximum capacity to minimum capacity during continuous operation.

TURNING BASIN Wider body of water at end of canal or narrow channel to turn around vessels.

TURNING LOOP Circular road segment at end of roadway to turning around.

TURNING MOVEMENT Designated turn of vehicular traffic at intersection.

TURNING PATH Smallest circular turn vehicle can make.

TURNING POINT Temporary point whose elevation is determined by additions and subtractions of backsights and foresights respectively.

TURNING TRACK WIDTH Radial distance between turning paths of outside of outer front tire and outside of rear tire nearest center of turn.

TURNOUT 1. Structure used to divert water from supply channel to smaller channel; 2. Point where control of water changes from irrigation district to customer.

TURNOVER When surface water replaces bottom water in impoundment.

TWENTY FOUR HOUR WARNING POINT Communication facility at state or local level, operating 24 hours day, which has capability to receive alerts and warnings, plus activate public warning system in its area of responsibility.

TWENTY-FOOT EQUIVALENT UNIT Cargo capacity used to describe capacity of container ships and terminals based on volume of 20-foot long container.

TWILIGHT Average time of civil twilight, that is time between civil dawn and sunrise in morning and between sunset and civil dusk in evening.

TWISTER Colloquial term for tornado.

TWO-DIMENSIONAL MODEL Model defined with two space coordinates (i.e., variables are averaged over third direction).

TWO-LANE HIGHWAY Roadway with only one lane in each direction.

TWO-RIBBON FLARE Flare that has developed as pair of bright strands (ribbons) on both sides of main inversion ("neutral") line of magnetic field of active region.

TWO-SIDED WEAVING SEGMENT Segment where one-lane on-ramp on one side of facility is closely followed by one-lane off-ramp on other side of facility.

TWO-WAY LEFT-TURN LANE Roadway that provides center lane exclusively for left turning vehicles coming from either direction.

TWO-WAY STOP-CONTROLLED Traffic control for vehicles on minor street to stop when they arrive at intersection of major street.

TWO-WAY SWITCH Two or more switches in different locations to control one circuit.

TWO-WAY VALVE Valve with two ports to control direction of fluid flow.

TYPE A JOINT Joint using wet concrete or epoxy between precast units.

TYPE A SOIL Cohesive soils having unconfined compressive strength of 1.5 $tons/ft^2$ or greater.

TYPE B JOINT Dry joint between precast units.

TYPE B SOIL Cohesive soil with unconfined compressive strength greater than 0.5 $tons/ft^2$ but less than 1.5 $tons/ft^2$.

TYPE C SOIL Cohesive soil with unconfined compressive strength of 0.5 $tons/ft^2$ or less.

TYPHOID Serious disease spread by contaminated food and water.

TYPHOON Tropical cyclone in Western Pacific Ocean where maximum 1-minute sustained surface wind is equivalent to hurricane or cyclone.

TYPHOON SEASON Part of year having relatively high incidence of tropical cyclones.

U

U BURST Fast radio burst spectrum of flare.

U.S. ARMY CORPS OF ENGINEERS Federal agency, part of Department of Army, has authority for approval of dredge and fill permits in navigable waters and tributaries thereof that enforces wetlands regulations and constructs and operates variety of water resources projects, mostly notably levee, dams and locks.

U.S. BUREAU OF RECLAMATION Federal agency whose mandate was to reclaim arid west of United States, operating in 17 western states, this agency builds, operates and maintains variety of irrigation, power and flood control projects.

U.S. GEOLOGICAL SURVEY Federal agency chartered to classify public lands and to examine geologic structure, mineral resources and products of national domain that provides information and data on Nation's rivers and streams that are useful for mitigation of hazards associated with floods and droughts.

UIC PROGRAM DIRECTOR Chief administrative officer of any state or tribal agency or EPA Region that has been delegated to operate approved UIC program.

ULTIMATE ANALYSIS Test to determine elements in compound.

ULTIMATE BEARING CAPACITY Load per unit of area (i.e., pressure) required to produce failure by rupture of supporting soil or rock mass.

ULTIMATE BEARING PRESSURE Pressure foundation will settle with no load increase.

ULTIMATE BIOCHEMICAL OXYGEN DEMAND Parameter that quantifies oxygen required for total biochemical degradation of organic matter.

ULTIMATE BIODEGRADABILITY Degradation level achieved when test compound is totally utilised by microorganisms resulting in carbon dioxide, water, mineral salts and biomass.

ULTIMATE CARBON DIOXIDE Specific amount of carbon dioxide by volume present in flue gas when exact amount of air is supplied to achieve correct air-fuel ratio.

ULTIMATE COMPRESSIVE STRENGTH Maximum amount of compressive stress that material can withstand before it breaks or fractures.

ULTIMATE STAGE BIOCHEMICAL OXYGEN DEMAND
Complete quantity of oxygen necessary to meet biochemical oxygen demand of sample.

ULTIMATE STRENGTH Maximum stress that material can sustain.

ULTIMATE STRENGTH LOAD Load where failure is expected.

ULTIMATE WATER DEMAND Water requirements (i.e., quantity, pressure and availability) for defined purpose at specified location.

ULTRA CLEAN COAL Coal washed, ground into fine particles, then chemically treated to remove sulfur, ash, silicone and other substances that's usually briquetted and coated with sealant made from coal.

ULTRA HIGH FREQUENCY Those radio frequencies exceeding 300 MHz.

ULTRA-FILTRATION Filtration using hydrostatic pressure forces and filter to retain colloidal particles, viruses, or large molecules.

ULTRA-THIN WHITETOPPING Construction technique that involves placement of 2 to 4 inch concrete overlay.

ULTRAFILTRATION Filtration using hydrostatic pressure forces and filter to retain colloidal particles, viruses, or large molecules.

ULTRASONIC TESTING Evaluation based on propagation of ultrasonic waves.

ULTRAVIOLET Electromagnetic radiation beyond (higher in frequency than) light visible to human eye.

ULTRAVIOLET ABSORPTION AT 254 OR 280 NANOMETERS
Measure of aggregate concentration of mixture of UV absorbing organic materials dissolved in analyzed water (e.g., lignin, tannin, humic substances, various aromatic compounds).

ULTRAVIOLET INDEX Index provides important information to help you plan your outdoor activities in ways that prevent overexposure to Sun's rays that's computed using forecasted ozone levels, computer model that relates ozone levels to UV incidence on ground, forecasted cloud amounts and elevation of forecast cities.

ULTRAVIOLET RADIATION Electromagnetic radiation of shorter wavelength than visible radiation but longer than x-rays.

ULTRAVIOLET RAYS Radiation from Sun that can be useful or potentially harmful.

UMBRA 1. Dark core or cores (umbrae) in sunspot with penumbra; 2. Sunspot lacking penumbra.

UNACCEPTABLE ADVERSE EFFECT Impact on aquatic or wetland ecosystem likely to result in significant degradation, significant loss or damage.

UNACCOUNTED FOR WATER Difference between volume of water delivered into network and volume of water consumed.

UNAMORTIZED CREDIT OR DEBIT Credit or debit accounted for in current interval.

UNBALANCED HEAD Condition where water pressure on upstream and downstream sides of object differ.

UNBRACED LENGTH Distance between structural member ends that are prevented from moving normal to member axis.

UNBUNDLING Electric service is traditionally provided on bundled basis, meaning that generation, transmission and distribution services are provided as single package.

UNBURNED HYDROCARBONS Hydrocarbons emitted after fuel is burned.

UNCERTAINTY Describes situations where potential outcomes cannot be estimated based on historical events.

UNCERTAINTY FACTOR One of several factors used in calculating reference dose from experimental data.

UNCLASSIFIED EXCAVATION Excavation paid for at fixed price.

UNCOMPACTED BACKFILL Material used in refilling excavation without material being compacted.

UNCONFINED AQUIFER Aquifer whose hydraulic head surface (i.e., water table) is at equilibrium with atmospheric pressure.

UNCONFINED GROUNDWATER Aquifer water with watertable.

UNCONFORMITY Break or gap in geologic record where rock layers were eroded or never deposited.

UNCONSOLIDATED DEPOSIT Deposit of loosely bound sediment that typically fills topographically low areas.

UNCONSOLIDATED FORMATION Loosely arranged or unstratified sediment whose particles are not cemented together.

UNCONTROLLED Intersection missing traffic control device.

UNCONTROLLED FILL Any fill without proper documentation.

UNCONTROLLED SPILLWAY Spillway that does not have gates and only governed by inflow.

UNCONVENTIONAL WASTE Hazardous, discarded material of no value.

UNDER-CONSOLIDATION In soil, when initial vertical effective stress is greater than pre-consolidated stress.

UNDERCURRENT Current below upper currents or surface of fluid body.

UNDERCUT BANK When water erodes soil from stream or river bank, leaving overhang.

UNDERCUT BEDDING Plane along where failure and sliding may occur because incline or free face is greater than bedding.

UNDERDRAIN To drain by forming underdrain.

UNDERFLOW Lateral motion of water through upper layers until it enters stream channel.

UNDERFLOW CONDUIT Permeable layer underlying surface stream channel.

UNDERFOOT CONDITIONS Relating to state of ground.

UNDERGROUND DRINKING WATER SOURCE Aquifer that supplies water for human consumption or contains less than 10,000 mg/l total dissolved solids.

UNDERGROUND INJECTION Placing fluid deep underground or below shallow soil layers.

UNDERGROUND INJECTION CONTROL Program under Safe Drinking Water Act that regulates use of wells to pump fluids into ground.

UNDERGROUND INJECTION WELLS Steel- and concrete-encased shafts into which hazardous waste is deposited by force and under pressure.

UNDERGROUND SERVICES Buried pipes, cables and equipment.

UNDERGROUND SOURCE OF DRINKING WATER Aquifer or portion of aquifer that supplies any public water system or that contains sufficient quantity of groundwater to supply public water system and currently supplies drinking water for human consumption or that contains fewer than 10,000 mg/l total dissolved solids and is not exempted aquifer.

UNDERGROUND STORAGE TANK Tank located at least partially underground and designed to hold gasoline or other petroleum products or chemicals.

UNDERGROUND WATER Subsurface water in unsaturated and saturated zones.

UNDERGROWTH Plants in forest that only reach relatively low height.

UNDERPASS Passage that crosses road, railroad, etc. in tunnel underneath it.

UNDERSATURATED FLOW Traffic flow unaffected by upstream or downstream conditions.

UNDERSTORY Foliage layer lying beneath and shaded by main canopy of forest.

UNDERSUN Optical effect seen by observer above cloud deck when looking toward Sun, as sunlight is reflected upwards off faces of ice crystals in cloud deck.

UNDERTAKING Any project, activity or program that could change character or use of historic properties.

UNDERTOW Relatively small-scale surf-zone current moving away from beach.

UNDIGESTED SLUDGE Sludge that has settled and is quickly removed from sedimentation tank before dissolved oxygen.

UNDISTURBED SAMPLE Soil sample that has been obtained by methods where every precaution has been taken to minimize disturbance to sample.

UNDULATION Wavelike motion.

UNFILTERED Pertains to constituents in unfiltered, representative water-suspended sediment sample.

UNFILTERED, RECOVERABLE Amount of given constituent in representative water-suspended sediment sample that has been extracted or digested.

UNFIT-FOR-USE TANK SYSTEM Tank system that has been determined through integrity assessment or other inspection to be no longer capable of storing or treating hazardous waste without posing threat of release of hazardous substances to environment.

UNIAXIAL TENSILE STRESS Tension stress acting in only one axis.

UNICOM Air-ground communication facility operated by non-air traffic control private agency to provide advisory service at uncontrolled airports.

UNIFIED PLANNING WORK PROGRAM Statement of work identifying planning priorities and activities to be carried out within planning area.

UNIFIED SOIL CLASSIFICATION SYSTEM Method of grouping and describing soils according to their engineering properties.

UNIFORM CHANNEL Channel with constant cross-section and constant roughness.

UNIFORM DELAY Delay based on assumption of uniform arrivals and departures.

UNIFORM FLOW Open channel flow that doesn't vary in magnitude and direction over space.

UNIFORMITY COEFFICIENT Ratio of diameter at 60 percent finer point to 10 percent finer point on gradation curve.

UNIMODAL Distribution having only one localized maximum (i.e., only one peak).

UNINTERRUPTED FLOW Traffic flow where there is no obstructions to movement of vehicles along road.

UNINTERRUPTED FLOW FACILITY Facility with no causes of interruption external to traffic stream.

UNION Joint or other connection uniting parts.

UNION BEND Bend with union at one end.

UNION SHOP Place of employment where newly hired workers join union within certain period of time.

UNIONIZED Neutral form of ionizable compound (i.e, acid, base).

UNIONIZED AMMONIA Neutral form of ammonia-nitrogen in water.

UNIT 1. Part of complete vehicle; 2. Volume of rock or ice of identifiable origin and age range defined by distinctive and dominant, easily mapped and recognizable features that characterize it; 3. Standard measure of quantity.

UNIT CONTROL POSITION WSR-88D radar operator uses this to control entire radar system.

UNIT COST Cost of producing unit of work, product or service.

UNIT DISCHARGE 1. Discharge per unit width; 2. Discharge through turbine when head on turbine is unity.

UNIT HYDROGRAPH Direct runoff (e.g., discharge) hydrograph resulting from unit (e.g., 1 inch) depth of excess rainfall produced by storm of uniform intensity over watershed for given time period.

UNIT HYDROGRAPH DURATION Time over which one inch of surface runoff is distributed for unit hydrograph theory.

UNIT HYDROGRAPH THEORY States that surface runoff hydrographs for storm events of same duration will have same shape and ordinates of hydrograph will be proportional to ordinates of unit hydrograph (e.g., hydrograph for $1/2$ inch of storm runoff will be half that of that from 1 inch hydrograph).

UNIT PRICE Price of single unit of measure or good.

UNIT PRICE CONTRACT Contract based on estimated quantities of unit prices.

UNIT SHEAR FORCE 1. Force per unit length; 2. Force or drag developed along channel bed by water.

UNIT TIME May be either hour or day, depending on incubation period.

UNIT TOW One to three barges pushed by towboat.

UNIT TRAIN Train where all cars carry same commodity and are shipped from same origin to same destination.

UNIT VENT Pipe that vents multiple traps.

UNIT WEIGHT Weight of substance in single unit of volume.

UNIT WEIGHT OF WATER Weight per unit volume of water.

UNIT WIDTH FLOW RATE Pedestrian flow rate calculated as pedestrians per minute per unit width.

UNITGRAPH Direct runoff (e.g., discharge) hydrograph resulting from unit (e.g., 1 inch) depth of excess rainfall produced by storm of uniform intensity over watershed for given time period that's sometimes called unit hydrograph.

UNIVERSAL COORDINATED TIME Local time at prime meridian that passes through Greenwich, England that's known as Coordinated Universal Time, Universal Time Coordinated (UTC), "Z time" or "Zulu Time."

UNIVERSAL GEOGRAPHIC CODE National Weather Service product that specifies affected geographic area of event, typically by state, county (or parish) or unique National Weather Service zone (land and marine).

UNIVERSAL TIME Local time at prime meridian that passes through Greenwich, England that's known as Coordinated Universal Time, Universal Time Coordinated (UTC), "Z time" or "Zulu Time."

UNIVERSAL TIME COORDINATED 1. Under international agreement, local time at prime meridian that passes through Greenwich, England that's known as "Z time," "Zulu Time" or prior to 1972, Greenwich Mean Time (GMT).

UNIVERSAL TYPE WEIGHTING AND RECORDING GAUGE Gauge which collects precipitation and then converts weight onto inked pen movement which traces on graph paper fixed to clock driven drum.

UNLINKED TRIP One trip on transit vehicle.

UNLISTED ITEMS Line item used in appraisal estimate for design changes and to estimate pay items that have little influence on total cost.

UNLOAD-RELOAD CYCLE Loading sequence resulting in rebound and recompression.

UNLOADING To remove or release (e.g., contaminant).

UNLOADING VALVE pressure-control Valve that dumps excess fluid to tank at set pressure.

UNMET DEMAND Queue at beginning of red phase of signalize intersection.

UNPAID FEDERAL INVESTMENT Year-end amount owed to Treasury which remains to be repaid.

UNRECLAIMABLE RESIDUES Residual materials of little or no value remaining after incineration.

UNSAFE CONDITION Situation in work place that may lead to property damage, incident or injury.

UNSATURATED Capable of dissolving more of solute at same temperature.

UNSATURATED ZONE Subsurface zone between land surface and groundwater where pores contain both water and air that may include capillary fringe.

UNSATURATED ZONE OF AERATION Zone between watertable and surface.

UNSEALING OF TRAPS Sewer gas escape due to reduction of trap's water seal.

UNSETTLED 1. Colloquial term used to describe condition in atmosphere conducive to precipitation; 2. With regard to geomagnetic levels, descriptive word specifically meaning that 8 is less than or equal to Ap Index that is less than or equal to 15.

UNSIGNALIZED INTERSECTION Road junction where right-of-way is not controlled by traffic signal.

UNSOCKETED PIPE Pipe joined using sleeve coupling.

UNSOUND Not whole, not solid or defective.

UNSTABLE AIR Air able to rise easily and has potential to produce clouds, rain and thunderstorms.

UNSTABLE SOIL Earth material that won't remain in place without additional support.

UNSTEADY FLOW Velocity at point varies with time.

UNSTEADY NON-UNIFORM FLOW Flow where velocity and water amount flowing per unit of time change at every point.

UNSTEADY-STATE MODEL Model where variables being investigated are time dependent.

UNSUITABLE MATERIAL Those soils that cannot be compacted in embankment or backfill or where excavated to finished grade result in unstable material.

UNWATERING Interception and removal of groundwater outside of excavations and removal of ponded or flowing surface water from within excavations.

UNWEATHERED No evidence of any mechanical or chemical alteration.

UP-VALLEY WIND Diurnal thermally driven flow directed up valley's axis, usually occurring during daytime that's part of along-valley wind system.

UPDRAFT Small-scale current of rising air.

UPFEED DISTRIBUTION Water fed to system by pump pressure.

UPFLOW COAGULATION Coagulation by passing liquid upward through settling sludge blanket.

UPFLOW CONTACT CLARIFIER Tank where water enters at bottom, is forced to up through filter bed and exits at surface.

UPGRADIENT Of or pertaining to place(s) from which groundwater originated or traveled through before reaching given point in aquifer.

UPLAND Elevated land above low areas along streams or between hills (i.e., elevated region from which rivers gather drainage) that's non-wetland.

UPLIFT 1. Upward pressure on structure; 2. Tectonic upheaval.

UPLIFT PRESSURE Internal hydrostatic pressure in embankment caused by level of water in reservoir acting through pressure-transmitting paths between soil particles in fill.

UPPER BANK Portion of streambank that extends from break in normal slope of surrounding land to high water line.

UPPER DETECTION LIMIT Largest concentration that instrument can reliably detect.

UPPER LEVEL Portion of atmosphere above lower troposphere, generally 850 hPa and above.

UPPER LEVEL DISTURBANCE Disturbance in upper atmospheric flow pattern that is usually associated with clouds and precipitation.

UPPER LEVEL SYSTEM General term for any large-scale or mesoscale disturbance capable of producing upward motion (lift) in middle or upper parts of atmosphere.

UPPER-AIR WEATHER CHART Weather maps that are produced for portion of atmosphere above lower troposphere, generally at and above 850 mb.

UPRATE Increase of greater than 15% in power plant unit output at existing facility through modifications or replacement of equipment.

UPRIGHT Vertical piece of any trench shoring setup.

UPRUSH Quick rush of water from breaking wave.

UPSET 1. Exceptional incident where there is unintentional and temporary non-compliance with permit limit because of factors beyond reasonable control of permittee; 2. To disturb, disrupt or adversely alter; 3. To thicken and shorten, tip or overturn.

UPSLOPE FLOW Air that flows toward higher terrain and hence is forced to rise.

UPSLOPE FOG Fog that forms when moist, stable air is carried up mountain slope.

UPSTREAM Located in area from which flow is coming (i.e., towards source of flow).

UPSTREAM BLANKET Impervious blanket placed on reservoir floor and abutments upstream of dam.

UPSTREAM FACE Inclined surface of dam in contact with reservoir.

UPSTREAM SLOPE Part of dam that is in contact with reservoir water.

UPSURGE Sudden strong rise or flow.

UPTAKE Absorption (e.g., nutrient by organism).

UPTAKE PHASE Period where evaluation organisms are being exposed to substance.

UPTIME Interval when machine can be utilized.

UPWARD-FLOW FILTER Where water flows, from bottom to top, through filter.

UPWELLING 1. Upward movement from lower source; 2. When strong winds push warm water away from coast and bring cold waters up to surface.

URANIUM Heavy silvery-white metallic element, with atomic number 92, highly radioactive and easily oxidized.

URANIUM MILL TAILINGS PILES Former uranium ore processing sites that contain leftover radioactive materials (i.e., wastes), including radium and unrecovered uranium.

URANIUM MILL-TAILINGS WASTE PILES Licensed active mills with tailings piles and evaporation ponds created by acid or alkaline leaching processes.

URBAN Related to city life.

URBAN AND SMALL STREAM FLOOD ADVISORY Advisory alerts public to flooding that is generally only inconvenience (i.e., not life-threatening) to those living in affected area.

URBAN AND SMALL STREAM FLOODING Flooding of small streams, streets and low-lying areas (e.g., railroad underpasses, urban storm drains).

URBAN AREA Predominantly cities, towns or developed areas where population is significant.

URBAN FLASH FLOOD GUIDANCE Guidance which estimates average amount of rain needed over urban area during specified period of time to initiate flooding on small, ungauged streams in urban area.

URBAN FLOODING Flooding of streets, underpasses, low lying areas or storm drains.

URBAN HEAT ISLAND Increased air temperatures in urban areas in contrast to cooler surrounding rural areas.

URBAN RENEWAL Process where older areas of city are redeveloped.

URBAN RUNOFF Storm water from city streets and adjacent domestic or commercial properties that carries pollutants of various kinds into sewer systems and receiving waters.

URBAN SITE Site that has greater than 50% urbanized and less than 25% agricultural area.

URBAN SPRAWL Unplanned, disorganized growth of housing on edge of city.

URBAN STREET Road in urban area with high number of driveways.

URBAN STREET SEGMENT Length between boundaries of urban street.

URBANIZATION To become urban in nature or character (i.e., residential, commercial and industrial development).

URBANIZED AREA Human population density area of 50,000 or more.

UREA-FORMALDEHYDE FOAM INSULATION Material once used to conserve energy by sealing crawl spaces, attics, etc.

URINAL Fixture used for urination.

USABLE STORAGE Quantity of groundwater of acceptable quality that can be economically withdrawn from storage.

USABLES Discarded waste usable in present condition.

USCS SOIL CLASSIFICATION SYSTEM Soil classification system to describe texture and grain size of soil.

USE ATTAINABILITY ANALYSIS Structured scientific assessment of factors affecting attainment of use which may include physical, chemical, biological and economic factors.

USE CLUSTER Set of competing chemicals, processes, and/or technologies that can substitute for one another in performing particular function.

USE VALUE Economic benefit associated with physical use of resource, usually measured by consumer surplus or net economic value associated with such use.

USED OIL Spent motor oil from passenger cars and trucks collected at specified locations for recycling.

USEFUL LIFE Estimated lifespan of depreciable fixed asset.

USEFUL STORAGE Reservoir water volume stored between minimum and normal levels.

USER DAY Participation in recreation activity at given resource during 24-hour period by one person.

USER FEE Fee collected from only those persons who use particular service, as compared to one collected from public in general.

USER PERCEPTION VARIABILITY When users experiencing identical conditions may perceive those conditions differently.

UTC 1. Under international agreement, local time at prime meridian that passes through Greenwich, England that's known as "Z time," "Zulu Time" or prior to 1972, Greenwich Mean Time (GMT).

UTILITY Service (i.e., electricity, gas, internet) provided to public.

UTILITY ACCESS HOLE Opening in underground system where workers enter for maintenance.

UTILITY CONNECTED SYSTEM Designed to provide portion of building's total electricity needs that connects with utility grid system.

UTILITY LOAD Total electricity demand for utility district.

UTILIZATION FACTOR Ratio of demand to availability.

V

V NOTCH Radar reflectivity signature seen as V-shaped notch in downwind part of thunderstorm echo.

V-NOTCH WEIR Weir that's V-shaped, with its apex downward used to accurately measure small rates of flow.

V-PLOW Plow with V-shaped blade that pushes material to both sides.

VACUUM Region of space that contains no matter.

VACUUM BREAKER Apparatus placed on valve to prevent back siphonage.

VACUUM DEAERATION Gear that removes dissolved gases from liquid in partial vacuum.

VACUUM FILTER Drum that rotates in tub of liquid to be filtered, used to dewater and extract out waste sludge.

VACUUM LOADER Truck, with street vacuum attachment, used for road sweeping, etc.

VACUUM PUMP Equipment used for extraction of gas or vapor from enclosed space, leaving container with partial vacuum.

VACUUM RELIEF VALVE Valve with vent to protect hot water tank against rupturing or imploding.

VAD WIND PROFILE Radar plot of horizontal winds, derived from VAD data, as function of height above Doppler Radar.

VADOSE WATER Water within zone between water table and ground surface.

VADOSE ZONE Subsurface extending from surface to regional groundwater table.

VADOSE ZONE FLOW SYSTEM Aggregate of rock, where both water and air enters and moves and that is bounded by rock that does not allow any water movement and by zones of interaction with Earth's surface, atmosphere and surface water systems.

VALDRIFT Air pollution transport and diffusion model developed to determine pesticide drift from aerial spraying operations in valleys.

VALID TIME Period of time during which forecast or warning, until it is updated or superseded by new forecast issuance, is in effect.

VALID TIME EVENT CODE Always is used in conjunction with and provides supplementary information to, Universal Geographic Code, to further aid in automated delivery of National Weather Service text products to users.

VALIDATION Comparison of model results with set of prototype data that were not used for verification.

VALIDATION MONITORING Determines if predictive model coefficients are adequately protecting targeted resources.

VALLEY Elongated depression between hills or mountains.

VALLEY DRAINAGE Water that gathers in depression.

VALLEY EXIT JET Strong elevated down-valley air current issuing from valley above its intersection with adjacent plain.

VALLEY FOG Fog that develops when mountains prevent dense, cooled air (by radiation) from escaping.

VALLEY VOLUME EFFECT Reduction in volume of valley (or basin) as compared to equal depth volume with horizontal floor.

VALUATION Estimation of somethings worth.

VALUE ENGINEERING Organized effort that analyzes selections for purpose of achieving essential functions at lowest life cycle cost.

VALUE PRICING Pricing based on how much customer believes product is worth.

VALUED ENVIRONMENTAL ATTRIBUTES/COMPONENTS Those aspects (i.e., components, processes, functions) of ecosystems, human health and environmental welfare considered to be important and potentially at risk from human activity or natural hazards.

VALVE Device fitted to pipeline or orifice where closure member is either rotated or moved in some way as to control or stop flow in conduit, pipe or tunnel.

VALVE ACTUATOR Device that uses power source to operate valve.

VALVE BANK Combination of valves for pump.

VALVE BEAT-IN Wear on valve face or seat caused by hammering of valve on seat.

VALVE BOX Enclosure in ground, set above valve(s), to allow operation.

VALVE BRIDGE Element that permits one rocker arm to control two valves.

VALVE CHAMBER Concrete portion of pressurized outlet works (i.e., room) containing gates between upstream and downstream conduits and/or tunnels in free-flow outlet works.

VALVE KEY Socket-wrench with T-head for valve spindle.

VALVE STEM Cylindrical portion of valve that moves up or down to open or close valve.

VALVE TOWER Reservoir tower that houses valve controls.

VANE Surfaces radially mounted along axis that are turned by or used to turn fluid or gas.

VANE PUMP Positive-displacement pump with rotor vanes that rotate inside.

VANE SHEAR TEST In-place shear test where rod with thin radial vanes at end is forced into soil and resistance to rotation of rod is determined.

VANEAXIAL FAN Fan consisting of disk-type wheel within cylinder.

VANPOOL Transport of pool of commuters in shared vehicle.

VAPOR Gas given off by substances that are solids or liquids at ordinary atmospheric pressure and temperatures.

VAPOR BLANKET Air layer, above water body, with moisture content greater that adjacent air.

VAPOR CAPTURE SYSTEM Any combination of hoods and ventilation system that captures or contains organic vapors so they may be directed to abatement or recovery device.

VAPOR DENSITY Unit volume weight of vapor divided by equal volume weight of gas (e.g., air, hydrogen).

VAPOR DIFFUSION Movement of vapor through vapor-permeable materials.

VAPOR DISPERSION Movement of vapor clouds in air due to wind, thermal action, gravity spreading and mixing.

VAPOR INCINERATOR Combustion apparatus for reducing emissions of organic compounds.

VAPOR LOCK When liquid fuel changes state from liquid to gas while still in fuel delivery system.

VAPOR LOCK DEVICE Device that eliminates or minimizes pipe vapor.

VAPOR PLUME Stack effluent visible due to water condensation.

VAPOR PLUMES Flue gases visible because they contain water droplets.

VAPOR PRESSURE 1. Measure of substance's propensity to evaporate; 2. Force per unit area exerted by vapor in equilibrium state with surroundings at given pressure.

VAPOR PROCESSING SYSTEM System designed to capture and process displaced organic compound vapors.

VAPOR RECOVERY SYSTEM System where volatile gases are captured instead of being released into atmosphere.

VAPORIZATION Conversion of solid or liquid into gas.

VARIABILITY Measurement of distance from mean of data range.

VARIABLE 1. Value that can be sensed and quantified and is liable to change.

VARIABLE AIR VOLUME HVAC system strategy where volume of air delivered to conditioned spaces is varied as function of ventilating needs, energy needs or both.

VARIABLE CAN RATE Charge for solid waste removal service based on volume or weight of waste disposed.

VARIABLE COST Cost that changes with change in volume of activity of organization.

VARIABLE COSTS Input costs that change as nature of production activity of its circumstances change.

VARIABLE FREQUENCY DRIVE Specific type of adjustable-speed drive that controls rotational speed of alternating current electric motor by controlling frequency of electrical power supplied to motor.

VARIABLE MESSAGE SIGN Mobile, electronic sign that gives travellers information about special events.

VARIABLE WIND DIRECTION Condition when wind direction fluctuates by 60° or more during 2-minute evaluation period and wind speed is greater than 6 knots or condition when wind direction is variable and wind speed is less than 6 knots during 2-minute evaluation period.

VARIANCE 1. Measure of variability; 2. Square of standard deviation; 3. Difference between what is expected and what happens; 4. Government permission for delay or exception in application of given law, ordinance or regulation.

VARIATE Measurable quantity capable of taking on number of values.

VARIATION Partial change in form, position, state or qualities.

VARVED Sedimentary bed or lamination deposited within one year's time.

VARVED SILT/CLAY Glacial lake deposit with silt or fine grained sand and clay layers.

VASCULAR PLANT Plant composed of or provided with vessels or ducts that convey water or sap.

VEBE TIME Time for concrete to fully consolidate in Vebe cylinder.

VECTOR 1. Quantity having both direction and magnitude; 2. Organism, often insect or rodent, that carries disease; 3. Plasmids, viruses or bacteria used to transport genes into host cell.

VEERING Clockwise shift in wind direction.

VEGETATED FILTER STRIPS Strips of grass or other close-growing vegetation that separates waterway (ditch, stream, creek) from intensive land use area (e.g., subdivision, farm).

VEGETATIVE CONTROLS Non-point source pollution control practices that involve plants (i.e., vegetative cover) to reduce erosion and minimize loss of pollutants.

VEHICLE 1. Device for carrying or transporting substances, objects or individuals; 2. Liquid content that acts as binding and drying agent.

VEHICLE CAPACITY 1. Ability of vehicle for carrying passengers and cargo; 2. Maximum number of vehicles that pass location over period of time.

VEHICLE CLASSIFICATION Classification based on height of vehicle over first two axles and total number of axles.

VEHICLE DETECTOR Device that detects vehicles based on presence of motor vehicles on roads.

VEHICLE MILES OF TRAVEL Product of multiplying amount of daily traffic on roadway segment by segment length.

VEHICLE MILES TRAVELLED Total number of vehicle miles travelled within specific geographic area over given period of time.

VEHICLE TRIP Vehicle movement between two locations.

VEIL OF CLOUD Thin layer of cloud that impedes seeing clearly.

VEIN Layer, seam or narrow irregular body of material different from surrounding formations.

VELOCITY 1. Vector quantity that has magnitude and direction; 2. Time rate of displacement of fluid particle from one point to another; 3. Rate of flow or occurrence.

VELOCITY AZIMUTH DISPLAY WSR 88-D product which shows radar derived wind speeds at various heights.

VELOCITY COEFFICIENT Ratio of actual velocity emerging from orifice to theoretical velocity.

VELOCITY CONVERSION FACTOR Variable that converts 3-second wind speed gust associated with 50-year mean recurrence interval to 3-second wind speed gust associated with other mean recurrence intervals.

VELOCITY CROSS-SECTION WSR-88D radar product displays vertical cross-section of velocity on grid with heights up to 70,000 feet on vertical axis and distance up to 124 nautical miles on horizontal axis.

VELOCITY HEAD Vertical distance that would statically result from velocity of moving fluid; 2. Pressure needed to increase fluid flow speed.

VELOCITY OF APPROACH Velocity flow approaches (e.g., weir, notch) before it flows past.

VELOCITY OF RETREAT Mean rate of flow just downstream of measuring weir.

VELOCITY PRESSURE 1. Due to motion, increase in moving fluid's pressure compare to its static pressure; 2. In flowing air, pressure due to velocity and density of air.

VELOCITY ZONES Areas within floodplain subject to potential high damage.

VELOCITY, AVERAGE INTERSTITIAL Average rate of groundwater flow to interstices expressed as product of hydraulic conductivity and hydraulic gradient divided by effective porosity.

VELOCITY, DARCIAN Rate of discharge of groundwater per unit area of porous medium measured at right angle to direction of groundwater flow.

VELOCITY-AREA METHOD Method to determine discharge using product of cross-sectional area of water times velocity.

VENT Opening that permits liquids and gases to pass into or out of confined space.

VENT PIPE 1. Conduit to remove gas and odors; 2. Conduit connecting vent stack to drain.

VENT STACK Vertical pipe that provides ventilation to building's drainage system.

VENTED Opening permitting discharge to pass to atmosphere.

VENTILATION Replacement or circulation of stale or noxious air with fresh air.

VENTILATION INDEX 1. Product of mixing depth and transport wind speed; 2. Measure of potential of atmosphere to disperse airborne pollutants from stationary source.

VENTILATION RATE Rate at which indoor air changes per unit of time.

VENTILATION/SUCTION Act of admitting fresh air into space in order to replace stale or contaminated air.

VENTING SYSTEM Pipes leading from waste pipes that release gases outdoors.

VENTURI Pressure jet that draws in and mixes air.

VENTURI EFFECT Speedup of air through constriction due to pressure rise on upwind side of constriction and pressure drop on downwind side as air diverges to leave constriction.

VENTURI FLUME Reduction in fluid pressure that results when fluid flows through constriction.

VENTURI METER Meter to measure increase in flow velocity and corresponding decrease in pressure in converging section of conduit.

VENTURI SCRUBBERS Air pollution control devices that use water to remove particulate matter from emissions.

VENTURI TUBE Tube with constriction to measure velocity or volume of fluid flow.

VERIFICATION Operation of testing equation of problem, to see whether it truly expresses conditions of problem.

VERMICULITE Hydrated silicate mineral that expands on heating, is used in insulation and as medium for planting.

VERNAL POOL Small lake or pond filled with water for only short time during spring.

VERNIER Device permitting finer measurement or control than standard markings or adjustments.

VERTICAL CLEARANCE Minimum vertical passage under overhead structures.

VERTICAL CURVE Meeting of different gradients in road or pipe.

VERTICAL DATUM Horizontal surface used as zero point for measurements of gauge height, stage or elevation for positions given in terms of latitude-longitude, State Plane coordinates or Universal Transverse Mercator coordinates.

VERTICAL DRAIN Drain installed to provide drainage paths for pore water in saturated soil.

VERTICAL LIFT BRIDGE Movable bridge where span rises vertically while remaining parallel with deck.

VERTICAL LIFT GATE All rectangular gates set in vertical guides within which gate moves vertically in its own plane.

VERTICAL PUMP Device that pumps water from well.

VERTICAL SERVICE SPACE Shaft oriented vertically to house building services.

VERTICAL VELOCITY Component of velocity (i.e., motion) in vertical.

VERTICAL VELOCITY CURVE Measuring velocities at enough vertical positions to plot velocity profile and calculate accurate mean velocity.

VERTICAL WIND SHEAR Change in wind's direction and speed with height.

VERTICALLY STACKED SYSTEM Low-pressure system, usually closed low or cutoff low, that is not tilted with height (i.e., located similarly at all levels of atmosphere).

VERTICALLY-INTEGRATED LIQUID WATER Property computed by RADAP II and WSR-88D units that takes into account three-dimensional reflectivity of echo.

VERY FINE GRAINED Grain diameter less than 0.004 inch.

VERY HARD Cannot be scratched with knife.

VERY HIGH FREQUENCY That portion of radio frequency spectrum from 30 to 300 MHz.

VERY LOW FREQUENCY That portion of radio frequency spectrum from 3 to 30 kHz.

VERY LOW-VOLUME LOCAL ROAD Road with average daily traffic volume of 400 vehicles/day or less.

VERY SOFT Can be deformed by hand.

VERY TIGHT Less than 0.004 inch.

VERY WIDE 0.394 inch to 0.984 inch.

VERY WINDY 30 to 40 mph winds.

VESICLES 1. Containing many small cavities formed by expansion of gas bubble or steam when rock solidifies; 2. Tiny holes in volcanic rock caused by gas bubbles trapped in lava when it cooled.

VESICULAR 1. Containing many small cavities formed by expansion of gas bubble or steam when rock solidifies; 2. Tiny holes in volcanic rock caused by gas bubbles trapped in lava when it cooled.

VHF OMNIDIRECTIONAL RANGE Short-range radio navigation system for aircraft to determine position and stay on course.

VIABLE Organism able to live and develop.

VIABLE POPULATION Population that has estimated numbers and distribution of reproductive individuals to ensure continued existence of species throughout its existing range in planning area.

VIBRATING SCREEN Screen that is vibrated to separate and move pieces resting on it.

VIBRATORY PLOW Plow that pulls blade as is moves, where motor vibrates blade.

VIBROTACTILE PEDESTRIAN DEVICE Device that communicates information about pedestrian timing by vibrating surface.

VICTAULIC COUPLING Grooved coupling pipe joining system that permits movement.

VICTAULIC PIPE Conduit with ends that receive victaulic couplings.

VICTOR AIRWAY Standard route for aircraft traffic between two VOR's or VOR intersections.

VIEWSHED View from particular vantage point.

VIGRA Streaks or wisps of precipitation falling from cloud but evaporating before reaching ground.

VINYL CHLORIDE Chemical compound, used in producing some plastics, believed to be oncogenic.

VIOLATION Failure to meet regulation.

VIOLATION RATE Percentage of vehicles in violation of prevailing regulation.

VIRGA Streaks or wisps of precipitation falling from cloud but evaporating before reaching ground.

VIRGIN COMPRESSION Compression corresponding to stresses greater than preconsolidation stress.

VIRGIN COMPRESSION CURVE Portion of compression curve corresponding to virgin compression.

VIRGIN COMPRESSION LINE Straight line approximating virgin compression curve.

VIRGIN MATERIALS Resources extracted from nature in their raw form (e.g., timber, metal ore).

VIRTUAL POTENTIAL TEMPERATURE Temperature parcel at specific pressure level and virtual temperature would have if it were lowered or raised to 1,000 mb.

VIRTUAL SLOPE Hydraulic gradient.

VIRTUAL TEMPERATURE Temperature parcel which contains no moisture would have to equal density of parcel at specific temperature and humidity.

VIRUS Submicroscopic, non-cellular structure consisting of core of DNA or RNA surrounded by protein coat, that requires living host cell to replicate and often cause disease in host organism.

VISCOMETER Instrument used to measure viscosity of liquid.

VISCOSITY Molecular friction within fluid that produces flow resistance.

VISCOSITY INDEX Unitless measure of change of viscosity with temperature.

VISCOUS FLOW Flow where there is continuous steady motion of particles.

VISIBILITY Distance at which given standard object can be seen and identified with unaided eye.

VISIBILITY PROTECTION PROGRAM Program specified by Clean Air Act to achieve national goal of remedying existing impairments to visibility and preventing future visibility impairment throughout United States.

VISIBILITY-LIMITED SIGNAL FACE Signal face that restricts visibility of signal indication from side or certain distance.

VISIBLE AREA What can be directly seen under favorable atmospheric conditions.

VISIBLE EMISSIONS Visible particulate matter that occurs due to emission process.

VISIBLE SATELLITE IMAGERY Uses reflected sunlight (i.e., reflected solar radiation) to see things in atmosphere and on Earth's surface.

VISITOR DAY Twelve visitor hours which may be aggregated by one or more persons in single or multiple visits.

VISITOR USE Visitor use of recreation and wilderness resource for inspiration, stimulation, solitude, relaxation, education, pleasure or satisfaction.

VISUAL ALARM Apparatus that serves as visual notification appliance.

VISUAL APPROACH SLOPE INDICATOR System of lights on side of airport runway that provides visual descent guidance information during approach.

VISUAL FLIGHT RULES Regulations that specify cloud and visibility limitations for aircraft operating with visual reference to terrain.

VISUAL IMPACT ASSESSMENT Analysis of potential visual impacts to landscape and views resulting from proposed development or action.

VISUAL SPECTRUM Portion of electromagnetic spectrum that eye is sensitive (i.e., light with wavelengths between 0.4 and 0.7 micrometers).

VITREOUS Of or resembling glass.

VITRIFIED CLAY PIPE Conduit made from clay and shale blend for sewers.

VOID 1. Space in soil or rock mass not occupied by solid mineral matter but rather occupied by air, water or other gaseous or liquid material; 2. Empty or vacant.

VOID RATIO Ratio of volume of void space to volume of solid particles.

VOIDLESS DENSITY Maximum theoretical density for material obtained by dividing mass of material unit volume by its specific gravity.

VOLATILE 1. Any substance that evaporates readily; 2. Explosive.

VOLATILE ACIDS Acids produced during digestion that are soluble in water.

VOLATILE FLAMMABLE LIQUID Flammable liquid whose temperature is above its flash point.

VOLATILE LIQUIDS Liquids which easily vaporize or evaporate at room temperature.

VOLATILE MASS OF LIVING SUBSTANCE Difference between dry mass and ash mass and represents actual mass of living matter.

VOLATILE MATERIAL 1. Components that easily evaporate at normal temperatures; 2. Components that are released as gas or vapor.

VOLATILE MATTER Unstable material that tends to not remain in one state and rapidly transitions to another.

VOLATILE ORGANIC COMPOUND Any organic compound that participates in atmospheric photochemical reactions, with high vapor pressure relative to their water solubility, that are emitted as gases from certain solids or liquids.

VOLATILE ORGANIC LIQUID Any organic liquid that can emit any volatile organic compound (VOC) into atmosphere.

VOLATILE SOLID Solids in water or other liquids that are lost on ignition of dry solids at 1,020 °F.

VOLATILE SOLIDS Those solids in water or other liquids that are lost on ignition of dry solids at 550 °C.

VOLATILE SYNTHETIC ORGANIC CHEMICALS Chemicals that tend to volatilize or evaporate.

VOLATILITY State of having low boiling point and evaporating readily.

VOLATILIZATION Entry of contaminants into atmosphere by evaporation from soil or water.

VOLCANIC Of, pertaining or produced by volcano.

VOLCANIC ASH Fine particles of mineral matter from volcanic eruption which can be dispersed long distances by winds aloft.

VOLCANIC BRECCIA Conglomerate-like rock made up of angular pieces of volcanic rock usually bound in volcanic ash.

VOLCANIC DUST Powder-size material of pulverized rock, minerals and volcanic glass created during volcanic eruptions.

VOLCANIC GLASS Natural glass produced by cooling of molten lava or liquid fraction of it, too rapidly to permit crystallization.

VOLCANIC ROCK Rock that forms from solidification of molten rock or magma at Earth's surface (e.g., extrusive igneous rock).

VOLCANISM Any natural phenomena and processes associated with action of volcanoes, geysers and fumaroles.

VOLCANO Vent or fissure on surface of planet, with magma chamber, periodically erupting forth lava and volcanic gases onto surface.

VOLCANOLOGY Study of volcanoes.

VOLT Unit of measurement of electromotive force that's difference of potential that would drive 1 ampere of current against 1 ohm of resistance.

VOLT PER METER Unit of electric field strength that's potential difference of 1 volt existing between two points that are 1 meter apart.

VOLT-AMP CURVE Output graph of welding power source.

VOLT-AMPERE 1. Unit of apparent power in AC circuit containing reactance that's equal to potential in volts multiplied by current in amperes, without taking phase into consideration; 2. One watt of direct current.

VOLT-AMPERES REACTIVE Unit of measure for reactive power.

VOLTAGE Electrical pressure (i.e., force which causes current to flow through electrical conductor).

VOLTAGE AMPLIFIER Device that puts out higher voltage than input voltage.

VOLTAGE DIP Reduction in voltage caused by short or sudden load.

VOLTAGE DRIFT Change in output voltage over period of time.

VOLTAGE DROP Voltage loss that occurs due to impedance.

VOLTAGE GAIN Ratio of output voltage to input voltage.

VOLTAGE MULTIPLIER Circuit that provides output voltage higher than input voltage.

VOLTAGE OPERATING BAND Generator voltage that can be adjusted and operated.

VOLTAGE REDUCTION Any intentional reduction of system voltage by 3% or greater for reasons of maintaining continuity of service of bulk electric power supply system.

VOLTAGE REGULATION Measure of change in voltage between sending and receiving unit.

VOLTAGE REGULATOR Apparatus designed to automatically maintain constant voltage level.

VOLTAGE-TO-GROUND Voltage between given conductor and grounded circuit point or conductor.

VOLTMETER Instrument for measuring electric potential in volts.

VOLTOHMMETER Combined voltmeter and ohmmeter.

VOLUME 1. Three-dimensional measure of space that comprises length, width and height; 2. Total number of vehicles on roadway that pass specific location.

VOLUME COVERAGE PATTERN Volumetric sampling procedure designed for surveillance of one or more particular meteorological phenomena.

VOLUME MEDIAN DIAMETER Statistical measure of average droplet size in spray cloud, such that 50% of volume of sprayed material is composed of droplets smaller in diameter.

VOLUME OF DAM Total space occupied by concrete forming dam structure computed between abutments and from top to bottom of dam sometimes called volume of concrete.

VOLUME REDUCTION Processing waste materials to decrease amount of space they occupy, usually by compacting, shredding, incineration or composting.

VOLUME SCAN Radar scanning strategy where sweeps are made at successive antenna elevations (i.e., tilt sequence) and then combined to obtain three-dimensional structure of echoes.

VOLUME VELOCITY PROCESSING Way to guess large-scale 2-dimensional winds, divergence and fall speeds from one-dimensional radial velocity data.

VOLUME-TO-CAPACITY RATIO Number of vehicles passing versus number of vehicles that could theoretically pass at capacity.

VOLUMETRIC Pertaining to measurement by volume.

VOLUMETRIC SHRINKAGE Decrease in volume, expressed as percentage of soil mass when dried, of soil mass when moisture content is reduced from given percentage to shrinkage limit.

VOLUMETRIC TANK TEST Volume of fluid in tank is measured directly or calculated from product-level changes.

VOLUNTARY OBSERVING SHIP PROGRAM International voluntary marine observation program under auspices of World Meteorological Organization.

VOLUNTARY SEPARATION Separation by originator from domestic refuse prior to collection.

VOLUNTARY STANDARD Industry best practices of consumer products.

VORTEX Whirlwind, whirlpool or similarly moving matter in form of spiral or column.

VORTEX SEPARATOR Device where liquid moves in swirling motion where particles are separated from stream, move towards center vortex and pulled down by gravity.

VORTICITY Measure of rotation of air in horizontal plane.

VORTICITY MAXIMUM Slang for center or maximum, in vorticity field of fluid.

VUGGY Voids 0.236 inch to diameter of core.

VULNERABILITY ANALYSIS Assessment of elements in community that are susceptible to damage if hazardous materials are released.

VULNERABILITY ASSESSMENT Evaluation of source quality and its vulnerability to contamination.

VULNERABLE ZONE Area over which concentration of chemical accidentally released could reach level of concern.

W

WAIT TIME Period of time waiting for vehicle or response.

WAIVER Legal document removing some requirement, right, claim or privilege.

WAKE Region of turbulence immediately to rear of solid body caused by flow over or around body.

WAKE TURBULENCE Disturbance in fluid or gas after object passes through it.

WALE 1. Horizontal timber used for supporting or retaining earth; 2. Ridge or low barrier.

WALER Horizontal brace used for support or to maintain required separation between components in order to help maintain form of construction under stress.

WALK INTERVAL Time period when WALKING PERSON signal is displayed.

WALKWAY Clearly defined path for pedestrians.

WALL CLOUD Localized, persistent, often abrupt lowering from rain-free base.

WALL FRICTION Frictional resistance mobilized between wall and soil or rock in contact with wall.

WALL HYDRANT Sill cock closed inside wall of building.

WALL-HUNG BASIN Wash basin fixed to wall.

WARM ADVECTION Movement of warm air into region by horizontal winds.

WARM AIR ADVECTION Movement of warm air into region by horizontal winds.

WARM AIR HEATING SYSTEM System that uses thermal energy to heat air and then forces warm air through system of ducts and registers.

WARM CORE LOW Low pressure area that is warmer at its center than at its periphery.

WARM FRONT Transition zone between mass of warm air and colder air it is replacing.

WARM OCCLUSION Frontal zone formed when cold front overtakes warm front and, finding colder air ahead of warm front, leaves ground and rises up and over this denser air.

WARM SECTOR Region of warm surface air between cold front and warm front.

WARM SPRING Spring produced by emergence of geothermally heated groundwater.

WARM-WATER FISHERY Water or water system that has environment suitable for species of fish other than salmonids.

WARNING Fourth of five Early Warning System components consisting of processes (including media) and equipment necessary to make public aware of potential, probable or imminent danger or risk that's issued when hazardous weather or hydrologic event is occurring, is imminent or has very high probability of occurring and used for conditions posing threat to life or property.

WARNING AREA Airspace of defined dimensions, extending from three nautical miles outward from coast of U.S., that contains activity that may be hazardous to nonparticipating aircraft.

WARNING BEACON Beacon that draws attention to increase awareness and enhancing safety.

WARNING DEVICE Apparatus that provides possible danger warning.

WARNING LIGHT Illuminated indicator that indicates either potentially serious problem or safety issue.

WARNING SIGN Sign that indicates potential hazard, obstacle or condition.

WARNING STAGE Depth of water in river at which National Weather Service reviews basin conditions for potential flooding.

WARNING TIME Amount of time between detection of failure or incipient failure and arrival of dam failure flood.

WARPING OR HINGED JOINT Joint where flexure is permitted without applicable horizontal or warping stresses movement.

WARRANT To give assurance or guarantee.

WARRANTY Guarantee that certain outcome or obligation will be fulfilled.

WARRANTY DEED Deed that guarantees clear title to buyer of property.

WASH 1. Move or erode by force of water in motion; 2. Normally dry bed of intermittent or ephemeral stream; 3. Shallow body of water.

WASH BORING Drilling in unconsolidated ground by use of water jet inside casing pipe.

WASH DOWN Wash something completely from top to bottom.

WASH LOAD That part of suspended load finer than bed material.

WASH WATER Water to flush or dissolve impurities.

WASH WATER GUTTER Trough that carries away water laden with suspended solids.

WASH WATER TANK Tank that holds water, above rapid sand filter beds, to clean them.

WASHBASIN Permanently installed sink to wash one's hands and face.

WASHLAND Any land flooded by river.

WASHOUT VALVE Opening in dam or conduit that can be opened to clear sediment accumulation.

WASHROOM Room intended to wash hands and face or laundry.

WASTE 1. Digging, hauling and dumping of valueless material to get it out of way; 2. Valueless material; 3. Unwanted materials left over from manufacturing process; 4. Refuse from places of human or animal habitation; 5. Plumbing drain line; 6. Large tract of uncultivated land.

WASTE BRANCH Pipe that carries liquid waste from sanitary fitting.

WASTE CHARACTERIZATION Identification of chemical and microbiological constituents of waste material.

WASTE DERIVED FUEL Combustible fuel created from conversion or managed wastes.

WASTE DISPOSAL UNIT Electronic garbage disposal connected to kitchen sink drain.

WASTE EXCHANGE Arrangement where companies exchange their wastes for benefit of both parties.

WASTE FACTOR Amount of material you purchase but cannot use.

WASTE FEED Continuous or intermittent flow of wastes into incinerator.

WASTE GAS Gases from incomplete combustion or chemical reaction.

WASTE GAS BURNER Apparatus that burns sludge digestion tank excessive gas.

WASTE GATE Gate in canal wall to remove excessive water.

WASTE GENERATION Weight or volume of materials and products that enter waste stream before recycling, composting, landfilling or combustion takes place (i.e., amount of waste generated by given source or category of sources).

WASTE HEAT Unused heat given to surrounding environment.

WASTE LOAD ALLOCATION Maximum load of pollutants each discharger of waste is allowed to release into particular waterway that's portion of stream's total assimilative capacity assigned to individual discharge.

WASTE MANAGEMENT Collection, transportation, disposal recycling and/or monitoring of waste.

WASTE MANAGEMENT UNIT Contiguous area of land on or where hazardous waste is placed or largest area where there is significant likelihood of mixing hazardous waste constituents in same area.

WASTE MANAGEMENT UNIT BOUNDARY Vertical surface located at hydraulically downgradient limit of unit and extending down into uppermost aquifer.

WASTE MINIMIZATION Measures or techniques (e.g., recycling) that reduce amount of wastes generated.

WASTE PILE Non-containerized solid, non-flowing hazardous waste.

WASTE PILES Non-containerized, lined or unlined accumulations of solid, non-flowing waste.

WASTE PIPE Pipe for carrying off liquid waste or excess liquid.

WASTE PROCESSING Changing physical or chemical properties of wastes.

WASTE RECEPTACLE Container for temporarily storing refuse.

WASTE REDUCTION Using source reduction, recycling or composting to prevent or reduce waste generation.

WASTE STACK Vertical pipe that carries discharge from any fixture.

WASTE STREAM Total flow of solid waste from homes, businesses, institutions and manufacturing plants recycled, burned or disposed of in landfills or segments thereof (e.g., residential waste stream, recyclable waste stream).

WASTE TREATMENT Procedures to ensure that waste has least practicable impact on environment.

WASTE TREATMENT LAGOON Impoundment made by excavation or earth fill for biological treatment of wastewater.

WASTE TREATMENT PLANT Facility containing series of tanks, screens, filters and other processes by which pollutants are removed from water.

WASTE TREATMENT STREAM Continuous movement of waste from generator to treater and disposer.

WASTE WATER Non-essential water.

WASTE-HEAT RECOVERY Recovering heat discharged as byproduct of one process to provide heat needed by second process.

WASTE-TO-ENERGY FACILITY/MUNICIPAL-WASTE COMBUSTOR Facility where recovered municipal solid waste is converted into usable form of energy, usually via combustion.

WASTEBANK Bank made of excess or unstable material excavated from construction site.

WASTELOAD ALLOCATION Proportion of receiving water's total maximum daily load allocated to one of its existing or future point sources of pollution.

WASTEWATER Water used in domestic or industrial activity that contains waste products.

WASTEWATER ANALYSIS Determination of composition of waste or effluent stream.

WASTEWATER CHARGE Fee for processing wastewater.

WASTEWATER INFRASTRUCTURE Plan or network for collection, treatment and disposal of sewage in community.

WASTEWATER IRRIGATION Reusing wastewater effluent for irrigation.

WASTEWATER OPERATIONS AND MAINTENANCE Actions taken after construction to ensure that facilities constructed to treat wastewater will be operated, maintained and managed to reach prescribed effluent levels in optimum manner.

WASTEWATER TREATMENT Processing of wastewater for removal or reduction of contained solids or other undesirable constituents.

WASTEWATER TREATMENT PLAN Facility containing series of tanks, screens, filters and other processes by which pollutants are removed from water.

WASTEWATER TREATMENT PLANT Plant for removing contaminants from wastewater.

WASTEWATER TREATMENT PROCESS Process that converts wastewater into effluent.

WASTEWATER TREATMENT TANK Tank designed to receive and treat influent wastewater through physical, chemical or biological methods.

WASTEWATER-TREATMENT RETURN FLOW Water returned to environment by wastewater-treatment facilities.

WASTEWATERS Any water that has been affected by human and animal use.

WASTEWAY 1. Waterway used to drain excess irrigation water dumped from irrigation delivery system; 2. Chute or other route for disposal of waste material.

WATCH Used when risk of hazardous weather or hydrologic event has increased significantly, but its occurrence, location, and/or timing is still uncertain that's intended to provide enough lead time so that those who need to set their plans in motion can do so.

WATCH BOX Slang for severe weather watch (e.g., Tornado, Severe Thunderstorm).

WATCH CANCELLATION Issued to let public know when severe weather watch (e.g., Tornado, Severe Thunderstorm) has been canceled early.

WATCH REDEFINING STATEMENT Tells public which counties/parishes are included in severe weather watch (e.g., Tornado, Severe Thunderstorm).

WATCH STATUS REPORTS Lets NWFO know of status of current severe weather watch (e.g., Tornado, Severe Thunderstorm).

WATER Substance found at room temperature and pressure as clear liquid and is present naturally as rain and found in rivers, lakes and seas.

WATER ABSORPTION Action of picking up water.

WATER ANALYSIS Chemical testing to measure water quality.

WATER BALANCE Accounting of inflow, outflow, storage of hydrologic unit.

WATER BALANCE MODEL Accounting of inflow, outflow and storage changes of water in hydrologic unit over fixed period.

WATER BAR 1. Continuous strip of waterproof material placed at concrete joints designed to control cracking and limit moisture penetration; 2. Interceptor dike used to prevent erosion by reducing flow length.

WATER BLAST Cleaning technique for cleaning internal and external surfaces with pressurized water.

WATER BREAK Obstacle that reduces flow velocity and scour.

WATER BUDGET Accounting of inflow, outflow and storage changes of water in hydrologic unit over fixed period.

WATER CHANNEL Ditch for directing water.

WATER CLOSET Flush toilet.

WATER COLUMN Imaginary vertical column of water extending through water body from surface to bottom.

WATER COLUMN STUDIES Investigations of physical and chemical characteristics of surface water, which include suspended sediment, dissolved solids, major ions and metals, nutrients, organic carbon and dissolved pesticides, in relation to hydrologic conditions, sources and transport.

WATER COMPENSATION Volume of water needed to meet rights holders requirements.

WATER CONDITIONER Process that changes condition of water passed through it by removing and/or altering minerals, chemicals and contaminants.

WATER CONSUMPTION Water withdrawals that includes evaporative losses, withdraw and product integrated.

WATER CONTENT 1. Ratio of weight of water in soil sample to weight of dry soil, expressed as percentage; 2. Water equivalent of snow on ground; 3. Amount of water contained in material (e.g., soil, rock or wood).

WATER CONTENT OF SNOW Amount of liquid water in snow at time of observation.

WATER CONVEYANCE EFFICIENCY Ratio of volume of irrigation water delivered by distribution system to water introduced into system.

WATER CONVEYANCE STRUCTURE Any structure that conveys water from one location to another.

WATER COOLING TOWER Structure that retreats heat from water by evaporation or conduction.

WATER CROP Runoff from drainage basin, including groundwater outflow that appears in stream plus groundwater outflow that bypasses gaging station and leaves basin underground (i.e., precipitation minus evapotranspiration).

WATER CUSHION Volume of water designed to reduce and control pressure drawdown applied to reservoir when flow valve is opened.

WATER CYCLE Processes (i.e., evaporation, precipitation, runoff, transpiration) by which water circulates between Earth's surface and atmosphere.

WATER DELIVERY SYSTEM Reservoirs, canals, ditches, pumps and other facilities to move water.

WATER DEMAND Water requirements for particular purpose (e.g., irrigation, power, municipal supply, plant transpiration, storage).

WATER DISTRIBUTION PIPE Internal conduit that carries water from potable source.

WATER DISTRICT Administrative region given task of supplying water and sewer needs to community.

WATER EQUIVALENT 1. Amount of water, in depth, obtained by melting solid precipitation (e.g., snow).

WATER EQUIVALENT OF SNOW Amount of water obtained by melting snow.

WATER EXPORTS Artificial transfer by pipes or canals of freshwater from one region or subregion to another.

WATER FILTER Apparatus to removal and/or reduce suspended solids in flowing water.

WATER FLOW FORMULA Mathematical methods to ascertain water discharge rate.

WATER FLOW SWITCH Device designed to send alarm to fire alarm panel and/or department when continuous flow of water occurs through sprinkler system.

WATER GAP Deep, narrow pass in mountain ridge where stream flows.

WATER HAMMER Pressure fluctuations caused by sudden increase or decrease in flow velocity.

WATER HAMMER ARRESTER Apparatus installed to remove water hammer by reducing hydraulic shock wave.

WATER HOLDING CAPACITY 1. Amount of soil water available to plants; 2. Smallest value that moisture content of soil can be reduced by gravity drainage.

WATER HOLE Depression where water collects.

WATER IMPORTS Artificial transfer (by pipes or canals) of freshwater to one region or subregion from another.

WATER INJECTION Procedure of recharging depleted water table by injecting water.

WATER INTAKE Structure that procures water from source.

WATER INVENTORY EQUATION Inflow equals outflow, plus change in storage.

WATER LANCE Pipe on end of water hose used to hydraulically remove solids.

WATER LEVEL Water elevation above datum.

WATER LEVEL GAUGE Apparatus to determine water surface level.

WATER LINE Line corresponding to surface of water touching any submerged object or body.

WATER LOSS 1. Collective term that includes water lost (discharged to atmosphere) through evaporation from soil and surface-water bodies and by plant transpiration in specific time period; 2. Amount of water that escapes system.

WATER MAIN Principal underground pipe for conveying water to residential and business properties.

WATER MANAGEMENT PLAN Plan developed during construction to help assure water quality compliance for both point and non-point pollution sources.

WATER MARK Point that represents maximum water body elevation over land.

WATER MASS BALANCE Inventory of difference source and sinks of water in hydrogeologic system.

WATER METER Device used to measure volume of water usage.

WATER PHONE Apparatus to find pressurized water pipe leaks by amplifying sound of escaping water.

WATER PLANT Treatment plant for water.

WATER POLLUTION Alteration of constituents of body of water by man (e.g., sewage effluent, fertilizer runoff, industrial chemical discharge) to such degree that water loses its value as natural resource.

WATER POLLUTION CONTROL PLANT Plant to treat domestic and industrial wastewater for pollution.

WATER POWER Mechanical or electrical energy derived from running or falling water.

WATER PURVEYOR Agency or person that supplies water (e.g., drinking water) to customers.

WATER QUALITY Describes chemical, physical, radiological and biological characteristics of water.

WATER QUALITY ASSESSMENT Evaluation of condition of water body using biological surveys, chemical-specific analyses of pollutants in water bodies and toxicity tests.

WATER QUALITY CRITERIA Levels of water quality expected to render body of water suitable for its designated use based on specific levels of pollutants that would make water harmful if used for drinking, swimming, farming, fish production or industrial processes.

WATER QUALITY LIMITED SEGMENT Any segment where it is known that water quality does not meet applicable water quality standards and/or is not expected to meet applicable water quality standards even after application of technology-based effluent limitations.

WATER QUALITY OBJECTIVES Water quality needed to protect beneficial values and uses.

WATER QUALITY PROGRAMS Programs that identify, restore and protect impaired waters.

WATER QUALITY STANDARD Law or regulation (i.e., state-adopted and EPA-approved ambient standards) that consists of beneficial use or uses of waterbody, numeric and narrative water quality criteria that are necessary to protect use or uses of that particular waterbody and anti-degradation statement.

WATER QUALITY STANDARDS Provisions of state, territorial, authorized tribal or federal law that describe desired condition of water body and means where that condition will be protected or achieved.

WATER QUALITY-BASED EFFLUENT LIMIT Value determined by selecting most stringent of effluent limits calculated using all applicable water quality criteria (e.g., aquatic life, human health, wildlife) for specific point source to specific receiving water for given pollutant.

WATER QUALITY-BASED LIMITATIONS Effluent limitations applied to dischargers when mere technology-based limitations would cause violations of water quality standards.

WATER QUALITY-BASED PERMIT Permit with effluent limit more stringent than one based on technology performance.

WATER RAM Device that forces flowing water to higher level using kinetic energy.

WATER RATE Charge for water supply.

WATER RECLAMATION Process that converts wastewater to water to be recycled for other purposes.

WATER REPELLENT Chemical used to make fabric water-repellent.

WATER REPELLENT PRESERVATIVE Liquid that penetrates surfaces to repel moisture.

WATER RESISTANT Able to resist water penetration to some degree.

WATER RIGHTS Legal rights to use of water.

WATER RUNOUT Runoff from drainage basin, including groundwater outflow that appears in stream plus groundwater outflow that bypasses gaging station and leaves basin underground (i.e., precipitation minus evapotranspiration).

WATER SATURATION That point at which material will no longer absorb water.

WATER SEAL Water contained in drain trap to prevent passage of odor.

WATER SERVICE Supply of potable water and fire protection water.

WATER SERVICE PIPE Conduit extending from potable water source to interior of building.

WATER SIDE ECONOMIZER Reduces energy consumption in cooling mode by allowing chiller to be turned off when cooling tower alone can produce water at desired chilled water set point.

WATER SOFTENER Device that reduces calcium and magnesium concentration of hard water by passing it through ion exchange resin.

WATER SOLUBILITY Maximum possible concentration of chemical compound dissolved in water.

WATER STANDARDS Provisions of state, territorial, authorized tribal or federal law approved by EPA that describe desired condition of water body and means where that condition will be protected or achieved.

WATER STORAGE POND Impound for liquid wastes designed to accomplish some degree of biochemical treatment.

WATER SUPPLIER Person who owns or operates public water system.

WATER SUPPLY Source or process of supplying water.

WATER SUPPLY FACILITIES System of infrastructure that collects, treats or distributes water.

WATER SUPPLY MAP Map showing location of water supplies readily available.

WATER SUPPLY OUTLOOK Seasonal volume forecast, generally for period centered around time of spring snowmelt (e.g., April-July).

WATER SUPPLY OUTLOOK PRODUCT Public product issued by Forecast Office which contains narrative and numeric information on current and extended water supply conditions.

WATER SUPPLY SOURCE Place where raw water can be withdrawn for need.

WATER SUPPLY SYSTEM Collection, treatment, storage and distribution of potable water from source to consumer.

WATER SURFACE ELEVATION Elevation of water surface above or below established reference level (e.g., sea level).

WATER SURFACE PROFILE Plot of flow depth changes along channel length.

WATER SURFACE STAGE Elevation of water surface above or below established reference level (e.g., sea level).

WATER TABLE Surface of level groundwater.

WATER TAX Charge for water often utilized to encourage less water usage.

WATER TEST Pressure evaluation on water pipes.

WATER THIEF Fitting that attaches to unthreaded faucet and garden hose.

WATER TOWER Large tank of water, in elevated position, acting as local reservoir.

WATER TRANSFER 1. Selling or exchanging water or water rights among individuals or agencies; 2. Artificial conveyance of water from one area to another.

WATER TREATMENT Any process that improves quality of water.

WATER TREATMENT LAGOON Impound for liquid wastes designed to accomplish some degree of biochemical treatment.

WATER TREATMENT PLANT Facility that improves water quality.

WATER TREATMENT SYSTEM System made up of several technologies to treat and condition water.

WATER TURBINE Device that converts energy of moving water to make electricity.

WATER USE Water actually used for specific purpose (e.g., domestic use, irrigation, industrial processing).

WATER USER Any individual, district, association, government agency or other entity that uses water supplied.

WATER VAPOR Water in gaseous state.

WATER VAPOR PLUME Appears in water vapor satellite imagery as plume-like object that extends from Intertropical Convergence Zone northward or southward into higher latitudes.

WATER VAPOR TRANSMISSION Speed of water vapor flow through unit area of material.

WATER VELOCITY Amount of water flowing at any given time.

WATER WASTE SURVEY Test to detect and remedy water supply leaks.

WATER WELL Excavation where intended use is for location, acquisition, development or artificial recharge of groundwater.

WATER WHEEL Wheel, propelled by running or falling water, used to power machinery.

WATER YEAR Time period form October 1 through September 30.

WATER YIELD Runoff from drainage basin, including groundwater outflow that appears in stream plus groundwater outflow that bypasses gaging station and leaves basin underground (i.e., precipitation minus evapotranspiration).

WATER-EFFECT RATIO Appropriate measure of toxicity of material obtained in site water divided by same measure of toxicity of same material obtained simultaneously in laboratory dilution water.

WATER-QUALITY CRITERIA Specific levels of water quality which, if reached, are expected to render body of water unsuitable for its designated use.

WATER-QUALITY DATA Chemical, biological and physical measurements or observations of characteristics of surface and groundwaters, atmospheric deposition, potable water, treated effluents and waste water and of immediate environment where water exists.

WATER-QUALITY GUIDELINES Specific levels of water quality which, if reached, may adversely affect human health or aquatic life.

WATER-QUALITY LIMITED SEGMENT Stretch or area of surface water where technology-based controls are not sufficient to prevent violations of water-quality standards.

WATER-QUALITY MONITORING Integrated activity for evaluating physical, chemical and biological character of water in relation to human health, ecological conditions and designated water uses.

WATER-QUALITY STANDARD Law or regulation that consists of beneficial designated use or uses of water body, numerical and narrative water-quality criteria that are necessary to protect use or uses of that particular water body and anti-degradation statement.

WATER-RESOURCE QUALITY Condition of water or some water-related resource as measured by : habitat quality, biological surveys, habitat-quality assessments, chemical-specific analyses of pollutants in water bodies, energy dynamics, hydrological regime, biotic factors and toxicity tests.

WATER-RESOURCES REGION Natural drainage basin or hydrologic area that contains either drainage area of major river or combined areas of series of rivers where in United States, there are 21 regions of which 18 are in conterminous United States and one each in Alaska, Hawaii and Caribbean.

WATER-RESOURCES SUBREGION Designated water-resources regions of United States that are subdivided into subregions where each subregion includes that area drained by river system, reach of river and its tributaries in that reach, closed basin(s) or group of streams forming coastal drainage system.

WATER-SOLUBLE PACKAGING Packaging that dissolves in water.

WATER-SOURCE HEAT PUMP Heat pump that uses wells or heat exchangers to transfer heat from water (i.e., groundwater) to inside of building.

WATERBORNE DISEASE Sickness caused by microorganisms or toxic substances in untreated or contaminated water.

WATERBORNE DISEASE OUTBREAK Significant occurrence of acute illness associated with drinking water from public water system deficient in treatment, as determined by appropriate local or state agencies.

WATERCOURSE Any surface flow (e.g., river, stream, tributary).

WATERLOGGED Soaked with water.

WATERLOGGED SOIL Soil so saturated that roots cannot respire due to excess water in soil.

WATERPROOF Unaffected or impervious to water in any state.

WATERPROOF MEMBRANE Substance that prevents penetration of water.

WATERPROOFING Treatment of something to make it waterproof.

WATERPROOFING COMPOUND Substance that prevents water penetration.

WATERS OF UNITED STATES All waters that are currently used, were used in past or may be susceptible to use in interstate or foreign commerce, including all waters subject to ebb and flow of tide.

WATERSHED Land area from which water drains toward common watercourse (e.g., natural basin, stream, river, reservoir).

WATERSHED APPROACH Coordinated framework for environmental management that focuses public and private efforts on highest priority problems within hydrologically-defined geographic areas taking into consideration both ground and surface water flow.

WATERSHED DIVIDE Boundary between catchment (i.e., drainage) areas.

WATERSHED PLANNING Assessment to protect water quality and quantity that focuses on watershed.

WATERSPOUT 1. Cyclone occurring over water; 2. Channel where water is discharged (e.g., gutter).

WATERSTAGE RECORDER Motor-driven (e.g., spring wound, electric) instrument for monitoring water surface elevation.

WATERSTOP Continuous strip of waterproof material placed at concrete joints designed to control cracking and limit moisture penetration.

WATERTIGHT So tight that water cannot enter or escape.

WATERWAY 1. Body of water that's navigable (e.g., river, channel or canal); 2. Conduit or watercourse to drain water.

WATERWAY OPENING Area of bridge opening at specified stage that's measured normal to particular flow direction.

WATERWAYS Spillways and outlet works.

WATERWORKS 1. Water supply system (i.e., reservoirs, pumps, pipes) of town; 2. Any single facility (e.g., filtration plant) within system.

WATT Unit of power equal to 1 Joule/s.

WATT PER METER KELVIN Unit of heat equal to watt/m x °K.

WATT PER SQUARE METER Unit of heat flux density equal to watt/m^2.

WATT PER STERADIAN Unit of radiant intensity equal to watt/steradian.

WATT-HOUR Electrical energy unit of measure equal to one watt of power supplied to or consumed from, electrical circuit steadily for one hour.

WATT-HOUR METER Meter that measures and registers, amount of electricity consumed in circuit.

WATTMETER Instrument for measuring electric power in watts.

WAVE Moving disturbance in level of liquid body or energy level of field.

WAVE CREST Highest part of wave.

WAVE EQUATION Second-order linear partial differential equation for wave description.

WAVE FRONT Surface representing corresponding points of wave that vibrate in unison.

WAVE HEIGHT Distance from wave trough to wave crest.

WAVE MODEL Froude number model where gravity waves are reproduced to provide driving mechanism for currents, impact forces, wave action, sediment transport, etc.

WAVE PERIOD Time between passage of consecutive wave crests or wave troughs past fixed point.

WAVE POWER Transport and harnessing of energy by ocean surface waves.

WAVE PROTECTION Riprap, concrete or other armoring on upstream face of embankment dam to protect against scouring or erosion due to wave action.

WAVE RUN-UP Extent of wave uprush landward that's measured vertically from still water level.

WAVE RUNUP Vertical height above stillwater level that water from specific wave will run up face of structure or embankment.

WAVE SPECTRUM Distribution of wave energy with respect to wave frequency or period.

WAVE SPEED 1. Distance wave travels over time period; 2. Speed where shock wave travels upstream or downstream (e.g., traffic).

WAVE STEEPNESS Ratio of wave height to wave length that's indicator of wave stability.

WAVE TROUGH Lowest part of wave.

WAVE VELOCITY Speed wave propagates at certain frequency.

WAVE WALL Solid wall built along top of dam (upstream and/or downstream edge) used for ornamentation, for safety of vehicles and pedestrians or to prevent overtopping caused by wave runup.

WAVE-BUILT TERRACE Gently sloping coastal feature at seaward or lakeward edge of wave-cut platform.

WAVE-CUT TERRACE Gently sloping coastal feature inclining away from base of eroded sea cliff.

WAVEBAND Range of electromagnetic wavelengths or frequencies.

WAVELENGTH Distance between crests or troughs of wave.

WAVELET 1. Small wave or ripple; 2. Fast-decaying oscillation.

WAVEWATCH III One of operational forecast models run at NCEP run four times daily, with forecast output out to 126 hours.

WAYBILL Document that lists details (e.g., final destination) of cargo.

WAYFINDING Process used by person or animal for orienting oneself and navigating.

WAYSIDE EQUIPMENT Train equipment located at grade crossing that's maintained and operated by railroad.

WAYSIDE HORN SYSTEM Stationary horns located at railroad crossings that warn motorists of oncoming trains.

WAYSIDE SIGNAL Fixed signal, outside train, along track.

WEAK ECHO REGION WSR-88D radar product that displays reflectivity for up to 8 elevation angles for radar operator selected location as set presentation of storm.

WEAR To erode, diminish or consume gradually.

WEAR AND TEAR Damage or depreciation resulting from ordinary use.

WEAR PLATE Liners used to safeguard against friction between two parts or materials that creates degradation.

WEARING SURFACE Layer placed on bridge deck to form road way surface.

WEATHER State of atmosphere at given time and place as regards to temperature, precipitation, wind velocity, direction and pressure.

WEATHER BALLOON Balloon that carries meteorological instruments aloft in order to measure, record and return data concerning atmospheric pressure, temperature and humidity.

WEATHER FORECAST OFFICE Responsible for issuing advisories, warnings, statements and short term forecasts for its county warning area.

WEATHER MAP Map that shows analysis of meteorological quantities at various levels of atmosphere.

WEATHERFAX Broadcasting graphic weather maps and other graphic images via HF radio.

WEATHERING 1. Process whereby earthy or rocky materials are changed in color, texture, composition or form (with little or no transportation) by exposure to atmospheric agents; 2. Slight inclination given to approximately horizontal surface to enable it to throw off water.

WEATHERING ZONE Layer of weathered material between ground surface and groundwater table.

WEATHERPROOF To make resistant to damage caused by weather.

WEAVING Unsteady motion back and forth (e.g., crossing traffic streams).

WEAVING FLOW Movement across path of another along length of facility without aid of signals or traffic control devices.

WEAVING SEGMENT Where two traffic streams cross paths at grade over length of highway.

WEAVING WIDTH Number of lanes between entry and exit areas within weaving segment.

WEBER Unit of magnetic flux that produces electromotive force of 1 volt in circuit of 1 turn when generated or removed in 1 second.

WEDGE Piece that tapers from thick end to chisel point.

WEDGE TORNADO Slang for large tornado with condensation funnel at least as wide (i.e., horizontally) at ground as it is tall (i.e., vertically) from ground to cloud base.

WEEP HOLE Drain embedded in concrete or masonry structure intended to relieve pressure caused by seepage behind structure.

WEFAX System for transmitting weather charts and imagery via satellite.

WEIGH Tire load measured by vehicle scale.

WEIGH SCALE Device to measure vehicle weight.

WEIGH STATION Highway checkpoint where trucks stop to get weighed.

WEIGH-IN-MOTION Devices that captures and records axle weights and gross vehicle weights as vehicles drive over measurement site.

WEIGHBRIDGE Platform for weighing vehicles.

WEIGHING-TYPE PRECIPITATION GAUGE Rain gauge that weighs rain or snow which falls into bucket set on platform of spring or lever balance.

WEIGHT Force on object due to Earth's gravitational attraction.

WEIGHT OF SCIENTIFIC EVIDENCE Considerations in assessing interpretation of published information about toxicity-quality of testing methods, size and power of study design, consistency of results across studies and biological plausibility of exposure-response relationships and statistical associations.

WEIGHTED MEAN Value obtained by multiplying each of series of values by its assigned weight and dividing sum of these products by sum of weights.

WEIGHTING OF SLOPE Additional material placed on slope of embankment.

WEIR Structure built across stream or channel for purpose of measuring flow.

WEIR FLOW Measuring structure where water flows over to calculate volumetric flow rate.

WEIR HEAD Weir water depth measured from bottom of weir notch to water surface upstream.

WEIR LOADING Tank discharge rate measure by what overflows weir.

WEIR PLATE Barrier across body of water that water flows over.

WEIR, MEASURING Device for measuring rate of flow of water where height of water above weir crest is used to determine rate of flow.

WELCOME CENTER Rest area with visitor center that's located after entrance from one state to another.

WELD To build up or fasten together metals by bonding on molten metal.

WELDED STEEL PIPE Conduit created from flat plates (i.e., skelp) that are bent and welded.

WELL Hole or shaft drilled into Earth to get water or other underground substances.

WELL CAPACITY Maximum speed well will produce (e.g., water, gas).

WELL CASING Conduit placed inside of drilled well.

WELL CONE OF INFLUENCE Depression in watertable due to well extraction.

WELL CURB Curb, parapet or enclosure around and above top of well.

WELL DRAIN Drain or vent serving to discharge water.

WELL DRILL Churn drill to bore well.

WELL FIELD Area containing one or more wells that produce usable amounts of water or oil.

WELL FLOODING Wells located in areas where flood waters reach or where groundwater is influenced by flood waters.

WELL GRADED Good representation in material of all particle sizes present from largest to smallest.

WELL HYDROGRAPH Graphical rendering of well water level through time.

WELL INJECTION Subsurface emplacement of fluids into well.

WELL INTAKE Structure for collecting water from well.

WELL INTERFERENCE When cones of depression for two or more wells overlap.

WELL LOG Record of events encountered during well drilling process.

WELL LOSS Well losses caused by damage to aquifer during drilling, head losses and friction losses.

WELL MONITORING Measurement by on-site instruments or laboratory methods of well water quality.

WELL PLUG Watertight, gastight seal installed in bore hole or well to prevent movement of fluids.

WELL POINT Hollow vertical tube, rod or pipe with driving point and fine mesh screen used to remove underground water.

WELL RECORD Information regarding hydraulic operation of well.

WELL SCREEN Filtering device on intake portion of well.

WELL SHOOTING Detonating explosives in well to boost water flow.

WELL STIMULATION Intervention performed on well to increase production by improving flow.

WELL STRAINER Screen before water entering well suction pump.

WELL VENT Conduit to allow gases to vent out.

WELL WATER Groundwater from well.

WELL WORKOVER Complex of operations aimed at restoring well production.

WELL-GRADED AGGREGATE Particle size gradation that fairly evenly spans in size from finest to coarsest.

WELLHEAD PROTECTION AREA Protected surface and subsurface zone surrounding well or well field supplying public water system to keep contaminants from reaching well water.

WELLPOINT Small diameter well for dewatering.

WELLPOINT PUMP Centrifugal pump with small-diameter pipes that are inserted into groundwater.

WELLPOINT SYSTEM Series of closely spaced small diameter shallow wells to remove groundwater.

WEST AFRICAN DISTURBANCE LINE Line of convection about 300 miles long, similar to squall line, that forms over west Africa north of equator and south of 15° North latitude.

WEST WALL Coast side boundary of Gulf Stream, typically south of Cape Hatteras.

WESTERLIES Prevailing winds that blow from west in mid-latitudes.

WESTERN AREA POWER ADMINISTRATION Markets and delivers reliable, cost-based hydroelectric power and related services within 15-state region of central and western U.S. WAPA's transmission system carries electricity from 55 hydropower plants, including those operated by Bureau of Reclamation.

WESTERN SYSTEMS COORDINATING COUNCIL Voluntary industry association created to enhance reliability among western utilities.

WET ADIABATIC Process where air, saturated with water, expands and compresses without external heat.

WET ADIABATIC INSTABILITY Condition in atmosphere where vertical movements of moist air typically increase.

WET ADIABATIC LAPSE RATE Reduction in temperature of moving air parcel per unit increase in height.

WET ALARM VALVE Check valve in fire sprinkler system.

WET ANALYSIS Method for determining effective diameters of particles.

WET BULB ZERO Height where wet-bulb temperature goes below 0 °C.

WET CONNECTION Connection made into under pressure water main for service line installation.

WET DEPOSITION Rain, sleet, snow or fog that has become more acidic than normal.

WET DIGESTION Sample decomposition by addition of liquid reagents to solubilize it.

WET FLOODPROOFING Approach to flood proofing which usually is last resort where flood waters are intentionally allowed into building to minimize water pressure on structure.

WET LABORATORY Laboratory where chemicals, drugs or other material or biological matter is tested and analyzed and which requires water, direct ventilation and specialized piped utilities.

WET MICROBURST Microburst accompanied by heavy precipitation at surface.

WET OXIDATION PROCESS Oxidation of soluble or suspended components in water using oxygen as oxidizing agent.

WET ROT Decay of timber by fungi that attack wood having high moisture content.

WET SCRUBBER UNIT Pollution control equipment that combines aqueous flow of slurry gasoline's from steam generating plant, designed to remove harmful gases and particulates from exhaust.

WET SOIL Soil with high moisture content, high expansive characteristics, high swelling and low shrinkage.

WET STORAGE DEBRIS Material that will not pass through 1-inch diameter round opening present in discharge from wet storage facility.

WET UNIT WEIGHT Unit weight of solids plus water per unit volume, irrespective of degree of saturation.

WET WEATHER GREEN INFRASTRUCTURE Infrastructure associated with stormwater management and low impact development that encompasses approaches and technologies to infiltrate, evapotranspire, capture and reuse stormwater to maintain or restore natural hydrologies.

WET WELL Sump for gravity-flow sewer systems.

WET-BULB DEPRESSION Difference between dry-bulb and wet-bulb temperatures.

WET-BULB TEMPERATURE Lowest temperature that can be obtained with evaporated water into air.

WET-BULB THERMOMETER Thermometer with bulb in wet reservoir.

WET-HEAD HYDRANT Fire hydrant with water valve at top.

WET-PIPE FIRE SPRINKLER SYSTEM Automatic sprinkler system that always has water in pipes.

WETLAND Area regularly wet or flooded and has water table that stands at or above land surface for at least part of year.

WETLAND BANK Restoration, creation or enhancement of wetlands to compensate unavoidable impacts to wetlands at another location.

WETLAND CREDIT Mitigation credit used to offset ecological losses in wetlands or streams that occur in waters of U.S.

WETLAND FUNCTION Process or series of processes that take place within wetland that are beneficial to wetland itself, surrounding ecosystems and people.

WETLANDS Any number of tidal and non-tidal areas characterized by saturated or nearly saturated soils, for long periods seasonally or continuously, that form interface between terrestrial and aquatic environments including freshwater marshes, swamps, ephemeral ponds, bogs and similar areas (e.g., wet meadows, river overflows, mudflats, natural ponds) and brackish and salt marshes, hydric soils and vegetation adapted for life in saturated soil conditions.

WETTABILITY Relative degree that fluid will spread into or coat solid surface in presence of other immiscible fluids.

WETTABLE POWDER Dry formulation that must be mixed with water or other liquid before it is applied.

WETTED PERIMETER Cross-sectional area of channel in contact with water.

WETTING 1. Liquid ability to maintain contact with solid surface; 2. Study of liquid deposited on substrate spreads out.

WETTING AGENT Substance (e.g., surfactant) that reduces surface tension of liquid.

WHALER Horizontal beam in bracing structure.

WHARF 1. Man-made landing place for ships on shore or river bank; 2. Bank of river; 3. Shore of sea.

WHARFAGE Fee charged for using wharf.

WHEEL DITCHER Large bucket trencher that excavates material onto conveyor belt that moves it aside of trench line.

WHEEL HORSEPOWER Power output measured at wheels.

WHEEL LINE 1. Portable irrigation systems on line of wheels; 2. Wheel grouping.

WHEEL LOAD Maximum load that wheel is designed to carry.

WHEEL SET Wheel-axle assembly of railroad car.

WHEEL TRACK Distance between axle from center of one tire to center of opposite tire.

WHEELCHAIR LIFT Machine used to transport wheelchair or other mobility device up or down stairs or across barrier.

WHEELING Transmission of electricity by entity that does not own or directly use power it is transmitting.

WHIRLWIND Small, rotating column of air.

WHITE GOODS Large household appliances (e.g., refrigerators, stoves, washing machines, driers).

WHITE LIGHT Sunlight integrated over visible portion of spectrum (4,000-7,000 angstroms) so that all colors are blended to appear white to eye.

WHITE LIGHT FLARE Major flare where small parts become visible in white light.

WHITE NOISE Random energy containing all frequency components in equal proportions.

WHITECAP Breaking crest of wave, usually white and frothy.

WHITTEMORE GAUGES Measure movements transverse and along joint or crack.

WHMIS CLASSIFICATION Hazard identification and product classification under WHMIS.

WHOLE EFFLUENT TOXICITY Aggregate toxic effect of effluent measured directly by toxicity test.

WHOLE-EFFLUENT TOXICITY Total toxic effect of effluent measured directly with toxicity test.

WHOLE-EFFLUENT-TOXICITY TESTS Tests to determine toxicity levels of total effluent from single source as opposed to series of tests for individual contaminants.

WHOLESALE CUSTOMERS Any entity that purchases electricity at wholesale level, including municipal utilities, private utilities, rural electric cooperatives or government-owned utility districts.

WHOLESALE POWER MARKET Purchase and sale of electricity from generators to resellers (who sell to retail customers) along with ancillary services needed to maintain reliability and power quality at transmission level.

WHOLESALE TRANSMISSION SERVICES Transmission of electric energy sold or to be sold, at wholesale in interstate commerce.

WHOLESALE WHEELING Process of sending electricity from one utility to another wholesale purchaser over transmission lines of intermediate utility.

WICKET GATE Gate which pivots open around periphery of turbine or pump to allow water to enter.

WIDESPREAD Areal coverage of non-measurable, non-convective weather and/or restrictions to visibility affecting more than 50% of forecast zone(s).

WIDTH OF DAM Width of dam as measured horizontally between upstream and downstream faces and normal to axis or centerline crest of dam.

WILD AND SCENIC RIVERS ACT Public law 90-542 that selects certain rivers possessing remarkable scenic, recreational, geologic, fish and wildlife, historic or other similar values, for preservation in free-flowing conditions.

WILDERNESS Tract or region of land uncultivated and uninhabited by human beings or unoccupied by human settlements.

WILDERNESS RESOURCE Resources identified in officially designated wilderness areas on Forest Service or Bureau of Land Management administered land.

WILDFIRE Any free burning uncontainable wildland fire not prescribed for area which consumes natural fuels and spreads in response to its environment.

WILDLANDS Any non-urbanized land not under extensive agricultural cultivation (e.g., forests, grasslands, rangelands).

WILDLIFE REFUGE Area designated for protection of wild animals, within which hunting and fishing are either prohibited or strictly controlled.

WILLINGNESS TO PAY Method of estimating value of activities, services or other goods, where value is defined as maximum amount consumer would be willing to pay for opportunity rather than do without.

WILLOW CARR Pool or wetland dominated by willow trees or shrubs.

WILLY-WILLY Slang for dust devil that was formerly used to denote tropical cyclone.

WILTING POINT Soil water content below which plants growing in that soil will remain wilted even when transpiration is nearly eliminated.

WINCH Drum that can be rotated so as to exert strong pull while winding in line.

WIND Horizontal motion of air past given point due to differences in air pressures where air moves from high to low pressure areas.

WIND ADVISORY Sustained winds 25 to 39 mph and/or gusts to 57 mph.

WIND CHILL ADVISORY National Weather Service issued product when wind chill could be life threatening if action is not taken.

WIND CHILL FACTOR Increased wind speeds accelerate heat loss from exposed skin.

WIND CHILL WARNING National Weather Service issued product when wind chill is life threatening.

WIND CONE Large, conical, open-ended tube designed to indicate wind direction and relative wind speed.

WIND COUPLET Area on radar display where two maximum wind speeds are blowing in opposite directions.

WIND DEPOSIT Material carried and deposited by wind.

WIND DIRECTION True direction from which wind is blowing at given location (i.e., wind blowing from east to west is east wind).

WIND ENERGY Process of creating electricity using wind.

WIND FARM Collection of wind turbines used to generate electricity.

WIND FIELD Three-dimensional spatial pattern of winds.

WIND FLOW Movement of air on large scale.

WIND GAP Dry valley, once occupied by stream, now occupied by another stream.

WIND GAUGE Device used for measuring wind force, speed and direction.

WIND GUST Rapid fluctuations in wind speed with variation of 10 knots or more between peaks and lulls.

WIND LOAD Measure of force exerted on structure surface by wind.

WIND MEASUREMENT Record of wind force, speed and direction.

WIND POWER Electric energy produced by harnessing wind.

WIND PROFILE Graphical representation of wind speeds with height.

WIND PUMP Windmill used for pumping water.

WIND RADII Largest radii of that wind speed found in that quadrant.

WIND ROSE Diagram, for given locality or area, showing frequency and strength of wind from various directions.

WIND SETUP Vertical rise in still water level at face of structure or embankment caused by wind stresses on surface of water.

WIND SHEAR Difference in wind speed, direction, and/or magnitude from point to point in given direction.

WIND SHEAR PROFILE Change in wind speed and/or direction usually in vertical.

WIND SHIFT Change in wind direction of 45° or more in less than 15 minutes with sustained wind speeds of 10 knots or more throughout wind shift.

WIND SHIFT LINE Long, but narrow axis across which winds change direction.

WIND SOCK Tapered fabric shaped like cone that indicates wind direction by pointing away from wind.

WIND SPEED Rate where air is moving horizontally past given point.

WIND TEE Large weather vane shaped like horizontal letter "T" at airfield to aid takeoffs and landings.

WIND WAVES Local, short period waves generated from action of wind on water surface (as opposed to swell).

WINDBLOWN 1. Dispersed by wind; 2. Growing in distorted shape produced by prevailing winds.

WINDBORNE Carried by wind.

WINDBREAK Hedge, fence or row of trees positioned to reduce wind force and damage.

WINDMILL Machine that translates linear motion of wind to rotational motion by means of adjustable vanes.

WINDOW PIPE Discharge pipe for dredge with openings in bottom.

WINDROW 1. Ridge of loose material; 2. Row of cut grain or hay allowed to dry in field.

WINDROW REVETMENT Deposit of fixed amount of erosion resistant material landward of streambank.

WINDWARD Side toward wind.

WINDY 20 to 30 mph winds.

WING DAM Barrier that only extends partway into waterway to deflect current.

WING WALL Wall that guides water into conveyance structure.

WINTER Typically coldest season of year during which Sun is farthest from overhead.

WINTER POOL Pool or height of water surface, of reservoir during winter.

WINTER SOLSTICE Time at which Sun is farthest south in Southern Hemisphere, on or around December 21.

WINTER STORM WARNING Product issued by National Weather Service when winter storm is producing or is forecast to produce heavy snow or significant ice accumulations.

WINTER STORM WATCH Product issued by National Weather Service when there is potential for heavy snow or significant ice accumulations, usually at least 24 to 36 hours in advance.

WINTER WEATHER ADVISORY Product issued by National Weather Service when low pressure system produces combination of winter weather (e.g., snow, freezing rain, sleet) that present hazard, but does not meet warning criteria.

WINTERS DOCTRINE Provides that establishment of Indian Reservation impliedly reserves amount of water necessary for purposes of reservation.

WIRE WEIGHT GAUGE River gauge comprised of weight that is lowered to water level.

WIRE-ENCLOSED RIPRAP Rock in wire enclosures.

WIRE-TO-WATER EFFICIENCY Efficiency of pump and motor together.

WITHDRAWAL Process of removing (e.g., taking water from ground source or diverted from surface-water source and conveying it to place for particular type of use).

WOLF NUMBER Historic term for Sunspot Number.

WOOD PACKAGING Wood products (e.g., pallets, crates, barrels).

WOOD PRESERVATIVE Chemical measures taken to ensure long life of wood.

WOOD TREATMENT FACILITY Industrial facility that treats lumber and other wood products for outdoor use.

WOOD-BURNING-STOVE POLLUTION Air pollution caused by emissions of particulate matter, carbon monoxide, total suspended particulates and polycyclic organic matter from wood-burning stoves.

WOODLAND Creature or object that grows, lives or exists in woodland.

WORK 1. Measure of energy expended in moving object; 2. Needed to complete contract; 3. Product of force and distance.

WORK DAY Day or part of day when work is done.

WORK HOURS Hours required that person spends working daily.

WORK IN PROGRESS Work that has been started, but not complete.

WORK ORDER Authorization to expend resources to perform task.

WORK PLAN Plans that are prepared which detail scope, direction and purpose of proposed plan.

WORK ZONE Segment along road where road construction is being performed.

WORK-ENERGY PRINCIPLE Increase in kinetic energy of rigid body is caused by equal amount of positive work done on body by resultant force acting on that body.

WORKABLE SLUDGE Digested sewage sludge that can be easily carried from drying bed.

WORKAROUND Response to unforeseen problem or risk.

WORKER Person who performs labor for living.

WORKERS RIGHT-TO-KNOW Legislation mandating communicating of information to employees.

WORKING CYCLE Machine operations.

WORKING DAY Days or part of days where work is done.

WORKING DRAWING Scale drawing that serves as guide for construction or manufacturing.

WORKING DRAWINGS Design sheet provided by designers to construction team that provide graphical and dimensioned information to construct project.

WORKING HORSEPOWER Horsepower delivered at flywheel.

WORKING LEVEL Unit of measure for documenting exposure to radon decay products, so-called "daughters."

WORKING LEVEL MONTH Unit of measure used to determine cumulative exposure to radon.

WORKING PRESSURE Pressure that something operates under and contains when working normally.

WORKING SHAFT Shaft used to excavate tunnel that's filled upon completion of task.

WORKING SPACE Void space around person or equipment.

WORKING WIDTH Distance between side of containment system closest to traffic flow before impact and furthest position of system section.

WORKPLACE HAZARDOUS MATERIALS INFORMATION SYSTEM Canada's national workplace hazard communication standard.

WORKSITE Site where work occurs.

WORM GEAR Toothed wheel worked by short revolving cylinder bearing screw thread.

WRAPPING GUST FRONT Gust front which wraps around mesocyclone, cutting off inflow of warm moist air to mesocyclone circulation and resulting in occluded mesocyclone.

WRONG-WAY ARROW Pavement markings placed on exit ramp to indicate wrong-way for traffic.

WROUGHT Having been worked or prepared by hammering.

WROUGHT PIPE Steel plate formed into tube and welded down seam.

WSPD Wind speed (m/s) averaged over 8-minute period for buoys and 2-minute period for land stations.

WSR-88D SYSTEM Summation of all hardware, software, facilities, communications, logistics, staffing, training, operations and procedures specifically associated with collection, processing, analysis, dissemination and application of data from WSR-88D unit.

WYE Y-shaped plumbing fitting.

WYE BRANCH Pipe fitting that joins main run of pipe at angle (i.e., shape of letter Y).

X

X-BAND Frequency band of microwave radiation where radars operate.

X-CRACK Crack pattern due to vibration.

X-RAY Short wavelength electromagnetic radiation used to create images of internal structure of objects.

X-RAY BACKGROUND Daily average background X-ray flux in 1 to 8 angstrom range.

X-RAY BURST Temporary enhancement of X-ray emission of Sun.

X-RAY FLARE CLASS Rank of flare based on its X-ray energy output.

X-RAY FLUORESCENCE Secondary radiation used to determine elemental composition of materials.

X-RAYS Very energetic electromagnetic radiation with wavelengths intermediate between 0.01 and 10 nanometers (0.1-100 Angstroms) or between gamma rays and ultraviolet radiation.

XENOBIOTA 1. Any biotum displaced from its normal habitat; 2. Chemical foreign to biological system.

XENON Heavy, gaseous chemical element with atomic number of 54.

XERISCAPE Landscaping that does not require lot of water.

XERISCAPING Landscaping based on native (i.e., adapted to local area), water-efficient plants to minimize need for irrigation (i.e., drought-resistant).

XEROPHREATOPHYTE Phreatophyte that's able to resist drought.

XEROPHYTE Plant adapted for growth under dry conditions.

XYLEM Vascular tissue in land plants primarily responsible for distribution of water and minerals taken up by roots.

Y

Y BRANCH Pipe fitting that joins main run of pipe at angle (i.e., shape of letter Y).

Y GROWTH RATE Experimentally determined constant used to estimate unit growth rate of bacteria while degrading organic wastes.

Y INTERSECTION When traffic approaches from stem of "Y" and continues along branch of "Y" with arrowhead.

Y-BRANCH Conduit fitting joining two other conduits at 45° angles.

YARD 1. Unit of length equal to 3 feet; 2. Area for storage and maintenance (e.g., open space on building lot); 3. Place where logs are accumulated.

YARD DRAIN Catch basin that gathers surface runoff water.

YARD HYDRANT Buried hardware that attaches to water supply pipe outside of well.

YARD RUBBISH Organic green-waste material (e.g., tree branches, brush, twigs, grass), but does not include earth or building materials.

YARD WASTE Part of solid waste composed of grass clippings, leaves, twigs, branches and other garden refuse.

YARDAGE Volume of material measured in yd^3.

YARDANGS Large wind-eroded mass of soft or poorly consolidated rock in desert region that lies parallel to prevailing winds.

YAW Rotation of vehicle, aircraft or vessel about its vertical axis.

YEAR Period during which Earth completes one revolution around Sun.

YEAR OF EXPENDITURE DOLLARS Project cost estimated in year to be built.

YEAST Single-celled fungus of wide variety of taxonomic families.

YELLOW CHANGE INTERVAL Interval that warns drivers of impending change in right-of-way assignment.

YELLOW SNOW Snow given golden or yellow appearance by presence in it of pine, cypress pollen or anthropogenic material or animal-produced material.

YELLOW-BOY Iron oxide flocculant (clumps of solids in waste or water) usually observed as orange-yellow deposits in surface streams with excess iron content.

YELLOWCAKE Concentrated form of uranium ore containing mixture of oxides.

YIELD 1. Quantity of water (expressed as rate of flow or total quantity per year) that can be collected for given use from surface or groundwater sources; 2. Mass per unit time per unit area; 3. To produce.

YIELD LINE Line instructing people to allow vehicles or people, where paths cross and have right of way, to move first.

YIELD STRESS Critical value of shear stress at yield point or yield strength.

YOCTA Prefix symbolizing 10^{-24}.

YOTTA Prefix symbolizing 10^{24}.

YOUNG RIVER Beginning of river where it flows quickly with lots of energy.

YOUNG VALLEY Valley in early period of growth.

YOUNG-OF-YEAR Refers to young (usually fish) produced in one reproductive year.

Z

Z TIME Local time at prime meridian that passes through Greenwich, England that's known as Coordinated Universal Time, Universal Time Coordinated (UTC), "Z time" or "Zulu Time."

Z/R RELATIONSHIP Empirical conversion relationship between radar reflectivity and precipitation rate.

ZEOLITE Minerals (i.e., aluminosilicates of sodium, potassium, calcium or magnesium) that have porous structure.

ZEOLITE SOFTENING Removal of calcium and magnesium hardness from water by base exchange using natural or synthetic zeolites.

ZERO AIR Atmospheric air purified to contain less than 0.1 ppm total hydrocarbons.

ZERO AIR VOIDS CURVE Curve showing relationship between dry unit weights and corresponding moisture contents, assuming that all of voids are completely filled with water.

ZERO DATUM Reference "zero" elevation for stream or river gauge.

ZERO MOISTURE INDEX Index of moisture when precipitation is adequate to supply all water needed for maximum evaporation and transpiration in year.

ZERO POPULATION GROWTH Population balance where number of people neither grows nor declines.

ZEPTO Prefix symbolizing 10^{-21}.

ZETTA Prefix symbolizing 10^{21}.

ZINC Chemical element with atomic number of 30.

ZINC EQUIVALENT Variable to evaluate effect on crop increase due to waste sludge inclusion to land.

ZIRCALOY Any set of specialist alloys of zirconium.

ZONAL FLOW Large-scale atmospheric flow where east-west component (i.e., latitudinal) is dominant.

ZONAL INDEX Measure of strength of middle-latitude westerlies, expressed as horizontal pressure difference between 35 $^{\circ}$ and 55 $^{\circ}$N latitude or as corresponding geostrophic wind.

ZONE Any given region or area of world.

ZONE FARE Fare proportional to length of trip.

ZONE OF AERATION 1. Comparatively dry soil or rock located between ground surface and top of water table where soil pores may either contain air or water; 2. Vadose zone.

ZONE OF CAPILLARITY Area immediately above water table where water may be drawn upward due to capillary action.

ZONE OF INTRUSION Area behind barrier that's free of fixed objects that vehicle may impact.

ZONE OF SATURATION 1. Zone where all interstices between particles of geologic material or all of joints, fractures or solution channels in consolidated rock unit are filled with water under pressure greater than that of atmosphere; 2. Area below watertable; 3. Phreatic zone.

ZONE TIME Time system where surface of globe is divided into 24 zones of 15 $^{\circ}$.

ZONED EARTHFILL Embankment dam that is comprised of zones of selected materials having different degrees of porosity, permeability and density.

ZONED EMBANKMENT DAM Embankment dam that is comprised of zones of selected materials having different degrees of porosity, permeability and density.

ZONING Legislative identification of areas of specified uses or restrictions.

ZONING ORDINANCE Written regulation pertaining to how people can use land and buildings within certain area.

ZONING PERMIT Document issued by regulatory agency permitting parcel of land to be used for specific purpose.

ZOOGLOEA Colony or mass of bacteria embedded in viscous gelatinous substance.

ZOOGLOEA MATRIX Floc formed mainly by slime producing bacteria.

ZOONOSIS Animal disease (e.g., rabies, anthrax) that can be transmitted to humans.

ZOOPLANKTON Small, free-floating aquatic animals (e.g., protozoa, crustaceans, eggs, larvae), found in lakes and reservoirs.

ZULU TIME Local time at prime meridian that passes through Greenwich, England that's known as Coordinated Universal Time, Universal Time Coordinated (UTC), "Z time" or "Zulu Time."

ZURICH SUNSPOT CLASSIFICATION Sunspot classification system that has been modified for NOAA Space Environment Services Center use.

CPSIA information can be obtained
at www.ICGtesting.com
Printed in the USA
BVHW040931200221
600461BV00005B/9

9 780983 908517